PERGAMON INTERNATIONAL LIBRARY
of Science, Technology, Engineering and Social Studies
The 1000-volume original paperback library in aid of education, industrial training and the enjoyment of leisure
Publisher: Robert Maxwell, M.C.

PLANKTON AND PRODUCTIVITY IN THE OCEANS

Second Edition

Volume 1 – Phytoplankton

By the same author

PLANKTON AND PRODUCTIVITY IN THE OCEANS, 2nd Edition,
Volume 2 Zooplankton

Other titles of interest

CORLISS
The Ciliated Protozoa, 2nd Edition

KEEGAN, O'CEIDIGH & BOADEN
Biology of Benthic Organisms

McLUSKY & BERRY
Physiology and Behaviour of Marine Organisms

NAYLOR
Cyclical Phenomena in Marine Plants and Animals

PARSONS, TAKAHASHI & HARGRAVE
Biological Oceanographic Processes, 2nd Edition

PURCHON
The Biology of the Mollusca, 2nd Edition

VAN WINKLE
Proceedings of a Conference on Assessing the Effects of
Power-plant Induced Mortality on Fish Populations

WOLFE
Fate and Effects of Petroleum Hydrocarbons
in Marine Organisms and Ecosystems

PLANKTON AND PRODUCTIVITY IN THE OCEANS

Second Edition

Volume 1 – Phytoplankton

JOHN E. G. RAYMONT
Department of Oceanography in the University of Southampton

with contributions from J. D. Burton and K. R. Dyer

PERGAMON PRESS
Oxford · New York · Toronto · Sydney · Paris · Frankfurt

UK	Pergamon Press Ltd., Headington Hill Hall, Oxford OX3 0BW, England
USA	Pergamon Press Inc., Maxwell House, Fairview Park, Elmsford, New York 10523, U.S.A.
CANADA	Pergamon of Canada, Suite 104, 150 Consumers Road, Willowdale, Ontario, M2J 1P9, Canada
AUSTRALIA	Pergamon Press (Aust.) Pty. Ltd., P.O. Box 544, Potts Point, N.S.W. 2011, Australia
FRANCE	Pergamon Press SARL, 24 rue des Ecoles, 75240 Paris, Cedex 05, France
FEDERAL REPUBLIC OF GERMANY	Pergamon Press GmbH, 6242 Kronberg-Taunus, Pferdstrasse 1, Federal Republic of Germany

First edition 1963

Second edition 1980

British Library Cataloguing in Publication Data

Raymont, John Edwin George
Plankton and productivity in the oceans. – 2nd ed. –
(Pergamon international library).
Vol. 1: Phytoplankton
1. Marine plankton
I. Title II. Burton, J D III. Dyer, K R
574.92 QH91.8.P5 79-40351

ISBN 0-08-021552-1 hardcover
ISBN 0-08-021551-3 flexicover

Printed and bound in Great Britain by
William Clowes (Beccles) Limited, Beccles and London

PUBLISHERS NOTE

Shortly before this book was sent for printing, the news of Professor Raymont's sad and untimely death was announced. The publishers would like to express their appreciation of the hard work and dedication shown by Professor Raymont in the preparation of this book and its companion volume.

Contents

* Contributed by K. R. Dyer.
† Contributed by J. D. Burton.

Chapter 6
Factors Limiting Primary Production: Light and Temperature

Chapter 7
Factors Limiting Primary Production: Nutrients*

Chapter 8
Factors Limiting Primary Production: Grazing

* Contributed by J. D. Burton.

Acknowledgements

Academic Press Inc. (London) Ltd., London
Academic Press Inc., New York
Akademie-Verlag, Berlin
George Allen & Unwin (Publishers) Ltd., London
American Association for the Advancement of Science, Washington, DC
American Geographical Society, New York
American Geophysical Union, Washington, DC (*Journal of Geophysical Research*)
Edward Arnold (Publishers) Ltd., London
Artemis Press Ltd., Horsham, Sussex
Blackwell Scientific Publications Ltd., Oxford
Cambridge University Press, Cambridge (for Marine Biological Association of the UK)
Conseil International pour l'Exploration de la Mer, Charlottenlund, Denmark
Duke University Press, Durham, North Carolina (for the journal *Ecology* – Ecological Society of America)
Elsevier/North-Holland Biochemical Press, Amsterdam
The Fisheries Research Board of Canada, Halifax, NS
Gauthier-Villars, Paris
Harvard University Press, Cambridge, Massachusetts
Johns Hopkins University Press, Baltimore
Institute for Marine Environmental Research, Plymouth
Indian Academy of Sciences, Bangalore, India
Liverpool University Press, Liverpool
Munksgaard, Copenhagen
Macmillan (Journals) Ltd., London
New York Academy of Sciences, New York
Oxford University Press (for *Journal of Experimental Botany*)
Prentice-Hall, Inc., New Jersey
Pergamon Press Ltd., Oxford
Ray Society, London
Springer-Verlag, Berlin
Skidaway Institute of Oceanography, Georgia
The Japanese Society of Scientific Fisheries, Japan
Charles C. Thomas, Springfield, Illinois
Idemitsu Shoten Co. Ltd., Tokyo
University of Miami (*Bulletin of Marine Science*)
University of Michigan (American Society of Limnology & Oceanography Inc.)
University – Ohio State (Botanical Society of America, *American Journal of Botany*)

University of California (Scripps Institution of Oceanography)
Yale University, Connecticut (Sears Foundation of Marine Research)
Universitetsforlaget, Oslo (Det. Norsk. Vidensk. Akad. Norway)
John Wiley & Sons, Chichester, England
John Wiley & Sons, Inc., USA
Woods Hole Oceanographic Institution, Massachusetts
Zeitschrift fur Naturforschung (Max-Planck Institut fur Chemie)

Introduction

The publication of the first edition of *Plankton and Productivity on the Oceans* arose from a need to attempt to give students, mainly advanced undergraduates in marine biology and those commencing research in biological oceanography, some appreciation of the variety of phytoplankton and zooplankton and of their relations with their environment. Some synthesis was also required of the course and significance of primary production, of its major controlling factors with their temporal and regional variations in the oceans, and of the inter-relationships between phytoplankton and zooplankton. In presenting an ecological approach to plankton studies an introduction to some of the physical and chemical characteristics of the marine environment was necessary. The discussion of secondary production was concentrated upon the zooplankton, but a brief account was included of the nekton and of benthic communities, chiefly in relation to the cycling of organic matter in the oceans.

Over the past 15 years the vast increase in research in marine biology, including planktonology, has created a need for updating the earlier text on plankton, despite the upsurge of excellent reviews dealing with particular aspects of biological oceanography. However, the great increase in the number and complexity of plankton studies, especially the advances in experimental studies and the more strictly quantitative approach to production, as well as the vast increase in literature, has necessitated a total reconsideration and complete rewriting of the text. The recent publication of books dealing specifically with benthos and with nekton has been a major influence on the decision for the rewriting of the present work to concentrate entirely upon the plankton. While separation of the benthos and nekton from the plankton is obviously artificial and precludes proper consideration of food-webs, the planktonic ecosystem, particularly in the open oceans, can be reasonably studied as a separate system.

Since the cycling of organic matter is a fundamental concept in biological oceanography, attention might be devoted initially to primary production as a convenient starting-point in the cycle. While contributions by benthic plants can be significant in inshore waters, the overwhelming importance of the planktonic algae in global primary production justifies a substantial study of phytoplankton production. The increasing use of the carbon isotope method of Steemann Nielsen since the late 1950s has led to a more precise quantification of primary production over the world ocean. Increasingly, however, the need for accurate assessment of the phytoplankton crop from different geographical regions and of temporal changes focuses the need for a detailed evaluation of the specific composition of phytoplankton communities. With our better appreciation of the importance of the nanoplankton in production there is also a demand for more detailed knowledge of the structure and life-cycles as well as the physiology of these minute algae. Estimates of chlorophyll, while a useful component in models of production, must be supplemented by detailed investigations of a whole range of algal taxa, including structure and physiology. Investigations over the past 15 or 20 years have demonstrated that growth rates, reproductive cycles, nutrient relationships, specific growth factors, rates of release of soluble organic substances, and light and temperature relationships may all show considerable species-specific variations. There is an increasing demand for such data relating to oceanic high and low latitude species as well as for the earlier-studied temperate neritic algae.

This volume therefore devotes considerable attention to the variety of marine phytoplankton, including nanoplankton forms, to the different types of, and temporal changes in, phytoplankton communities, and also attempts to relate experimental studies on planktonic algae especially to primary production and algal crop in the seas. In discussing the major factors controlling production, the review pays particular attention to primary production with reference to latitude, season, climate and ocean region. The ecological approach implies the need for some knowledge of those physical and chemical characteristics of the marine environment of particular significance to planktonic existence. Dr. Dyer contributes an introduction to pertinent physical aspects. Dr. Burton follows with a chapter on the chemical composition of seawater; in relation to factors influencing production, he deals later in some detail with nutrients and uptake kinetics. To some extent the present volume attempts to integrate the area dealt with in the excellent study of production and its controlling factors by Drs. Parsons and Takahashi in "Biological Oceanographic Processes" with a broader study of planktonic organisms and their community temporal and spatial variations.

While this volume is devoted to the phytoplankton, consideration of grazing in particular leads logically to some discussion of the zooplankton and to secondary production. An account of the taxonomy, horizontal and vertical distribution of zooplankton and of communities, and experimental studies on the physiology and biochemistry of zooplankton is left to Volume 2.

This work does not attempt a complete literature coverage but it is intended that the bibliography be sufficiently wide, both with regard to various oceanic regions and to the active centres of research of the many nations, to encourage students to commence a serious study of original biological oceanographic literature.

My sincere thanks are due to my publishers for their assistance and to the authors, including many personal friends, of books and scientific papers for permission to include a number of original figures and other illustrations in this volume. My thanks are also due to the publishers of these books and journals; these permissions are separately acknowledged.

Apart from the contributed chapters by Drs. Dyer and Burton, I have received substantial help and guidance in preparing this book from several colleagues (Professor Charnock, FRS, Dr. Williams, Dr. Robinson, Mr. Phillips) who have read certain sections, as well as from past and present students for informal discussions concerning plankton. Dr. Burton has also undertaken the laborious task of reading the manuscript and proofs. Above all, I acknowledge the most valuable and unstinting assistance of my wife, who has helped in every aspect of the writing, searching the literature and preparing the bibliography, selecting illustrations and constantly critically reviewing the manuscript. For the errors which remain I accept full responsibility. It is a pleasure to thank Miss Mallinson and Miss Brady for help in preparing certain illustrations. Finally, I am deeply in debt to my secretary, Mrs. Gathergood, for cheerfully accepting the laborious task of repeatedly retyping the manuscript and for much other assistance.

Chapter 1
Physical Aspects of the Oceanic Environment

The main source of natural energy on the earth's surface is derived from the sun's radiant energy. Even though only one part in 2×10^9 of the sun's radiant energy is received by the earth, it supports physical and biological processes, is the main cause of the wind system in the atmosphere, and of the currents in the ocean. Because the axis of the earth's rotation is inclined to the plane of its rotation round the sun, the amount of energy falling at a particular place on the earth's surface varies seasonally as well as diurnally and with latitude. The resulting differences in temperature at the surface of the sea and the land cause pressure differences which drive the atmospheric circulation and this in turn is responsible for driving the more vigorous of the ocean circulation systems. However, differences in the temperature and salinity of seawater due to direct heating and the resulting evaporation, or to freezing, effect the smaller scale mixing processes.

Thus, broadly speaking, it is the seasonally averaged conditions which drive the large-scale, global or ocean-wide, features. Shorter-term variability in both the atmosphere and the ocean is largely due to inherent instabilities. In the atmosphere we are familiar with the instabilities in the form of depressions which pass over in a few days. The corresponding instabilities in the deep sea take typically weeks to pass a given place. In shallow seas there is a more direct coupling between atmosphere and ocean and consequently a fair degree of correlation of oceanographic variations with storms. Also, of course, tides are more apparent in shallow waters and these give large, though fairly regularly fluctuating motions. All these shorter term effects make it difficult to see long-term trends and this is one of the hazards of oceanographic studies.

The density of seawater will be influenced by temperature change, but since water has a high specific heat and takes a finite time to heat up and cool down, the fastest and most extreme fluctuations are ironed out. Density will also be affected by the concentration of dissolved salts and seawater is a very complex solution of many dissolved ions. However, in the open ocean the proportional composition remains remarkably constant even though the total dissolved salts can vary significantly. The average ocean salinity is about 35‰ (parts per thousand). The effect of temperature and salinity on water density is shown in Fig. 1.1. It is normal to consider density in terms of specific gravity, the ratio of the water's density to that of pure water at 4°C. Also oceanographers have adopted the notation $\sigma_{s,t,p} = (\text{specific gravity} - 1) \times 10^3$. The pressure term is involved because of the compressibility of water, but since in general we compare water masses at the same depth (pressure) or over the same range of depth, its effect can be neglected for descriptive purposes. However, if one were to raise a mass of water it would expand and adiabatically cool. The original temperature is called the *in situ* temperature and its value after cooling on raising to the surface, is the potential temperature. σ_t is the

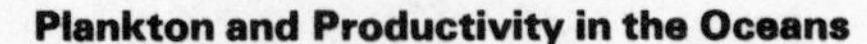

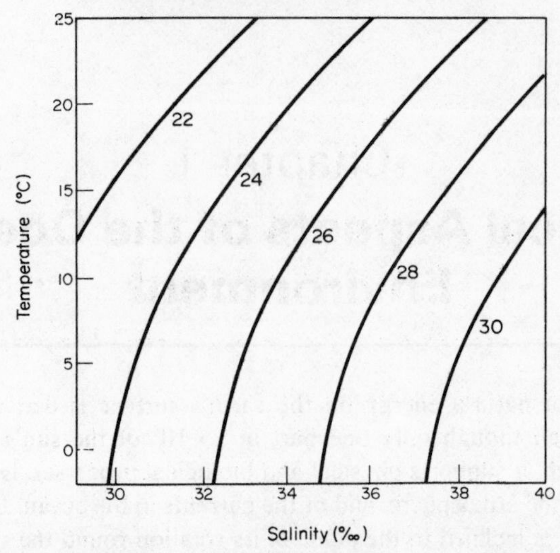

Fig. 1.1. Relation between density σ_t and temperature and salinity.

specific gravity using *in situ* salinity and temperature, but at atmospheric pressure. σ_θ is the value using potential temperature. Thus water of salinity 34.67‰ at a depth of 6450 m and an *in situ* temperature of 1.93°C has a potential temperature of 1.25°C, a σ_t value of 27.74 and σ_θ of 27.79.

In the open ocean where the salinity changes are fairly small, the main control on the water density is temperature. In estuaries and coastal regions where the salinity differences are greater, temperature differences play a lesser role. An important feature caused by salinity is the depression of the freezing point. Thus at a salinity of 35‰ the freezing point is about −1.8°C. Coupled with this is a depression in the temperature of maximum density. In fresh water the maximum density is at 4°C and it is unusual for the bottom of a deep body of water to have a temperature lower than this. Above a salinity of about 25‰ the maximum density is below the freezing point and consequently it is possible to have the whole water column at a temperature very close to freezing point.

Temperature also affects the molecular viscosity (Fig. 1.2). Here it is obvious that

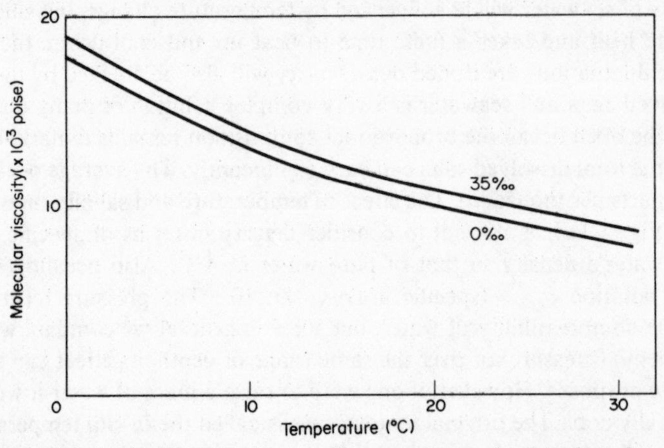

Fig. 1.2. Variations of molecular viscosity of water with temperature and salinity (data from Defant, 1961).

salinity has a much smaller effect than temperature. The combined effect of density and of viscosity variations appears in the variations of fall velocity with temperature. Figure 1.3 shows the fall velocity of quartz spheres of density 2.65 g/cm^3 in stationary pure water. For the smaller grains the fall velocity is about tripled for a temperature rise from

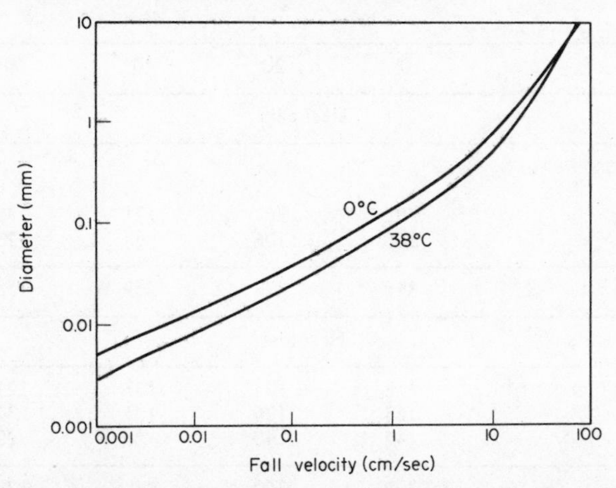

Fig. 1.3. Fall velocity of quartz spheres in still water (from Raudkivi, 1967).

0°C to 38°C. This effect has been used to explain the complexity of the shape of the tropical plankton, as being an attempt to reduce their fall velocity in the warmer water. However, the occurrence of turbulence effectively hampers the settling of particles in the ocean and would cause fall velocities considerably lower than those shown in Fig. 1.3. It is thus likely that the variation in form of the plankton may well be a response to differences in the levels of turbulence at different latitudes.

Heat Balance of the Ocean

Provided a long enough average is considered, taking the earth as a whole the inflow of energy from the sun is balanced by re-radiation into space. The inflow is of high temperature radiation, 50% being in the visible part of the spectrum, whereas the re-radiation is of longer infra-red wavelengths.

Consideration of the heat budget of the ocean surface shows that there is a nett inflow into the sea at the tropics and a nett outflow at higher latitudes. There thus has to be a transport of heat from the tropics towards the poles both in the atmosphere and in the ocean, about 50% of the poleward heat transport being by the ocean currents.

The upper atmosphere receives on average about 700 langley/day (1 ly = 1 g cal/cm^2). The actual amount reaching the surface obviously depends on latitude as well as being reduced by the atmospheric path and its content of dust and water vapour. At a latitude of 20° the amount reaching the sea surface is about 370 ly/day. Of this about two-thirds is direct solar radiation and one-third diffuse radiation from the sky. These proportions again depend on latitude, and on cloudy days the proportion of direct radiation will be reduced. Part of this solar energy will be reflected at the sea surface, depending on the angle of incidence and on the sea state. Some of the energy absorbed is quickly re-radiated into the atmosphere. The re-radiation from the sea surface depends on the temperatures of the lower layer of the atmosphere and the water surface and on the vapour pressure in the near-surface air. There is also a small exchange of heat by conduction and convection (Table 1.1).

Table 1.1. Heat budget of the total ocean (ly/day) (from Defant, 1961)

Latitude	0°	20°	40°	60°	80°
		Heat gain			
Direct solar radiation after allowing for cloudiness	202	267	171	80	44
Diffuse radiation	166	106	98	73	41
Total heat gain	368	373	269	153	85
		Heat loss			
Effective back-radiation	118	144	133	121	131
Evaporation heat	163	176	125	36	6
Convection	45	40	20	20	20
Total heat loss	326	360	278	177	157
Gains–losses	+41	+13	−9	−24	−72

The rest of the radiant energy is available to heat the surface water, though some will be used to evaporate some of the water. About 60 ly are required to evaporate 1 mm of water. Evaporation is greatest in the tropics and sub-tropics (Fig. 1.4). At mid latitudes there is, consequently, a small energy surplus available during daylight hours to heat the near surface water. Under steady state conditions this would all be re-radiated back at night. During the summer there is greater gain during the day than loss at night. This results in both diurnal and seasonal temperature fluctuations.

Consideration of the yearly averaged energy balance shows that the loss at the poles

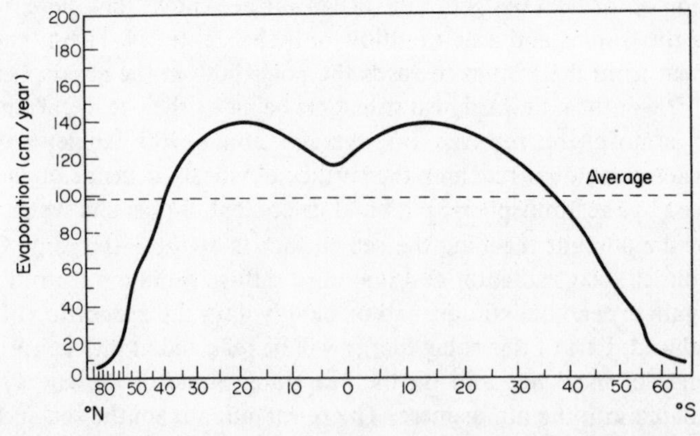

Fig. 1.4. Average annual evaporation from the ocean at various latitudes (after Wust *et al.*, 1954).

in the winter outweighs the gain during the summer (Fig. 1.5). Multiplication of the values by the areas for each latitude should give an approximate balance.

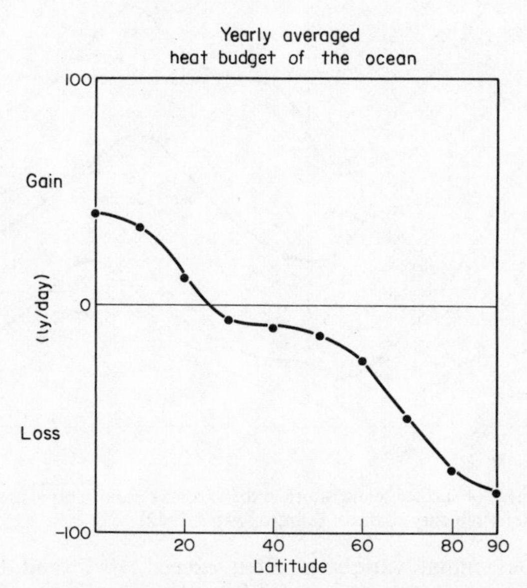

Fig. 1.5. Variation with latitude of the yearly averaged heat budget of the ocean (data from Defant, 1961).

Temperature Cycles

Diurnal cycle

In the open ocean the diurnal variation is generally less than 1°C and the variation decreases with increased latitude and with depth, being generally negligible at 10 m. Turbulence can cause a fairly quick mixing of the heat through the top few metres of the water and this tends to minimize the magnitude of the temperature variations. The maximum value occurs at about 1400 hours and the minimum at about 0400 to 0500 hours. In near-shore regions the diurnal variation can be larger, but this is often caused by covering and uncovering of tidal flats, sometimes producing 5°C variation. However, the magnitude is affected by whether the high tide occurs in the early afternoon or not.

Annual cycle

In the tropics and at the poles the annual changes of temperature are small. The highest variation is in the sub-tropical regions where the clear skies give the maximum heat gain in summer and heat loss in winter (Fig. 1.6). The largest ranges are in the northern hemisphere because of the influence of the cold continental winter winds. In the tropics the maximum and minimum occur at the equinoxes when the sun is overhead, whereas in extra-tropical regions they occur in August and February. At some positions, such as off the Grand Banks of Newfoundland, seasonal fluctuations in the current systems can give large variations. The annual cycles in the Bay of Biscay and south of Japan are shown in Fig. 1.7. The maximum temperature fluctuation occurs later with increasing depth and does not reach as high a magnitude as at the surface.

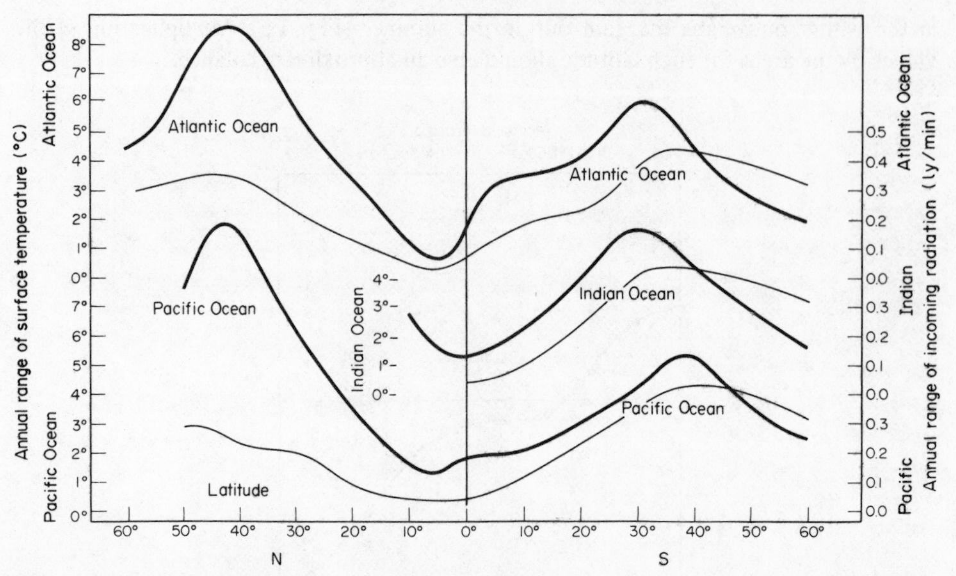

Fig. 1.6. Average annual ranges of surface temperature in the various oceans plotted against latitude (thick curves) and ranges in radiation (thin curves) (from Sverdrup *et al.*, 1942).

In coastal areas the annual variations often exceed 10°C and these are often associated with warm summer and cold winter run-off from the land.

Vertical Temperature Distribution

The time lag between the heat rise at the surface and that at depth is due to the time taken for the turbulence to carry the heat downwards. The rate at which this transfer occurs depends on the temperature gradient and on a vertical eddy-diffusion coefficient. The value of this coefficient is several orders of magnitude larger than the coefficient of thermal conductivity. Thus heat transfer which by conduction would take years, by turbulent exchanges takes days. Unfortunately it is difficult to calculate the vertical eddy-diffusion coefficient because it is not a basic physical property of the fluid, but more a measure of the amount of turbulence present. Surface heating will produce a temperature gradient and consequently also a density gradient which will inhibit the turbulence and which confines the heat to a surface layer bounded by the thermocline beneath. In many respects a slow increase in temperature will be mixed more readily downwards than a fast one. However, there is an additional factor already mentioned and that is surface waves. Waves cause an orbital motion of the water to a depth about equal to their wavelength. The amplitude of the motion decreases with depth. By themselves these motions do not cause any mixing, but in the presence of a density gradient they can cause large velocity gradients. Additionally, the action of wind on the water will produce a velocity gradient in the surface layers. These velocity gradients cause turbulence which can promote fairly vigorous mixing, but obviously of a fairly local and spasmodic occurrence. Detailed measurements in the thermoclines in the ocean have shown step-like temperature layers which may have been the result of this mixing process.

The development of the vertical distribution of temperature can be demonstrated with reference to Fig. 1.7A. In June, July and August the surface is heated and a zone of high temperature gradient formed. This thermocline gets gradually deeper and less steep by downward mixing. In autumn the surface water cools. As it becomes cooler the winds cause enough turbulence to mix the water column to some depth. If the surface cooling is sufficiently intense an additional mixing mechanism comes into effect. This is thermohaline convection. The denser water will sink to a level where it is in equilibrium carrying its temperature and salinity with it almost unchanged and the lighter water beneath will rise towards the surface. This is a comparatively efficient mixing mechanism which is particularly effective at high latitudes during the autumn and early winter cooling. In polar areas it can cause the surface water to be carried down through almost the whole water depth. Even in the Mediterranean thermohaline convection is important in the winter, when the cold Mistral winds cool the surface water, and in the Gulf of Lyons this generates the comparatively cold Mediterranean Bottom Water.

In shallow water, thermocline development is restricted because of the additional mixing caused by tidal movement of the water. The turbulence arising from the flow over the rough bottom mixes the water, whereas the heat flux downwards from the surface tends to stabilize it. The balance between these two opposing effects depends on the tidal velocities (u) and on the water depth (h), since the tidal mixing processes are most vigorous near the sea bed. It has been shown that the extent of the summer

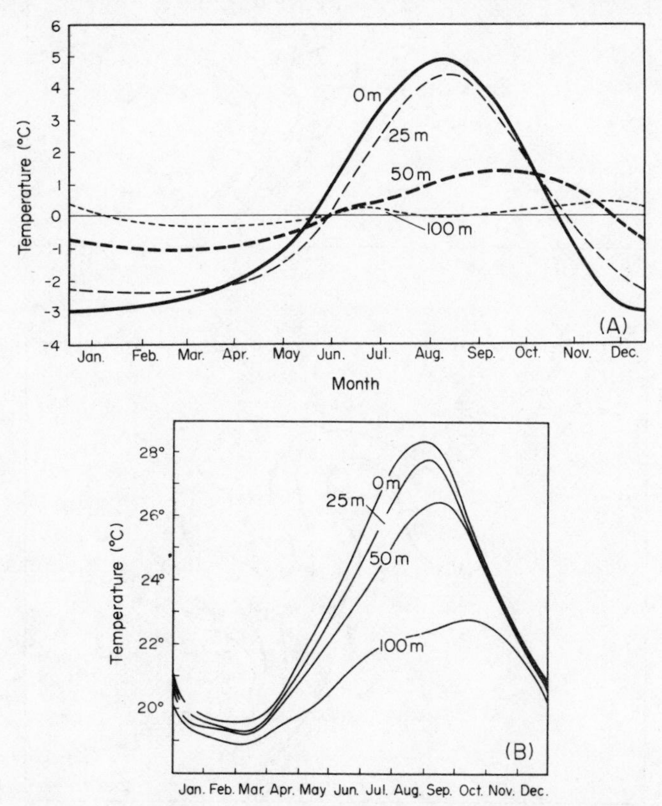

Fig. 1.7. (A) Annual variation of temperature about the mean, at different depths off the Bay of Biscay. (B) Annual variation of temperature at different depths in the Kuroshio off the south coast of Japan (from Sverdrup *et al.*, 1942).

stratification is essentially determined by the parameter h/u^3, the critical value being in the range 50 to 100 for the shelf seas around north-west Europe, h being measured in metres and u in m/sec. Figure 1.8 shows that at its maximum extent the thermocline in the English Channel is restricted to the western end. During spring there is pronounced build-up but its destruction in autumn is slow because of the stability of the thermocline.

At the boundary between the fully mixed and stratified areas a front is formed. The tidal motions carry the front backwards and forwards, but the density differences cause residual currents resulting in slight convergence and sinking at the front. A temperature section through a front is shown in Fig. 1.9A and the residual currents in Fig. 1.9B. The front is generally marked by an accumulation of debris, by colour differences and by differences in wave amplitude across the front. The frontal area is often one of high

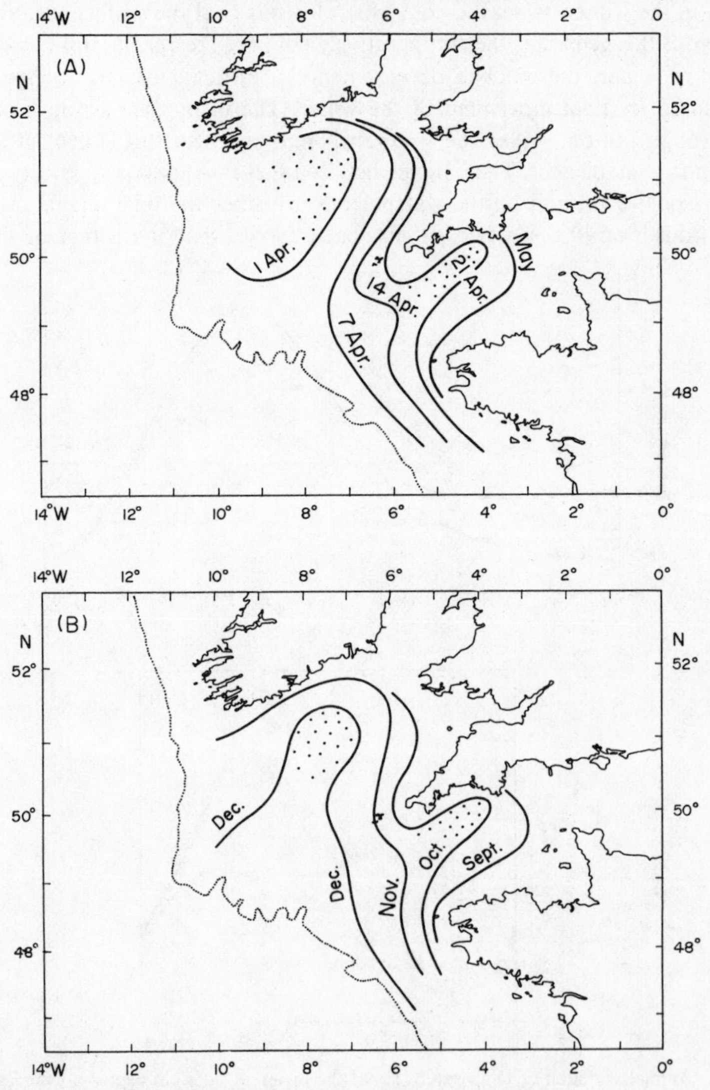

Fig. 1.8. Summer thermocline in the English Channel and Celtic Sea: (A) Development by weekly contours. (B) Retreat by monthly contours (from Pingree, 1975).

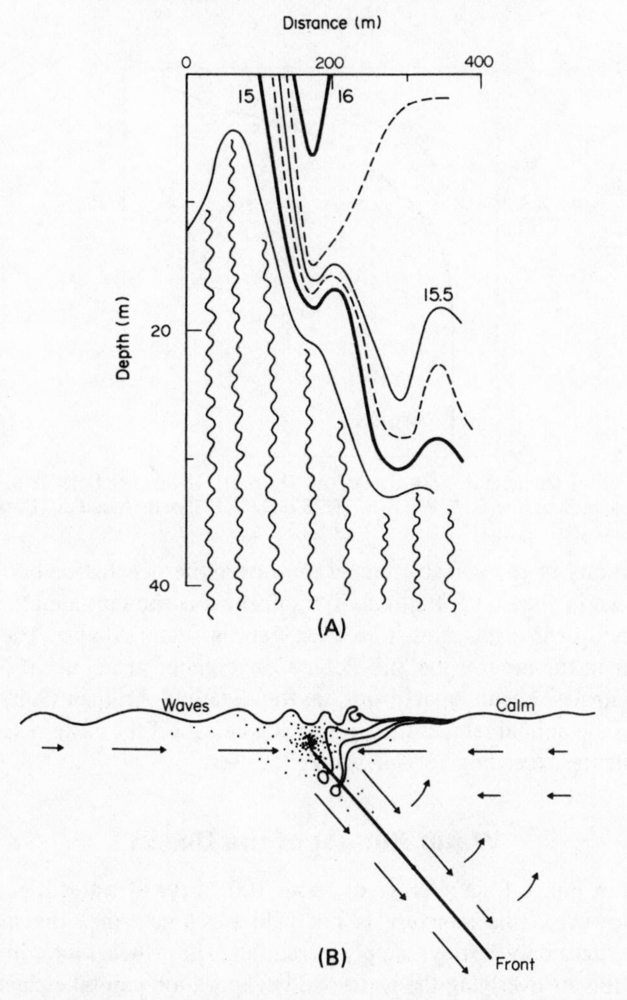

Fig. 1.9. Section across a front near Jersey: (A) Temperature. (B) Schematic picture showing components of circulation normal to the front (from Pingree *et al.*, 1974).

productivity because of the good light penetration in the clearer warmer water and the high nutrient supply in the bottom water brought to the surface.

In the open ocean there is a permanent thermocline. This is restricted to tropical and sub-tropical areas, between about 40° north and 40° south of the equator. The water above it is a maximum of about 200 m thick, and is subject to the direct effects from the atmosphere, to diurnal and annual temperature variations, and to wind mixing. This surface layer is virtually homogeneous. In the thermocline the temperature decreases with increasing depth at a maximum rate of about 5°C per 100 m. However, because salinity decreases at the same time, the thermocline does not produce a particularly steep density gradient. Below about 1000 m depth the temperature becomes more uniform. Typical temperature profiles are shown in Fig. 1.10. Again, where the thermocline outcrops at the surface there is a front called the Sub-Tropical Convergence. The vertical section in the Atlantic (Fig. 1.11) shows how relatively thin the surface mixed layer is and that the bulk of the oceanic water has a temperature less than 5°C.

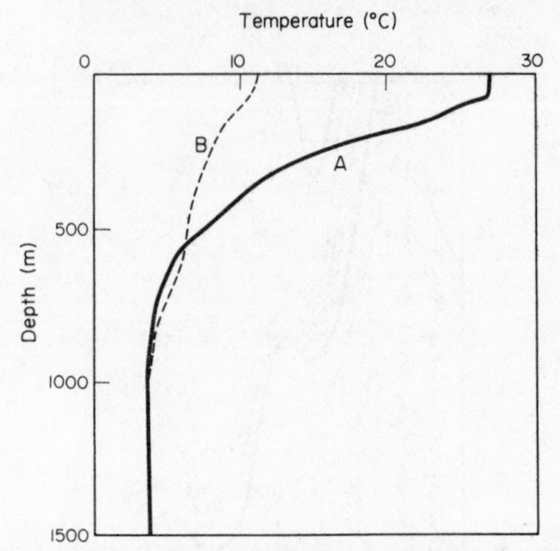

Fig. 1.10. Variation of temperature with depth: (A) *Discovery* II Stat. 684. 15°38′S, 29°50′W (South Atlantic). (B) William Scoresby Stat. WS 438. 39°18′S, 1°59′E (South Atlantic). (Data from *Discovery Reports*, Vol. 4, 1932.)

The distributions of the surface temperature over the oceans for both summer and winter are shown in Fig. 1.12. Particularly noticeable is the latitudinal arrangement of the isotherms and their displacement between February and August. The isotherms are closest together in the area of the Sub-Polar Convergence at about 50°N and S of the equator. The warmest ocean, apart from the Red Sea and Arabian Gulf, is the western Pacific Ocean. The annual temperature range is greatest off the eastern seaboards of the northern continents, exceeding 15°C in the Yellow Sea.

Water Budget of the Ocean

As is shown in Fig. 1.4, an average of about 100 cm/yr of water is evaporated from the oceans. However, this moisture is not held for long, since the moisture in the atmosphere averages only 10 days supply of rainfall. The critical factor in evaporation is the dryness of the air overlying the water, and evaporation can take place by night and by day. The centres of highest evaporation are off the east coasts of the main continents in high latitudes, and off the west coasts in the tropics where the dry winds blow off the continents.

Over the open oceans there is an overall excess of evaporation over precipitation, whereas over the land rainfall is 50% in excess of evaporation. There is obviously a balance with river run-off providing the link. This is shown diagrammatically in Fig. 1.13. As most of the precipitation occurs at high latitudes, particularly in the northern hemisphere, and most evaporation in the tropics, the local difference between precipitation and evaporation will determine the surface salinity (Fig. 1.14). Between 50°N and 50°S of the equator the comparison is most striking.

The evaporation causes a cooling and an increase in salinity. Consequently, unless there is rapid heating at the same time, there is an increase in density and a fairly rapid mixing. As a result the diurnal variation of salinity is small and as large a variation can occur through local precipitation.

The annual variation is largest in the coastal areas near river mouths and in the polar regions where the freezing and melting of ice can cause a 25% variation in the thin

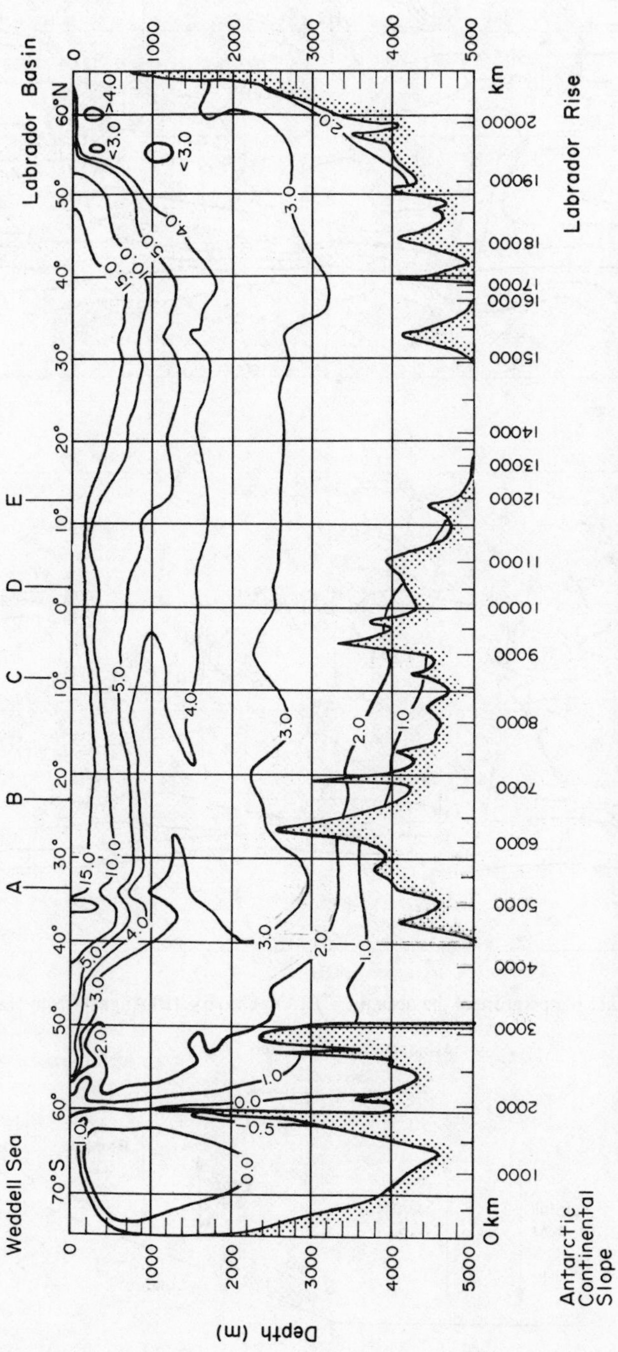

Fig. 1.11. Longitudinal temperature (°C) section through the western trough of the Atlantic (from Wust, 1930).

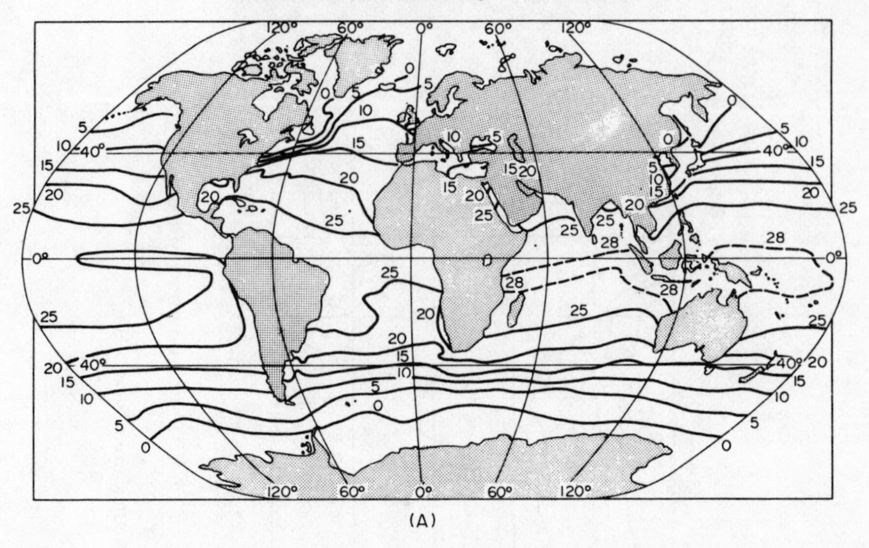

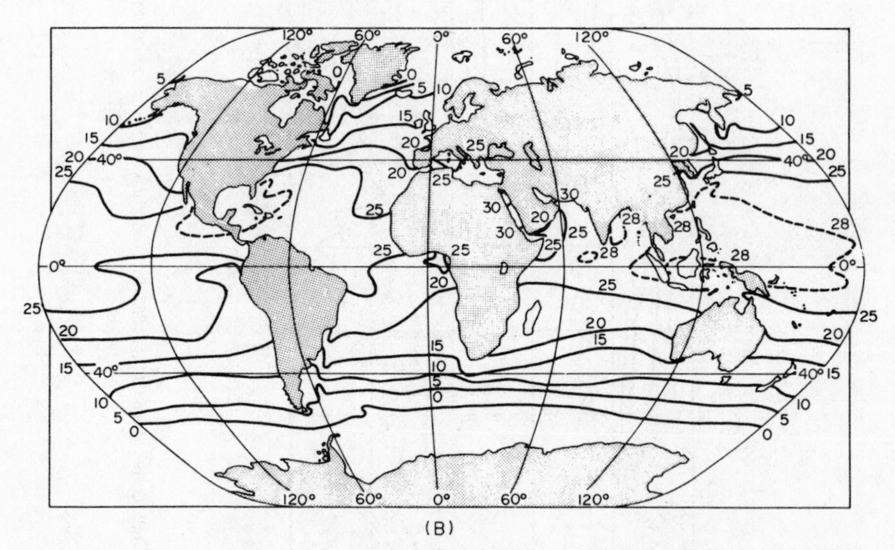

Fig. 1.12. Surface temperature of the oceans (°C): (A) February. (B) August (from Harvey, 1976).

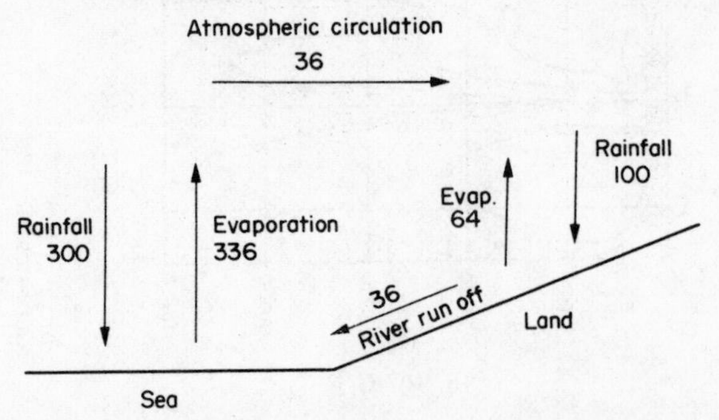

Fig. 1.13. Diagrammatic water budget of the ocean. Units 10^{15} kg/yr (after Harvey, 1976).

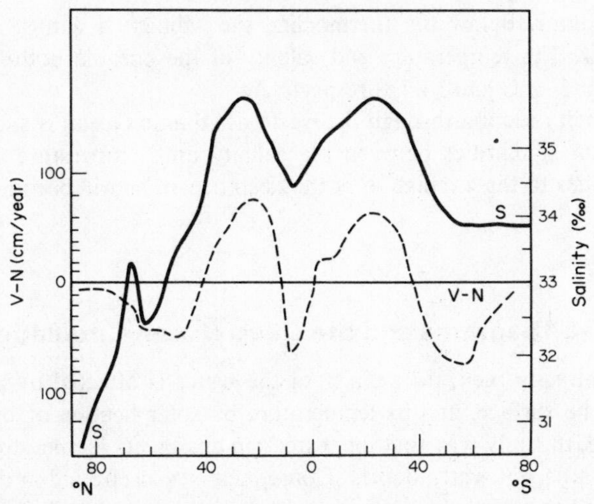

Fig. 1.14. The relation between surface salinity in the Atlantic Ocean and the excess of evaporation over precipitation (V–N) (from Wust *et al.*, 1954).

surface layer. In the open ocean the annual variations are generally less than 1%.

The surface salinity of the oceans is shown in Fig. 1.15. The sub-tropical highs are quite distinctive, with lower salinities at the poles and near to land. The Atlantic is the most saline of the major oceans, because of the high evaporation under the dry winds blowing off the continents, and the easy dispersal of moisture over the continents. Additionally, there is a large area of monsoon rainfall which lowers the salinity in the Indian Ocean and western Pacific.

In the open ocean an increase in salinity with depth is not essential for stability as the temperature generally decreases fast enough for density to increase with depth. The highest salinity values normally occur in the surface layers and decrease with depth. However, in sub-tropical areas there is often a salinity maximum associated with the top of the thermocline, where high-salinity water is brought in laterally by currents flowing

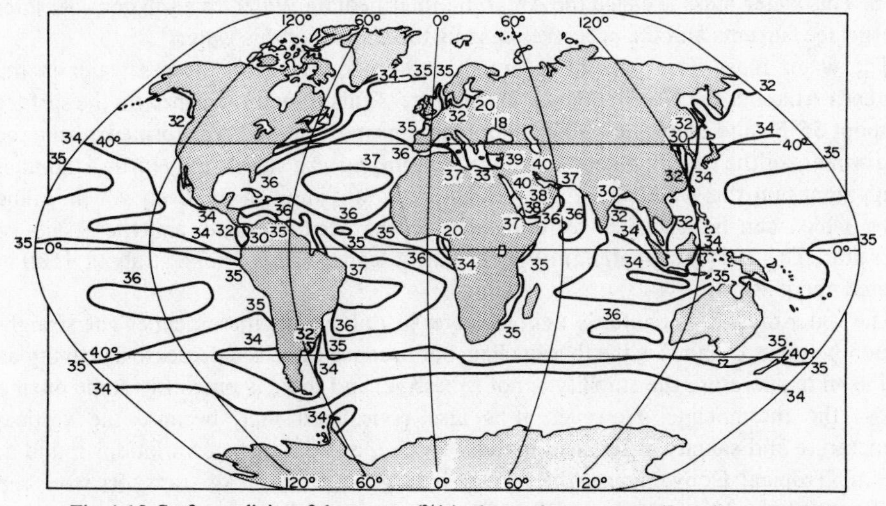

Fig. 1.15. Surface salinity of the oceans (‰) in the northern summer (from Harvey, 1976).

towards the equator. Below the thermocline the salinity is almost entirely between 34.5 and 35.0‰. The temperature and salinity of the oceanic bottom water is fairly constant at about 1–2°C and 34.7‰ respectively.

A vertical salinity section through the western Atlantic Ocean is shown in Fig. 1.16. There are certain similarities between the salinity and temperature distribution (Fig. 1.11) that are clues to the circulation of the deep ocean, as will become apparent in the next section.

T-S Diagrams and the Deep Ocean Circulation

As we have already seen, the salinity of the water is affected by precipitation and evaporation at the surface, and its temperature by solar heating or by cooling. Away from the surface, the only way that the water can change its temperature and salinity is by mixing with adjacent water bodies. Consequently a useful aid in distinguishing the various water bodies and studying their sources and mixing is the T-S diagram. This has the added attraction of indicating the density structure of the ocean as well, showing those depths where there is a large or a small density gradient.

On a T-S diagram a water mass with a unique temperature and salinity will be represented as a point. Mixing between two water masses will take place on the straight line joining them. When there are three water masses, then the middle one will mix with that above and below. This process is illustrated in Fig. 1.17. The T-S curve (a), taken at about 33°S in the western South Atlantic, shows three straight-line segments, the possible result of mixing of four basic water masses. Looking at the sequence of curves (a) to (e) shows that mixing of the water mass Z with that of T causes an increase in both salinity and temperature towards the north. The characteristics of the overlying water mass U and the deeper layer B change little over the distance. The changes indicate that the source of Z is in the southern hemisphere and that of T in the northern. Comparison with Figs. 1.11 and 1.16 shows that the water mass Z is a tongue of low salinity cold water penetrating below the thermocline from the sea surface at about 55°S. This water mass is called the Antarctic Intermediate Water. We will consider later the surface currents and the processes causing the sinking of this water.

The water mass T is part of a large volume occupying the deeper water in the northern Atlantic; the North-Atlantic Deep Water. This is mainly formed on the surface at about 55°N, and there is evidence that a large proportion of it was formed during the cold winters of the early nineteenth century. On the eastern side of the North Atlantic it is apparent that this water mass receives an additional input of relatively warm, saline water which can be traced to an outflow from the Mediterranean and the Straits of Gibraltar. This is even apparent in Fig. 1.16 as a higher salinity zone at about 1500 m located about 30°N.

The water mass U is the fairly uniform layer just above the thermocline. The straight section between U and Z is the thermocline, but because there is a decrease in salinity as well as in temperature the stability is not excessive, and there is obviously some mixing across the thermocline. However, it is also considered that, because the vertical temperature and salinity differences match so well the zonal surface variations found at the Sub-Tropical Convergence, the former is largely the result of transport from the surface at about 30° to 40°N and S of the equator towards the equator along lines of

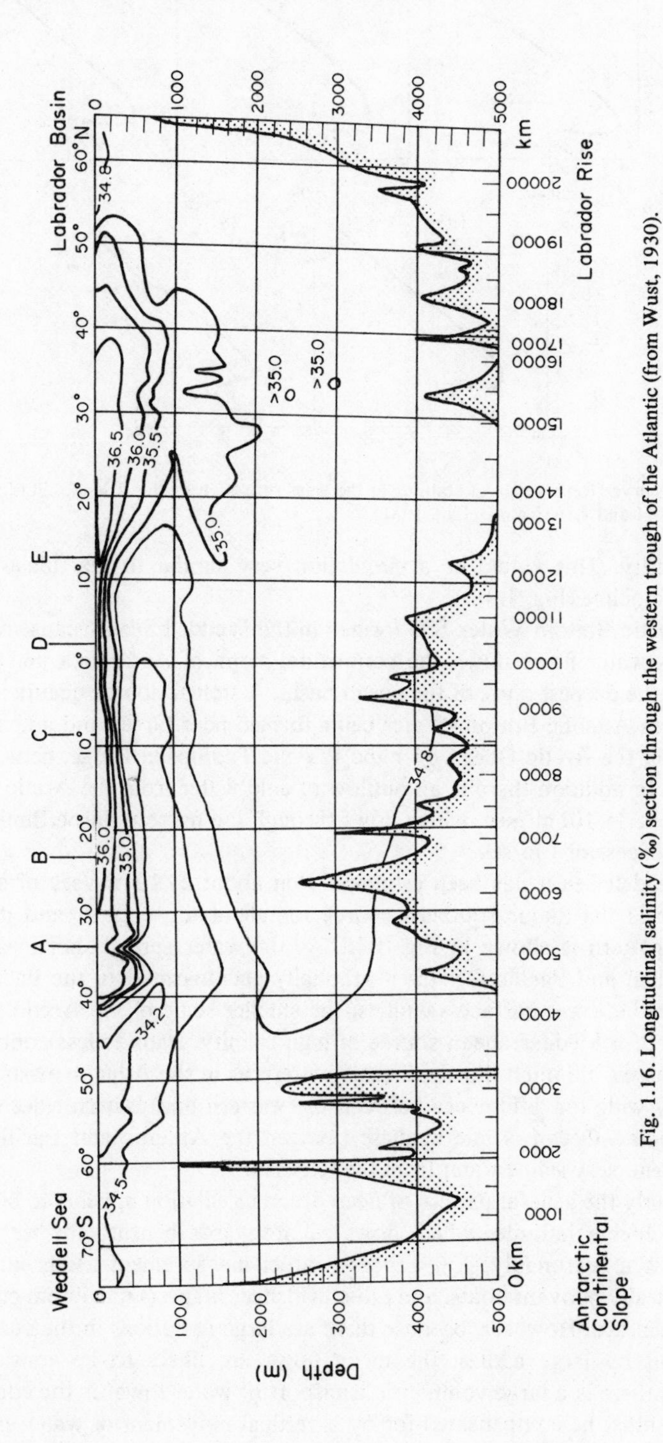

Fig. 1.16. Longitudinal salinity (‰) section through the western trough of the Atlantic (from Wust, 1930).

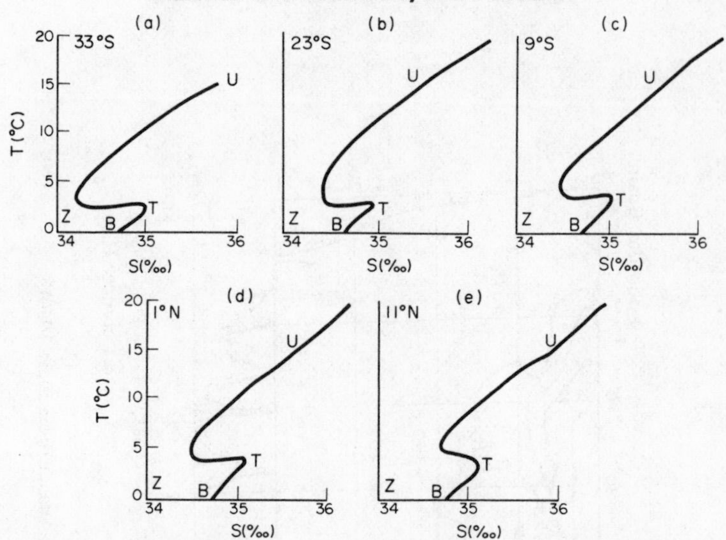

Fig. 1.17. T-S curves for a series of stations in the western trough of the Atlantic. Station positions are shown in Fig. 1.11 and 1.16 (from Defant, 1961).

constant density. This would be a circulation very similar to that illustrated for the summer thermocline (Fig. 1.9).

The Antarctic Bottom Water B is formed in the Weddell Sea. Because of its extreme coldness this water flows down the continental slope of Antarctica and flows northwards along the deepest parts of the ocean basin. A similar source occurs in the Arctic, the cold North Atlantic Bottom Water being formed near Greenland and spilling intermittently from the Arctic Ocean over the Wyville-Thompson Ridge, between Scotland and Iceland. In addition there is an outflow of cold water from the Arctic Ocean, contributing about 1×10^6 m^3/sec, which flows through the narrow Faroe Bank Channel at a velocity in excess of 1 m/sec.

In the Weddell Sea it has been estimated that about 2×10^7 m^3/sec of deep water is formed. This is the major southern source for all of the oceans, and its postulated distribution pattern is shown in Fig. 1.18. As the water spreads northwards into the Atlantic, Indian and Pacific Oceans it gradually gets warmer. In the Pacific, this fact coupled with the lower surface salinities, the smaller scale of the Arctic contribution, and the lack of a Mediterranean source of high salinity, leads to less contrast between the water masses, although the same basic pattern as in the Atlantic exists. The Pacific is sufficiently wide for differences between the western and eastern sides to be noticeable. The Indian Ocean is intermediate between the Atlantic and Pacific and has a source of warm, very saline water from the Red Sea.

Consequently the general pattern of deep ocean circulation appears to be a sinking of water in the higher latitudes which flows equatorwards beneath the warmer tropical water. This is a reasonably slow process; direct measurement using current meters, tracking neutrally buoyant floats, and other evidence, suggests maximum currents of the order of 20 cm/sec. However, because there are large variations in the currents caused by tides, and by large eddies, the mean flows are likely to be considerably less. Nevertheless there is a large volumetric transport of water towards the equator in deep water which must be compensated for by a vertical movement of water exchanged by mixing across the thermocline. It has been estimated that the upward movement averages 4×10^{-5} cm/sec over the whole area of the thermocline, and this water must then be discharged polewards in the warmer surface currents, with only about 10% of this volume being lost by evaporation and transported towards the poles in the atmosphere.

Surface Ocean Circulation

The circulation of the surface of the ocean is fairly closely coupled to that of the atmosphere. They both interact on each other by exchanging heat, moisture and momentum. Both transport heat towards the poles. It has been estimated that the Gulf Stream would take 3 years to decay in the absence of winds. This is illustrative of the fact that, in general, it is the mean atmospheric circulation that affects the ocean currents rather than the day-to-day situation.

When the wind blows across an open ocean far from any boundaries, the resulting movement of the water will be greatest at the surface and will decrease in magnitude with depth. However, because of the rotation of the earth a deflecting force affects the movement which causes the water to move *cum sole* towards the right in the northern hemisphere and towards the left in the southern. This force is the Coriolis Force and its magnitude depends on latitude, being zero at the equator and maximum at the poles. The resultant of the wind stress on the surface and the Coriolis Force causes the surface water to move at 45° (*cum sole*) to the wind direction, this angle increasing gradually, and the current velocities diminishing, with increasing depth below the surface (Fig. 1.19). At a depth called the Ekman Depth the current is opposite to that on the surface and is very small. The Ekman Depth varies with latitude and with the character of the turbulent mixing. This spiral of current beneath the wind is called the Ekman Spiral and

Fig. 1.18. Model of the circulation of deep water from sources in the North Atlantic and Weddell Sea (from Stommel, 1957).

it has the property that the depth integrated transport of the water (the Ekman transport) is at right angles to the wind direction. This is obviously of great importance in considering the wind-driven ocean circulation. It is only when the movement of water is restricted by land that it coincides with the direction of the wind.

In the atmosphere there is a mean transfer of heat towards the poles. In a simplified system there would be cold dense air giving high pressure at the poles and warm less dense air at the equator giving low pressure. The cold air would flow near the ground towards the equator where it would rise on being warmed. The warm air would then return towards the poles at a higher level. Because of the rotation of the earth the moving air would be deflected so that the surface wind would have an easterly component. In a single-cell circulation this would lead to a nett frictional force on the earth's surface tending to reduce the speed of the earth's rotation. The diagrammatic representation of the actual pressure distribution and the wind system is shown in Fig. 1.20. One consequence of this wind system is that it gives equal westward and eastward components of the frictional force. The high-pressure areas at the tropics cause clear skies, high temperatures and high evaporation rates. The low-pressure area at the equator results in more cloud and heavy convection rain. The boundary between the warm westerly airstream and the polar masses at about 55°N and S is particularly unstable with the development of large eddies (depressions) which travel along the polar front. Warm air is moved polewards in front of the depressions and cold air moved equatorwards behind them, providing a good north–south exchange of heat. As we shall see later, similar processes act in the ocean. There are also seasonal differences in the pressure distributions over the continents which cause the monsoons in Asia.

We can now consider how the surface waters react to this mean atmospheric circulation. As can be seen in Fig. 1.20, because of Ekman transport there will obviously be a tendency for water to build up at the sub-tropical high-pressure areas at the expense of the low-pressure areas. Also there will be a component of water movement towards the east beneath the westerlies and towards the west beneath the trade winds. The con-

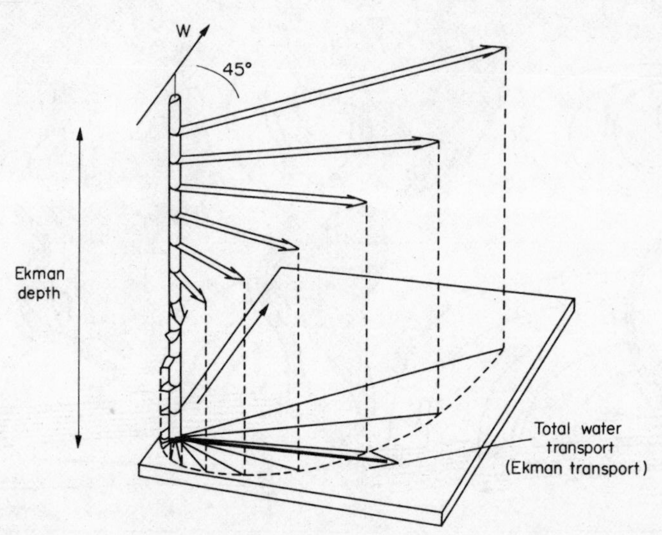

Fig. 1.19. The Ekman Spiral showing the wind direction (W) and the direction and speed of currents from the surface downwards, for the northern hemisphere (after Sverdrup *et al.*, 1942).

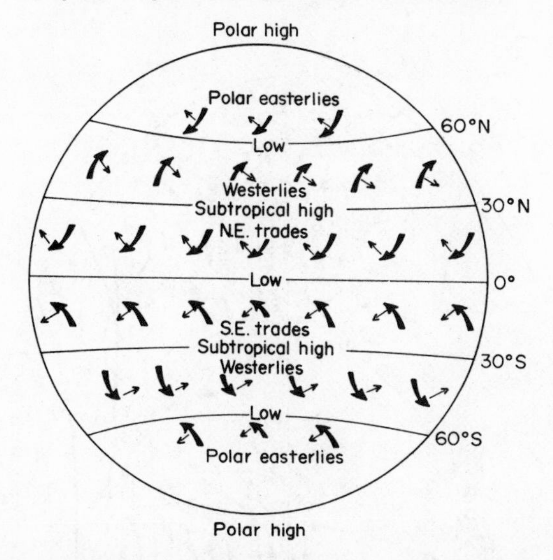

Fig. 1.20. Schematic diagram of zonal atmospheric pressure system and winds near the earth's surface. Thick arrows wind systems. Thin arrows wind-driven water movement.

tinents have a more or less north–south orientation and consequently water would build up on the western side of the oceans at the equator and on the eastern side at high latitudes. These effects give a couple, or a vorticity, which will tend to create gyres in areas between the equator and about 50°N and S of the equator.

In the northern hemisphere these gyres would be in a clockwise sense. Because of water movements, the Coriolis Force acts towards the centre of the gyres. This is balanced by an outward density-induced force produced by the presence of high temperature, high-salinity water in the centre of the gyre. The balance between the density distribution and the Coriolis Force is called the Geostrophic Balance and describes fairly well the main features of the flow. Consequently from knowledge of the density distribution the currents can be calculated reasonably well, despite the approximations involved. One assumption made in this balance concerns the force created by the sloping water surface. Because the waters in the centre of the gyre are relatively light they give a sea surface gradient outwards from the centre of the ocean. The sea level difference between the centre of the North Atlantic and the coasts, for instance, is about 1 m. Even though the actual sea surface gradient is small, it helps to balance the Coriolis Force created by the water motion. Thus the circulation of the surface layer of the sub-tropical ocean is the result of the wind stress being applied over the whole of that part of the ocean and is balanced by the zonal variations in sea surface density (mainly the result of temperature). Though the circulation is balanced over the whole ocean, locally the wind and the currents may not be balanced so that there is often a component of transport radial to the gyre. This leads to convergences and divergences, downwelling and upwelling of water, respectively.

The currents at the ocean surface are summarized in Fig. 1.21. The gyres are quite distinctive in all of the sub-tropical areas except the northern Indian Ocean where the monsoons cause seasonal variation in the current patterns. Near the equator where there are comparatively strong westward currents a considerable build-up of water occurs on the western side of the oceans. This leads to a narrow and shallow counter-

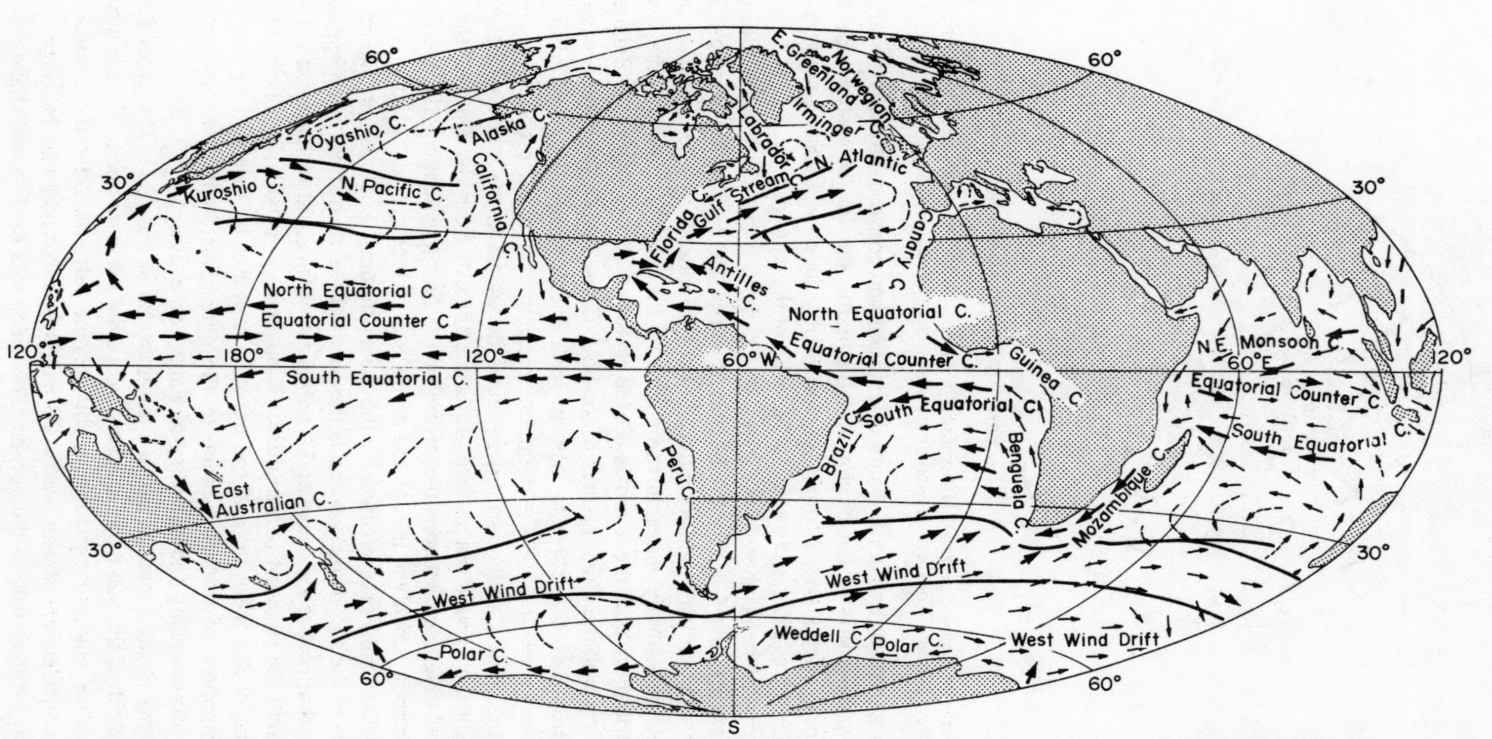

Fig. 1.21. Ocean surface currents in the northern winter with sub-tropical and sub-polar convergences (after Harvey, 1976, with additions).

current flowing eastwards. Coriolis Force in this area is small so that there are no forces to deflect the current.

Over the sub-tropical ocean the water acquires a clockwise rotation tendency in the northern hemisphere due to the wind system. Coriolis Force also produces a clockwise rotation tendency in the same area, but this varies with latitude. In order to prevent the rotation increasing indefinitely some braking by friction is required. As the friction is proportional to the square of the current speed the balance is achieved by having the circulation within the gyres asymmetrical and having faster currents on the west sides of the oceans than on the east sides. Thus the Gulf Stream is of higher velocity than the Canaries Current and the areas of highest current velocity do not necessarily coincide with the areas of highest winds. Where the warmer central portions of the gyres meet the eastward flowing limbs at 30° to 40° latitude, the Sub-Tropical Convergences are formed. These are not clear-cut boundaries, often covering several degrees of latitude, and varying in position seasonally. The Sub-Polar Convergences which are formed where the warmer water meets the cold polar waters are generally better defined, particularly in the southern hemisphere. Around the Antarctic continent the predominant easterly winds lead to a westward water circulation. This leads to a build up of cold water against the Antarctic Peninsula particularly in the winter and the formation of the Antarctic Bottom Water. In the North Atlantic and North Pacific there are also gyres formed by the interaction of the westerlies and the polar easterly winds which rotate in the opposite sense to the sub-tropical gyres.

It is instructive to look at a number of particular areas in rather more detail.

The Gulf Stream

The Gulf Stream is initially the edge of the higher temperature, higher salinity, yet lower density, sub-tropical water mass and does not become a wind drift until it has progressed further into the Atlantic, and into the area of maximum westerly winds, in the region 40° to 50°N. It is formed from the Florida Current and the Antilles Current which join north of the Bahama Islands. The former is the result of the North Equatorial Current flowing into the Caribbean and the Gulf of Mexico. The Antilles Current is the arm of the Equatorial Current which flows northwards outside the West Indies. The combined transport off Chesapeake Bay is in the region of 8×10^7 m^3/sec. The temperature and salinity sections in Fig. 1.22 show that the landward side of the Gulf Stream, where there is a southward-flowing colder coastal current, is fairly sharply defined both in terms of temperature and salinity. In contrast, there is little difference between the Gulf Stream and the oceanic water to the east. Also it is only the top few hundred metres that form the Gulf Stream. From synoptic oceanographic surveys and from recent remote sensing surveys it is apparent that the Gulf Stream is far from a continuous stream in a fixed path. The current meanders over quite a wide area, occupying at any time only one-third of a 150-km-wide zone. The meanders sometimes become detached and large masses of cold water become enveloped in the warmer sub-tropical water. These structures are known as "cold core rings". They have a cyclonic, anticlockwise rotation and appear mainly to form west of about 60°W. Anything between 8 and 14 can be visible at any time and they wander slowly south-westwards, at about 2 km/day, remaining recognizable for up to 2 years. Their paths are very variable, some rejoin the Gulf Stream off Florida, while others disappear into the Sargasso Sea

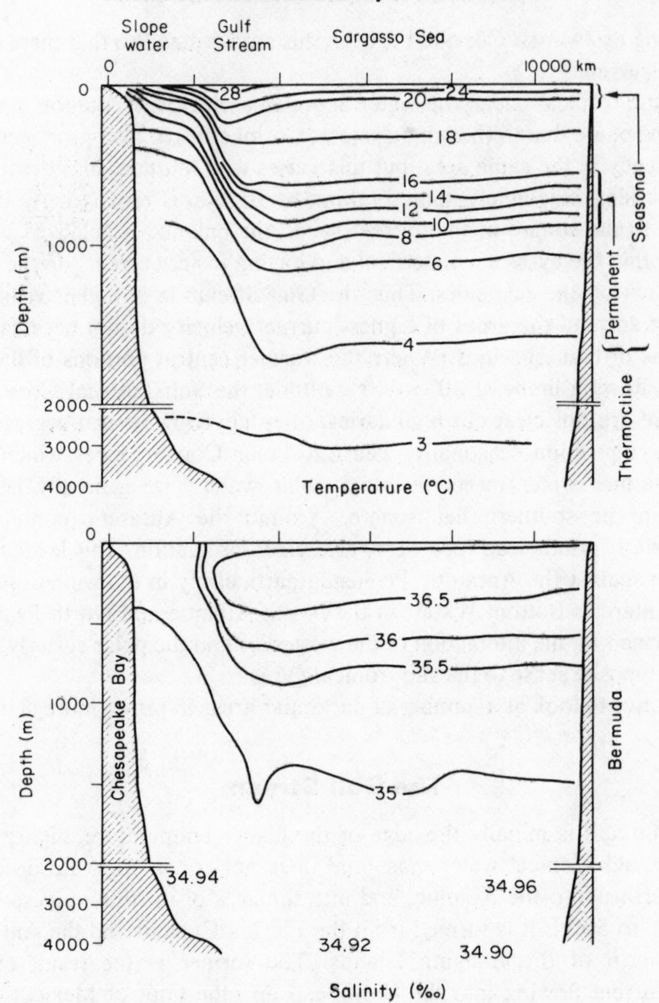

Fig. 1.22. Temperature and salinity sections across the Gulf Stream from Chesapeake Bay to Bermuda, August/September 1932. (According to Iselin, Pickard, 1963.)

(Richardson, 1976). The rings are 150 to 300 km in diameter, have peak tangential velocities of 100 cm/sec, and reach to depths of 3000 m. Consequently they raise the thermocline by 500 m or so and have considerable potential as well as kinetic energy. The rings have a biomass considerably higher than that in the surrounding sea, and the biological contrast is especially marked in the distribution of slope forms such as *Nematoscelis megalops* (Wiebe and Boyd, 1978).

Anticyclonic or warm core rings also form on the coastal side of the Gulf Stream. They are believed to form as frequently as the cold ones, but only last up to 6 months because of the restricted area. Most appear to rejoin the Gulf Stream near Cape Hatteras.

The meanders are generated by a general instability in the current and may also be triggered by pulsations in the strength of the Florida Current. Some of these pulsations are related to changes in the strength of the trade winds, others to the weather effects within the Gulf of Mexico. A useful account of the Gulf Stream's characteristics has been written by Stommel (1958).

North Pacific Ocean

It is apparent from Fig. 1.21 that there are major similarities between the circulation of the North Pacific and the North Atlantic. The surface distribution of salinity (Fig. 1.15) and temperature (Fig. 1.12) have already been commented on. The centre of the region is occupied by high salinity, warm water being driven in a clockwise gyre by the trade winds at low latitudes and the westerlies at high latitudes. On the western side, the Kuroshio Current has a transport of between $6–7 \times 10^7$ m^3/sec, with maximum velocities of over 150 cm/sec, very similar to the Gulf Stream. This current keeps close to the continental shelf of Japan until about 36°N, 140°E; then it flows east in a remarkable meandering path up to about 160°E. The meanders may well be associated with the topographic changes of the seabed of the South Honshu Ridge and the Japan Trench. Further east, the current is transformed into the weak, broad North Pacific Current which undergoes many seasonal and spatial variations due to mixing and incursion of cold sub-arctic water across the normal boundaries of the warm sub-tropical gyre. This effect is shown in Fig. 1.23, a temperature section across the North Pacific at 35°N. In the east, the California Current is the relatively cold southward-flowing current which, coupled with offshore winds, produces upwelling of colder water to the surface and high nutrient and plankton levels; there is a roughly inverse relationship of zooplankton volume to temperature in the California Current.

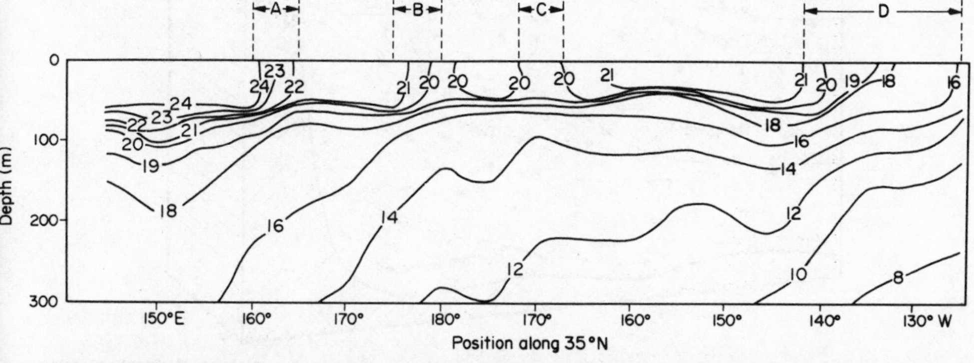

Fig. 1.23. Distribution of temperature (°C) on a section at 35°N across the North Pacific Ocean. A, B, C, D mark the zones of influx of cold subarctic water (from Wong *et al.*, 1974).

To the north there is a cyclonic gyre, with the Alaska Current circulating in the Gulf of Alaska and a current flowing westwards outside the Aleutian Islands. Between this current and the North Pacific Current there is consequently a divergence of surface waters bringing cold deeper water to the sea surface. There is a complex and variable system of currents between the Aleutian Islands, possibly seasonally related to the presence of ice in the eastern Bering Sea in winter. In the western Bering Sea the currents are also complex, but with a tendency to a further anticlockwise gyre. The western side of this gyre, the Kamchatka Current, which continues southward as the Oyashio Current, is strongest in winter, and varies seasonally by a factor of 3. These variations obviously will affect the location and extent of the Sub-Polar Convergence in the North Pacific.

The pattern of circulation observed at the surface appears to extend downwards to depths as great as 3000 m, with only moderate modification. At increased depths the sub-tropical anticyclonic gyre retreats polewards and an easterly flow occurs in the lower latitudes.

The Antarctic Convergence

The Antarctic Convergence in the South Atlantic is found at about 50°S and the Sub-Tropical Convergence at about 40°S. The former is fairly well marked by a rise in the sea surface temperature towards the north and it varies in position by only about a degree of latitude during the year. The Sub-Tropical Convergence is more variable in its position and more diffuse. As can be seen from Fig. 1.24, there is a rise in the sea surface by about a metre across the convergence and a change in water density. This is balanced by the Coriolis Force associated with the strong West Wind Drift that flows around the Antarctic continent between 40° and 65°S. The strongest winds, however, occur in the zone between 50° and 60°S and this results in a small northwards component in the flow in these latitudes. Thus there is a convergence of water at the

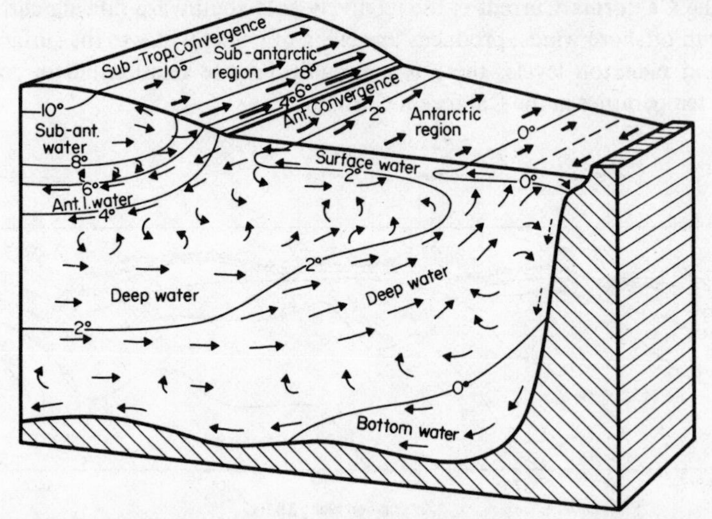

Fig. 1.24. Schematic representation of the currents, temperatures and water masses of the Antarctic regions with the Antarctic and Sub-Tropical Convergences (from Sverdrup *et al.*, 1942).

northern limit of this zone. The deep water to the south which has originated in the North Atlantic, rises towards the surface because of the buoyancy created by the mixing with cold but less saline water below and the warmer more saline water above. Near the surface it is cooled slightly and mixed with less saline water formed by the melting of the ice. As it is still slightly less dense than the water beneath, it stays at the surface but gradually moves northwards. In flowing northwards this surface water comes up against the less dense warmer, but more saline, water towards the north and flows beneath it as the Antarctic Intermediate Water. In the area of mixing with the surface water the currents are complex. The differences in the currents over the top

100 m or so, coupled with the vertical migration of plankton, have been used to explain the distribution of krill near South Georgia (cf. Hardy, 1967). The presence of the Convergence is the result of a complex inter-relationship between zonal wind variation, surface cooling and alterations of salinity due to melting of ice.

Upwelling

The major areas of upwelling occur where the current and wind are in the same direction, parallel to a coast, lying on the left-hand side in the northern hemisphere and the right-hand side in the southern. In a sense the upwelling of deep water just described south of the Antarctic Convergence is an example. However, the Canaries Current is a more typical one. In that case the Canaries Current flows southwards in the area of the North-east Trade Winds, with the African coast on the left-hand side. The Coriolis Force associated with this current balances the effect of the sea surface slope and the density gradients. The effect of the Ekman transport beneath the wind, however, is to move the surface water towards the right. Because of the presence of the coast this water is then replaced by water which has the same characteristics as that occurring at 100 or 200 metres depth further offshore, by a slow overturning (Fig. 1.25). These waters are relatively nutrient-rich and, coming into the photic zone, are highly productive. Upwelling areas thus support active fisheries. However, the upwelling water is

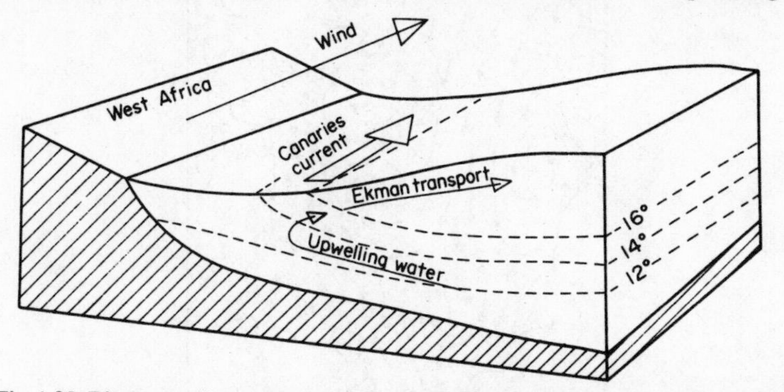

Fig. 1.25. Diagrammatic representation of upwelling produced by the Canaries Current.

relatively cold, resulting in frequent fogs, particularly famous in California. The upwelling is not a continuous phenomenon, varying in its occurrence locally and in some cases seasonally with drastic consequences on the fisheries. This is well documented in the area of the Peru Current (Wyrtki, 1966) where the anchovy fishery shows large fluctuations (Longhurst, 1971). Seasonal upwelling also occurs in the monsoon regions. The summer monsoon blowing towards India creates upwelling off the coast of Somalia, whereas in winter upwelling can occur off the western coast of India under the offshore north-easterly winds.

Coastal Areas: The North Sea

Each coastal area has its own peculiar characteristics which are related to the nearby ocean currents and water masses, the local weather, tides and river run-off. The North Sea is unusual in that water enters both from the Atlantic in the north and the English

Channel in the south. The general circulation throughout the year is from the north of Scotland southwards down the east coast of Britain. Water flowing through the Dover Strait flows north-eastwards into the German Bight. The merging of the two streams leads to a series of anticlockwise gyres in the central North Sea. After joining water flowing out of the Baltic Sea, the mixture flows northwards along the Norwegian coast (Fig. 1.26). Thus the residual circulation is a large anticlockwise one. There are seasonal variations in the circulation: the maximum inflow of North Atlantic water is between September and February with little occurring during spring and summer. The magnitudes of the residual currents are also affected by the local weather, particularly in the central part, and the flow through the Dover Strait has been related to wind direction and barometric pressure differences between the English Channel and North Sea. Because of the density gradients and the wind stress at the surface, the residual flows near the bottom may not show the same patterns as at the surface. The residual flows are those remaining after averaging out quite large tidal currents. In the northern North Sea the maximum tidal currents average about 0.25 m/sec, but this rises gradually towards the south and towards the headlands and river mouths. In the Dover Strait the velocities reach 1.5 m/sec.

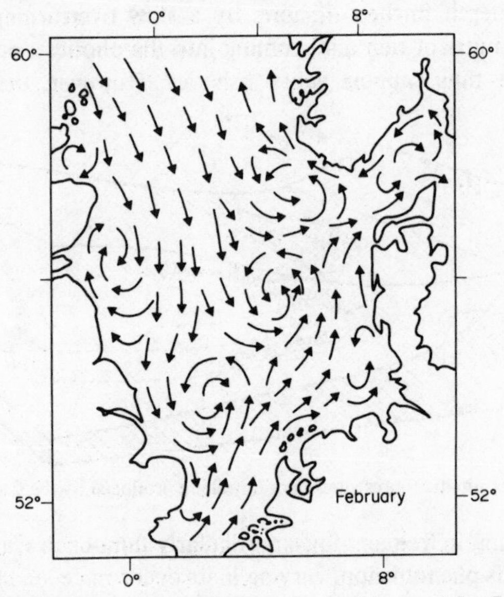

Fig. 1.26. Surface residual currents in the North Sea in February (after Bohnecke, 1922).

The temperature and salinity distributions reflect the interaction of the oceanic and the river water. Water of high salinity enters the North Sea between the Shetlands and Norway and through the Dover Strait. The river run-off produces low salinity water in a fairly narrow coastal zone. The salinity variations in this zone are related to the seasonal fluctuations in run-off. Particularly noticeable (Fig. 1.27) is the lower salinity close to the Norwegian coast. This is due to the summer outflow from the Baltic and from the Scandinavian mountains as a result of snow melting. Longer term fluctuations in salinity have been detected in the North Sea and these have been correlated with periodic changes in the rainfall and mixing processes in the North Atlantic.

The temperature distributions (Fig. 1.28) show in winter a general decrease in temperature towards the coast, while in summer the temperature increases towards the

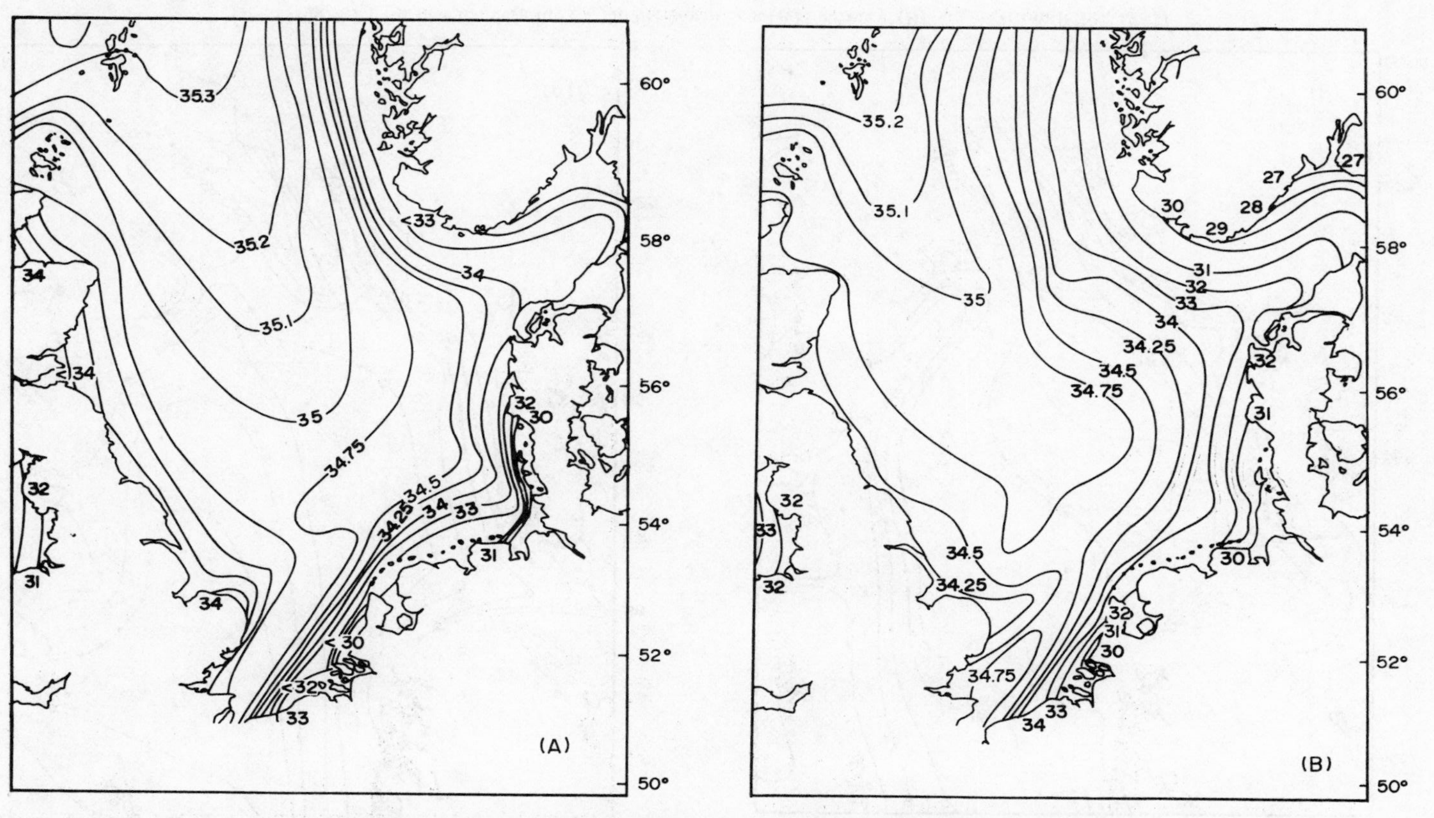

Fig. 1.27. Surface salinity (‰) in the North Sea: (A) February. (B) August (from ICES, 1962).

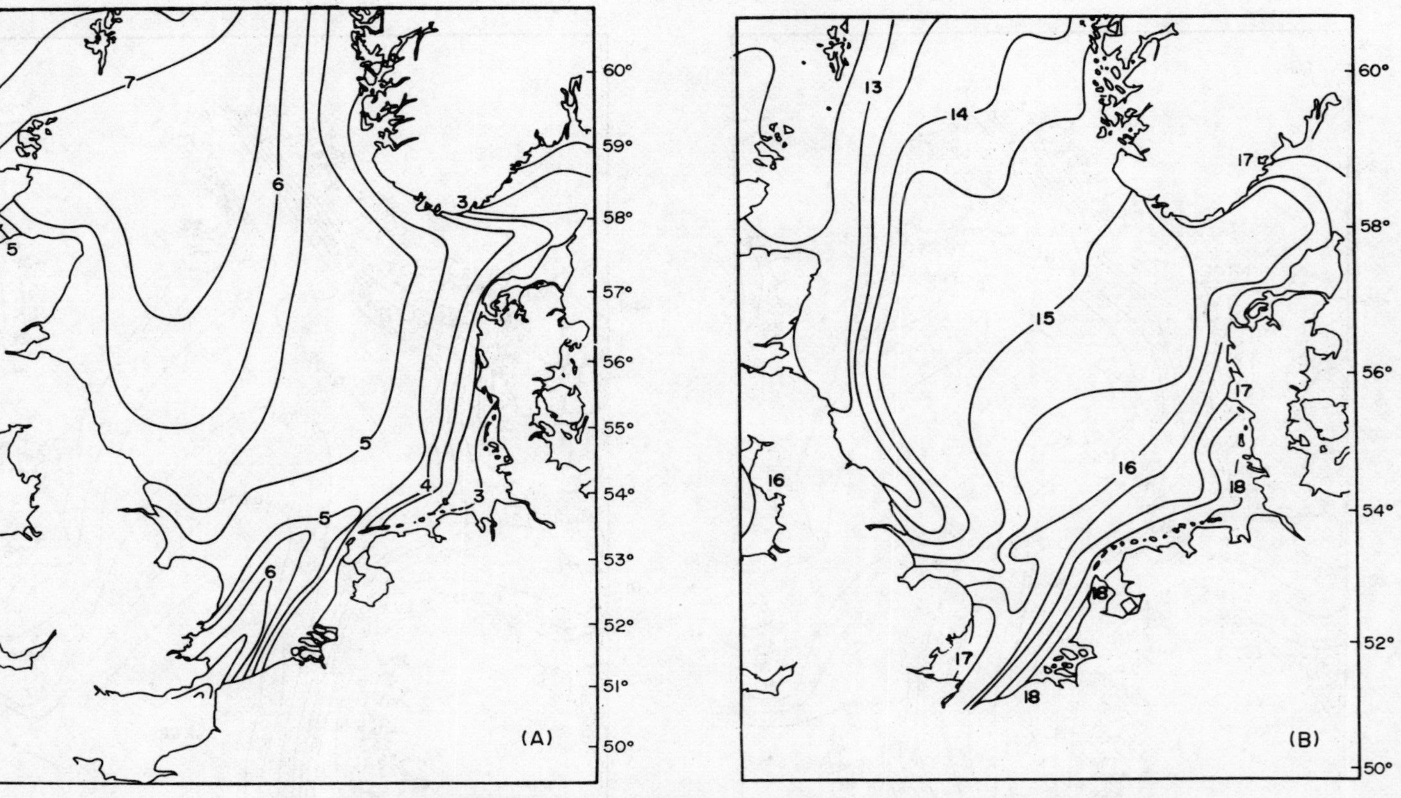

Fig. 1.28. Surface temperature (°C) in the North Sea: (A) February. (B) August (from ICES, 1962).

coast. This reflects the temperature of the river run-off. The temperature of the water in the central North Sea is related to those prevailing in the North Atlantic and in the English Channel. In February the temperature distribution shows well the penetration of the North Atlantic water down the British coast and of English Channel water along the continental coast. In the summer the central and northern North Sea is thermally stratified. Fluctuations in mean temperature have also been detected, with a minimum in the early 1920s and a maximum around about 1940. There appears to have been a gradual decline in temperatures during the 1950s. These variations are probably associated with climatic fluctuations. Lee (1970) has reviewed knowledge of the currents and watermasses of the North Sea.

Estuaries

In estuaries the river water interacts directly with the coastal water and there will be a strong longitudinal salinity gradient from the river to the sea (Fig. 1.29). Additionally, because of the density differences, there will be a tendency for the river water to flow out over the sea water. This leads to a vertical salinity gradient and the formation of a wedge-shaped salt intrusion on the bottom. Because of tidal motion the water mass moves up and down the estuary and the water motion creates turbulence which tends to mix the water column more thoroughly. A result of the interaction of the vertical stratification with the tidal flow is to cause a residual two-layer motion. This process is illustrated in Fig. 1.30. Near the sea bed the flood current is stronger than the ebb, and near the surface the ebb is stronger than the flood. This results in a residual downstream flow of fresher water on the surface which can be many times larger than the river discharge. There is a compensating inflow near the estuary bed. A classification series has been established for estuaries ranging from highly stratified to vertically

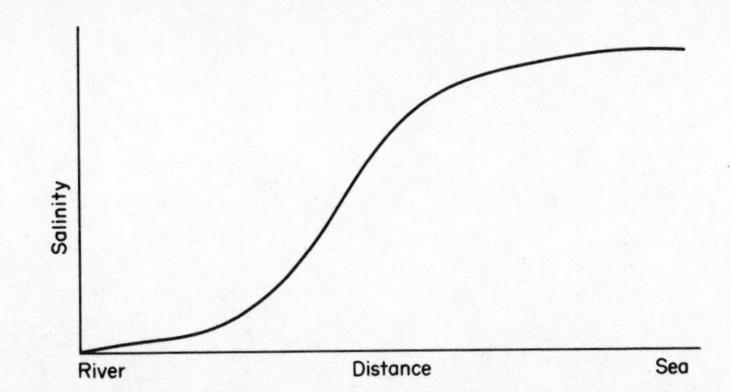

Fig. 1.29. Variation of tidal and depth mean salinity along an idealized estuary.

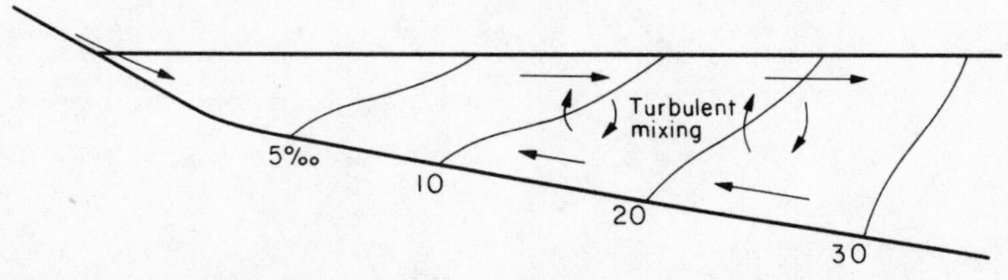

Fig. 1.30. Diagrammatic representation of the mean salinity distribution and circulation in a partially mixed estuary.

homogeneous, and related to river discharge and tidal flow. The understanding of estuarine circulation, however, is made more difficult by the fact that wind has been shown to be particularly effective in altering the residual water flow. Summaries of estuarine characteristics and circulation patterns are available in Dyer (1973) and Officer (1976).

Chapter 2

Chemical Composition and Characteristics of Seawater

The average composition of seawater reflects the composition of the various inputs to the ocean and the differences in the rates of removal of individual components from the ocean. Variations in composition within the ocean arise in part through the exchanges of material at the boundaries with other parts of the hydrosphere (such as rivers and waters of glacial origin), the atmosphere, the sea-bed and the continental lithosphere. These exchanges lead to localized gradients in concentration which may extend for hundreds of kilometres in the case of some river inputs or may be confined to a microlayer as in some atmospheric exchange processes. Inputs at the boundaries of the oceanic reservoir are not uniform. There are, for example, large differences in the composition of individual river waters. For some components, removal occurs on time scales which are short relative to the internal mixing time of the reservoir and thus variations in composition can be maintained between various major bodies of water within the ocean. A further important cause of spatial variation arises through the uptake and scavenging of material by organisms and detritus in surface waters, with at least partial release into bottom waters following the downward transport of material. This can lead to major concentration gradients when the flux of material is large relative to the reservoir concentration and the return of intermediate and deep water to the surface by physical mixing processes is relatively slow.

Biological processes have played a major role, on geological time-scales, in determining the chemical conditions which characterize the earth's surface. They also exert important influences on the composition of seawater on shorter time-scales, through their part in the transport of material in the ocean and its removal to sediments and their effects on the chemical forms (species) which are present. At the same time, chemical characteristics have an important influence on biological activity, exemplified to differing degrees by the contrast in productivity between oligotrophic and eutrophic waters and in life processes between oxic and anoxic waters. Those features of the present-day composition of seawater which show marked temporal and spatial variations are of particular interest from the marine biological standpoint, since they are often closely related to the cycles of organic production and decomposition and have immediate relevance in relation to productivity.

In the development of marine chemistry, much emphasis has been placed on understanding the processes by which the composition of the ocean is regulated. An approximate steady-state balance between input and removal processes, when considered on appropriate time-scales, may be assumed for most constituents. Some features of the composition of seawater resemble those predicted for an equilibrium system. From the standpoint of biological oceanography the geochemical control of the composition of seawater is of interest mainly in the context of the evolution and adaptation of organisms in the sea. It seems certain that the remarkable stability of the main compositional variables over long periods of time has been relevant to ecosystem development. There is no evidence that, excluding marginal regions, the main chemical characteristics of seawater have shown major evolutionary change over at least the past 6×10^8 yr (Mackenzie, 1975).

The emphasis of this chapter is on those chemical characteristics of seawater which are of specific biological significance and the broad features of their distribution in the ocean. A summary of processes of input and removal, and of the consequences of the time-scales of these processes in terms of compositional features, is followed by accounts of the main groups of constituents and of important properties related to composition. The account given is primarily for waters containing free oxygen; the main chemical differences between oxic and anoxic water bodies are outlined on p. 58. Fuller accounts, and information on aspects which lie beyond the immediate scope of this work, can be found in several recent works, particularly the comprehensive treatise edited by Riley and Skirrow (1975) and Riley and Chester (1976, 1978).

Processes of Supply and Removal of Constituents in the Ocean

Products of weathering of the continental lithosphere are carried to the ocean with freshwater drainage both in solution and as particulate material. Particles, from arid regions particularly, are carried also by wind (eolian) transport. Eroded material enters with glacial inputs. Particles from continental volcanic activity may also be deposited directly, or washed with run-off, into the ocean, but such activity is more significant in relation to volatile materials. A variety of surface processes, including outgassing from porous materials such as soils, also contribute volatile compounds. Such material may become associated with particles and aerosols in the atmosphere. While large particles are mainly deposited gravitationally, scavenging by precipitation is important for smaller particles which may be transported on a global scale.

Exchange of material across the ocean-atmosphere interface is the main route of entry for some important dissolved gases. For certain of the less abundant gases the ocean acts as a nett source for the atmosphere.

At the sea-bed, local injections of material to overlying waters occur in regions of volcanic and geothermal activity, and more extensive areas of oceanic crust are exposed to submarine weathering. Considerable exchanges of material between water and rocks of essentially basaltic composition can occur in some of these interactions which are now thought to be significant in the control of the concentrations of some constituents (for example, magnesium) in seawater. Other inputs of material at the sea-bed can arise through diffusional fluxes set up in the pore waters of bottom sediments as a result of post-depositional (diagenetic) changes.

The quantitative importance of all of these sources has not been well established. Goldberg (1971) has compared the global fluxes of riverborne and eolian material:

	Input (10^{14} g/yr)
Riverborne particulates	180
Riverborne dissolved material	40
Eolian transport of continental debris	1–5

A greater mass of riverborne material thus enters the ocean as solid phases than in a dissolved form. Much of this material, however, enters in a relatively small number of highly turbid rivers, particularly in south-eastern Asia, and in some systems, draining well-weathered terrains, the dissolved load predominates. Much of the lithogenous particulate material deposits in the coastal and shelf zones, but finer fractions undergo long-range transport and contribute to pelagic deposits. In abyssal and oceanic ridge regions, the importance of eolian inputs and local submarine sources becomes enhanced because of the low rates of accumulation of continentally derived solids.

Geochemical mass balance calculations (Horn and Adams, 1966) are consistent with the hypothesis that igneous rock weathering has supplied the total material present in the ocean, sediments and sedimentary rocks, in the case of most elements. An important group of elements, including chlorine and sulphur (the dominant anion-forming elements in seawater), bromine and boron are present, however, in amounts which cannot be explained on this basis. They may have been residual in a primitive ocean but the case was cogently argued by Rubey (1951) that these volatile elements have accreted, together with water and carbon dioxide, by release from the earth's interior, continuously over the ocean's history.

Dissolved material is removed from the ocean by a variety of sedimentary processes. Some becomes incorporated in pore waters but for the majority of constituents removal in solid phases is most important. Such phases may be formed by organisms (biogenous deposits), the most notable examples being the calcareous (calcitic or aragonitic) and siliceous (opaline) skeletal remains, which constitute, particularly the former, abundant pelagic sediments. They may also arise by inorganic processes, such authigenic phases being exemplified by the regionally abundant ferromanganese concretions. For many minor constituents the principal route of removal is through incorporation by lattice substitution in, or adsorption on, the major sedimentary components. In addition to sedimentary processes, submarine weathering processes, as indicated above, can also remove material by exchange reactions, while hydrothermal phases may scavenge some material from overlying waters.

The removal processes act differentially upon the input material, the rates of removal reflecting the geochemical reactivity of individual elements. Thus, much of the iron measured as a dissolved fraction in river water is present as a colloid which becomes destabilized in contact with seawater (Boyle, Edmond and Sholkovitz, 1977), so that where iron-rich rivers enter the sea, large-scale removal of the element occurs in the early stages of estuarine mixing. In contrast, currently operating sedimentary sinks for sulphate, adequate to balance the present-day natural river input, have not been clearly identified (Berner, 1972) and it is possible that the present geological period is one of temporary imbalance for this ion. For many elements estimates have been made of the mean reservoir life-times (residence times) of the elements, relative to river input or sedimentary removal, using a simple one-box model (Barth, 1952; Goldberg and Arrhenius, 1958). The magnitudes of the mean reservoir life-times range from high values (e.g. 10^8 yr for sodium, 10^7 yr for magnesium, potassium and lithium, 10^6 yr for calcium) to values which are formally as low as 10 to 100 yr for elements such as iron, aluminium and thorium which are dominantly associated with particulate material. From these figures it is apparent that most elements have been supplied to seawater in amounts greatly exceeding the content of the total oceanic reservoir, during the ocean's history.

The significance of these contrasting geochemical reactivities is apparent from a comparison of the residence times with the time-scales of renewal of the ocean water in the hydrologic cycle and the internal mixing of the ocean reservoir. The oceanic residence time of water relative to circulation in the cycle of evaporation and continental precipitation is about 4×10^4 yr. Internal mixing is comparatively rapid, the maximum residence time of deep water, relative to the surface mixed layer, not exceeding the order of 10^3 yr. Elements which are supplied in high concentrations and which tend to accumulate in seawater relative to the input composition can thus become well mixed. This is the basis for the observed uniformity of proportions of the major dissolved constituents. Geochemically reactive elements, by contrast, show greater variability in concentration. The major vertical internal flux, as discussed above, can lead to substantial concentration differences between surface, intermediate and deep waters. With elements of moderate or long residence time, the consequences of this flux depend upon its magnitude relative to the reservoir concentration. This is exemplified by the contrasts in the distribution of, for example, calcium and phosphate, which are discussed in subsequent sections.

Particulate and Dissolved Material in Seawater

Particulate material in seawater consists of a mixture of phases of diverse origins, principally lithogenous (e.g. clay minerals, quartz), biogenous (living and detrital organic and skeletal material) and inorganic authigenic (e.g. hydrous iron oxide), with additional components held by adsorption. The complexity of composition of solid phases in seawater can greatly influence chemical behaviour. For example, the solubility behaviour of particles of calcium carbonate or opal in seawater is greatly modified by the presence of organic coatings.

The amounts of suspended particulate material vary widely between different environments. In the Yellow River, which carries the largest sediment load of any river in the world, suspended material may account at times for as much as 40% by weight of the material transported (Todd and Eliassen, 1938, as cited by Holeman, 1968). The global average concentration of particulate material in rivers, estimated as some 500 mg/l, is considerably weighted by such turbid rivers. This value may be compared with a mean concentration of 50 mg/l in rivers entering the North Sea (McCave, 1973). In estuaries the concentration are highly variable, ranging from less than 1 mg/l to the range of fluid muds; concentrations of 10 to 100 mg/l appear to be most typical for temperate estuaries. The suspended loads in shallow estuaries are frequently influenced by extensive resuspension of bottom sediments. In near-shore waters, an average concentration of 1 mg/l has been reported, while the average concentration in open surface waters is of the order of 100 μg/l (Chester and Stoner, 1972).

On the basis of the GEOSECS survey, it appears that the concentrations of particulate material in surface open Atlantic Ocean waters show a pronounced latitudinal variation with minimal values in equatorial waters (Lal, 1977). The overall range is 10 to 600 μg/l. Waters below 200 m generally contain 10 to 20 μg/l. Surface-water concentrations in the Pacific Ocean are in a lower range than in the Atlantic. Low concentrations in deep waters are not necessarily characteristic of the entire water column. A significant feature is the occurrence of nepheloid layers, with higher concentrations in some regions; these layers may extend for some hundreds of metres above the bottom sediments.

In open-ocean environments the terrestrially derived material includes a significant fraction with particle sizes in the clay range, i.e. <2 μm, this material consisting of riverborne particles which have not been deposited by gravitational settling (which may be biologically accelerated) in environments nearer shore, and long-range eolian transported material. The lithogenous fraction is accompanied by variable amounts of biogenous material. Because of the complex origin of the suspended material in any specific environment, it is difficult to arrive at representative data for its composition. However, particulates dominated by lithogenous contributions are enhanced in silicate and aluminosilicate phases. The mineralogy of these particulates reflects the weathering processes in source areas and the pathways of transport to the ocean, as evidenced by the composition of the bottom sediments to which they contribute (Griffin, Goldberg Windom, 1968).

Variations in the concentrations of dissolved material are indicated by the distribution of salinity, which has been discussed in Chapter 1. In open-ocean surface waters these variations are produced mainly by differences in the balance between evaporation and precipitation. In coastal waters dilution with continental drainage has a major local influence. In estuaries variations in salt content occur down to the river water concentrations, which globally average 120 mg/l (Livingstone, 1963). In some regions, and in isolated lagoons, high rates of evaporation from relatively shallow water bodies can markedly increase the salinity. Ocean waters have salinities in the range 33 to 37‰ and 75% of the total volume of ocean water has a salinity in the rather narrow range of 34 to 35‰.

In considering the concentrations of individual elements dissolved in seawater the operational nature of distinctions between dissolved and particulate material must be considered. Chemical species in true solution have sizes ranging up to 1–10 nm. The distinction between dissolved and particulate material employed by marine chemists is based upon separations using filters of about 0.5 μm average pore diameter. The cellulose ester membranes most generally used have actual cut-offs which are ill-defined, in that they vary according to filter and sample characteristics, and which are lower than the specified pore size (Sheldon, 1972). The use of such filters represents a compromise between demands of practicality for routine analysis and clear definition of fractions on a size basis. The results reported for dissolved elements include macromolecular and colloidal fractions, as well as forms in true solution. The significance of this varies greatly according to the chemistry of the individual elements. For many elements the implications are not significant, but for an element such as iron a negligible part of the analytically defined "dissolved" fraction may be in true solution. In subsequent sections, which are concerned mainly with the dissolved forms, the above facts should be borne in mind.

Salinity and Chlorinity

Concentrations of individual constituents are often referred to either salinity or chlorinity as an index of the bulk composition of a seawater sample. Salinity has been employed as a major parameter in physical oceanographic work, being used in conjunction with temperature and pressure to estimate the *in situ* density of seawater, and, in conjunction with temperature, to characterize water masses (cf. Chapter 1). In oceanographic usage it is not, however, an actual property of the water but a defined quantity which approximates to the total dissolved solids, being about 0.05‰ lower. Measurement of the total dissolved material is difficult in practice, because of changes in the composition of the residual solids on drying, and is in any event unsuited for routine shipboard measurements. In the early work of the Knudsen Committee, salinity was defined gravimetrically and the relationship between salinity, defined in this way, and chlorinity was established by analysis of a series of samples of varying salinity. The relationship between salinity and sigma-o was also established. Measurements of salinity were thereafter made, for five decades, almost always through determinations of chlorinity. This procedure assumes essentially constant proportions between the major constituents in different samples of seawater.

The chlorinity of seawater, a parameter which is measurable both accurately and routinely by titrimetry with silver nitrate, approximates to the chlorine equivalent of the chloride and bromide content of the water. It is rigorously defined (Jacobsen and Knudsen, 1940), in terms of the atomic weight values accepted in 1938, as the number, expressed in per mille of a seawater sample, identical with the number giving the mass with unit gram of atomic weight silver just necessary to precipitate the halogens in 0.3285234 kg of the seawater sample; weighings are *in vacuo*. This established the definition in terms of a given sample of pure silver, as used in the redetermination of atomic weights in 1938, avoiding any changes due to subsequent redeterminations of atomic weights. Chlorosity, a quantity occasionally used instead of chlorinity, is defined in the same way except that the litre at a stated temperature replaces the kilogram.

In the late 1950s and early 1960s, advances in electronic technology led to the introduction of reliable conductivity bridges for measurement of conductivity relative to given reference solutions. For routine work on board ship, the development (Brown and Hamon, 1961) of instruments using an inductively coupled cell, with a temperature-compensation circuit, was particularly important. Instrumentation was also developed for *in situ* measurement of conductivity, together with temperature and pressure, at closely spaced intervals throughout the water column.

The rapid change from the use of chlorinity to that of conductivity, as a means of estimating salinity, led to the need for reappraisal of concepts relating to salinity. The questions arising have proved to be complex and the work of an expert panel, established by UNESCO, ICES and other international organizations, which has considered them, is not fully complete. Accurate measurements of the proportions of the major dissolved constituents in representative samples of seawater confirmed, as is discussed in more detail below, that deviations from constancy of ionic ratios were small. The relationship of values of salinity, as derived from conductivity measurements, and density shows somewhat less scatter than that shown when salinities are derived from chlorinity measurements (Cox, Culkin and Riley, 1967). This is attributable to the fact that conductivity reflects the concentration of all ions present and there is thus a reduced effect arising from the small variations which do occur in the relative proportions of those ions which contribute significantly to the salinity.

The work of the international panel has led to several changes in definitions. Chlorinity remains satisfactorily defined as by Jacobsen and Knudsen (1940) and is related to salinity by the expression

$$S = 1.80655 \text{ Cl}. \tag{1}$$

Both salinity and chlorinity are always expressed on a per mille (‰) basis.

The empirical relationship between salinity, derived on the basis of equation (1), and conductivity was established by Cox *et al.* (1967) and is

$$S = -0.08996 + 28.29720R_{15} + 12.80832(R_{15})^2 - 10.6789(R_{15})^3 + 5.98624(R_{15})^4 - 1.32311(R_{15})^5 \quad (2)$$

In this equation R_{15} is the conductivity ratio at 15°C, i.e. the ratio of the conductivity of the sample at 15°C to that of water notionally of salinity exactly 35‰ at the same temperature. Equation (2) has been used to produce International Oceanographic Tables (UNESCO, 1966) for use with conductivity instruments and has been recommended for adoption as the current definition of salinity (Wooster, Lee and Dietrich, 1969). Wilson (1975) has pointed out that the above relationship does not provide a new basic definition but is a restatement in terms of conductivity of the definition of chlorinity, since it is from values of chlorinity that those of salinity were derived. The reference solutions (IAPSO Standard Seawater) used with salinometers are currently standardized in terms of chlorinity. Work is in progress to enable them to be standardized also in terms of absolute conductivity.

The conversion of chlorinity or conductivity ratio to salinity adds no information of interpretative value to that provided by the original data but the widespread usage of salinity in the past has made it desirable to retain the concept. The more detailed aspects of the salinity problem, reviewed recently by Wilson (1975), are of most concern in physical oceanography, where the maximum precision attainable in measurement of the parameters influencing density is of practical value.

Major Dissolved Constituents

The highest precision currently attainable in the measurement of salinity is about 0.001‰. Elements with dissolved concentrations at or above 1 mg/l thus contribute significantly to the salinity and those with concentrations in this range throughout the open ocean are described as major constituents. These constituents, by virtue of their long residence times and relatively high oceanic reservoir concentrations, mostly occur in rather closely constant proportions, and their changes in concentration within the ocean are dominantly a result of water-mixing processes. They are thus often referred to also as conservative constituents. The concentrations of the major constituents and their ratios to chlorinity are shown in Table 2.1. The most recent wide-scale investigations of the constancy of the relative composition of seawater from different regions have been those of Culkin and Cox (1966), Morris and Riley (1966) and Riley and Tongudai (1967), who examined samples representative of the major oceans and water masses and important marginal seas. These studies showed that for sodium, magnesium, potassium, sulphate and bromide, variations are close to or within the analytical precision. Calcium shows a significant enrichment, relative to chlorinity, in deep water, amounting to between 0.5 and 1%, this enrichment being a consequence of the substantial flux of calcium from surface waters to the deep ocean, created by the formation, transport and dissolution of calcium carbonate.

The concentration given in Table 2.1 for bicarbonate ions includes the carbonate ions which are present. Considerably greater variations occur in the ratio of the concentration of bicarbonate (with carbonate) to chlorinity than for other major components. The chemistry of inorganic carbon species in seawater is more complicated than that of the other major constituents, primarily because it involves also the behaviour of the free carbon dioxide in the system. These questions are considered in a subsequent section (p. 52).

Table 2.1. Major dissolved constituents of seawater[a]

Constituent	Concentration (g/kg) at $S = 35‰$	Ratio of concentration (g/kg) to chlorinity (‰)
Na^+	10.765	0.5557
Mg^{2+}	1.294	0.0668
Ca^{2+}	0.412	0.0212
K^+	0.399	0.0206
Sr^{2+}	0.0079	0.00041
Cl^-	19.353	0.9989
SO_4^{2-}	2.712	0.1400
HCO_3^-	0.142	0.00735
Br^-	0.0674	0.00348
F^-	0.0013	0.000067
H_3BO_3	0.0256	0.00132

[a] Values are taken or derived from Millero (1974) and refer to the total concentration of each constituent, i.e. no account is taken of complexation or, in the case of boric acid, dissociation; those for HCO_3^- include the CO_3^{2-} present.

The magnitude of variations in the ratio of the concentration of strontium to salinity has been a matter of controversy, reflecting analytical uncertainties (Wilson, 1975). Recent data suggest that significant variations exist, particularly between surface and deeper waters. They do not, however, exceed 3% and since the concentration of the element is low, this variation is of minor significance in terms of its effect on salinity relationships. Variations in the ratio of boron to salinity also appear to be relatively small (Wilson, 1975) and similar considerations to those raised for strontium indicate that their significance for salinity relationships is marginal.

The contribution of the fluoride concentration to salinity is close to the limits of significance in terms of the precision of measurement of the latter quantity. In most waters the ratio of the concentration of fluoride to salinity is uniform within the precision of measurement (Warner, 1971) but some exceptions have been reported. The most notable were found in deep waters, mainly from the North Atlantic, which showed an excess of fluoride, up to 30% above the normal ratio (Riley, 1965; Brewer, Spencer and Wilkniss, 1970). Interpretation of these findings has been complicated by the fact that the excess fluoride is not detected by electrochemical sensing of the activity of the fluoride ion (Brewer *et al.*, 1970).

It should be noted that among the non-conservative constituents in the ocean, dissolved silicon, present mainly as undissolved silicic acid, and the total dissolved organic carbon, exceed concentrations of 1 mg/l in some water masses; for silicon the average concentration is some 2 mg/l. At the level of precision now often attained in the measurement of salinity for hydrographic purposes the variability in the proportions of several constituents which could hitherto be ignored becomes significant in terms of the relationship between conductivity, salinity and density. This has been examined in detail for dissolved silicon, total carbon dioxide and alkalinity (Brewer and Bradshaw, 1975).

It follows from the largely conservative behaviour of the constituents discussed in this section that their spatial and temporal variations in seawaters not subject to major local influences, are closely related to those of salinity. Their distributions may be predicted on this basis with generally good precision and it is thus only in studies where there is specific interest in small variations and their causes that their concentrations are directly measured in seawater; the carbonate species are exceptions to this statement for reasons which are discussed subsequently. Deviations from the constancy of ionic ratios occur in some marginal environments, however, such as in low salinity waters in estuaries. The global average river water contrasts markedly in composition with seawater, most notably in the much higher proportions of calcium and bicarbonate ions, and this contrast is reflected by changes in proportions of ions, as well as in their concentrations, along salinity gradients. For major constituents, however, the higher absolute concentrations in seawater lead to the general dominance of the seawater characteristics at relatively early stages of mixing. Accounts of estuarine chemical processes are given in Burton and Liss (1976). In anoxic waters of the Black Sea, a decrease in the ratio of concentration of sulphate to chlorinity has been reported by Skopintsev, Gubin, Vorob'eva and Vershinina (1958). Evaporation of isolated bodies of seawater in lagoons leads to alterations in the ionic ratios, as a result of the differential removal of components into solid phases. Isolated brines of atypical composition occur also in some deep-water basins of the Red Sea rift valley as a feature associated with tectonic processes (Degens and Ross, 1969).

The chemical composition of seawater is most simply expressed in terms of the concentrations of particular elements or ions, these being the quantities normally measured in analysis. In complex aqueous media, however, ions tend to associate in a variety of complexes and such associations, together with more general effects of increases in the ionic strength of a solution, mean that there are considerable differences between the analytical concentrations and the activity of a given ion. The activity is a thermodynamic quantity which may be considered as a measure of the effective concentration in terms of chemical equilibria. For many purposes it is important to evaluate the activity and to consider the nature of the associated forms of an ion. This is the case not only in physicochemical studies themselves but also in relation to the bioaccumulation and toxicity of elements in seawater (Burton, 1979).

The differences between activity and concentration, which are most marked for charged forms, arise because of the non-ideal behaviour of dissolved species, particularly through the interactions of a non-specific nature which can occur between ions, on account of their charges, and in many cases, through complexing effects specific to particular ions or other ligands. These differences are taken into account quantitatively by the activity coefficient (f) defined by the relationship

$$f = \frac{a}{c} \tag{3}$$

where a is the activity and c the concentration of a given species. Two main conventions have been adopted by solution chemists. The first makes use of the infinite dilution scale on which f approaches unity as the ionic strength approaches zero. The second approach, using an ionic medium scale, defines f as unity in an ionic medium of specified ionic strength. A non-interactive electrolyte at a relatively high ionic strength forms a suitable ionic medium, enabling equilibria for other species to be examined under conditions where their activity coefficients always remain close to unity. Such a scale has major advantages for marine chemistry, but since in practice many of the basic data, such as stability constants, required for physicochemical calculations on seawater, are available for the infinite dilution scale, much use has been made of extrapolations from the latter.

Activity coefficients, as defined by equation (3), normally refer to single species, relating, for example, the concentration and activity of a particular ion, e.g. Ca^{2+}. In this example the coefficient may be described as the free ion activity coefficient. In marine chemistry it is often convenient to use total activity coefficients which relate the activity of the free ion (e.g. Ca^{2+}) to the total concentration of the dissolved element which will be made up by a number of different significant species (in this example, Ca^{2+} and its ion pairs such as $CaHCO_3^+$, $CaSo_4^\circ$ and $CaCO_3^\circ$). Thus

$$a_{Ca^{2+}} = f_{Ca^{2+}(total)} c_{Ca^{2+}(total)}. \qquad (4)$$

The ratio $f_{Ca^{2+}}$ (total)/$f_{Ca^{2+}}$ (free) represents the fraction of the total calcium present in unassociated forms. For calcium ions in seawater of 35‰ at 25°C and 1 atmosphere total pressure, the free activity coefficient has been estimated as 0.26 (Berner, 1971). About 87% of the calcium ions behave as free ions, the remainder behaving as if ion-paired. The value of $f_{Ca^{2+}}$ (total) is thus 0.23.

One approach, pioneered by Garrels and Thompson (1962), to the modelling of chemical speciation in seawater, and the derivation of total activity coefficients, is based on Bjerrum's treatment of associations in terms of ion-pairs formed between ions of opposite charges. Estimated values for free ion activity coefficients in seawater are used in conjunction with association constants for the relevant ion-pairs to evaluate the extent of specific associations; the negligible tendency of chloride ions to associate with the major cations considerably simplifies this evaluation. The speciation of the element is thus derived and total activity coefficients can be derived as above. The general success of ion-pair modelling in this field does not necessarily imply the existence of the ion-pairs as actual chemical species, although for some there is independent evidence that they do occur (Millero, 1974). Other approaches (Leyendekkers, 1973; Whitfield, 1973, 1975a), based on the more general Bronsted-Guggenheim treatment of specific ion interactions, lead directly to estimates of total activity coefficients.

Pytkowicz and his colleagues (Kester and Pytkowicz, 1969; Pytkowicz and Hawley, 1974) have determined experimentally the association constants for the most relevant ion-pairs at the ionic strength of seawater (such constants are variously referred to as stoicheiometric, conditional or apparent). This approach amounts to the adoption of a specialized ionic medium scale and gives a direct attack on the problem of chemical speciation. It does not require the assignment of individual free ion activity coefficients but assumes that these are identical in solutions of the same ionic strength, independently of composition. The main features of the model of Pytkowicz and Hawley (1974), for the speciation of the major constituents studied, are shown in Table 2.2. While there are differences in detail, the speciation broadly resembles that obtained in the original ion-pair model of Garrels and Thompson (1962). Extension of the model to take account of effects of temperature and pressure over the oceanic range (Kester and Pytkowicz, 1970) has led to controversy (Millero, 1971, 1974; Pytkowicz, 1972) concerning the assumptions involved. Clarification of this question is particularly needed in relation to the occurrence of $MgSO_4^\circ$ since this has important effects on ultrasonic absorption (Fisher, 1972). From Table 2.2 it is seen that the major cations occur mainly as the free

Table 2.2. Ion-pair model of speciation of some major constituents[a] in seawater at 25°C and 1 atmosphere total pressure

Constituent	Free ion	Ion-paired with			
	(%)	SO_4^{2-} (%)	HCO_3^- (%)	CO_3^{2-} (%)	
Na^+	98	2	<0.1	<0.1	
Mg^{2+}	89	10	0.2	0.2	
Ca^{2+}	88	11	0.3	0.4	
K^+	99	1	—	—	
Constituent	Free ion	Ion-paired with			
	(%)	Na^+ (%)	Mg^{2+} (%)	Ca^{2+} (%)	K^+ (%)
SO_4^{2-}	39	37	19	4	0.4
HCO_3^-	81	11	6.5	1.5	—
CO_3^{2-}[b]	8	16	44	21	—
F^-	51	—	47	2	—

[a] After Pytkowicz and Hawley (1974).
[b] Significant amounts of carbonate are associated as $Mg_2CO_3^{2+}$ (7%) and $MgCaCO_3^{2-}$ (4%). The fractions of Mg and Ca involved in these associations and in fluoride ion-pairs are negligible (<0.1%).

ions but that formation of ion-pairs with sulphate is significant for the alkaline earths, calcium and magnesium. Complexing as ion-pairs has a greater effect on the major anions, with the notable exception of chloride. For carbonate, the free ion accounts for less than 10% of the total concentration.

In Table 2.3 values are given for the free ion activity coefficients, as modified by Berner (1971) from those assigned in the Garrels-Thompson model, and for the total activity coefficients obtained by alternative approaches and by experimental measurement. Agreement between the total activity coefficient values is generally good. It is probable that the low values for sulphate, given by the ion-pair models, mainly reflect incorrect assignment of the free ion activity coefficient.

Dissolved and Particulate Organic Material

The relative importance of various pools of organic carbon in the ocean is illustrated by the data for coastal Pacific Ocean water given in Table 2.4. The organic carbon is a small fraction of the dissolved inorganic carbon, present mainly as bicarbonate. The dominance of dissolved organic carbon (DOC) over the particulate organic carbon (POC) is a general feature in the ocean. Although values an order of magnitude higher may arise in inshore and estuarine areas, the levels characteristic of coastal waters are about 1 mg DOC/l and 0.1 mg POC/l (Head, 1976). Menzel (1974) has reviewed information on the concentrations of organic carbon in the open sea. Surface waters typically have concentrations of dissolved organic carbon of about 1 mg C/l, rarely exceeding 1.5 mg C/l, and the concentration decreases below the euphotic zone with

Table 2.3. **Activity coefficients of some major ions in seawater at 25°C and 1 atmosphere total pressure**

		Na^+	Mg^{2+}	Ca^{2+}	K^+	Cl^-	SO_4^{2-}	HCO_3^-	CO_3^{2-}
Free ion activity coefficients[(a)]		0.71	0.29	0.26	0.63	0.63	0.17	0.68	0.20
Total activity coefficients									
Ion pair models[(b)]	(1)	0.70	0.25	0.23	0.62	0.63	0.07	0.51	0.02
	(2)	0.69	0.26	0.23	0.62	0.63	0.066	0.55	0.016
Specific interaction models[(c)]		0.67	0.23	0.21	0.63	0.68	0.12	0.59	0.03
Measured values[(d)]		0.68	0.23	0.21	0.64	0.68	0.11	0.55	0.02

(a) From Berner (1971).
(b) Values obtained by two approaches: (1) from estimates of individual ion activity coefficients and ion-pair association constants at the ionic strength of seawater (Garrels and Thompson 1962, modified by Berner 1971); (2) from estimated free ion activity coefficients, shown above, and the extent of ion pairing as shown in Table 2.2.
(c) At ionic strength 0.7, except for HCO_3^- and CO_3^{2-} ($I = 0.5$); average estimates from Whitfield (1973) and Leyendekkers (1973).
(d) From Millero (1974).

Table 2.4. **Amounts of organic carbon in various pools in surface water of the Pacific Ocean**[(a)]

Pool	Equivalent μg C/l
Dissolved organic carbon	1000
Particulate organic carbon	125
Phytoplankton	20
Zooplankton	2
Fish	0.02

(a) Data taken from Riley and Chester (1971).

relatively uniform concentrations in the range 0.5 to 0.8 mg/l in waters below 500 m. A profile of DOC, which is probably characteristic of situations with a well-developed pycnocline, is given in Fig. 2.1. Concentrations of POC are, as expected, more variable than those of DOC, generally amounting to 0.01–0.1 mg C/l in surface layers, with the lower values being more typical, and 0.003 to 0.01 mg C/l in deeper waters (Menzel, 1974).

An example of a vertical profile of POC is given in Fig. 2.1. The particulate organic material (POM) usually forms a significant part of the total particulate matter even in coastal waters (Manheim, Meade and Bond, 1970). Lal (1977) gives a value of 25% for the fraction of POC in total particulate material in open Atlantic Ocean waters. Oceanic POM is mostly derived from phytoplankton ultimately and there is a general correlation of high concentrations of POC with regions of high primary productivity (Hobson, Menzel and Barber, 1973). Detrital POM generally forms, however, the dominant part, even in near-surface waters, so that this relationship does not arise in a simple, direct way. As discussed by Parsons (1975) the visual appearance of much of the POM does not conform to that of the remains of organisms. It seems probable that considerable amounts of POM are formed from dissolved organic matter (DOM), by aggregation through the action of bacterial growth. Furthermore, inorganic particles can adsorb organic material on their surfaces and there is evidence to suggest that interfacial processes related to bubble formation at the sea surface may lead to the formation of POM. Limited information exists on the chemical composition of POM and on the compositional differences between surface and deep-water material. The study of POM in deep waters is analytically difficult because of its low concentration. From the data summarized by Parsons (1975) it appears that the weight ratio of carbon to nitrogen in deep-water POM is characteristically above 10 and thus significantly higher than the typical value of about 6 found in phytoplankton. Parsons (1975) suggests that the ratio in deep-water POM is probably greater also than that in POM in the euphotic zone. It is possible that such a difference might reflect the removal of the more readily utilized organic material.

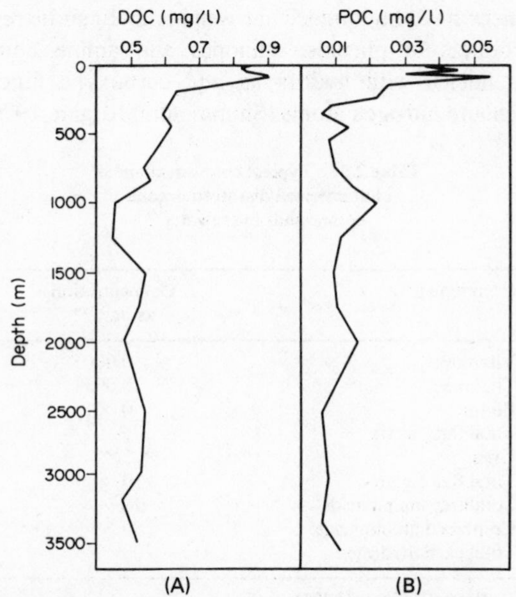

Fig. 2.1. Vertical distribution of (A) dissolved organic carbon and (B) particulate organic carbon, in the Gulf of Mexico at 24°N, 93°W (from Maurer, 1976).

Much of the carbon fixed in the biosphere passes eventually to the pool of dissolved organic matter, by excretion, autolysis and microbial activity. The riverborne input of DOC amounts to less than 1% of the rate of carbon fixation in the ocean (Duce and Duursma, 1977). The importance of the sources in contributing to the oceanic reservoir of DOC depends also, however, upon the rapidity with which various components are mineralized. While DOC is a convenient analytical entity, its use as an environmental parameter is apt to be misleading unless it is borne in mind that it embraces a wide range of compounds which are being mineralized at widely differing rates. Thus a substantial flux of rapidly mineralized component may contribute little to the integrated reservoir of DOC; furthermore, major changes in the concentrations of biologically important organic compounds may be reflected to only a small, even a negligible, degree in measurements of DOC.

Information on the average concentrations of particular groups of organic compounds is summarized in Table 2.5. From this it is apparent that labile organic compounds, such as amino acids and fatty acids, account for only a small part of the DOC, most of which has not been characterized in terms of its molecular composition (the fraction described as total carbohydrate is of uncertain significance in this respect). The organic composition is best understood in terms of two major fractions, that which is readily utilized and recycled by microbial processes and that which is resistant to such biodegradation (Bada and Lee, 1977). The resistant material may be broadly equated with the dominant uncharacterized fraction of DOC. Kalle (1937) used the term Gelbstoff to describe the organic material which shows characteristic absorption at short wavelengths in the visible spectrum (see Chapter 6) and this includes at least part of the uncharacterized material. The latter is frequently referred to also as marine humus or humic material, by somewhat uncertain analogy with the polyelectrolytic and macromolecular humic substances of terrestrial origin which occur in fresh waters. These humic substances have high molecular weights and can be regarded structurally as condensation products of phenols, quinones and amino compounds. In these structures, aromatic nuclei, with hydroxylic and carboxylic functional groups, are linked through oxygen and nitrogen atoms (Stumm and Morgan, 1970).

Table 2.5. Typical concentrations of characterized dissolved organic compounds in seawater[a]

Component	Concentration as μg C/l
Vitamin B_{12}	0.0005
Thiamine	0.005
Biotin	0.001
Total fatty acids	5
Urea	5
Total free sugars	10
Total free amino acids	10
Combined amino acids	50
Total carbohydrates	200

[a] From Williams (1975).

More information is available on the humic material in marine sediments than on the dissolved substances. It appears that the marine humic materials contain similar functional groups to the terrestrial materials, but that they are notably higher in nitrogen content and lower in oxygen content (Rashid and King, 1970; Stuermer and Harvey, 1974). Fulvic acids from seawater show a more aliphatic character than their terrestrial equivalents (Stuermer and Payne, 1976). Stuermer and Harvey (1974) found that most of the humic material in waters in oceanic and coastal environments was in the fulvic acid fraction (that which is soluble in dilute mineral acid) and that the bulk of the material was below 700 in apparent molecular weight. Data for the Gulf of Mexico (Maurer, 1976) indicated that up to 15% of the DOC occurs as compounds with apparent molecular weights above 1000 and that material of molecular weight greater than 10^5 comprises not more than 5% of the total DOC. A study on the coastal seawater of Tokyo Bay (Ogura, 1974) showed that for waters of salinity greater than 33‰, containing about 1 mg DOC/l, from 35 to 50% of the DOC was accounted for by material in the apparent molecular weight range 10^4 to 10^5; significant fractions (17 to 30%) showed apparent molecular weights exceeding 10^5 and material of low apparent molecular weight (<500) was also significant (8 to 34%).

While more work is needed to clarify fully the characteristics of the DOM even in terms of molecular-weight distribution, it does appear that the marine DOM contains a higher proportion of low molecular weight material than is characteristic of fresh water. The marine DOM shows also significant differences from terrestrial material in the ratio of the stable carbon isotopes ^{12}C and ^{13}C (Williams and Gordon, 1970; Stuermer and Harvey, 1974). Together with the chemical differences already noted, these features point to an authigenic origin for the DOM. It seems probable that material resembling marine DOM can be formed by condensation reactions from compounds exuded from macroalgae, such as polyphenols, carbohydrates and amino compounds (Sieburth and Jensen, 1970). Similar reactions acting on material from planktonic algae may account for the formation of humic material in the ocean but sites of formation have not been identified. The less aromatic character of fulvic acids in seawater, as compared with their terrestrial counterparts, could reflect the low abundance of lignin in oceanic environments (Stuermer and Payne, 1976).

The dominance in marine DOM of compounds which are resistant to mineralization is shown by the fact that DOC in deep water of the Pacific Ocean has a content of ^{14}C corresponding to an apparent age of 3400 yr (Williams, Oeschger and Kinney, 1969). It is thus considerably older than the inorganic carbon present in the same water mass. Such "ages" cannot be interpreted unequivocally because a given apparent age may arise from various combinations of younger and older material. The value does, however, show the existence of a significant microbially resistant fraction.

Analyses of dissolved organic nitrogen and phosphorus have been summarized by Williams (1975). From some investigations of the former there appears to be a decrease in concentrations from 60 to 110 μg N/l in the mixed layer to values of 30 to 60 μg N/l in deeper waters, but in some regions higher and less variable concentrations have been reported. Analytical problems may underlie some of these differences. Dissolved organic phosphorus also decreases in concentration with depth, although recent analyses using photo-oxidative breakdown of the organic material suggest a smaller difference between surface and deep waters than was indicated by earlier studies. The values obtained by this method range overall from the limit of detection to 12 μg P/l.

Dissolved Gases other than Carbon Dioxide

The gases which are generally classed as non-variable in the atmosphere are listed in Table 2.6. With these gases, exchange across the air–sea boundary tends to equalize the partial pressure of a gas in the atmosphere and in solution, so that the surface water is saturated under the prevailing conditions. The rate of molecular diffusion in the aqueous boundary layer is the most important factor controlling the rate of exchange for sparingly soluble gases (Liss, 1973), although processes in the gas phase become important for very soluble gases. The thickness of the boundary layer varies according to sea surface conditions but characteristically is of the order of 100 μm. Below the boundary

Table 2.6. Abundance of gases in the atmosphere[a]

Gas	Partial pressure[b] (atm)
Nitrogen	0.78080 ± 0.00004
Oxygen	0.20952 ± 0.00002
Argon	$(9.34 \pm 0.01) \times 10^{-3}$
Carbon dioxide[c]	$(3.3 \pm 0.1) \times 10^{-4}$
Neon	$(1.818 \pm 0.004) \times 10^{-5}$
Helium	$(5.24 \pm 0.004) \times 10^{-6}$
Methane	2×10^{-6}
Krypton	$(1.14 \pm 0.01) \times 10^{-6}$
Carbon monoxide	$(0.1 - 0.2) \times 10^{-6}$
Nitrous oxide	5×10^{-7}
Xenon	$(8.7 \pm 0.1) \times 10^{-8}$

[a] After Kester (1975).
[b] The limits indicate the variability or the uncertainty of the values. The gases N_2O and CH_4 are variable but the extent of variations has not been well defined.
[c] The partial pressure of carbon dioxide has shown a long-term increase from a value of about 2.90×10^{-4} atm in the mid-nineteenth century.

layer dissolved gases are transported by physical mixing of water bodies represented by diffusion coefficients which are much greater than those for molecular diffusion. The mixed layer generally has fairly uniform concentrations of dissolved gases as a result of turbulent mixing.

The mass of gas which is dissolved in a given body of surface water at equilibrium depends upon a number of factors, namely, the atmospheric pressure, temperature and humidity, the temperature and salinity of the water body, and the surface conditions. The atmospheric boundary layer will normally be saturated with water vapour and at a closely similar temperature to the sea surface. Atmospheric pressure variations can cause significant changes in the saturation concentrations, since the solubility is directly proportional to the partial pressure in the gas phase (Henry's Law). When bubbles are carried below the surface, exchange of gas will occur under significant additional hydrostatic pressure. The presence of dissolved salts reduces the solubility of gases in aqueous solution, but over much of the ocean the range of surface salinities is narrow and the major factor influencing dissolved gas concentrations is temperature. To a first approximation contours of surface dissolved gas concentrations thus follow isotherms.

A useful first-order model, for consideration of the distribution of dissolved gases throughout the ocean, is provided by the assumption that a sample of water collected anywhere in the ocean may be assigned a saturation concentration, on the basis of its temperature and salinity, using experimentally determined gas solubilities. These saturation values refer only to surface equilibration conditions and should not be confused with values for *in situ* pressures. Since the entry of gases is in general only significant at the interface with the atmosphere, the effects of higher pressures on gas solubility are not of major oceanographic interest. Values for the solubilities of oxygen and nitrogen under various conditions of temperature and salinity are shown in Tables 2.7 and 2.8,

Table 2.7. **Solubility of oxygen (ml/l) in pure water and seawater[a]**

Temperature (°C)	Salinity (‰) 0	15	30	35	38
1	9.94	8.97	8.11	7.84	7.68
5	8.93	8.09	7.33	7.09	6.95
10	7.89	7.17	6.52	6.32	6.20
15	7.05	6.43	5.87	5.69	5.58
20	6.35	5.81	5.32	5.17	5.07
25	5.77	5.30	4.86	4.73	4.65
30	5.28	4.86	4.47	4.35	4.28

[a] Values refer to equilibration of seawater with an atmosphere of normal composition, at 100% relative humidity and 1 atm total pressure. Data from UNESCO (1973).

Table 2.8. **Solubility of nitrogen (ml/l) in pure water and seawater[a]**

Temperature (°C)	Salinity (‰) 0	20	30	35	38
1	17.95	15.48	14.38	13.86	13.55
5	16.26	14.09	13.11	12.65	12.38
10	14.51	12.64	11.80	11.40	11.17
15	13.08	11.45	10.73	10.37	10.17
20	11.90	10.47	9.82	9.51	9.33
25	10.91	9.64	9.06	8.78	8.63
30	10.06	8.93	8.41	8.16	8.02

[a] Values refer to equilibration of seawater with an atmosphere of normal composition, at 100% relative humidity and 1 atm total pressure. Data from Weiss (1970) based on measurements by Murray, Riley and Wilson (1969).

respectively. The assignment of these values in the way described assumes that equilibration was completed at the sea surface at an atmospheric pressure of 760 torr. On the basis of this model, useful interpretations may be made by comparing the observed concentrations in a sub-surface sample with the saturation concentrations corresponding to the temperature and salinity of the sample. The relationship between temperature and solubility is non-linear, so that if two volumes of water are independently equilibrated with the atmosphere at different temperatures and then mixed, the concentrations of dissolved gases will not be identical with the concentration which would be attained if the mixed water were itself equilibrated. This effect, however, and the use of *in situ* temperature values, rather than the strictly appropriate potential temperature, can be neglected in obtaining a general view of the behaviour of dissolved gases by comparing observed and saturation values. For such comparisons use is made of the per cent saturation of a gas in a water sample, given by

$$\% \text{ saturation} = \frac{100G}{G^1} \qquad (5)$$

where G is the observed concentration of the gas and G^1 is the saturation value corresponding to the temperature and salinity of the water.

Comparison of the concentrations of dissolved gases such as oxygen and nitrogen, measured in samples of surface seawater, with the saturation values, shows that there is a reasonable approximation to equilibrium over much of the ocean. For gases which behave conservatively when dissolved in seawater, the variations from 100% saturation values are generally within the limits accountable for by the variability of atmospheric pressure and surface conditions.

In sub-surface waters concentrations may be modified by the introduction of material from non-atmospheric sources but such effects are usually negligible. Significant anomalies in helium concentration in bottom waters, however, are related to the input of the gas, released by mantle outgassing, in tectonically active regions. More generally, variations from 100% saturation concentrations reflect the involvement of gases in biological processes. The dissolved gases may thus be divided into those which behave conservatively, such as the noble gases and, under most conditions, nitrogen, and those which behave non-conservatively, of which oxygen and carbon dioxide are most important. The behaviour of carbon dioxide is considered in a separate section since its chemistry involves also several ionic species.

Nitrogen is a good example of a gas which generally behaves conservatively. Although the gas is utilized by nitrogen-fixing organisms and released by denitrifying bacteria (cf. Chapter 7), the influence of these processes is small in the present context. Most sub-surface samples show concentrations which conform with 100% saturation, within the limits of the uncertainties associated with possible variations in conditions of equilibration. These uncertainties can be much reduced by considering ratios of nitrogen to argon, rather than concentrations of nitrogen. Argon can be regarded as completely inert and differences in surface equilibration conditions will affect each gas similarly. Ratios of nitrogen to argon in samples of oxic seawater from various regions and depths have been shown to be essentially constant (Benson and Parker, 1961). The isotopic composition of dissolved nitrogen in oxic waters is also essentially constant and is closely similar to that of atmospheric nitrogen, features which would not be expected if the concentrations of the gas were significantly influenced by biochemical reactions.

In contrast, the concentrations of oxygen in the sub-surface ocean show extensive and marked deviations from the 100% saturation values. While the mixed layer often shows rather uniform concentrations of oxygen, production of the gas during photosynthesis can lead to significant supersaturation in parts of the euphotic zone, when productivity is high and the water column is stable. Under these circumstances some of the excess oxygen will tend to be lost to the atmosphere and diurnal variations, as well as longer term seasonal variations, in oxygen concentration can arise. Below the euphotic zone, concentrations are generally significantly below the 100% saturation values, as a result of respiratory demands on the gas. Except in areas with rapid vertical mixing, an oxygen minimum zone is commonly found in vertical profiles, with the concentrations of oxygen increasing again at depth because of the thermohaline circulation of cold water which equilibrated with the atmosphere at high latitudes. The degree of depletion in the minimum zone varies considerably. Very marked depletion is seen in the eastern tropical North Pacific, where concentrations below 0.1 ml/l occur in a minimum zone of considerable thickness. The sub-surface Atlantic Ocean generally has higher percentage saturation values than the Pacific Ocean. Vertical profiles of dissolved oxygen in several oceanic regions are shown in Fig. 2.2.

The difference between the saturation concentration and the observed concentration

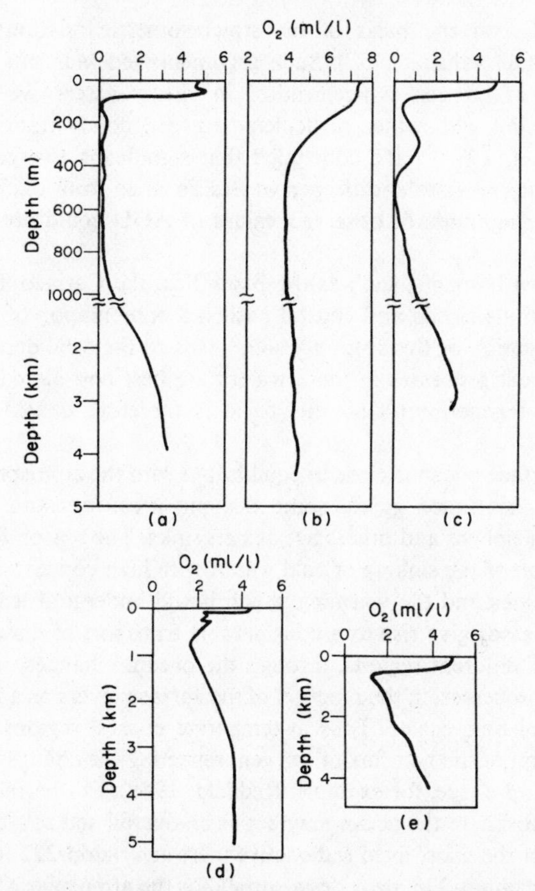

Fig. 2.2. Vertical distribution of dissolved oxygen in various oceanic regions: (a) eastern tropical Pacific Ocean (12°N, 114°W); (b) Antarctic Convergence (30°S, 36°W); (c) eastern tropical Pacific Ocean (6°N, 104°W); (d) Indian Ocean (13°S, 75°E); (e) western tropical Pacific Ocean (21°N, 17°W) ((a) to (c) from Richards, 1965; (d) from Kester, 1975; (e) from Broecker, 1974).

of oxygen for a particular sample is a measure of the oxygen consumption which has taken place in the water since surface equilibration. It is generally described as the apparent oxygen utilization (AOU), following Redfield (1942). The relationship of the AOU to the concentration of regenerated nutrients in the water body is considered in a subsequent section (p. 55). The amount of oxygen consumed in a given water sample can be treated as a consequence entirely of the respiratory demand, dominated by heterotrophic microbial activity. The oxygen concentration in the sample has been altered, however, not only by respiratory utilization but also by mixing with water bodies of different oxygen concentration. The main uncertainties in present understanding of the way in which these factors interact to give rise to the observed distribution of dissolved oxygen arise in relation to the sites where oxidative processes occur. Menzel (1970) has argued, on the basis of the distributions and characteristics of DOC and POC (p. 41), that oxidative processes occur dominantly in the upper 200 m of the ocean. Menzel and Ryther (1970) were able to show that changes in oxygen content along a section of the oxygen minimum zone in the South Atlantic Ocean were about an order of magnitude greater than would correspond to the decrease in the concentration of DOC, on the basis of the stoicheiometric relationships used in the interpretation of AOU values (p. 57). Such arguments led Menzel (1970) to conclude that the changes in oxygen concentration in these waters were brought about dominantly by mixing with water of depleted oxygen concentration. Other workers (Ogura, 1970; Craig, 1971) have concluded that significant changes in oxygen concentration in the intermediate and deeper waters do arise from oxidative processes *in situ*, although the magnitude of observed values of AOU could not be accounted for fully on this basis.

In specialized environments, such as the Black Sea, the Cariaco Trench, part of the Baltic Sea, and various fjords and coastal basins, a combination of oxidative demand and restricted circulation of the bottom waters leads to the total depletion of dissolved oxygen. The chemical processes in these waters are best considered in relation to the cycles of nutrient regeneration and this topic is therefore considered subsequently (p. 57).

Although the surface ocean is close to equilibrium with the atmosphere as regards the concentrations of the major gases, some oceanic areas constitute permanent nett sources for the atmosphere and others act as nett sinks. The reason for this is apparent from a consideration of the sinking of cold water, with high concentrations of dissolved gases, at high latitudes and the warming which it will undergo if it upwells at a lower latitude. Such processes give rise to a considerable transport of dissolved gas between the atmospheres of different regions, through the ocean. Changes, including those related to biological processes, in the function of the surface layers as a nett source or sink can arise on several time scales. Thus in temperate coastal regions seasonal changes occur in the direction of the nett flux of oxygen, reflecting the changes in biological activity and in temperature (see, for example, Redfield, 1948). For certain gases which are transient in the atmosphere the ocean may act as an overall nett source. Thus the standing concentration of the short-lived radioactive inert gas radon-222 in the ocean, maintained by decay of radium-226, provides a supply for the atmosphere (Broecker, 1965).

Recent investigations of a number of non-conservative gases at low concentrations in seawater have shown several notable phenomena (Seiler and Schmidt, 1974). Carbon monoxide is often highly supersaturated in surface waters. Below 100 m concentrations usually correspond to slight supersaturation. The gas is produced in surface waters mainly by algae. Supersaturation of hydrogen is also found in surface waters and it is marked also near the sea-bed, features which are explicable in terms of bacterial production. Bacteria also produce methane in surface waters; very high concentrations occur in anoxic waters and some polluted areas. The role of the ocean in relation to nitrous oxide has received special attention, because of its importance for the atmospheric chemistry of ozone. The flux of nitrous oxide from the ocean to the atmosphere, related probably to bacterial production, exceeds any other known source. The behaviour of this gas is discussed further in Chapter 7.

Redox Potential

The oxidation-reduction (redox) potential (Eh) of a natural aqueous system, which provides an index of the oxidizing or reducing tendency of the medium, is a major determinant of chemical conditions and is closely linked with biological processes. This potential gives a quantitative measure of the energy change involved in the addition or removal of electrons. Since oxidation and reduction reactions involve the transfer of electrons, the redox condition may also be expressed in terms of the activity of electrons (a_{e^-}) using the parameter, pE, which is analogous to pH and is defined by the equation

$$pE = \log a_{e^-} \tag{6}$$

A number of elements can occur in more than one oxidation state in aqueous solution; the most important in seawater re H, O, C, N and S. Other redox elements whose chemistry is of particular interest in relation to geochemical processes in the sea include Fe, Mn and I. The redox potential of a solution containing several such components is affected by the activities of all the redox species present but in seawater free oxygen is usually present in concentrations which make it the dominant electron-accepting species. In evaluating the redox potential theoretically, it is also necessary to identify the reduced form which completes the redox couple.

Until recently it was assumed, following Sillen (1961), that the redox potential was determined by the equilibrium

$$O_{2(g)} + 4H^+ + 4e^- \rightleftharpoons 2H_2O. \tag{7}$$

From this it can be calculated that for seawater at the characteristic pH of 8.1, the pE is 12.5, which represents a strongly oxidizing condition. The true value for the pE is currently a matter of controversy. Conventional methods for measurement of the redox potential in oxic waters do not give absolute values. Breck (1972), on the basis of experimental data, concluded that the pE was as low as 8.5 and considered that it may be determined by the O_2/H_2O_2 couple, depending upon the presence of a very low steady-state activity of hydrogen peroxide. He has shown (Breck, 1974) that use of this pE gives an equilibrium redox model for seawater which resembles the real system more closely than does the Sillen model. It seems probable that the redox potential is a mixed potential, not a single value corresponding to equilibrium. Its effective magitude remains to be definitely established. The agreement or disagreement of the observed redox speciation of particular elements with predictions of equilibrium models provides little basis for assessment of this question because of the presence of metastable species and steady-state concentrations of thermodynamically unstable species, continuously produced by biological processes. This is discussed in Chapter 7 for the case of nitrogen. The pE of seawater shows only small changes with a decrease in the concentration of free oxygen until very low concentrations are reached. With the disappearance of free oxygen, other oxygen-containing species are utilized as electron acceptors in respiratory processes, first nitrate, and, when that is exhausted, sulphate. In anoxic waters containing sulphide there is a marked reduction in the redox potential, which is then controlled by the SO_4^{2-}/HS^- couple, to about −4.1 (Brewer, 1975); this leads to changes in the speciation of elements such as Fe, Mn and I. The redox conditions in sulphate-reducing and oxic regimes are distinctly separated. In the marine environment intermediate conditions are restricted to very narrow zones, because the intermediate processes afford poor poising and low reductive capacity (Breck, 1974). There are, however, limited zones of sub-oxic conditions, such as those in the tropical eastern Pacific, with permanent oxygen deficiency to a level at which partial reduction of nitrate occurs (see Chapter 7). Boundaries between oxic and anoxic waters occur in the water column only where conditions of stratification and bottom topography lead to permanent or temporary isolation of bottom waters receiving inputs of organic material (examples are given on pp. 58–9). Such redox boundaries are more common within bottom sediments, and in near-shore sediments they are often situated close to, or even at, the interface between the sediment and the overlying water.

Carbon Dioxide and the Carbonate System

The factors which have already been discussed (p. 46) as affecting the concentrations of dissolved gases which are in equilibrium at the air–sea interface, influence in a similar way the concentrations of carbon dioxide molecules. The behaviour of this gas is, however, much more complex than that of, for example, oxygen, since in addition to its participation in biological processes it is involved in complex equilibria with combined forms, principally bicarbonate and carbonate ions, which enter and leave the system by non-atmospheric pathways. These equilibria can be expressed simply as

$$H_2O+CO_{2(g)} \rightleftharpoons H_2CO_3 \rightleftharpoons H^+ + HCO_3^- \rightleftharpoons 2H^+ + CO_3^{2-}. \quad (8)$$

Because CO_2 reacts with water, the transfer rate from the gas phase into solution is increased. Despite this the rate remains lower than those of the ionization stages so that equilibrium in the distribution of molecular CO_2 between the surface water and the atmosphere is not usually fully attained (Skirrow, 1975). Since CO_2 is utilized in photosynthesis and released in respiration, diurnal and seasonal variations occur in its partial pressure in surface waters, related to biological activity. Variations on these time-scales are, however, also brought about through changes in temperature (Takahashi, 1961; Gordon, Park, Hager and Parsons, 1971) and through seasonal changes in vertical mixing (Gordon *et al.*, 1971). A summary of the large-scale variations in partial pressure of CO_2 in surface oceanic waters, and their relationship to patterns of oxidation is given by Skirrow (1975).

The reaction of CO_2 with water gives rise to protons, bicarbonate and carbonate ions (equation (8)). The entry of carbon dioxide to the ocean from the atmosphere thus reduces the pH ($pH = -\log a_{H^+}$). Most of the total carbon dioxide present in the ocean, however, exists as bicarbonate and carbonate ions which have entered in river water, their charges being balanced mainly by the abundant cations Ca^{2+}, Na^+, Mg^{2+} and K^+. Bicarbonate and carbonate ions entering thus, are derived from continental weathering processes involving CO_2. These anions act as proton acceptors (bases) and thus tend to increase the pH. In oxic seawater there are no comparably abundant species which tend to associate with hydroxyl ions. The combined effect of the carbonate equilibria gives surface seawater a pH characteristically in the range 7.8 to 8.4. The presence of anions, such as bicarbonate, formed from weak acids, produces a buffering effect in seawater, damping the variations in pH which are mainly caused by differences in the concentration of free CO_2, arising through differences in surface solubility and changes due to photosynthesis and respiration. Such variations are reflected by alterations in the equilibria between the carbonate species, as illustrated in Fig. 2.3. Increase in pH favours the formation of carbonate ions and decrease in pH favours the formation of free CO_2. In the pH range of natural seawater bicarbonate is the dominant species. For most of the ocean, less than 1% of the inorganic carbon exists as free CO_2 and the fraction as H_2CO_3 is negligible.

The total alkalinity (A) of seawater is the amount of hydrogen ion required to convert all the anions of weak acids to the unionized acids. In oxic waters of natural pH, significant contributions to the total alkalinity are made by only three species, namely HCO_3^-, CO_3^{2-} and $H_2BO_3^-$. Thus,

$$A = c_{HCO_3^-} + 2c_{CO_3^{2-}} + c_{H_2BO_3^-}. \tag{9}$$

The value of A is characteristically about 2.4 meq/l. Borate usually contributes less than 10% of the total alkalinity. For some purposes the carbonate alkalinity (CA) may be required:

$$CA = c_{HCO_3^-} + 2c_{CO_3^{2-}}. \tag{10}$$

The alkalinity of seawater shows greater variations than do the concentrations of the major constituents in general. Nonetheless, the ratio of total alkalinity (eq/l) to chlorinity (expressed on a per mille basis) is sometime used in the form:

$$\text{Specific alkalinity} = \frac{A \times 10^3}{Cl}. \tag{11}$$

Values for specific alkalinity are usually in the range 0.12 to 0.13.

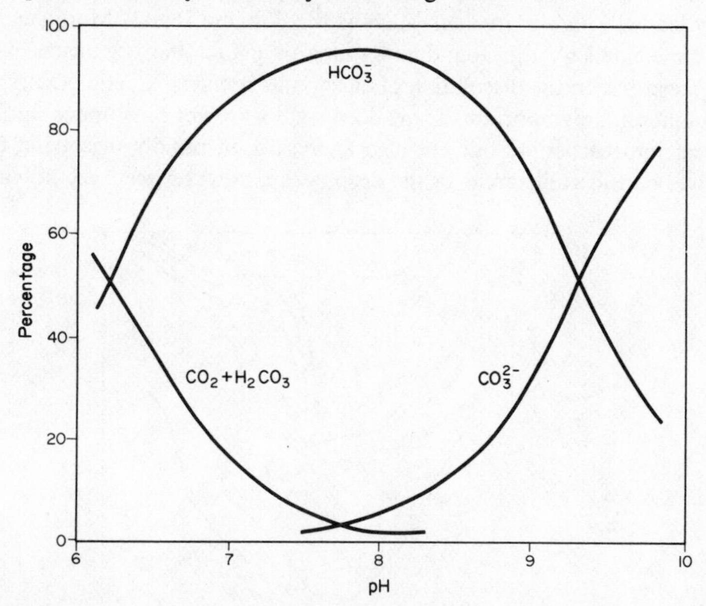

Fig. 2.3. Variations with pH in the proportions of carbon dioxide (CO_2+H_2CO_3), bicarbonate and carbonate ions in seawater (Cl = 19‰) at 0°C (from Skirrow, 1975).

Substantial effort has been devoted to the evaluation of equilibrium constants for the carbonate equilibria and their variations with temperature and salinity. From such values equations can be derived which inter-relate the main analytical parameters which can be used to characterize the system, namely the pH, total alkalinity, partial pressure of carbon dioxide and total carbon dioxide (the sum of the concentrations of all the inorganic carbon species). These relationships have been widely used to obtain a full analytical description of the carbonate system from a minimum of analytical measurements. A detailed review is given by Skirrow (1975). Until quite recently, the partial pressure of carbon dioxide was always calculated from other parameters. Since this quantity is a small fraction of the total, calculated values are subject to considerable uncertainty, as a result both of uncertainties in the relationships used and of errors in measured parameters. Direct measurements of the partial pressure can now be made relatively easily and this approach is much to be preferred.

The carbonate system affords not only a short-term mechanism for buffering pH changes in the ocean but also a partial regulation of the pH on geological time-scales. The global average composition of river waters is dominated by bicarbonate ions and thus the continuous input of river material tends to increase the pH. An increase in pH leads, however, to the formation of more carbonate ions (see Fig. 2.3) and this process is restricted by the limited solubility of calcium carbonate. The degree of saturation of ocean waters with respect to this phase is shown for a section of the Pacific Ocean in Fig. 2.4. Considerations of the solubility of calcium carbonate must take into account the existence in marine environments of two polymorphic crystalline phases, calcite and metastable aragonite. Inorganic precipitation from seawater leads to the formation of aragonite. Organisms may produce either form as skeletal material depending primarily upon the species concerned. The surface ocean is supersaturated with respect to both polymorphs. Direct precipitation arises only in shallow, warm environments, where high productivity leads to high pH values, such as the Bahama Banks. More generally, precipitation is mediated by the abundant groups of planktonic organisms which form calcareous skeletons. In the dissolution of calcareous remains, kinetic factors, as well as solubility equilibria, are important. Some deep waters are not only under-saturated with respect to calcium carbonate but are also aggressive in rapidly dissolving calcareous particles. Over considerable areas of the deep ocean, however, nett accumulation takes place.

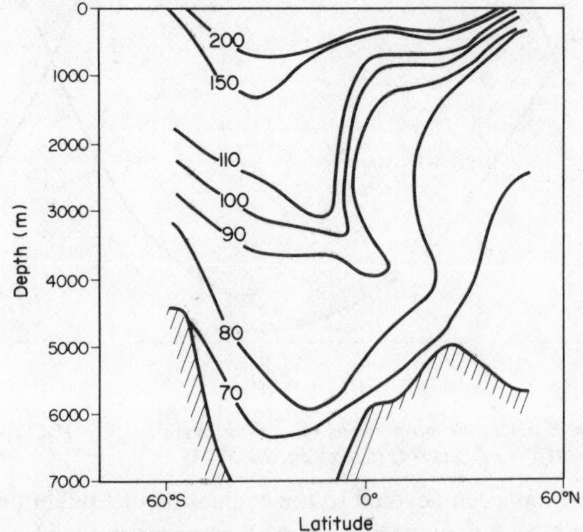

Fig. 2.4. Variations in the degree of saturation of calcium carbonate (calcite) in a vertical section of the Pacific Ocean at 170°W (from Hawley and Pytkowicz, 1969).

The formation and sedimentation of calcium carbonate does not operate as an equilibrium control on alkalinity but it corresponds coarsely to the operation of such a system by biological mediation. The non-ideal solubility behaviour of calcium carbonate in the real system is associated with the presence of other substances. Magnesium ions, and probably dissolved organic matter, inhibit precipitation from supersaturated solutions, and surface coatings on particles, including organic layers, appear to reduce rates of dissolution. Considerations of mass balance (Mackenzie and Garrels, 1966a, b) indicate that some alkalinity must be removed by mechanisms other than carbonate deposition and there are substantial arguments in support of the hypothesis that reactions with aluminosilicates play an important role in the long-term stabilization of oceanic pH and composition (Sillen, 1961, 1967; Garrels, 1965).

The ocean plays a central role in the global cycle and budget of carbon. The major reservoirs of the exchangeable carbon system are shown in Table 2.9 with their carbon contents. The inorganic carbon in the deep ocean forms a massive reservoir, relative to the atmospheric and surface ocean reservoirs in which short-term changes exert their maximum effect. At the present time particular attention is focused on the capacity of the ocean to mitigate the consequences of man's increased mobilization of carbon dioxide by fossil-fuel consumption. This involves consideration of the rates of exchange between the major reservoirs and the new distributions corresponding to modified steady-state systems. There have been many discussions of the complex considerations involved (see, for example, Machta, 1972, and Keeling, 1973). Substantial fractions of fossil-fuel carbon dioxide and anthropogenic ^{14}C have already passed into the ocean.

Table 2.9. Exchangeable carbon system[(a)]

Reservoir	Content (10^{19}g C)
Atmosphere[(b)]	0.062
Ocean	
Inorganic above pycnocline	0.075
Total inorganic	3.8
Organic	0.1
Terrestrial biosphere	0.16

(a) Data from Skirrow (1975).
(b) Pre-industrial value. Concentration increased by about 10% to 1973.

Micronutrient Elements and their Relationships with Oxygen: Anoxic Conditions

The important micronutrients, nitrate, phosphate and dissolved silicon, have received particular attention because of their importance in relation to primary productivity. This aspect is considered in more detail in Chapter 7, where their utilization, the processes of their regeneration from organic material, and their major spatial and temporal variations are discussed. The general ranges of concentrations for these constituents are 1 to 500 $\mu g\ NO_3^-{-}N/l$, <0.1 to 90 $\mu g\ PO_4^{3-}{-}P/l$, and 10 to 5000 μg Si/l.* The dominant feature of their distribution, on the oceanic scale, is the marked depletion in the mixed layer characteristic of areas with a permanent pycnocline. Examples of vertical distributions which show this feature, are given in Fig. 2.5. Broecker (1974) has described how

* For some purposes it is more useful to express the concentrations of constituents in terms of, for example, μmole/l or μg-atom μg-at)/l, rather than μg/l. This arises especially with the principal micronutrient elements. In this volume both conventions are used, depending upon the context. Conversions are given in Chapter 7.

the extremely effective removal of the nutrients from the euphotic zone leads to concentrations in the deep waters of the ocean which increase in the direction of the principal deep-water circulation. Thus for the major oceans, the concentrations are greatest in Pacific deep waters and lowest in Atlantic deep waters. Much more uniform vertical concentrations of the nutrients are found in areas of rapid vertical mixing. The widespread depletion of surface waters reflects the utilization of the micronutrient elements in organic production and their subsequent vertical transport in association with detrital organic material. The cycle of dissolved silicon, present mainly as undissociated silicic acid, involves its uptake in the formation of biogenous silica (opal) and the dissolution of much of this material in intermediate and deeper water. The cycles of nitrogen and phosphorus are more complex. Nitrate and phosphate are taken up and

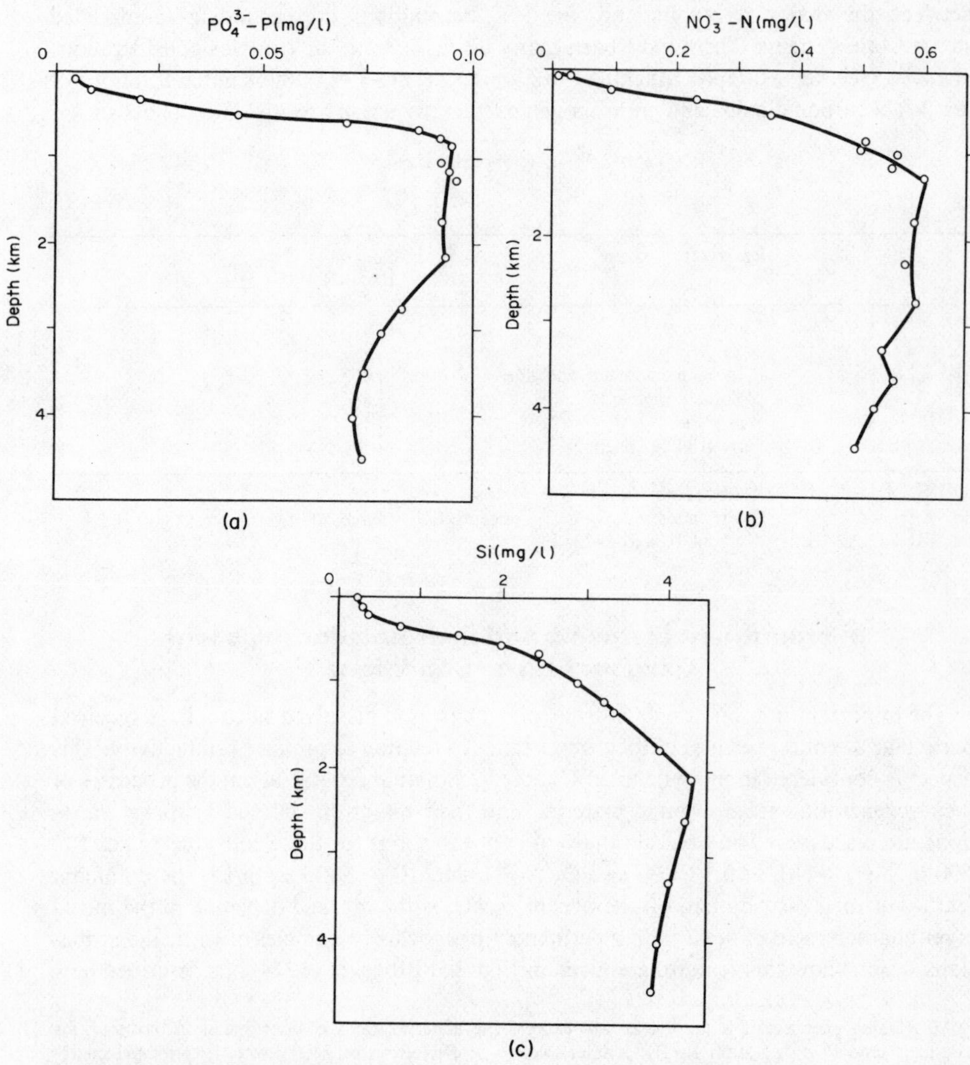

Fig. 2.5. Vertical distributions of dissolved phosphate, nitrate and silicon in the Pacific Ocean at 21°N, 170°W (from Broecker, 1974).

the elements are incorporated into a variety of organic molecules, such as amino acids, proteins, nucleic acids, phospholipids and aminophosphonic acids. The ions are regenerated by hydrolytic and oxidative processes, with the formation of various intermediates, primarily dissolved organic nitrogen and phosphorus compounds, ammonia and nitrite. These intermediates may be directly utilized in the production of organic material but in intermediate and deeper waters the cycle has been substantially completed by the reformation of phosphate and nitrate.

In the production of organic material under environmental conditions, carbon, nitrogen and phosphorus are used in ratios which vary only rather narrowly, when considered over extended areas and time-scales. Thus, the average composition of the organic matter formed can be represented by a quasi-stoicheiometric formula. On this basis the formation of the material may be represented by the equation (Redfield, Ketchum and Richards, 1963):

$$106CO_2 + 122H_2O + 16HNO_3 + H_3PO_4 \longrightarrow (CH_2O)_{106}(NH_3)_{16}H_3PO_4 + 138O_2 \quad (12)$$

The microbial processes responsible for the decomposition of the organic material and the regeneration of the nutrients, i.e. the reversal of the above reaction, are complex (see Chapter 7). For example, nitrifying bacteria, as discussed more fully on p. 301, are involved in the oxidation of ammonia to nitrite and thence to nitrate; this reaction is thermodynamically favoured under appropriate conditions of pE and pH but proceeds mainly by biological mediation. Nevertheless, considerable success has been obtained (Redfield *et al.*, 1963) in modelling the relationship between concentrations of nutrients and oxygen in sub-surface waters on the basis that the overall process of regeneration can be represented by the reverse reaction to that given in equation (12). In this way the apparent oxygen utilization (see p. 49) can be used, in conjunction with the quasi-stoicheiometric relationship shown, to obtain estimates of the amounts of the nutrients in a given water sample which are of oxidative origin. As pointed out previously, there is uncertainty as to how much of the oxidation has actually occurred *in situ* in the intermediate and deep waters.

When free oxygen is not available, the next thermodynamically favoured process for the decomposition of organic material is by denitrification in which nitrate is reduced to molecular nitrogen. This provides only about 10% of the energy available through the use of free oxygen as a terminal electron-acceptor. Nitrite is produced as an intermediate in denitrification which is carried out by specialized types of bacteria such as *Pseudomonas*. Several reactions are possible (Redfield *et al.*, 1963) including:

$$(CH_2O)_{106}(NH_3)_{16}H_3PO_4 + 84.8HNO_3 \longrightarrow 106CO_2 + 42.4N_2 + 16NH_3 + H_3PO_4 + 148.4H_2O \quad (13)$$

in which ammonia is formed, as well as nitrogen, and

$$(CH_2O)_{106}(NH_3)_{16}H_3PO_4 + 94.4HNO_3 \longrightarrow 106CO_2 + 55.2N_2 + H_3PO_4 + 177.2H_2O \quad (14)$$

in which the ammonia is oxidized. It appears that both reactions occur so that some ammonia accumulates.

Following the exhaustion of nitrate and nitrite, sulphate reduction is the next thermodynamically favoured means for the oxidation of organic matter, a process carried out by bacteria such as *Desulphovibrio*. This reaction may be conventionally represented as:

$$(CH_2O)_{106}(NH_3)_{16}H_3PO_4 + 53SO_4^{2-} \longrightarrow 106CO_2 + 53S^{2-} + 16NH_3 + H_3PO_4 + 106H_2O. \quad (15)$$

This is consistent with the observed accumulation of ammonia in anoxic waters. The decomposition of organic matter is less efficient when nitrate and sulphate are used as terminal electron-acceptors so that organic matter tends to accumulate in anoxic environments.

As discussed above (p. 51) the denitrification stage occurs within a very limited zone between oxic and sulphate-reducing regimes. Some overlap between the processes must be expected under environmental conditions and mixing processes create transient situations where reacting species coexist. When anoxic water mixes with oxic water, hydrogen sulphide is rapidly oxidized by dissolved oxygen, giving mainly thiosulphate and sulphate (Cline and Richards, 1969).

The distributions of certain parameters in the Black Sea, as shown in Fig. 2.6, illustrate the consequences of these processes. The marked pycnocline, in association with the shallow sill at the Bosphorus, produces an effective isolation of the bottom water. The oxic zone forms a relatively shallow surface layer. Concentrations of nitrate show a characteristic initial increase with depth and then decrease gradually to an undetectable level in the zone where the oxygen concentrations approach zero. The concentration of phosphate, by contrast, continues to increase in the deeper water. Concentrations of hydrogen sulphide increase markedly with depth in the anoxic zone.

There are numerous differences in the chemistry of anoxic and oxic regimes. The conservative behaviours of sulphate and dissolved nitrogen are modified by sulphate reduction and denitrification. Alkalinity has to be redefined to take into account the unusually

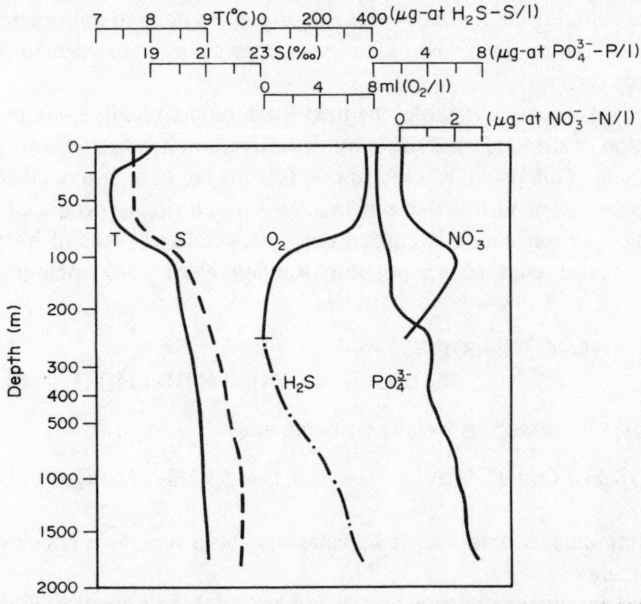

Fig. 2.6. Vertical distribution of temperature, salinity, dissolved oxygen, hydrogen sulphide, phosphate and nitrate at a station in the Black Sea (from Grasshoff, 1975).

high concentrations of proton-acceptors such as phosphate, and the presence of ammonium ions; values exceeding 12 meq/l have been reported (Gaines and Pilson, 1972). The speciation of minor redox elements is greatly modified as a result of the marked change in pE which, as already noted, decreases in the sulphate-reducing zone to about −4.1. This aspect is illustrated in Fig. 2.7, for dissolved manganese and iodine in the Cariaco Trench. The marked increase in concentration of manganese(II) in the anoxic waters is attributable to the stability of this oxidation state at the lowered pE. In the overlying oxic zone, manganese(II) becomes oxidized to manganese(IV) which has a low solubility under the prevailing conditions. The concentrations of dissolved iodine are more uniform throughout the water column but there is a pronounced change in speciation from iodate to iodide at the lower pE prevailing in the anoxic zone.

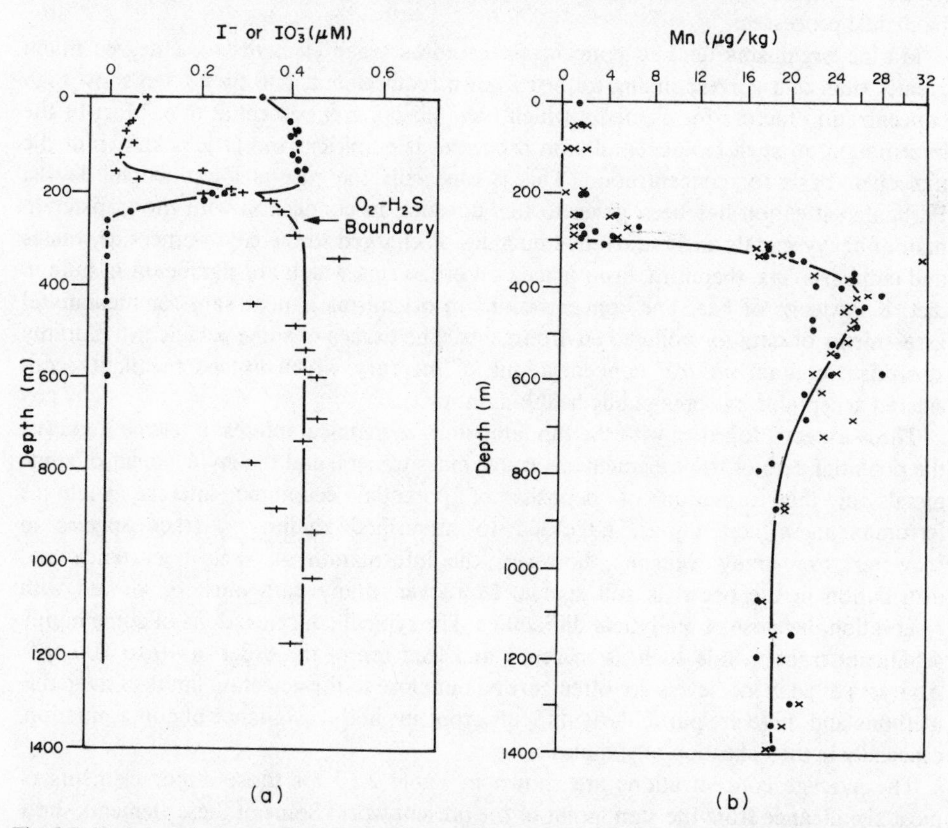

Fig. 2.7. Variations in the chemical speciation of dissolved iodine and in the concentration of manganese (II) with changes in depth and redox conditions in the Cariaco Trench. (a) Profiles of iodate (●) and iodide (+) (from Wong and Brewer, 1977). (b) Profile of manganese(II) (from Brewer, 1975); the two sets of data shown were obtained using different analytical procedures.

Minor Elements

The majority of the elements present in the earth's crust have now been detected in seawater. In addition to the principal micronutrients discussed above, some of the less thoroughly studied elements are of known significance for biological processes, either because they are essential for growth and development (e.g. iron and zinc) or because they have toxic properties (e.g. cadmium and mercury). Essential elements may themselves become toxic at certain concentrations. The assessment of sub-lethal effects of chemical species is difficult and there is no reason to suppose that the typical environmental concentrations of, for example, trace metals are optimal for organic production. Experiments by Steemann Nielsen and Wium-Andersen (1970) indicate that an increment of 1 μg Cu/l in culture solution has at least an initial effect on growth in freshwater algae. Other studies on this element are discussed in Chapter 7. For most elements, the difference between concentrations used in order to produce measurable effects in experimental studies and those characteristic of natural environments is so great that no valid deductions can be made from these studies as to the implications of normal environmental characteristics, and their variations, for productivity or other biological processes.

Marine organisms tend to concentrate essential trace elements to a degree much greater than that corresponding to their known requirements and they often show high concentration factors for elements which have no apparent essential role. Most of the information on such bio-accumulation processes is empirical and little is known of the molecular basis for concentration. This is especially the case at lower trophic levels. Particular attention has been given to this question in connection with the transfer to man of heavy metals and radioactive nuclides discharged to the environment as wastes and emphasis has, therefore, been placed on organisms which are significant in human diet. Knowledge of base-line concentrations in organisms is necessary for meaningful assessments of data for polluted environments. The tissues of some pelagic fish, notably swordfish, contain *natural* concentrations of mercury which exceed the limits considered acceptable by some public health authorities.

These aspects, together with the fundamental geochemical interest in element cycles, the potential uses of trace elements as water mass tracers, and the involvement of some metals in the formation of deposits of potential economic interest, such as ferromanganese concretions, have led to intensified studies on trace species in seawater. For many elements, however, the information on their occurrence and distribution in the ocean is still sparse. Moreover, many data must be viewed with reservation, because of analytical difficulties. The typical concentrations of some highly significant trace metals such as mercury and lead are of the order of 10 to 100 ng/l. Analyses at such low levels are often carried out close to the detection limits of available methods and there are particularly difficult problems in the avoidance of contamination, especially in the collection of samples.

The average concentrations are shown in Table 2.10 for those minor elements of most significance from the standpoint of the present work. Some of these elements show essentially conservative behaviour in the ocean and thus occur at fairly uniform concentrations. Thus, two of the rarer alkali metals, lithium and rubidium, show only narrow variations in their ratios of concentration to salinity. Other examples of trace metals which show relatively little variability in concentration over a wide range of oceanic environments are caesium, molybdenum, vanadium and uranium.

For elements with relatively short residence times, which tend to be depleted in oceanic waters, relative to river input concentrations, marked concentration gradients often arise in marginal waters. Such effects have been accentuated in some cases by polluted inputs. Concentrations of zinc in western British coastal waters are shown in Fig. 2.8. Some of the major concentration gradients are associated with waste inputs

Table 2.10. **Concentration and inorganic speciation of some minor elements in seawater**[a]

Element	Concentration[b] ($\mu g/1$)	Dominant dissolved species[c]
Lithium	180	Li^+
Beryllium	0.006	$BeOH^+$
Aluminium	<1	$Al(OH)_4^-$
Vanadium	2	$H_2VO_4^-$, HVO_4^{2-}
Chromium	0.3	$Cr(OH)_2^+$, CrO_4^{2-}
Manganese	0.02–0.1	Mn^{2+} , $MnCl^+$
Iron	2	$Fe(OH)_2^+$, $Fe(OH)_4^-$
Cobalt	0.05	Co^{2+}
Nickel	0.1–1	Ni^{2+}
Copper	0.2	$Cu(OH)_2^\circ$, $CuCO_3^\circ$
Zinc	0.01–1	$Zn(OH)_2^\circ$, Zn^{2+}
Arsenic	2	$HAsO_4^{2-}$, $H_2AsO_4^-$
Selenium	0.1	SeO_4^{2-}, SeO_3^{2-}
Rubidium	120	Rb^+
Molybdenum	11	MoO_4^{2-}
Cadmium	0.005–0.1	$CdCl_2^\circ$, $CdCl^+$
Tin	0.01	$SnO(OH)_3^-$
Antimony	0.2	$Sb(OH)_6^-$
Iodine	60	IO_3^- , I^-
Caesium	0.4	Cs^+
Barium	10	Ba^{2+} , $BaSO_4^\circ$
Mercury	0.01–0.1	$HgCl_4^{2-}$
Lead	0.001–0.02	$PbCO_3^\circ$, $PbCl_2^\circ$
Radium	8×10^{-8}	Ra^{2+}
Thorium	$<1 \times 10^{-4}$	$Th(OH)_4^\circ$
Uranium	3	$UO_2(CO_3)_3^{4-}$

[a] Based upon Brewer (1975) with additions and modifications.

[b] Typical open oceanic concentrations or ranges are shown. Ranges are not given for elements which commonly vary by only about a factor of 2 from the typical values. As discussed in the text some elements are markedly depleted in the euphotic zone.

[c] The probable inorganic species for oxic conditions are shown. The uncertainties referred to in the text should be borne in mind as should the existence of colloidal and polymeric forms, within the conventional dissolved fractions, for some elements.

(for example, in Liverpool Bay) whilst in other areas (for example, off the coast of mid-Wales) effects of mineralization in the drainage area explain the observed enhancement in the coastal zone.

Information reported for the oceanic distributions of some of the non-conservative elements shows a high degree of scatter, and it is not possible to account for such distributions in terms of the known oceanographic processes which explain the distribution of, for example, the principal micronutrient elements. Boyle, Sclater and Edmond (1977) have argued cogently that incompatibility of distributions with known processes implies inaccuracy of the analytical data. This situation is well exemplified by studies on copper. Recent work (Boyle and Edmond, 1975; Moore and Burton, 1976; Bender and Gagner, 1976; Boyle *et al.*, 1977) suggests that the concentration of this widely studied element is significantly lower (at about 0.1 to 0.2 μg/l) than earlier work had indicated and that vertical profiles of copper show systematic variations, probably related mainly to the same transport mechanisms that produce the vertical gradients in concentrations

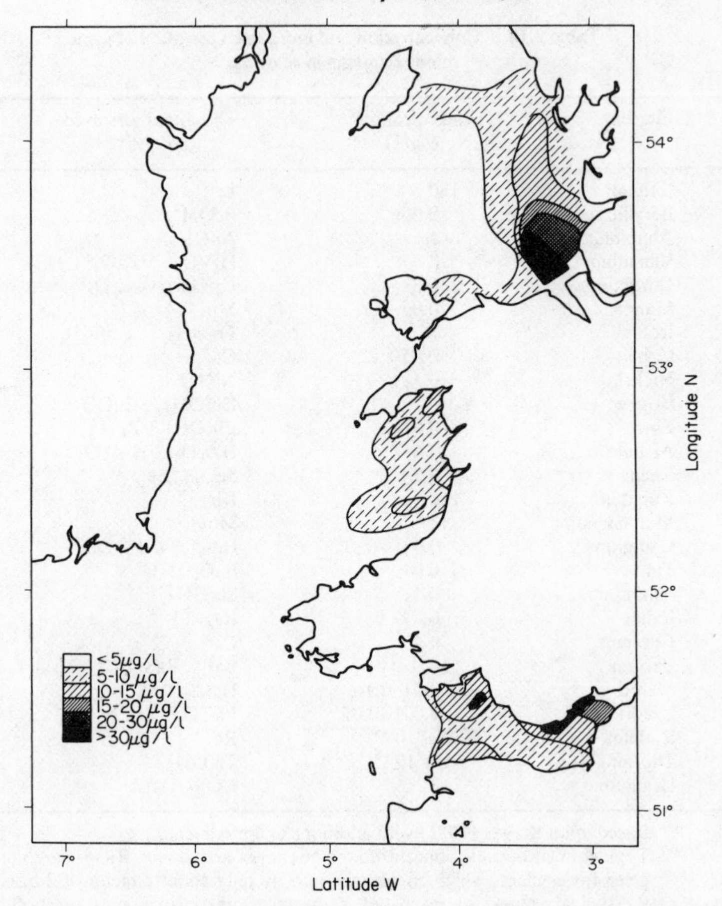

Fig. 2.8. Concentrations of dissolved zinc in western British coastal waters in 1971 (from Abdullah, Royle and Morris, 1972).

of the micronutrients. Other examples of trace elements which show distributions which appear to be related to such biological transportation processes include barium and radium (Wolgemuth, 1970), cadmium (Boyle, Sclater and Edmond, 1976; Bruland, Knauer and Martin, 1978a), nickel (Sclater, Boyle and Edmond, 1976; Bender and Gagner, 1976), zinc (Bruland, Knauer and Martin, 1978b) and selenium (Sugimura, Suzuki and Miyake, 1976; Measures and Burton, 1978). As examples of this type of distribution, vertical profiles are given in Fig. 2.9 which show the close coherence of barium and the chemically similar radioactive nuclide, radium-226, as well as the resemblance to the distribution of the nutrient elements, particularly silicon (cf. Fig. 2.5). The movement of trace metals into deeper water may be related to their associations with either skeletal materials or the organic fraction of biogenous material. In the case of nickel, Sclater *et al.* (1976) concluded from a multiple regression analysis of the concentrations of nickel, silicon and phosphorus that the vertical distribution of the element was related to its incorporation in both types of material.

In Table 2.10, information is given also on the principal inorganic chemical species of the elements, as present in oxic waters of normal pH at 25°C and 1 atmosphere total pressure. Information of this kind has been obtained by the application of equilibrium ion association models, analogous to those used successfully for the major constituents (see p. 40). Such applications have been made by Zirino and Yamamoto (1972). Dyrssen and Wedborg (1974) and Ahrland (1975). Whitfield (1975b) has extended a specific interaction model to minor constituents, to derive total activity coefficients. It is apparent from such models that many metal ions are dominantly associated under seawater conditions. An example of the results of equilibrium modelling is given in Fig. 2.10 which shows the distribution of copper between its various inorganic dissolved forms in seawater, as a function of pH.

The inclusion of environmentally relevant organic ligands in equilibrium models has been precluded by the lack of relevant data on the association constants for complexes formed between metals and the humic materials which form the dominant metal-binding dissolved organic components in natural waters. Apparent constants for such associations have recently been evaluated by Mantoura and Riley (1975) and these have been incorporated into an equilibrium model (Mantoura, Dickson and Riley, 1978). Since it is possible that metals may form inert (kinetically stable) complexes with organic ligands of this kind, the equilibrium model provides a description of a boundary condition for the system. In the environment non-equilibrium features may be found. There is evidence for some elements, notably copper, that significant fractions in coastal waters are available analytically only after the decomposition of organic material (Williams,

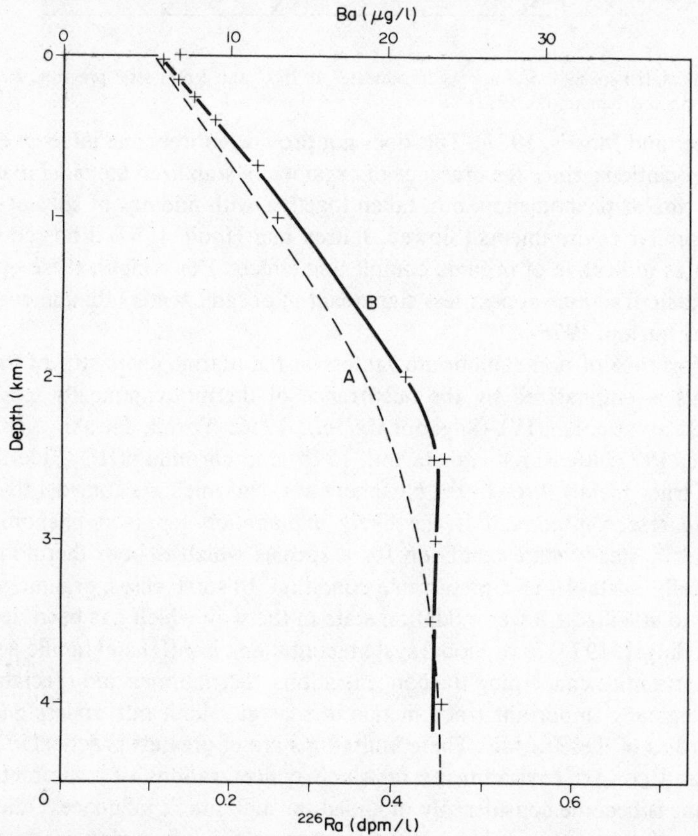

Fig. 2.9. Vertical distributions of radium-226 (A) and barium (B) at the 1969 North Pacific Ocean GEOSECS station (28°N, 122°W) (from Wolgemuth, 1970).

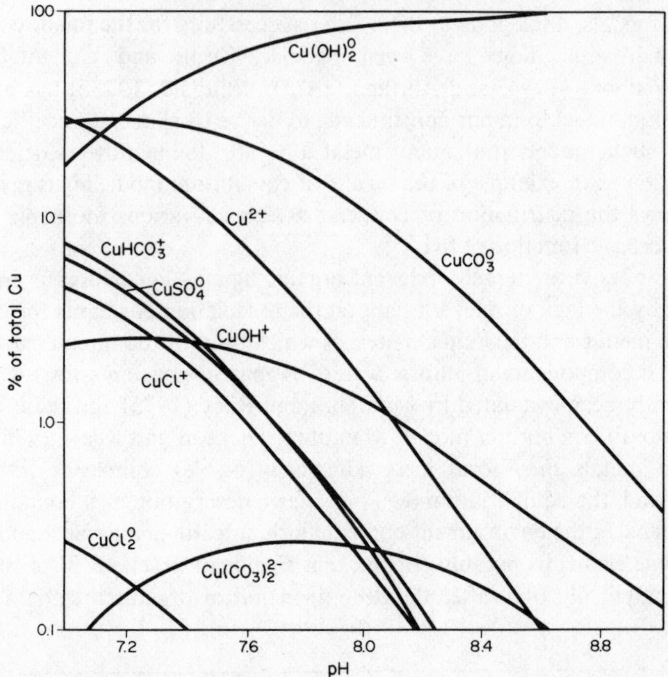

Fig. 2.10. Calculated speciation of copper in seawater at 25°C and 1 atm total pressure, as a function of pH (from Zirino and Yamamoto, 1972).

1969; Foster and Morris, 1971). This does not provide entirely unequivocal evidence for organic associations, since the presence of organically stabilized colloidal material could produce a similar phenomenon, but, taken together with findings of solvent-extractable copper in similar environments (Slowey, Jeffrey and Hood, 1967) it has generally been interpreted as indicative of organic complexing effects. For copper, these operationally defined organic fractions appear less significant in oceanic waters than in coastal waters (Moore and Burton, 1976).

The importance of non-equilibrium factors in the marine chemistry of the dissolved trace metals is emphasized by the occurrence of thermodynamically unstable redox states, such as selenium(IV) (Sugimura *et al.*, 1976; Yoshii, Hiraki, Nishikawa and Shigematsu, 1977; Measures and Burton, 1978) and chromium(III) (Elderfield, 1970). Cycling of trace metals through the biosphere at rates which are considerable relative to the oceanic reservoir content is the likely explanation for such phenomena, which reflect either a steady-state condition for a species which is both thermodynamically and kinetically unstable, or a metastable condition. In some cases, organic associations may serve to stabilize a lower oxidation state in the way which has been demonstrated (Theis and Singer, 1974) for a model system containing iron(II) and tannic acid.

The uncertainties concerning the concentrations, distributions and speciation of some of the biologically important trace metals in seawater limit our understanding of the biological roles of these metals. These limitations are of greatest practical importance in estuarine and coastal environments in which concentrations of toxic metals have in some instances become considerably modified through man's influence. Other questions which arise relate to the possible roles of trace metals in influencing productivity; these are considered in Chapter 7.

Chapter 3
Primary Production

The most essential factor which binds an ecosystem together is the constant interchange of matter. The system is in a sense cyclic since energy is transferred through an ecosystem but is eventually lost, being degraded as heat.

A convenient starting-point in the cycle is the synthesis of organic matter of higher potential chemical energy from substances of lower potential chemical energy. This synthesis in the marine ecosystem involves the reduction of carbon dioxide, producing organic materials – *autotrophic* production. While, as in the terrestrial environment, carbohydrates are frequently a product of the autotrophic process, many substances including amino acids, proteins, lipids and a variety of other organic compounds forming the substance of the living cells result from autotrophic production.

By contrast, a large array of bacteria and other micro-organisms in the marine environment, in addition to all the diversity of animal species, can synthesize the complex materials of their body substance only by using organic substances as a carbon source; they cannot reduce CO_2. Moreover, they require organic substrates as a source of chemical energy, obtained by oxidative phosphorylation, to effect the synthetic processes. Such production is termed *heterotrophic* and is, of course, equivalent to the secondary production typical of the animal kingdom. It is well to emphasize that both autotrophic and heterotrophic production provide high-energy phosphate compounds, hydrogen donors and a range of carbon intermediates from which the building of more complex molecules is achieved (cf. Fig. 3.1).

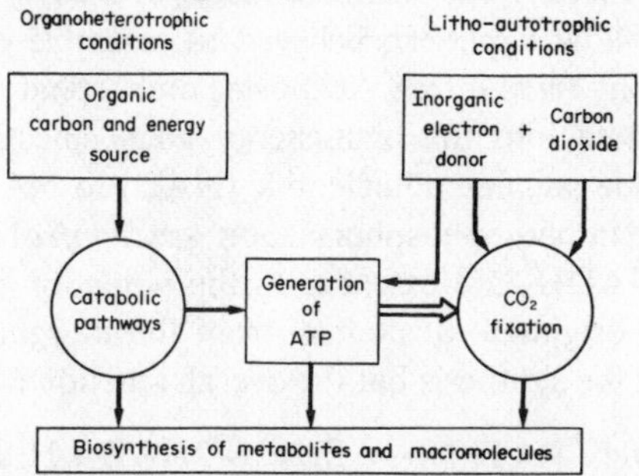

Fig. 3.1. Autotrophic and heterotrophic synthetic pathways (Schlegel, 1975).

Autotrophy demands an input of energy. The overwhelmingly important form of autotrophic production in the sea involves the utilization of solar energy by green plants in photosynthesis. A second very minor type of autotrophic production is due to certain species of bacteria which obtain the energy for the synthetic process by first effecting the oxidation of comparatively simple inorganic substances. The chemical energy thus released is subsequently utilized in the reduction of CO_2 to higher energy, relatively complex, organic compounds – chemo-autotrophy or chemolitho-autotrophy.

Chemo-autotrophic Production

The distribution of chemo-autotrophic bacteria responsible for this type of production in the sea is comparatively little known, though it appears that some species are limited to environments where organic matter is relatively rich, a somewhat unusual condition in the oceans. It is generally believed that few chemo-autotrophic bacteria occur in the seas. The concentration of organic matter is low in ocean water; therefore autotrophic bacteria requiring rich environments are very limited in their distribution. They may be present in association with the comparatively larger quantities of organic matter in littoral areas, and perhaps in deeper areas isolated from general circulation. They occur in the deoxygenated zone of the Black Sea but even there they are mainly limited to the boundary between the oxygenated and deoxygenated layers (Sorokin, 1964).

The types of substances used as electron donors in the oxidations by autotrophic bacteria include reduced or incompletely oxidized forms of hydrogen, sulphur (S, H_2S, $S_2O_3^{2-}$), iron (Fe^{2+}) and nitrogen (NH_4^+, NO_2^-). The final hydrogen acceptor is normally oxygen, then reduced to water, but some bacteria can utilize nitrite or nitrate. Examples include, with molecular H_2:

$$2H_2 + O_2 = 2H_2O$$

with sulphur bacteria using H_2S or S:

$$2H_2S + O_2 \longrightarrow 2S + 2H_2O$$
$$2S + 2H_2O + 3O_2 \longrightarrow 2SO_4^{2-} + 4H^+$$

with iron bacteria converting ferrous to ferric iron:

$$4Fe^{2+} + 4H^+ + O_2 \longrightarrow 4Fe^{3+} + 2H_2O$$

With these obligate chemo-autotrophic bacteria, CO_2 is the sole source of carbon in the synthesis of organic matter. The chemical energy is used by the bacteria in pathways reducing CO_2 which follow a pattern believed to resemble closely that employed in photosynthetic autotrophy (*vide infra*). Adenosine diphosphate (ADP) is converted with the addition of phosphate into the high-energy adenosine triphosphate (ATP). The reduction of nicotinamide adenine dinucleotide (NAD) to $NADH_2$ (comparable to the production of $NADPH_2$ in photophosphorylation – *vide infra*) is also achieved. The two substances ATP and $NADH_2$ are essential to the reduction of CO_2 to carbohydrate. The electrons required originate, of course, from the inorganic donor. A number of enzymes are involved in the synthesis but the overall reaction is:

$$CO_2 + 2NADH_2(+ATP) \longrightarrow CH_2O + 2NAD + H_2O(+ADP + P_i)$$

The pathway of CO_2 fixation appears to involve the Calvin reductive pentose phosphate cycle as in green plants (*vide infra*), with the CO_2 accepted by ribulose-diphosphate, its conversion to phosphoglyceric acid, reduction to triose phosphates and subsequent cyclical interconversions. The role of various enzymes in this cycle will not be described but some discussion is given by Schlegel (1975).

Nitrifying bacteria include several autotrophic forms. Apart from the ammonia-oxidizing *Nitrosomonas* found in soil, *Nitrosocystis* is now known from the marine environment. Nitrite-oxidizing bacteria also occur in the sea and include *Nitrococcus mobilis* and *Nitrospina gracilis*. These latter species are obligate chemo-autotrophs, i.e. they cannot use organic substrates as a carbon source. There are a number of marine bacteria, however, which apparently are facultative chemo-autotrophs, i.e. though they have some ability to perform organic synthesis using CO_2, they can also utilize organic substrates as a carbon source (cf. Schlegel, 1975).

If chemo-lithotrophic bacteria need only the chemical energy from oxidations of simple inorganic compounds to reduce CO_2, they could be regarded as preceding photosynthetic organisms, but most of the species are aerobic and they are probably therefore of more recent evolution.

Chemo-autotrophic bacteria not only have a limited distribution in the ocean. They require comparatively rich organic matter, or ammonia, hydrogen sulphide or other simple compounds which have been derived from the decomposition of previously formed organic matter, even though its formation may have been considerably separated in space and time. This dependence applies equally to all heterotrophic organisms. While they may be of the greatest significance in cycling organic matter and in providing particulate matter suitable as food for higher trophic levels, because they have an absolute requirement for organic matter, both as energy and carbon source, they are entirely dependent on the previous bio-synthesis of such organic material.

Non-photosynthetic organisms cannot, therefore, make a contribution to the synthesis of organic matter from CO_2 and water *ab initio* in the absence of earlier formed matter or its products. Photosynthetic organisms are able to perform this synthesis – photo-autotrophic production – and this is the only real source of primary production in the oceans. (No consideration is given here to the possibility that primitive early heterotrophs may have utilized organic materials, chemically synthesized under the influence of radiation, before CO_2 and O_2 were present in the earth's atmosphere.)

Photo-autotrophic Production

Primary production is dependent in the seas on the photosynthesis of green plants, with a possible very minor contribution of a few species of "green" (i.e. truly photosynthetic) bacteria. The energy necessary for the process and accumulated as chemical energy in the organic matter is derived from light (photoautotrophy). The conversion of light energy to chemical energy is not in itself unusual, but the products of most photochemical reactions are extremely unstable, and when such products decompose, the energy is lost, being degraded as heat. It is a unique feature of green plants that in their conversion of radiant energy the resulting chemical energy is fixed in relatively stable materials which can be accumulated in organisms until they are required, though they may subsequently be degraded with release of energy to the organism. The photosynthetic process is, therefore, the basis of primary production in the marine ecosystem.

Photosynthesis in the marine environment is almost confined, except in exceedingly shallow waters, to production by algae, and even in shallow waters algae play an important role (cf. p. 127 *et seq.*). There is no reason to believe that the photosynthetic process in algae is dissimilar from that of land plants. In photosynthesis it is possible to distinguish between a series of reactions which are entirely dependent on light energy and involve the conversion of radiant energy into chemical energy (the "light reaction") from a further series of reactions where a multiplicity of compounds are formed and interconverted to produce a fairly stable array of carbohydrates and other organic compounds. This latter phase, which can proceed without light energy, is known as the "dark reaction".

Normally the light and dark reactions are linked together; indeed the photochemical reactions proceed so rapidly that algae with even the briefest exposure to illumination show a whole range of carbon intermediates. Nevertheless, it is helpful to separate the two reactions and to understand something of the biochemistry of each process in order to appreciate the factors which affect photosynthesis in the marine environment.

Photosynthetic Pigments

In the photosynthesis of algae the conversion of light to chemical energy is intimately dependent upon chlorophyll. The chlorophyll in the great majority of algae is contained in specific plastids (chloroplasts) inside the cells which also contain other pigments, carotenes and xanthophylls, and in some algae – phycobilins. Such plastids are sometimes known then as chromatophores or chromoplasts (cf. Chapter 4). Chloroplasts vary in size, shape and number, but electron-microscope studies show that each is bounded by a double membrane and that stacks of membranes occur inside the chloroplast. The paired membranes form discs between which is a less dense matrix containing granules, lipid droplets, etc. The discs are the chlorophyll-containing structures and it appears that the regular arrangement of the chlorophyll molecules on the membranes is of the utmost importance in the efficient conversion of light to chemical energy. Even in the blue-green algae, where the plant pigments are unique amongst algae in not being contained in plastids (cf. Chapter 4), there appears to be evidence that the chlorophyll is arranged in a network of lamellae inside the cell.

Chlorophyll exists in several closely allied forms in algae (cf. Chapter 4). Chlorophyll *a* occurs in all groups and shows the basic structure of four pyrrole groups forming a porphyrin ring with a magnesium atom at the centre. To one of the pyrrole groups is attached a long unbranched hydrocarbon chain (phytol "tail") (Fig. 3.2). The different chlorophylls (*a*, *b*, *c* and *d*) show small differences in the side groups. The phytol "tail", common to all chlorophylls, is associated with lipid and the magnesium "centre" with protein. Of the other accessory pigments, the carotenes are long hydrocarbon chains, ending in a ring structure, and with short side groups and a series of double bonds (cf. ß-

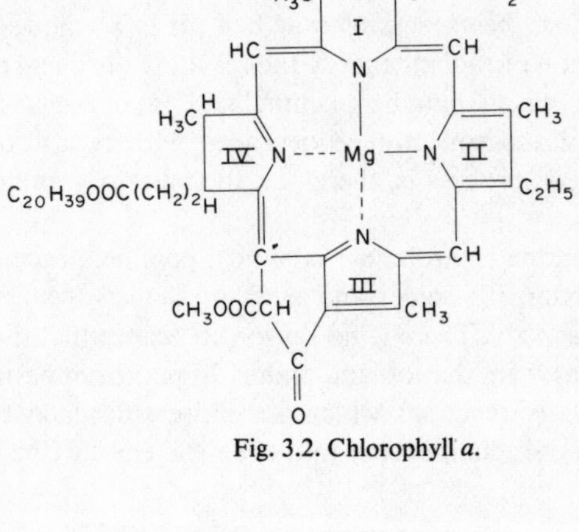

Fig. 3.2. Chlorophyll *a*.

carotene – Fig. 3.3). Xanthophylls (e.g. lutein of higher plants) are oxy-derivatives of carotenes. The phycobilins (biliproteins) are protein-associated pigments with a basic structure of four open pyrrole rings, as illustrated in Fig 3.4. They are divisible into phycoerythrins and phycocyanins. The distribution of these various pigments amongst the algae is described in Chapter 4.

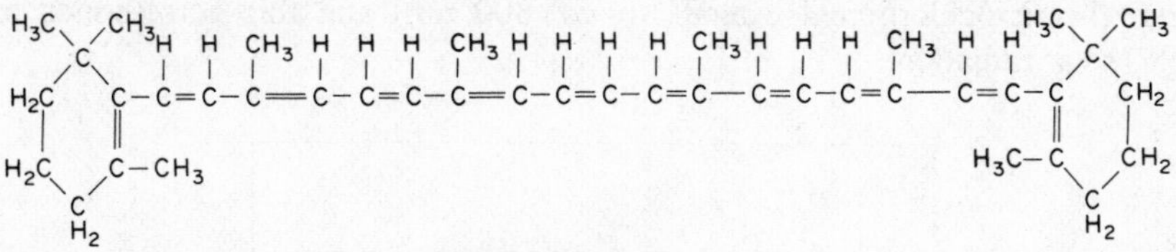

Fig. 3.3. *β*-Carotene.

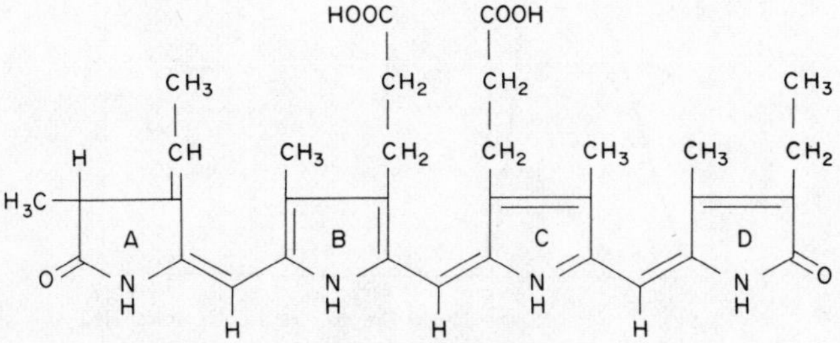

Fig. 3.4. Phycocyanobilin.

Any pigment to be photosynthetically active must absorb light. Chlorophylls have their maximum absorption in the red (650 to 700 nm) and in the blue-violet (*ca.* 450 nm) ranges of the spectrum (cf. Fig. 3.5); the various chlorophylls differ in the precise wavelengths showing maximal absorption (cf. Bogorad, 1962). Carotenoids show the

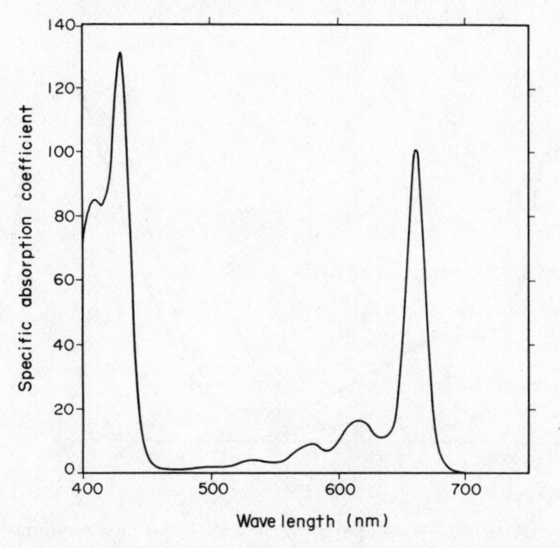

Fig. 3.5. Absorption spectrum of chlorophyll *a* (after Bogorad, 1962).

greatest absorption in the blue-green, e.g. fucoxanthin (Fig. 3.6). Maximal absorption varies between the two groups of phycobilins – phycoerythrins absorb in the blue, green and yellow and phycocyanins in the yellow-red (cf. Fig. 3.7). The absorption peaks for all these pigments show some variation with the solvent employed, and also probably differ slightly between *in vitro* and *in vivo* measurements (cf. Halldal, 1974). The effective range for all the photosynthetic pigments, however, is approximately from 400 to 700 nm (bacteriochlorophyll extends to *ca.* 800 nm) and this corresponds to the "visible" solar radiation.

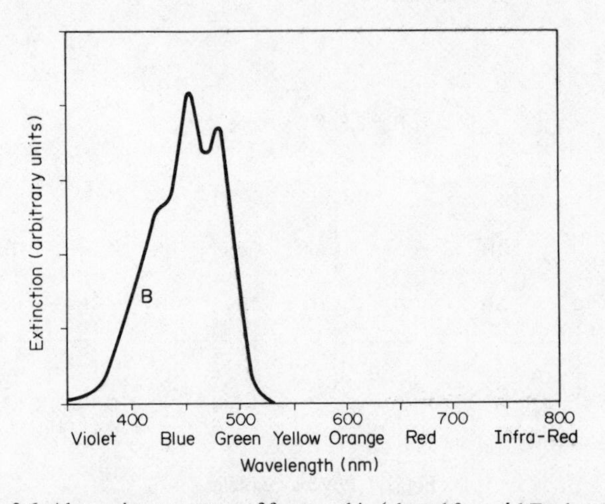

Fig. 3.6. Absorption spectrum of fucoxanthin (altered from ó hEocha, 1960).

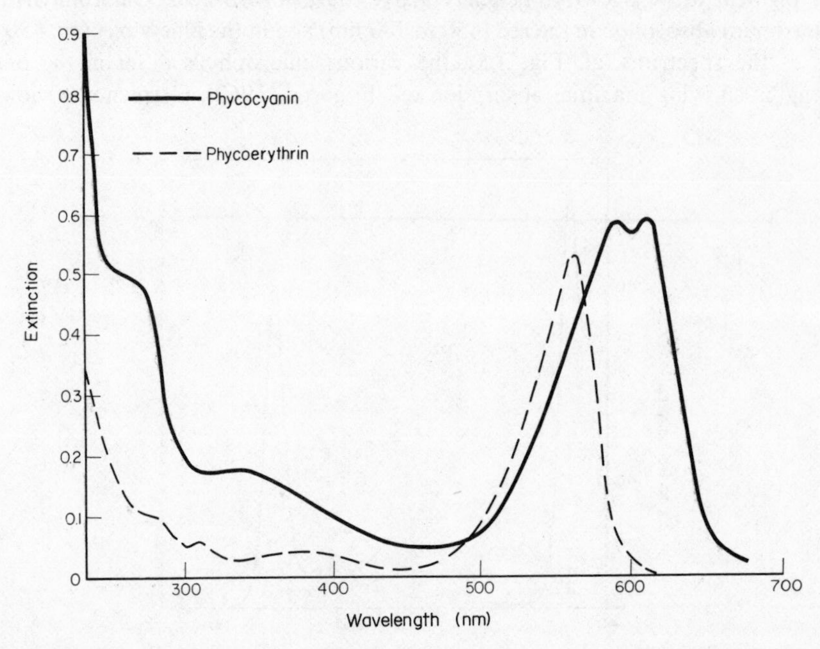

Fig. 3.7 Absorption spectra of (———) phycocyanin and (- - -) phycoerythrin in *Hemiselmis* (altered from ó hEocha, 1960).

The absorption of visible radiation in the sea is described in Chapter 6. It is generally accepted that the universal chlorophyll *a* is the photosynthetically active pigment, but with the rapid attenuation of light in the ocean, especially of the longer wavelengths, the role of the accessory pigments in trapping other wavelengths could be of the greatest significance. Fucoxanthin, which is relatively abundant in diatoms and some other algae (cf. Chapter 4), has been shown to be able to transfer light energy to chlorophyll. The demonstration involves the extent of fluorescence by chlorophyll when diatoms are stimulated by light of different wavelengths. With blue-green light, when about 75% of the energy must be absorbed by the fucoxanthin of diatoms, the fluorescence is about equal to illumination with red light, when the absorption will be by chlorophyll alone. Similarly, the efficiency of photosynthesis by *Navicula minima* was about equal with green light (fucoxanthin only effective) and red light (chlorophyll active only). Indeed, the experiments of Tanada (1951) with this diatom suggest that the quantum yield of photosynthesis, negligible beyond 700 nm, is almost constant from wavelengths of 680 to 520 nm, and falls only slightly at shorter wavelengths, to rise again to a smaller peak at about 430 nm (cf. Fig. 3.8). Cell suspensions may be illuminated by monochromatic light of different wavelengths to investigate the relative photosynthesis at each wavelength. An action spectrum may be obtained showing the quantum yield of photosynthesis as a function of wavelength. Action spectra would be expected to parallel absorption spectra but they do not always follow each other precisely. Strickland (1965) believes that discrepancies are mainly due to illuminating algae by one wavelength at a time; the simultaneous use of two wavelengths gives better agreement. Absorption and action spectra of the various pigments demonstrate that while carotenoids such as fucoxanthin and peridinin are active, with chlorophyll, in photosynthesis, some of the carotenoids do not appear to play a significant part. They may have other important functions such as shading the active pigments against intense

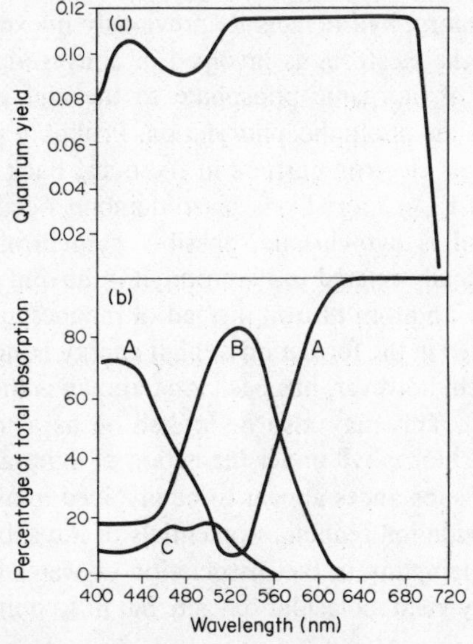

Fig. 3.8. (a) Quantum yield of photosynthesis of the diatom *Navicula minima* as a function of wavelength. (b) The estimated distribution of light absorption among pigments in living cells of *Navicula minima* as a function of wavelength; A, chlorophylls *a* and *c*, B, fucoxanthins, and C, other carotenoids. (Redrawn from T. Tanada, 1951).

radiation (cf. Halldal, 1974). The phycobilins have been proved to be active in utilizing and transferring solar energy to chlorophyll (cf. Haxo, 1960). The reduced photosynthetic activity at *ca.* 470 nm in Fig 3.9 is due to the absorption of light by the inactive carotenoids.

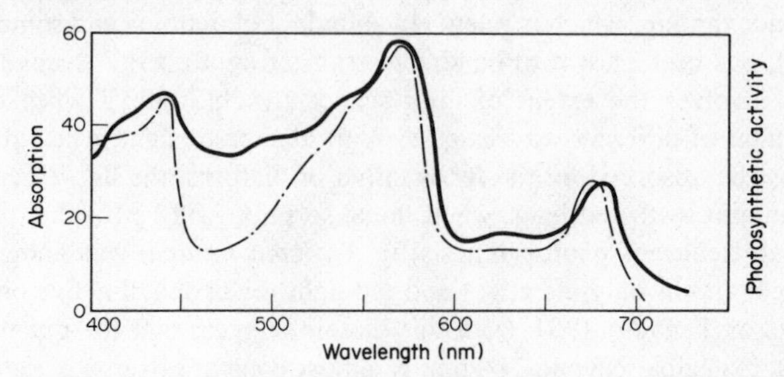

Fig. 3.9. Absorption (———) and photosynthetic action spectra (—·—) for *Phormidium ectocarpi* (after Haxo, 1960).

Before attempting to outline the course of photosynthesis it is essential to emphasize that exact details are still open to debate. Fortunately a number of algae have been used in investigations as well as higher plants. A fuller discussion is given by Fogg (1968). Though the role of chlorophyll in photosynthesis is stressed, the chloroplasts contain a number of other substances which are essential to the process. Many enzymes have been identified; several cytochromes, quinones, flavoproteins and an iron-containing protein, ferredoxin, are amongst those known to be active in electron transport.

When light is absorbed by chlorophyll the molecule appears to become "excited"; electrons acquire a higher energy level. The different energy levels are attained at the expense of the absorbed quanta of light energy. With the return of the electrons to a lower energy level the chlorophyll regains its previously unexcited state. In this return, however, the energy of the electrons is involved in converting adenosine diphosphate (ADP) in the presence of inorganic phosphate to the high-energy bond ATP. This process is thus cyclic–cyclic photophosphorylation. Probably at least two intermediate substances are involved as electron carriers in the cycle back to a lower energy level. The first may be vitamin K, or more likely plastoquinone, occurring in the chloroplasts of algae, and the second is cytochrome, possibly cytochrome B_6 or cytochrome f. (Cytochromes are chemically related to chlorophyll in having a similar porphyrin ring system, but they contain an atom of iron instead of magnesium.) So far the light reaction has, therefore, resulted in the formation of high-energy bond ATP.

Another reaction which, however, has been long known is the reduction of NADP to $NADPH_2$ (cf. Fig. 3.10). This may also be looked on as a transfer of electrons first expelled from "excited" chlorophyll under the action of light, and is coupled with ATP formation. A number of substances appear to be involved in an electron transfer chain which depends on the oxidation-reduction potentials of the substances. It is possible to envisage the chain as originating in the dissociation of water into $(H)^+$ ion and $(OH)^-$ ion; the $(OH)^-$ ions finally yield molecular oxygen, but in so doing, donate electrons.

$$H_2O \xrightarrow{\text{Light}} (H)^+ + (OH)^-$$
$$4(OH)^- \longrightarrow 2H_2O + O_2 + e^-$$

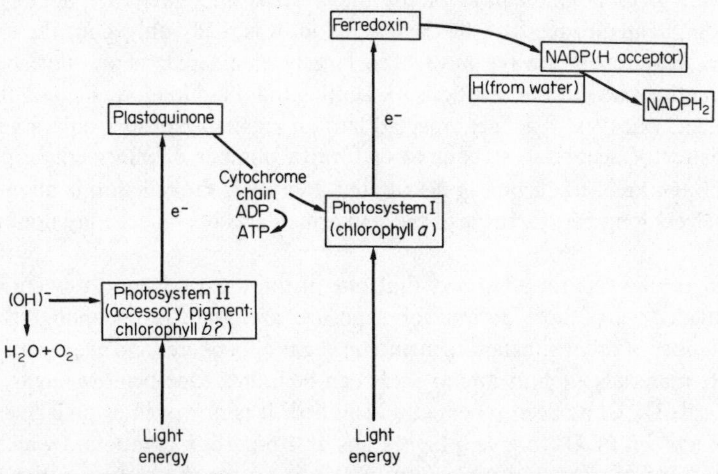

Fig. 3.10. Diagrammatic representation of the main course of photosynthesis.

This is in contrast to the mechanisms in chemo-autotrophic bacteria where incompletely oxidized forms of sulphur, iron or hydrogen are used as hydrogen donors (cf. p. 66). In the algae the transport of electrons from $(OH)^-$ ions is via a photosynthetic system, possibly involving an accessory pigment excited by light. In some algae this could be fucoxanthin, or phycobilins; in algae, such as the Chlorophyceae, which do not possess these pigments, it may by chlorophyll *b*. The energy is transferred to chlorophyll *a*, the electron chain including cytochromes (B_6, f) and probably plastoquinone and a flavin. ATP formation occurs during the transfer (Fig. 3.10). The enzyme systems involved apparently require chloride and manganese ions. The electrons passed to chlorophyll *a*, at a higher energy level due to the light quanta, are passed to nicotinamide adenine dinucleotide phosphate (NADP) (cf. Fig. 3.10). It appears that the iron-containing substance ferredoxin is critically involved at this stage, probably in conjunction with another enzyme – a flavoprotein. Ferredoxin can accept electrons at the redox potential and pass these to NADP which is thereby reduced (cf. Round, 1965; Fogg, 1968).

The overall equations for these light reactions may therefore be crudely expressed as follows:

$$ADP+P_i \xrightarrow{\text{Light}} ATP$$

(Cyclic photophosphorylation)

$$2NADP+2H_2O+2ADP+2P_i \longrightarrow 2NADPH_2+O_2+2ATP$$

(Non-cyclic photophosphorylation)

Although the precise course of even the light reaction in photosynthesis is not fully accepted, in sum ATP and $NADPH_2$ have been produced; water has been split with the evolution of oxygen, but so far there has been no participation by carbon dioxide.

The dark reaction involves the reduction of CO_2 by $NADPH_2$ already formed in the light reaction, with the formation of carbohydrate. $NADPH_2$ thus acts as the hydrogen donor but the potential chemical energy of ATP is required for the reaction. The basic equation is:

$$CO_2+2NADPH_2+nATP \longrightarrow (CH_2O)+H_2O+2NADP+nADP+nP_i$$

This, however, gives no indication of the many steps and different pathways in the extremely complicated process of the dark reaction. It is really only since the use of ^{14}C that the dark reaction pathways have been largely elucidated; algae have been used particularly in the study of the reactions. Assuming that illumination or pre-illumination has taken place, once ^{14}C has been injected into an algal suspension, only a very short period (a matter of seconds) is required to form a number of intermediate products, among which the labelled carbon is distributed. Indeed, if the reaction is allowed to go for an excessively long period, some of these intermediate substances may not be readily recognized.

Chromatographic separation shows that one of the most constantly occurring substances, provided very short periods of exposure are used, is phosphoglyceric acid (PGA). A number of other substances including sugar phosphates and even traces of non-carbohydrate materials such as amino acids can be found. One pentose sugar, ribulose diphosphate (RuDP), has been repeatedly identified. It is now well established that this substance reacts with CO_2 to give a highly unstable 6-carbon compound which almost immediately splits to give two molecules of PGA. So far the stages in the reaction, though enzyme controlled, have not required energy or any source of hydrogen; there has been no reduction of CO_2. The most essential reaction is that under the influence of ATP, PGA adds on another "active" phosphate group and can then be reduced by $NADPH_2$ (produced in the light reaction) forming two molecules of glyceraldehyde phosphate, a well-known triose sugar. This pathway is usually termed the Calvin reductive pentose phosphate cycle (cf. p. 66). Though there are a number of pathways in the subsequent formation of hexoses and pentoses, fundamentally two molecules of glyceraldehyde may be regarded as condensing to give one molecule of a hexose sugar.

The complicated paths which may be followed in the photosynthetic carbon cycle are illustrated in Fig. 3.11. The main pattern observed is that fructose-diphosphate is

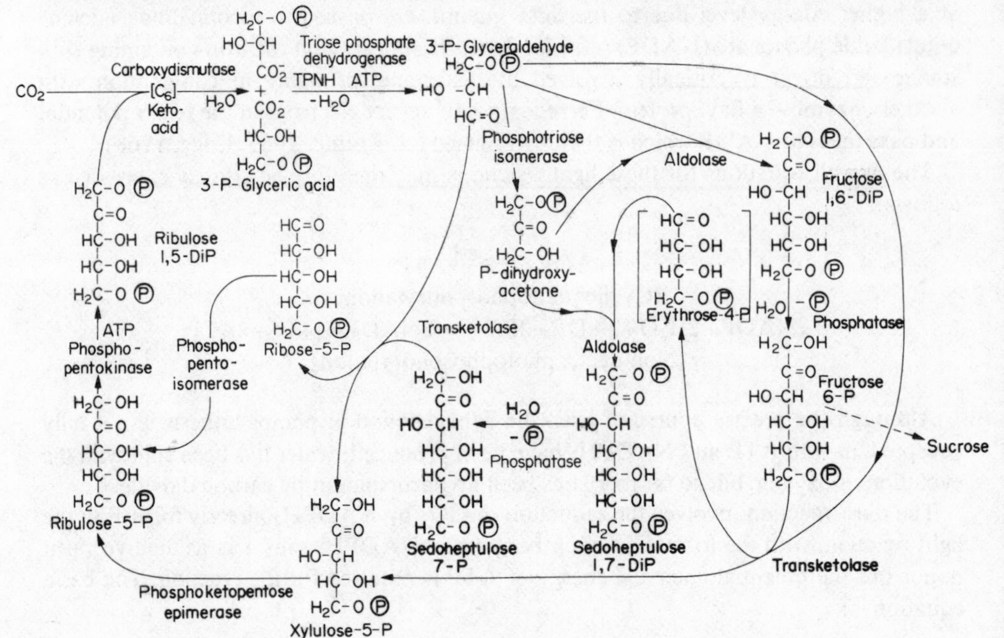

Fig. 3.11. Photosynthetic carbon reduction cycle (Lewin, 1962).

produced, perhaps from glyceraldehyde and dihydroxyacetone under the influence of aldolase enzyme. Fructose-phosphate can react with glyceraldehyde phosphate to produce 4– and 5–carbon carbohydrates. The 4–carbon erythrose-phosphate and another triose, dihydroxyacetone, can react to give a 7–carbon carbohydrate sugar, sedoheptulose diphosphate, which has often been observed in $^{14}CO_2$ experiments. This can transform with glyceraldehyde to give two pentoses, ribose and xylose, and both of these pentoses can then be converted into ribulose 5–phosphate. It is particularly important to see the position of this pentose in the cycle, because ribulose 5–phosphate, under the action of ATP, can add on a further phosphate bond giving ribulose diphosphate (RuDP). This might be regarded as the completion of the cycle commenced when RuDP reacts with CO_2. For simplicity, only one enzyme, aldolase, which catalyses the formation of fructose diphosphate and also the production of sedoheptulose diphosphate, has been mentioned, but a considerable number of enzymes are involved in the variety of pathways (cf. Holm-Hansen, 1962 and Fig. 3.11).

The precise form of photosynthesis in the truly photoautotrophic bacteria is somewhat different from the pattern described for algae. Bacteria capable of performing photosynthesis ("purple" and "green" sulphur bacteria) do not possess the chlorophyll *a* typical of plants, but have similar substances, bacteriochlorophyll and chlorobium chlorophyll. These bacteria are reasonably common where decomposition yields hydrogen sulphide and other reducing substances. They use hydrogen sulphide, molecular hydrogen, or even organic substances, essentially as hydrogen (electron) donors, but no oxygen is evolved during their photosynthetic activity.

Photophosphorylation occurs in photosynthetic bacteria much as in green plants; ATP is produced under the influence of light energy absorbed by bacterio-chlorophyll. Light energy also effects the reduction of NAD to $NADH_2$, through the activation of bacteriochlorophyll and an associated enzyme chain. As in chemo-lithotrophic bacteria, $NADH_2$ is produced rather than $NADPH_2$, but a reduction of CO_2 then follows to carbohydrate, the pathways apparently following the same Calvin pentose reductive cycle.

Dark (Heterotrophic) Fixation of Carbon Dioxide

Most plants, including algae, as well as non-photosynthetic organisms are able to fix CO_2 by metabolic paths other than the photosynthetic "dark reaction". Metabolic reactions involving liberation of CO_2 are usually reversible, and though de-carboxylation is favoured under normal conditions carboxylation can occur, though the amount would appear to be relatively small and require the expenditure of metabolic energy. Labelled ^{14}C entering Krebs cycle intermediates during dark fixation may also pass into several amino acids after transamination. Thus Holm-Hansen (1962) notes that alanine with labelled ^{14}C often occurs in dark fixation experiments. Among other pathways he quotes the formation of pyruvate from acetylphosphate:

$$CH_3CO.OPO_3^{2-}+H_2+CO_2 \rightleftharpoons CH_3COCOOH+HOPO_3^{2-}$$

The pyruvate may then partake in transamination reactions.

Yentsch (1974) points out that while in eutrophic waters dark fixation may be commonly found to amount to 5 to 10% of the carbon fixation in light, there is great variability in the extent of dark fixation in oligotrophic areas. The level of dark fixation also appears to be related to population size (cf. Morris, Yentsch and Yentsch, 1971a; also Chapter 9). There is considerable difficulty therefore in measuring primary production in oligotrophic tropical oceans, since there is little uniformity amongst workers in assessing dark fixation; some investigators subtract the variable values for dark fixation from the light carbon fixation measurements, some assume a constant dark fixation value, and some disregard it.

Zajic and Chiu (1970) point out that some algae are facultative heterotrophs, i.e. they are able to utilize certain organic materials as a carbon source in cellular photosynthesis instead of CO_2; some species are obligate heterotrophs, i.e. at least part of their carbon *must* come from organic compounds. In some, organic nitrogen may be utilized in addition to organic carbon. In *Chlorella* apparently glycine can be used as nitrogen source, but is useless as a carbon source. Heterotrophic growth is accelerated by light, though obviously it proceeds also in darkness.

Products of Photosynthesis

During normal photosynthesis hexose sugars are frequently formed and they may be regarded as some of the relatively stable products, but disaccharides can be produced, and even better known are the condensations of hexose units which give a variety of polysaccharides. Starch, though perhaps the best known polysaccharide, occurs plentifully only in certain algae, particularly Chlorophyceae. Laminarin and closely related substances are constantly occurring polysaccharides in certain algae, and a whole variety of other carbohydrate materials are also found (cf. Percival, 1968).

Observations (Meeuse, 1962; Kreger, 1962; Handa, 1969; Handa and Tominaga, 1969) have suggested that the carbohydrates produced by algae can be divided approximately into the somewhat more stable "wall" polysaccharides, generally more resistant to degradation, and "reserve" carbohydrates which are called upon more immediately during periods of respiration. Analyses by Parsons, Stephens and Strickland (1961) demonstrated a variable quantity of crude fibre, more or less equivalent to wall carbohydrate, in different algal species. Apart from cellulose, which, though occurring in some algae (e.g. Chlorophyceae), is not universal, many other polysaccharides may be constituents of "wall" material, especially xylans and mannans (Kreger, 1962), as well as uronic acids. Mucilages may also make an important contribution. Their exact composition appears to vary in different algae, but when broken down they yield varying proportions of hexoses (especially glucose, mannose, galactose), pentoses (xylose, arabinose, etc.), as well as uronic acids; a considerable proportion of the polysaccharide in mucilages is sulphated (cf. Chapter 4). Condensed forms of uronic acids may also occur. Diatoms have cell walls (valves) which contain silica intimately related to such a kind of organic material, usually referred to as pectin. Pectin is a polygalacturonid acid in which the (COOH) group is methylated. However, according to Lewin (1962) the "pectin" of diatoms has not been completely identified.

Handa (1969) found that when *Skeletonema* was placed in the dark after a period of active photosynthesis a loss of over 40% of carbohydrate occurred and the materials left, which are relatively resistant, were mostly mannans and pentosans. The amount of reserve carbohydrate suggested by Handa is to some extent confirmed by Ricketts (1966) who obtained a loss of about 37% of the carbohydrate when algae were placed in the dark. Reserve carbohydrates, apart from starch, include laminarin, chrysolaminarin, and related substances, such as leucosin. There is considerable discussion in the literature as to the relationship of chrysolaminarin, chrysose, leucosin and laminarin. Some of these carbohydrates apparently only differ by the groups at the end of the long polysaccharide chains and the first three may be identical. The important reserve polysaccharide, laminarin, is typical of the Phaeophyceae (cf. Chapter 4). As Meeuse (1962) has indicated, laminarin may be a mixture of nearly identical polysaccharides (cf. Percival, 1969), but essentially the main structure is glucose units with linkage through $\beta-1{:}3$ bridges (see Fig. 3.12). Some of the chains have mannitol as end units. Leucosin and chrysolaminarin (= chrysose?) appear to be similar, except that mannitol units are absent from the ends of the chains. It seems likely that in some algae the reserve carbohydrate is not entirely standard in that a mixture of very similar carbohydrates may be present. They may differ, for instance, in whether any of the chains exhibit branching.

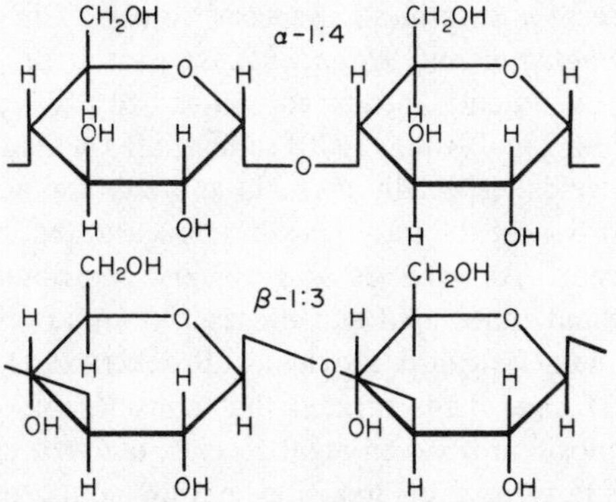

Fig. 3.12. α1–4 and β1–3 linkages in algal polysaccharides (after Morris, 1967).

Although starch is not so commonly found in algae, apart from the Chlorophyceae, a similar compound is found in the Cryptophyceae and Euglenophyceae and in dinoflagellates (cf. Chapter 4). In the red algae a special product occurs which differs only slightly from normal starch ("floridean starch"); this may be identical with the starch found in blue-green algae. In "true" starch, "floridean" starch and "cyanophycean" starch the glucose units of the chains are united through $\alpha-1{:}4$ linkages (cf. Fig. 3.12).

In addition to polysaccharides, sugar alcohols also occur as reserve materials in algae. Among the variety observed in different groups, glycerol is occasionally found, but the most important is mannitol which forms a conspicuously large proportion of dry weight in such brown weeds as *Laminaria*.

Although "simple" sugars, such as galactose, mannose, fructose, glucose, arabinose and ribose, have been recorded for most groups of algae, the amounts of these products would in general appear to be relatively very small. A variety of organic acids have been found in algae and some of these may be important as extracellular products. Glycollic acid is regarded as particularly important by Fogg (1966) and may be excreted as an extracellular product.

Among the other products of photosynthesis are lipid materials. Some algae maintain only a relatively small reserve of lipids but they are common in diatoms, the Chrysophyceae, and dinoflagellates (cf. Chapter 4). Parsons *et al.* (1961) claim that for many species of phytoplankton the gross biochemical composition is similar (cf. also Handa,

1969). Many phytoplankton species, however, show distinct variations in lipid (and other components) with age of culture; it is generally accepted, for instance, that "older" cultures, especially those experiencing nitrogen deficiency, tend to have increased total lipid.

Lipid synthesis almost certainly starts from acetyl coenzyme A, most probably formed from pyruvic acid. Phospho-pyruvic acid is relatively easily derived from PGA (cf. Fogg, 1968)

$$CH_2O \sim P.CHOH.COOH \longrightarrow CH_2{:}CO \sim P.COOH + H_2O$$

From acetyl coenzyme A, fatty acids may be produced, though the reduction which occurs in the formation of fatty acids almost certainly involves $NADPH_2$. Triose sugars may also produce glycerol so that neutral fats may be produced. It is somewhat contradictory, however, that when stimulated by addition of glycerol in light, although growth rate is enhanced for several algal species (mostly Chrysophyceae and Crytophyceae: also *Phaeodactylum* and *Nannochloris*), starch or carbohydrates such as leucosin appear to be mainly produced. Thus any effect appears to be on gluconeogenesis (cf. Cheng and Antia, 1970) rather than on lipid synthesis.

Although saturated acids, especially myristic and palmitic acid ($C_{14:0}$; $C_{16:0}$),* occur in algae, a large proportion of the fatty acids are unsaturated and some relatively long chain compounds occur. As well as fatty acids contributing to glycerides and phospholipids, sterols and other lipid substances are found (cf. Chapter 4). Detailed investigations by Ackman, Jangaard, Hoyle and Brocker (1964) (e.g. Ackman, Tocher and McLachlan, 1968) have demonstrated that considerable differences exist in the precise fatty acid composition between algal species, but that age of culture and other factors may influence fatty acid composition, a view supported by Pugh (1971) (cf. Chapter 4). Changes also occur in the fatty acid constituents of phytoplankton collected from the field. Jeffries (1970) has demonstrated seasonal changes in both the phytoplankton and microzooplankton.

Even a relatively short exposure of $^{14}CO_2$ to algae under illumination results in the formation of ^{14}C labelled amino acids. The synthesis of amino acids is markedly reduced when combined inorganic nitrogen is present in relatively low concentration. The formation of amino acids and their ultimate polymerization to form proteins is greatly accelerated whenever photosynthesis occurs. The same substance PGA occupies an important position in the formation of amino acids. Phospho-pyruvic acid seems to be derived first from PGA. This can react with ammonium directly, or with other nitrogen sources which are reduced to ammonium, and in the presence of $NADPH_2$ gives rise to an amino acid such as alanine (Fogg, 1968).

$$CH_2{:}CO \sim P.COOH + NH_4^+ + NADPH_2 \longrightarrow CH_3.CH(NH_2).COOH + NADP + P_i$$

Pyruvic acid may also act with CO_2 to give the double carboxyl acid, oxaloacetic (Fogg, 1968). Thus with $^{14}CO_2$:

$$CH_2{:}CO \sim P.COOH + {}^{14}CO_2 \longrightarrow HOO^{14}C.\ CH_2CO.COOH + P_i$$

In the presence of ammonium, oxaloacetic acid may be converted into aspartate. Although alanine and aspartic acid may be identified among the first amino acids, others may be produced. Presumably extensive transamination takes place in plants

* For explanation of fatty acid notation, see Chapter 4.

which finally gives the assortment of some twenty or so common amino acids. An example is the conversion of aspartic to glutamic acid (and vice versa), under the influence of glutamic-oxaloacetic transaminase:

$$\begin{array}{lllllll}
\mathrm{COOH} & & \mathrm{COOH} & & \mathrm{COOH} & & \mathrm{COOH} \\
| & & | & & | & & | \\
\mathrm{CH_2} & & \mathrm{CH_2} & & \mathrm{CH_2} & & \mathrm{CH_2} \\
| & & | & & | & & | \\
\mathrm{CH_2} & + & \mathrm{C=O} & \rightleftharpoons & \mathrm{CH_2} & + & \mathrm{CH(NH_2)} \\
| & & | & & | & & | \\
\mathrm{CH(NH_2)} & & \mathrm{COOH} & & \mathrm{C=O} & & \mathrm{COOH} \\
| & & & & | & & \\
\mathrm{COOH} & & & & \mathrm{COOH} & &
\end{array}$$

glutamic+oxaloacetic $\rightleftharpoons$ 2-oxoglutaric+aspartic.

Various amino acids condense to form the array of proteins found in tissues. With these complex transformations, which to a large extent stem from the formation of PGA, the three major biochemical components, carbohydrates, lipids and proteins, are formed, from which the cellular protoplasm of the algae is ultimately constituted.

Holdsworth and Colbeck (1976) have recently examined pathways of CO_2-fixation in the diatom *Phaeodactylum*. The initial product in short-term fixation experiments was confirmed as phosphoglyceric acid (PGA), but ^{14}C-labelled aspartate was also rapidly produced; the likely pathway would appear to be via phosphoenolpyruvate and oxaloacetate. A variety of carbohydrates (free sugars, phosphate esters, polysaccharides including chrysolaminarin or a very similar substance and "wall" polysaccharides) and a substantial proportion of lipid materials (fatty acids, carotenoids) and of amino acids, all carrying the ^{14}C label, were also formed in somewhat longer CO_2-fixation exposures. Apparently wavelength of light could influence the photosynthetic product (cf. p. 103).

Much is known about the synthesis of particular components of plant tissues: the various plant pigments (chlorophylls, carotenoids, etc.), certain vitamins, and those proteins constituting the enzyme systems of algal cells. Nevertheless, the essence of primary production is the building up of relatively stable high-energy organic materials from low-energy substances.

Factors influencing Photosynthesis

A number of substances are essential to the photosynthetic process, some of the compounds such as water, carbon dioxide (mainly as bicarbonate) and several other substances being always present in excess in the sea (cf. Chapter 2). One of the essential requirements of algae, in common with other green plants, is for nitrogen in the synthesis of amino acids and their products. They also have an absolute requirement for phosphorus, for example, in the formation of nucleic acids. The provision of phosphate ions which play such a significant rôle in phosphorylations is also essential, and although the majority of phosphorylation reactions are cyclic, some phosphate is probably lost. Other phosphorus compounds in living organisms include the important, mainly structural, phospholipids. The cell sap of algae contains a variety of cations and anions, usually giving a remarkably constant chemical composition. Some of these (e.g. Na^+, K^+, Mg^{2+}, Ca^{2+}, Cl^-, SO_4^{2-}) are abundantly present in seawater; other elements may occur only at great dilution in the sea, though algae may have an absolute requirement for them. In addition to the requirements for the delicately balanced cellular fluid, some metals are essential for the formation of particular compounds, for example, manganese, copper, and zinc in certain enzymes, iron in the cytochromes and ferredoxin, as well as other compounds. Magnesium, which is present in relatively large concentrations in the sea, forms part of the chlorophyll molecule.

Clearly certain elements in seawater are essential to the photosynthetic process and for the whole metabolism of algal cells, and some nutrients could be limiting factors in photosynthesis. A detailed analysis of the role, and of the supply and demand of essential nutrients, is given in Chapter 7. The other most obvious requirement for photosynthesis is light. Marine plant life, whether phytoplankton, macro- or microbenthic algae, must be limited by the depth to which light can penetrate, and light is absorbed very rapidly in water (cf. Chapter 6). The rate of photosynthesis will fall as the light intensity decreases and at a point where the light intensity is very low, the photosynthetic activity over a 24-hour period will be balanced by the catabolic processes due to respiration. The depth at which this balance is achieved is known as the "compensation depth" and the limiting light as the "compensation intensity".

Early studies on photosynthesis to estimate compensation intensity used the oxygen production technique in which water samples with their natural plankton populations were suspended at various depths, shading being avoided by the string of bottles being arranged on a slanting rope. The change in oxygen (Δ O) from the initial concentration was determined for each bottle. If photosynthesis is represented as *P*, then the change in oxygen is:

$$\Delta \mathrm{O} = P - R_p - R_n - R_{(z+b)}$$

where R_p = respiration of plant tissue originally present;
R_n = respiration of newly formed algal tissue;
$R_{(z+b)}$ = respiration of zooplankton and bacteria.

The rate of photosynthesis measured in this way may be regarded as equivalent to nett photosynthesis, or nett production. If a parallel series of bottles be placed alongside the existing ones, but every such alternate bottle is blackened to exclude light, the black ("dark") bottles will measure the total respiration due to original plant tissue and to zooplankton and bacteria. If the amount of oxygen consumed in these "dark" bottles is added to the oxygen change in the "light" bottles at each depth, the gross or total production at each depth can be approximately evaluated. The respiration of the newly formed plant tissues (R_n) has not been included, but the error will be small in short-term experiments. Gaarder and Gran (1927) carried out one of the earliest experiments using this technique, in a Norwegian fjord, and found a compensation depth of approximately 10 m for experiments during March.

For each species of phytoplankton there is a minimum light intensity (compensation intensity) which is necessary to achieve minimal overall primary production. In view of the varying dark and light periods during a day, it is usually more satisfactory to calculate the compensation intensity for a species over a 24-hour period, but it is perfectly feasible to estimate a compensation intensity over some shorter time scale. Above the compensation intensity, provided other factors remain optimal, production increases more or less linearly with increase in light energy. With further rise in light intensity, the rate of increase in production tends to fall, though production continues to rise, until a level is achieved at which the response is maximal; any additional increase in energy is not followed by a concomitant increase in photosynthesis. The quantity of light necessary for maximal photosynthesis is known as the saturation intensity. The rate of carbon assimilation in most phytoplankton species tends to remain more or less constant with considerable further increase in light energy, but at a much higher light energy value, photosynthetic activity begins to decrease in the majority of species, i.e. an inhibitory effect is apparent at high light intensities (cf. Fig. 3.13).

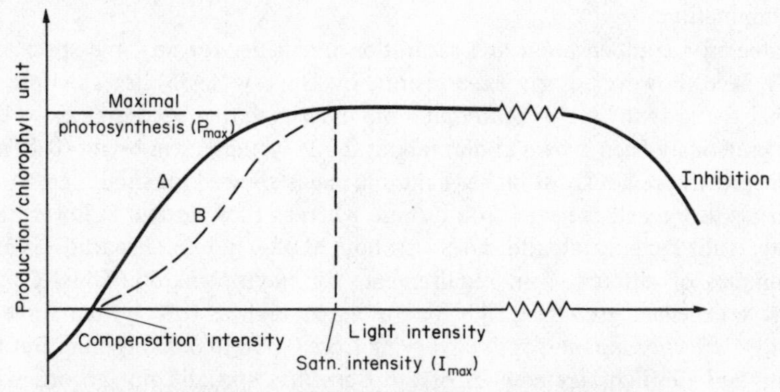

Fig. 3.13. Relation between photosynthetic rate and light intensity.

The early part of the increase in photosynthesis with increasing illumination is more or less linear, but there has been considerable discussion over the precise shape of the production/light intensity curve. The type B curve (Fig. 3.13) gives a relatively poor fit to most of the data. Although there may be some species variation, most phytoplankton tends to give a rectangular hyperbola type of relationship (approximately type A, Fig. 3.13). Photosynthetic productivity must be related to biomass; usually productivity is normalized to chlorophyll content (cf. Fig. 3.13). At saturation intensity the maximum production can then be expressed as mg C/mg chl *a*/hr – often termed the assimilation number. The *P*/*I* relationship essentially describes the photochemical ("light") reaction and, not surprisingly, is more or less independent of temperature. Platt, Denman and Jassby (1977), describing the problem of the precise form of the *P*/*I* curve for phytoplankton algae, point to small variations from the rectangular hyperbola. A total *P*/*I* curve should also take into account any photo-inhibition effect.

At near saturation intensity the "dark" reaction with related enzymes will play the major part; P_{max}, in the absence of limiting factors, will reflect the enzymatic activity level of the "dark" reaction. The initial slope of the *P*/*I* curve, however, is independent of any photo-inhibitory factor and essentially is dependent on the light reaction. Both the assimilation number and the initial slope value show a considerable range (0.1–20 mg C/mg chl *a*/hr, and 0.01–0.20 mg C/mg chl *a*/hr, respectively). Although the initial slope is much less variable, Platt *et al.* (1977) show that it is not constant but may vary, for example, seasonally. For a variety of phytoplankton species it can also be demonstrated that one quantum of light activates (or is absorbed by) the same number of chlorophyll molecules (8), within reasonably narrow limits. The amount of "negative production" at zero light intensity (Fig. 3.13) which is probably equivalent to basic cell respiration, is a matter of dispute; probably in part it is species specific.

While the *P*/*I* relationship gives a general pattern for phytoplankton, the compensation intensity and the saturation intensity as well as P_{max} and the initial slope vary for different species. In Gaarder and Gran's early experiments, the species of phytoplankton with which they were mainly concerned were *Lauderia borealis, Thalassiosira gravida* and *T. nordenskioldii*. While *T. gravida* and *Lauderia* multiplied most rapidly in the upper 2 m and possibly right at the surface, *T. nordenskioldii* showed best growth at 2 to 5 m, or possibly deeper, indicating lower light requirements. An average value for compensation intensity has been accepted as equivalent to 1% surface illumination.

The values for compensation and saturation intensities for any one species are not absolutely fixed, however. Early experiments, by Harvey (1955), for example, demonstrated that the growth rate of *Biddulphia mobiliensis* varied according to whether the cells had previously been grown in dim (about 0.024 ly/min) or in bright (0.108 ly/min) light. Cells grown previously at higher light intensities showed maximum growth in subsequent trials at intensities near 0.108 ly/min, whereas those grown at lower intensities grew better subsequently at intensities at about 0.024 ly/min. Braarud (1961) gives some examples of different light requirements for phytoplankton. Thus, *Coccolithus huxleyi* gave excellent growth at light intensities as high as 0.30 ly/min while, on the other hand, *Ceratium* spp. had optima varying from 0.015 to 0.030 ly/min. But Braarud points out that the light preferences of the *Ceratium* spp. did not accord with their vertical distribution. There are often substantial differences between the light requirements of phytoplankton species as determined in the laboratory and as observed in the field. Such differences may reflect the physiological state of the cells, and this is partly influenced by their past history including light acclimation, a view earlier expressed by Harvey (1955) (cf. also Maddux and Jones, 1964). Not only is acclimation experienced to light intensity, acclimation to temperature affects subsequent light and temperature optima. This is not surprising since the "dark" (non-photochemical) reaction in photosynthesis is a series of essentially enzyme-controlled reactions, temperature and substrate concentrations being two obviously important factors in such transformations. How far adaptation to light intensity involves more than variation in chlorophyll content (cf. p. 87) may be tested by normalizing production to chlorophyll. The role of accessory pigments is a possible factor in adaptation (*vide infra*). Since, however, environmental factors such as light, nutrients and temperature can affect pigment concentration and the whole cellular metabolism, the physiological history of algae must affect the *P*/*I* curve. Moreover, in addition to longer-term adaptations, short-term rhythms in cell physiology can cause variations in productivity in relation to light intensity, even though other factors are constant and optimal.

While such factors as light, temperature and nutrient level and to a lesser extent, pH, and other variables affect production, these are not independent but interact so that the effects of previous acclimation to a particular set of conditions can affect subsequent growth in a complex manner. Curl and McLeod (1961) investigated some of the interrelationships in the diatom, *Skeletonema costatum*. At temperatures between 5 and 18°C the photosynthetic rate increased with temperature, and the light-saturation value was fairly stable at intensities of between 0.072 and 0.096 ly/min. At temperatures from 20–30°C photosynthesis was diminished but the saturation intensity was also reduced to a value of about 0.030 ly/min. Provided nutrients (nitrate and phosphate) were present in maximal amounts, the temperature optimum approached 20°C, but when nutrients were in limited supply, not only did the photosynthetic rate decline but the temperature optimum was lowered (cf. also Smayda, 1963b).

Jones in Halldal (1966) reports a somewhat comparable relationship between light intensity and nutrient level. Continuous cultures of *Nitzschia closterium* and *Carteria* sp. were grown at constant temperature (16°C) at different light intensities ranging from approximately 0.0048 to 0.024 ly/min. By markedly increasing the nutrient concentrations, the algae not only showed a greatly enhanced growth rate but were able to grow at much higher light intensities. Thus for *Carteria*, whereas maximal growth occurred in relatively low nutrient levels at about 0.0072 ly/min, when nutrients were increased about a thousand times, maximal growth occurred at about 0.018 ly/min (cf. Fig. 3.14). Temperature/growth relationships were also modified by nutrient level. *Nitzschia* showed an optimum temperature of 15°C at low nutrient concentrations; the optimum rose to 23°C at the increased nutrient level. Ignatiades and Fogg (1973) found that the response of *Skeletonema* to total carbon dioxide concentration depends on the light intensity. At comparatively high intensities photosynthesis reached carbon saturation at much higher total CO_2 concentrations.

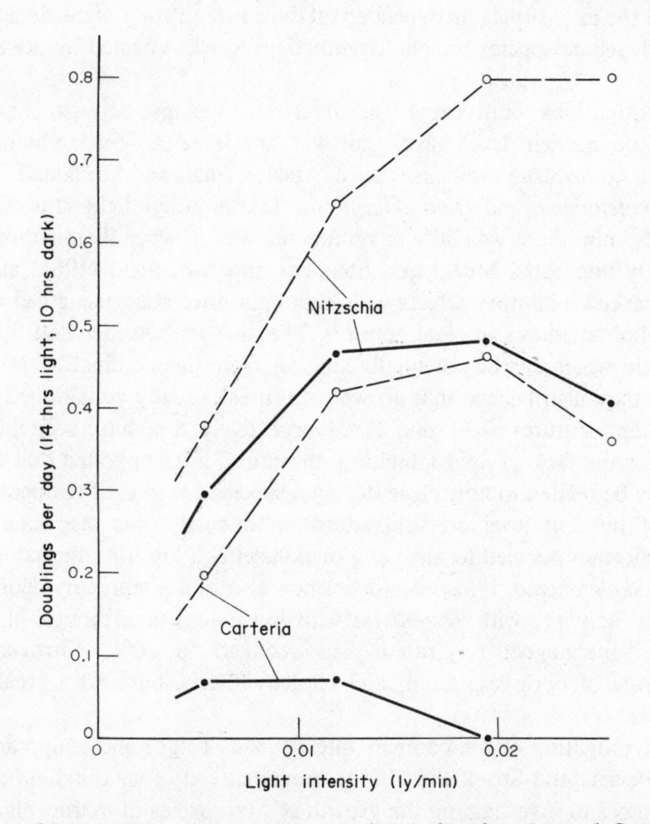

Fig. 3.14. Effect of light intensity on the growth rate of *Nitzschia closterium* and *Carteria* sp. at two nutrient levels. At 140 mg N/l and 15.5 mg P/l (dashed line) and 125 μg N/l and 12.5 μg P/l (solid line). Temperature constant at 16°C (from Jones in Halldal, 1966).

Algal cultures, maintained under standard conditions, may show varying photosynthetic rates depending on their state of growth (age), which may be accompanied by changes in enzyme concentrations and other cellular components. Thus optimal conditions for temperature, light and nutrients may change during growth and their complex inter-relationships are not unexpected. Moreover, the factors promoting nuclear division, membrane and wall formation, and other aspects of cell division will not be identical in their effects, nor necessarily have comparable effects on photosynthetic rate. The experiments by Akinina (1966), dealing with two dinoflagellates, *Prorocentrum* and *Gymnodinium*, illustrate some of these complex effects in relation particularly to light-saturation intensities. For *Prorocentrum micans* light saturation occurred at intensities of about 0.096 to 0.120 ly/min but the precise level varied with the duration of experiment and age of culture. Thus after 10 days' growth the light saturation intensity was less that half. Even more striking results suggested that whereas in very young, actively dividing, cultures saturation intensities reached as high as 0.36 ly/min, for very old, slowing dividing, cultures the saturation intensity might be a tenth of the value. The level of light saturation also depended on the nutrient available; the intensity was more than halved with phosphate impoverished cells. Fewer experiments were carried out on *Gymnodinium covolesci*, though a relatively high light saturation intensity approaching 0.096 to 0.120 ly/min was obtained, and the intensity again depended on the concentration of nutrient (phosphate). In both dinoflagellate species the photosynthetic rate was affected by pre-experimental illumination.

Some investigations concerning the inter-relationships between light intensity, temperature and nutrient level have dealt with the level of light inhibition as well as saturation and compensation intensities. McAllister, Shah and Strickland (1964), using cultures of *Skeletonema* and *Dunaliella*, found that although light saturation occurred at about 0.1 ly/min, there was little or no inhibition even when illumination reached as much as 0.4 ly/min. Jitts, McAllister, Stephens and Strickland (1964) also noted an absence of marked inhibitory effects with high light intensities, using cell division as a criterion of photosynthesis in algal cultures. The authors consider that the absence of ultra-violet light which can be potentially limiting, from the illumination, may have been a factor. But they also believe that growing cultures, already conditioned to relatively high optimal temperatures of 20° and 25°C, respectively, may have been the more likely explanation for the lack of light inhibition. In nature, it is suggested that the extent of inhibition may be related to how close the algal species are to their temperature optima. The effect of nutrient level on light-saturation intensity was also examined. Only phosphate deficiency seemed to have any obvious effect, lowering the saturation value, especially in *Skeletonema*. Phosphate-deficiency also had a markedly depressing effect on respiratory activity; with *Dunaliella* with low phosphate, though photosynthesis was virtually unchanged, respiration was reduced to 20%. Nitrogen deficiency lowered the rate of both respiration and photosynthesis, but had a greater effect on photosynthesis.

Another investigation of the complex interactions of light and temperature by Jitts, McAllister, Stevens and Strickland (1964) employed sixty-four different combinations of the two factors in investigating the growth of five species of marine algae. The data on *Thalassiosira nordenskioldii* might be of particular interest since the diatom has been shown to grow well at low temperatures, but only at relatively low light intensities.

These conditions, low temperature and low light intensity, are typical of early spring in temperate latitudes when *T. nordenskioldii* flourishes in the field. The diatom may also appear, however, in considerable quantities in the summer plankton. Smayda and other authors have suggested that the changed light requirements may be dependent on nutrient conditions. The results of Jitts *et al.* (1964) suggest that the reappearance of the diatom in the summer is consistent with its proven ability to grow at much higher light intensities, provided temperature is raised also.

One of the problems in evaluating saturation intensities and other light characteristics is to separate species specific differences from conditioning effects of the environment. Thus Ryther (1956a) suggests saturation intensities for several diatoms, including *Skeletonema*, slightly exceeding 0.1 ly/min, agreeing with Jenkin's (1937) value for *Coscinodiscus* and with Talling's (1960) data for *Chaetoceros*. Jitts *et al.* (1964) give a saturation intensity for *Skeletonema* which is only slightly lower, but these latter investigators find a considerably lower intensity (0.02 ly/min) for *Thalassiosira*. While this may be a genuine specific difference, it might be attributed to environmental conditioning which may modify light preferences considerably.

In this connection, the light and temperature characteristics of tropical phytoplankton species, especially those living in the upper part of the euphotic zone, might be expected to show marked differences from species from higher latitudes. There appear to be rather few data for warm water forms.

With regard to light, Qasim, Bhattathiri and Devassy (1972a) tested eleven species of marine tropical algae and found that saturation intensities (as I_k – cf. p. 93) for most species ranged between 0.066 and 0.090 ly/min* but *Dinophysis miles* and *Rhizosolenia styliformis* had exceptionally high saturation intensities (0.21 and 0.14 ly/min, respectively). These intensities are much higher than those observed for most temperate algal species; Steemann Nielsen and Jensen were also quoted as having found a high value (0.18 ly/min) for tropical phytoplankton. Qasim *et al.* (1972) found that for each species the relationship between illumination and photosynthesis was approximately linear at lower light intensities (Fig. 3.15). At suprasaturation intensities, many species (e.g. *Rhizosolenia styliformis, Planktoniella sol, Ceratium furca*) showed obvious inhibition; with others (e.g. *Triceratium favus, Coscinodiscus radiatus*) no inhibition was observed, at least up to intensities of about 0.18 ly/min, and apparently, for these two species, even at full sunlight. For all species photosynthesis was related to the total radiant energy within the spectral range, 700–430 nm, though *Dinophysis* showed a peak response between 700 and 538 nm. The phytoplankton showed no obvious dependence of photosynthesis in saturating light conditions on wavelength. Despite the variability of carotenoids and chlorophylls in marine phytoplanktonic algae, Qasim *et al.* (1972) found that the response of these tropical algae to variation in the quality and intensity of the light was generally rather similar. Findlay (1972), working with a tropical species of diatom, *Coscinodiscus pavillardii*, found that saturation occurred at 0.036 ly/min; no inhibition was observed at intensities up to 0.090 ly/min. The optimum temperature, though not very accurately determined, appeared to be high (*ca.* 25–29°C). Findlay compared these data with those from Thomas (1966c) who obtained a saturation intensity of 0.036 to 0.039 ly/min and a high optimum temperature (23–33°C) for a tropical species of *Chaetoceros*. Thomas found some growth with this species even at 37°C; on the other hand, growth was inhibited at moderate temperature

* The values have been re-calculated from the original data.

(11°C). The optimum temperature for growth varied somewhat with light intensity, recalling earlier results on light/temperature effects. Other tropical species isolated by Thomas were a *Nannochloris* sp., with a high temperature optimum (27–37°C) and a saturation intensity of 0.048 ly/min and a *Gymnodinium* sp. – optimum temperature 23–29°C; saturation intensity 0.045 ly/min. In these two species, growth ceased at lower temperatures of 11° and 16–17°C, respectively. The saturation intensities for all three species are of the same order as those quoted earlier (p. 84) for temperate diatoms. Thomas also points out that his values are only slightly greater than that for *Asterionella*, another temperate species. Although saturation intensities for tropical plankton would not, therefore, appear to be substantially different, the optimal temperatures for photosynthesis in the warm-water forms are very different from those for temperate phytoplankton. The surface temperature in Thomas' area of collection was *ca.* 27°C, so that the phytoplankton may truly be regarded as tropical flora. The high-temperature optima and the very moderate lower temperature limits at which growth ceased would suggest warm-water, environmentally adapted species. It is also remarkable that an attempt at adaptation, involving growing these species for a prolonged period at only 21°C, did not modify the temperature optima. Thomas points out that these tropical algae exhibited a relatively high rate of cell division under optimal conditions – 6 divisions/day *Chaetoceros*, 4.5 divisions/day *Nannochloris* but less for *Gymnodinium*.

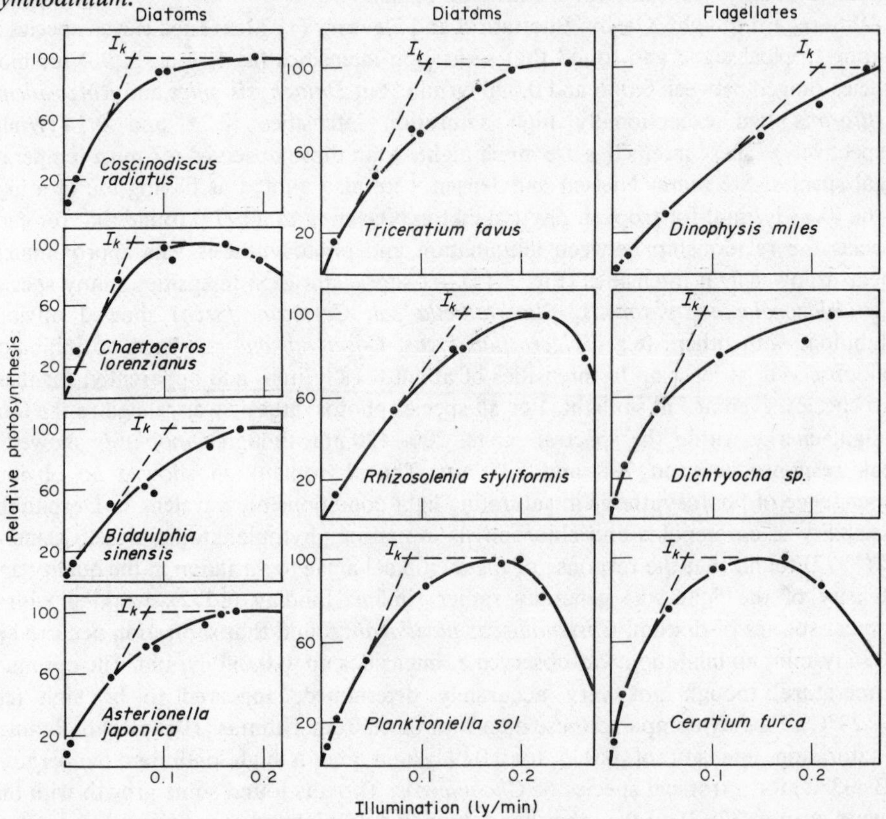

Fig. 3.15. Photosynthesis in different algae as a function of light intensity. I_k values are indicated (Qasim *et al.*, 1972).

The complex inter-relationships between light, temperature and concentration of various nutrients, and other less obvious factors, must together determine the relative success of the different phytoplankton species, so realizing the floral composition of the population in an area. It is widely acknowledged that many species are able to exist under a considerable range of environmental conditions so that species sometimes exhibit a broad geographical distribution. But more usually, where competition between species is operative, species will be more restricted to those areas where near-optimal environmental conditions prevail. In many areas, particularly those experiencing pronounced seasonal changes, the interaction of environmental factors will play a most important role in the succession of phytoplankton species (cf. Chapter 5).

Smayda (1969) has examined the effects of light intensity, temperature and salinity on the growth of certain diatoms. With *Detonula confervacea*, increase in temperature from 2° to 12°C resulted in better growth, though the difference between 7° and 12° was slight. No growth occurred at 16°C – a result not surprising for an Arctic diatom; on the other hand, good growth occurs in nature at temperatures of <1°C. Within the optimal salinity range of 15 to 30‰, a relationship was established between optimum light intensity and temperature: at 2°C optimum light intensity was 0.013 to 0.039 ly/min; at 7°C it ranged from 0.039 to 0.078 ly/min; at 12°C from 0.078 to 0.117 ly/min. Negligible growth occurred at 5‰. The range in growth rate was very wide: the maximum was about 1.5 divisions/day; low rates approaching 0.1 to 0.2 divisions/day were recorded, at salinities exceeding 10‰. *Rhizosolenia fragilissima* also showed a considerable range in division rate in response to the combined effects of temperature, light intensity and salinity (cf. Ignatiades and Smayda, 1970a). Good growth occurred at 12°, with the optimum ranging from 18°C to 25°C in culture after preconditioning; growth was slower at 9°C. Satisfactory growth occurred at salinities between 10 and 35‰, with an optimum at 20 to 25‰ which appeared to be independent of temperature. The optimum light intensity for growth was about 0.039 ly/min and at lower salinities this appeared to be largely temperature independent. At 30 to 35‰ salinity, however, optimal light intensity was related to temperature; the optimum was 0.078 to 0.117 ly/min for temperatures of 18 to 25°C. The maximum daily division rate in culture was 1.2 divisions/day. Investigations in the field (Ignatiades and Smayda, 1970b) involving enrichment suggested that apart from the combined effects of temperature, light and salinity, the chemical "water quality", including trace metals and other unknown factors, all affected growth rates. Maximum growth rate in field experiments was 0.7 to 0.8 divisions/day, i.e. somewhat less than in laboratory culture.

It has long been known that one of the effects of light intensity is on the chlorophyll content of algal cells. Acclimation to reduced light intensities markedly increases chlorophyll content. Experiments by Steemann Nielsen, Hansen and Jorgensen (1962) demonstrated that when the same species of alga was grown in dilute culture, to avoid shading, at two widely different light intensities, 0.018 ly/min and 0.180 ly/min, the cells acclimatized to the lower light intensity exhibited greater photosynthetic efficiency, but became much more rapidly light saturated (cf. Harvey – *vide supra*). The greater efficiency was due to increased chlorophyll content of the cells acclimatized to the lower light intensities. Thus when the photosynthetic activity for the two groups of algae was plotted against chlorophyll content, the curves were identical, though the cells previously adapted to lower intensities became light saturated at lower energy levels (see Fig. 3.16). Other factors affect chlorophyll content, above all nutrient concentration (cf.

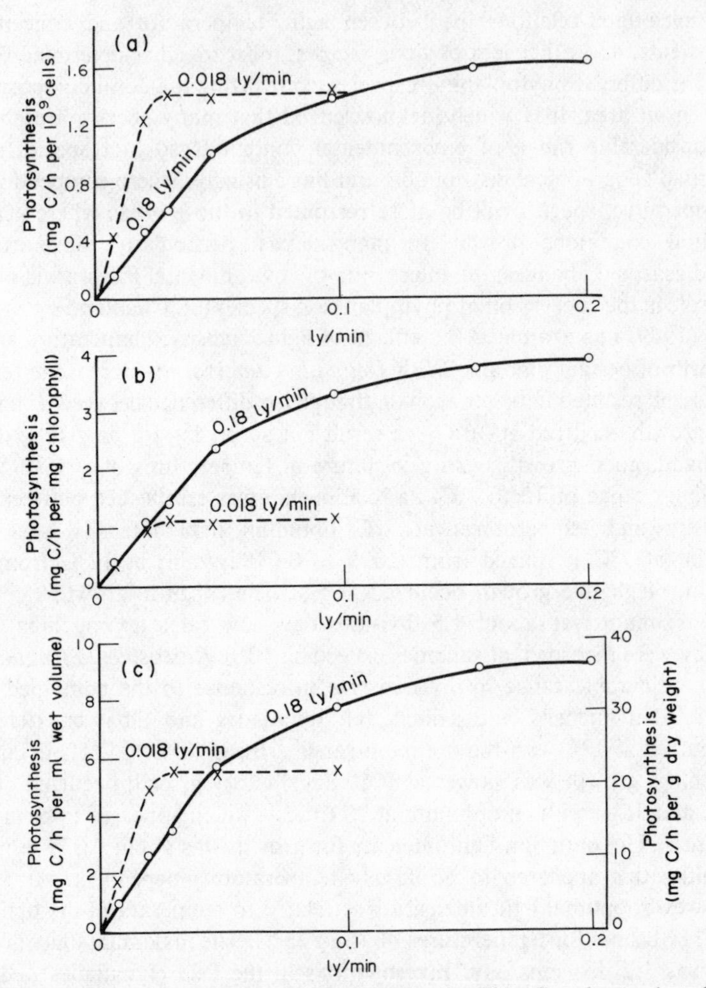

Fig. 3.16. Photosynthetic rate as a function of light intensity, for *Chlorella vulgaris* adapted at two light intensities (0.018 and 0.18 ly/min) a, b and c illustrate photosynthesis calculated per unit cell number, per unit of chlorophyll, per unit wet volume or dry matter, respectively (Steemann Nielsen *et al.*, 1962).

Steele and Baird, 1962, 1965). Nitrogen limitation markedly reduces chlorophyll content (cf. Yentsch and Vaccaro, 1958; Thomas, 1970).

The interactions of such factors as light intensity, temperature and nutrient level on photosynthetic efficiency do not, however, depend on chlorophyll content alone. The accessory pigments of marine algae play a very significant role in photosynthesis, even though their precise modes of action may not be fully understood. Apart from the possible direct participation of certain of these pigments in the photosynthetic mechanism (cf. p. 71) it seems that the function of some is to absorb light energy of particular wavelengths to which chlorophyll *a* is relatively insensitive; this will be of particular significance with the pronounced differential attenuation of wavelengths in the sea (cf. Chapter 6). Halldal (1966) has emphasized the importance of the accessory pigments also at lower intensities. From experiments on *Anabaena* he claims that at high intensities (0.060 ly/min), which will be experienced only in the uppermost surface layer, chlorophyll *a* and phycocyanin are both relatively important in absorbing light energy, but that at lower intensities (e.g. 0.0048 ly/min) phycocyanin becomes of greater significance and phycoerythrin then also plays a part (cf. Fig. 3.17). Since in the sea, even at very moderate depths, the light is attenuated and also shifted to maximum penetration at shorter wave lengths (*ca.* 500 nm for coastal waters), the importance of some of the accessory pigments as an energy "trap" becomes obvious (cf. Fig. 3.18). Despite doubts about the function of some accessory pigments, a more or less constant chlorophyll/carotenoid ratio seems to be essential for optimal photosynthesis, but this ratio can vary even in the same species with external factors. Bunt, Owens and Hoch (1966) showed a marked destruction of carotenoids, especially fucoxanthin, in addition to a somewhat less obvious deleterious effect on chlorophyll, when the Antarctic diatom, *Fragilaria sublinearis*, was exposed to comparatively high temperature (24°C), intense light increasing the destructive effect.

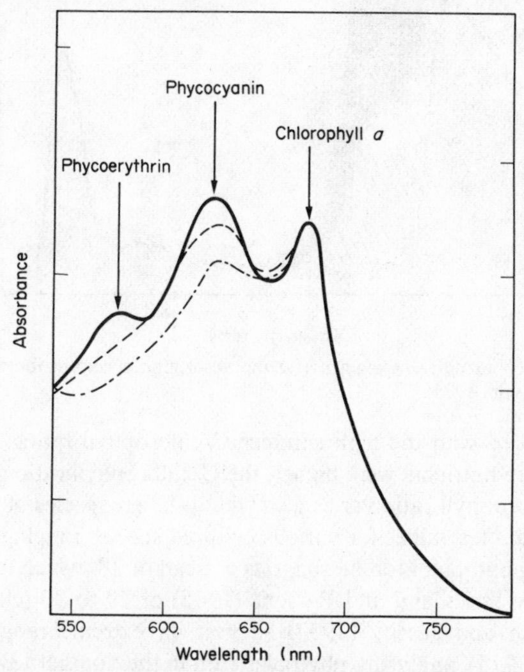

Fig. 3.17. Absorption curves of *Anabaena* grown at different light intensities: (———) 0.0048 ly/min, (–––) 0.036 ly/min, (–·–·–) 0.060 ly/min (Halldal, 1966).

The overall variability of the amount of photosynthetic pigment may be expressed as the ratio of cell carbon to chlorophyll concentration (C/chlorophyll). Steele and Baird (1962, 1965) demonstrated very considerable fluctuations in the C/chlorophyll ratio for natural phytoplankton populations. For a sea-loch (L. Nevis, Scotland) the mean value for summer phytoplankton was 74; lower ratios (*ca.* 35) were typical of spring and autumn. On the Fladen Ground (North Sea), over the period of the spring increase, the C/chlorophyll ratio was as low as approximately 20. Low values were typical also of autumn, but over the summer, peak values, even three times that in Loch Nevis, were found. Steele and Baird quote a ratio generally exceeding 100 over the summer. Presumably the reduced nutrient levels during summer with comparatively high light

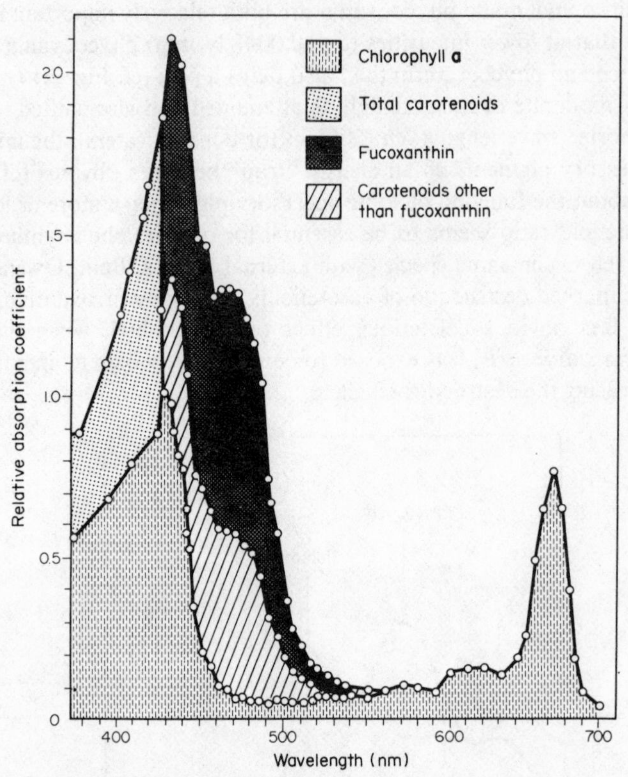

Fig. 3.18. The contribution of different pigments to the absorption spectrum for the diatom *Nitzschia closterium* (after Whittingham, 1976).

intensities are correlated with the high summer C/chlorophyll ratios. Over the Faroe-Shetland plateau where nutrients were higher, the C/chlorophyll ratio over summer was quite low. The C/chlorophyll ratio varies also with different species of algae; Zeitzschel (1970), for example, quotes values for various cultured species ranging from 25 to more than 100. For mixed phytoplankton he suggests a mean of 38, which may be compared with a range given by Strickland and Parsons (1965) of 20 to 70 (cf. also Strickland, 1965). Platt, Denman and Jassby (1977) suggest an extreme range of 20 to 200. Gieskes and Kraay (1975), analysing phytoplankton in the Southern Bight of the North Sea, found that the chlorophyll content was too variable to be used as an index of phytoplankton concentration. Apart from variations in the C/chlorophyll ratio due to species composition, the ratio varies considerably with nutrient concentration, light and temperature, as well as with pre-conditioning (*vide supra*).

Yentsch and Vaccaro (1958) suggest that N-depletion reduces both chlorophyll and tissue nitrogen but that chlorophyll is affected more rapidly. There is a marked variation in the carotenoid/chlorophyll ratio with lack of nitrogen. Steele (1962) also emphasizes the effect of nutrient level and considers that chlorophyll levels are relatively higher, for example, near the base of the euphotic zone where nutrient levels might be expected to be somewhat raised, provided that the depth of the euphotic zone is reasonably large. In latitudes where seasonal effects are marked, Steele suggests that lowered chlorophyll values associated with lower nutrients may be expected during summer. The

C/chlorophyll ratio may be of significance in relation to the observations of Ketchum, Ryther, Yentsch and Corwin (1958), that in open warmer oceans, apart from gross photosynthesis being lowered, the nett/gross ratio of photosynthesis was lower. Ketchum suggests that the phytoplankton may be less "healthy"; this may be interpreted as in part related to a lowered photosynthetic pigment ratio.

Light intensity and temperature may affect the C/chlorophyll ratio, although nutrient concentration is perhaps the most important factor. Bunt (1967) has grown the Antarctic diatom *Fragilaria* at temperatures ranging from −1.6° to 7.5°C and at two light intensities (0.0006 and 0.0194 ly/min). The C/chlorophyll *a* ratio varied from 20 to approximately 60, with a mean of 30. The range for this one species, grown under different conditions of temperature and light, is almost as great as that given for different species under "normal" conditions. Moreover, the data obtained by Bunt and Lee (1970) from field samples from McMurdo Sound show remarkably wide variations in the C/chlorophyll *a* ratio. While some of the values obtained outside the usual growth season are abnormally high and may be due to the inclusion of large amounts of organic detritus, even during November/December (the period of major algal growth) values ranged from 24 to 59. Bunt cautions against the acceptance of carbon standing stock data derived from chlorophyll values in the Antarctic.

Comment has already been made on the dependence of algal photosynthesis on light intensity in relation to the varying proportions of various photosynthetic pigments (cf. p. 71). Ryther (1956) emphasized the varying light requirements between different taxa of phytoplankton. He suggested that Chlorophyceae (*Dunaliella, Platymonas, Nannochloris*) were light saturated at 0.032 to 0.048 ly/min, diatoms (*Nitzschia, Navicula, Skeletonema*) at somewhat higher values (0.065 to 0.13 ly/min) and dinoflagellates (*Gymnodinium, Exuviaella, Amphidinium*) at even higher intensities (0.161 to 0.194 ly/min). He further proposed that for phytoplankton species generally, inhibition is apparent at about 0.065 ly/min above the saturation value. At intensities of 0.52 to 0.65 ly/min (equal to about full noon sunshine) photosynthesis in green algae and diatoms was only 5–10% of that at the saturation intensity; in dinoflagellates it amounted to 20–30%. This can be regarded, however, only as a very broad generalization. For example, some phytoplankton species in Antarctic waters, or near the base of the euphotic zone at lower latitudes, have very low light requirements; some tropical diatoms live successfully near the sea surface at very high intensities (cf. Thomas; Findlay – *vide supra*).

Steemann Nielsen and Hansen (1959) have also drawn attention to the adaptation to light intensity shown by groups of phytoplankton from various oceanic habitats, rather than to the different light requirements being related to taxonomic distinctions. To avoid the complication due to variations in the amounts of photosynthetic pigment, the rate of photosynthesis calculated as the amount of carbon assimilated was normalized to chlorophyll. Steemann Nielsen and Hansen found that although photosynthesis increases at lower light intensities at very approximately the same rate for a variety of marine phytoplankton, light saturation (P_{max}) occurs at very different light intensities between, for example, surface Arctic, surface temperate and surface tropical populations; moreover, for each of these broad latitude groupings the phytoplankton assemblages occurring deep in the euphotic zone show different characteristics. Both light saturation values and photosynthetic values at light saturation (P_{max}) show very wide differences between the various habitat groups (cf. Fig. 3.19). Many marine

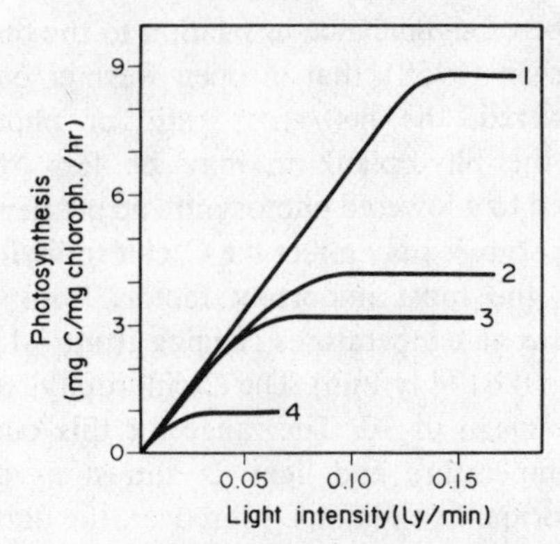

Fig. 3.19. Light intensity and the rate of gross photosynthesis for marine phytoplankton from different habitats (from Steemann Nielsen, 1959). (Cf. also Steemann Nielsen and Hansen, 1959.)

algalogists therefore distinguish what may be termed a "sun" plankton, occurring near the surface and adapted to relatively high light intensities, from a "shade" plankton living near the base of the euphotic zone and adapted to lower intensities. On the whole the ratio of carbon to chlorophyll tends to be lower in the "shade" forms.

The existence of "shade" and "sun" types, especially in highly stratified tropical waters, is widely recognized. Ryther and Menzel (1959) investigating an area of the Sargasso Sea during winter, when the water was virtually isothermal to 400 m, and when the euphotic zone approached 150 m in depth, found that during that season the phytoplankton at three depths, corresponding respectively to 100%, 10% and 1% of the surface light, was of a similar "sun" type, becoming light saturated at about 0.30 ly/min. In summer, however, when marked stratification occurred, with a thermocline at a depth of 25–50 m, while the surface plankton adhered to the "sun" type, below the thermocline, at about 100–150 m, corresponding to about 1% of the surface radiation, the phytoplankton was of the "shade" type; it became light saturated at an intensity of less than 0.060 ly/min. At depths between the surface and 100 m, the plankton was rather intermediate in character. For Japanese waters of the Kuroshio Current, Saijo and Ichimura (1962) demonstrated that while photosynthesis varied more or less directly with light energy at relatively low light intensities for plankton from the surface, 20 m and 50 m depth respectively, there were clear differences in the light saturation values, reflecting the "sun" (surface) and "shade" (deeper) types of phytoplankton.

Thomas (1970) investigated photosynthetic activity in phytoplankton from the eastern tropical Pacific. He also emphasized the importance of calculating photosynthetic rate per unit of chlorophyll, in view of the variability in content of photosynthetic pigment. Thomas found somewhat lower values for photosynthesis at light saturation per unit of chlorophyll (i.e. assimilation index or assimilation number, according to different authors) in natural phytoplankton populations characteristic of nutrient-poor water as compared with those from richer nutrient waters of the tropical east Pacific.

As we have already indicated, however, not only do light-saturation values and the assimilation number (index) vary amongst phytoplankton, particularly with "sun" and "shade" species, but the initial slope of the *P*/*I* curve differs between species, indicating differences in photosynthetic efficiency.

Talling (1957) proposed a function which is a measure of the energy at light saturation. This function (I_k) is defined as the light intensity corresponding to the intersection of the extension of the initial slope of the *P*/*I* curve with P_{max}. To some degree I_k thus describes the ratio between the photochemical and enzymatic ("dark reaction") parts of the photosynthetic process. Figure 3.20 indicates that two algal species (1) and (2) may have similar I_k values but that the photosynthetic efficiency can none the less be higher in one of the species (species (2) in the examples). Species (3), on the other hand, has a higher I_k ("sun" type) but the photosynthetic efficiency is lower, even slightly less than species (1). The "sun" type alga can utilize comparatively high light intensities efficiently, but a "shade" species (2) shows a higher photosynthetic rate at low light values. (For simplicity, the compensation intensity for all three species is regarded as identical.)

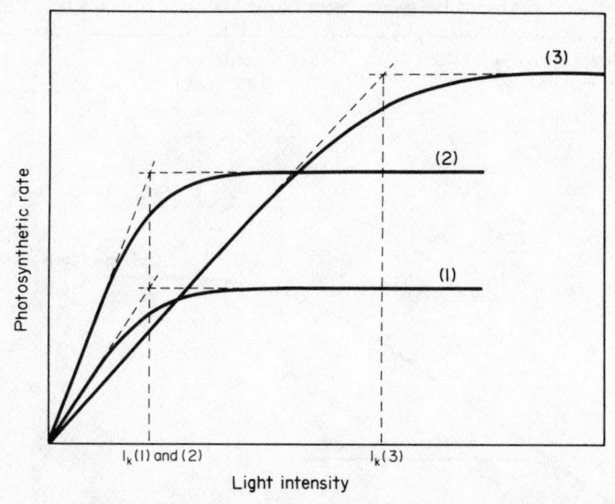

Fig. 3.20. Three types of *P* vs *I* curves. (1) and (2) shade type algae showing similar I_k values but with higher photosynthetic efficiency in (2) than (1). Sun-type community (3) showing lower photosynthetic efficiency than (1) or (2) at lower light intensity (Parsons and Takahashi, 1973).

Parsons and Takahashi (1973a) quote values for the initial slope of the *P*/*I* curve and for P_{max} for a number of species in culture and for several natural populations. The variations are considerable: initial slopes vary between 0.1 and 0.65 mg C/mg Chl *a*/hr per klux for cultures and from 0.05 to 0.8 for natural populations; P_{max} ranges from 1.1 to 62.0 mg C/mg Chl *a*/hr for cultures and from 0.1 to 6.0 mg C/mg Chl *a*/hr in natural phytoplankton samples. Some of the variations have already been noted as corresponding to different geographical ranges and to vertical distribution.

Yentsch and Lee (1966) have referred photosynthetic rate to the ratio: P_{max}/P. If this ratio is plotted against light intensity, "light" and "dark" reactions can to some extent be separated, and each may be related to environmental parameters. Apart from the obvious effect of light intensity on the light reaction, environmental factors affect mainly the dark reaction (cf. p. 81). Decreased temperature, for example, has a marked effect on the dark reaction in *Nannochloris* cultures, and the authors believe this is true of natural phytoplankton populations, especially if accompanied by nutrient deficiency. Acclimatization to lower light intensities also affects the dark reaction more obviously. Although high light intensity, and possibly high temperature, has an influence on the light reaction, this is probably due to photo-oxidation of chlorophyll. The light reaction is, however, not entirely independent of temperature (cf. Yentsch, 1974). Steemann Nielsen (1974) emphasizes that while light adaptation is obviously related to chlorophyll content and thus to the light reaction, intense illumination reduces cell size and this can result in increased enzyme concentration per cell, affecting the dark reaction. Enzyme concentration is mainly related to temperature, adaptation to lowered temperature increasing enzyme (and cell protein) content. Although "shade" forms exhibit lower I_{max} and lower P_{max} per unit of chlorophyll (assimilation number) than "sun" forms, the P/I curves are fairly similar at lower light intensities (i.e. the initial slopes are not widely different). Strickland (1965) points out that if, however, photosynthetic rate is calculated per cell or per carbon unit, the P_{max} values are much closer but initial slopes, at lower light intensities, show much greater differences, presumably reflecting the marked variations in photosynthetic pigment between "sun" and "shade" species (cf. Fig. 3.21).

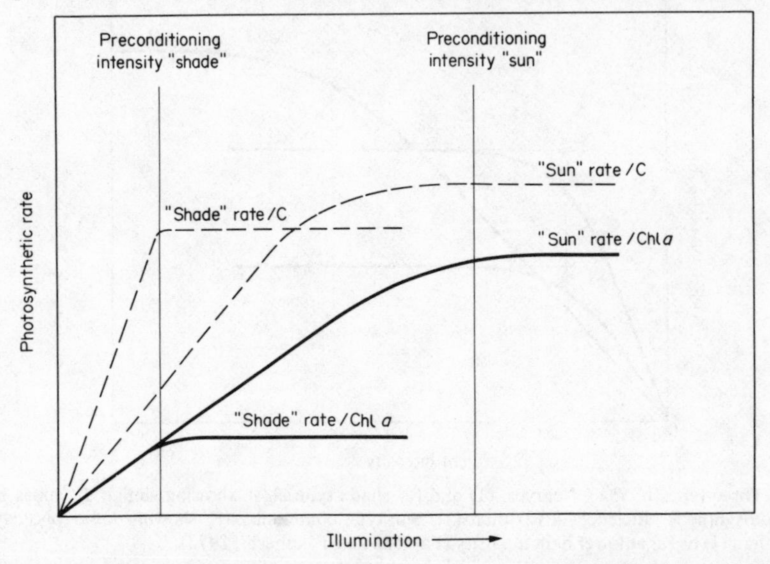

Fig. 3.21. Effect of light and preconditioning on rate of photosynthesis (Strickland, 1965).

While different algae may be characterized as "sun" or "shade" type, a species may become adapted to changed light intensities. Steemann Nielsen and Park (1964) demonstrated that this adaptation can be quite rapid. Phytoplankton from an area in Friday Harbour (Washington State), with a marked lack of stratification, and where the whole phytoplankton was adapted to relatively high light intensities, was isolated in bottles and transferred to a depth where the light was only 5% of the surface value. The phytoplankton cells showed a change to the dark-adapted type within 3 days, the main change being a marked increase in chlorophyll content. However, not all species adapt easily; Steemann Nielsen reports that the common diatom *Skeletonema costatum* did not adapt to low light intensities, despite varying temperature and nutrient levels; unexpectedly the species appears to be adapted to relatively high intensities (cf. Halldal, 1966).

Although surface ("sun") plankton generally has a lower chlorophyll content, inhibition of photosynthesis in plankton exposed to high light intensities does not necessarily nor solely involve an immediate destruction of chlorophyll. Steemann Nielsen (1962a) demonstrated that *Chlorella* grown in low light intensities and transferred to high intensities showed a depression in photosynthetic rate (apparently both photochemical and enzymatic processes), irrespective of chlorophyll changes. A small amount of photo-oxidation during inhibition can reduce the concentration of enzymes involved in the dark processes. After a short period of return to darkness, these processes were completely reactivated. The chlorophyll and carotenoid content of phytoplankton cells must clearly be related to the period required for the destruction and reactivation of pigments when exposed to darkness, and other environmental factors. Moreover, variations in the pigment content of phytoplankton cells on a 24-hour cycle have been recognized. A recent review of periodicity in phytoplankton, including changes in pigment content, is due to Sournia (1974).

Interest in the persistence of chlorophyll in plant cells below the euphotic zone has partly stemmed from the discovery of phytoplankton at great depths. Although deep-living plankton of various taxa has been observed by Bernard, Wood and other workers, including the observation of "olive green" cells by Fournier (1971), it remains to be proved whether chlorophyll in these cells is photosynthetically active. Even in algae taken just below the euphotic zone in the Sargasso Sea, Menzel and Ryther (1960) found a lack of photosynthetic activity. Algae may sink or be carried by turbulence to depths below the compensation point and it is ecologically important to ascertain whether the chlorophyll remains active for a reasonable time so that if the cells were returned to the euphotic zone they could become productive again.

The problem of algae maintaining their position in the euphotic zone is intensified over winter in temperate and high latitudes. Though, particularly over oceanic depths, survival for some species may depend on exceedingly few healthy cells, distributed in the very shallow uppermost layers ("hidden flora"), the ability of algal species to remain viable over periods of darkness obviously increases the species' survival value. Some neritic forms can produce resting spores, but this is certainly not universal even among coastal species (e.g. *Asterionella, Skeletonema*) and does not hold for oceanic forms. Smayda and Mitchell-Innes (1974) tested dark survival for nine coastal diatoms, subjected to total darkness for up to 90 days, at a temperature of 15°C. Tests during the period showed that seven species, including *Asterionella, Lithodesmium, Chaetoceros didymus*, *C. curvisetus*, *Thalassiosira* and *Ditylum*, retained their photosynthetic powers for the whole period, though their division rate fell with time and no growth or division occurred in darkness. *Skeletonema* retained its photosynthetic capacity for only about 7 weeks, though to some extent the retention may be temperature dependent – at 20°C viability was only 1–4 weeks, lengthening to 24 weeks at 2°C. Smayda and Mitchell-Innes also believe that very brief exposure to light as opposed to continued darkness may greatly extend the period of viability. This is borne out by experiments of Umebayashi (1972) who found that certain algae would survive for very long periods (*Skeletonema, Cyclotella nana, Chaetoceros calcitrans* – 9 months; *Phaeodactylum* – 25 months; *Nitzschia* – 34 months) in darkness, but with fairly frequent very short exposures to light, whereas survival was reduced in continuous darkness. Survival under ice is relevant in this context and is of considerable ecological significance. Bunt and Lee (1972) demonstrated that of four species of algae from sea ice, two diatoms and a flagellate survived 3 months in total darkness, following some 3 months of growth in light. Survival in the dark was apparently not aided by the addition of various organic substrates, suggesting that heterotrophy did not play an important role in dark survival, at least in these species. Low temperature (−1.8°C) may be very significant. Horner and Alexander (1972) found a complex phytoplankton community under sea ice; diatoms (especially *Nitzschia frigida*, with *N. closterium, Amphiprora, Fragilariopsis, Navicula* and unidentified pennate species), flagellates (*Dinema, Eutreptiella*, cryptomonads, etc.) with colourless flagellates, ciliates, and even some Metazoa were present. Although the diatoms probably did not act as "seed" for the open water blooms which were mainly due to different centric species, the production due to the ice flora was important. Survival of the diatom flora under ice, though critical, was not obviously related to heterotrophy; uptake of a variety of organic substrates was negligible. Fogg (1977) suggests, from experiments by Morgan and Kalff on *Cryptomonas erosa*, that low temperature may promote carbohydrate accumulation while restricting cell division, and, by reducing metabolic demands, may enable Antarctic phytoplankton to use reserve carbohydrate conservatively during the long period of survival in darkness.

Whatever the habitat, species undoubtedly show marked differences in their ability to retain the photosynthetic function in darkness. Smayda and Mitchell-Innes quote findings of Antia and Cheng (1970) that the survival of thirty-one species of marine algae varied with species (at 20°C) from 1 week to 6 months, and that no obvious taxonomic pattern could be discovered. Precise conditions of the experiment, in addition to the type of culture employed, may all modify results – thus *Dunaliella* was found to survive for 7 weeks by Antia and Cheng in contrast to about 5 days by Yentsch and Reichert (1963), and to the results of Hellebust and Terborgh (1967). Earlier work by Ignatiades and Smayda (1970a) demonstrated that *Rhizosolenia fragilissima* in total darkness retained good viability for 11 days and a more limited survival for 23 days, in contrast to the prolonged survival of seven other diatoms species examined by Smayda and Mitchell-Innes.

The effect of low temperature in prolonging the viability of resting spores is potent at least for some diatoms, normally exposed to such conditions. Thus Durbin (1978) found that resting spores produced by two diatoms, *Thalassiosira nordenskioldii* and *Detonula confervacea* found in Arctic waters and in winter/spring in boreal areas, would survive for long periods in darkness at low temperatures, especially about 0–5°C. At 15°C *Thalassiosira* would not form resting spores, and spores produced at lower temperatures by both species survived for only a very brief time (< 7 days) in darkness when exposed to a temperature of 20°C. At 0°C *Thalassiosira* spores maintained in the dark remained viable for even 576 days! "Unfavourable conditions", represented by low nutrient concentration, promoted spore formation in both *Thalassiosira* and *Detonula*, but only at comparatively low temperatures at which the species flourish in their natural environment. In Narragansett Bay, for example, the two diatoms can bloom in late winter but disappear over the time when temperatures exceed 10° or 12°C. No vegetative growth can occur for either diatom at temperatures greater than 15°C, but resting spores cannot carry the species over the summer; re-inoculation must come from cooler, deeper, neighbouring areas. On the other hand, these high-latitude diatoms can survive the long dark Arctic winter as resting spores at the prevailing low environmental temperatures.

Apart from the effects of prolonged or irregular periods of darkness on photosynthetic capacity, it has been suggested that the regular alternation of the normal dark/light diurnal cycle might affect phytoplankton, not only through the *direct* effect of light energy on photosynthesis but through changes in pigment content and the photosynthetic mechanism. While such changes are often attributed to diurnal light variations, other explanations, including cellular rhythms, must be considered. A comprehensive review is given by Sournia (1974).

Diurnal changes in the amount of phytoplankton pigments have been investigated by, amongst others, Yentsch and Scagel (1958) who found that chlorophyll *a* in particular exhibited marked diurnal variations. For surface phytoplankton the concentration of chlorophyll *a* per cell was high in the morning but fell to a minimum about midday; some rise occurred in the afternoon with chlorophyll reaching a maximum in the early evening. The variations depended, however, on depth; for example, the midday depression in cell chlorophyll was not true of deeper-living plankton. Yentsch and Scagel relate the variations to intensity and duration of light; the reduction in chlorophyll near midday in surface phytoplankton was held to be due to decomposition at excessively high light intensities; conversely, in the deepest part of the photic zone, where light was limiting, decomposition was minimal. The diurnal changes in chlorophyll were thus held to be a result of both synthesis and decomposition.

The changes in chlorophyll content in natural phytoplankton populations do not seem to be paralleled by such obvious changes in carotenoids. Yentsch and Scagel (1958) claim that rates of synthesis and decomposition of chlorophyll *a* were approximately five times greater than rates for carotenoids. The ratio of chlorophyll to carotenoid may change accordingly even diurnally in algal cells, presumably modifying photosynthetic capacity, and presenting another difficulty in relating quantities of chlorophyll to carbon content.

It has long been known that natural populations of phytoplankton frequently show periodic variation in their photosynthetic ability (cf. Doty and Oguri, 1957; Shimada, 1958). At constant illumination, conveniently measured at light saturation, there tends to be a maximum in photosynthesis in the late morning, somewhat lower values in the afternoon and a minimal period at night or early evening – a diurnal rhythm in photosynthesis. Yentsch and Reichert (1963) believed that such variations were a persistent feature of many areas. The effect could be due to an increase in chlorophyll after a period of darkness and a reduction in chlorophyll during the day at relatively high light intensities; alternatively, darkness might promote photosynthetic ability apart from chlorophyll content.

Few areas of phytoplankton research have been so bedevilled by confusion of thought as the "diurnal rhythm" of photosynthesis. The confusion stems from several causes, chiefly from differences in methodology, including estimates of production, pigments, quantity and quality of illumination, "adaptation" or "pre-conditioning" of phytoplankton cultures employed, and so forth. Sournia (1974) refers to a number of diurnal (diel) variations in algal cells, such as ATP content, cell volume, cell carbon and other chemical constituents – phosphorus, nitrogen, etc. Regarding diurnal changes in chlorophyll, he emphasizes the difficulties involved in estimating daily changes in pigment quantities accurately in the field, though, quoting from the researches of Glooschenko and colleagues, it appears that genuine diurnal variations can be demonstrated, irrespective of the effects of migration or passive sinking and accumulation of algal cells, and of differential grazing activities by zooplankton. Ryther and Menzel (1961) showed fluctuations between day and night in nutrients and also diurnal changes opposite in phase, in chlorophyll and in ^{14}C assimilation, during studies in the Sargasso. Sournia draws attention to variations in the diurnal rhythm with latitude, with depth in the euphotic zone, and in phasing, and warns about confusion with the "pseudo-rhythm" obtained from ^{14}C-deck-simulated experiments – an effect due essentially to a differential loss of labelled carbon through nocturnal algal respiration. His most important conclusion is that a diurnal rhythm in photosynthetic activity ("afternoon" or "midday depression") can be recognized for marine phytoplankton, independent of biomass or pigment changes (cf. Fig. 3.22), and that this probably reflects internal biochemical mechanisms.

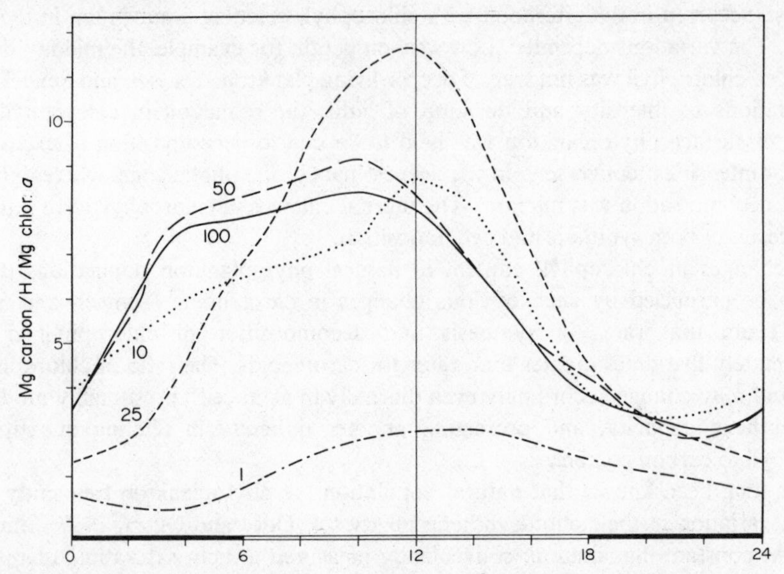

Fig. 3.22. Photosynthetic activity in relation to sampling time. Samples taken at different times of the day at five depths, corresponding to 100, 50, 25, 10 and 1%, respectively, of incident light, were incubated under identical conditions (after Sournia, 1967).

Yentsch and Reichert (1963) were interested in the possible effects of retaining algae for varying periods in darkness on their subsequent photosynthetic ability. The photosynthetic capacity was measured per unit of chlorophyll to obviate the effect of changes in chlorophyll content. The results showed a rapid rise in the rate of photosynthesis with increasing previous periods in darkness, with a maximum after about 20 to 24 hours of darkness; after about 50 hours darkness there was a decline in photosynthetic rate. After 100 hours of darkness photosynthesis only reached compensation level. The pattern held whether the nitrogen source for photosynthesis was nitrate or ammonium, so that the effect could not be attributed only to nitrate-reductase activity (cf. Chapter 7). McAllister, Shah and Strickland (1964) might be considered as dealing in part with the same phenomenon when they describe *Dunaliella* cells, reared at low light intensities, as having higher constants for gross photosynthesis at moderate intensities than cells reared at high light intensities. They claim that apart from the effects of high and low light intensity pre-conditioning and of darkness on chlorophyll content, darkness is associated with the production of enzymes concerned in photosynthesis. Despite the adaptation in photosynthetic rate in "shade" phytoplankton, partly related to chlorophyll content, *Dunaliella* cultures did not show a clear correlation between gross photosynthetic rate and chlorophyll *a* at low light intensities. Yentsch and Reichert's observation that *Dunaliella* kept in darkness, after a brief increase in photosynthetic capacity, showed a decline, which after more than about 100 hours was a total and apparently irreversible loss, is generally confirmed by Hellebust and Terborgh (1967). They found an increase in photosynthetic rate (both light and dark reactions) and in the activities of RuDP carboxylase and aldolase, two enzymes involved in the Calvin cycle, when cultures of *Dunaliella tertiolecta*, grown under continuous illumination, were transferred to darkness for some 24 hours. But continued maintenance of the alga in darkness resulted in a rapid decrease in photosynthetic rate and enzyme activity. When *Dunaliella* was exposed to continuous but low light intensities, considerably below the compensation point, instead of total darkness, however, the algae retained their photosynthetic capacity and enzyme activities for a much longer period. Lowered temperature also appeared to favour retention of photosynthetic capacity; at a temperature of 5°C, under continuous illumination of only 0.0003 ly/min, the cells remained photosynthetically active for 3 weeks. Hellebust and Terborgh found a decline in the rate of respiration with time; this will tend to lower the compensation intensity.

The ability of phytoplankton to survive at light intensities below the compensation depth is related in part to the problem of deterioration of chlorophyll. Our knowledge of partly decomposed chlorophyll in the seas is limited. Yentsch (1965a and b) points to the difficulty of determining the total phaeo-type pigment which may be phaeophorbide or phaeophytin in marine phytoplankton (but cf. Lorenzen, 1976). In either case the chlorophyll has lost its central magnesium atom; phaeophytin retains the phytol but in phaeophorbide the phytol is lacking. Chlorophyllide results from the loss of phytol from chlorophyll. Yentsch, using as a criterion the ratio of the fluorescence of a normal chlorophyll sample to the fluorescence of the same sample after acidification, finds that the fraction of pigment as chlorophyll decreases, and the fraction as phaeophytin increases, with depth. Thus in the Indian Ocean, where the absolute amount of chlorophyll is highly variable, ranging over most of the region between 25–150 mg/m^2, throughout the upper 50 m phaeophytin is hardly measurable at some stations and very rarely exceeds 40% of the total. On the other hand, below 50 m, chlorophyll and phaeophytin are approximately equal, and at depths greater than 100 m the percentage of phaeophytin exceeds that of chlorophyll (Table 3.1). El-Sayed and Jitts (1973), however, report for the south-eastern Indian Ocean that generally the vertical distribution of phaeopigments was similar or only slightly deeper than that for chlorophyll. The total pigment maximum was subsurface ranging from 10 m to even 120 m but averaging about 25 m.

Table 3.1. The mean percentage of chlorophyll to total pigment for nine depths at a station in the Indian Ocean. (Data from Yentsch, 1965)

Depth (m)	%
0	81
10	81
25	86
50	64
100	58
125	30
150	36
175	30
200	27

Jeffrey (1974), examining the distribution of photosynthetic pigments with depth off New South Wales, found that the upper 35 m had comparatively large amounts of active photopigments. A difficulty was the occurrence of what appeared to be chlorophyll *a* which was not active. At deeper levels (> 40 m), no chlorophyll *a* was recovered, but small quantities of carotene and of xanthophylls were present. At greater depths (*ca.* 100 m), degradation products, phaeophorbide and phaeophytin occurred, with some chlorophyll *c*.

Undoubtedly much of the phaeophytin present in natural waters arises from the grazing activities of zooplankton with the subsequent decomposition of phytoplankton cells; how far bacteria promote phaeophytin formation is uncertain. It is widely suggested that chlorophyll is converted also to phaeophytin in nutrient deficient cultures and when phytoplankton is kept in darkness. Yentsch (1965a) examined the effect of darkness using *Phaeodactylum tricornutum*. Laboratory cultures in the light showed the expected increase in chlorophyll, with little phaeophytin present. When the cultures were placed in darkness, chlorophyll first increased to reach a maximum after about 70 hours of darkness; thereafter there was a decline in total pigments, particularly chlorophyll, with a rise in phaeophytin. Continuing periods in darkness led to a fall in total pigment but the amount of phaeophytin increased substantially; after about 270 hours of darkness, chlorophyll was virtually absent and only phaeophytin was present (Fig. 3.23). The possibility of recovery was tested; *Phaeodactylum* cultures, previously 25 days in darkness, when exposed to light showed a slow production of chlorophyll. After about 250 hours exposure to light the amount of phaeophytin progressively declined whereas the amount of chlorophyll rose substantially and after about 500 hours all pigment present was chlorophyll (Fig. 3.24). With natural populations of phytoplankton, taken from about 250 m depth in the Straits of Florida, over 75% of the pigment was phaeophytin. When a sample was placed in the light there was a slow reduction in phaeophytin content and, after a lag period, an increase in the amount of chlorophyll which continued with time of exposure (Fig. 3.25).

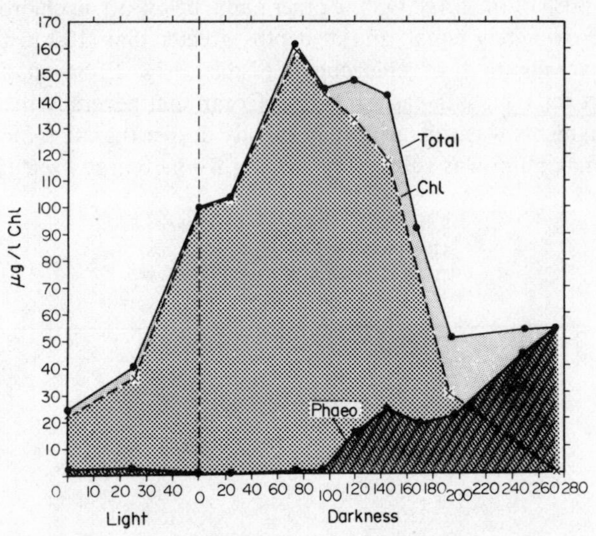

Fig. 3.23. The effect of darkness on the proportions of chlorophyll and phaeophytin in a culture of *Phaeodactylum tricornutum* (after Yentsch, 1965).

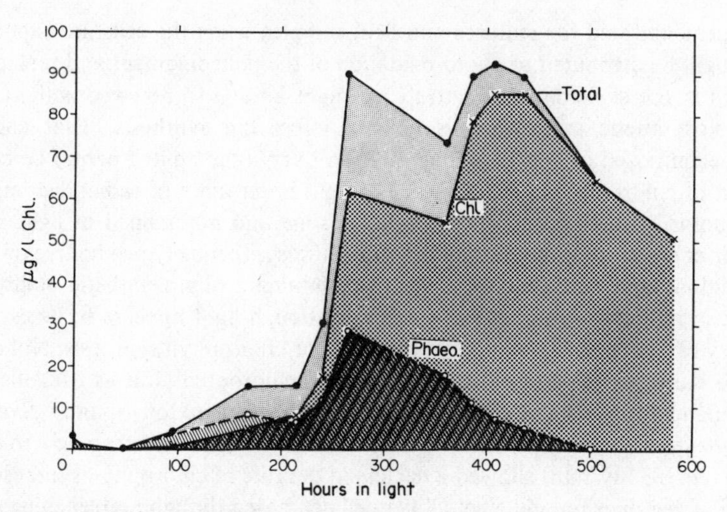

Fig. 3.24. The proportions of chlorophyll and phaeophytin in a darkened culture of *Phaeodactylum tricornutum* on re-exposure to light (after Yentsch, 1965).

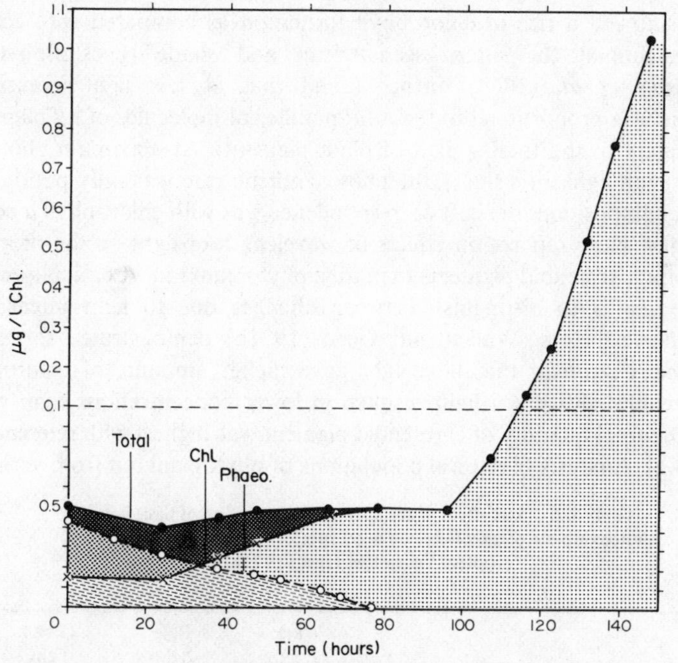

Fig. 3.25. The proportions of chlorophyll and phaeophytin in a sample of seawater from 200 m in the Straits of Florida when irradiated at 0.030 ly/min at 18°C (after Yentsch, 1965).

Some experiments with *Dunaliella* and *Skeletonema* attempted to test the effect of nutrient deficiency. Although there was the expected marked decline in production and, after some time, a fall in chlorophyll, there was no obvious effect of nitrogen deficiency on the chlorophyll/phaeophytin ratio. Nutrient deficiency, therefore, does not appear to promote phaeophytin production, but darkness is a major factor in converting chlorophyll to phaeophytin. The recovery of chlorophyll content in cells with degraded pigment, demonstrated for cultures and field samples when the cells are exposed again to light, might be attributed to photo-oxidation of the phaeopigment and replacement of magnesium in the skeleton; alternatively, it might be due to renewed activity of some intact photosynthetic unit which is able to effect the synthesis. That chlorophyll, partially decomposed in darkness, may at least over some limited period be capable of conversion to chlorophyll in the presence of light is certainly of ecological importance. Cells temporarily sinking below the euphotic zone and maintained at light intensities below their compensation point may become photosynthetically productive on return to light conditions, but specific differences in maintaining photosynthetic ability may be significant in the relative success of species. Although light appears to be essential for the stability of chlorophyll, light is not required for chlorophyll synthesis. Not only does transfer to darkness stimulate the production of chlorophyll, but as McAllister, Shah and Strickland (1964) have shown, cultures of phytoplankton (mainly *Skeletonema* and *Dunaliella*; also *Monochrysis* and *Amphidinium*) transferred from low to high light intensities (*ca.* 0.35 ly/min) showed a decline in the rate of chlorophyll synthesis relative to the rate of cell division; chlorophyll per cell decreased though no bleaching occurred. When kept at the standard pre-conditioning intensity (0.075 ly/min), cell division and chlorophyll synthesis kept pace with each other. Cells transferred to lower intensities (0.02 ly/min) showed a rise in chlorophyll formation as compared with cell division. These changes simulate the patterns seen in "sun" and "shade" types of phytoplankton cells. McAllister *et al.* (1964) further found that at low light intensities gross photosynthesis was proportional to the *total* number of molecules of *all* plant pigments (or approximately to the total weight of plant pigment). At maximum photosynthesis (i.e. relatively high light intensities), the photosynthetic rate was only poorly related to plant pigment composition; the best correspondence was with chlorophyll *a* content.

There are few clear reports on effects of wavelength of light on the proportions of chlorophylls and carotenoid pigments in marine phytoplankton. A difficulty with experiments in the sea is to distinguish between changes due to light intensity and to wavelength. Nevertheless, Wallen and Geen (1971b) demonstrated for cultures of *Dunaliella* and *Cyclotella* that blue light gave higher amounts of chlorophyll than exposure to white light; green light resulted in lower concentrations than white light. Conversely, the total quantity of carotenoid pigment was highest with reference to green light (Table 3.2). Analysis of natural populations of phytoplankton from various depths

Table 3.2. The effect of the spectral composition of light in the photosynthetic pigments in two algal species (all units as $\mu g/10^6$ cells) (after Wallen and Geen, 1971b)

	White light	Blue light	Green light
Cyclotella nana			
Chlorophyll *a*	1.23	1.48	1.07
Chlorophyll *c*	0.27	0.34	0.19
Total carotenoids	1.36	1.07	1.62
Dunaliella tertiolecta			
Chlorophyll *a*	2.68	2.98	2.36
Chlorophyll *b*	1.34	1.69	1.14
Total carotenoids	1.19	0.96	1.33

in stratified waters also showed differences in the carotenoid/chlorophyll ratio which paralleled the differences observed in laboratory cultures. Since in their culture experiments Wallen and Geen had used similar light intensities, the authors attribute the differences in carotenoid/chlorophyll ratio, even in the natural environment, to wavelength rather than to intensity.

The results of Wallen and Geen (1971a, b and c) also indicated that algal cultures and natural phytoplankton could exhibit considerable differences in the proportions of the type of photosynthate, dependent on light conditions, and there was considerable evidence for attributing the effects to differences in wavelength rather than in intensity. Blue and green light, for example, increased protein synthesis relative to carbohydrate formation, while white light favoured carbohydrate synthesis; lipid seemed to vary little. Growth and division rate of the algae, chlorophyll formation and RNA and DNA production were higher for blue than for white light, but green was less effective. Blue light also appeared to favour the production of some free amino acids, especially aspartic, glutamic, serine and alanine, and certain organic acids, malic and fumaric.

The effect of light on photosynthesis in the sea is usually regarded as the product of light intensity and length of day; this has been stressed, for example, in the initiation of the spring outburst in temperate latitudes (cf. Chapter 6). Photo-period *per se* has not been so generally regarded as an important factor. However, evidence from culture experiments suggests that different photo-periods may affect growth in certain species of diatoms and dinoflagellates (e.g. *Gonyaulax tamarensis*). Castenholz (1964) showed for two littoral diatoms, *Fragilaria striatula* and *Synedra tabulata*, that growth was day-length dependent, division rates being significantly lower during short day periods. *Melosira moniliformis* and *Biddulphia aurita*, by contrast, showed little dependence of growth on day length. *Fragilaria* grew well under continuous illumination even at high light intensities, whereas *Biddulphia* exhibited some growth inhibition under these conditions. *Melosira* and *Synedra* showed some decrease in growth rate at high intensities, with prolonged illumination. For example, with *Melosira* such light intensities over long (15-hr) periods caused marked inhibition but the compensation intensity, which was low, was little affected by day length.

Jorgensen (1966) obtained a considerable measure of synchrony in *Skeletonema*, with 12-hour dark and light periods, division occurring during the light period. There was a rhythmic variation in the photosynthetic pigments, chlorophyll *a* and *c* and fucoxanthin, with increasing amounts occurring only in the light (cf. Sournia, 1974). With a higher light intensity (0.060 ly/min), on the other hand, division occurred in the dark period.

Pasche (1966a) followed the uptake of ^{14}C and calcium in light and darkness in the coccolithophorid *Coccolithus huxleyi*. Although coccolith formation in the alga occurred in light and darkness, the rate was very much higher in the light. Cell growth of the coccolithophorid with varying photo-periods showed that increasing day length had a marked effect on growth up to a 16-hour day; there was very little difference between this day length and continuous illumination, and no inhibition with continuous light (Paasche, 1967). The saturation intensity found (*ca.* 0.03 to 0.05 ly/min for most day lengths) was roughly equivalent to that for the diatom, *Nitzschia turgidula*, an observation that does not give support to the theory that coccolithophorids are especially adapted to high light intensities. However, laboratory culture experiments are difficult to interpret in relation to natural plankton growth. Another indication of such discrepancies was that growth was optimal at *ca.* 21°C in *Coccolithus* cultures, whereas growth occurs readily under natural conditions at *ca.* 7–10°C. Paasche, investigating the increase in chlorophyll in algae when light falls below saturating levels, found for *Coccolithus huxleyi* that chlorophyll *a* (per unit cell volume) increased with shorter day length and decreasing intensities.

Under a suitable combination of day length and intensity to give *ca.* 1 division/day, cell division in *Coccolithus*, although only roughly synchronous (over about 6 hours), occurred invariably in darkness. Discussing periodicity of cell division in other algae, Paasche (1968a) points out that whereas Chlorophyceae and Euglenophyceae seem usually to divide in darkness, no generalization appears to hold for diatoms or dinoflagellates. Species differences appear also with regard to the effect of continuous illumination; some algae need a dark period, others do not. Thus Boutry, Barbier and Ricard (1976) found evidence from experiments on culturing the diatom *Chaetoceros simplex calcitrans* for the requirement of a dark period ("scotophase") for continued active photosynthesis. They also demonstrated an effect of a short period of ultra-violet irradiation, though this may be partly due to effects on bacteria. Photo-period (and ultra-violet radiation) markedly influenced the proportion of total sterols, and the composition of the sterols synthesized. For two planktonic diatoms, Paasche (1968a) found species specific reactions to photo-periods. *Ditylum* showed improvement of growth with higher intensities and increasing day length to a 16-hour period which was equivalent to continuous illumination. Marked inhibition occurred at high intensities (> 0.03 ly/min) with constant light. For *Nitzschia turgidula*, no increase in saturation intensity beyond *ca.* 0.03–0.05 ly/min occurred with increasing day length, and although, as with *Ditylum*, increasing day length had a positive effect up to 16 hours light (equivalent to continuous illumination), no inhibition occurred with constant light. Presumably in some phytoplankton species continuous light interferes with certain enzymatic reactions, perhaps particularly those concerned in RNA and DNA synthesis, so "upsetting" normal cell division.

In both *Nitzschia* and *Ditylum*, Paasche achieved, by suitable combinations of photo-period and intensity, a considerable measure of synchrony of cell division, especially in *Ditylum*. With approximately 1 division/day in both diatoms, division occurred in the light, and division in darkness was markedly reduced, especially over short light periods for *Ditylum* (Fig. 3.26). This is in contrast to *Coccolithus* which divided in the dark period. It is significant, however, that Richman and Rogers (1969)

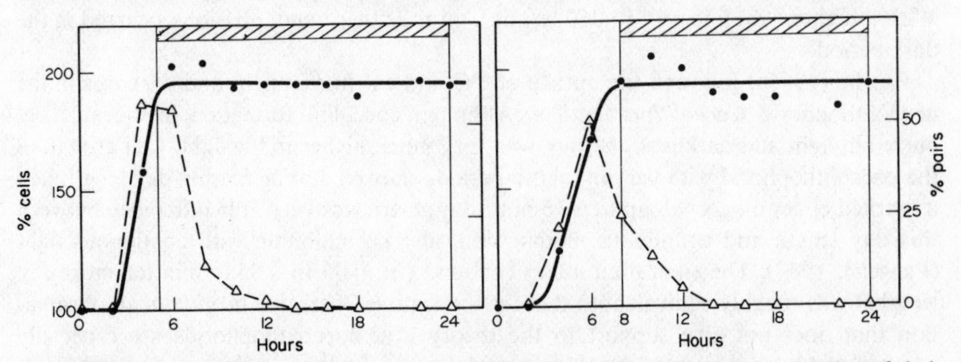

Fig. 3.26. Synchronized cell division in *Ditylum brightwellii*. Hatched areas indicate duration of dark periods, (———) relative cell concentration, (– – – – –) percentage paired cells (after Paasche, 1968).

obtained synchronous division in the same diatom, *Ditylum*, in darkness, with adjusted photo-period and a temperature of 15°C. They believe that temperature may be an important modifying factor, though possibly clonal differences exist inside a species. Sournia (1974) includes an impressive list of instances of synchrony in cell division of algae. For both *Nitzschia* and *Ditylum*, Paasche found that the chlorophyll per cell (or per cell volume) increased, as in *Coccolithus*, with shorter photo-periods and with decreasing intensities, but that chlorophyll synthesis could continue in darkness. Another effect of continuous bright light which has been reported is pronounced carbohydrate (polysaccharide) formation; conversely it is claimed that in darkness there is an increased rate of conversion to protein.

Day length may, therefore, be of significance *per se* in various aspects of phytoplankton growth and metabolism. Apart from the direct effect of alternating periods of light and darkness on photosynthesis, fluctuations are induced in pigment content and enzyme concentration and thus in photosynthetic capacity. Eppley, Holmes and Paasche (1967), using a photo-period to adjust *Ditylum* cultures to approximately 1 division/day, showed that not only did the photosynthetic rate increase to a maximum during the light period, but that synthesis of chlorophyll, carotenoids, cell carbon and carbohydrate increase in light, and that cell division occurred then. The separation of the daughter cells was preceded sequentially by the division of the protoplast and formation of new cell walls; apparently the earlier division of the chromatophores must have occurred in darkness. Other changes in the dark period are a decline in carbohydrate, possibly associated with increased production of other products, and buoyancy changes which may result in sinking (cf. Chapter 6).

Growth and Photosynthesis

Cell growth and division in autotrophic algae and photosynthetic activity are often regarded as equivalent. While these activities are closely related, Hellebust (1970) points out that light requirements for photosynthesis and growth are sometimes not identical. The growth rate in two marine coccolithophorids, for example, was saturated at light intensities far below those necessary for maximum photosynthesis. Halldal (1966), discussing the effects of light intensity on growth and photosynthesis, includes a statement by Myers that when algal growth and photosynthesis are related to light intensity, using the same dimensions, the curves are not identical; light-saturated rates, for example, are frequently dissimilar. Rates of photosynthesis over short-term experiments are often much greater than growth rates. On the other hand, the work of Jitts *et al.* (1964) suggested that the light requirements for maximal growth rates, with optimal temperature conditions, were high and very similar to those for maximum photosynthesis. Thomas (1966c) found growth-saturation intensities for a number of tropical marine phytoplankton species which were generally considerably lower than those reported for growth and photosynthesis by Jitts *et al.* for more temperate species. McAllister, Shah and Strickland (1964), using mainly bacterial free cultures of *Skeletonema* and *Dunaliella*, investigated growth constants using several parameters (cell carbon, chlorophyll, carotenoids, cell numbers), as well as estimating photosynthetic rate. For *Skeletonema* the rate of cell division and production of cell carbon was the same, but was about 40% below the nett photosynthetic rate; the discrepancy may be partly due to release of dissolved organic matter during photosynthesis (*vide infra*). Gross photosynthesis, however, also appeared to be substantially independent of cell division.

To some extent the different effects of such environmental factors as light intensity and nutrient concentration on cell division and photosynthetic rate may be related to diurnal (diel) periodicity in life processes of phytoplankton. A daily rhythm in cell division may not be identical in intensity and timing with that of photosynthetic activity. As already indicated, various cell processes may be differently affected by dark/light periods; species may exhibit different patterns of responses and other environmental factors may modify the results (cf. Sournia, 1974). Certainly in laboratory experiments photosynthetic activity and cell division are not always equal in their response to varying conditions. Photosynthesis may exhibit a more obvious direct response to temperature increase (within limits) than cell division. Jitts *et al.* (1964) remark on the relatively low Q_{10} for cell division in the algae tested. Holdway (1976) found with mixed cultures of the two diatoms *Asterionella japonica* and *Stauroneis decipiens*, that maximal growth occurred at a lower temperature (14°C) than photosynthesis (22°C). Exposure to sublethal concentrations of some toxic substances can also be followed by differential effects on production and growth. Thus, cultures of *Asterionella* and *Stauroneis* in the presence of low copper concentrations showed cell division to be more depressed than photosynthesis. Nutrient concentration and salinity are among other factors which do not necessarily affect cell multiplication and photosynthesis equally. Tentative explanations have included the suggestion that division is controlled in part by the attainment of a critical volume (possibly related to osmotic change) and that rates of synthesis of photosynthetic products and of RNA and DNA are not necessarily identical. In broad terms, however, photosynthesis and growth in predominantly nearshore or estuarine phytoplankton would appear to be similarly affected by lowered salinity. Qasim, Bhattathiri and Devassy (1972b) demonstrated a remarkable adaptability by tropical species to changes in salinity in inshore Cochin waters. Maximum photosynthetic rates occurred at low salinities, mainly in the range 10‰ to 20‰; large populations indicating active division were also found when salinities were well below full seawater. The positive effect of low salinity is not confined to tropical species.

In relation to rhythms in natural populations of phytoplankton, the results of shipboard experiments involving nitrogen enrichment of populations (Eppley, Carlucci, Holm-Hansen, Kiefer, McCarthy, Venrick and Williams, 1971) are relevant in that strong evidence was adduced for periodicity in a number of cell processes. These included N-assimilation, involving apparently both nitrate- and nitrite-reductase activity, the latter being a photosynthetic activity with the dependence on light being species specific. Other functions exhibiting periodicity were C-assimilation, phosphate assimilation, cell division and possibly certain aspects of chemical composition. Although chlorophyll synthesis did not show periodicity in the experiments, it has already been widely recognized in some species of phytoplankton. These shipboard investigations established probably the first recorded example of approximately "synchronous" cell division in a natural population; the diatoms divided mainly in the afternoon and early evening. With such a variety of cell processes and with species differences, the apparently conflicting reports of the effects of photo-period (as one environmental factor) on division rate and photosynthesis are not surprising.

There is little information on the influence of different wavelengths of light on growth rate as on the synthesis of pigments and other compounds (cf. p. 102), for phytoplankton species. Hellebust (1970) refers to a few investigations indicating that there are changes in growth rate in some species; for example, Kain and Fogg's work on *Prorocentrum micans* demonstrated differences in growth rate of the dinoflagellate depending upon whether incandescent or fluorescent lighting was used. Wallen and Geen (1971a) showed that photosynthetic activity, growth rate, and I_k for *Cyclotella* and *Dunaliella* were higher in blue light, and lower in green light, than in white of the same intensity. This was true whether experiments were carried out at high or low light intensities (cf. Fig. 3.27). They believe that the dark reaction was more affected than the light reaction by changes in wavelength. In view of the rapid changes in spectral distribution of light under water and of differences between coastal and oceanic depths, it is essential that further information is obtained on the effects of different wavelengths (cf. Platt *et al.*, 1977). Paasche (1966b) found that while coccolith formation and photosynthetic activity were closely linked in *Coccolithus huxleyi*, blue light had a greater effect on coccolith formation (cf. Table 3.3).

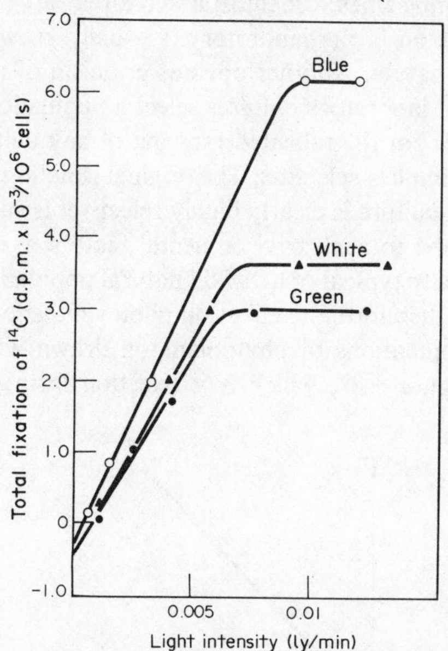

Fig. 3.27. Fixation of ^{14}C in relation to light intensity for *Cyclotella nana*, using blue, white and green light (after Wallen and Geen, 1971a).

Table 3.3. The ratio of photosynthesis (*P*) at two wavelengths (442 and 668 nm) and the ratio of coccolith formation (*C*) at the same wavelengths; light energy in equal quanta in region of linear response (from Paasche, 1966b)

Expt. no.	P_{442}/P_{668}	C_{442}/C_{668}
1	2.99	4.15
2	3.06	4.21
3	2.45	3.07
4	2.05	2.72
5	2.96	4.15
Mean	2.70	3.66

The Applicability of Culture Experiments

Data on the precise effects of light on growth and photosynthetic activity are often drawn from experiments on cell cultures. While acknowledging the value of experiments on pure cultures, Motoda and Kawamura (1963), Braarud (1961), Smayda (1963b, 1970) and others have pointed out that species in culture can react differently from phytoplankton in the natural environment. There is general recognition today of this difficulty in relating laboratory results to field conditions. Among the complicating features is the general use in the laboratory of uni-algal or axenic cultures. In the field, bacteria may modify the reaction of algae; moreover, a number of algae have been shown to have antagonistic or stimulating effects on other phytoplankton species (cf. Aubert and Aubert, 1969; and other references, p. 124). Other complications involve the usually high concentrations of nutrients in experimental media and the presence of glass/water interfaces which may modify algal behaviour. The rapidly changing environmental conditions in batch culture is another feature, and although chemostat methods may avoid some of these complexities, chemostat cultures are also undoubtedly artificial. Moreover, phytoplankton in the laboratory is usually grown at densities far greater than those observed in nature. Another obvious criticism of most cultures, whether batch or continuous, is that laboratory cultures select a population of a species. Sometimes indeed clones are used, but the repeated growing of any culture, with transfer from laboratory to laboratory, implies selection. The original isolation of a phytoplankton cell for the establishment of a culture is clearly highly selective. Is such a population, often subsequently highly adapted to such environmental factors as light intensity, temperature and nutrient media, really typical of a "wild" natural population?

Motoda and Kawamura (1963) followed, therefore, the effect on photosynthetic rate of varying light intensities using *natural* populations of phytoplankton drawn widely over the northern North Pacific region. Figure 3.28, which is a selection from their

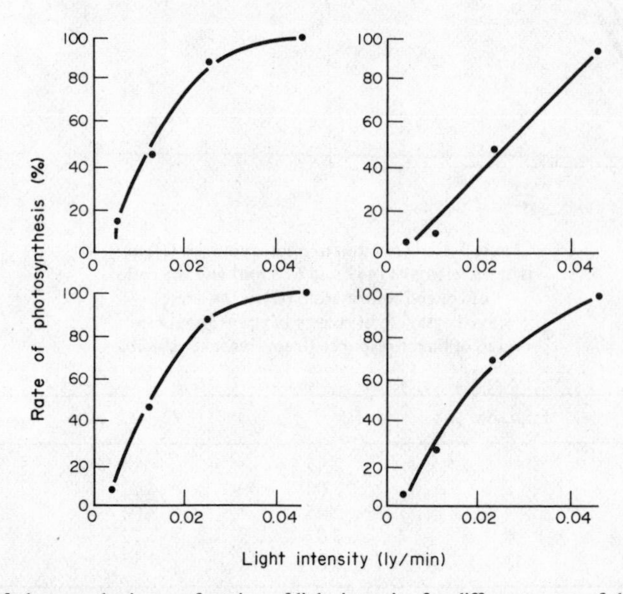

Fig. 3.28. Rate of photosynthesis as a function of light intensity for different areas of the northern North Pacific (after Motoda and Kawamura, 1963).

results, demonstrates the expected effect of increase in photosynthetic activity with increasing light, and further suggests that light saturation was achieved at different light intensities with different populations. Light saturation occurred at intensities of only 0.030 ly/min at some stations, at about 0.048 ly/min at other stations, and at much higher intensities in other areas. An attempt is made to correlate the results with the dominant plankton species, essentially diatoms, at each station. Thus, whereas a species such as *Chaetoceros debilis* may be light saturated at 0.024 to 0.030 ly/min, *Fragilaria striatula* and *Denticula marina* are saturated at higher intensities (0.048 ly/min) and other species (*Nitzschia seriata*, *Coscinodiscus subtilis*, *Chaetoceros atlanticus* and *Asteromphalus hepactis*) only at intensities much in excess of 0.048 ly/min. These values are all distinctly lower than those suggested for diatoms generally by Ryther (1956a), but Motoda and Kawamura believe that the difference may be related to the particular climatic conditions of the northern North Pacific. Mist in those areas leads to reduced light intensities over considerable periods of time; the phytoplankton may be to some extent adapted to lower light intensities. The results also suggest that inter-specific variability may be much greater than variability between larger taxa.

Compensation Intensity

Although specific differences have been widely recognized for light-saturation values of phytoplankton, little is known about differences in compensation intensity; an early value quoted is *ca.* 0.002 ly/min for *Coscinodiscus eccentricus* (Jenkin, 1937). The acceptance of the depth of the euphotic zone as equivalent to the level receiving 1% of the surface light to some extent assumes that species have similar compensation intensities. However, many authors have pointed out that those species which normally live near the base of the euphotic zone, and even more, those species which live in shallow coastal waters, close to the bottom or as tychopelagic species, must probably be adapted to much lower intensities and presumably have correspondingly lower compensation intensities. The work of Bunt and his colleagues (e.g. 1966, 1968, 1970) has drawn attention to the existence of epontic diatoms. These live below the marine ice, either attached to the base of the ice or in the brash layers immediately below (cf. Chapter 6). Growth for these epontic diatoms appears to be possible at extraordinarily low light intensities, as also would appear to hold for the flora noted by Horner and Alexander (1972). Very little light can penetrate Antarctic ice, especially with snow cover, but Bunt and Lee (1970) showed that growth could start as early in the Antarctic spring as September/October. Some Antarctic algae must presumably have very low compensation intensities. This was confirmed by Bunt (1968) for a few Antarctic species successfully maintained in culture. *Stauroneis membranacea* and *Synedra*, for example, were found to have compensation intensities of *ca.* 0.0001 ly/min at the lowest temperature possible, the freezing point of seawater; there was some rise in compensation intensity with temperature, presumably reflecting some increase in metabolic demands. For *Fragilaria sublinearis* the intensity was slightly higher – *ca.* 0.004 ly/min – but this is approximately an order of magnitude below the compensation intensity suggested by Jitts *et al.* (1964) for experiments on the diatoms *Skeletonema* and *Thalassiosira nordenskioldii*. These very low Antarctic values correspond to about 0.1 to 0.06% of full sunlight rather than the 1% normally reckoned as the lower limit of the euphotic zone.

Bunt refers to the work of English (1961) who claimed higher compensation intensities of 0.003 to 0.0129 ly/min for Arctic microalgae under pack ice. He questions whether the presence of leads in the Arctic pack could have affected the light requirements of the algae. If English's data are confirmed, a very great difference exists between the light demands of Arctic and of, at least, some Antarctic species. Andersen (1977) describes algae trapped in ice at Godhavn (Greenland) which appear to have compensation intensities also far in excess of those described by Bunt. They probably represent truly planktonic algae with higher light requirements, but they contribute substantially after the ice breakup to primary production.

That some of the algae in the hypophotic zone (with its lower limit equivalent to 10^{-4} of surface illumination) (cf. Chapter 6) in Antarctic waters are photosynthetically active suggests that the compensation intensities of some of the species must be at least an order of magnitude below usually quoted values (Mandelli and Burkholder, 1966). Anderson (1969) has also found active algal cells in Oregon offshore waters, with compensation intensities estimated at one-tenth that given earlier by Jenkin (cf. Chapter 9).

Strickland (1965) suggested that the few accurate determinations of instantaneous compensation intensities gave values certainly below 0.01 ly/min and probably less than 0.005 ly/min for natural phytoplankton populations near the bottom of the euphotic zone. Algae below pack ice or trapped below pycnoclines have exceptionally low compensation intensities. Over a 24-hour period, however, he considered that these compensation intensities should be approximately doubled. To some extent algal cells may be transported across the compensation depth so that intensities can be approximate only. Flagellate cells may actively maintain their level above the compensation depth in some species by phototactic movements. Those phytoplankton cells that are specially adapted to shade conditions will tend to have lower compensation values; intensities less than 0.002 ly/min seem likely.

Paasche (1967, 1968a) found minimum light intensity requirements of 0.004 ly/min for *Coccolithus huxleyi*, and for *Ditylum* and *Nitzschia turgidula* about 0.003 ly/min, for a 6-hour light period. Values were slightly less for longer photo-periods. Values given by Jitts *et al.* (1964) are slightly greater (0.003 to >0.01 ly/min); Castenholz's (1964) data for littoral species are in the same range (0.001 to 0.01 ly/min). However, there are differences inside this range for various species. For example, the compensation intensities and light-saturation levels given by Castenholz for four diatoms are shown in Table 3.4.

Table 3.4. The compensation and saturation intensities for four littoral diatoms (after Castenholz, 1964)

	Compensation (ly/min)	Saturation
Fragilaria	0.008–0.010	0.04
Biddulphia	0.004–0.006	0.03
Synedra	0.003	*ca.* 0.02
Melosira	0.001–0.002	*ca.* 0.02

McAllister *et al.* (1964) determined compensation intensities for bacteria-free cultures of marine algae; nett photosynthesis was estimated from O_2-evolution (cf. Table 3.5).

Table 3.5. Compensation intensities for different algae (from McAllister *et al.*, 1964, slightly altered)

	Compensation intensity (ly/min)
Dunaliella (low light pre-conditioning)	0.013
Dunaliella (high light pre-conditioning)	0.013
Skeletonema	0.006
Monochrysis	0.01
Amphidinium	0.01
Dunaliella (low P)	0.002
Dunaliella (N-deficient)	0.015
Skeletonema (P-deficient)	0.006
Skeletonema (N-deficient)	0.012

The values were much the same as those obtained, using cell division as a criterion, by Jitts *et al.* (1964). Some differences in compensation intensity appear even between these few species; notably the intensity for *Skeletonema* is approximately half that of other species. Some effect of nutrient level is apparent, especially the markedly lower compensation intensity for phosphate-deficient *Dunaliella* as compared with cells in normal media. It is suggested that this reduction is due to the lowering of respiration with reduced phosphate. Pre-conditioning to light intensity does not appear to affect the compensation intensity, at least in so far as results with *Dunaliella* are typical. On the other hand, there is some indication from the results of Hellebust and Terborgh (1967) with *Dunaliella* cultures that compensation intensity may be appreciably lowered with prolonged culturing at very low light levels and lower temperature.

Jitts *et al.* (1964) do not quote precise values for compensation intensities since they were obtained by extrapolation. Nevertheless, the data indicate specific differences, with *Skeletonema* and *Thalassiosira nordenskioldii* (at low temperatures) having lower compensation intensities than *Monochrysis*, *Amphidinium* and *Dunaliella* (Table 3.6), a result in agreement with that of McAllister *et al.* (1964). The values for the two diatoms are only slightly higher than Jenkins' compensation intensity for *Coscinodiscus* (0.002 ly/min).

These latter data for compensation intensities mostly refer either to littoral algae or to temperate, mainly neritic species. Thomas (1966c), investigating certain tropical

Table 3.6. Minimum light requirements for growth in different algae (Jitts *et al.*, 1964)

	Minimum light for cell division (ly/min)
Dunaliella	0.015–0.025
Amphidinium	0.01–0.05
Monochrysis	0.01–0.02
Skeletonema	0.003–0.006
Thalassiosira	0.005–0.01

phytoplankton forms living at relatively high ambient temperatures (*ca.* 27°C), found that although temperature optima for photosynthesis were high, compensation light intensities were more or less normal, approximately 0.002 and 0.0006 ly/min for *Gymnodinium* and *Chaetoceros*, respectively. Findlay (1972) obtained a higher compensation point for the tropical diatom *Coscinodiscus pavillardii* (0.006 ly/min), but his value is similar to the findings of Jitts' *et al.* for diatoms (cf. Table 3.6).

Phytoplankton Respiration

A major factor determining the compensation intensity for a phytoplankton species is the level of respiration of the cell. The ratio of photosynthesis to respiration (PS/R) is of the greatest importance in productivity. Yentsch (1962) showed that Ryther's data for the Rhode Island area, assuming a reasonable level of photosynthesis and a low nett production confined to the winter months, could be interpreted only if the PS/R ratio was of the order of 10:1 or 15:1. With a PS/R ratio of only 5:1, for example, production would occur only during June and July.

Akinina (1966), from experiments on *Prorocentrum*, found a relatively constant level of respiration over a wide range of light intensities. The compensation light intensity was calculated as approximately 0.0024 ly/min. At near saturation light intensities, the respiration:photosynthesis ratio was about 15%, or a PS/R ratio of 7:1.

Many authorities have assumed a PS/R ratio of approximately 10:1 for natural populations when photosynthesis was maximal. Steemann Nielsen and Hansen (1959b) state that in most of the Dana experiments the level of respiration was generally some 6–10% of maximal photosynthesis, but there was considerable variation between different times and areas (cf. Fig. 3.29). As Yentsch (1962) points out, however, over the whole 24-hour period and throughout the whole euphotic zone, respiration may be of the order of 40% of photosynthesis. Qasim, Wellershaus, Bhattathiri and Abidi (1969) found in the tropical waters of Cochin Backwater very considerable variation in the level of respiration over a 24-hour period from day to day. Total daily respiration ranged from 20 to 45%, though some of this high value may be attributed to bacterial activity. Qasim *et al.* assumed the same level of respiration over the 24-hour period when calculating nett production. Considerable discrepancies occurred in attempting to

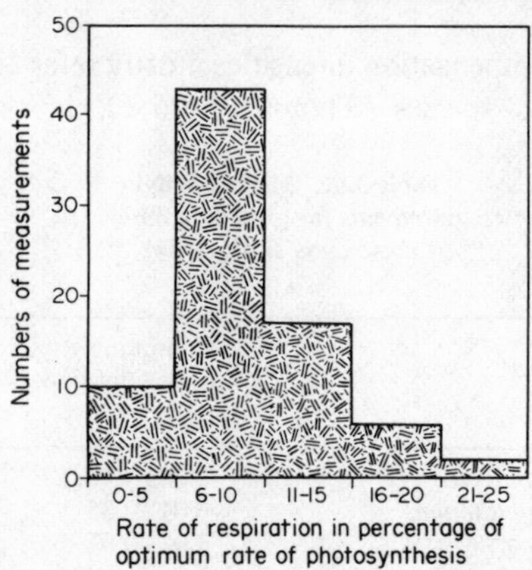

Fig. 3.29. The scattering of seventy-eight measurements of respiration made during the *Dana* cruises 1956–7 (Steemann Nielsen and Hansen, 1959).

compare gross production (as O_2 production) and near nett production (as ^{14}C estimates). There has been considerable debate as to whether ^{14}C-fixation represents nett photosynthesis, gross photosynthesis or some value between these. In short-term ^{14}C experiments it has been suggested that a substantial proportion of the assimilated ^{14}C lost in respiration is again photosynthesized, so that the average respiratory loss is less than 10%. In 24-hour experiments the ^{14}C technique measurements may estimate near nett production. Probably the variable physiological state of algal cells, in addition to the effects of environmental factors and to specific floristic differences, means that the level of production measured is not absolute. Fogg (1975) believes that provided the concentration of heterotrophic organisms is not unusually high, fixation of ^{14}C in light should not be corrected by subtracting values for dark fixation. Different intermediates in the TCA cycle become labelled with $^{14}CO_2$ in light conditions, although the cycle continues apparently at the same rate in light and darkness.

The respiratory rate of algae is not fixed; temperature and respiratory rate are directly related and the precise effects of other environmental factors may be of considerable significance. With nutrient deficiency, for example, Ryther (1954b) found that *Dunaliella* cultures showed a more rapid decline in photosynthesis than in respiration; when the cells ceased division the two processes were believed to be about equal in value.

Investigations dealing specifically with the respiratory rate of algae are few; early experiments on *Chlorella* suggested that O_2 uptake might be unaffected by light, at least at moderate and low intensities, though certain tests with other algae suggested that high light intensities could cause either a decrease or an increase in O_2 uptake. Gibbs (1962) describes some early experiments on the production of CO_2 and consumption of O_2 simultaneously, using a chlorophycean and a chrysophycean. The results suggested that below the compensation point light had very little effect on O_2 consumption, that O_2 consumption increased at higher light intensities, but that light inhibited the production of CO_2. Eppley and Sloan (1965) tested a number of phytoplankton species in culture and suggested that respiration was enhanced in light as compared with darkness only in *Dunaliella*.

Estimates of respiration have normally involved measuring O_2 uptake during darkness. The assumption has usually been made that respiration is the same in darkness and in light; on this assumption the addition of the amount of O_2-uptake in respiration to the quantity of O_2 production in nett photosynthesis has been reckoned as indicating gross photosynthesis. However, the assumption is not necessarily well founded. Certainly the interactions between respiration and photosynthesis are not fully understood. Many of the intermediates in the photosynthetic and respiratory processes are common.

Fogg (1968) points out that ATP and $NADPH_2$ in particular, products of the photochemical reaction, are also well recognized as intermediates in respiration. NAD admittedly functions more importantly than NADP in respiration, but probably hydrogen is transferred between NAD and NADP. As Fogg points out, if as a result of photophosphorylation the amount of ADP in the cell is lowered due to its conversion in the light to ATP, this will limit respiration. Equally, if any $NADPH_2$ produced in light transfers hydrogen to NAD, then the level of NAD may limit respiratory activity.

Isotopic oxygen studies have not entirely decided the question whether respiratory and photosynthetic paths are separate. In some plants, for example, *Chlorella*, respiration seems to be unchanged by light and darkness so that photosynthetic O_2 and respiratory O_2 are apparently not exchanged. On the other hand, some experiments suggest that an exchange of O_2 can occur, and that the interaction of the processes can be substantial leading to an apparent sharp decrease of respiration in light. Probably the degree of independence of the two processes is not only species specific, but may depend on the physiological state of the cell. Fogg (1975) points out that the recognition of a process, photorespiration, in algal tissue has further complicated our views. The process is dependent on light intensity. Uptake of O_2 and the release of O_2 with photosynthesis may be followed using O_2 isotopically labelled. According to Fogg the substrate is glycollate and the oxidation does not result in ATP production; possibly the process acts as an overflow mechanism when, with high light intensity but lack of nutrients, photosynthate is produced in excess of the amount converted to cell materials. Burris (1977) confirmed the occurrence of photorespiration in several algal species including some planktonic forms. He also observed high dark respiration rates in relation to the rates of photosynthesis in *Glenodinium* and in zooxanthellae algae; this could be a significant factor in their growth rates.

For most algae the respiratory quotient appears to be close to unity suggesting that carbohydrate is used as the main substrate for respiration. EMP glycolytic pathways, HMP shunt and TCA cycles are apparently all involved in respiratory pathways in algae; a number of labelled-carbon intermediates can be identified, although only some respiratory enzymes in these pathways have been recognized in algae. Whereas photosynthesis occurs in the chloroplast, the EMP and HMP enzymes are believed to be distributed in the cytoplasm, and the TCA and the electron-transport system in the mitochondrion. Although, then, many authorities believe that the processes of photosynthesis and respiration interact to some degree, there is possibly some physical separation between the chloroplast intermediates of photosynthesis and those intermediates in the respiratory process. The exchange of substances like ADP and NAD which might limit respiration during active photophosphorylation may therefore be restricted. Although photosynthesis and respiration in the algal cell both provide high-energy phosphorus substances and hydrogen donors, together with a number of carbon intermediates, so that the synthesis of more complex molecules becomes possible, many algal species appear to be obligate phototrophs, i.e. they can grow only in light and cannot grow in darkness on organic substrates (cf. Fig. 3.1).

Among factors which appear to be likely to depress the respiratory rate is the depletion of substrate reserves. Frequently there is a marked rise in respiratory rate with the addition of substrate, especially carbohydrate. It is possible by the addition of a ^{14}C-labelled substrate to demonstrate the oxidation of the material and subsequent evolution of the labelled CO_2. On the other hand, the addition of substances such as glucose to normal cells sometimes appears to suppress the indigenous respiration level. The effect apparently varies again from species to species and with the physiological state of the cell.

The concentration of certain nutrients appears to affect respiration; for example, the addition of ammonium to nitrogen-starved cells causes a much more obvious increase in respiration than addition of nitrate or nitrite. Photosynthetic rate is also raised by the addition of ammonium to nitrogen-starved cells, but the stimulation is not so marked. Yentsch and Reichert (1963), using *Dunaliella*, suggested that the level of respiration in darkness varied with the nitrogen source and was equivalent to approximately 27–37%

of the initial photosynthetic rate, depending on whether nitrate or ammonium was added. Respiration gradually decreased when cultures were kept for longer periods in darkness. Akinina (1966), using dinoflagellate cultures, suggested that phosphate impoverished cells had a somewhat raised level of respiration though respiration was virtually constant over a very wide range of light intensities. These findings may be compared with those of Jitts *et al.* (1964) which have been referred to earlier.

Little has been said so far on the effect of temperature on the level of respiration and on nett production. Interaction between temperature, light and nutrient conditions, some of the factors important in phytoplankton growth, might play a part in species succession, but at first sight temperature might be expected to be of lesser direct importance in determining production. Increased temperature, while having a positive effect on photosynthesis at relatively high light intensities, would increase respiratory requirements. The anabolic and catabolic effects of temperature on production might cancel, provided respiratory and photosynthetic relationships with temperature were similar, and optimal temperatures for respiration and photosynthesis more or less coincided. Such a condition may hold for some species of temperate phytoplankton. Thus Jitts *et al.* (1964) found several temperate species relatively insensitive to temperature over a range of more than 10°C. An exception was *Thalassiosira nordenskioldii*; near the extreme limits of temperature the species became more sensitive to changing light conditions.

This similarity of relationship between temperature and photosynthesis and temperature and respiration may not always be true particularly in cold-water plankton. Antarctic diatoms seem especially well adapted to photosynthesizing at extremely low light intensities (*vide supra*) and at permanently low temperatures. Bunt (1964) shows that not only are epontic diatoms able to photosynthesize at remarkably low light levels, but that this holds for *Phaeocystis*, an alga of worldwide distribution.

A more detailed analysis of temperature effects on photosynthesis by Bunt (1967) included experiments with several diatom species and one flagellate, isolated from Antarctic waters and grown under laboratory conditions. Bunt suggests that the diatom, *Fragilaria sublinearis*, is an obligate cryophilic species. It grows well in its natural habitat at temperatures below 0°C, though at such temperatures it demands higher light intensities than those quoted earlier, a range of about 0.006 to 0.015 ly/min being required for best growth. Some increase in primary production follows an increase in temperature up to about 7°C, but a temperature of 10°C appears to be lethal. Optimum growth of this species was found with light intensity 0.02 ly/min and temperature 5–7°C. There are differences between the species isolated; for example, *Stauroneis membranacea* appears to have a slightly lower temperature optimum than *Fragilaria*, but for all species compensation intensities were extremely low, of the order of 0.0005 ly/min.

In these Antarctic species the responses of the photosynthetic and respiratory processes to temperature appear to be somewhat separated, so that the optimum temperature for photosynthesis is appreciably lower than that for respiration. The diatom is better adapted to achieve a good level of primary production at relatively low temperatures, but may be restricted because of these characteristics to Antarctic waters.

In temperate phytoplankton, photosynthesis appears to vary directly with temperature, provided fairly high light intensities are available. Experiments growing phytoplankton under strictly controlled conditions of temperature, light intensity and photo-period (e.g. Paasche, 1967, 1968a; Eppley *et al.*, 1967; Jorgensen, 1966) indicate that optimal temperatures in the laboratory are frequently higher than might be supposed from environmental temperatures (e.g. *Ditylum* 20–27°C; *Nitzschia turgidula* about 20°C). Braarud (1961) found for eight dinoflagellate species that the optimum temperature was in the region of 15–20°C, though differences existed in the precise relationship. For example, *Gonyaulax* had a fairly sharp fall in growth at temperature below 20°C, whereas for *Exuviaella* the optimum appeared to be above 20°C; *Ceratium tripos* in contrast to *C. fusus* and *C. furca*, grew fairly well at temperatures below 20°C, whereas for *Exuviaella* the optimum appeared to be above mainly in the summer in Oslofjord so that a temperature optimum of 15–20°C would be in reasonable agreement, *C. lineatum* was found in abundance at lower temperatures, and *C. tripos* was predominant at colder seasons. For the coccolithophorid, *Coccolithus huxleyi*, the temperature optimum was about 20°C, although in the field growth occurs at temperatures as low as 7°C (cf. Paasche, 1966a). Braarud observed for two widely distributed dinoflagellates, *Prorocentrum micans* and *Peridinium trochoideum*, that a difference existed in temperature response, depending on geographical area (cf. Fig. 3.30). With diatoms, Braarud obtained reasonable agreement for *Biddulphia aurita* and *B. sinensis* between field and experimental optimal temperatures, but results with *Asterionella japonica* and *Thalassiosira nordenskioldii* posed problems (Table 3.7). As suggested earlier, marked growth of *T. nordenskioldii* at

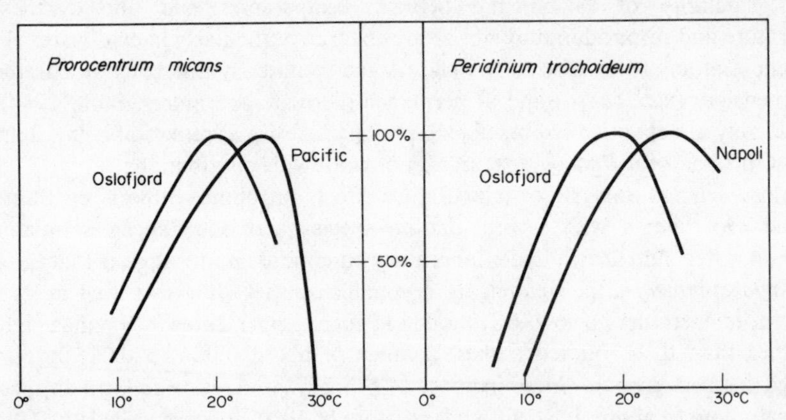

Fig. 3.30. Temperature/growth response in strains of two dinoflagellates from different geographical regimes (Braarud, 1961).

Table 3.7. Temperature—growth relationships in marine plankton diatoms (from Braarud, 1961)

Species	Field temperature during periods of abundance	Optimal temperature in culture
Biddulphia aurita	1°C	5°C
Biddulphia sinesis	13°C	16°C
Asterionella japonica	8°C	20–25°C
Thalassiosira nordenskioldii	2–3°C	Excellent growth at 10–11°C

low temperatures in early spring may be in part correlated with high nutrient concentration. Some of the difficulties in relating experimental findings on laboratory cultures to field conditions have already been discussed.

Although in some seas photosynthetic rate may not be maximal at environmental temperatures, throughout the world ocean species of phytoplankton are present capable of effective photosynthesis under the temperature regimes. Adequate light energy is, however, an absolute requirement, and regional and seasonal variations in incident light and the rate of attenuation of light with depth are of paramount importance in primary production. The indirect effects of temperature in relation to stratification may be highly significant (cf. Chapter 6).

Excretion by Algae

The variety of dissolved organic substances found in seawater is largely derived from the decomposition of marine plants and animals; near land this may be reinforced by organic matter derived from the terrestrial environment (cf., for example, Prakash, 1971). Part of the dissolved organic matter may arise, however, from materials released from phytoplankton cells during life, substances frequently described as excretory products. Putter (1909–1925) commented on the significance of relatively large amounts of dissolved organic matter in seawater; he believed that part of the material was leached from phytoplankton cells during growth.

The extent to which excretion occurs in phytoplankton is a matter for debate. Duursma (1961), for example, contended that practically all the dissolved organic matter in seawater was derived from decomposition processes. The increasing weight of evidence from laboratory cultures and examination of natural phytoplankton blooms leads, however, to the conclusion that excretion is a normal part of phytoplankton growth, though the extent may be extremely small (cf. Thomas, 1971). The term excretion is perhaps not entirely warranted; under some circumstances algae not only release dissolved organic material to the medium but absorb organic substances. Eppley and Sloan (1965) observed that although *Dunaliella* excreted much glycerol, *ca.* 70% of this could be reabsorbed even during an hour in darkness.

The measurement of excretory processes by algae sometimes leads to conflicting results. While some of the differences may be species specific and often reflect varying external conditions as well as changing physiological condition of the phytoplankton population, some differences may be ascribed to imperfect technique such as rupture of algal cells during filtration. Arthur and Rigler (1967), Wallen and Geen (1971c) and Williams, Berman and Holm-Hansen (1972) are among several investigators who have called attention to experimental errors in phytoplankton excretion measurements. Sharp (1977) has recently discussed the reality of excretion by healthy algae and has referred to some of the errors in its measurement.

Dissolved organic matter may be directly estimated by analysis of the filtrate of a quantitative sample of phytoplankton from the field or from culture. Another often employed method is a modification of the ^{14}C-assimilation technique described more fully in Chapter 9. Following the addition of $NaH^{14}CO_3$ to an algal suspension or phytoplankton sample and incubation, and after the removal of the algae by filtration, the filtrate is acidified and any remaining $NaH^{14}CO_3$ removed by bubbling with CO_2. The radioactivity of the filtrate is subsequently determined. Possible errors, apart from faulty filtration, include the removal of some of the organic material by heterotrophic processes; the limitation that only products synthesized during the incubation period with ^{14}C can appear as excreted matter; and the fact that the bicarbonate added may be contaminated by radioactive organic material.

Eppley and Sloan (1965) estimated the dissolved organic matter excreted by phytoplankton by the two methods: the increase in soluble organic carbon and the amount of ^{14}C appearing as dissolved organic material in ^{14}C assimilation experiments. Any increase in soluble organic matter cannot be ascribed entirely to excretion and the authors believe that the ^{14}C method gave the better estimate of excretion, though some excretory products may have been unlabelled. Quantities of carbon excreted by a variety of phytoplankton species in culture and by a natural "red tide" showed a considerable range, dependent on species, and on method: 0.13–0.50 mg C/l for *Dunaliella*; <0.01–0.60 mg C/l for the four diatoms, *Ditylum*, *Thalassiosira*, *Cyclotella* and *Skeletonema*; 0.01–0.07 mg C/l for *Peridinium* and two coccolithophorids. *Dunaliella* excreted a constant fraction, about 13%, of the carbon synthesized, irrespective of cellular physiological activity, but with diatoms, with the possible exception of *Thalassiosira*, excretion was inversely proportional to physiological activity. Excretion in *Ditylum* could even exceed 30% with reduced activity. *Thalassiosira* appeared to show the same excretory rate in light and in darkness, and possibly as much during the exponential growth phase as when growth was reduced.

Eppley and Sloan (1965) questioned whether the commonly observed discrepancy between a higher carbon assimilation (nett) rate deduced from O_2-production experiments, and a lower rate as observed in ^{14}C assimilation experiments (and from direct measurements of increase in particulate carbon) could be entirely explained by the amount of excreted soluble carbon. Higher estimates were obtained from O_2-measurements even when due allowance was made for excretion. A photosynthetic quotient

$$\text{P.Q.} = \frac{O_2 \text{ produced}}{CO_2 \text{ absorbed}}$$

of 1.25 is frequently employed for determinations with diatoms, but Eppley and Sloan found that the P.Q. showed some variation with environmental conditions; for example, it tended to be lowered by high light intensity. It is possible that, diatoms, in contrast to other groups of algae have a peculiar metabolic path releasing "extra" O_2, so that their P.Q. would be somewhat raised. There was no confirmation in their experiments, comparing "healthy" and "incipient senescent" cells of *Dunaliella*, of a relationship between the nett/gross photosynthesis ratio and the proportion of chlorophyll/carotenoids as suggested by Ketchum, Ryther, Yentsch and Corwin (1958).

Ketchum and Corwin (1965) estimated the increase in dissolved organic phosphate during a 10-day bloom of phytoplankton in the Gulf of Maine. They attributed a small part of the increase to excretion by the phytoplankton, but recognized that some dissolved organic material could have been derived from decomposition, especially from grazing by zooplankton. Estimates of excreted organic matter were also obtained by comparing nett production measurements from O_2 production and from CO_2 changes, with the amount of particulate material synthesized over the same period estimated directly as particulate carbon, particulate phosphorus, or from ^{14}C assimilation. This approach has been employed by McAllister, Parsons, Stephens and Strickland (1961) and Antia, McAllister, Parsons, Stephens and Strickland (1963) using the plastic sphere technique with natural phytoplankton populations. Such estimates of dissolved organic matter excreted must be influenced by errors inherent in the various methods employed (e.g. O_2 determination, CO_2 measurement, ^{14}C counting, etc.). Banoub and Williams (1973) calculated from observations at E1, off Plymouth, that about half the products of algal photosynthesis were transformed into dissolved organic matter. They quote the results of Williams and Yentsch for Florida waters, which indicated that 5–20% of photosynthate may be excreted (cf. p. 123). While accepting that not all dissolved organic matter at E1 results from excretion (one obvious source of dissolved organic matter is through zooplankton grazing), they imply that a considerable portion of the photosynthate is "lost" by algae.

Estimates of primary production with laboratory cultures of *Dunaliella* and *Skeletonema*, using several methods, were carried out by McAllister, Shah and Strickland (1964); some of the measurements by the ^{14}C method were appreciably lower than the nett values found by the O_2 method, the differences being attributed to excretion. Ryther, Menzel, Hulburt, Lorenzen and Corwin (1970), measuring primary production in upwelled waters of the Peruvian Current, found that estimates based on nutrient and O_2 changes, and to a somewhat less extent, direct measurements of particulate carbon and particulate phosphorus, all gave higher values for production than ^{14}C assimilation determinations. While some of the difference could be due to excretion of organic matter, there might have been some loss of organic material from the euphotic zone. Some direct measurements of excretion were also made (*vide infra*). Strickland, Holm-Hansen, Eppley and Linn (1969) showed that relatively little organic matter was excreted in tank cultures of phytoplankton. For *Ditylum* the maximum amount was 10% of photosynthate estimated by direct organic analysis, and about 5% measured by ^{14}C estimates.

Many factors may influence algal excretion. Hellebust's (1965) investigations, which involved twenty-two species of phytoplankton, suggested that when algae were grown in the log phase under similar conditions the majority of the species excreted only a relatively small amount (3–6%) of the carbon photosynthesized. A few algae, notably *Olisthodiscus*, *Amphidinium*, two species of *Chaetoceros*, and *Chlorococcum*, excreted much larger amounts, up to 25% of the assimilated carbon, under the same conditions. Some studies indicated that excretion of dissolved organic matter tended to occur only when algae were becoming somewhat "unhealthy", especially towards the end of a bloom. Fogg (1963), on the other hand, advanced arguments for the belief that excretion was typical during normal healthy growth of algae, and Hellebust's experiments showed some excretion during the log phase of algal development. Antia *et al.* (1963) also commented on the release of soluble organic matter while algae were in vigorous growth. The percentage of dissolved organic matter may tend to rise with declining physiological activity of algae (cf. Eppley and Sloan's results), but with *Dunaliella* and possibly with *Thalassiosira*, excretory rate remained more or less constant. Not all investigators have found the suggested inverse relationship between excretion and physiological activity, though Ryther *et al.* (1970) confirmed the suggestion, and Anderson and Zeutschel (1970) observed that photosynthetic rate was inversely related to the amount of dissolved organic matter released, when this was expressed as percentage of production. Hellebust (1965) also found small amounts (4–16%) of assimilated carbon excreted by natural healthy phytoplankton, but a much larger fraction (17–38%) excreted at the end of a bloom when a number of empty frustules were found.

Light is one of the main factors believed to influence algal excretion. McAllister, Shah and Strickland's (1964) data indicated that pre-conditioning to light could affect the capacity for excretion in *Dunaliella*. The intensity of light is the most usually quoted factor affecting excretion, however. High light intensities, especially those which may cause some inhibition of photosynthesis, have often been correlated with increased excretion. Hellebust's experiments demonstrated that when a number of algae were grown at two moderate light intensities (0.018 and 0.06 ly/min) there was relatively little difference in the percentage of carbon excreted; there was a slight indication of increased excretion at the lower light intensity. Even with widely different intensities (0.018 and 0.15 ly/min), changes in the amount of carbon excreted were slight, with the exception of *Exuviaella*, which released approximately three times the amount of carbon at the higher intensity, and also of *Skeletonema*, which exhibited quite constant excretion rates over a remarkably wide range of light intensities (0.084–0.39 ly/min), but showed a considerable increase in percentage excretion at lower intensities. Only excessively high light intensities resulted generally in a large rise in the amount of carbon released. For example, twelve species of algae, incubated for five hours in full sunlight (0.72 ly/min), excreted organic carbon amounting to between 8.5% and 52% of the photo-assimilated carbon. This may reflect some damage to the cells by the intense sunlight. Watt (1966) claimed from observations on freshwater species that the extent of carbon released from algae increased sharply with marked inhibition of photosynthesis. In tropical marine waters, high percentages of carbon released by phytoplankton are believed by Thomas (1971) to arise partly from surface light inhibition. Samuel, Shah and Fogg (1971) obtained little evidence for increased excretion in surface tropical waters at Cochin, but there was little inhibition, probably due to the high turbidity of these backwaters.

Horne, Fogg and Eagle (1969) found in inshore Antarctic seas usually a low percentage (*ca.* 1%) of carbon liberated as extracellular product at the depth of maximum photosynthesis, but percentage release tended to rise at both lower and higher light intensities. With very high intensities and consequent inhibition of photosynthesis, the percentage release could amount to 38%.

There is evidence that extracellular organic matter is released in darkness and that *percentage* release may increase in the dark (cf. Samuel, Shah and Fogg, 1971) but differences between species may be significant. Eppley and Sloan (1965) found for *Ditylum* and *Skeletonema* that excretion increased during the first period of darkness, though it was uncertain whether this continued with residence in darkness; for *Thalassiosira*, by contrast, excretion was similar in dark and light. Eppley, Holmes and Paasche (1967) kept cultures of *Ditylum* under 8/16-hour light/dark cycles and found that only small amounts (2–3%) of assimilated carbon were excreted, but sometimes larger amounts were excreted during the "early" period of darkness.

There have been few experiments on the possible effect of light quality on algal excretion. Wallen and Geen (1971c), while admitting some possible artefacts due to filtration pressure, found that whereas in the lower photic zone only *ca.* 10% of the ^{14}C fixed by algae was excreted as soluble photosynthate, in surface/subsurface layers from 13% to as much as 34% of ^{14}C fixed was excreted. They attributed the differential effect to the wavelength of light at the two depths rather than to intensity differences, although light intensity was almost certainly an additional factor. McAllister *et al.* (1964) also found differences in the fraction of soluble organic matter believed to be excreted, according to the type of light source employed (filtered tungsten or cool-white fluorescent lamps). They considered that the effect might be attributable to differences in the spectral distribution of light energy.

Another factor affecting the amount of dissolved organic matter released by algae is distance from shore. This appears to be mainly a reflection of the usually more eutrophic areas inshore and more oligotrophic regions offshore, especially in tropical seas. Table 3.8 gives some values for the percentage of carbon released by algae from various areas. Although the wide differences in percentage of dissolved organic material excreted reflect in part particular conditions and precise methods of determination, there is a fairly clear relationship between the quantity and type of water. Anderson and Zeutschel (1970) found that release of organic matter generally declined with increasing depth. The highest values within the euphotic zone were found in the most productive waters, and the lowest values in oligotrophic areas, though the radioactivities measured were generally low. However, they showed that excretion reckoned as a percentage of total production ranged from 1% for surface waters in the most productive area to 49% for surface oligotrophic waters; at several stations percentage excretion also increased with depth.

Table 3.8. The excretion by phytoplankton in different regions. The extracellular material released is expressed as a percentage of estimated production

Region		Percentage extracellular release
North Sea		7
Inshore Antarctic (Eutrophic)		1–2
Inshore Indian Ocean (Oligotrophic)		1–20
Southern Indian Ocean		5–32
Inshore Pacific		35–40
Inshore Tropical (Oligotrophic)		<1–23 Av: 6.6
Coastal Pacific	Eutrophic surface	>1
	Oligotrophic surface	<49
Eutrophic Estuary		<7
Atlantic Coastal surface		<13
Atlantic Coastal below surface		<21
Westernmost Sargasso (Oligotrophic)		<44
Peruvian	surface coastal	1–3
	coastal water column	4–12
	offshore column	45
Pacific oceanic tropical		30

Ryther *et al.* (1970) also found that whereas surface samples incubated at full sunlight released only 1–3% of carbon fixed, those samples collected from the bottom of the euphotic zone and incubated at *ca.* 1% incident light, released some 25% of the carbon assimilated. These findings applied to stations in markedly eutrophic Peruvian waters, but at one offshore station where organic production was much lower the release of organic soluble matter amounted to 45%. Differences in the amounts of excreted substances must thus be related to rates of photosynthesis in eutrophic and oligotrophic waters. However, there are some problems in interpretation of the data of

Ryther *et al.*; in some experiments a fairly constant amount of excretion was associated with a considerable range of values for photosynthetic rate. It is worth noting that Smith, Barber and Huntsman (1977) found that in the eutrophic waters of the upwelling region off north-west Africa, excretion amounted on average to only 8.7% of the total carbon fixed, and that the amount varied directly with the quantity of carbon fixation and was little influenced by variations in light or nutrient levels.

Though there does not appear to be clear evidence for a close relationship between photosynthetic rate and percentage excretion, Thomas' (1971) investigations suggest that an increase in percentage of organic substances released by algae occurred in proceeding to very oligotrophic areas. Thus, in a highly productive inshore area such as a Georgia estuary, the rate of release of dissolved organic matter is considerably higher (<1 to 40 mg C/m^3/day) than offshore (e.g. 0 to 2 mg C/m^3/day in western Sargasso), but the percentage of extracellular material excreted rises in proceeding offshore into relatively "poor" areas – <7% in Georgia estuary; <13% in surface south-east coastal waters; <44% in western Sargasso. High light intensity may be a factor in increasing percentage excretion in open waters.

Some of the difference between percentage excretion in oligotrophic and eutrophic waters has been related to cell density. Certain experiments with algal cultures have demonstrated an inverse relationship between cell density and percentage soluble organic matter released. Samuel *et al.* (1971) demonstrated in experiments in tropical waters at Cochin that variations in chlorophyll *a* concentrations of 117 to 0.54 mg/m^3 were accompanied by differences in excretion of soluble organic matter of from 5.0% to 18.9%, respectively. Apart from some inhibition of photosynthesis due to high light intensity in tropical open oceans which might raise the percentage of dissolved organic matter excreted, there may be other factors. Earlier studies have indicated that the "healthy" condition of algal cells was considerably influenced by nutrient levels, and since there is some evidence for increased excretion being associated with reduction of physiological activity, the algae in oligotrophic areas might be considered as less active with the generally lower nutrient level. Stress certainly appears to increase excretion. Thus, concentrations of toxic substances may affect the percentage of dissolved organic material excreted. Holdway (1976) described an increase in extra-cellular ^{14}C products excreted by diatom cultures when these were subjected to sublethal copper concentrations and quotes supporting evidence by Steemann Nielsen and Kamp-Nielsen (1970). Temperature and salinity may also modify excretion and may act synergistically.

Ignatiades (1973) and Ignatiades and Fogg (1973) have reaffirmed their belief that excretion is a normal process in healthy phytoplankton, though the amount varies with many factors, senescence, preconditioning to darkness, low nutrient level and low population densities, for instance, leading to an increased release of extracellular substances. Field work in Loch Etive, Scotland, demonstrated that *Skeletonema* showed differences in the amount of organic carbon excreted with season and depth. During March extracellular excretion formed 3.5% of the total ^{14}C fixed at a depth of 1 m. At 10 m, approximately the limit of the euphotic zone, there was increased excretion (13.9%). In July, when photosynthetic activity was less than during the spring, excretion at 1 m depth accounted for 15.9 to 18.5%. The most obvious effect was a very marked increase in the relative amount of photosynthate excreted at the 1 m level by October, when it amounted to 46%. Ignatiades found a significant inverse correlation between crop, as chlorophyll or as carotenoid concentration, and percentage excretion. Experiments carried out by Ignatiades and Fogg demonstrated that *Skeletonema* excreted very little (2 to 4%) of the carbon assimilated when in the exponential growth phase and when exposed to limiting or saturating light intensities, but with light intensities causing inhibition the amount of ^{14}C excreted increased substantially. Increased bicarbonate concentration in the medium resulted in a rise in the relative amount of extracellular substances released; ageing cultures showed both an absolute and relative increase in excretion. Although *Skeletonema* could retain its viability in darkness for up to 37 days, the amount of ^{14}C fixed on re-exposure of the alga to light fell with the period in darkness, but the percentage of carbon excreted rose to very high values. Nutrient deficiency similarly caused a very clear increase in the relative amount of carbon excreted as the particulate ^{14}C fixation declined (cf. Fig. 3.31). The authors also suggest that with reduced cell population although the relative amount of carbon released declined, the absolute amount excreted per cell was inversely related to the cell population (Fig. 3.32).

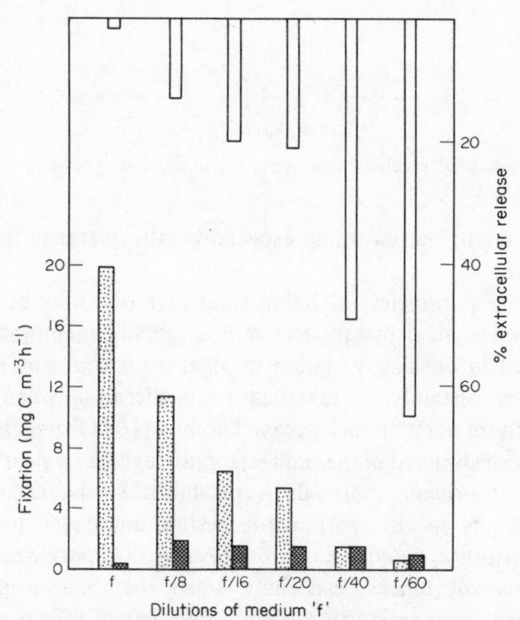

Fig. 3.31. Amounts of particulate matter (light hatching) and soluble extracellular material (solid shading) fixed as a function of different concentrations of culture medium "f" by *Skeletonema*. Extracellular release is also shown as percentage (after Ignatiades and Fogg, 1973).

Williams and Yentsch (1976), however, found generally low values for algal excretion at a station near the Bahamas, where primary production was low and fairly typical of an oligotrophic area, and where, therefore, high values for excretion might have been expected. Excretion ranged from <1% to 23% (mean 6.9%) of the total carbon fixation. Excretion was also comparatively low for three axenic cultures of algae (*Skeletonema*, *Dunaliella*, *Monochrysis*) during logarithmic growth. Williams and Yentsch failed to confirm the suggestion of an effect of cell density on excretion. Neither the field data nor the culture experiments gave any indication of an increase in percentage excretion with low cell populations or with low photosynthetic rates which could vary a hundred-fold. Further investigations on excretion, particularly for open-

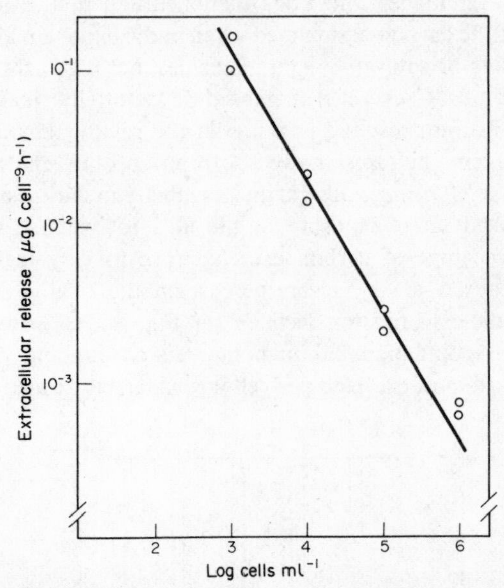

Fig. 3.32. Absolute amount of extracellular release per cell in relation to cell concentration of *Skeletonema* (after Ignatiades and Fogg, 1973).

ocean populations, might be rewarding especially with reference to cell density effects and productivity level.

Although the list of parameters affecting algal excretion may be still debatable, any factors which influence algal metabolism and especially membrane permeability are likely to be involved in causing variation in algal excretion. The release of dissolved organic matter must certainly be taken into consideration when assessing primary production especially in very "poor" areas. Thomas (1971) suggests that productivity may have been under-estimated in these oligotrophic regions by nearly 50%.

A wide variety of organic materials is produced in the extracellular excretions of phytoplankton. Some species, particularly certain dinoflagellates including *Gymnodinium breve*, *Gonyaulax catenella*, *G. tamarensis*, *G. polyedra* and *Prymnesium*, liberate highly toxic substances, especially when they bloom giving rise to "red tides", and these can have seriously adverse, even lethal, effects on a wide range of animals (cf. Gunter, Williams, Davis and Walton Smith, 1948; Abbott and Ballentyne, 1957; Brongersma Sanders, 1957; Lo Cicero, 1975). Oguri, Soule, Juge and Abbott (1975) emphasize that particular conditions (nutrient levels, hydrographic factors, etc.) are responsible for such red tides rather than abnormally accelerated division rates. Massive populations reaching 16 to 20 × 10⁶ cells/l can be built up from seed stock (Steidinger, 1975). Some of the toxins, which include haemolytic and neurotoxic agents, are heat stable (cf. Lo Cicero, 1975).

Many algal species release material which is not obviously toxic but which can promote or discourage the growth of other species (Jorgensen and Steemann Nielsen, 1961; Williams and Yentsch, 1975). Aubert and Aubert (1969), Aubert, Pesando and Gauthier (1970), Aubert and Pesando (1971) and Aubert and Gambarotta (1972) found that about a quarter of some eighty species of diatoms had antibiotic effects on certain bacteria. Extracts of some species had growth-promoting effects. An investigation of

Asterionella japonica indicated that thermolabile materials in the form of a fatty acid and a nucleoside were excreted. Certain species of peridinian could counteract the antibiotic effect; for example, *Prorocentrum* secreted an extracellular substance of a protein nature which acted in this manner.

Little is known of the detailed composition of algal excretions. Collier (1953, 1958) described carbohydrate-like excretory substances produced by *Prorocentrum* and by *Gymnodinium breve*. Guillard and Wangersky (1958) demonstrated the production of soluble extracellular carbohydrates by a variety of flagellates, the amount released being highest in old cultures. The same authors (Wangersky and Guillard, 1960) showed the production of an organic base by a marine dinoflagellate, *Amphidinium carteri*. The release of carbohydrates by *Isochrysis* and *Prymnesium* under such varied conditions as increased and decreased salinity, reduced light, and low nitrogen has also been followed by Marker (1965). Fogg, Stewart, Fay and Walsby (1973) report on the liberation of extracellular nitrogenous compounds, both amino acids and peptides, by Cyanophyceae. Fogg (1963, 1966) emphasized the importance of glycollic acid in freshwater and marine phytoplankton. Appreciable amounts released by phytoplankton may form an important reserve; the dissolved material may be used by bacteria; to some extent it may be taken up by the algae and used in metabolism. Fogg considers that the relatively small amounts of glycollate obtained under normal conditions of analysis is a reflection of the considerable amount which is rapidly utilized. Other organic acids, apart from glycollate, and a variety of organic compounds (polysaccharides, amino acids and peptides) may be produced as extracellular products, in addition to growth-promoting and growth-inhibiting substances whose biochemistry is largely unknown.

Fogg suggests that there are two major types of extracellular product. Intermediate substances of relatively low molecular weight may be released into the sea before being converted into more permanent metabolic products; to some extent they may be reabsorbed from the medium under certain conditions (e.g. glycollic acid). Substances of higher molecular weight, polysaccharides, polypeptides, etc., are released to the medium mainly in relation to the amount being metabolized by the cell (i.e. in relation to growth); little uptake of these materials occurs from the medium.

While the nature of growth-promoting and growth-inhibiting substances is largely unknown, the release of extracellular nicotinic acid by *Chlamydomonas* has been reported by Nakamura and Gowans (1964). Some of the substances released may act as growth-promotion agents due to their functioning as chelators (Johnston, 1964).

Parsons and Strickland (1962) commented on the limited heterotrophy of algae, though some uptake of carbon compounds can occur in some algal groups. Lewin and Lewin (1960) and Lewin (1963) showed marked differences in the heterotrophic abilities of various diatom species; some could utilize organic carbon in the form of glucose and lactate, but generally heterotrophic powers were rather small. For chrysomonads, Pintner and Provasoli (1963) demonstrated that an array of organic acids and carbohydrates were used as carbon sources, though specific differences were evident; for example, *Coccolithus huxleyi* had very little heterotrophic ability. Chrysomonads used adenylic acid and a range of amino acids as nitrogen source, though the substances were inferior to nitrate. Guillard (1963) found that some diatoms and unicellular flagellates could make use of organic nitrogen (amino acids, urea, uric acid) in bacterial-free culture, and Provasoli and McLaughlin (1963) demonstrated limited utilization of dissolved organic nitrogen by dinoflagellates.

The role of glycollic acid is not entirely resolved. Hellebust (1965) found glycollic acid as an extracellular excretion in all twenty-two species of algae which he investigated, but it formed a major proportion (9–38%) in only four species, *Olisthodiscus*, *Chaetoceros pelagicus*, *Chlorococcum* and *Skeletonema*. Fogg believes that alternative culture conditions might have stimulated larger quantities in all the algae. With *Skeletonema*, low light intensities favoured the release of glycollic acid. Hellebust could not demonstrate uptake of glycollate by *Skeletonema* but *Cyclotella nana* absorbed considerable quantities.

An examination of the substances isolated from the extracellular products of algae by Hellebust showed that relatively small amounts of protein were liberated by most algae. Chloroform-soluble (i.e. lipid-like) materials were liberated in somewhat larger amounts by most species – about 5% of the carbon excreted. A considerable fraction of excreted material appeared as organic acids of relatively low molecular weight. Few, apart from glycollic and possibly malic and citric acid, were identified. Most of the material consisted of amino acids of comparatively small molecular weight, and sugar alcohols (mannitol, glycerol, etc.). Some higher molecular weight polypeptides and carbohydrates were apparently also excreted. A few species excreted one substance to a very large extent, e.g. *Chlorella*; proline; *Dunaliella tertiolecta*: glycerol; *Olisthodiscus*: mannitol.

The variety of organic substances listed refers to excretion by algae in the log phase of growth. Just as the total amount of extracellular excretion may change in some species in ageing cultures, so the products of metabolism liberated may change with the phase of growth. Whatever the range of substances, there is little doubt about the significance of phytoplankton excretion. Even if it be proved that heterotrophic uptake is generally of very minor importance in primary production, the effects of algae in conditioning the seawater by their excretions may be paramount in determining the floristic assemblage of plankton communities, favouring or depressing the growth of other algal species, especially in species succession (cf. Johnston, 1964). Moreover, functioning as chelators and other growth-promoting agents, extracellular excretions may directly influence the productivity of an area (cf. Chapter 7).

Primary Production and Depth in the Marine Environment

Since primary production must be limited by the compensation intensity, and light is so rapidly attenuated in the sea, production over the world ocean is due solely to the phytoplankton population inhabiting the surface and near-surface layer. No growth of plants on the sea bottom is possible in open sea. In very shallow inshore waters, on the other hand, sufficient light may reach the bottom for plants to grow on the substratum. Moreover, plants in the inter-tidal areas may add to production in shallow adjacent waters.

Typical flowering plants inhabiting sandy shores and shingle beaches are mostly above high-tide mark and cannot contribute much to the marine environment. On muddy shores, however, the salt marsh type of environment is well known and salt marsh flowering plants may add significantly to overall production. Salt marshes are typical of sheltered environments such as estuaries and bays, but in certain geographic regions cover large areas. In tropical regions a particular type of this environment is the mangrove swamp which can be very extensive and highly productive.

In temperate regions the inter-tidal muddy areas have a few flowering plants (*Spartina* and *Salicornia*); as mud accumulates, with very slightly higher levels and better drainage, a considerable variety of flowering plants may be present. Jefferies (1972) summarizes the range of plants in British salt marsh communities, and comments on their remarkable phenotypic plasticity. The more typical salt marsh may be succeeded landwards by other communities leading to a fresh-water type of marsh environment with such plants as rushes (*Phragmites*, *Scirpus*, *Juncus*, etc.). A generalized picture is given by Eltringham (1971).

The more seaward areas of mud flats have other flowering plants, restricted to a few genera, which live completely submerged in the shallow waters. These include *Zostera*, *Ruppia*, *Posidonia*, *Cymodocea* and *Thalassia*. According to Dawson (1966) most of these marine grass-like flowering plants are essentially warm-temperate species, though some penetrate to high latitudes, in particular species of *Zostera*. The first four genera occur as eel grasses in the Mediterranean (cf. Peres, 1967). Thomas, Moore and Werk (1961) stress the importance of the turtle grass *Thalassia testudinum* in warm sheltered areas from Florida to Brazil; the species is an extremely important part of the benthic flora in shallower waters of Florida bays (Pomeroy, 1960). The remarkable uptake and release of phosphate from sediment and from the water by *Zostera* and *Thalassia* communities is notable (cf. McRoy, Barsdate and Nebert, 1972; Patriquin, 1972; Patriquin and Knowles, 1972). Substantial nitrogen fixation by turtle grass communities in waters where nutrient levels are low can be an important factor in the high productivity of turtle grass beds. Stengel (1976), discussing the importance of flowering plants in the primary production of shallow seas, points to the significant part played by *Zostera* in transporting phosphate in either direction between sediment and water. Phosphorus released to the water may be carried to considerable distances, enriching the water in the pelagic zone. The fixation of nitrogen by *Thalassia* beds through associated bacteria and blue-green algae also contributes to nutrients. These grasses have been described as acting either as a source or a sink for the two nutrients.

Very few animals graze directly on sea grasses, apart from a few herbivorous fishes such as parrot fishes and surgeon fishes. Other grazers are geese, manatees, turtles and, in the Caribbean, some sea urchins. The grasses contribute to the general food supply, however, mainly through detritus, but their productivity is among the highest in the world.

Though tidal flats and shallow waters appear to be dominated by eel and turtle grasses, certain algal species may be present in considerable abundance. Green algae such as *Enteromorpha*, *Ulva* and *Ulothrix*, and in warm waters such as the Mediterranean, *Caulerpa* are found, as well as filamentous blue-green algae. Occasionally, unattached brown algae may be present in salt marsh communities, but typically a very considerable assemblage of algae grows on the eel grasses and turtle grass, adding much to the overall production of the region (cf. Peres, 1967). Qasim and Bhattathiri (1971) point to the importance of *Thalassia* and *Cymodocea*, often found abundantly near coral reefs, in binding the loose coral sand and debris. In Kavaratti Lagoon (Laccadives) they reported a high nett production equivalent to 5.8 g C/m^2/day, due chiefly to *Thalassia hemprichii*, with algae such as *Ulva* and *Cladophora*.

Standing crop and production from the marsh grasses, sea grasses, mangroves and macro-seaweeds in shallow muddy waters may be extremely high. Jefferies quotes values for *Spartina* marshes which may exceed 1000 g/m^2/year (cf. Table 3.9). Schelske and Odum (1962) estimated 2000 g organic dry matter/m^2/year for Georgia salt marsh estuaries dominated by *Spartina*. Petersen (1918), dealing with *Zostera* beds in Danish fjords, which can extend to 10 or 15 m depth, emphasized much longer ago the vast importance of flowering plants and of seaweeds growing on them. Typically the *Zostera* was broken down, but the detritus formed the major food in the ecosystem, though possibly Petersen under-estimated the contribution of the phytoplankton.

Table 3.9. Estimated annual mean nett productivity (Jefferies, 1972)

Location	Principal species	Estimated annual nett productivity g (dry wt)/m^2/year
Georgia, USA	*Spartina alterniflora*	3700
North Carolina, USA	*Spartina alterniflora*	650
Bridgwater Bay, UK	*Spartina anglica*	960
Norfolk, UK	*Spartina anglica*	980
Norfolk, UK	*Limonium vulgare*	1050
Norfolk, UK	*Salicornia* spp.	867

Other major contributions to production in shallow water areas are microbenthic algae. These are mostly represented by bottom-living diatoms; a few other algae, dinoflagellates and blue-green algae, may be found. Large populations of benthic diatoms may occur not only on sands and muds but on the fronds of seaweeds and on eel and turtle grasses. The total production of salt marshes may, therefore, approach the highest found in marine and terrestrial environments. Georgia estuaries rank "among the most productive natural ecosystems in the world" (Schelske and Odum, 1962).

Another contributor to primary production is the phytoplankton, present only during high tide in inter-tidal areas. However, tidal movements markedly affect light penetration in muddy areas. The stirring of the sediment and, at high tide, the depth of water will reduce effective light penetration for the fixed flowering plants, seaweeds and benthic micro-flora; reduced light will also limit production by phytoplankton. Some species may be able to photosynthesize effectively at very low light intensities.

Benthic diatoms if buried deeply in mud deposits would be unable to photosynthesize, but these algae have some powers of movement and can migrate surfacewards. Studies by Aleem and more recently by Hopkins on the migration of diatoms on mud flats are summarized by Eltringham (1971). The diatoms are normally under the surface of the mud when the tide is full but come to the surface during low tide. They migrate just beneath the surface immediately before the tide reaches them. Though light is a factor in the movement of diatoms, Hopkins demonstrated that the vertical migration persists in the absence of stimuli. The basic diurnal rhythm is related to day length but with a superimposed tidal rhythm. Mechanical disturbance and wetting of the mud may be additional stimuli. The movements of the diatoms would appear to achieve maximum photosynthesis during exposure to the greatest light. The great majority of algae are in the near-surface; Eltringham quotes data showing that generally 60% of the diatoms live in the top 2 mm of mud; a greater proportion live in the top millimetre in fine mud than in coarser muds (cf. also Pomeroy). Micro-algae on sand flats show remarkable survival. Pamatmat (1968) found live chlorophyll-bearing algae to a depth of 10 cm; when brought into full light they photosynthesized as rapidly as algae on the surface. He suggests that the algae may survive heterotrophically when buried. Grontved (1960), working in the shallow Danish fjords, suggested that over the productive season, March to October, phytoplankton production was only a quarter that of the bottom micro-flora, though a strong seasonal variation in production was observed (Grontved, 1962). Pamatmat calculated from Grontved's figures that 115–178 g C/m^2/year was produced by micro-algae. Pomeroy (1959) found an annual gross production of 200 g C/m^2. Pamatmat's own data for an inter-tidal flat are of the same order of magnitude. He suggested that the contribution by phytoplankton to total photosynthesis over tidal mud-flats was negligible, but it is likely that the phytoplankton production was somewhat under-estimated. W. E. Odum (1970) also demonstrated that detritus arising from sea grasses, macro-algae, mangroves and benthic micro-flora was of much greater significance than the phytoplankton in shallow waters. Detritus makes a large contribution to the total food production; mullet graze extensively on the detritus, for example, though they will preferentially exploit living benthic algae, if these are available.

Marshall (1970) found that the total primary production in shallow marine areas of southern New England was high, approximating to 300 g C/m^2/year. Table 3.10 suggests that the macro-flora contributed the largest proportion; benthic micro-algae were also important; phytoplankton made a smaller contribution. Pomeroy (1960) stated that production in a turtle grass area in Florida was divided between phytoplankton, marine flowering plants, essentially *Thalassia*, and the benthic micro-flora, but that the relative importance of the three varied with depth. In water less than 2 m depth, which made up 75% of the total area, all three were about equal in importance. In deeper water the fixed plants became unimportant and primary production depended mainly on the phytoplankton. The compensation depth for the phytoplankton was about 5–7 m. *Thalassia* is not inhibited by strong illumination as is the phytoplankton, but no *Thalassia* was found below about 2 m depth (see Fig. 3.33). Schelske and Odum (1962) comment on the relatively high nutrient level and the rapid turnover in Georgia marshes. Production continues all the year round though, with the turbidity, light must be a limiting factor. *Spartina*, with two crops per year, was considered to constitute about two-thirds to three-quarters of total production; benthic diatoms gave a more or

Table 3.10. Estimated contributions from different sources to annual production of organic carbon in shallow waters of southern New England (Marshall, 1970)

	g C/m^2/yr	
Macroflora, including *Zostera* and its aufwuchs, also macroscopic algae	125	
Benthic microflora	90	
Deposition of phytoplankton	50	
Allochthonous matter	0–10	
Conversion of dissolved organic matter (aggregates)?	No estimate	
	265–275	plus unknowns

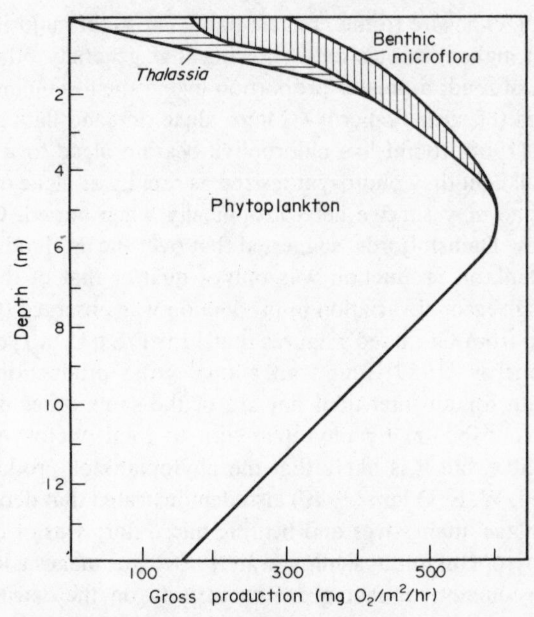

Fig. 3.33. Variation with depth of the relative primary production of the plant populations in Boca Ciega Bay. Most of the bay is less than 2 m deep (after Pomeroy, 1960).

less constant contribution with the flood-tide periods producing maximally in summer and the low-tide-exposed periods producing maximally over winter. Over the year the benthic micro-flora was estimated as producing about one-quarter to one-third of the total. While the contribution of the phytoplankton was originally thought to be small (*ca.* 10%), it has undoubtedly been under-estimated. Estimates gave as much as 5 g organic matter/m^2/day in summer. If such production lasted only 100 days, the contribution would be about one-quarter of the whole.

Coastal shallow areas characterized by a hard rocky substratum may have an extensive flora of macroscopic seaweeds. Though of relatively simple algal structure, some seaweeds are large plants and can form dense forests in shallow regions, contributing considerably to primary production. Seaweeds usually reach their maximum abundance in temperate and boreal regions of the world. Though generally less evident in warm waters, they can be plentiful in association with coral reefs, especially the calcareous *Lithothamnion* type of seaweed. In inter-tidal zones they occur abundantly in colder waters, except on those Arctic and Antarctic shores which are barren of algae, due to the scouring action of the ice. Below low-tide mark even very cold waters may show dense seaweed crops. Twenty-three species are reported at depths exceeding 37 m from the Antarctic Ross Sea and many more species from shallower areas. In the Arctic Chukchi Sea brown seaweeds, including *Phyllaria* and *Laminaria*, as well as some red seaweeds, are abundant at depths exceeding 35 m. While in tropical areas, especially on coral reefs, there is generally a strong emphasis on calcareous forms, some sandy and gravel shores may include species such as *Caulerpa* and *Halimeda* and especially *Sargassum*. The warm Mediterranean has a fairly rich algal flora (cf. Peres, 1968). The far western Caribbean and the east coast of central America, by contrast, are virtually destitute of fixed algae (Prescott, 1969). Among areas of abundant algal growth are parts of Australia and south-west Africa.

Although many species of red and green algae are found on coasts, the most obvious are the large brown seaweeds. These form the typical kelp forests of temperate and high latitudes. The largest seaweed is *Macrocystis*, some 50 to 70 m in length; some records suggest 100 m.

Several species of *Macrocystis* form great kelp beds, typical of Pacific coasts of South America, also off Kerguelen and the Australasian islands of Tasmania and New Zealand, and along the coast of the extreme south of Africa. *Macrocystis* beds are characteristic also off the North American Pacific coast as far south as California, and in north-west Mexico, including Baja California. Other major kelp species include *Nereocystis*, at least 25 m in length, and particularly abundant along the Pacific coast of North America. The two genera of brown algae, *Laminaria* and *Fucus*, though not individually of such great size as *Macrocystis*, contribute greatly to luxuriant seaweed growth. In the sub-Antarctic, kelp beds can be considerable; species of *Durvillea* form an important constituent.

Despite their real contribution to overall production of organic matter in shallow seas, seaweeds, because of their necessity for anchorage, must be confined to a fairly narrow layer of coastline. A very few species have the capacity for abundant vegetative growth drifting in open oceans, notably "Sargassum weed". Levring, Hoppe and Schmid (1969) indicate the crop of floating *Sargassum* in the North Atlantic to be 4 to 11×10^6 tons (fresh weight). Ultimately some of the older weed, especially when weighted with epizoontic species, may sink to great depths and contribute to the food supply of benthic organisms (cf. Schoener and Rowe, 1970). The vast majority of seaweeds, however, are fixed to the substratum. Like other plants, they require light as an energy source for the photosynthetic process and must be limited in depth.

Prescott gives on average a depth limit of about 50 m for macrobenthic seaweeds in northern seas, as compared with some 100 m for tropical seas. The Mediterranean with its remarkably clear water has macrobenthic seaweeds living at about 130 m maximum depth, with an extreme record of 180 m off Minorca. Wave action on shore increases turbidity so that light penetration is very often much reduced; depths so far given are the *maximum* depths of light penetration. For many temperate species of seaweeds the depth limit is much less; the maximum depth for *Macrocystis*, for instance, is about 30–35 m; for *Nereocystis* about 25 m. In British waters *Laminaria* beds do not exceed *ca.* 20 m. For many Arctic regions the brown kelp beds have a similar maximum depth but in some areas algae live more deeply. Levring, Hoppe and Schmid (1969) cite algae living to 80 m off the east coast of Greenland, and to 50 m under ice (occasionally deeper) in Arctic Canada. Some even more remarkable records come from the Antarctic; in the Ross Sea area benthic algae are quoted as living down to 150 m. Possibly some Arctic and Antarctic benthic algae have very low compensation intensities so that they grow at greater depths than in other regions of the world. The limitation of light even in the clear waters of the Mediterranean is real. According to Peres (1967) the luxuriant vegetation zone is limited to about 40 m.

Production and standing crops of benthic macro-algae have not been very widely studied, though considerable emphasis has been given to some of the brown algae with their importance in the alginic acid industry. Scagel (1966) quotes data from Aleem that *Macrocystis* on the west coast of North America may have a standing crop of between 5 and 9 kg/m^2. Some estimates by Towle and Pearse (1973) give *ca.* 6 kg/m^2 as standing crop for the densest part of a kelp bed of *Macrocystis pyrifera* in Monterey Bay, a value which compares well with data by McFarland and Prescott and by Aleem. Production, according to Towle and Pearse, is about 7 g C/m^2/day, agreeing with measurements quoted by McFarland and Prescott (5 to 6 g C/m^2/day), though Clendenning's estimates are less than half, and Aleem's only *ca.* 0.1 g C/m^2/day. A high productivity is also quoted for other brown algae (from 4 to 20 g C/m^2/day).

Kanwisher (1966), referring to the surprisingly rapid growth of macroscopic seaweeds, states that in the absence of self-shading they can double their mass in 2 to 5 days. At Woods Hole he estimates standing crop of *Fucus* and *Enteromorpha* may reach 2 kg and 0.3 kg/m^2, respectively, and under the most favourable conditions, photosynthetic production could amount to 20 g/m^2/day. Kain (1977) recently recorded the standing crop of *Laminaria hyperborea* in Britain as 10–20 kg (fresh weight)/m^2 decreasing with depth; production was 0.4 kg C/m^2/yr (= 1.1 kg dry matter) between 4 and 11.5 m depth. A rather similar estimate of production (0.14–0.43 kg C/m^2/yr) was made by Hatcher, Chapman and Mann (1977) for a kelp bed of *L. longicruris*.

Benthic seaweeds with submerged sea grasses, emergent flowering plants and benthic micro-algae, therefore, contribute together to the enormous productivity of the narrow coastal waters. Reinforced by the phytoplankton crop in these areas, the total amount of organic matter produced is overwhelmingly greater than in off-shore waters. The balance of contributions between benthic macro-algae, flowering plants, micro-algae and phytoplankton varies greatly from area to area. In general, phytoplankton becomes progessively more important with increasing depth.

The continental shelf with a nominal depth limit of 200 m and, in some regions, the upper areas of the continental slope, receive contributions of organic matter derived ultimately from the intense primary production of shallow coastal areas. Much of the material is in the form of organic detritus. Since, however, the average depth of the world ocean is *ca.* 3800 m and benthic plants of whatever kind are restricted to very shallow coastal waters, their total contribution to primary production in the world ocean must be very small. Wood (1963), quoting Braarud, suggested that macrophytes contributed about 2%, Zenkevitch estimated *ca.* 1% and Ryther (1963) indicated a rather larger contribution (*ca.* 5%). The importance of production by benthic macrophytes and micro-algae in the usually comparatively rich environments close inshore is beyond question. Their enrichment of many continental shelf areas is also significant, but the overwhelmingly important agent in primary production of the world ocean is the phytoplankton.

Chapter 4
The Algae of the Phytoplankton

A Review of the Major Classes

Although in shallow waters a restricted number of genera of flowering plants contribute very significantly to primary production (cf. Chapter 3), the algae are generally dominant in the marine environment. In shallow waters both benthic and planktonic algae are significant, but over deeper waters and throughout the open oceans only the phytoplankton contributes, apart from the drifting masses of *Sargassum* weed which occur in some areas of the ocean.

Among the algae, both benthic and planktonic, the variety is considerable. This great group of plant organisms includes the macroscopic seaweeds, as well as unicellular forms and very small colonies of varied shape.

The phytoplankton is composed of single cells or of relatively simply organized, small colonies, but a considerable diversity of algal groups is represented. One particular type of colony is found in nearly all these groups: numbers of non-motile cells are embedded in a mucilaginous matrix (the palmelloid stage).

Where thalloid, filamentous or palmelloid forms constitute the dominant stage in the life history, a considerable contribution to the marine phytoplankton may come from the production of flagellated zoospores or non-motile aplanospores. Certain algal groups (Cyanophyceae and Rhodophyceae), however, never produce motile stages.

Detailed morphological characteristics are not easy to describe accurately in such small organisms as phytoplankton, but more recent electron-microscope studies, elucidating fine structure of such cell constituents as scales, cell wall (cf. Dodge, 1973), flagella, etc., have contributed to specifying group features. Biochemical characters including photosynthetic pigments, especially the variety of xanthophylls (cf. Nakajama, 1962), and storage products, also show a remarkable diversity throughout the algae. Algalogists will readily admit, however, that many species, particularly amongst the more minute marine algae (nanoplankton), are but poorly known, the lack of knowledge applying to the cell morphology and often to the life cycle.

Significant revisions of the classification of algae have resulted from these studies of the fine structure of cells, in addition to the elucidation of life histories and to biochemical studies by algalogists—revisions which are still in progress.

An example of recent detailed studies of the fine structure of algae that has yielded results which may be of taxonomic value comes from the investigations of Dodge (1973) on the eye-spot (stigma). This is an orange or red-pigmented structure, present in many motile algal cells and of considerable morphological complexity. While apparently lacking in Haptophyceae and relatively uncommon in Dinophyceae and Cryptophyceae, the eye-spot, according to Dodge, is almost universally present in

Chlorophyceae, Prasinophyceae, Euglenophyceae, Xanthophyceae and Chrysophyceae. Dodge has classified the types of stigma in algal cells:

(1) The eye-spot is not obviously associated with flagella but is part of the chloroplast (Chlorophyceae, Prasinophyceae).
(2) The eye-spot is part of the chloroplast and closely associated with a flagellum (Chrysophyceae, Xanthophyceae).
(3) The eye-spot is independent of the chloroplast but adjacent to flagella (Euglenophyceae).

He puts forward the idea that three evolutionary lines may exist among algae, according to this range of eye-spot structure, and draws attention to the fine structure of other organelles (chloroplast, flagellum, cell wall, nucleus, ejectile organelle) as being valuable criteria in algal phylogeny. Scagel (1966), reviewing some of the arguments for a new classification, included the condition of nucleus, flagella and plastids, especially the presense or absence of chlorophylls *b* and *c* in addition to chlorophyll *a*, and the identification of carotenoid pigments, as of importance in delimiting major taxa. Boney (1970) also discusses some of the problems in the classification of certain algal groups. One difficulty is that it has long been recognized that in various algal taxa, species are known which are colourless and saprophytic or holozoic; some may even become facultative saprophytes on loss of chlorophyll.

An example of suggested major groupings by Christensen (1962) separates the blue green algae (Cyanophyceae) from the others, largely on the lack of a fully formed nucleus. The red algae (Rhodophyceae) are frequently separated from the remainder largely on account of their lack of flagellated cells. As Prescott (1969) points out, the Rhodophyceae and Cyanophyceae also have in common the possession of phycobilin pigments. Christensen then distinguishes those algae which possess only chlorophyll *a* and *b*, namely Chlorophyceae, Euglenophyceae, Prasinophyceae and Loxophyceae; the group is given the rank of a phylum (Chlorophyta). In a comparable way the blue-green and red algae, each of which has only a single class, are elevated to phyletic rank (Cyanophyta and Rhodophyta). The rest of the algae are grouped by Christensen and many other algalogists as the Chromophyta, a major phylum with some nine classes of algae, sometimes spoken of as the brown or yellow-brown algal groups. They include such mainly unicellular or simple colonial forms as the Dinophyceae, Chrysophyceae and Bacillariophyceae but also one major group of macroscopic seaweeds–the Phaeophyceae. While there is general agreement that these groups and the Cryptophyceae possess chlorophyll *c* in addition to *a*, the investigations of Jeffrey (1965, 1968, 1972) have demonstrated clearly the presence of two types of chlorophyll *c* (c_1 and c_2) in some diatoms, dinoflagellates, chrysomonads and a cryptomonads, as well as in brown seaweeds.

Although the quantities of c_1 and c_2 are frequently more or less equal, some algae show marked differences (cf. Table 4.1). The total amount of chlorophyll *c* is always less than *a*, though the chlor *a*/chlor *c* ratio can show wide variations.

It would be inappropriate in this text to attempt to comment critically on the phyletic origins of algal groups. An outline working classification with comments on the general biology of marine representatives of each algal class, however, may be helpful. The comments apply essentially to planktonic forms; macroscopic seaweeds are mentioned only briefly.

Table 4.1. Relative proportion of chlorophylls a, c_1 and c_2 in marine algae (after Jeffrey, 1972)

Marine algae	Chlorophylls (% of total chlorophyll by weight)		Ratio chlorophyll $c_1:c_2$
	a	$c_1 + c_2$	
Diatoms			
Phaeodactylum tricornutum	78	22	1:1
Thalassiosira decipiens	69	31	1:1
Thalassiosira aestivalis	77	23	1:1
Thalassiosira sp. (P.H.)	85	15	2:1
Chaetoceros didymum	74	26	1:3
Skeletonema costatum	83	17	1:5
Nitzschia closterium (NC31)	85	15	1:1
Navicula sp.	84	16	1:1
Cylindrotheca closterium var. *californicum*	85	15	1:1
Coscinodiscus centralis	—	—	2:1
Dinoflagellates			
Amphidinium carteri	63	37	0:1
Gymnodinium simplex	67	33	0:1
Prorocentrum micans	77	23	0:1
Chrysomonads			
Monochrysis lutheri	84	16	1:1
Isochrysis galbana	71	29	1:1
Cryptomonads			
Chroomonas sp.	80	20	0:1
Phaeophyceae			
Sargassum fallax	75	25	1:1
Colpomenia sinuosa	89	11	1:1
Eklonia radiata	81	19	1:2
Hormosira banksii	85	15	1:1

Reference should be made to a series of papers covering check lists of British Marine Algae (e.g. Parke and Dixon, 1964, 1968, 1976) which include many valuable comments on problems of algal classification.

1. Cyanophyceae (The "blue-green algae")

The most characteristic feature of these algae is that the nucleus of the cell is not clearly defined, and there does not appear to be a discrete nuclear membrane or nucleolus. The DNA is not combined with histones. Mitochondria are not recognizable. The photosynthetic pigments are distributed through the periphery of the cell and are not enclosed in obvious plastids (chloroplasts or chromatophores),* but are associated with membranes ramifying through the outer cell region and arranged in flattened vesicles. The pigments include phycobilins, phycocyanin being dominant, in addition to chlorophyll a, and various carotenoids including β-carotene and xanthophylls, some of which, such as myxoxanthin, oscillaxanthin and myxoxanthophyll, are apparently found only in Cyanophyceae. The cells, therefore, may be of varied colour in different species: from blue-green to olive, or sometimes yellow, red or violet. The cell membrane

* See footnote on p. 137.

may be indistinct or very thick and mucilaginous. Reproduction is by simple division, by fragmentation or by non-motile spores; sexual stages are unknown.

The cell walls have been described as containing cellulose, with pectin and mucilaginous layers. Chapman and Chapman (1973) point out that the true cell envelope of at least two layers is composed largely of mucopeptides (built up from amino sugars and amino acids) and thus differs from other algae. Fogg, Stewart, Fay and Walsby (1973) refer to a lipopolysaccharide component in addition to the mucopolymer of the cell wall. The outer mucilaginous sheath has been reported as partly of cellulose; it appears to be composed, however, of pectic acids and mucopolysaccharides and possibly is variable in composition. It surrounds the cell proper, the colony or filaments.

Many filamentous types develop enlarged cells with relatively thick walls (heterocysts). These may be concerned with nitrogen fixation; in some filamentous species they appear to play a part in vegetative reproduction or spore formation (cf. Fogg *et al.*, 1973).

Some planktonic blue-green algae possess gas vacuoles which presumably aid flotation, but generally, the cyanophycean cell is not provided with a plasma vacuole. Flagellated cells do not occur but the colonies can exhibit motility with a type of gliding movement which appears to be associated with mucilage production; the trichome (*vide infra*) in the filamentous types may show movement inside the mucilage sheath. Assimilation products include a special type of starch, which appears to be virtually identical to glycogen, and sometimes trehalose amongst carbohydrates, as well as glyco-proteins and sometimes lipids. These algae are generally of rather simple construction; they may be unicellular, or colonies of varied shapes. In the filamentous forms the living structure within the sheath (the trichome) is divided into cells by cross walls; the filaments may be simple or branched and there may be several filaments. There are many marine species, some large enough to be classed as benthic seaweeds, though they never attain the large size characteristic of some other seaweed groups. Cyanophyceae are especially found in salt marshes.

A number of species, unicellular, small colonial forms or filaments, are included in the phytoplankton, but the blue-greens are not a dominant phytoplankton group (cf. Wille, 1908). Cyanophyceae, often as part of the nanoplankton, are encountered normally in low densities in almost all seas, including inshore Danish waters, (cf. Steemann Nielsen, 1940). Occasional red tide outbursts are caused by blue-green algae, though these are more usually encountered in brackish areas. Brongersma-Sanders (1957) quotes several examples of mass algal blooms due to blue-green species. While such outbreaks are rather exceptional, one species, *Trichodesmium* (*Oscillatoria*) *erythraea*, normally appears in extraordinarily dense patches in almost all warm seas, usually with a water temperature exceeding 25°C and salinity near 35‰. *Trichodesmium* may form a most discontinuous pattern, huge patches being irregularly distributed over the ocean. These develop comparatively rapidly, and may be so dense that at first sight little other plankton seems to be represented (cf. Ramamurthy, 1970).

Some species of blue-green algae can fix elemental nitrogen, although this seems to be less common among marine forms; Fogg *et al.* (1973) state that *Trichodesmium* is the only planktonic marine species capable of nitrogen fixation. Though the evidence is not certain, the fixation, which may be at a high rate, is probably due to the algae and not to associated bacteria. The alga may thus thrive in waters low in combined nitrogen but cf. Kimor and Golandsky (1977) *Trichodesmium* also assimilates nitrate, nitrite, ammonia, urea and glycine. There is considerable evidence that substantial N-fixation occurs in the central N. Pacific gyre (Mague, Mague and Holm-Hansen, 1977) due to *Trichodesmium* and to *Richelia*. This latter organism is a blue-green alga, endophytic in *Rhizosolenia* spp. Mague, Weare and Holm-Hansen (1974) and Venrick (1974) suggested that blooms of *Rhizosolenia* in the region were associated with increased nutrients due to N-fixation by *Richelia* rather than by bacteria. Sournia (1970) suggests that the importance of a number of Cyanophyceae other than *Trichodesmium* and *Richelia*, including several small species, has perhaps been underestimated. The possibility of nitrogen fixation by some of the forms is of particular interest.

2. Rhodophyceae (The "red algae")

The great majority of these are thalloid; in addition to chlorophyll and carotenoids, the chloroplasts have phycobilin pigments of which phycoerythrin is the more abundant. The general pigmentation typically is some shade of red, and these are often called the red seaweeds. Of assimilation products they store a particular form of starch known as floridean starch (cf. Chapter 3). A detailed review by Percival (1968) includes an account of other mucilaginous polysaccharides. The red algae never have motile stages though the life cycles are complex. This class is essentially marine.

3. Phaeophyceae (The "brown seaweeds")

The great majority of these marine seaweeds have a thalloid macroscopic structure which is relatively complex; they include the important kelps. Reproduction involves a complex alternation of generations and includes motile reproductive cells with two lateral flagella. The chloroplasts (chromatophores)* are brown, the chlorophyll pigments being masked by a number of carotenoids, particularly fucoxanthin. The products of assimilation include laminarin and mannitol as well as lipid.

4. Chlorophyceae (The "green algae")

These algae include a range of body form from unicellular species or microscopic colonies to filamentous forms and thalloid species which are benthic macroscopic green seaweeds. Such forms, however, do not approach the complexity of the larger brown kelps. The planktonic microscopic colonies may consist of flagellated cells, or of non-flagellated cells embedded in mucilage but reproducing usually by flagellated zoospores (Tetrasporales). Unicellular phytoplankton species may be non-motile (e.g. *Nannochloris*); reproduction may be by flagellated zoospores (*Chlorococcum*) or by non-motile aplanospores (*Chlorella*). The unicellular species possess two similar flagella (e.g. *Chlamydomonas, Brachiomonas*); a few genera (e.g. *Carteria*) have four flagella of similar length ("isokont"). The flagella are typically smooth and emerge from pores in the cell walls at the apical end. A cell wall is usually present and is of a specialized type with a number of layers, including cellulose microfibrils frequently arranged in differently orientated layers; other materials including glycoprotein and pectin are often present. In some genera the cell wall appears to be very thin (e.g. *Dunaliella*); in certain species of *Chlamydomonas* the wall may be quite thick but different species are characterized by cell walls of different thickness.

* Some authors reserve the term chloroplasts for plastids which are green; they use the term chromatophores or chromoplasts for those plastids in which the green colour is not predominant.

Chlorophyceae produce as an assimilation substance true starch which appears to be very similar, if not identical, to the starch of higher green plants. Some distinction has been attempted between this substance and the starch formed by Prasinophyceae (*vide infra*) but Boney (1970) points out that the distinction appears to be based on colour reactions alone. Some species also produce sucrose and glycerol. The "green algae" characteristically possess pyrenoids; these structures are present in some members of most algal classes. They consist mainly of a protein matrix, and are always associated with the chloroplast, though not necessarily embedded in it. In Chlorophyceae, the pyrenoid is attached to the chloroplast and usually has a starch sheath. Chlorophylls in the chloroplasts, usually single, include only chlorophylls *a* and *b*. Of the carotenoids β-carotene is generally more abundant than in other algal groups; some Chlorophyceae apparently possess small amounts of other carotenes. Riley and Wilson (1967) found that carotenes were represented by β-carotene alone. The xanthophylls include a considerable amount of lutein; neoxanthine and violaxathine are found in smaller amounts. The investigations of Riley and Segar (1969) confirmed these findings. In general the photosynthetic pigments are more similar than in other algal groups to those found in higher green plants. Sexual reproduction is commonly exhibited amongst the Chlorophyceae.

5. Prasinophyceae

Typically these are motile monads which have four equal flagella though some have only a single flagellum (e.g. *Micromonas*). There are reports that a second flagellum can be seen in species of *Micromonas* at certain stages as a short internal structure. Other Prasinophyceae may have two unequal flagella as in *Heteromastix*. The flagella originate from an apical pit or groove, sometimes from a lateral depression. The cell wall is absent; instead mineral scales cover the body surface, in some cases more than one type of scale being present. There is evidence that the formation of scales appears to be associated with the Golgi (cf. Manton and Parke, 1965; Manton, Rayns, Ettl and Parke, 1965). In genera such as *Platymonas* (*Tetraselmis*) (Fig. 4.1 and Fig. 4.2) a theca is developed from the scale-like structures but, in contrast to the Chlorophyceae, cellulose appears to be absent (Manton and Parke, 1965). Scales fre-

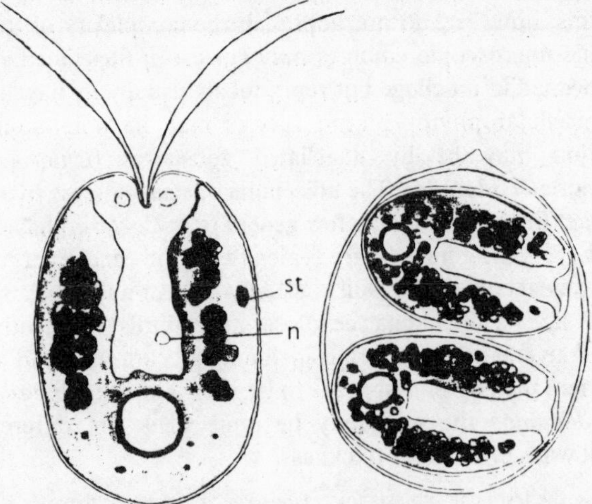

Fig. 4.1. *Platymonas apiculata* and cyst stage. n = nucleus; st = stigma (from Butcher, 1952).

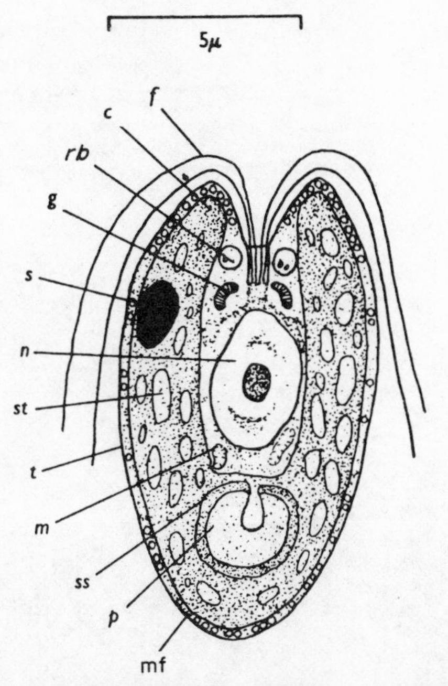

Fig. 4.2. *Platymonas convolutae*: c = chloroplast, f = flagellum, g = golgi body, m = mitochondrion, mf = muciferous body?, n = nucleus, p = pyrenoid, rb = refractive body of unknown nature, ss = starch shell, s = stigma, st = stroma starch, t = theca (from Parke and Manton, 1967).

quently cover the flagella; remarkably fine hairs which are very easily detached may also occur on the flagella (cf. Parke and Manton, 1965 and Fig. 4.3).

Typically there is a single chloroplast usually with a pyrenoid, with starch "shell" and protein core, though as Boney (1970) points out, the precise structure of the pyrenoid can differ even in two genera such as *Prasinocladus* and *Platymonas*. The chloroplasts appear green, since as in Chlorophyceae, of the chlorophylls only *a* and *b* are present. Of other pigments, β-carotene is important, with some α-carotene; Boney describes a protochlorophyll-like pigment also in Prasinophyceae. Riley and Wilson (1965, 1967) and Riley and Segar (1969) confirmed in general that the carotenoid composition resembled that of Chlorophyceae, with lutein as the chief xanthophyll and with minor amounts of neoxanthin and violaxanthin. However, Riley and Segar's work showed that while four species of Prasinophyceae showed this pigment array, *Heteromastix* had four unusual xanthophylls in place of lutein. Boney (1970) also points to the variation in the assortment of xanthophylls among Prasinophyceae.

The presence of chlorophylls *a* and *b* is one of the several resemblances between monads belonging to this class and those assigned to the Chlorophyceae, but there are a number of characters which separate the two groups (cf. Boney, 1970; Peterfi and Manton, 1968).

Sexual processes are not known in Prasinophyceae but cyst formation may occur. A palmelloid stage may be included as, for example, in *Halosphaera*, where indeed the non-motile phase is the obvious and well-recognized vegetative stage in the life history. However, as Parke and Hartog-Adams (1965) have demonstrated (cf. Fig. 4.4), the reproductive cycle of *Halosphaera* includes motile stages which are very similar to the genus *Pyramimonas*, another member of the Prasinophyceae.

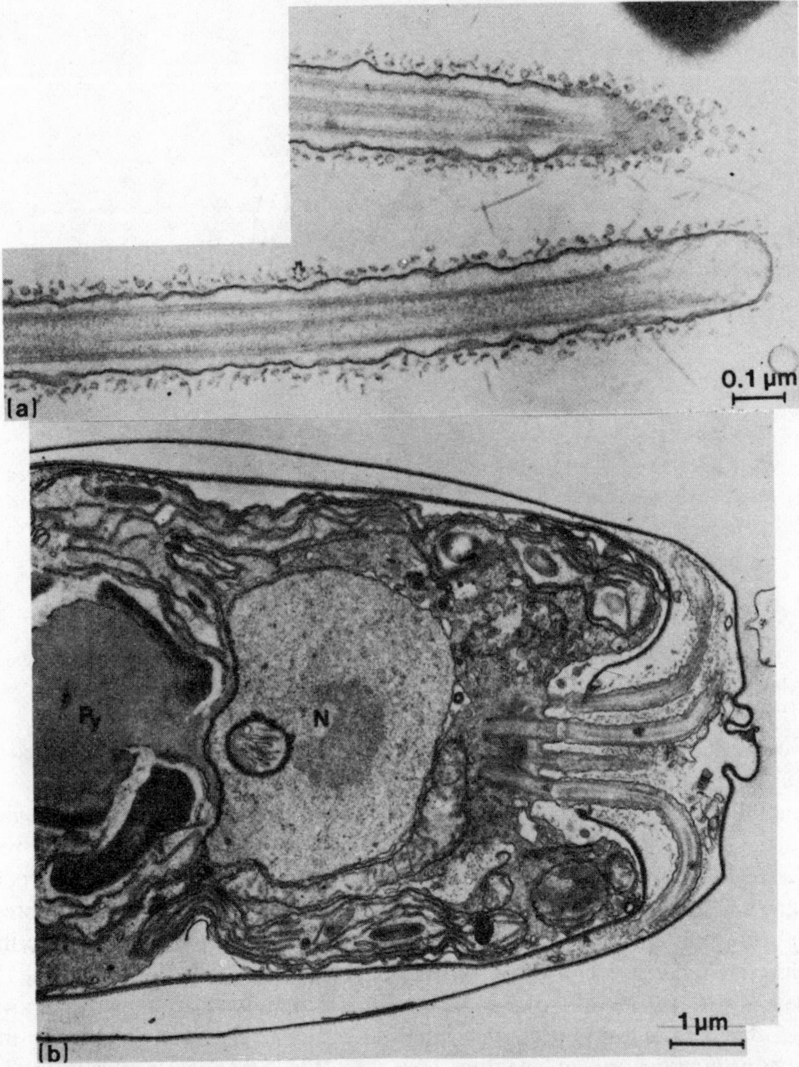

Fig. 4.3. (a) *Platymonas tetrathele*, two flagella in longitudinal section; note scales and hairs on surface (×30,000). (b) *Prasinocladus marinus*, longitudinal section through flagellar end of cell. Four flagella enclosed between inner and outer theca; N = nucleus, Py = pyrenoid (×12,000) (after Manton and Parke, 1965; and Parke and Manton, 1965).

Although other genera of Prasinophyceae may include a non-motile phase, this may not be palmelloid; for example, in *Prasinocladus* a non-motile stage occurs as a small dendroid colony. The genus *Micromonas* includes two species, *M. pusilla* and *M. squamata*, the first of which, originally known as *Chromulina pusilla*, is a minute flagellate less than 2 μm in diameter. Despite its small size this may be a very important nanoplankton organism; Knight-Jones (1951) records it as being widely distributed around the British Isles and very abundant. Indeed it is probably the commonest organism in inshore waters round our coasts and in the North Sea. A detailed description of this and related species by Manton and Parke (1960) suggests that the species has an even wider distribution, occurring in oceanic waters. Descriptions of other

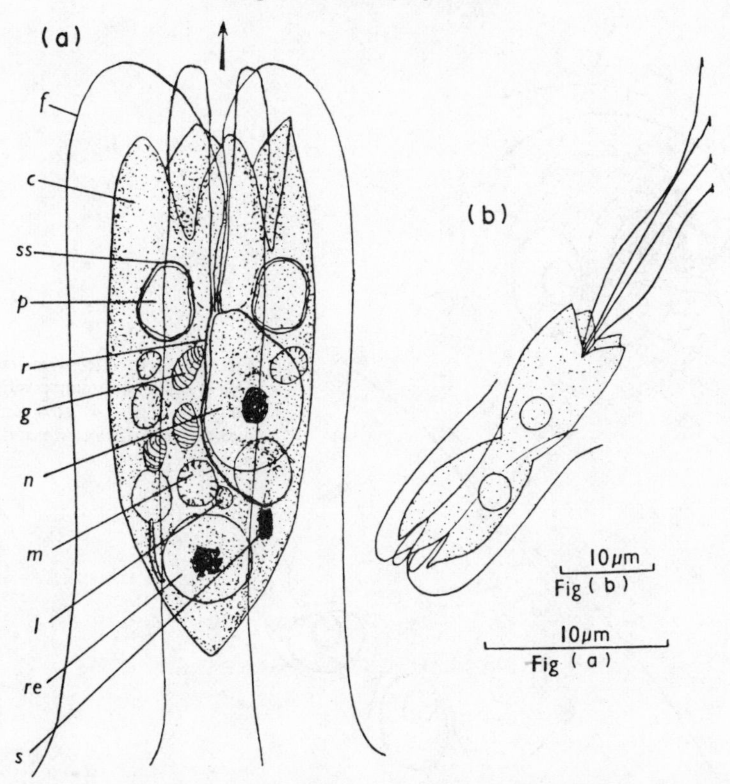

Fig. 4.4. *Halosphaera* in motile phase: (a) *H. viridis*, typical position in swimming in arrowed direction; c = chloroplast, f = flagellum, g = golgi body, l = lipid body, m = mitochondrion, n = nucleus, p = pyrenoid, r = rhizoplast, re = reservoir, s = stigma, ss = starch sheath. (b) *H. russellii* late fission stage (from Parke and Hartog-Adams, 1965).

species of Prasinophyceae include that of *Heteromastix* by Manton, Rayns, Ettl and Parke (1965); of *Prasinocladus* by Parke and Manton (1965); and by Manton and Parke (1965) for *Platymonas*.

Manton *et al.* (1965) in their account of *Heteromastix* describe the cell as undergoing binary fission while motile (Fig. 4.5). They also refer to an observation of Korshikov that the life history includes a sexual process and an encystment stage, though they have not confirmed this observation. In *Prasinocladus* cell division takes place in the non-motile stage (the predominant phase), but cyst formation is known (Parke and Manton, 1965). In *Platymonas* cell division also occurs in non-motile cells; two daughter cells are completely formed inside the old theca before release (cf. Fig. 4.6).

Recent investigations have clearly demonstrated that many prasinophyceans include a motile stage (which may be dominant) in the life history. *Pachysphaera* is a genus with a non-motile cyst, which can exceed 100 μm in diameter according to species, and a motile stage (Parke, 1966). The cyst first contains a single nucleus and chloroplast, but following their repeated division, numerous small cells are formed. These are then liberated as the motile stages which are much smaller (of the order 10μm) and have four very long flagella. The flagella and the body of the cell are covered with scales. The motile stage can itself undergo fission (Fig. 4.7).

To some extent the reproduction of *Halosphaera* is similar. The immature cyst is

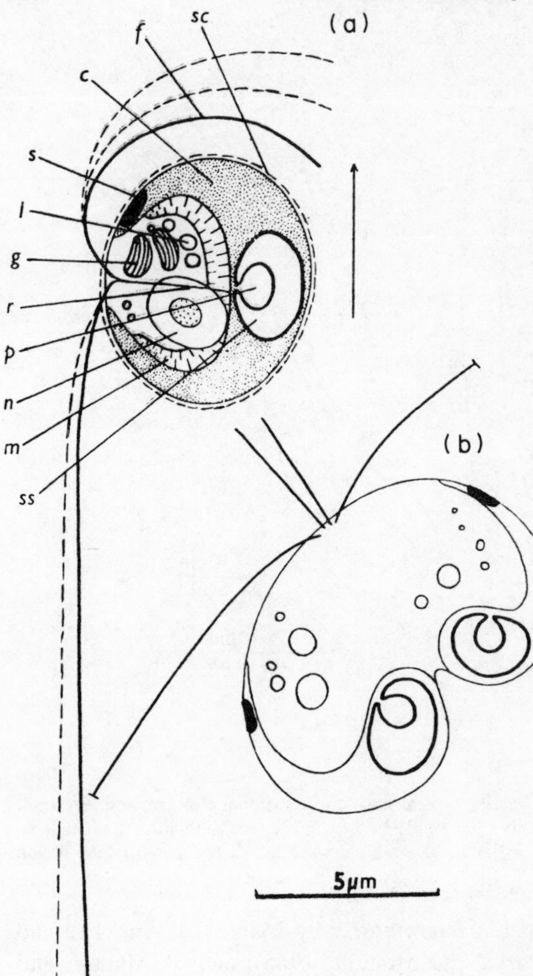

Fig. 4.5. Left. *Heteromastix rotunda.* (a) Cell in rapid swimming position. Abbreviations as Fig. 4.4; sc = scales. (b) Late fission stage (from Manton, Rayns, Ettl and Parke, 1965).

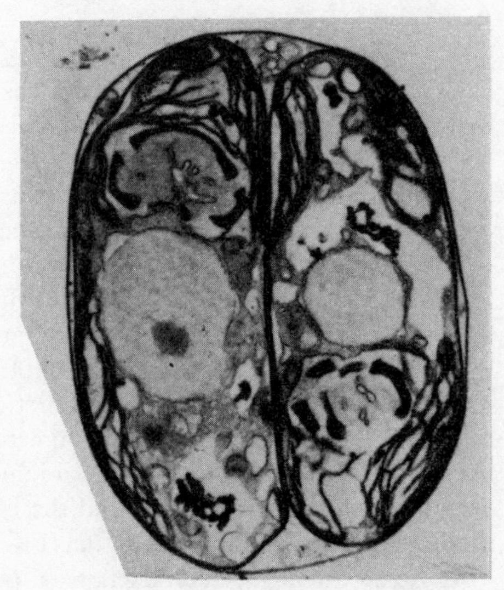

Fig. 4.6. Right. *Platymonas tetrathele*, cell division inside theca (×5000) (from Manton and Parke, 1965).

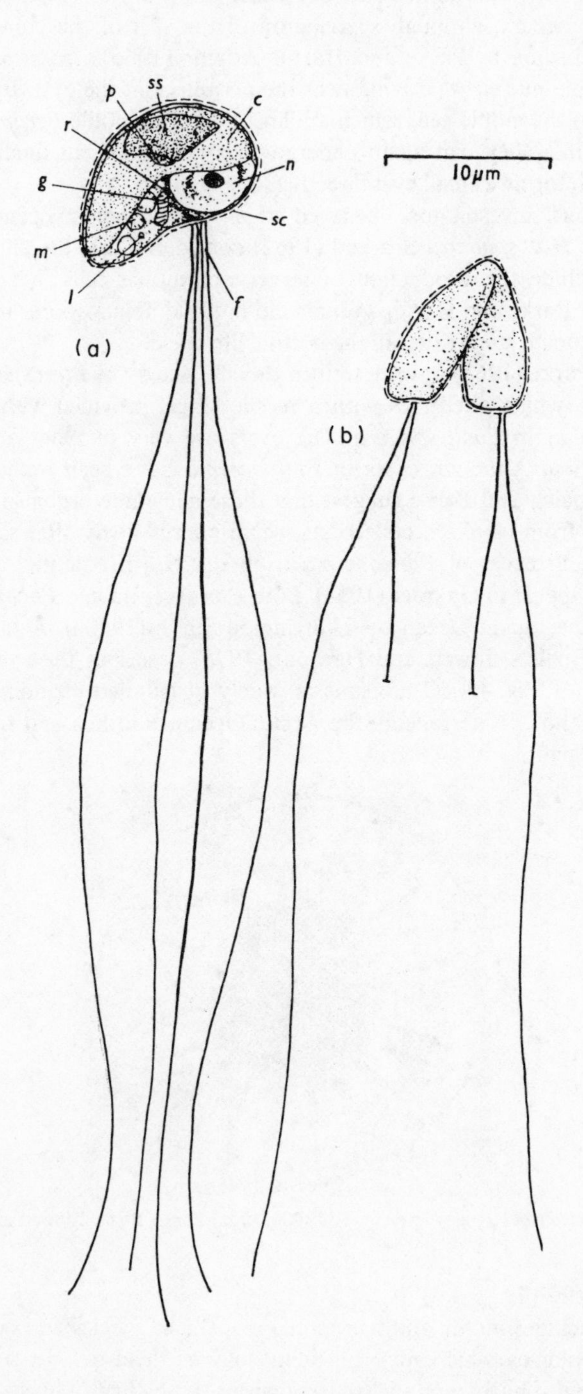

Fig. 4.7. *Pachysphaera pelagica*, (a) motile phase. Abbreviations as Fig. 4.4. (b) Late fission stage (×2500) (from Parke, 1966).

small and has a single large nucleus, but when the cyst has reached its full diameter nuclear divison occurs, although segregation of the rest of the contents lags behind. Eventually, according to Parke and Hartog-Adams (1965), "rosettes" are produced, each with a single nucleus and lying near the periphery of the cyst. By further division these give rise to the motile cells which are liberated. The motile *Pyramimonas*-like cells have four flagella. They can again undergo asexual fission but finally they lose their flagella and each forms a small cyst directly (see Fig. 4.4).

Some previous investigators believed that aplanospores occurred during the reproduction of *Halosphaera*. Braarud (1962) considered that the annual reproductive cycle might include the production of larger non-motile cells in addition to motile swarmer stages. Parke and Hartog-Adams did not find aplanospores in their study, and doubt whether such a stage exists in the normal life cycle.

Boalch and Parke (1968) have described the life history of *Pterosperma*. The known phase is a cyst, which alternates with a motile stage, provided with four long scale-covered flagella as in *Pachysphaera*. The cysts are very buoyant and remain at the surface of the sea. Although cysts of *Pterosperma* have been found in the Western Approaches, Boalch and Parke suggest that these neustonic organisms (cf. Volume 2) are usually lost from plankton collections, being poured away after samples have been sedimented. Occurrences of *Pterosperma* from the North Atlantic, and references to earlier records, appear in Gaarder (1954). Both *Pterosperma* and *Pachysphaera* are also reported from the Indian Ocean by Thorrington-Smith (1970a). A full account of the genus (Parke, Boalch, Jowett and Harbour, 1978) describes the cysts (phycoma) of several species (cf. Fig. 4.8). The genus is widely distributed including the North and South Atlantic and Mediterranean, the Arctic Ocean, Red Sea and Indian Ocean and the north-east Pacific.

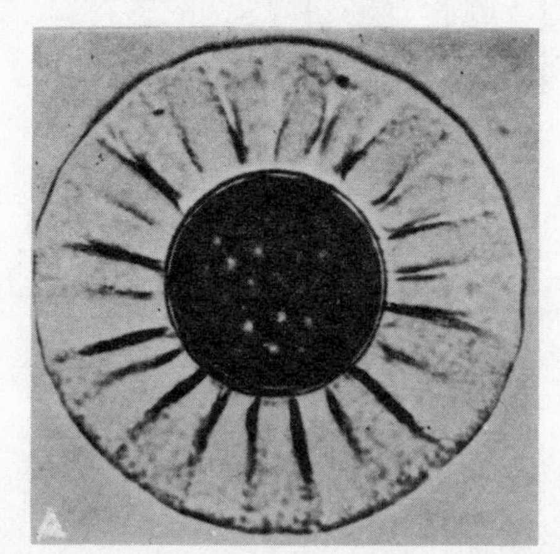

Fig. 4.8. *Pterosperma moebii* – phycoma phase (×300) (from Parke, Boalch, Jowett and Harbour, 1978).

6. Chrysophyceae

This group includes motile and non-motile unicellular species and occasionally small rather simple colonies which may be palmelloid or dendroid. In some the cell can become amoeboid; phagotrophy is well known in some chrysophyceans. Indeed a few species such as *Paraphysomonas* (Lucas, 1967, 1968) are without chloroplasts and maintain themselves entirely holozoically. The suggested classification takes account of the view of Parke (1961) that two series of chrysophycean genera exist, now separated as the Haptophyceae and Chrysophyceae (cf. Christensen, 1962; Parke and Dixon, 1968).

Typical Chrysophyceae are biflagellate, one of the flagella being furnished with hairs (pleuronematic), the other smooth and usually distinctly shorter. Uniflagellate forms, however, are also well known; apparently it is the "hairy" flagellum which is retained. In *Mallomonas*, a uniflagellate form, Parke (1961) indicates that the smooth flagellum, gradually reduced in size, may be represented by the photoreceptor peduncle. The genus *Chromulina* is included in the Chrysophyceae, apart from the two species, *C. pusilla* and *C. squamata*, now referred to the genus *Micromonas* and included in the Prasinophyceae. The chrysophycean species possess only a single flagellum (Fig. 4.9) though Parke (1961) suggests that *C. psammobia* has a second very short, internal flagellum.

The chrysophyceans appear golden yellow or golden brown in colour, though a few species are more or less green. The chloroplasts number one, or more usually two, and are peripheral in position. Many genera lack pyrenoids. The chloroplasts are characterized by possessing chlorophyll *a* and chlorophyll *c*; in addition β-carotene and an abundance of xanthophylls occur. There seems general agreement that the xanthophylls include especially fucoxanthin; diadinoxanthin is also important. Although Prescott (1969) lists chlorophylls *d* and *e* as also occurring in some Chrysophyceae (cf. also Bogorad, 1962) the majority of investigators report chlorophyll *d* in the Rhodophyceae only. Chlorophyll *e* is even more restricted in distribution, being identified in certain Xanthophyceae. Certainly only chlorophylls *a* and *c* are present generally in Chrysophyceae. Riley and Wilson (1967) mention only these two pigments, though they point out that their relative proportions are very variable in different species. The stored photosynthetic products of Chrysophyceae include chrysolaminarin, frequently referred to as leucosin (cf. Chapter 3), and lipid.

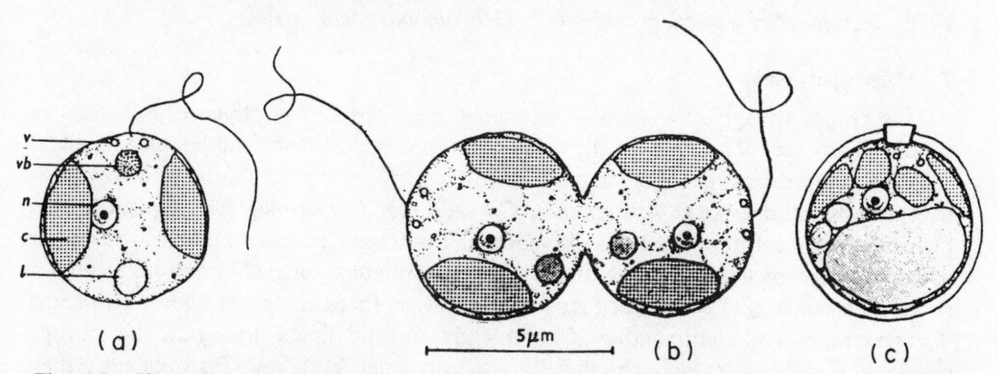

Fig. 4.9. *Chromulina pleiades* (a) young motile cell: c = chromatophore, l = leucosin, n = nucleus, v = vacuole, vb = vibrating body. (b) Late division stage. (c) Mature cyst (×5000) (from Parke, 1949).

Characteristically the cell surface of Chrysophyceae contains minute scales typically composed of silica; cysts may be formed. The silicoflagellates, included in the Chrysophyceae, form an internal skeleton of relatively conspicuous siliceous spicules, and they differ from most other Chrysophyceae in having numerous chloroplasts (Fig. 4.10). The scales of chrysomonads show a wide range of structure and complexity. They may appear more or less smooth, or as Parke (1961) describes, they may be ornamented with all kinds of systems of papillae, folds and ribs. The scales are exceptionally small so

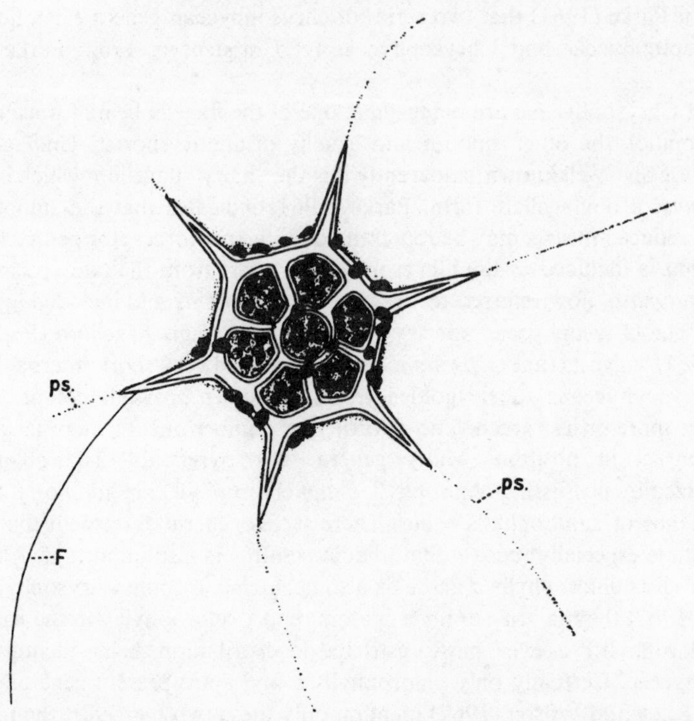

Fig. 4.10. A silicoflagellate, *Distephanus speculum*, F = flagellum; ps = pseudopodia (from Marshall, 1934).

that the fine structure can be elucidated only by electron-microscope studies (cf. Dodge, 1973). A number of genera, however (e.g. *Ochromonas*), lack scales.

7. Haptophyceae

This group, though now usually separated from the Chrysophyceae, is similar in certain characters. The form of the body may be unicellular and motile, of palmelloid colonial form, at least in some stage of the life history, or it may consist of filamentous microscopic colonies (e.g. *Apistonema*, *Chrysonema*, *Chrysotila*). The typical haptophycean motile cell resembles the chrysomonad in its golden-brown colour, the one or usually two (sometimes four or more) chloroplasts containing chlorophylls *a* and *c*, some β-carotene, and abundant fucoxanthin and neofucoxanthin, in addition to some diadinoxanthin and diatoxanthin. Chrysolaminarin and lipids are again the storage products of photosynthesis. The flagella are paired and both smooth; frequently they are equal in length but they may be unequal. In addition, typical Haptophyceae such as *Prymnesium, Phaeocystis* and *Chrysochromulina* are characterized by having a third filiform structure, the haptonema, capable of coiling, and which can be used for temporary anchorage. Considerable variability exists, however; in *Prymnesium* and *Phaeocystis* the haptonema is short and apparently does not coil, while genera such as *Isochrysis* and *Dicrateria* apparently do not possess a haptonema (Fig. 4.11) though Parke (1961) suggests that when cells are subject to very careful investigation this structure might prove to be present. *Pavlova*, a tiny monad, was formerly included in the Chrysophyceae, but studies by Green and others (e.g. Green, 1976) have proved the

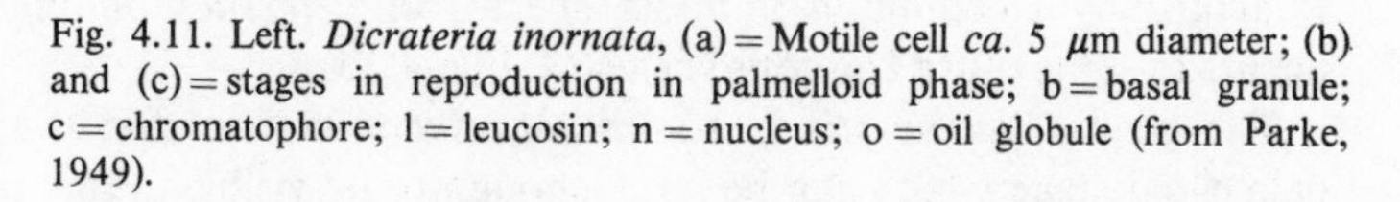

Fig. 4.11. Left. *Dicrateria inornata*, (a) = Motile cell *ca*. 5 μm diameter; (b) and (c) = stages in reproduction in palmelloid phase; b = basal granule; c = chromatophore; l = leucosin; n = nucleus; o = oil globule (from Parke, 1949).

Fig. 4.12. Below. *Pavlova lutheri* (Droop) comb. nov.: *af*, anterior flagellum; *g*, Golgi body; *ha*, haptonema; *m*, striated metabolite bodies with small refringent caps; *n*, nucleus and nucleolus; *p*, plastid; *pf*, posterior flagellum; *s*, stigma (from Green, 1975).

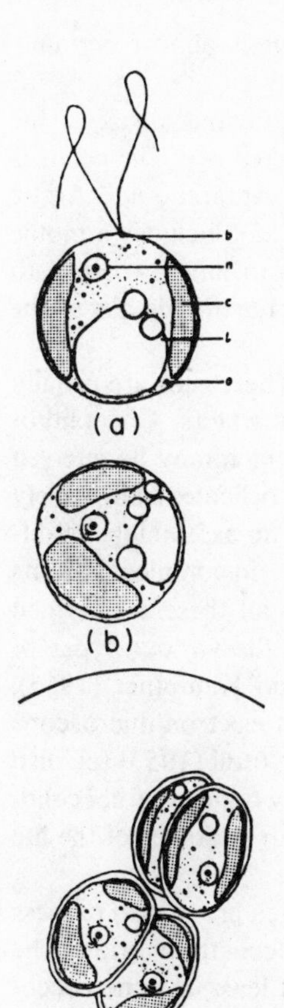

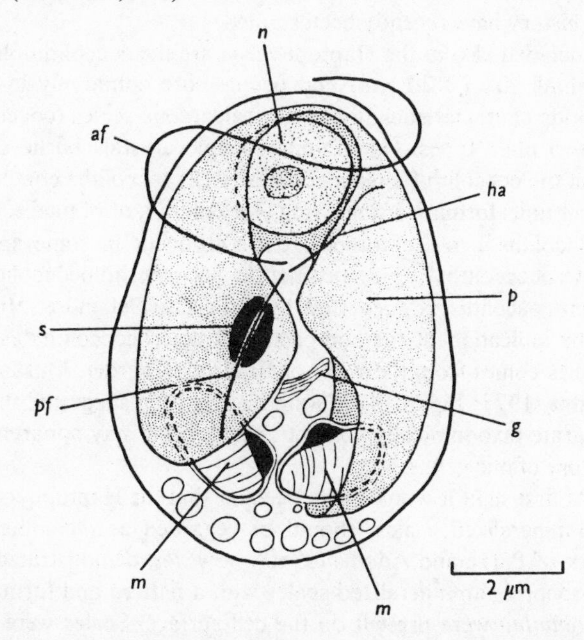

presence of a small haptonema and two very unequal flagella. The genus has now been transferred to the Haptophyceae. *Monochrysis lutheri* has also been shown to possess a second very short posterior flagellum and a slender haptonema (Fig. 4.12) and has been included by Green (1975) in the genus *Pavlova*. Some indication of the great variability in structure and movement of the haptonema in different species is given by Leadbeater (1971).

Although some genera including *Isochrysis* and *Diacrateria* are naked, i.e. they do not possess body scales, these are typically developed, and though they are then very minute and delicate and can be examined only by electron-microscopy, they serve to distinguish Haptophyceae from Chrysophyceae in that the scales are unmineralized. They are of an organic nature only, possibly largely carbohydrate. They are usually round or oval, and with a rim enclosing a plate composed of a lattice of cross striations on the outerface; the inner face has radiating ridges. Despite their delicacy and their uniform basic structure, there is considerable variation between species. Studies such as those of Parke, Manton and Clarke (1955, 1956, 1958 and 1959) show that two kinds of scale may be present even in the same species of *Chrysochromulina*. Further investigations by Manton and Parke (1962) and Manton (1967) demonstrated the development of the scales in relation to the Golgi cisternae. Species such as *Chrysochromulina ericina* and *C. pringsheimii* have scales with very long spines.

Phagotrophy is common in most species of *Chrysochromulina*; it almost certainly occurs in some other Haptophyceae (cf. Fig. 4.13).

The genus *Phaeocystis* is somewhat exceptional in that the obvious phase is the palmelloid stage which can be large enough to be visible to the naked eye. The genus is very widely distributed in warm and temperate waters as well as Antarctic and Arctic Seas. The life history, described among others by Kornmann (1955), includes a motile stage which perhaps undergoes division. The motile cell appears to have two smooth flagella with a short haptonema of the typical haptophycean cell. Further details of the life history have recently been studied.

Included also in the Haptophyceae are the coccolithophorids. These cells are usually of small size (<20 μm) and occur more commonly in warmer waters. Coccolithophorids characteristically possess calcareous scales (coccoliths) which may be referred to two main types: (1) sub-microscopic crystals borne on very delicate scales solely form the coccolith ("holococcoliths"); (2) coccoliths composed of an assemblage of different units forming a single ring ("cricolith"), or of plates, ribs, etc., in complex patterns ("placoliths"); other complex patterns occur in some species – all these are termed "heterococcoliths". This distinction between holococcoliths and the various types of heterococcoliths is suggested by Braarud, Deflandre, Haldall and Kamptner (1955). Some indication of the complexity of heterococcoliths as seen in electron-microscope studies comes from the work of Braarud, Gaarder, Markali and Nordli (1953) (cf. also Dodge, 1973; Fig. 4.14). Though it has been suggested that the two kinds of coccolith separate taxonomically distinct genera, both may apparently occur in stages of the life history of one genus.

At first sight it would seem peculiar that the Haptophyceae, which are said to possess non-mineralized scales, should be regarded as including the coccolithophorids. The work of Parke and Adams (1960), however, demonstrated that at least in some coccolithophorids unmineralized scales with a pattern and form extremely similar to *Chrysochromulina* were present on the cell surface. Scales were of different sizes; moreover, although unmineralized at first, some of the scales began to develop a deposit of calcite crystals to produce the typical coccolith. An internal origin for coccoliths was also indicated. Manton and Leedale (1969) studied the origin of coccoliths in two coccolithophorids, *Coccolithus pelagicus* and *Cricosphaera carterae.* In both species a coccolith is attached to the margin of an unmineralized scale; the scales and coccoliths arise internally in cisternae of the Golgi, and smaller unmineralized scales, independent of coccoliths, are also present. A comparison between the unmineralized scales of such haptophyceans as *Chrysochromulina* and *Prymnesium* and those of coccolithophorids might indicate that the coccolith represents the rim of the scale.

Paasche (1964) had earlier drawn attention to an internal origin of coccoliths, though there was a suggestion that such a development might not necessarily be true for all types of coccolith. The investigations of Outka and Williams (1971) for *Hymenomonas carterae*, however, clearly demonstrated coccoliths arising in Golgi cisternae and in sequential order. The coccolith was shown to be derived from two precursors — a single scale base to which was added multiple electron-dense "coccolithosomes" forming an outer rim sheath. Calcium carbonate was then added, filling the rim sheath, to form the definitive coccolith (cf. Dodge, 1973).

Paasche (1964) showed that in *Coccolithus huxleyi*, as well as the "normal" coccolithophorid cell with a layer of some fifteen coccoliths, "naked" cells exist, lacking

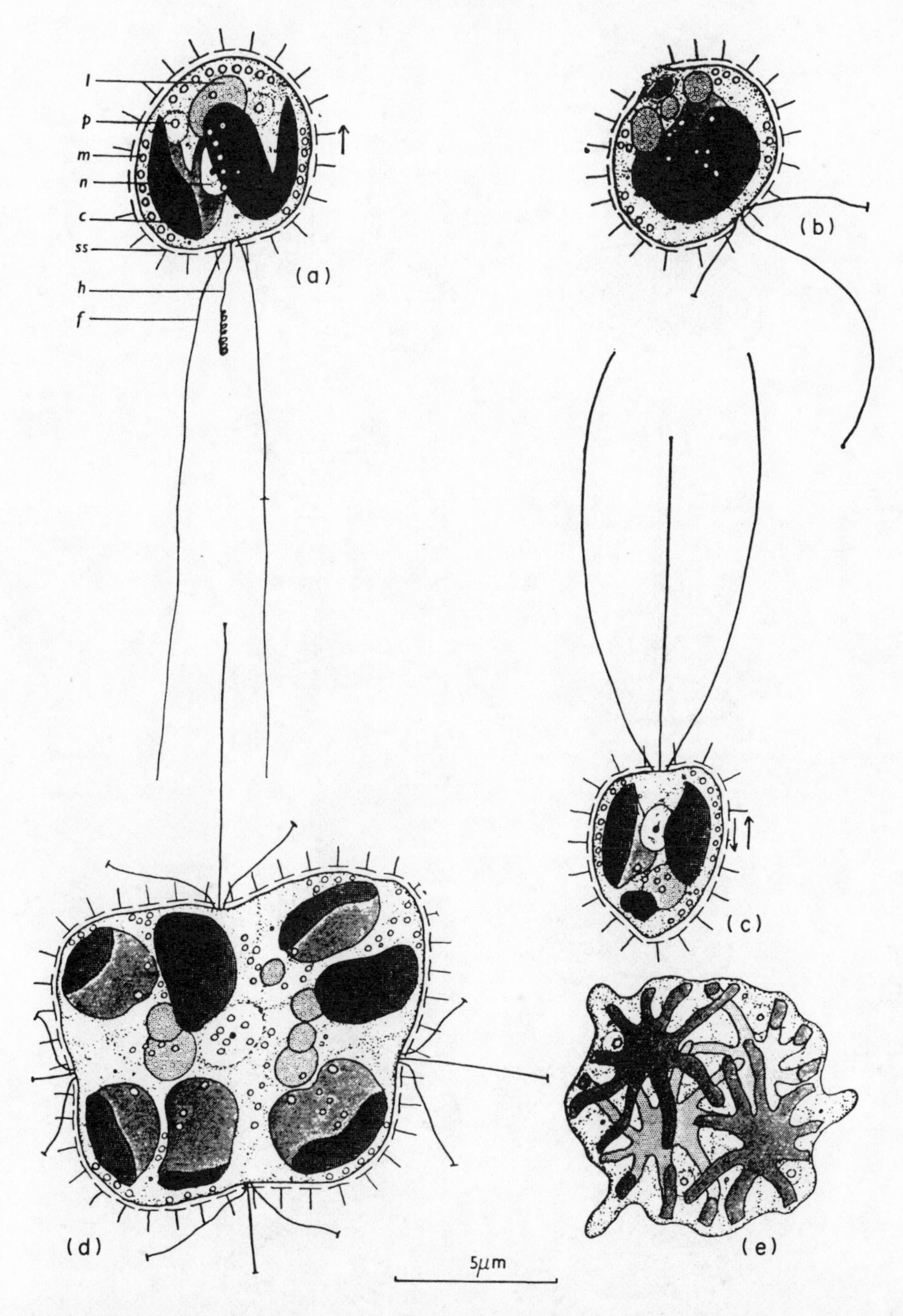

Fig. 4.13. *Chrysochromulina brevifilum*. (a) Swimming cell, chromatophores dividing; c = chromatophore, f = flagellum, h = haptonema, l = leucosin vesicle, m = muciferous body, n = nucleus, p = pyrenoid-like body, ss = spined scales. (b) Cell anchored on extended haptonema; note graphite just ingested at non-flagellar pole. (c) Individual swimming with flagella and haptonema in front. (d) Stage with four pairs of flagella and four haptonemata prior to double-fission. (e) Large amoeboid individual (×5000) (from Parke, Manton and Clarke, 1955).

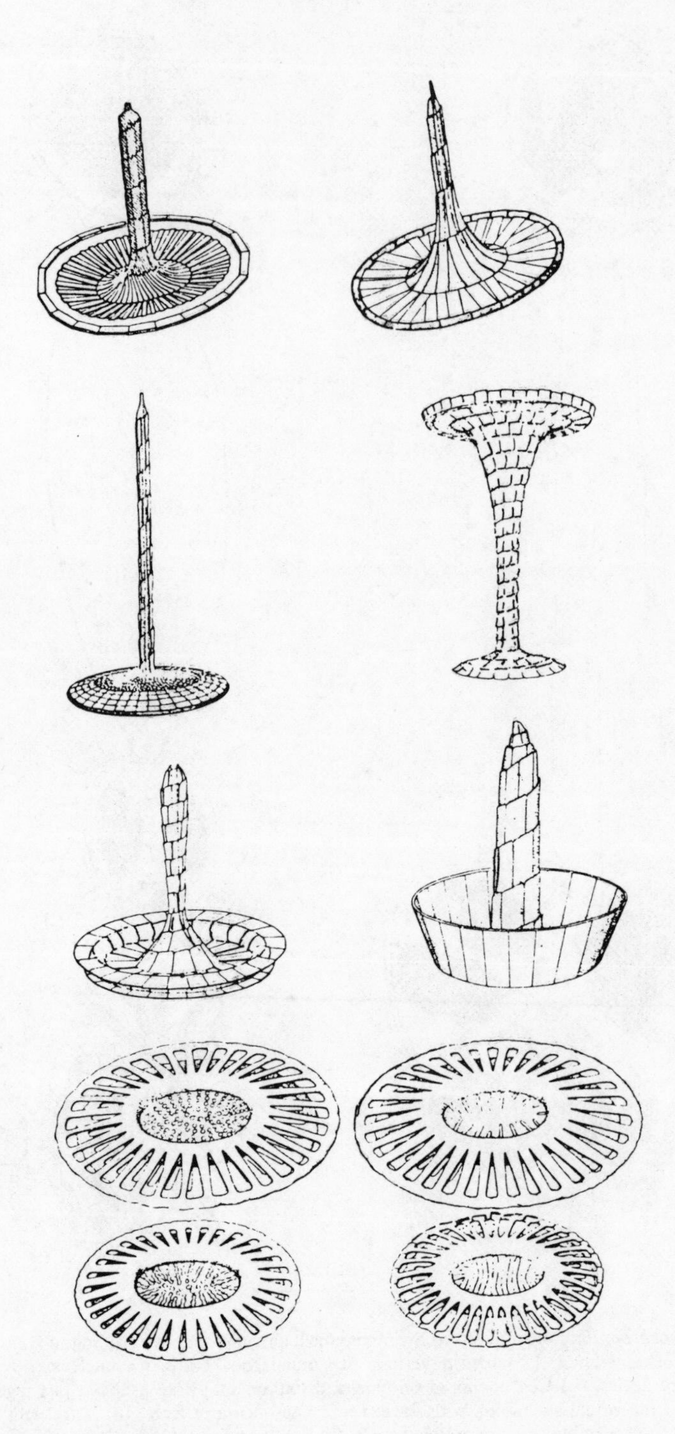

Fig. 4.14. Some examples of the variety of form of coccoliths (modified from Dodge, 1973).

coccoliths. Braarud believed that these may be stages in the reproduction cycle, naked forms undergoing binary fission (often with an amoeboid stage intervening) and then developing coccoliths. Paasche demonstrated that normal cells could undergo binary fission and that clones of normal and naked cells could occur in culture. Cells could lose the ability to develop coccoliths; equally, naked strains could acquire coccoliths. Coccolith formation is essentially light dependent and normally occurs with photosynthesis, though the two processes are independent. Other factors include nutrient level, pH and the amount of CO_2 in solution (cf. Paasche, 1966a and 1968b). Though most of the information on naked strains is derived from cultures it is very probable that naked and normal coccolith-bearing cells occur in nature. The variability in coccolith formation led Paasche to emphasize that there is very little evidence for the view that coccolithophorids are essentially tropical or sub-tropical in distribution because of their ability to live successfully at high light intensities with protection from the excessive radiation from their dense coccoliths! Pienaar (1976) has described a new species, *Hymenomonas lacuna.* In culture the cells quickly lose their flagella and form non-motile clumps. Although the two subequal flagella are well developed, no emergent haptonema is seen but the base is visible in electron-microscope sections. The coccoliths of unusual pattern (cf. Fig. 4.15) form a single layer, inside which are several layers of small unmineralized scales, immediately outside the cell membrane. The Golgi body is an active area of coccolith and scale production. Blackwelder, Weiss and Wilbur (1976) have demonstrated that *Cricosphaera (Hymenomonas) carterae*, which is unusual amongst coccolithophorids in tolerating a wide range of salinities, can lay down calcium carbonate as coccoliths after considerable de-calcification has been induced. With a calcium-deficient medium, calcification is possible provided strontium is substituted in the medium, although the coccoliths are composed of calcite with only trace strontium.

The life histories of coccolithophorids have been the subject of considerable research. Parke and Adams (1960) showed that the motile *Crystallolithus hyalinus* is a phase in

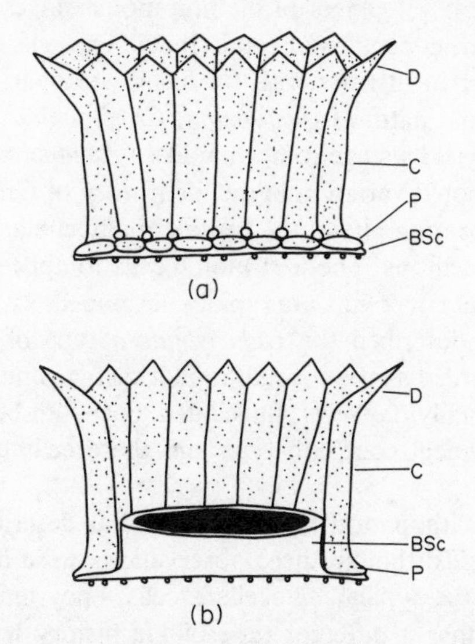

Fig. 4.15. Reconstruction of coccolith of *Hymenomonas lacuna*: (a) seen inside elevation; (b) in section. BSc = basal plate; P = small proximal flange; C = vertical central region; D = distal flange (after Pienaar, 1976).

the life history of *Coccolithus pelagicus* which is non-motile. The motile phase (*Crystallolithus*) has the haptonema and two flagella typical of Haptophyceae and possesses holococcoliths; it exhibits phagotrophy. Reproduction in the motile phase is typically by binary fission. Cell division can continue, probably for a very prolonged period, but eventually the motile cells begin to lie on the bottom, gradually increasing in size to form the non-motile phase. This non-motile cell, which bears placoliths (heterococcoliths), typically divides twice to produce four daughter cells, although apparently two daughter cells are sometimes liberated after only a single division. This production of *Coccolithus pelagicus* type non-motile cells can continue, perhaps almost indefinitely (cf. Paasche, 1968c) but eventually the four daughter cells develop flagella to be liberated as the motile phase. An alternation of generations thus occurs, and though this is known from culture, both phases have been observed from the same area in the sea. Rayns (1962) further confirmed that in the alternation of generations in *Cricosphaera carterae* there was a chromosomal change; the motile phase was diploid whereas the non-motile phase, which has a considerable range of form, was haploid.

Boney and Burrows (1966) found that cultures of *Cricosphaera* sp., a motile coccolithophorid, could give rise to benthic stages similar to benthic algae generally known under such generic names as *Apistonema* (filamentous and coccoid forms), *Chrysonema* and *Ochrosphaera*. Motile coccolithophorids were repeatedly liberated in culture.

Parke (1961) had earlier shown that a range of forms of bottom-living small algae, coccoid, filamentous and dendroid types, belonging to such genera as *Cricosphaera*, *Apistonema*, *Chrysonema* and *Chrysotila*, might occur in the life histories of various motile coccolithophorids. These non-motile stages could be found as benthic algae in inshore environments. Parke points out that the motile coccolithophorid stage has the typical two smooth flagella and that usually a haptonema may be observed, but often only in the "naked" stage of the cell before the coccoliths are developed. She describes as for *Crystallolithus* the motile coccolithophorids continuing division by binary fission almost indefinitely. In the development of the non-motile stage, however, a tetrad stage occurs, which may cast the coccoliths or retain them. The cells can then remain as non-motile coccoid groups, or form rows with further division; in some cases, they form colonies even of a dendroid nature (*Apistonema*, *Chrysonema* types). Repeated secretion of a membrane may yield a stage somewhat like *Prasinocladus*.

Paasche (1968c) also noted variation in the life history of *Cricosphaera* spp. Following the motile coccolith-bearing phase, the benthic multicellular stage shows variability of form and may be filamentous. The first motile cells to appear, following the benthic phase, may be "naked" but coccoliths are rapidly developed.

Paasche (1968c) also described the *Ochrosphaera* type of life history (cf. Parke, 1961). What may be regarded as the coccolithophorid stage undergoing binary fission is here non-motile. Occasionally, however, flagellated, coccolith-bearing, swarmers occur; appear to be of fairly typical coccolith type and these cells develop into non-motile coccolithophorids again.

Another type of coccolithophorid life history is that described by Bernard (1948) for *Cyclococcolithus fragilis*, though these observations were not made on cultures of the species. As well as the typical biflagellate cells, encysted and palmelloid stages occur. Bernard believes that a different type of life history holds for near-shore and more open-sea populations. Offshore, the encysted and orange-coloured spore stages are common, apart from the typical flagellated coccolithophorid. Inshore, palmelloid stages are very abundant in addition to the bi-flagellate stage, and dark-coloured spores occur. But repeated division, apparently for both types of population, can produce the comparatively very large cell masses ("plates") with very considerable numbers of cells that sink relatively rapidly (see Fig. 4.16). Palmelloid stages can apparently reach great depths in the Mediterranean (cf. Bernard and Elkaim, 1962). The suggestions of Bernard (1953) that in warm seas the density of coccolithophorids is generally low and that the reproduction rate is slow (doubling in 6 to 10 days) may be contrasted with Paasche's (1968c) results for cultures of more widespread species; *Cricosphaera elongata*, 2.25 divisions/day; *Coccolithus huxleyi*, 1.85 divisions/day. Undoubtedly details of many coccolithophorid life histories have still to be elucidated; the life history of even the very widespread species, *C. huxleyi*, is not known precisely. It has been suggested that in the life history of some coccolithophorids the swarmers could undergo sexual fusion, but sexual reproduction is not known for certain in any species.

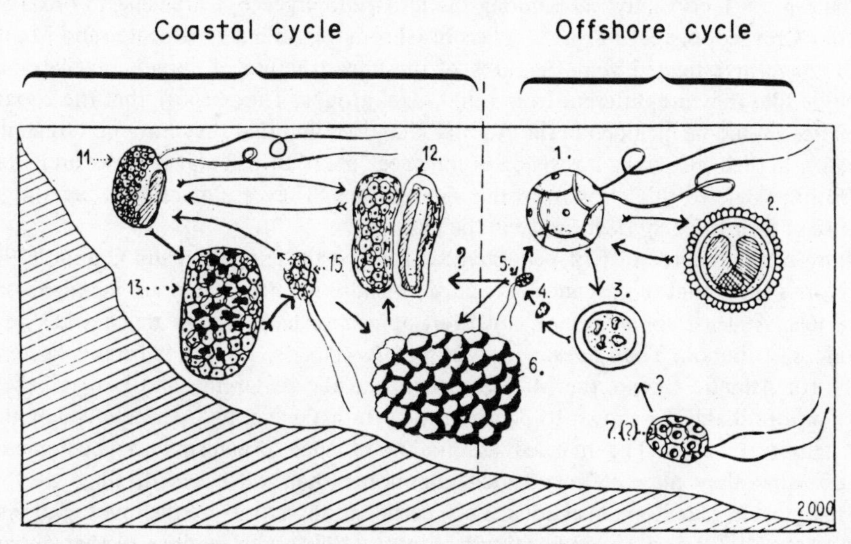

Fig. 4.16. *Cyclococcolithus fragilis* – life histories in coastal and in offshore waters. Offshore: (1) typical free swimming cell; (2) cyst; (3) and (4) orange spore formation and liberation; (5) young orange flagellated stage; (6) mass of cells lacking flagella arising from (5); (7) olive-green cells common at depth in Mediterranean. Inshore: (11) dominant stage inshore, flagellated but coccoliths smaller than in (1); (12) palmelloid stage; (13) large individual sunk to near bottom, and with sixteen black spores, one of which has released a flagellated cell (15) similar to stage (5) (after Bernard, modified, 1948).

8. Craspedophyceae

A group of little-known nanoplankton organisms is sometimes separated from the Chrysophyceae under the name of Craspedophyceae. Although the number of species was thought to be not very large, some of these organisms are neustonic flagellates which hang from the surface film and therefore are rather infrequently sampled with ordinary plankton methods.

Typically the cells possess a collar made up of a ring of very fine tentacles and surrounding a flagellum. While the flagellum with the collar appear to be concerned intimately with food collection, the ingestion of food particles occurs mainly at the sides of the cell. Obviously these organisms are essentially holozoic and phagocytic, taking in particulate food. They are, therefore, often grouped among zooflagellate taxa as the choanoflagellates. Some of the families possess a lorica surrounding the cell, variously described as composed of chitin or cellulose, but in one family, the Acanthoecaceae, the lorica is composed of silica fibres united together to form a basket-like structure which may be of quite complex pattern (cf. Norris, 1965; Leadbeater, 1972). Leadbeater and Manton (1974) confirmed the presence of silica in the lorica of *Stephanoeca*.

Though there would appear to be no good reason for including holozoic flagellates among the phytoplankton, one genus, *Stylochromonas*, has golden-brown chloroplasts and is photosynthetic. Norris (1965) also mentions that another genus, *Microsportella*, is faintly pigmented and may be photosynthetic. Until its photosynthetic ability has been confirmed, however, this organism is retained in the genus *Diaphanoeca*. Norris argues that the choanoflagellates are related to photosynthetic Chrysophyceae. In this connection he also includes the family Pedinellaceae (which Parke and Dixon (1968) retained in the Chrysophyceae) among the Craspedophyceae. Christensen (1962) also includes Craspedophyceae as a sub-class of Chromophyta but Leadbeater and Manton (1974) have investigated some features of the fine structure of choanoflagellates and conclude that they are different from other algal groups. They believe that the choanoflagellates should be included in the Animal Kingdom and that they have no direct algal ancestry. In their most recent revision of the check list of British algae, Parke and Dixon (1976) have deleted this class from the algal groups. They are mentioned here largely because of their frequent occurrence in the neuston.

Many of the records of Craspedophyceae such as those of Boucaud-Camou (1966) and Norris show that the organisms occur in inshore habitats, often among algae or in tide pools. At least some species, however, appear to have a very much wider, even oceanic, distribution. Thus *Parvicorbicula socialis* (Norris, 1965) has been collected in the North Atlantic Ocean, the Mediterranean and the Antarctic, and Norris believes that it will probably be shown to occur in the North Pacific, and perhaps widely in at least temperate seas. The marked adaptation of such organisms to neustonic life suggests that they may be very much commoner than we believed, since they are generally lost in plankton collections. Very pertinent, therefore, are the investigations of Leadbeater (1972) and Leadbeater and Manton (1974) who emphasize that choanoflagellates (Craspedophyceae) are world wide in distribution, and that the number of species is very large amongst the nanoplankton of temperate seas of both northern and southern hemispheres, as well as in the Arctic.

9. Xanthophyceae

Typically these are motile flagellates, but non-motile cells exist, including palmelloid species; some are stalked forms; some Xanthophyceae occur as filaments. Non-motile species may give rise in asexual reproduction to zoospores of the typical flagellate structure, but aplanospores occur in some species. A few forms such as *Rhizochloris* are amoeboid, or can become amoeboid (e.g. *Heterochloris*). The cells are characteristically biflagellate, one flagellum being markedly longer than the other and having well-developed lateral fibrils, the other being smooth; a few species are uniflagellate. The colour of the cells is commonly a light yellowish green, owing to an unusual abundance of xanthophylls, especially according to some authorities "flavoxanthin"; some species, however, appear bright green. In addition to chlorophyll *a*, chlorophyll *c* is reported by some workers. However, Chapman and Chapman (1973) list chlorophyll *a* only; carotenoids include β-carotene. There are discrepancies in the literature regarding the precise list of the more minor photosynthetic pigments. For example, Xanthophyceae are said by some authorities to possess chlorophyll *e*; Jackson (1976) and also Bogorad (1962) quote chlorophyll *e* for one species of Xanthophyceae only. Riley and Wilson (1967) point out that β-carotene is usually the dominant carotenoid and although xanthophylls are abundant, their precise identification, especially that of lutein, is doubtful. They further suggest that although relatively abundant fucoxanthin occurs in *Olisthodiscus*, that is anomalous for the group and perhaps this genus is incorrectly included in Xanthophyceae (cf. also Parke and Dixon, 1976). A small alga provisionally placed with Xanthophyceae has, however, been described by Riley and Segar (1969) as possessing fucoxanthin and diadinoxanthin, and substantial quantities of an unknown xanthophyll. Chapman and Chapman (1973) state that only diadinoxanthin and heteroxanthin are identified with certainty.

Some Xanthophyceae appear to have no cell walls, e.g. Heterochloridaceae. Pectin is commonly found in those species having cell walls and some of them possess silica. The chloroplasts may be few or, more frequently, numerous; pyrenoids when present are embedded in the chloroplast and have no starch sheath. Lipid or chrysolaminarin may be stored by the cells.

Xanthophyceae are not common in the marine environment, especially among the phytoplankton, but some small species may have been overlooked. *Meringosphaera* appears to be a widespread planktonic form originally included in the Xanthophyceae (cf. Parke and Dixon, 1968). Records for European waters occur in Apstein (1908). Thorrington-Smith (1970a) states that *M. mediterranea* occurred commonly in Indian Ocean samplings.

10. Bacillariophyceae

This group, the diatoms, frequently forms the dominant phytoplankton of temperate and high latitudes. A fuller account of the distribution and relative abundance of diatoms appears in a later chapter.

Many genera are unicellular (e.g. *Coscinodiscus*, Fig. 4.20) but some form loose colonial associations of characteristic shape, most usually chains (e.g. *Chaetoceros*) (Fig. 4.17) but sometimes of other patterns (e.g. *Asterionella*, *Eucampia*) (see Figs. 4.18 and 4.19). The association in any event seems to be largely mechanical since the individual cells are apparently capable at all times of an independent existence; no obvious special benefit appears to derive from association except in so far that some forms may improve their flotation capacity (cf. p. 165).

Perhaps the most characteristic feature of diatoms is that the cell has the ability to form an external skeleton, largely of silica, called the frustule, and composed of two overlapping halves (valves), with a fairly distinct region known as the girdle lying between (Fig. 4.20). The girdle usually has a different structure from the valves proper and may have minute teeth which to some extent hold the valves together; it may consist of a single band or may be made up of numerous segments. In some diatoms such as *Corethron* and *Rhizosolenia* the valves are less easily seen and the girdle is greatly developed with intercalary bands giving rise to a more or less tubular type of cell (Fig. 4.21). In many diatoms the upper valve or epitheca and the lower valve or hypotheca fit together in a manner somewhat similar to the two halves of a petri dish. Each valve consists of a flattened or convex plate and can vary in shape between species, being circular, elliptical, triangular, square, polygonal, or even irregular. Typically, when examined through a microscope, a diatom may present one of two views according to how it is lying; it may show the valve or the girdle, and the appearance may be very different (cf. Fig. 4.22). When one of the axes of the cell is greatly hypertrophied, however, as in *Rhizosolenia*, one view of the cell is much commoner.

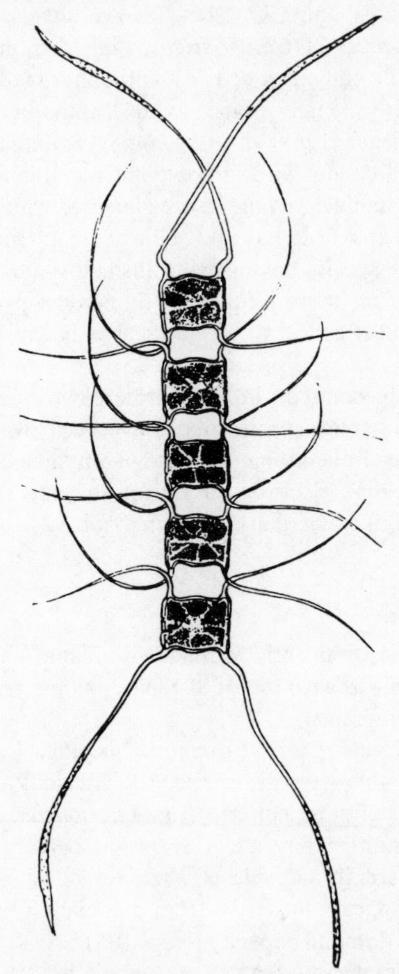

Fig. 4.17. *Chaetoceros laciniosus*, 25 μm across (from Lebour, 1930).

The valves may show the most remarkable complexity of ornamentation, which the use of the electron-microscope (cf. Hendey, 1954, 1971; Desikachary, 1957) has done much to reveal. Hendey suggests that the ornamentation may take the form of four major categories: (1) *Puncta*: small dot-like markings which are now known to be small holes in the frustule; they may be irregularly scattered, or regularly arranged in lines called striae. (2) *Areolae* or *alveoli*: larger structures consisting of chamber-like areas of varied outline within the thickness of the valve walls; they may be closed by membranes on either the inner or outer surface. (3) *Canaliculi*: fine tubular channels running through the valve wall. (4) *Costae*: solid rib-like structures which appear to act as

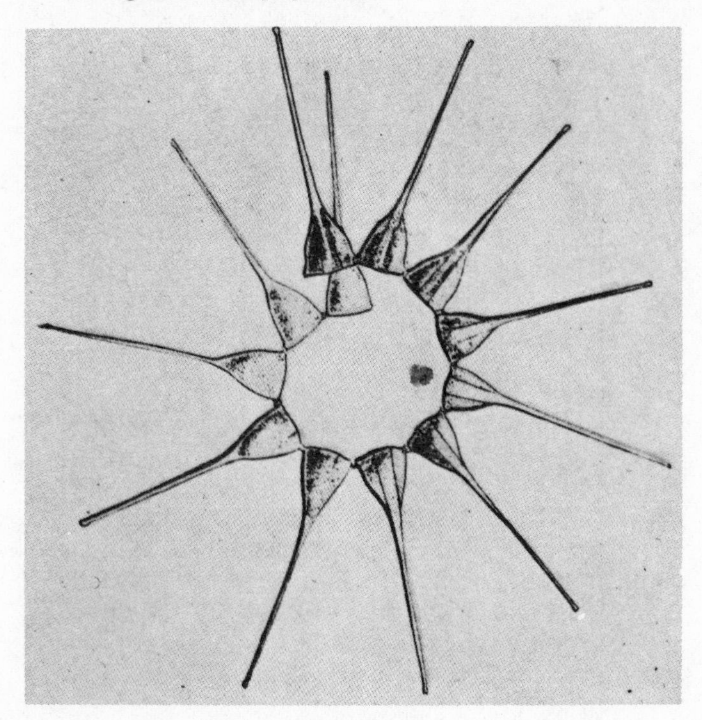

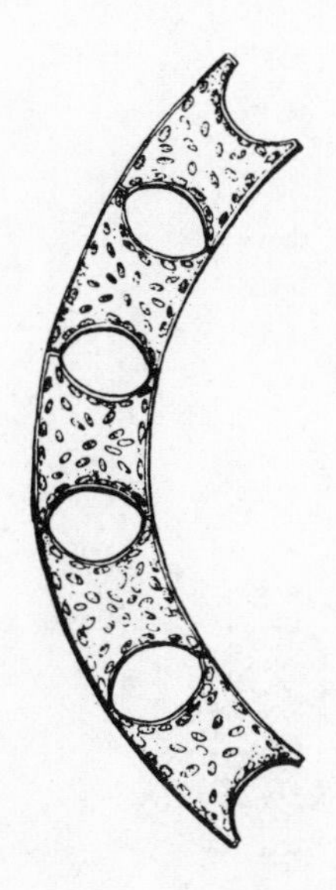

Fig. 4.18. Above. *Asterionella japonica*, 58 μm broad (from Lebour, 1930).

Fig. 4.19. Left. *Eucampia zoodiacus*, 25 μm across (from Lebour, 1930).

Fig. 4.20. Below. *Coscinodiscus eccentricus*, 70 μm across (from Lebour, 1930).

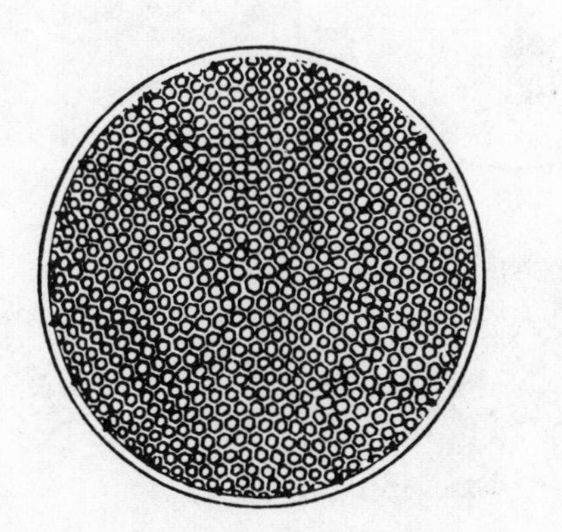

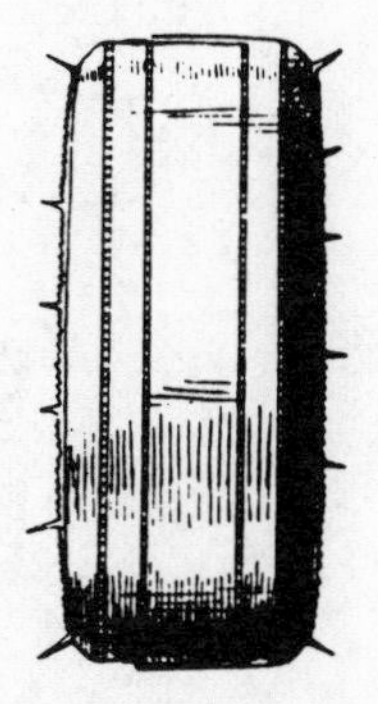

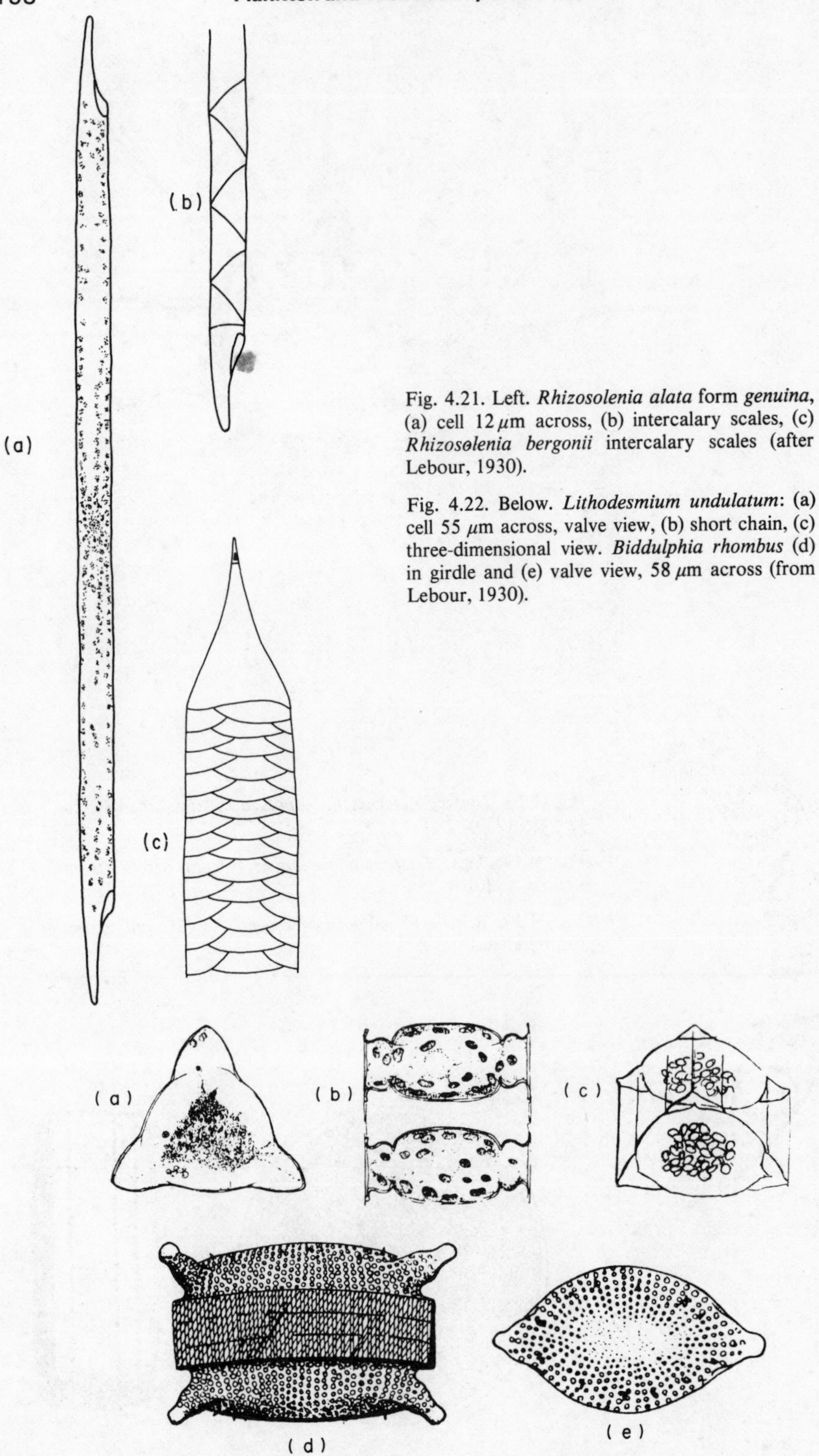

Fig. 4.21. Left. *Rhizosolenia alata* form *genuina*, (a) cell 12 μm across, (b) intercalary scales, (c) *Rhizosolenia bergonii* intercalary scales (after Lebour, 1930).

Fig. 4.22. Below. *Lithodesmium undulatum*: (a) cell 55 μm across, valve view, (b) short chain, (c) three-dimensional view. *Biddulphia rhombus* (d) in girdle and (e) valve view, 58 μm across (from Lebour, 1930).

strengtheners for the valve surface; essentially they are heavy depositions of silica. These various valve structures may be arranged in different patterns, according to species. For example, the patterns may be centric, radiating, or concentrically arranged with reference to a point or sometimes to more than one point; sometimes the pattern radiates from the angles of the cell or there may be a pinnate arrangement where the patterning is more or less symmetrically arranged on each side of a central line (axis). A further complexity is that the wall structure of diatoms may consist of either a single layer of silica, or of a double layer of siliceous material, separated by vertical walls which Hendey has described as usually arranged in a hexagonal pattern. This type of structure may lead to extremely complex forms; the two surfaces, for example, almost invariably possess different detailed structure and the alveoli may exhibit considerable complexity (Figs. 4.23, 4.24 and 4.25).

Apart from ultra-microscopic ornamentation the valves in various genera may be provided with a variety of small teeth, shorter or longer spines, or siliceous projections of different shapes. Since the shape of the valve itself varies also greatly with the genus, it is not surprising that the precise form and ornamentation of the valve is largely used in identification.

Although taxonomists such as Hendey (1954, 1964) do not believe that the pennate and centric diatoms form a natural division it is convenient to distinguish the two

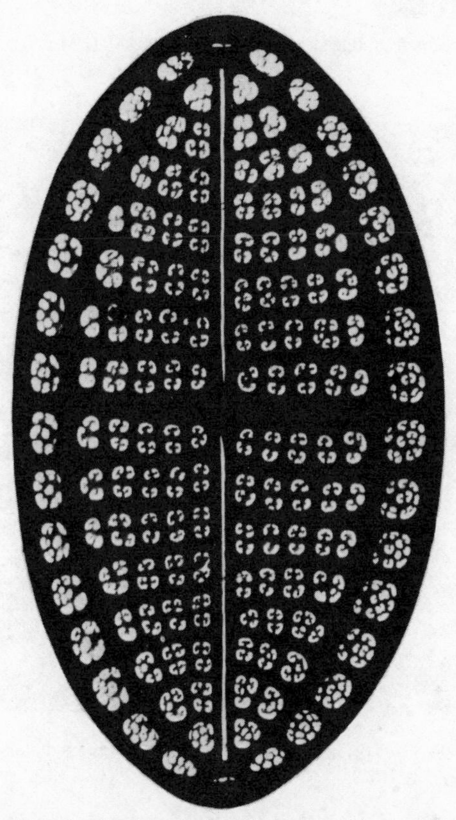

Fig. 4.23. One valve of the frustule of *Cocconeis* (Bacillariophyceae). Organic components of the cell have been cleaned away leaving only the silica (×8000) (from Dodge, 1973).

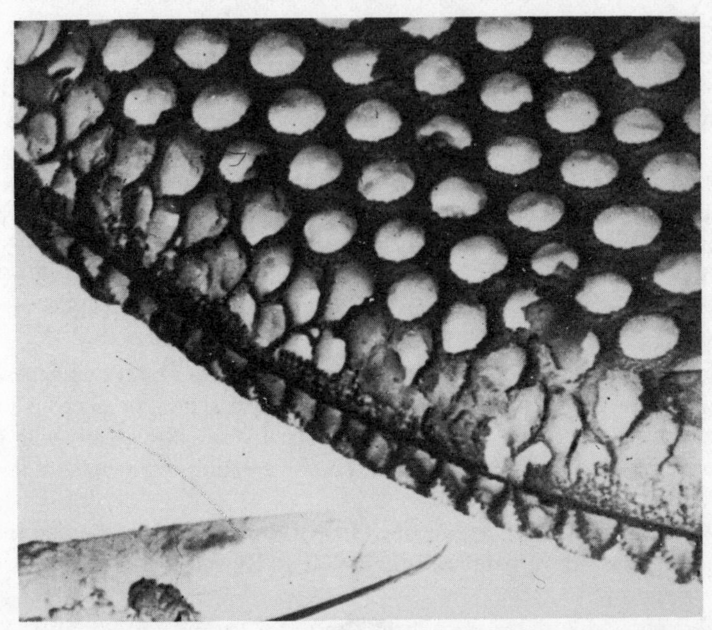

Fig. 4.24. *Thalassiosira eccentricus*, fine structure of valve (×6000) (EM photograph by Holdway, 1976).

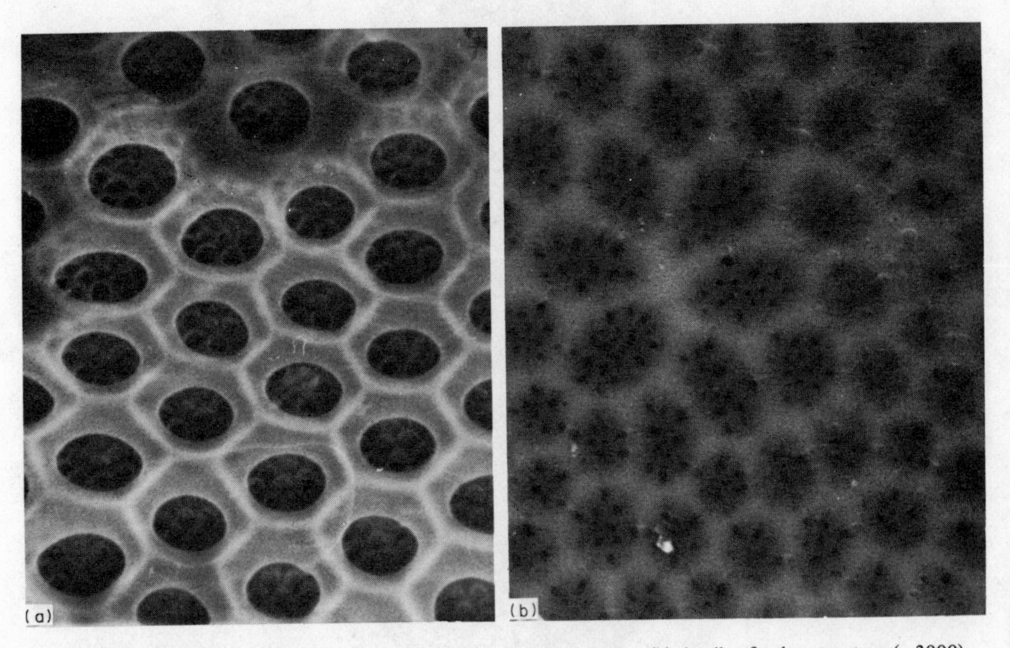

Fig. 4.25. *Thalassiosira eccentricus* (a) and *Coscinodiscus concinnus* (b) details of valve structure (×3000) (EM photograph by Holdway, 1976).

groups. In the pennate forms the cells are more or less elongated in one direction and the valve structure is arranged with reference to this apical axis: typically the structure and ornamentation are thus feather-like in orientation. In many species there is a bilateral symmetry in the valve structure. Along the long axis is a narrow clear space known as the axial area (Fig. 4.26). Sometimes this is structureless and is then known as the pseudoraphe; the middle is usually enlarged to form a clear so-called central area. In many pennate diatoms the central axial area is furnished with a narrow slit, the raphe (cf. Fig. 4.26) interrupted in the middle by a thickening called the central nodule, and at each end of the raphe, towards the edge of the valve, by a further thickening known as the polar nodule. According to Hopkins and Drum (1966) there is a continuation groove extending towards the edge of the valve, and apparently ending in an anterior pore. Along the course of the slit in species which possess a raphe the protoplasm may come in contact with the surrounding water. In a large number of pennate diatoms the valves are similar in that the raphe or pseudoraphe occurs on both valves, but some forms have only a single raphe or pseudoraphe. While in typical linear pennate diatoms, the puncta, costae or other patterns of ornamentation are disposed symmetrically on each side of the raphe and the diatoms have regular, more or less elongated, frustules (Fig. 4.26), in other pennate diatoms the shape is very variable – fusiform, oval, sigmoid or roughly circular, – and the raphe is carried along the *margin* of the valve, typically on a wing-like extension. Although many pennate diatoms are benthic in sand and mud, some species are planktonic.

A number of marine diatoms, including many plankton species, belong to the so-called centric group which do not possess a raphe. The structure of the valves is arranged radially or concentrically with reference to a central point. In very approximate terms centric diatoms have often been likened to a pill box as compared with the date box or boat-like pennate forms, but this is a very crude representation.

Hendey (1974) in a discussion of the problems of classifying diatoms has criticized the suggestion that the mode of reproduction might be used as a criterion, centric diatoms exhibiting sexual reproduction by oogamy (cf. p. 172); he emphasizes that sexual reproduction has not been studied in many species. A classification on the possession of resting spores (*vide infra*) has been proposed, but Hendey comments that this could be a more recent evolutionary development, largely an adaptation to planktonic existence, and thus to some extent convergent. While therefore avoiding a classification into Centricae (Centrales) and Pennatae (Pennales), the terms centric and pennate diatoms will be retained, largely for convenience.

The amount of silica in the cell walls of diatoms varies greatly with species as well as with culture conditions. Diatoms have an absolute requirement for silicon. According to Lewin and Guillard (1963) the usual range for silicon is between 4 and 50% of the dry weight though data outside this range have been quoted, especially for species under culture. An exceptional form is *Phaeodactylum*; Lewin and Guillard point out that the silica content is less than 1% of the dry weight. In the form of *Phaeodactylum* with oval cells only one valve is silicified, while in the fusiform form the cells normally do not have silica. It is frequently stated that planktonic species tend to have thinner valves which presumably help flotation; while this may be broadly true, some planktonic species (e.g. *Coscinodiscus*) can have considerable amounts of silica. The silica appears to be laid down as amorphous hydrated silica but this is intimately connected to an organic constituent of the cell wall which is usually described as "pectin". Lewin (1962) has

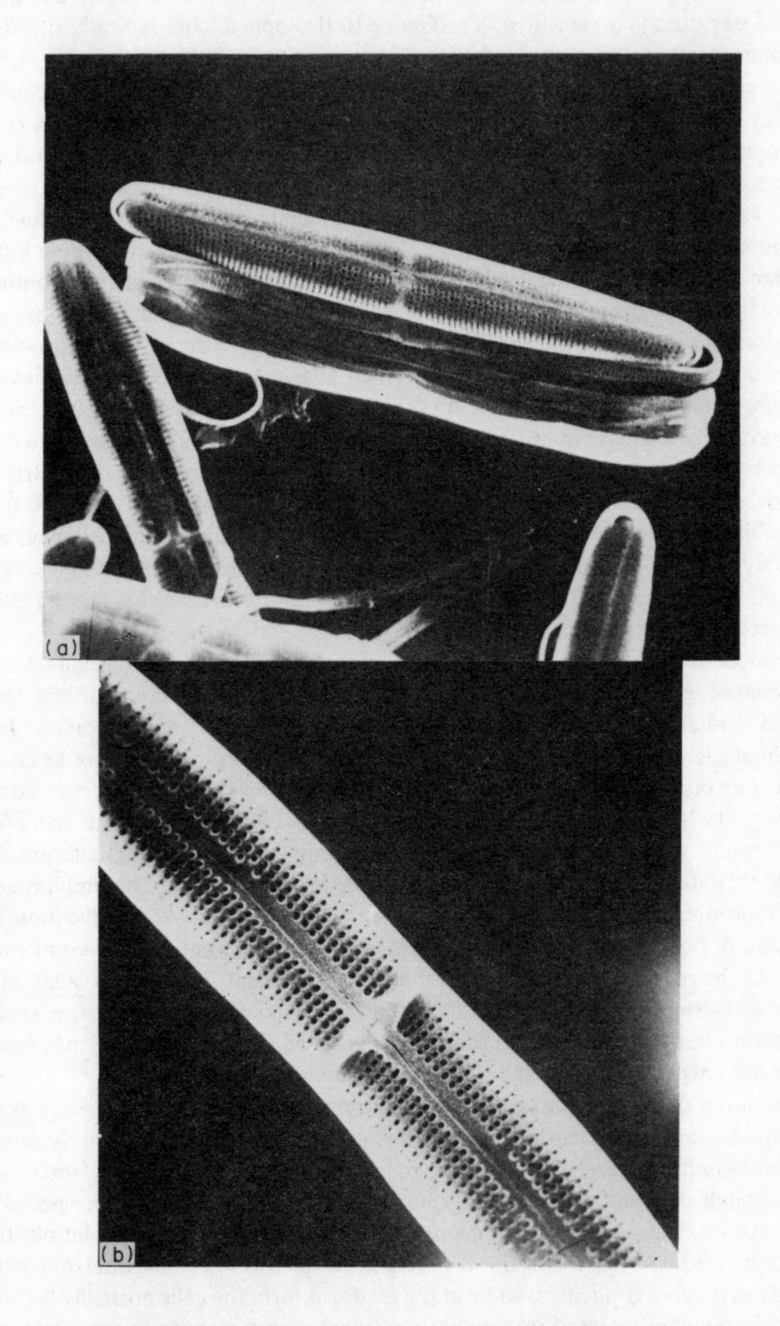

Fig. 4.26. *Stauroneis decipiens*: (a) cell showing valves and girdle (×2750); (b) details of valve showing axial area, raphe, etc. (×4750) (EM photograph by Holdway, 1976).

expressed doubts on the precise chemical nature of the organic material (cf. also Hecky, Mopper, Kilham and Degens, 1973; Haug and Myklestad, 1976).

Inside the lining membrane of the cell the general protoplasm forms a comparatively thin lining surrounding a large vacuole filled with cell sap. The nucleus may be more or less central in position as in many pennate diatoms; in others, particularly in many centric forms, it is displaced to one side, but in any event, cytoplasmic strands may extend across the vacuole. The chloroplasts (chromatophores) may be small and fairly numerous as in many centric species; typically in pennate forms they are reduced in number, often to two, and are relatively large and often lobed. They are usually brownish in colour, owing to the abundance of carotenoid pigments. The chloroplasts possess chlorophylls *a* and *c*; β-carotene occurs and there is a record in one species of ε-carotene. An abundance of xanthophylls, above all of fucoxanthin, is present. The xanthophylls appear to be fairly specific; apart from fucoxanthin with neofucoxanthin, only diadinoxanthin and diatoxanthin normally occur and then in fairly small amounts. Riley and Segar's (1969) analysis of four planktonic diatoms confirmed that fucoxanthin is overwhelmingly important of the carotenoids, but diadinoxanthin was the only minor xanthophyll identified. Neofucoxanthin is certainly present, but there is some doubt as to whether it is in part an isomerization product (cf. Riley and Wilson, 1967; Jeffrey, 1974).

The characteristic carbohydrate product of photosynthesis is a polysaccharide, chrysolaminarin, similar to laminarin, but probably lacking mannitol in the molecule (cf. Lewin and Guillard, 1963; Morris, 1967, and Chapter 3). Diatoms characteristically also produce lipids as a product of photosynthesis. Oil globules can sometimes be conspicuous in diatom cells, and aged, particularly nitrogen-deficient cultures, can sometimes be very high in lipid. Grontved (1952) describes unusual conditions where a rich patch of *Coscinodiscus* in the North Sea produced so much oil that a slick was evident. However, it must not be assumed that lipid is the major or most characteristic product of photosynthesis. Lipid content varies with species and with conditions; normally protein and often carbohydrate is much greater in amount (cf. pp. 192–202).

Diatoms do not possess cilia, flagella or other organs, giving locomotory powers, though uniflagellate spermatozoids occur in some species. Although marine planktonic diatoms typically do not have motility, a number of species, especially those living in benthic deposits, possess this ability, though it is restricted to certain pennate species which possess a true raphe.

The precise mechanism of movement is still not fully known. Arguments have developed concerning whether movement can go on without contact with the substratum and whether a particular part of the diatom must be in contact. Early theory connected movement with the cytoplasmic streaming in the raphe; mucilage, which is widely produced and often extruded by diatoms, was thought to be intimately connected with motility. Lewin and Guillard doubt whether mucilage extrusion is the true basis; they speak about an energy transfer near the raphe and discuss transverse waves of small amplitude in the raphe protoplasm, which are transmitted to the exterior medium. Other investigators who have analysed movement in diatoms include Jarosch (1962), Hopkins and Drum (1966) and Drum and Hopkins (1966). Although movement is exhibited mainly by species living in mud or sand, in *Bacillaria paxillifer*, which can occur in neritic plankton, the narrow elongated cells adhere to form a colony, and the cells are able to slide over one another, often going into reverse movement.

In apparently all species of diatoms, when a cell divides, the two daughter cells can move apart and lead separate existences. Many diatoms indeed remain as solitary individuals (e.g. *Ditylum*, *Coscinodiscus* and *Nitzschia*) but some species form colonies. Colony formation is frequently associated with the liberation of mucilage, so much of which may be secreted that an irregular mass can be produced in which the diatom cells are embedded (Fig. 4.27); sometimes even unrelated species are embedded together. A fairly common pattern is that the cells unite to form a more or less flat ribbon (e.g. *Fragilaria*) (Fig. 4.28). In *Grammatophora* the cells partly separate but adhere together by a corner to form a sort of zig-zag chain; *Thalassiothrix* and some species of *Biddulphia* are somewhat similar (Fig. 4.28). Many of these diatoms are essentially bottom-living forms which are swept up into the over-lying waters to become temporarily planktonic – "tychopelagic". In *Licmophora*, another tychopelagic diatom, the colonies form fan-shaped groups of wedge-shaped cells. The most typical planktonic diatom colonies are chain-like, the cells of the chain united by mucilaginous threads. In *Thalassiosira* the cells have only a single thread in the centre of each valve which holds them together; *Coscinosira* has several threads (Fig. 4.29). Spines on the valves may help to hold the cells together; for instance, *Skeletonema* has a ring of marginal spines. In *Corethron* the cells may be joined by a central cluster of spines ending in small hooks. The numerous species of *Chaetoceros* and the genus *Bacteriastrum* have cells equipped with elongated spines (often called setae), which twist to some degree so that they interlock, holding the cells together to form the chain-like colony (Figs. 4.17 and 4.30).

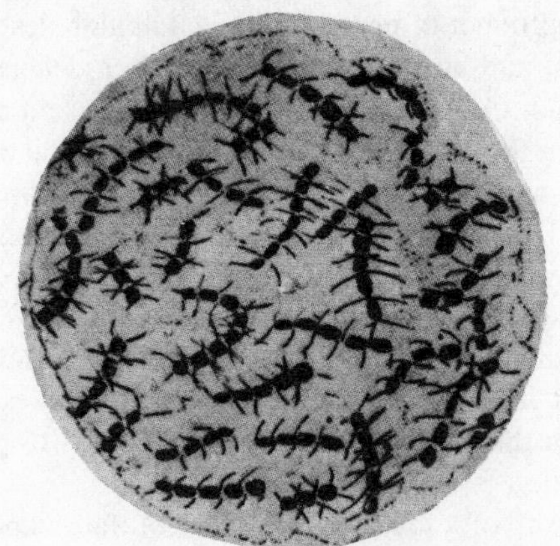

Fig. 4.27. *Chaetoceros socialis*, 15 μm across. Chains embedded in a slimy colony (from Lebour, 1930).

Investigations by Schone (1970) suggest that the motion of the seawater, especially the degree of turbulence, can have a pronounced effect on certain colony-forming diatoms. For four chain-forming species, *Chaetoceros curvisetum*, *Cerataulina bergonii*, *Guinardia flaccida* and *Skeletonema costatum*, Schone found a marked shortening of chain length with increasing sea force. This relationship, established first in the field, Schone attributed to a direct mechanical effect; in particular to trapped air bubbles due to wave action. With *Skeletonema*, whereas a sea force of three arbitrary units caused a decrease in chain length of 10%, a force of seven units decreased chain length by about 40 to 50%.

The effects of turbulence on chain breakage and cell number are complex, however,

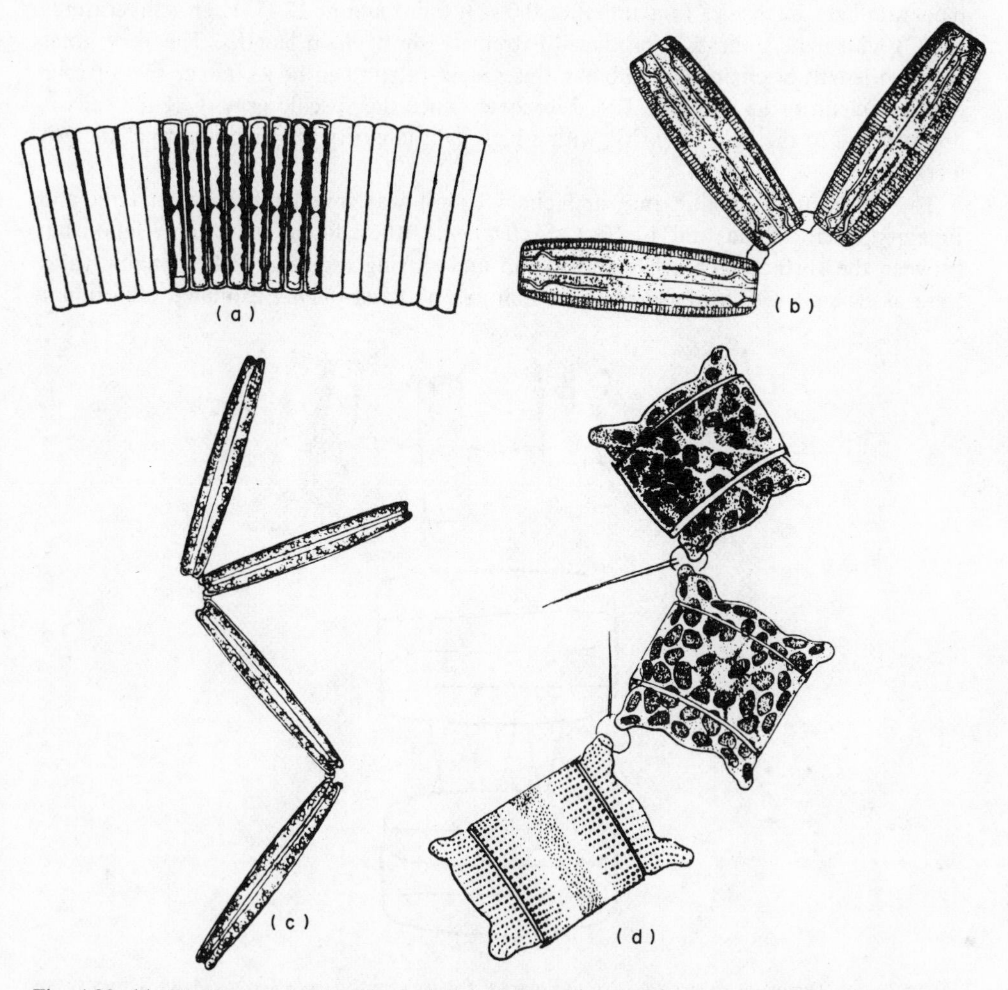

Fig. 4.28. (a) *Fragilaria striatula*, 55 μm broad; (b) *Grammatophora marina*, valve 58 μm long; (c) *Thalassiothrix* (*Thalassionema*) *nitzschioides*, valve 80 μm long; (d) *Biddulphia obtusa*, 35 μm across (from Lebour, 1930).

at least in *Skeletonema*. In the field, increased turbulence and cell numbers appeared to be directly correlated, so that a marked increase in water movement was accompanied by a prolongation of the exponential phase of cell division and a substantial increase in overall cell density followed. The population, however, on reaching the plateau phase, very rapidly declined; the cells became senescent and sank. Chain length is also partly dependent upon the age of the diatom population. The effect of turbulence may be very different in non-colonial forms. Schone finds that cell division rate in a solitary species of *Chaetoceros* is reduced by strong turbulence.

Turbulence in any event is only one factor influencing chain length in diatoms. In general, conditions which are sub-optimal, such as poor light intensity and reduced nutrient, cause decreased chain length. Post-bloom populations tend to show reduced chain length. Schone (1972) suggests that the diatom *Thalassiosira rotula* "prefers" relatively low light intensities and moderate temperatures; greater chain lengths occur at moderate light intensities (maximum at 0.072 ly/min) and at 12°C. High temperatures (22°C) with light saturation produced extremely short chain lengths. The very great range in length of chains is notable in this species; there can be as few as two or four cells per chain or as many as 128. Moreover, since the specific gravity of the cells is proportional to the light intensity, with a high light intensity the sinking rate of the cell increases.

The problem of sinking rate and chain length was investigated by Smayda and Boleyn (1965; 1966a and b). For *Skeletonema*, they found no clear relationship between the surface area/cell volume ratio and sinking rate, but higher sinking rates were associated with senescence. Thus cultures of *Skeletonema* exhibited only minor

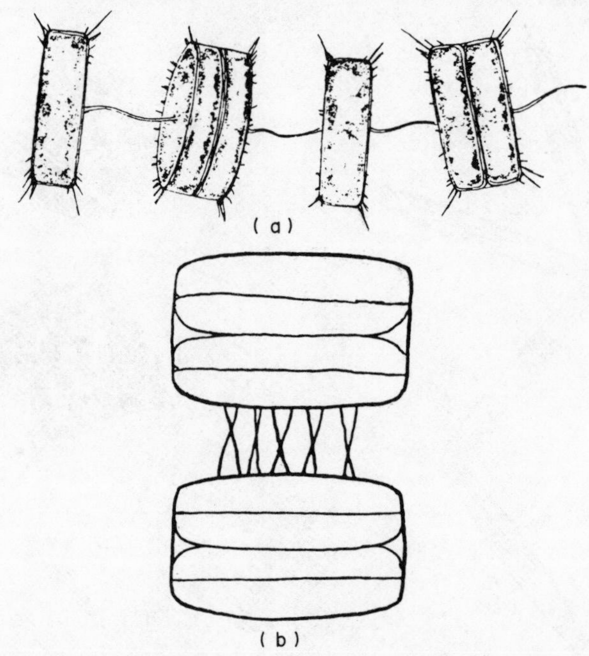

Fig. 4.29. (a) *Thalassiosira gravida*, 20 μm across; (b) *Coscinosira polychorda*, 24 μm across (from Lebour, 1930).

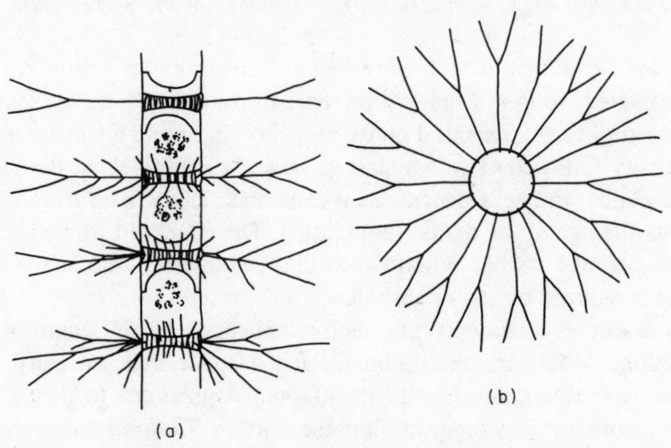

Fig. 4.30. *Bacteriastrum hyalinum*, 25 μm across: (a) chain, (b) valve view (from Lebour, 1930).

variations in sinking rate over the first 9 days of growth, but older (11-12 day) cultures showed a much higher sinking rate; the increase was not linear with age. Renewal of growth was accompanied by a decrease in sinking speed. Irrespective of culture age, however, sinking rate was also inversely correlated with colony size. Further, in natural populations, increasing age of diatom population was accompanied by a reduction in colony size. *Skeletonema* therefore tends to sink through the euphotic zone as the population ages, due both to senescence itself and to a shortened mean chain length. The decreased sinking rate with greater chain length may be associated with the silica rods which occur between the cells. Elongation of the rods may increase the surface area of the chain, without any significant increase in volume.

The relationship between sinking rate and senescence holds for other diatom species (e.g. *Thalassiosira* spp., *Bacteriastrum*), where the rate increased fairly steadily with age, but this relationship is not universal – *Nitzschia seriata* and *Chaetoceros lauderi* showed no obvious effect of senescence. Moreover, in the chain-forming diatoms, *Thalassiosira rotula*, *Nitzschia seriata*, *Bacteriastrum*, *Asterionella* and *Leptocylindrus* (Smayda, 1970), in contrast to the findings for *Skeletonema*, the greater the number of cells per chain, the faster the sinking (cf. Table 4.2). Possibly the greater number of cells per chain reduces the effective surface area owing to the spikiness of the cells in *Bacteriastrum* and *Chaetoceros*; in *Nitzschia seriata* the effect may result from the cell apices overlapping in chain formation.

Table 4.2. The different influence of colony formation on sinking behaviour in two species of diatoms (from Smayda, 1970)

Bacteriastrum hyalinum		Skeletonema costatum	
Colony size (cells)	Mean sinking rate (m/day)	Colony size (cells)	Mean sinking rate (m/day)
1–5	0.79	2–5	0.73
6–10	1.31	6–10	0.55
11–15	3.21	11–20	0.32
		>20	0.13

Smayda and Boleyn believe that cell size and shape, as well as age and colony formation, all influence diatom sinking rates. The mean sinking rate of such species as *Thalassiosira nana* and *T. rotula* which increased with the age of culture is probably partly attributable to an increase in cell size in the older cultures.

The ratio of cell surface area to volume must be an important factor in sinking. Apart from the differences in sinking rates with chain length, the post-auxospore cells (*vide infra*) of *Rhizosolenia setigera* sank faster than the pre-auxospore cells. Similarly in *Chaetoceros lauderi*, the post-auxospore cells sank faster possibly reflecting the decrease in surface area/volume when auxospores develop. In *Chaetoceros* the volume was estimated to have increased sevenfold with auxospore formation, from 3400 μm^3 to 24,400 μm^3; the surface area/cell volume ratio declined consequentially from 0.41 to 0.19.

Whatever the effect of precise cell shape (morphology) in flotation, it must play a comparatively minor role, since dead cells without change in form sink at a much faster rate. Size itself may also not be all that important in sinking rate since during active growth similar sinking rates occur in diatoms of greatly different volumes. Nevertheless, laboratory experiments with single-celled phytoplankton, all in logarithmic growth, indicated an increased sinking rate with cell size (cf. Smayda, 1970) though large cells appear to sink proportionately less rapidly than small ones. For example, the increase in cell size between *Cyclotella nana* and *Coscinodiscus wailesii* represents a volume increase of nearly 10^4, and although *Coscinodiscus* sinks much more rapidly the increase in rate is only fourteenfold. The most significant factor is the ratio of cell surface area to volume which determines frictional drag.

For diatoms the silica content is also significant; too great a quantity of silica may increase the density several times, thus nullifying the advantage of increased surface area. Smayda insists that lipid content cannot be regarded as important in flotation; even 40% lipid would decrease the cell density by approximately only 3%. It is also a common observation that although aged populations of phytoplankton tend to sink faster, it is these very cells at the end of their rapid vegetative growth which tend to have higher lipid levels.

Apart from the effects of cell size and shape, and of lipid or other cell constituents, some observers believe that diatoms may show neutral or positive buoyancy. Gross and Zeuthen (1948), working with *Ditylum*, showed that healthy cells have positive buoyancy. This was due to an active ionic regulation; a lower concentration of bivalent ions (sulphate, calcium and magnesium) was maintained in the cell sap, though the cell sap was isotonic with the seawater. Many observers believe that neutral buoyancy, actively maintained, is a common feature of diatoms, though direct evidence is slight. Eppley, Holmes and Paasche (1967) imply that *Ditylum* has neutral buoyancy which changes with time, darkness inducing sinking. Strickland, Holm-Hansen, Eppley and Linn (1969) also claimed that a large-scale culture of *Ditylum* showed neutral buoyancy at a certain stage of growth. Another diatom, *Stephanopyxis*, has also been described as exhibiting buoyancy. More direct evidence comes from the study by Beklemishev, Petrikova and Semina (1961) who found a total lack of magnesium ion in the cell sap of the large diatom, *Ethmodiscus rex*, and a reduced concentration of other ions (calcium, sulphate and the monovalent ions, sodium and potassium) as compared with seawater. Certainly, active neutral buoyancy may not be universal in diatoms; a difficulty arises for freshwater species since ion concentration in fresh-water is so low.

Smayda (1970) doubts whether neutral buoyancy is genuine; perhaps diatoms are slowly sinking. If sinking is the rule, it is extremely slow. For entirely undisturbed waters several months might be required for some species to sink 100 m! The observed range in sinking rates of living vegetative cells varies from 0 to 30 m/day. Dead cells descend at a much higher rate; for instance, a large *Coscinodiscus* at 60 m/day and even higher rates for the very large diatom, *Ethmodiscus rex* (Smayda, 1970). Palmelloid stages of certain algae, with their relatively large diameter, may also descend very rapidly (cf. Bernard and Elkaim 1962; Bernard, 1963). Smayda points to a marked increase in sinking rate in many diatom species depending on whether they are in the so-called neutrally buoyant stage, the actively dividing stage, or the senescent stage. Boleyn (1972) described in cultures of *Ditylum* normal vegetative cells, approximately 40 μm in diameter; enlarged cells, 70 μm in width, termed "pseudo-auxospores", since these were not apparently the product of zygotic fusion (see p. 171); and "spheroidal" cells which were plasmolysed and equivalent to the resting spores of Gross (1937). Boleyn found that sinking rates were lowest in the "pseudo-auxospores", senescent cells which had passed the logarithmic growth stage had a higher sinking rate, and resting spores an even faster rate, approaching five times that of normal cells. Nutrient availability had no effect on the sinking rate of cells in the logarithmic phase of growth; senescent cells, given nutrient enrichment, tended to have a lower sinking rate providing light was available. It is generally accepted that the physiological condition of diatoms is a potent factor in flotation. Despite the paucity of direct evidence, it would appear reasonable also to continue to accept active buoyancy mechanisms, at least for some species.

Although "healthy" phytoplankton has a low sinking rate, retention within the euphotic zone may not be the only consideration. Munk and Riley (1952) have suggested that diatoms when sinking may achieve an accelerated absorption of nutrients, and that to maintain reproduction a minimum sinking rate is essential, even though in nutrient-rich waters the rate may be extremely low. Smayda further argues that the various behavioural and morphological characters which modify sinking rate are not so much concerned to achieve suspension but rather to allow vertical movement *within* the euphotic zone. During their time in the euphotic zone, the cells are twisting and turning, increasing their absorption of nutrients. Circulation patterns, especially convection cells in the upper layers of the ocean, favour phytoplankton being retained in the euphotic zone and the various spiky projections characteristic of many phytoplankton cells may render the path of descent of the diatom very complex, the cell zig-zagging and twisting to and fro instead of sinking in a more or less direct path. Chain formation may help in the complex movement of diatoms, the pattern facilitating rotation of the colony. The complex paths of single cells and colonies thus promote the exploitation of the whole euphotic zone. The cell projections may have other minor functions; in some species they contain chloroplasts so that the photosynthetic activity of the cell as well as nutrient absorption is increased, e.g. *Chaetoceros* species of the sub-group *Phaeoceros*. There is very little evidence for the view that spines on phytoplankton act as protective devices against the grazing activity of zooplankton.

Some diatom cells are relatively large. *Ditylum* can exceed 100 μm in length and may approach 150 μm; several species of *Rhizosolenia* have lengths greater than 500 μm; some species (e.g. *Rhizosolenia styliformis*) exceed 1 μm (1,000 μm). A number of species of *Coscinodiscus* are also of considerable size: *Coscinodiscus oculus-iridis* often exceeds 200 μm in diameter, *C. asteromphalus* even 300 μm. The tropical diatom *Ethmodiscus* may reach 2 mm in diameter. At the opposite extreme some diatoms are less than 5 μm in size. Just as we have tended to overlook many of the smaller species of other algal groups making up the nanoplankton, so some of the very small diatoms have only recently been described. Collier and Murphy (1962) draw attention to some exceptionally small species; Thorrington-Smith (1970b) also lists some small, little known, planktonic diatoms friom the Indian Ocean.

As Harvey (1950) has described, the volume of diatom species shows a very great range corresponding to the variation in cell dimensions. He quotes the smallest species as having a cell volume of 20 μm^3 and the largest $2 \times 10^7\ \mu m^3$ – a range of six orders of magnitude. Even this may not take full account of the most extreme sizes of diatoms which have been more recently discovered; Smayda (1970) suggests that the diameter of cells may vary from about 2 to 200 μm (cf. also Chapter 9). The variation in size does not apply to species alone; individuals of one species can vary greatly in volume. Harvey (1950) suggests that *Ditylum brightwellii* may show a thirty-fold increase in size. The major cause of the change in cell size is the method of reproduction. The normal method is by vegetative cell division, with the production of two daughter cells. The two valves of the frustule normally part slightly, the nucleus and protoplasm divide and new valves are formed within the valves of the parent cell (cf. Boleyn, 1972; Stosch and Drebes, 1964). At each division, therefore, when the valves separate, one of the original valves passes to each daughter cell (cf. Fig. 4.31). The hypotheca of the original diatom thus forms the epitheca of one daughter cell: the average size of a continuously dividing population of diatoms must therefore decrease. Hendey points out that the progressive decrease in cell size may have an important influence on the precise shape of the frustule, affecting the coarseness of striation and the precise outline; this can introduce difficulty in taxonomic classification.

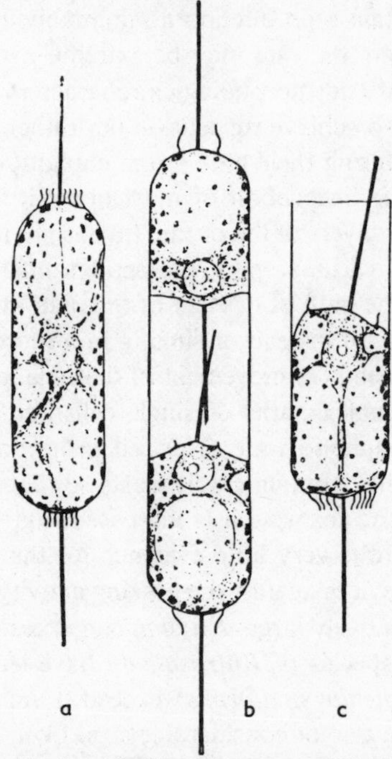

Fig. 4.31. *Ditylum*: (a) a normal vegetative cell, (b) cell in process of division, (c) a recent daughter cell from such a division (from Gross, 1937).

Reduction in size with repeated cell division is common to all but a very few diatoms, but according to Lewin and Guillard (1963), a very few species apparently maintain constant size in culture and presumably can do so in nature. The plasticity of the cell wall and the particular construction of the girdle in diatoms that have relatively little silica allow the cells to resume their original size and shape. A culture of *Nitzschia pallia* is cited in which, after the cells had decreased in size under poor culture conditions, they maintained a constant reduced size when more favourable conditions obtained. However, in general the average size of a continuously dividing diatom population decreases. Wimpenny (1936a, 1946) and Lucas and Stubbings (1948) followed the decline in cell size of natural populations of a diatom (*Rhizosolenia styliformis*). Indeed, stocks of diatoms in the North Sea distinguished by size differences have been identified and the history of the populations followed. Cushing (1953, 1963) has used the decline in average size of a diatom population as an indication of division rate.

The restoration in mean size of a diatom population seems frequently to be consequent upon the formation of auxospores. In this process the old cell valves are thrown off, the protoplast subsequently increasing considerably in size. In species such as *Ditylum brightwellii*, no obvious changes precede the swelling of the cell, although there are suggestions of nuclear changes; perhaps a type of autogamy occurs (Gross, 1937). Following the increase in size of the cell, a membrane consisting largely of pectic substances together with a little silica is formed round the protoplast (the perizonium); subsequently new valves of normal kind, but of larger size, are developed. A sudden restoration of the cell size of the species to its characteristic level thus occurs.

In all species where auxospore formation has been identified it appears to be associated with reduction of cell size to a minimum due to continued vegetative division. Lewin and Guillard (1963) point out that some species can apparently "go below" this minimum, but then can continue reproducing only by vegetative division until the clone dies out. (There are claims that very rarely some clones can continue division at more or less constant size). Hendey (1964), in emphasizing that cell size has a profound influence on auxospore formation, quotes Geitler who found three categories of cells distinguishable on a size basis: (a) those of maximum size, capable only of vegetative division, so that the cells are gradually decreasing in average dimensions; (b) cells of small size, which under particular conditions are capable of forming auxospores, though they may continue binary fission; (c) very small cells, incapable of forming auxospores, but which may continue vegetative division until death apparently ensues.

While auxospore formation is conditioned by cell size, other internal physiological conditions may play a part, and external conditions such as nutrient concentration and light can "trigger off" the process. Dilution of culture was often said to initiate auxospore formation. Gross (1937) concluded that once a minimal size of cell had been reached, increasing density of culture, which would involve reduction of nutrients as well as possible release of metabolites and perhaps reduced light intensity, promoted the production of auxospores. The complex mechanisms controlling auxospore production may vary with the species.

Our knowledge of auxospore formation is mainly derived from cultures, where with some species auxospores can be produced frequently. For example, Soli (1963) quotes 8-14 days for auxospore formation in *Chaetoceros*. On the other hand, Subramanyan believes that the process may be long delayed in nature, and Hendey holds that for many species it occurs relatively infrequently. Differences in frequency with culture condition are indicated by Ignatiades and Smayda's (1970) results with *Rhizosolenia fragilissima*.

In some diatoms, in contrast to *Ditylum*, more complicated changes in cell morphology are seen in auxospore formation. For example, Seaton (1970), describing this process in *Rhizosolenia hebetata*, states that cell diameter is again important; a minimum cell diameter of $<5\,\mu$m is necessary, as compared with a normal cell diameter of *ca.* 20 μm. In the development a bladder-like formation is seen at the side of the cell with two protoplasmic masses which apparently fuse together. The bladder tends to round off and a weakly silicified pseudocalyptra appears. Later at the opposite end, a heavily silicified calyptra develops which finally separates the auxospore bladder from the parent. Much later, with an increase in cell length, a fully silicified calyptra replaces the membraneous pseudocalyptra. Two forms, *Rhizosolenia hebetata* forma *semispina* and *R. hebetata* forma *hiemalis*, both seem capable of developing in this manner; there appears to be some plasticity of the species. The diameter of the new cell formed from the auxospore may vary from 19 to 27 μm. Seaton points out that very small cells, capable of forming auxospores, often appear to have the protoplast divided into 16 or 32 "packages". These lack chloroplasts and may represent microspores which could develop into flagellated swarmers, acting as male spermatozoids. In this event, the formation of the auxospore could follow a form of presumably oogamous sexual reproduction (*vide infra*).

The question of sexuality and auxospore formation has been discussed by many investigators, including among more recent authors, Hendey (1964), Stosch and Drebes (1964), Takano (1967), Migita (1967a, b) and Drebes (1966, 1967). The form of sexual reproduction is generally accepted as being variable, although precise observation is difficult. It is much easier to observe the formation of auxospores than to follow precisely the intracellular events which lead to their development. However, at least in several species of diatoms, intracellular changes have been described.

There is agreement that the vegetative cells of all diatoms are diploid. Many authorities suggest that centric diatoms are oogamous, whereas pennate diatoms do not produce oogonia or flagellated male gametes, and sexual reproduction is characteristically isogamous. Hendey (1974), however, points out that investigations indicate only that *some* species of centric diatoms follow a type of oogamous sexual reproduction.

The oogamous reproduction of centric species is highly variable. Young oogonia are usually distinguished by some degree of enlargement of the cell, and normally after each of the two reducing divisions, one nucleus disintegrates so that only a single haploid nucleus and a single oogonium remain. However, in species such as *Biddulphia mobiliensis*, two egg cells are produced. A greater degree of variability applies to the male cells. The spermatogonia may arise directly as vegetative cells as in *Cyclotella tenuistriata*, or from one cell mitotic division may produce a variable number of smaller spermatogonia, each of which apparently undergoes meiotic division to produce four spermatozoids. In some species with small spermatogonia (*Biddulphia mobiliensis*, *Lithodesmium*, *Streptotheca*) all the material including the chloroplasts is used in the formation of spermatozoids while in forms with larger spermatogonia (*Biddulphia rhombus*, *B. granulata*, *Melosira varians*) some cytoplasm and the chloroplasts remain behind. Each sperm cell typically possesses a single flagellum. In some centric species, separate gametes have not been discovered, despite careful investigation; perhaps reproduction is by intracellular autogamy.

Sexual reproduction anticipates the formation of auxospores in *Stephanopyxis palmeriana* (Drebes, 1966). Production of auxospores is as usual partly a function of cell size: cultures of the diatom showed cell diameters varying from 19 to 156 μm (cf. Fig. 4.32), but only cells between *ca.* 19 and 60 μm were capable of undergoing sexual reproduction. Increase in light intensity (0.024 ly/min) and in temperature (21°C) were important in promoting sexual reproduction. In the formation of spermatozoids, the male mother cell produces four, six or eight spermatogonia, each of which then undergoes meiosis to form finally four uniflagellate colourless spermatozoids. Future oogonia may be distinguished by the elongation of the mother cell, the swollen nucleus and the increased number of chloroplasts. Only a single oogonium is formed, one nucleus disintegrating at each meiotic division (Figs. 4.33 and 4.34).

Auxospore formation in the related species, *Stephanopyxis turris*, is very similar

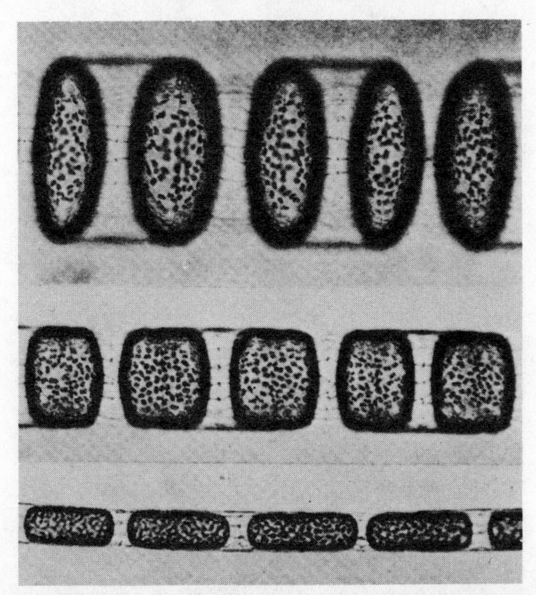

Fig. 4.32. *Stephanopyxis palmeriana*, three chains with cells of different sizes (×200) (from Drebes, 1966).

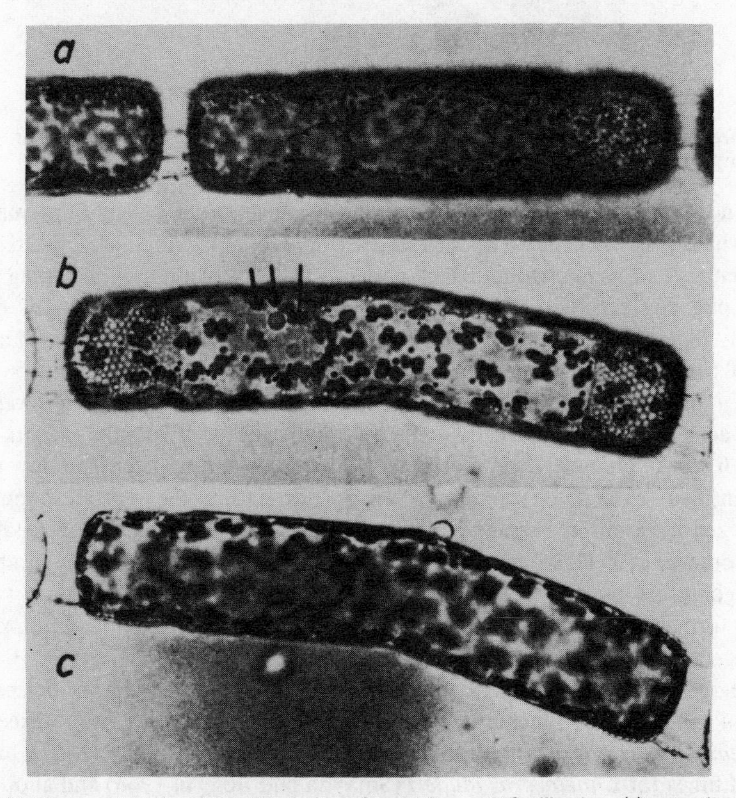

Fig. 4.33. *Stephanopyxis palmeriana*. Formation of oogonium and fertilization: (a) oogonium in meiotic prophase; (b) mature egg, one haploid nucleus (right arrow) with central nucleolus; two pycnotic nuclei (left arrows); (c) slightly bent oogonium, sperm on surface (×450) (after Drebes, 1966).

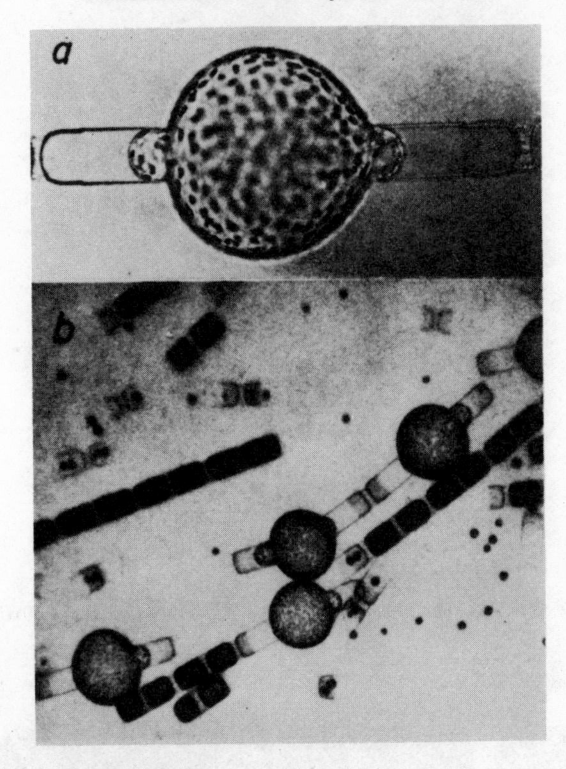

Fig. 4.34. *Stephanopyxis palmeriana*. (a) Zygote inflated to form auxospore (×216); (b) four auxospores and vegetative cell chains (×55) (after Drebes, 1966).

(Stosch and Drebes, 1964). The male mother cell forms four or eight spermatogonia, each of which produces four uniflagellate spermatozoids lacking chloroplasts. Vegetative cells enlarge to form oogonia which undergo meiosis, two nuclei becoming pycnotic to leave one egg cell with a haploid nucleus. Apart from reduced cell diameter, moderately high temperature (15 to 21°C) and light intensity (0.012 ly/min) appear to be optimal for auxospore formation in this species.

Sexual reproduction also accompanies formation of auxospores in *Skeletonema costatum* according to Migita (1967 a and b). Small cell size is essential (Migita suggests cells 5 to 6 μm in diameter) and increased temperature and transfer from low to higher light intensities appear to promote sexuality. In nature the diatom population is characterized by a fall in average cell diameter in spring, with marked vegetative division; restoration of cell size is achieved with auxospore development in summer. The spermatogonia are produced after one or two cell divisions of an ordinary vegetative cell; each forms four colourless uniflagellate spermatozoids. Each oogonium, produced from a vegetative cell, forms a single egg cell (Fig. 4.35).

The degree of cell enlargement after auxospore formation varies with species (cf. Figs. 4.34 and 4.35). Whereas the increase in cell diameter was threefold for *Skeletonema*, it was only twofold for *Melosira moniliformis* (Migita, 1967), about two and a half times for *Chaetoceros lauderi* (Smayda and Boleyn, 1966) and about fivefold for *Rhizosolenia hebetata*, according to Seaton's measurements. The increase in cell volume is obviously greater – for *Chaetoceros lauderi* Smayda and Bóleyn estimate

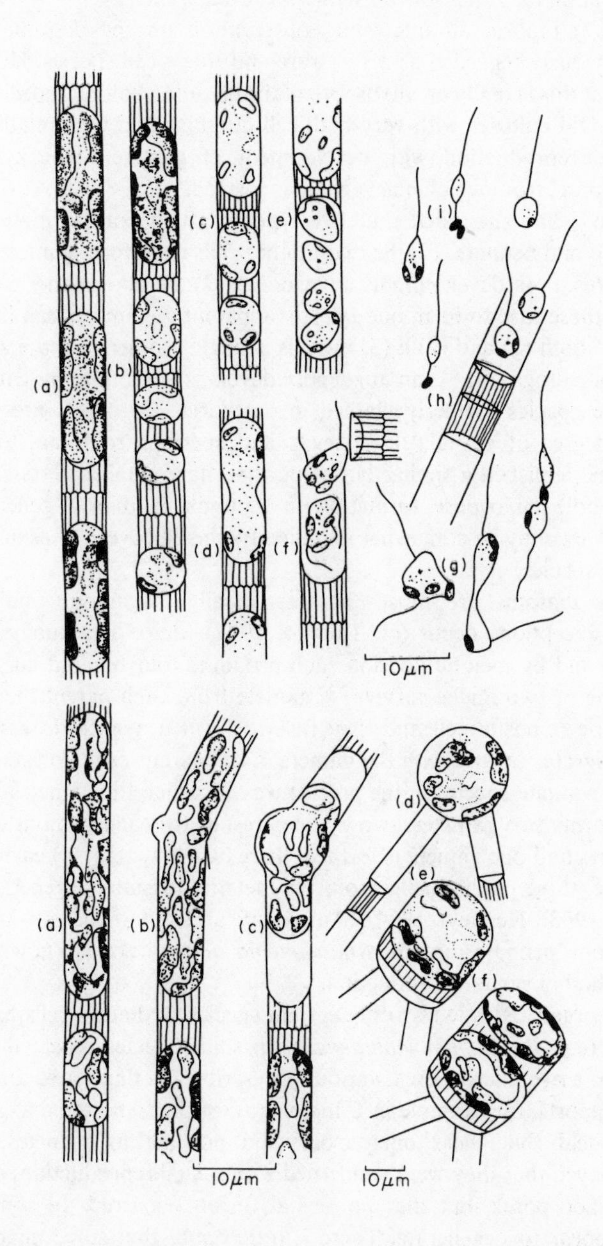

Fig. 4.35. Sexual reproduction in *Skeletonema costatum*. Upper figure – formation of spermatozoids. (a): Narrow vegetative cells which will later develop into spermatogonia. (b), (c): Spermatogonia produced by division of narrow cells. (d), (e): Formation of spermatozoids in each spermatogonium. (f), (g): Unseparated spermatozoids. (h): Liberation of spermatozoids. (i) Liberated spermatozoids. Lower figure – fertilization of oogonium and auxospore formation. (a): Stretched oogonium (egg cell). (b): Fertilization of egg cell. (c), (d): Development of zygote. (e): Fully grown auxospore. (f): Vegetative cell division of auxospore (after Migita, 1967).

seven times, Harvey's earlier estimate for *Ditylum* is thirty times, and Smayda and Boleyn's calculation for *Rhizosolenia setigera* is even greater.

Drebes (1967) reports an interesting observation on the diatom, *Bacteriastrum solitarium*, previously regarded as a very rare and single-cell species. His investigations demonstrate that this is really an auxospore of the common colonial species, *B. hyalinum* (cf. Fig. 4.30). Old cultures with very high cell densities and with relatively small cells undergo sexual reproduction with development of gametes. The zygote forms an enlarged auxospore, from which a new chain is formed.

Takano (1967) has suggested that five types of auxospore formation occur in all diatoms, centric and pennate: (1) the two mother cells each form gametes which fuse to form two zygotes, both developing as auxospores; (2) the two mother cells each form a single gamete; these fuse to form one zygote (apparently, gametes can be isogamous or anisogamous in both (1) and (2)); (3) there is a single mother cell; the zygote develops by automixis or autogamy; (4) an auxospore develops without apparently any trace of sexuality (some species formerly classed in categories (3) or (4) are now known to exhibit some degree of sexuality); (5) eggs and spermatozoids are formed. Hendey (1964), who has published a similar but somewhat more detailed classification, believes that on the whole auxospore formation in diatoms exhibits a general decrease in sexuality; this view may be somewhat modified by the discovery of sexual reproduction in other centric species.

The pennate diatoms are most characteristically isogamous and their gametes amoeboid but exceptions occur (cf. Takano, 1967). Pairs of sexually mature diploid cells conjugate and by meiotic division each produces four haploid nuclei, but in most species only one or two nuclei survive. A gamete from each parent then approaches a gamete from the opposite cell and they fuse to form a zygote; in some species one gamete only migrates to fuse with the gamete in the parent cell. Considerable variation exists between pennate species in the precise way in which the gametes are formed: (a) each gamont forms two gametes, two zygotes result; (b) each gamont cuts off a small cell which aborts and one gamete is formed, there is one zygote; (c) each gamont forms a single gamete, three of the nuclei from the meiotic divisions degenerating (cf. Lewin and Guillard, 1963; Hendey, 1964). Autogamy and automixis are known in a few species. Only one pennate diatom, *Rhabdonema adriaticum*, is known to exhibit the oogamous method of sexual reproduction.

Numbers of organized bodies within certain species of diatom cells have been called "microspores" (e.g. *Biddulphia mobiliensis*). In some species repeated division of the cell gives rise to these microspores, variously described as flagellated and as amoeboid, and there are reports, for example in *Chaetoceros*, of their liberation as swarming cells. While Gross held that these microspores did not act as gametes, Braarud and Wimpenny believed that they were concerned with sexual reproduction; others held that they acted as zoospores, but that no sexual fusion occurred. In some diatoms the microspores appear to degenerate. There is little doubt that some microspores act as spermatozoids, but as Hendey (1964) has emphasized, their function in general is uncertain and a number of different phenomena have probably been included under the general term of microspore production.

Under optimal conditions diatoms can divide exceedingly rapidly; Spencer (1954) has shown a maximum doubling rate for *Phaeodactylum* (*Nitzschia*) *tricornutum* of 10 to 12 hours. Soli (1963) gives a similar maximum (11 hours) for the doubling time in

Chaetoceros didymus. Some discussion of the growth and division rates of marine phytoplankton and their dependence on complex relationships between light, temperature, nutrient conditions and other factors, has been included in Chapter 3. Lanskaya (1963) emphasizes that even when optimum conditions exist, diatom species show considerable variation in rates of division; the great range in division rate for any one species was also confirmed. Some diatoms show a maximum doubling rate of <10 hours, though Lewin and Guillard (1963) question whether these excessively high reproductive rates are maintained, or whether they result from a period of suppressed growth. The highest division rates obtained by Lanskaya in mass blooms were *Skeletonema*, with a doubling in 3 hours, and *Chaetoceros socialis*, doubling in 8 hours. Comparatively slow rates under optimal conditions were typical of *Coscinodiscus janischii* (30 hours) and *Rhizosolenia calcar-avis* (28 hours). In culture the range for any one species could be considerable, e.g. *Skeletonema* 3 to 56 hours; *Chaetoceros socialis* 8 to 84 hours, the rate varying widely in a species with the time of year.

Raymont and Adams (1958) obtained a wide range in division rate in large-scale cultures of *Phaeodactylum*. Although a maximum rate of one division in *ca.* 24 hours could be achieved, under especially poor light conditions the diatoms appeared to divide only once in 28 days or more, though the culture was still healthy. Further work on mass culture of this species by Ansell, Raymont and Lander (1963) and Ansell, Raymont, Lander, Crowley and Shackley (1963) confirmed the great range in doubling time; whereas frequently a doubling time of less than 24 hours was achieved, at the other extreme cultures divided only once in 5 or 6 days or – very exceptionally – in 10 days. There seems no reason to suppose that *Phaeodactylum* is unusual in this characteristic. Indeed, though less attention appears to have been devoted to minimal rates, data such as those of Lanskaya and Smayda tend to confirm the results.

A maximum reproductive rate in the natural environment of a doubling in 1 to 2 days would appear to be a reasonable estimate (cf. one division in 36 to 38 hours – Harvey, 1950) and would agree with the more recent estimate by Ignatiades and Smayda (1970a) of a maximum rate in the field of 0.8 division per day. Cushing (1975) believes, however, that the maximum field rate has been under-estimated. He obtained a rise in division rate during a spring increase in the North Sea reaching more than one division per day (cf. Chapter 8).

For many diatoms the reproductive rate is exceedingly reduced or almost halted during winter months in higher latitudes, though the cells appear to be healthy. The ability of diatoms to survive over unfavourable periods, as, for example, over winter in darkness, has already been discussed (Chapter 3). For many diatom species, although only a few cells survive in healthy vegetative condition, and these display little growth, as soon as conditions become favourable, rapid reproduction occurs leading to a bloom. For oceanic diatoms this would appear to be the means of survival of the species. Some of the diatoms usually inhabiting neritic areas, however, are known to form resting spores. In typical resting spores the protoplasm forms a more or less compact body, usually dark in colour, and with comparatively thick walls. Such cells, being dense, sink into deeper layers in a neritic area and await a new season of favourable conditions before they "germinate" and give rise to a new bloom. Resting spores are probably important among the diatoms inhabiting polar regions since they remain dormant in ice. Whereas auxospores apparently do not act as resting stages, resting spores tide some neritic diatoms over unfavourable periods. Apart from the formation of a dense protoplast, including the chloroplasts and nucleus, but with loss of the vacuole, there is usually in a resting spore a special type of siliceous valve different from that characteristic of the species. Indeed, the structure of the resting-spore valve may appear so different and they may be so resistant that many of them, discovered in fossil deposits, have been listed as separate genera and species. Gross (1937, 1940) described the formation of resting spores with the loss of cell sap and strong silification of the cell valves in *Ditylum*, and also germination of the spores, with the gradual increase in cell volume, expansion of cytoplasmic strands and, finally, cell division.

The formation of resting spores in *Stephanopyxis turris*, described by Stosch and Drebes (1964), shows that the spores develop in pairs from the division of adjoining vegetative cells in the chain. During cell division, one new valve develops in each dividing cell of a different type ("heterovalve"), being of a thicker heavier construction, with differences in the pores and in the spines. Division is prolonged, and during the process polar streaming of the protoplast occurs towards the modified heterovalve. Ultimately most of the cell contents are included in the half-cell with the heterovalve, leaving only a few chloroplasts and a nucleus in the other half. As cell division is completed, the unmodified valve is separated with its very reduced protoplast from the major part which forms a compact spore (Fig. 4.36). In germination, elongation and division occurs with the formation of a new valve typical for the species.

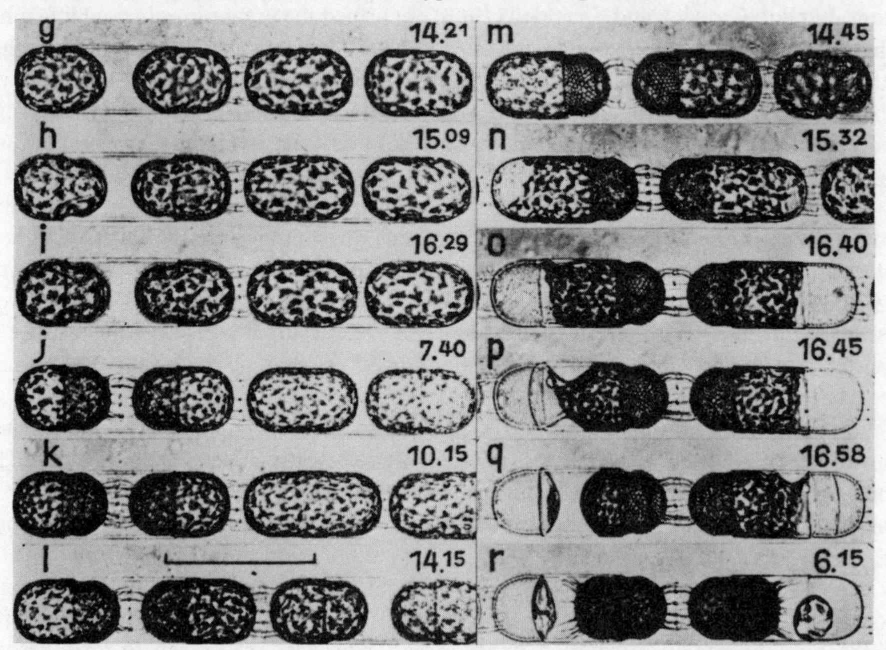

Fig. 4.36. Resting spore formation in *Stephanopyxis turris*. Note the formation of heterovalves in the left-hand pair of cells (k, 1). Subsequent loss of most of contents in half-cell not forming spore (m-r) (after Stosch and Drebes, 1964).

Drebes (1966) describes a similar process of resting-spore formation in *Stephanopyxis palmeriana*, with the development of a modified heterovalve. Size does not appear to be significant in inducing resting-spore formation in *S. palmeriana*; spores developed from cells ranging from 19 to 90 μm. Temperature was not a major factor, but nutrient depletion, especially phosphate lack, was important. With nutrient deficiency, a temperature of *ca.* 12°C seemed most effective. *S. turris* also produced resting spores over a very wide range of cell size (30 to 115 μm), but in this species somewhat lower temperature and moderate light intensity (0.006 ly/min) appeared to be optimal for spore production. Durbin (1978) has confirmed the different morphology, especially the heavy sculpturing of one valve in resting spores of the diatoms, *Thalassiosira nordenskioldii* and *Dentonula confervacea.* Only a single spore was produced from one vegetative cell.

According to Gross (1937), over-crowding appears to promote spore formation in *Ditylum* cultures; in the sea important factors are poor light intensity, reduced nutrient conditions and possibly low temperature. Hendey points out that in the germination of resting spores, particularly as described by Gross, considerable enlargement of the protoplast occurs before the formation of new valves, so that spore formation might be a method of restoring cell size in certain species, but their main function is to tide a "seed" stock over unfavourable periods (cf. Chapter 3).

11. Pyrrophyceae

This is an important marine phytoplankton group of algae, often known as the dinoflagellates or, less accurately, as the peridineans. There are very few coccoid forms; the luminescent planktonic *Pyrocystis* with very large spherical cells is a well-known warm-water genus. In the reproduction of some species of less typical form, swarmers resembling free living dinoflagellates have been recognized (*vide infra*). Amoeboid forms are very rare, but *Dinamoebidium* is one holozoic marine genus lacking chloroplasts; some other species may also become temporarily amoeboid. One or two filamentous genera (e.g. *Dinothrix*), reproducing by dinoflagellate-like swarmers, are known. The vast majority of species of Pyrrophyceae, however, are typically unicellular and biflagellate. Many of the symbiotic algae (zooxanthellae) of corals, sea anemones and other Coelenterata, and also of Radiolaria, have been identified as dinoflagellates. As with other algal groups, a number of Pyrrophyceae do not possess chloroplasts and cannot perform photosynthesis; many are holozoic, some saprozoic. It seems likely that even some normally photosynthetic species practise holozoic nutrition to some degree. The Blastodinaceae are parasitic; the cells are colourless, and the majority are ecto- or endoparasites of Metazoa; *Blastodinium* and *Oodinium*, for example, are well-known parasites of copepods. Reproduction is by means of gymnodinioid (dinoflagellate-like) zoospores. Very occasionally, in a few dinoflagellates chain formation occurs, for instance in *Polykrikos* and one or two species of *Ceratium*; normally the cells remain unicellular.

The usual pyrrophycean cell has two flagella; one ("transverse") is band-like forming a type of helical ribbon with a row of fine long hairs on one side. The other ("longitudinal") is frequently said to be simple, though Chapman and Chapman (1973) report that it is pleuronematic, with short fine lateral hairs; these are absent in *Amphidinium.* Leadbeater and Dodge (1966) state that the longitudinal flagellum of *Woloszynskia micra* has two rows of short fine hairs. The dinoflagellate cell shows other characteristic features; trichocysts are frequently present (cf. Leadbeater and Dodge, 1966). Typically also there are structures known as pusules, resembling contractile vacuoles but showing no contractile activity (cf. Dodge, 1973). The nucleus is somewhat peculiar, being prominent in the cell, with the chromosomes typically visible and remaining in a condensed state even in the interphase condition; a mitotic spindle appears to be absent (cf. Dodge, 1960, 1963, 1973). In some species, such as *Prorocentrum*, the nuclear membrane is not obvious. A further peculiar feature is that while DNA has been identified in the nucleus, histones are apparently absent.

The food reserves include starch and lipid material. Some dinoflagellates (e.g. *Gonyaulax*) liberate toxins which can be highly injurious to both invertebrates and vertebrates (cf. Chapter 3). The chloroplasts, typically few and plate-like in the Desmophyceae, are normally small and numerous in the Dinophyceae. Most characteristic, however, is the contained pigment. Chlorophylls *a* and *c* are present; Jeffrey (1972) points out that the proportion of *c* to *a* is highest in dinoflagellates (cf. Table 4.1 on page 135). Also occurring are β-carotene and a group of xanthophylls which appear to be peculiar to Pyrrophyceae, including usually dinoxanthin, peridinin and diadinoxanthin. The cells tend therefore to have a golden-brown colour, though this can vary considerably from species to species. Riley and Segar (1969), studying two species of Pyrrophyceae, confirmed the great importance of peridinin among the xanthophylls. Fucoxanthin is usually stated to be absent in Pyrrophyceae, but there appears to be considerable variability between species. Riley and Wilson (1967) demonstrated that whereas *Prorocentrum micans* had a dominance of peridinin and lacked fucoxanthin, *Gymnodinium veneficum* had a considerable amount of fucoxanthin but peridinin was apparently absent. Mandelli (1968) has also identified fucoxanthin in the dinoflagellate, *Glenodinium foliaceum*; Leadbeater and Dodge (1966) found no peridinin in the newly described marine dinoflagellate, *Woloszynskia*.

One of the common distinctions in dividing the Pyrrophyceae relates to the external cell membrane. It is usually held that some dinoflagellates have only a firm periplast ("non-thecate" or "naked" forms) while the rest have a comparatively thick cellulose cell wall ("thecate" or "armoured" types). Leadbeater and Dodge (1966), however, have drawn attention to the presence of very fine plate-like structures in the external cell membranes of *Woloszynskia micra*; these would not normally be visible by ordinary light microscopy, and this organism would previously have been classed as "non-thecate". The authors raise the question whether the distinction between thecate and non-thecate species is valid; the thickness of the wall is essentially a matter of degree. Dodge (1973) describes the remarkable increase in thickness of plates in the middle ("thecal vesicle") layer of the cell wall in dinoflagellates. In the thecate species, *Aureodinium pigmentosum*, an apparently non-motile phase occurs, which lacks the theca (Dodge, 1967).

The Pyrrophyceae are fairly obviously divisible into two groups, the Desmophyceae and Dinophyceae.

(1) *Desmophyceae*. This is much the smaller group but at least two genera, *Exuviaella* and *Prorocentrum*, are important and widespread phytoplankton forms. In these algae the two flagella arise from the anterior of the cell, the thread-like flagellum being directed forwards, and the band-like one bending round at right angles after emerging from the cell. (Fig. 4.37). There are no transverse or longitudinal furrows as in the Dinophyceae. The cell wall is not composed of separate plates but has a longitudinal suture which divides it into two valves. Reproduction occurs by longitudinal division while the cell is motile; during division the suture dividing the two valves separates so that after fission each daughter cell retains one valve from the parent.

(2) *Dinophyceae*. A prominent transverse groove (the girdle) divides the cell into an

anterior part (the epicone) and a posterior part (the hypocone). The girdle houses the band-like transverse flagellum which arises through a pore; the wave-like beating of the flagellum may cause the cell to spin to some extent on its axis. There is also a longitudinal furrow (sulcus) which runs from the posterior end of the cell part way forwards. From a pore in this sulcus arises the other (longitudinal) flagellum which runs back and usually extends beyond the cell, trailing behind in the water (Fig. 4.38).

Fig. 4.37. Left. 1a: *Exuviaella marina*, 36 μm long; 3a: *E. perforata*, 22 μm long and 5a: *Prorocentrum micans*, 37 μm long (from Lebour, 1925).

Fig. 4.38. Below. Diagram to show structure of a naked dinoflagellate: e = epicone; h = hypocone; g = girdle; s = sulcus; tf = transverse flagellum; lf = longitudinal flagellum; afp = anterior flagellar pore; pfp = posterior flagellar pore (from Lebour, 1925).

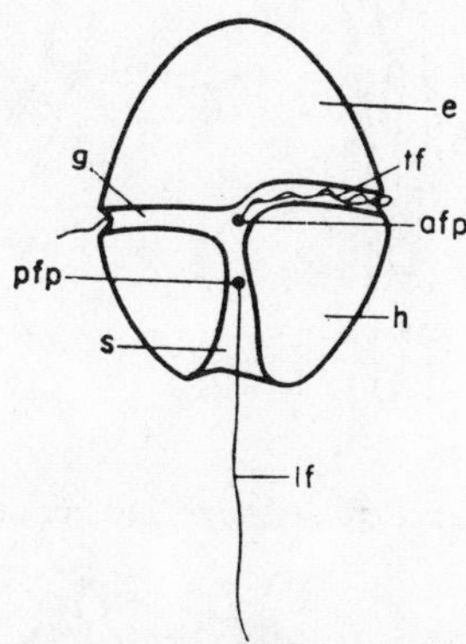

In a number of genera (e.g. *Gymnodinium*, *Gyrodinium*, *Amphidinium*) the cell is furnished only with a firm periplast. Although the investigations of Leadbeater and Dodge, already referred to, throw some doubt on any sharp distinction between such non-thecate and thecate species, these genera contrast with the majority of dinoflagellates which possess an obvious cellulose wall. In typical thecate forms this is divided into a number of separate plates, frequently arranged in three series on both epicone and hypocone. The plates are sculptured with minute pores and frequently with small spines. The precise arrangement of plates is typical of the species. One of the most widely known genera is *Ceratium* (with a very large number of species) (Fig. 4.39); *Peridinium* and *Gonyaulax* are other important genera (Fig. 4.40); *Glenodinium* is unusual in having small indistinct plates. *Dinophysis* has numerous though regularly arranged plates; the genus is also characterized by the girdle being far forward, so that the epicone is small and the cell possesses typical wing-like expansions (Fig. 4.41).

Reproduction in Dinophyceae is by oblique cell division, either with the cells in motion or resting. In some species the protoplast leaves the theca before division so that the whole theca is renewed; in others, half of the original theca is retained by the daughter cell and the remainder regenerated. In *Dinophysis* division is apparently longitudinal. Sexual reproduction is known in *Ceratium* and *Glenodinium*. In addition,

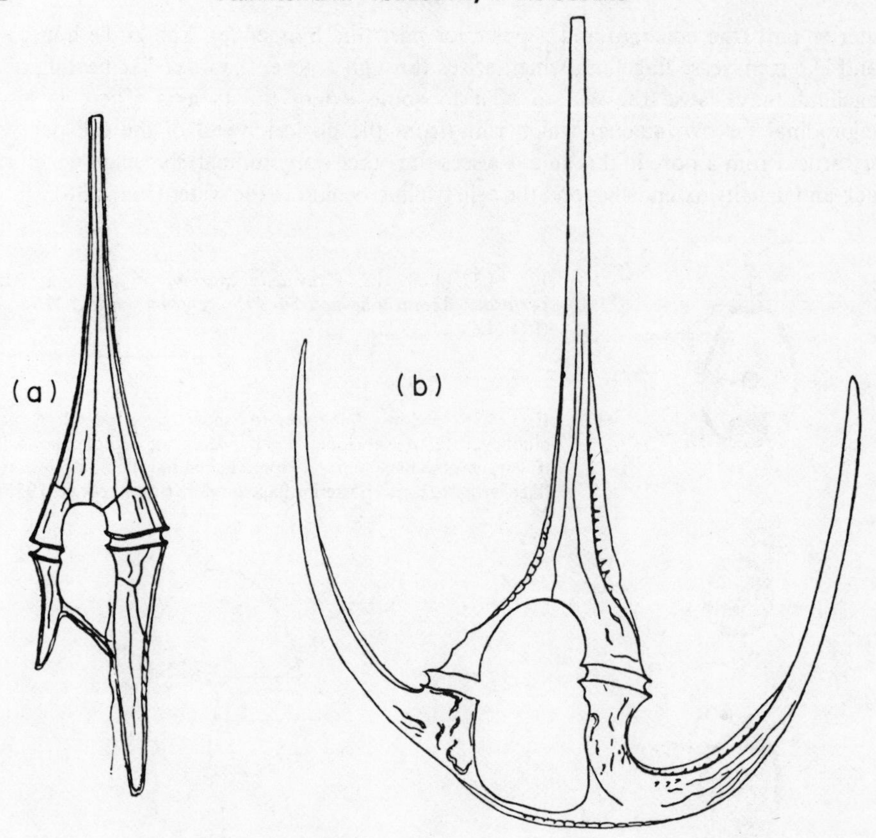

Fig. 4.39. (a) *Ceratium furca*, 150 μm long. (b) *Ceratium tripos*, 270 μm long (from Lebour, 1925).

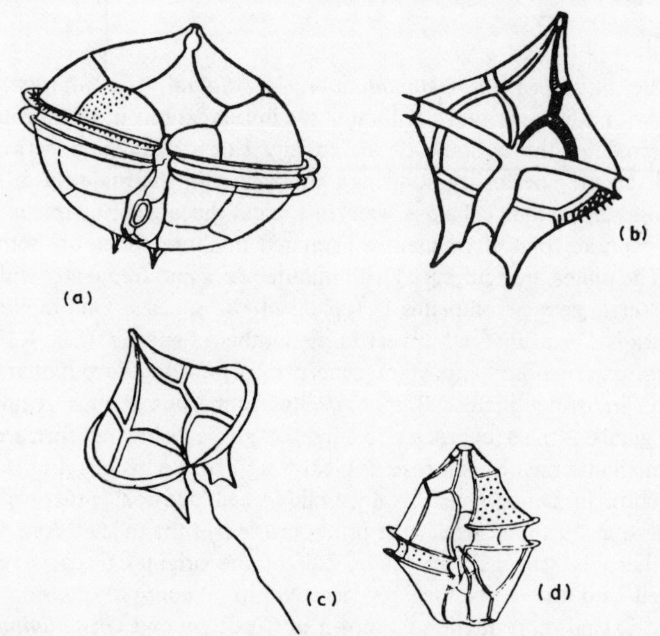

Fig. 4.40. *Peridinium*: (a) *P. oratum*, 64 μm across; (b) *P. divergens*, 56 μm across; (c) a living cell of the same species; (d) *Gonyaulax polyedra*, 42 μm long (from Lebour, 1925).

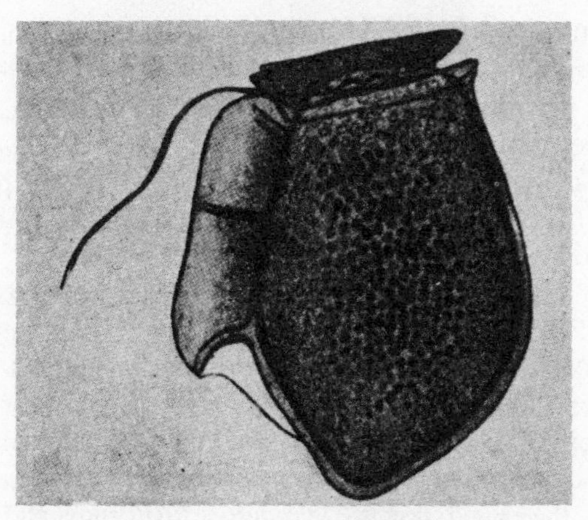

Fig. 4.41. *Dinophysis* – The two flagella are visible, the transverse flagellum is lying in the girdle (after Gran, 1912, in Murray and Hjort).

reproduction by zoospores generally resembling *Gymnodinium* is known for some coccoid, filamentous and parasitic species. The detailed life cycle of some of these Pyrrophyceae is still, however, a matter for dispute.

For example, in the well-known species *Pyrocystis lunula* the dominant vegetative phase is a relatively large globose cell, uni-nucleate, with many chloroplasts, a large vacuole and thin cellulose wall. This stage, usually described as the primary cyst stage, is said to divide to produce 8, 16 or 32 typically crescent-shaped cells which, when released from the parent, are usually termed "secondary cysts". These in turn divide to form 1–8 gymnodinioid flagellated swarmers of small size (Fig. 4.42). These are presumed to develop into the large cyst stage, but this final development has apparently

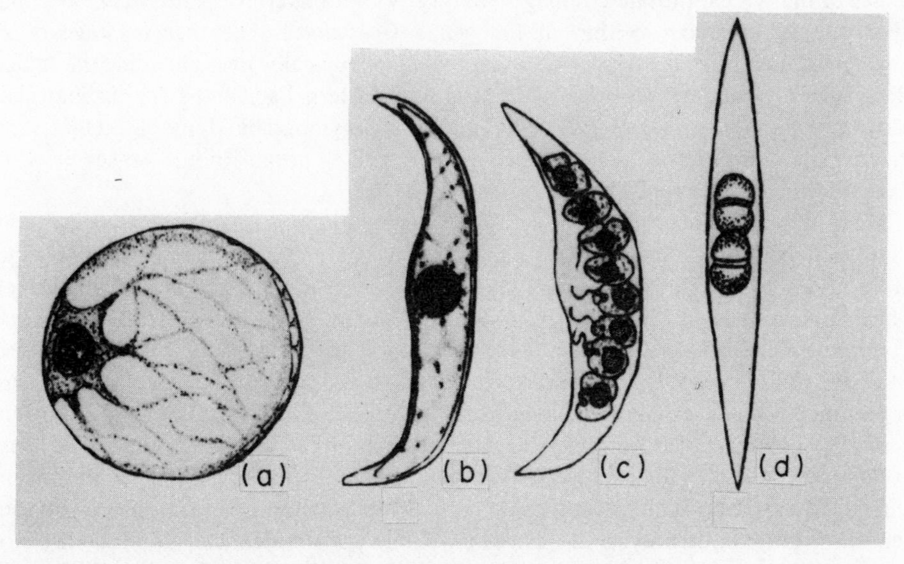

Fig. 4.42. *Pyrocystis lunula*: (a) primary cyst; (b) and (c) stages in life cycle; secondary cysts, 80–150 μm; (d) *Pyrocystis fusiformis*, 600–1600 μm (after Tregouboff and Rose, 1957).

never been proved; there are indications that the swarmers may undergo binary fission first. Some suggestions have been made of a parasitic stage in the life history. This might gain support from the findings of Drebes (1969) of a new dinoflagellate species, *Dissodinium pseudocalani*, ectoparasitic on the eggs of *Pseudocalanus elongatus*. After consuming the egg material, this dinoflagellate forms a special primary cyst, which divides to produce 8, 16 or 32 secondary cysts. Each secondary cyst forms 16 or 32 colourless, flagellated gymnodinioid swarmers.

Swift and Durbin (1971) cultured the crescent-shaped cyst cells of *P. lunula* and observed that they divided to produce 1 or 2 (rarely 4) "aplanospores"; these swelled rapidly during or after release from the parent cyst, and though gymnodinioid, they lacked the transverse flagellum, and the girdle was relatively poorly developed. Apparently, after a brief free swimming existence they formed the crescent-shaped non-motile stage again; no typical large globose primary cysts were obtained in culture. The authors found that the allied forms, *Pyrocystis fusiformis* and *P. noctiluca*, also produced aplanospores which swelled to produce the large vegetative cells again. Bonquaheux (1972) showed that cyst stages of *P. fusiformis* and *P. elegans* produced one or two biflagellate gymnodinioid swarmers which after a short free swimming period, rounded up to form the cyst stage; this in turn divided again to form swarmers. This cycle, repeated every 2 or 3 days, apparently continued indefinitely in culture. It is uncertain how far such differences in reproductive cycles observed in culture may occur in nature.

The investigations of Taylor (1972) dealing largely with a re-examination of unpublished reports by Kofoid and Michener, have shed light on distinct differences in the free swimming stages of *P. lunula* and two other species, *P. pseudonoctiluca* (= *noctiluca*) and *P. fusiformis*. All three have primary and secondary cysts, the primary cysts being the large, planktonic, vacuolated cells with thin cellulose walls. In the latter two species, however, the secondary cysts give rise after division to motile stages of the typical thecate dinoflagellate type, with a pattern of plates in epicone and hypocone very similar to that in the genus *Gonyaulax*. The thecate stage of *P. pseudonoctiluca* is particularly widely distributed in tropical waters including the Indian Ocean, North Atlantic, Caribbean, Central and Eastern Pacific. Taylor distinguishes between *Dissodinium* (*Pyrocystis*) *lunula*, having gymnodinioid motile stages, and *Pyrocystis* (*sensu strictu*) which has *Gonyaulax*-like thecate motile stages (e.g. *P. pseudonoctiluca*) (Fig. 4.43).

Many dinoflagellates are known to undergo encystment. Some species may temporarily surround themselves with a gelatinous coating, presumably a resting stage. More typically encystment occurs with the formation of a thick wall around the cell which corresponds to a resting spore. Smayda (1958) quotes species of *Gonyaulax*, *Ceratium* and *Peridinium* amongst those capable of producing resting spores. De Sousa and Silva (1962) report that in culture *Glenodinium* and *Gonyaulax* can produce more or less thick-walled resting cysts. The work of Wall and his colleagues (Wall, 1970) in particular has expanded our knowledge of resting spores in dinoflagellates. At least twenty-five species are capable of encystment, and the cysts of some recent dinoflagellate species such as *Gonyaulax digitale* can be recovered from shallow bottom deposits. Many modern cysts are extremely similar, if not identical, to fossil stages of dinoflagellates, known for many years, but classified under different genera. Such fossil specimens are known as hystrichospheres. Modern cysts have a thick wall and are fre-

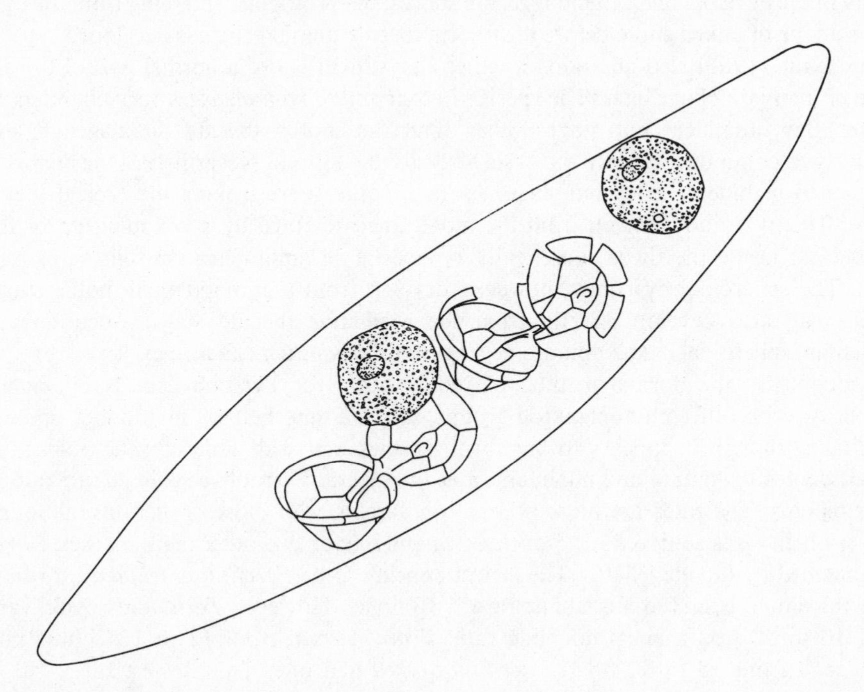

Fig. 4.43. *Pyrocystis fusiformis* forma *fusiformis*, containing two daughter cells with dissociated thecae (after Taylor, 1972).

quently extremely spiny; they lack the complete arrangement of plates of the thecate dinoflagellate. Wall and Dale (1967) emphasize that there may be differences in the appearance of the cyst in apparently the same species, though whether these differences are genetic or due to environmental factors is uncertain.

A large number of cysts belonging to several species of dinoflagellates were recovered from shallow water marine bottom detritus during winter by Wall and Dale (1967) and Wall, Guillard and Dale (1967). Raising the temperature to 16–25°C caused excystment of these resting spores which developed into cultures of such motile thecate dinoflagellates as *Gonyaulax scrippsae*, *G. spinifera* and *G. digitale*. A culture of *Peridinium trochoideum* produced abundant calcareous cysts; this formation of calcareous cysts has been observed in two other genera, although it is uncommon for dinoflagellates. A long period, up to 9 months of encystment, appears to be true of *P. trochoideum*; factors inducing encystment are not obvious since the species has very wide temperature/salinity tolerance limits (Wall, Guillard, Dale, Swift and Watabe, 1970). Possibly prolific spore formation enables this species to colonize inshore marine areas. There is a suggestion that encystment in dinoflagellates may not necessarily be a reaction to adverse conditions in all species; it might be a normal part of the reproductive cycle.

In the excystment of resting spores, Wall and Dale (1970) point out that some dinoflagellates such as *Gonyaulax spinifera* first develop as a motile stage, uniflagellate but later forming two flagella, equipped with a girdle, but lacking thecal plates. This gymnodinioid stage gradually transforms into a typical thecate dinoflagellate. Commonly in the development of dinoflagellate spores, the protoplast emerging from the cyst forms a type of naked spore before the typical thecate dinoflagellate is developed.

Buchanan (1968) also questions whether encystment is not a normal part of the life cycle of many dinoflagellates. The species *Pyrodininium bahamense* is very abundant in Oyster Bay, Jamaica, but stages other than the motile thecate dinoflagellate are relatively uncommon; possibly the cysts settle to the bottom. Nevertheless, the life cycle appears to include an encysted stage – a non-motile spore lacking the typical thecal plates. The cyst may develop from the motile thecate stage by a rounding up of the protoplasm inside the theca; apparently no special cell multiplication follows excystment. The spheroidal cyst may, however, develop from a gymnodinioid motile stage, which may also develop directly from the vegetative thecate stage. According to Buchanan, spheroidal naked non-motile stages of *Pyrodinium* can also exist.

Undoubtedly the normal method of reproduction for Pyrrophyceae is by binary fission. As with other phytoplankton groups, the rate may be very high under optimal conditions, though it appears to vary with species and with time of year. Lanskaya (1963) quotes maximum and minimum rates for several Pyrrophyceae in culture and in mass blooms. The rates are more or less comparable with those of diatoms, although the very high rates achieved for *Skeletonema* and one or two other diatom species were not matched by dinoflagellates. The fastest appears to be *Peridinium triquetrum* which at its maximum achieved a doubling rate in 10 hours. However, *Peridinium* could vary from 10 to 50 hours in its doubling rate; *Prorocentrum* from 12 to 127 hours and *Exuviaella* from 15 to 90 hours. *Ceratium furca* had one of the lowest rates with a maximum doubling time of 48 hours. Lanskaya suggests maximum densities for common dinoflagellate species in the sea are lower than maximum populations of diatoms, and she attributes this to selective grazing, since the division rates are claimed as comparable. There are several indications, however, that rates of multiplication of dinoflagellates in the sea often do not achieve such high values as diatoms. An experiment studying the growth of natural phytoplankton with nutrient enrichment of the seawater showed that *Prorocentrum*, the only common dinoflagellate, had a lower division rate than all the common diatom species, whatever the nutrient addition (cf. Eppley, Carlucci, Holm-Hansen, Kiefer, McCarthy, Venrick and Williams, 1971).

It is difficult to elucidate from laboratory culture the complete life histories of some Pyrrophyceae as they occur in the sea. Although the complexity of the life cycles of some species is well recognized, the role of some phases of the life history in nature is in dispute.

12. Cryptophyceae

These algae consist of an ovoid, slipper-shaped or bean-shaped biflagellate cell, in which there appears to be no cell wall, although a periplast, more complex at least in some species than the plasma membrane, gives the body its typical shape. The two flagella are slightly unequal, but are unusual in that both are pleuronematic (i.e. equipped with lateral hairs). The flagella arise from an apical depression or furrow lined by trichocysts. The body also possesses a contractile vacuole. Only one, or more frequently two, chloroplasts are present; they are comparatively large for the cell and lobed in outline (Fig. 4.44). They vary in colour very greatly among species, from olive-green to yellow, brown, red or even bluish.

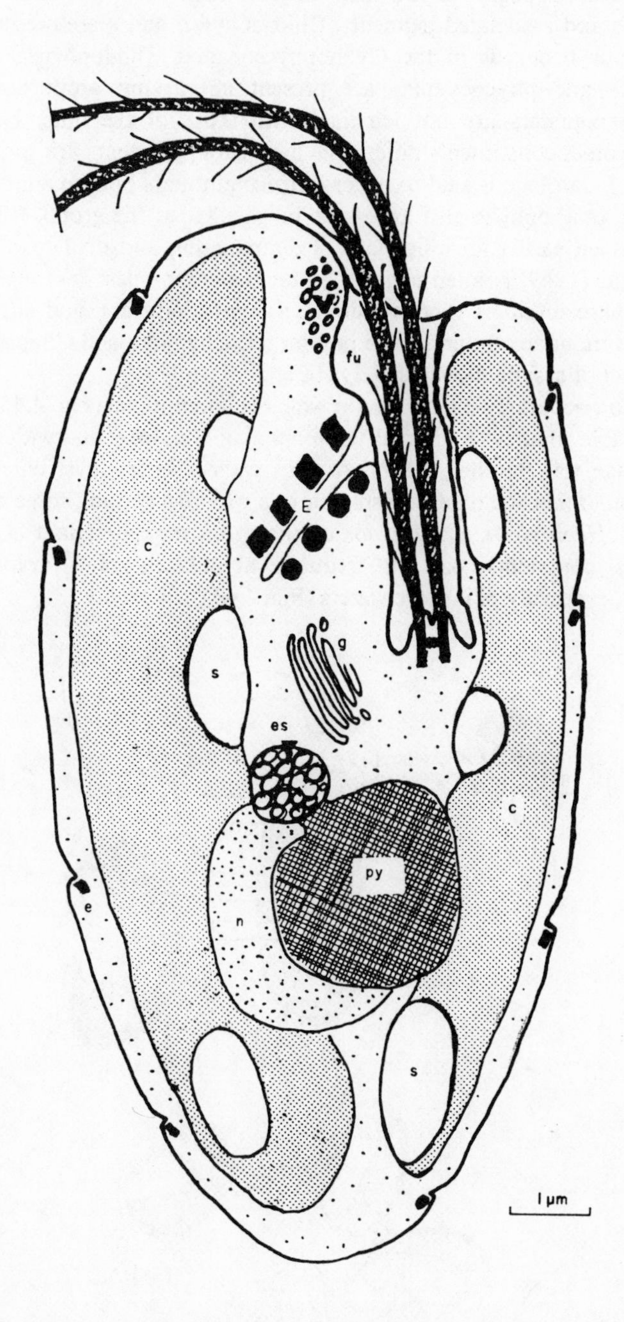

Fig. 4.44. Diagrammatic longitudinal section of a member of the Cryptophyceae showing the anterior furrow (fu) and the position of chloroplasts (c), pyrenoid (py), eye-spot (es), starch (s), Golgi (g), large and small ejectosomes (E) and (e), contractile vacuole (V) (after Dodge, 1973).

This variation appears to be due to the somewhat peculiar assortment of photosynthetic and associated pigments. Chlorophyll *a* and *c* are present, and in addition, biliproteins occur as in the Cyanophyceae and Rhodophyceae. But although phycoerythrins and phycocyanins are present in varying proportions in different species, these pigments are not identical with those of the other two major algal groups; the protein constituents differ. The carotenoid pigments are somewhat unusual among algae: β-carotene is said to occur in trace amounts but the major carotene is -carotene. The xanthophylls also appear to be peculiar to the group. Riley and Wilson (1967) questioned earlier identifications of diatoxanthin and diadinoxanthin, and later Riley and Segar (1969) pointed out that by far the commonest xanthophyll throughout the class is alloxanthin. Crocoxanthin is a minor pigment and in one family of cryptomonads monodoxanthin is also present (cf. Chapman and Chapman, 1973). The storage product of assimilation is mainly starch.

Reproduction occurs by binary fission which is longitudinal (Fig. 4.45); some species appear to be able to enclose the body in a mucilaginous coat and with division form a palmelloid stage (cf. Butcher, 1967). A thick-walled resting cyst characterizes some species. Although this is a relatively small group with few species, some of the species of *Cryptomonas*, *Hemiselmis* and *Rhodomonas* can be very abundant at times. Knight-Jones (1951) comments on the frequent abundance of *Cryptomonas* in the nanoplankton, especially of inshore waters (Fig. 4.46).

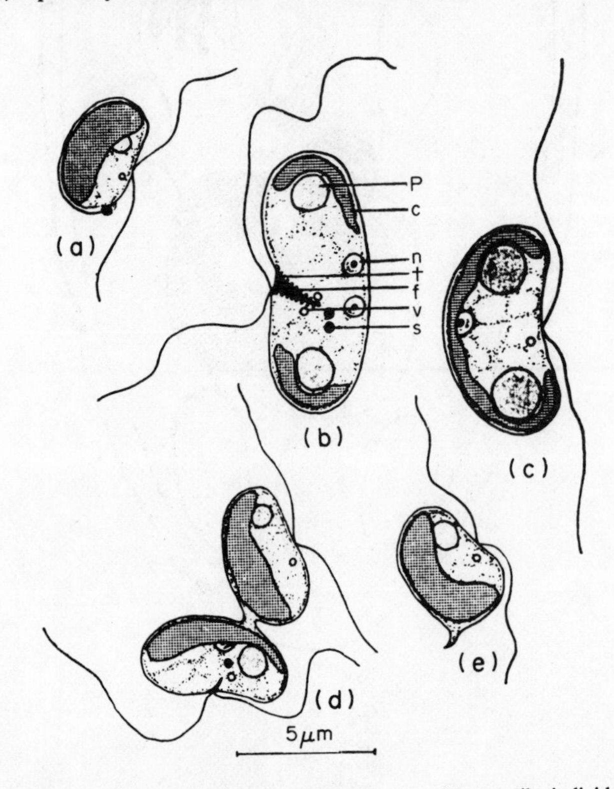

Fig. 4.45. *Hemiselmis rufescens*, a = Young motile cell; b = older motile individual; c, d = stages in fission; e = daughter-cell just after fission; c = chromatophore; f = furrow; p = pyrenoid; s = stigma; t = trichocysts; v = contractile vacuole (from Parke, 1949).

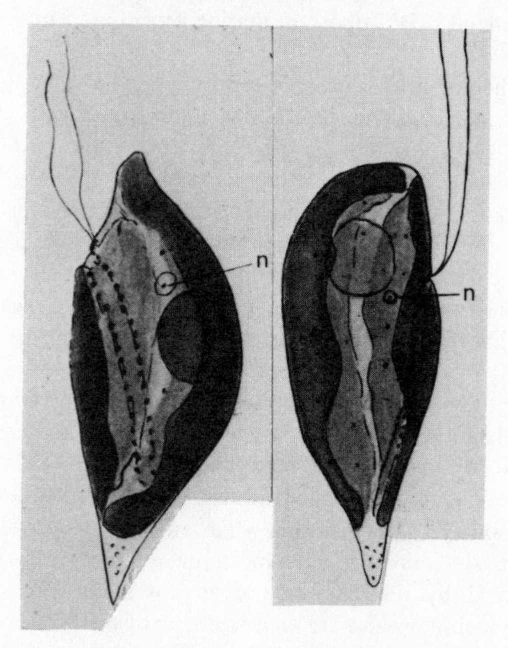

Fig. 4.46. *Cryptomonas acuta*, lateral views (×4500), n = nucleus (after Butcher, 1952).

13. Euglenophyceae

There are unicellular forms, typically fusiform or cylindrical in shape, flagellated, and lacking a cellulose cell wall. The outermost cytoplasmic layer forms a very firm or somewhat elastic layer, often showing complex striations; a specially differentiated protein layer consisting of strips with helical symmetry has been demonstrated. Warts may appear on the exterior and in species which show a particularly firm, well-defined outline, there may be ridges on the surface. When the pellicle is somewhat elastic, the body is capable of flexible, more or less regular changes of shape (metaboly), sometimes described as "euglenoid movement".

The flagella arise anteriorly from an invagination of the cell which consists of a tubular portion called the canal and an enlarged chamber, the reservoir. Typically one or sometimes more contractile vacuoles are present, discharging into the reservoir. In some species a pair of flagella is present, but frequently only a single one emerges from the invagination as a locomotory flagellum (Fig. 4.47b), the second flagellum in this case ends inside the invagination. In one or two genera, where only a single flagellum apparently exists, it is probable that a small second flagellum is present but obscured in the anterior end of the body. A very few species have more than two locomotory flagella. Whether one or more flagella emerge from the anterior end, each bears a single row of lateral hairs.

As in other algal groups, some members are colourless, but in the Euglenophyceae the number of colourless species is very large. Numerous euglenoids live in freshwaters, especially in highly polluted environments with a large organic content. The relatively few marine species are sometimes common in brackish ditches and ponds (cf. Butcher, 1961); others may be found in inshore waters, especially in somewhat polluted estuaries (cf. Palmer and Round, 1965). Important marine genera are *Eutreptia* and *Eutreptiella* (Fig. 4.47a).

The nutrition in the colourless forms is saprophytic, and many species, though they possess chlorophyll and can photosynthesize under good light conditions, have the ability of living saprophytically in darkness. Leedale (1967) states that all photosynthetic euglenoids need at least one vitamin, and that organic substances may enhance primary production, although they have not been shown to have an absolute demand for organic substrates. A few colourless species can take particulate food.

The shape of the chloroplasts varies in different genera. While some have numerous disc-shaped chloroplasts with or without pyrenoids, in other species the chloroplasts are fewer, but larger. The form is very variable: ribbon-shaped, or plate-like, with the margin entire, lobed or even finely dissected (cf. Leedale, 1967). The chloroplasts appear grass-green; they contain chlorophylls *a* and *b*. In addition to chlorophyll, β-carotene and several xanthophylls including antheraxanthin and neoxanthin are present (Leedale, 1967). According to Chapman and Chapman (1973) diadinoxanthin is present in the Euglenophyceae, but the identification of other xanthophylls is doubtful.

The product of photosynthesis is normally paramylum (paramylon), a starch-like carbohydrate but with β1–3 linkage. Unlike starch, paramylum is insoluble in boiling water, is not affected by diastase, and does not stain with iodine. Granules of paramylum are very common and are characteristic of the cells of euglenoids. Oil may also be found.

Euglenoids reproduce typically by longitudinal binary fission. The nucleus is fairly prominent. According to Leedale, mitosis does not involve a spindle; the nucleolus

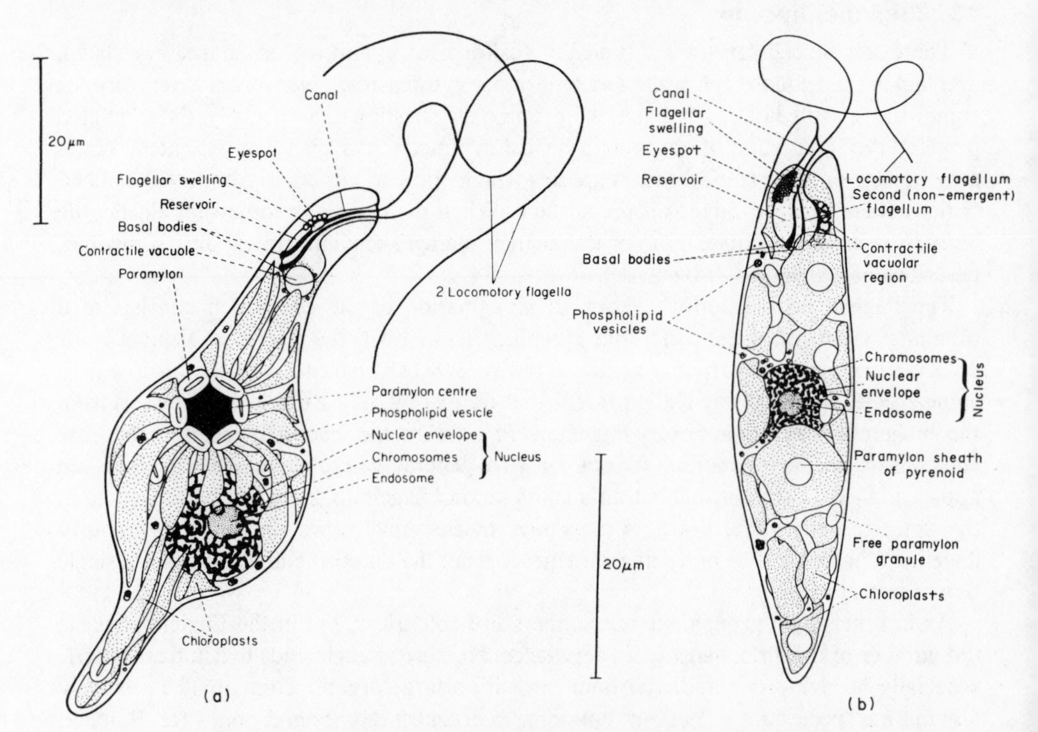

Fig. 4.47. (a) *Eutreptia pertyi*, (b) *Euglena gracilis* (after Leedale, 1967).

persists and divides during cell division. Green photosynthetic species typically have an eye-spot which is independent of the chloroplasts. A photoreceptor may be present as a swelling on the base of the main flagellum.

Euglenoids, though not normally prominent in marine phytoplankton, may be common in inshore waters, at least sporadically and their importance may have been somewhat under-estimated in more off-shore waters. Hulburt (1975) comments on the unusual occurrence of *Eutreptia marina* as a dominant phytoplankton species during April/May 1972 in an area off Grand Banks.

The Biochemical Composition of Algae

In describing the classes of algae, reference has frequently been made to chemical characteristics of cell walls and to the inclusion of particular chemical elements in types of algal cell. Earlier work on chemical composition, summarized by Vinogradov (1953), dealt mainly with the different amounts of particular elements, for instance, the large amounts of silicon in diatoms, and the appreciable quantities of other elements, notably iron and iodine, as contrasted with the relatively small amounts of calcium, magnesium and sulphur. There were also data, including those from Brandt's earlier investigations, for total carbon, nitrogen and phosphorus content of phytoplankton. Lewin and Guillard (1963) pointed to the great range of silicon in diatoms, of the order of 4 to 50% dry weight. Variations were intraspecific as well as interspecific, and the authors refer to Strickland's findings that rate of growth as well as availability of silicon and other factors affect the silicon and the organic content, thus influencing carbon, nitrogen and phosphorus. Table 4.3 indicates the very different silicon contents of diatom species grown in clonal culture. Various phytoplankton species may exhibit unusual ash contents. Boney (1970) quotes the coccolithophorid *Syracosphaera* as containing 36.5% dry weight as ash of which 24% is calcium carbonate. *Tetraselmis maculata* has a relatively high ash content (23.8%) – mainly chloride.

Table 4.3. Silica content of some diatoms grown in clonal culture (from Lewin and Guillard, 1963).

Species	Silica content (SiO_2 as % dry weight)
Navicula pelliculosa (grown with Si limiting)	4.3
Navicula pelliculosa (unlimited Si)	21.6
Skeletonema costatum	30.6
Coscinodiscus sp.	47.3
Phaeodactylum tricornutum	0.5 (mean value)

One of the chief interests in the composition of the phytoplankton, however, as primary producers lies in the proportions of protein, lipid and carbohydrate and the ratio of organic material to ash. Earlier work suggested that the amount of protein in marine phytoplankton was higher than in land plants. Brandt (1898), analysing plankton hauls dominated by diatoms, gave a mean value for protein of 30% of the dry organic matter; Moberg (1926) found a range of from 24% for hauls near shore to 60% for off-shore nettings, though such natural phytoplankton collections are to a variable degree contaminated by zooplankton and more especially by detritus. Ketchum and Redfield (1949), analysing six species of unicellular algae from cultures, confirmed that the amount of protein was large (40–55% of the dry organic weight) and that the amount of lipid (20–25%), though smaller, was also great as compared with the majority of land plants. Such studies with algal cultures avoid the problems of contamination inherent in the analyses of natural plankton collections, but it is now widely recognized that the composition of a single species of alga may be greatly affected by the age and vigour of the culture and by physical and chemical factors of the environment (cf. Chapter 3). Analyses must therefore be made under strictly similar conditions of growth, if comparisons are to be made between the composition of various phytoplankton species, and the chemical composition of a selected species should be investigated under different growth parameters.

Parsons, Stephens and Strickland (1961) investigated the composition of eleven species of marine phytoplankton grown under identical conditions. Their results agreed with those of earlier workers in emphasizing the importance of protein as the principal organic constituent; in four species of Chlorophyceae and Chrysophyceae the amount of protein approached or even exceeded 50% of the total dry weight. In two dinoflagellate species (*Amphidinium* and *Exuviaella*) carbohydrate amounted to 30.5 and 37%, respectively, of dry weight, and only in these two species thus exceeded the protein content (28% and 31% dry weight, respectively). The amounts of lipid were appreciable but varied in the different species from as little as $<2\%$ of the dry weight to concentrations exceeding 10%, with the two dinoflagellates reaching even higher values (18 and 15% dry weight). Only the diatom *Chaetoceros* showed lipid in excess of carbohydrate.

Table 4.4 shows the quantities of the three major biochemical constituents calculated as percentages of the total organic matter.

Parsons *et al.* (1961) claim a strong similarity in chemical composition of the algae investigated, bearing in mind the diversity in size and the different evolutionary origin of the species, and they suggest that analysis of net phytoplankton would give a reasonable picture of the composition of total phytoplankton. They indicate an average ratio of

Table 4.4. Biochemical composition of eleven species of algae, calculated as percentages of total organic matter (data from Parsons *et al.*, 1961)

Species	Protein	Carbohydrate	Lipid
Tetraselmis	74	21	4
Dunaliella	60	33	7
Monochrysis	53	34	12
Syracosphaera	71	23	6
Chaetoceros	72	14	15
Skeletonema	59	33	8
Coscinodiscus	74	18	8
Phaeodactylum	52	38	10
Amphidinium	37	40	24
Exuviaella	37	45	18
Agmenellum	45	39	16

protein:carbohydrate:lipid of 4:3:1 and believe this is especially suitable as a food source. It is somewhat difficult to accept these views without reservation. The ratio holds for *Monochrysis* but as Corner and Cowey (1968) have remarked, *Phaeodactylum* has a ratio of 5:4:1 and *Skeletonema* 8:4:1. Both algae are very acceptable foods for zooplankton, and even though there are some suggestions (cf. Ferguson, 1973) that *Skeletonema* has a lower assimilation ratio for zooplankton than some other algae, this may not be associated with the different chemical composition. Even wider differences are seen in the ratio of some of the other algae, e.g. *Syracosphaera* 12:4:1, *Tetraselmis* 18:5:1, *Chaetoceros* 5:1:1 (Table 4.4.).

Analysis of cultures of *Phaeodactylum*, *Skeletonema* and *Coscinodiscus* in exponential growth by Chong (1970) also indicated considerable differences in the ratio protein:carbohydrate:lipid. While the amounts of protein agree reasonably well with the data obtained by Parsons *et al.* for all three diatoms, fairly substantial differences appear in respect of ash (*Phaeodactylum*) and of lipid for these diatoms.

Data from Ricketts (1966) for marine flagellates are less easy to compare since carbohydrate was not always determined but there are indications of fairly substantial differences in the major organic fractions, although Ricketts comments on an overall similarity of biochemical composition. A comparison by Myklestad (1974) of nine species of marine planktonic diatoms also indicates that the basic biochemical composition can vary considerably. While all nine species tended to accumulate carbohydrate (glucan) in the stationary phase in batch culture, and the proportion of protein to carbohydrate fell with time, there were substantial specific variations; for example, in the logarithmic growth phase *Thalassiosira gravida* had 52% and *Skeletonema costatum* 11% of the dry organic weight as protein. In the stationary growth phase, *Skeletonema* had 90% and *Corethron hystrix* 13% of dry organic matter as carbohydrate.

To a considerable degree the differences in composition of the same species, all grown in the log phase, which are suggested by various authors may be ascribed to differences in culture media. Undoubtedly the physical/chemical environment as a whole plays a major role in the composition of algae. In general, high nitrogen level in the medium promotes high protein; on the other hand, nitrogen depletion tends to increase the lipid content. The tendency for older algal cultures to show higher lipid levels may often be associated with nitrogen depletion of the medium (cf. Collyer and Fogg, 1955; Spoehr and Milner, 1949). To some degree effects of nitrogen deficiency and age of culture on the proportions of protein, fat and carbohydrate are illustrated by the variations seen in a number of diatom species harvested under different conditions of culture (cf. Table 4.5). The data include some of those already reviewed from Parsons *et al.* (1961). It is sometimes difficult to appreciate the variations in proportions of organic constituents critically, since in many instances the amounts of protein, fat and carbohydrate do not add to 100%. Perhaps some constituents were missed or proper drying procedures not completed. In general, however, the high level of protein was confirmed except where marked nitrogen deficiency obtained; some lipid values were very much higher than those found by Parsons *et al.*

The experiments of McAllister, Parsons, Stephens and Strickland (1961) and Antia, McAllister, Parsons, Stephens and Strickland (1963), using the plastic sphere technique, also supply evidence for nitrogen depletion affecting the chemical composition of phytoplankton. Phosphate depletion was not obvious in the experiments; indeed there was evidence for a relatively rapid release of phosphorus from cells as well as for some regeneration of phosphate. Nitrate depletion, however, was marked and was accompanied by reduction of cell protein, but an increase in the phytoplankton biomass still followed and was accompanied by substantial increases in carbohydrate and to some extent in lipid. The authors point out that the classic concept of diatoms having overwhelming amounts of lipid was not substantiated, even though nutrient-poor conditions tend to accompany higher lipid contents. Strickland, Holm-Hansen, Eppley and Linn (1969), employing experiments in a deep tank, found nitrogen deficiency affecting levels of chlorophyll and protein in phytoplankton species, but the most marked change was a very sharp six-fold increase in cell carbohydrate with nitrogen lack in a population of the dinoflagellate *Cachonina*. Carbohydrate fell with the addition of nutrient and renewed cell growth followed. Another observation by Strickland *et al.*, apparently not recorded elsewhere, was a comparison of the composition of algal cells grown on ammonium or nitrate as nitrogen source. The dinoflagellate *Cachonina* showed only relatively small differences in chemical composition. On the other hand, for the diatom *Ditylum*, while protein, carbohydrate and DNA were more or less unchanged, chlorophyll and ATP content were higher in cells reared on nitrate, and lipid was distinctly greater in cells on a nitrate medium. In experiments involving the preferential use of ammonium over nitrate and the induction of nitrate reductase in *Ditylum*, *Cachonina* and *Gonyaulax*, it was noticeable that the two dinoflagellates resembled each other in basic biochemical composition and were much higher in lipid and carbohydrate and lower in protein than the diatom *Ditylum*.

Paasche (1968a) noted changes in overall chemical composition of phytoplankton consequent on exposure to light. Cultures of *Ditylum* exhibited a relative increase in carbohydrate when exposed to continuous illumination. Light may thus retard protein synthesis; earlier work on *Cyclotella cryptica* showed that excess carbohydrate was formed during photosynthesis but was converted to protein during darkness.

Ricketts (1966), describing wide variations especially in lipid and carbohydrate in algae, ascribes these mainly to the state of nutrition of the cells. He believes, however,

Table 4.5. Biochemical constituents of some diatoms expressed as percentages of ash-free dry weight (data taken from Lewin and Guillard, 1963)

Species	Lipid	Carbohydrate	Protein
Rhabdonema adriaticum (from plankton)	44	14	13
Chaetoceros decipiens (from plankton)	28	13	22
Chaetoceros sp. (unialgal culture in exponential growth)	10	9	49
Skeletonema (unialgal culture in exponential growth)	8	34	61
Skeletonema (unialgal culture grown 2–4 weeks)	22	35	44
Phaeodactylum tricornutum (unialgal culture in exponential growth)	7	26	36
Phaeodactylum tricornutum (fusiform cells grown 16-day pure culture)	39	2	46
Phaeodactylum tricornutum (oval cells grown 16-day pure culture)	27	21	38
Navicula pelliculosa (pure culture in low N. medium)			
8 days	28	29	37
33 days	28	29	10

that much of the carbohydrate which is produced with cultures grown under alternating light and dark periods is not available as a food reserve, although carbohydrate is the first food material to be drawn on and this is followed by use of lipid reserves.

Increase in lipid in older algal cultures is described by Chong (1970). Lipid content of static cultures of *Phaeodactylum* nearly doubled (46% of the organic matter) as compared with cultures in the log phase of growth (25%). Protein showed a corresponding fall (27% in static as against 51% in log phase cultures) while carbohydrate was practically unchanged. "Older" (17-day) cultures of *Coscinodiscus* also showed a small increase in lipid over 7-day cultures (44% as against 37% organic matter). The 17-day cultures were not static, however.

Miller (1962) points out that a number of factors, including high light intensity and nitrogen deficiency, are well recognized as promoting lipid synthesis. Phosphorus deficiency does not have this effect. Aged cultures are associated with a high lipid content, but it is not entirely clear whether this is solely a nutrient (nitrogen) deficiency effect.

Age of culture was found to be a factor in carbohydrate and protein synthesis, using *Coscinodiscus eccentricus* cultures, by Pugh (1969; 1975) who also examined the effects of salinity. During the initial lag phase carbohydrate increased with dry weight, but protein synthesis was retarded and there was no rise in lipid. During the subsequent exponential growth phase, Pugh found a decline in carbohydrate, and it was not stored in the stationary phase. These changes were similar for cultures at all salinities. An increase in protein as a percentage of dry weight occurred during the exponential or stationary phase of *Coscinodiscus*, but the timing appeared to relate to salinity. At higher salinities (30‰ and 35‰) the peak value for protein was achieved fairly rapidly, before the stationary phase and there was then a marked decline. At 20‰ salinity the increase was slow, and the peak coincided with the maximum in cell number. At 25‰ the increase in protein was fairly early and the peak occurred at the end of the exponential growth phase. The decrease in nitrogen and phosphorus in the medium was presumably related to the levelling off in protein synthesis.

Some details are available of the carbohydrate constituents of phytoplankton algae. Information has already been given on reserve carbohydrates, and attention drawn to the fact that free sugars usually occur only in very small quantities in phytoplankton (Chapter 3). Hydrolysates of the whole cells of marine diatoms, however, have been found to yield glucose, galactose, mannose, xylose, ribose, rhamnose and glucose; hexuronic acids have been identified. Other phytoplankton species analysed by Parsons *et al.* (1961) showed glucose, galactose and ribose on hydrolysis, with hexuronic acids in all except dinoflagellates. Glucose was predominant in all species. Not all polysaccharide is reserve material, however. Glucuronic acid has been identified from the capsular material of *Navicula pelliculosa* and the mucilage from *Phaeodactylum* yielded xylose, mannose, fucose and galactose. The presence of polysaccharides in extra-cellular material thus appears fairly certain. Polysaccharides also participate in wall formation of diatoms; hemicellulose and a cellulose-like material yielding mixtures of hexoses and pentoses on hydrolysis have been found. Parsons *et al.* (1961) found acid- and alkali-insoluble carbohydrate which they termed "crude fibre" in all species of phytoplankton which they examined. This "crude fibre" they interpreted as comprising the more resistant carbohydrate constituents of the cell wall as contrasted with the stored (reserve) carbohydrate. The proportion of crude fibre was variable. In

Monochrysis, *Syracosphaera*, *Phaeodactylum* and *Amphidinium* the amount was small (<5% of total carbohydrate). It amounted to almost 10% in *Dunaliella*, and exceeded 10% in *Tetraselmis* and especially in *Exuviaella*; presumably these algae have tough carbohydrate cell walls. In two diatoms, *Chaetoceros* and *Coscinodiscus*, crude fibre exceeded 20% of total carbohydrate.

In more recent studies of carbohydrates, Myklestad and Haug (1972) identified a glucan with $\beta 1-3$ type of linkage (leucosin=chrysolaminarin, possibly) as the cellular reserve carbohydrate in batch cultures of the diatom, *Chaetoceros affinis*. Low nutrient media yielded high carbohydrate but the protein/carbohydrate ratio fell at the stationary phase of the culture, whatever the original nutrient level. Extracellular polysaccharide was also produced. Myklestad (1974) and Haug and Myklestad (1976) showed that extracellular carbohydrate could be produced by other diatom species, but the quantity was variable; for instance, it was high in cultures of *Chaetoceros curvisetus* and *C. affinis*, and low in *Skeletonema* and *Thalassiosira gravida* cultures, grown under identical conditions. The polysaccharide in *Chaetoceros* yielded rhamnose, fucose and galactose as monosaccharides on hydrolysis. It thus differed from the "wall" polysaccharides found by the same investigators in cultures of five species of *Chaetoceros* and *T. gravida* and *Corethron hystrix*. These polysaccharides were obtained as alkali-soluble material following extraction of other carbohydrates with dilute acid, and probably constitute the amorphous matrix of the cell wall, associated with silica. (Resistant fibrous material also was present.) The alkali-soluble carbohydrates formed from 5% to 15% of the total organic matter and the quantity was not markedly affected by the phase of growth of cultures. On hydrolysis these polysaccharides yielded almost no glucose, but a mixture of other monosaccharides, especially rhamnose and fucose, with galactose, mannose, xylose and ribose. The proportions varied, however, with species; *Chaetoceros* spp. were all rich in rhamnose, *Thalassiosira* had none and *Corethron* very little. Haug and Myklestad (1976) confirmed that the cellular reserve carbohydrate was an acid-soluble polysaccharide of the glucan type, which yielded 95% of monosaccharides as glucose on hydrolysis.

Another study on the carbohydrate (and amino acid) composition of algae relates particularly to silica deposition in diatom cells (cf. Hecky, Mopper, Kilhan and Degens, 1973). The cell walls of several diatom species exhibited increased concentrations of xylose, mannose and, in three species, of fucose, and markedly reduced glucose concentrations in comparison with the cell contents. Of amino acids, glycine, serine and threonine were relatively enriched in the cell walls and glutamic and aspartic acids, the aromatic amino acids (tyrosine and phenylalanine), and the sulphur-containing amino acids were reduced as compared with the cell contents. Protein increase in the cell walls had earlier been found to be associated with increase in silica, and the further studies suggested that a protein rich in serine and threonine might act as a template for Si deposition. Possibly the hydroxyl groups of serine and threonine condense with silicic acid, with the elimination of water. The frequency of reacting OH groups may thus determine the degree of polymerization (and hydration) of the deposited silica ranging from $SiO_2.2H_2O$ to $SiO_2.nH_2O$. The somewhat variable polysaccharide coating (mainly of resistant mannans and xylans) serves possibly as an external protection against the solution of silica; an intact organic coating may decrease the solubility of diatom frustules. Figure 4.48 suggests a hypothetical arrangement for the organic and silica layers.

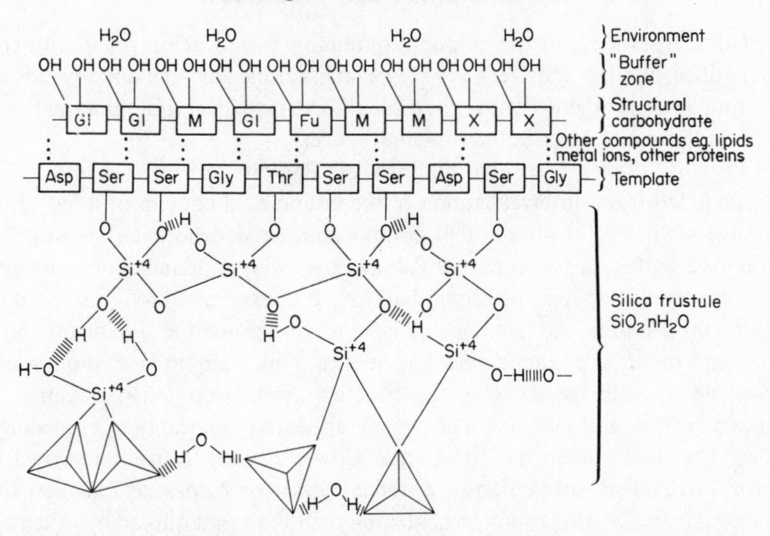

Fig. 4.48. Hypothetical arrangement of organic layers in the diatom cell wall. Outer layer of polysaccharides consisting of various sugars: *Gl* glucose; *M* mannose; *Fu* fucose; *X* xylose. Outward directed hydroxyl groups of sugars depict the hydrophilic buffer zone. Residues in the protein template are: *Ser* serine; *Gly* glycine; *Thr* threonine; *Asp* aspartic acid. Hatched lines represent hydrogen bonds. Tetrahedra emphasize the three-dimensionality of the silicic acid (and resulting silica) with Si in four-fold coordination with oxygen atoms at the points of the tetrahedra (from Hecky *et al.*, 1973).

Concerning free carbohydrates in phytoplankton, Boney (1976) makes an interesting distinction between the two groups of "green" algae with regard to their production of "simpler" carbohydrates. Prasinophyceae (*Micromonas*, *Tetraselmis*, *Prasinocladus*) synthesize mannitol as the main product, whereas Chlorophyceae produce sucrose or glycerol. Wallen and Geen (1971 a and c), investigating ^{14}C uptake in *Dunaliella*, also noted that glycerol was a substantial proportion of the ethanol-soluble fraction, whereas the diatom *Cyclotella* formed mainly glucose as soluble product. Ethanol-insoluble polysaccharide in both algae yielded mainly glucose on hydrolysis. With natural phytoplankton, ^{14}C uptake experiments showed a variety of soluble carbohydrates, chiefly glucose, but including fructose, mannose, galactose, sucrose, etc. Organic acids, especially malic and fumaric acids, were formed both by the cultured species and by natural phytoplankton.

Investigations on the amino acid composition of marine phytoplankton by Parsons *et al.* (1961) indicated considerable differences between species of algae, though certain amino acids – aspartic, glutamic, glycine, alanine and lysine – occurred in comparatively large amounts in all species investigated. Arginine was also important. Aspartic acid was frequently the chief amino acid but the amounts varied between species from 14.8% to 35.5% of the total amino nitrogen. Whereas aspartic was virtually equivalent to glutamic in *Agmenellum*, *Phaeodactylum* and *Coscinodiscus*, the ratios for *Syracosphaera*, *Monochrysis*, *Tetraselmis* and *Skeletonema* were 7:1, 4:1, 3:1 and 2:1, respectively. Other major amino acids showed considerable variations (e.g. glycine 8.6–21.2%, alanine 7.5–28.5%). The ratio of alanine to glycine was nearly 1:1 for a number of the algae including *Syracosphaera*, *Coscinodiscus*, *Phaeodactylum* and *Agmenellum*, but was approximately 3:1 for *Monochrysis* and 0.5:1 for *Skeletonema*. The less abundant amino acids were also variable in different species though the data indicated that histidine and the sulphur-containing amino acids generally occurred in small quantities. Wallen and Geen (1971a) also list some eighteen amino acids identified in ^{14}C uptake experiments with phytoplankton. Aspartic and glutamic acids and alanine were the major acids in the protein fraction.

Somewhat in contrast to the findings of Parsons *et al.*, Cowey and Corner (1966) concluded from an investigation of the amino acid content of a few algae that a considerable similarity of amino acid composition existed between species. This view was supported by Chau, Chuecas and Riley (1967) who conducted a much wider study involving twenty-five species of phytoplankton. For example, though some differences were apparent, aspartic acid, alanine and leucine were present in significant and more or less constant amounts; glutamic acid was another major amino acid, though somewhat more variable between species; glycine, serine and lysine were fairly abundant and constant in proportion, and histidine and cystine appeared invariably in exceedingly small quantities. The data quoted by Corner and Cowey (1968) from unpublished work by Walne for *Tetraselmis suecica* and *Chlamydomonas coccoides* also suggest that these species have generally similar amino acid composition to that quoted by other authors.

Chau, Chuecas and Riley (1967), however, included only one diatom, *Phaeodactylum tricornutum*, in their analyses; this species has frequently been shown to be somewhat aberrant amongst diatoms. In a later investigation involving five diatom species Chuecas and Riley (1969b) found some major differences in amino acid composition between "typical" diatoms and other marine algae. In diatoms, they maintain, in contrast to Cowey and Corner's (1966) results, that glycine is more abundant than glutamic or aspartic acid, serine is often the major amino acid, and alanine and arginine are usually abundant, with lysine relatively low. The authors emphasize that this pattern of amino acids differs sharply from that of other marine phytoplankton and that *Phaeodactylum* does not conform to the typical diatom pattern.

Preliminary investigations by Ferguson (unpublished) on the amino acid composition of laboratory cultures of four diatoms, *Coscinodiscus eccentricus*, *Skeletonema costatum*, *Chaetoceros septentrionale* and *Ditylum brightwellii*, harvested after 20 days, suggested that the composition of the four diatoms was similar and generally agreed with the analyses of Cowey and Corner. Thus major amino acids were aspartic, glutamic, leucine and alanine; lysine and glycine were slightly less abundant; serine with valine and others were next in importance. It would be of interest to have more analyses of natural phytoplankton crops, dominated by different taxa and at different stages of flowering (e.g. bloom and post-bloom periods) to compare with laboratory cultures. Although some information on the amino acid composition of phytoplankton might suggest relatively little variation, few taxa have yet been studied and most of the species analysed are from coastal and temperate regions.

The total lipids of phytoplankton may be divided into several fractions: glycerides, including mono-, di- and tri-glycerides, phospholipids, free fatty acids, sterols and hydrocarbons. There are also glyco-lipids (cf. Kates and Volcani, 1966) and lipo-protein complexes. Triglycerides are often regarded as the typical storage fat material, but they may not represent the major lipid fraction. Thus Miller (1962) quotes *Monodus* having about 61% triglycerides, 30% mono- and di-glycerides and 9% free fatty acids making up the total lipid content, whereas *Phaeodactylum* in culture might have even as much as 80% free fatty acids. Lee, Nevenzel and Paffenhofer (1971) also found that phospholipids, mainly the "structural" lipids, were the dominant fraction (*ca.* 50% of the total) in three diatoms (*Lauderia*, *Chaetoceros* and *Skeletonema*); triglycerides accounted for only 12–16%. Wax esters, a very significant fraction of the total lipid in some zooplankton (cf. Volume 2), were undetectable in these diatoms. Though hydrocarbons are more usually regarded as a minor fraction, the importance and apparently widespread occurrence of hydrocarbons in phytoplankton algae is now recognized, largely from such investigations as those of Clark and Blumer (1967), Lee, Nevenzel, Paffenhofer, Benson, Patton and Kavanagh (1970), Lee and Loeblich (1971) and Blumer, Mullin and Guillard (1970). A list of sterols identified in several marine algae is given in Miller (1962); fucosterol, cholesterol and sitosterol would appear to be the major sterols.

The more detailed analysis of lipid components of phytoplankton is concerned chiefly with the fatty acids, although more recently hydrocarbons have received attention. Early work emphasized the importance of $C_{16:0}$,* and of the C_{16} and C_{18} unsaturated acids; this was supported by Kayama, Tsuchiya and Mead (1963) who quoted the major saturated acids for the diatom *Chaetoceros simplex*: $C_{14:0}$ (13%); $C_{16:0}$ (18%), and the chief unsaturated acids: $C_{16:1}$ (48%); $C_{18:1}$ (9%).

Apart from variations in fatty acid constitution in different algal taxa, the composition appears to be considerably affected by the culture technique and age of culture. Thus Miller (1962), commenting on analyses in *Chlorella* states that nitrogen deficiency tends to lower the trienoic and tetraenoic C_{16} and C_{18} fatty acids, while increasing the dienoic and monoenoic acids. Chuecas and Riley (1969a) also summarize the results of some earlier investigations by stating that $C_{16:0}$ and $C_{18:1}$ are relatively richer in the more abundant lipid produced in nitrogen-deficient cultures under light saturation, whereas nitrogen-rich media tend to increase the proportion of polyunsaturates while reducing the amount of total lipid.

Ackman, Jangaard, Hoyle and Brockerhoff (1964) point out for *Skeletonema* that nitrate concentration and age of culture may affect fatty acid composition; not only does a marked decline in the C_{18} series (except $C_{18:4}$) occur with age, but there is a very marked accumulation of $C_{14:0}$ (10%–>30%), a decline of $C_{22:6}$ and some reduction of $C_{20:5}$. In contrast to some other investigations, they state that the ratio of $C_{16:0}$ to $C_{16:1}$ remained almost constant (about 0.5). Whether the lipid metabolic changes associated with age of culture are solely a reflection of nitrogen depletion or whether they are related to changes in pH, exocrine concentration, etc., is open to question. Although Ackman *et al.* (1964) found a general resemblance between the fatty acids of natural populations of phytoplankton (essentially *Biddulphia sinensis*) and of *Skeletonema* cultures, a strict comparability between natural phytoplankton and cultured algae is doubtful. Apart from taxonomic differences, nutrient conditions and age of culture introduce variability. Temperature may also affect lipid composition; at least in some species, increase in temperature may result in a rise in the proportion of the less unsaturated fatty acids (cf. also Ackman, Tocher and McLachlan, 1968). Pugh (1971) has examined the effect of salinity on fatty acid composition of cultures of *Coscinodiscus eccentricus*. In general Pugh found that the analyses of Ackman *et al.* (1964) on *Skeletonema* agreed with results for *Coscinodiscus* and that there was little overall difference in fatty acid composition for *Coscinodiscus* cultured at the four salinities: 20, 25, 30 and 35‰. Considerable variations occurred, however, during the

* The number preceding the colon gives the number of carbon atoms in the chain; the number of double bonds is indicated by the number following the colon.

growth cycle of the diatom (from early log phase to senescence) and some of the variations appeared to be related to salinity. The C_{16} series of fatty acids ($C_{16:0}$, $C_{16:1}$, $C_{16:2}$, $C_{16:3}$, $C_{16:4}$) was dominant, forming 44% of the total fatty acid and rising to 76% in old cultures; $C_{16:0}$ itself varied little but the ratio of $C_{16:0}/C_{16:1}$ fell from *ca.* 1.0 (1.36 in the 35‰ culture) to *ca.* 0.5 on senescent cultures. As with other diatoms, the other important fatty acids were $C_{14:0}$, $C_{20:5}$ and $C_{22:6}$, and the C_{18} series was present only in small amounts. Though $C_{14:0}$ varied in its proportions with growth phase and to some extent with salinity, there was no marked accumulation with age as Ackman had found. Some rise occurred with cell division, notably at 20‰ salinity. $C_{20:5}$ was the only acid in the C_{20} series in significant quantity. The average concentration was 11% but the amount varied and was maximal at cell division. The average quantity was far greater than the other highly unsaturated acid, $C_{22:6}$. The $C_{13:0}$ acid, found usually in very small amounts in diatoms, increased in early growth stages and amounted to significant concentrations at 25‰ and 35‰ salinities.

One of the main problems in analysing fatty acids is the extraction technique. Chuecas and Riley (1969a) emphasize the considerable differences seen in total lipid and in the proportions of saturated and polyunsaturated fatty acids when the extraction employs polar and non-polar solvents. The now more usually employed chloroform-methanol mixture extracts structural lipid and polyenoic acids, not readily extracted by the relatively non-polar solvents (e.g. ether). This might explain, for instance, Kayama's failure to detect fatty acids higher than C_{18}, although the C_{20} series is usually a major contributor in diatoms. Another source of error is the danger of auto-oxidation of the unsaturated acids.

Chuecas and Riley (1969a) cultured twenty-seven species of marine phytoplankton under conditions which gave rapid growth. The algae were harvested in the exponential growth phase to avoid changes due to physico-chemical conditions and to age of culture. They emphasized the great variety of fatty acids in algae while accepting that overall patterns may follow taxonomic groupings (cf. Ackman *et al.*, 1968) (Fig. 4.49). Of the saturated acids, C_{14} and C_{16} predominate in almost all groups; unsaturated acids of the C_{16}, C_{18}, C_{20} and C_{22} series are very abundant and may together exceed 80% of total fatty acids; $C_{16:1}$, $C_{16:4}$, $C_{18:3}$, $C_{18:4}$, $C_{20:5}$ and $C_{22:5}$ are generally well represented. However, some taxonomic patterns appear to hold. For example, in agreement with other workers, they find that diatoms appear to have comparatively little $C_{18:3}$ and $C_{18:4}$ (indeed the whole C_{18} series is poorly represented), and relatively little $C_{22:5}$, but substantial amounts of the C_{16} series and $C_{20:5}$. Cryptophyceae have considerable amounts of the C_{18} series among the polyenoics and are especially rich in the monounsaturated fatty acid, $C_{20:1}$. Only one other algae (*Chlamydomonas*) had a substantial quantity of $C_{20:1}$. Dinoflagellates appear to show some remarkable specific differences in the proportion of the C_{16} series, e.g. *Peridinium* had a large amount of $C_{16:0}$ and little of the unsaturated C_{16} fatty acids; *Prorocentrum* had mostly $C_{16:1}$ and reasonable quantities of polyenoic acids. Both species have substantial and similar amounts of $C_{20:5}$ and of total quantities of the unsaturated C_{18} fatty acids.

Chuecas and Riley report that although two species of *Chlamydomonas* differed in the amount of unsaturated C_{18} acids and of $C_{20:1}$, both species lacked $C_{20:5}$ and $C_{22:5}$, a character unique among the algae analysed. They found reasonable quantities of the two polyenoic acids in *Dunaliella*. Ackman *et al.* (1968), however, found that *Dunaliella* also lacked C_{20} and C_{22} polyunsaturates. If these polyenoics are generally

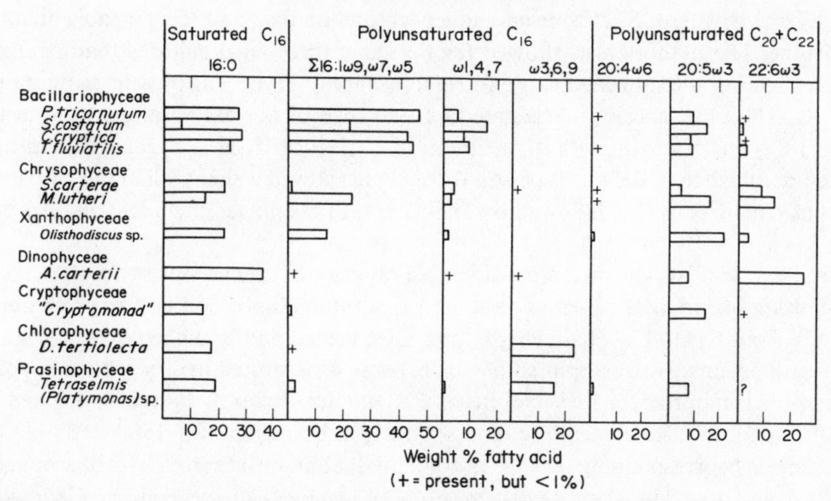

Fig. 4.49. The C_{16} chain length fatty acids and the principal C_{20} and C_{22} chain length fatty acids in eleven species of marine phytoplankton species (slightly modified from Ackman *et al.*, 1968).

lacking in the Chlorophyceae it would separate the class from other marine phytoplankton. The results of analyses of other algal taxa mentioned by Chuecas and Riley include a general resemblance between two species of Chrysophyceae and the Bacillariophyceae, the dominant unsaturated fatty acids being $C_{16:1}$, $C_{16:3}$ and $C_{20:5}$, and a peculiar pattern of fatty acids shown by *Olisthodiscus*, sometimes regarded as a member of the Xanthophyceae. The predominant acid was $C_{18:4}$, $C_{24:0}$ was also unusually plentiful and $C_{22:5}$ abundant; the unsaturated C_{16} acids were very poorly represented. An analysis of *Olisthodiscus* by Ackman *et al.* (1968) had indicated a more normal pattern. Ackman *et al.* (1968) point to a distinction in fatty acid pattern between Chlorophyceae (*Dunaliella*), lacking C_{20} and C_{22} polyunsaturates, and Prasinophyceae (*Tetraselmis*) (cf. Fig. 4.49).

The synthesis of polyunsaturated long-chain fatty acids is of interest in view of the large quantities of polyenoic C_{20} and C_{22} acids in most phytoplankton species. The production of these polyenoic acids is probably via $C_{18:2}$ and $C_{18:3}$, in turn probably synthesized from $C_{18:1}$. Polyunsaturates of the C_{16} series, Ackman (1964) suggests, may be produced from the abundant $C_{16:1}$ rather than from C_{18} polyunsaturates. The accumulation of $C_{14:0}$ reported in aged cultures by Ackman may be the result of the blocking of synthetic pathways to C_{16} and higher acids. There is a suggestion that a certain level of polyunsaturates is required by algae, for example for their synthesis of structural phospholipids, after which "depot" fats are mostly produced, consisting mainly of saturated or monounsaturated acids.

Limited information is now becoming available on some of the hydrocarbons found in marine phytoplankton. This subject has a wider interest, since there is some belief that the fossil hydrocarbons present as oil deposits may represent accumulations of plankton.

Pristane, identified by Blumer, Mullin and Thomas (1964) in zooplankton and in higher links of the marine food chain, is a branched saturated hydrocarbon, thought to be derived from the phytol ($C_{20}H_{40}OH$) part of the chlorophyll molecule of phytoplankton. Clark and Blumer (1967) found that cultures of three phytoplankton algae (*Syracosphaera*, *Skeletonema* and a cryptomonad) all had recognizable quantities of pristane. The authors also showed for the same three algal cultures that a range of hydrocarbons of the saturated straight chain (n-paraffin) type was present as minor constituents. The three species investigated had a range of normal paraffins from at least C_{14} and C_{32} chain length, with a maximum of $C_{15}H_{32}$ or $C_{17}H_{36}$. *Syracosphaera* had a marked dominance of the C_{17} type. All three algae showed a slight fall in the amounts of paraffins from C_{17} to C_{21}, followed by an increase in the higher chain lengths, especially C_{29}.

Patterson (1967) confirmed a considerable range of hydrocarbons in different species of *Chlorella* including a series of saturated n-paraffins, both odd and even numbered, certainly from C_{17} to C_{36} chain length, and unsaturated and branched hydrocarbons in minor proportions. However, in some algae, some unsaturated hydrocarbons appear to be of special importance. Lee, Nevenzel, Paffenhofer, Benson, Patton and Kavanagh (1970) found in *Skeletonema* cultures considerable quantities (80% of the total hydrocarbons corresponding to 14% of total lipid) of an unsaturated hydrocarbon, later identified as $C_{21:6}$. The same hydrocarbon was identified in the diatom, *Chaetoceros septentrionale*, but not in Chlorophyceae (*Dunaliella tertiolecta*). Blumer, Mullin and Guillard (1970) independently identified the same hydrocarbon in diatoms, Chrysophyceae, Dinophyceae (*Gymnodinium*) and Cryptophyceae (*Cryptomonas*). The authors stated that diatoms generally appeared to have $C_{21:6}$ as a major hydrocarbon; its occurrence in Xanthophyceae was debatable, and it was absent in Chlorophyceae. It remained for Lee and Loeblich (1971) to demonstrate that $C_{21:6}$ was the major hydrocarbon (80–90%) in a variety of algal classes (diatoms, dinoflagellates, cryptomonads, phaeophytes and chrysophytes). It occurs also in *Platymonas*, quoted as a chlorophyte, but now usually included in the Prasinophyceae. The hydrocarbon is absent in most cyanophytes, rhodophytes, xanthophytes and chlorophytes. Although straight-chain saturated and unsaturated hydrocarbons from C_{15} to C_{36} are reported for all algal taxa, the quantities are very small. Dinoflagellates and euglenoids may be exceptional in having significant amounts of $C_{21:4}$ and $C_{21:5}$. $C_{21:6}$ may be confined to photosynthetic species, since two non-autotrophic dinoflagellates, *Noctiluca* and *Oxyrrhis*, though possessing substantial amounts of total hydrocarbon, lacked $C_{21:6}$. Lee and Loeblich suggest that a metabolic pathway might be from the fatty acid $C_{22:6}$, relatively widespread in algae, which is decarboxylated. Diatoms, which are unusual amongst phytoplankton in possessing so little $C_{22:6}$, are especially active in this decarboxylation process; this is particularly true of older cultures.

There is need for continued detailed analyses of biochemical components of different phytoplankton taxa. The results, taken in conjunction with experiments on algae, particularly those employing isotope techniques, may lead to a fuller understanding not only of primary production, but of the synthesis of particular substances which may be of especial significance in marine food chains.

Chapter 5
Phytoplankton Crop

Methods of Assessment

The crop (standing crop) is the biomass of phytoplankton and may be measured in various ways, for example, as the weight of matter (wet or dry) per unit volume, or under a unit of surface (usually 1 m^2) throughout the length of the euphotic column. Since inorganic detritus may be mixed with phytoplankton in significant amounts, particularly in turbid areas, it is usually preferable to determine the weight of organic matter, or of carbon. In the estimation of carbon a useful and accurate technique is to measure carbon dioxide using an infra-red analyser. Organic detritus will be included in such assessments, and the amount of particulate non-living organic matter may at times be large, for instance towards the end of a phytoplankton bloom. Direct estimates of phytoplankton cell density have been, therefore, most usually made to determine phytoplankton crop. These direct counts were especially employed in earlier surveys, but since phytoplankton algae vary greatly in size interspecifically and intraspecifically (cf. Chapter 4), cell numbers may be misleading in estimating biomass. Total algal cell volume has been used as an index of crop, but cell volume may not give a proper indication of plant tissue, and cell surface area or plasma volume has been employed to determine crop and productivity. Some further discussion has been included in Chapter 9. Protein estimation and ATP measured as luminescence in luciferin/luciferase reactions, have also been employed for crop assessment. The latter method is sensitive and distinguishes living material from detritus. All methods, however, have certain disadvantages, with reference to the speed of the technique or the degree of accuracy. A particular problem is the proper conversion of the parameter measured to the quantity of organic matter.

One of the most frequently employed techniques in assessing standing crop is an estimation of cell pigment, usually chlorophyll *a*, since this is universally present in marine algae. The method, described later, is relatively rapid and convenient but suffers from certain limitations, in particular that different algal species, and even the same species under different growth conditions, do not have a constant chlorophyll:carbon ratio. The example of "sun" and "shade" phytoplankton species demonstrates that light intensity affects cell chlorophyll concentration. Nutrient levels, particularly nitrogen, and other factors can also modify chlorophyll:carbon ratios. The range of values is indicated in Chapter 3.

The method of collection of phytoplankton samples may introduce other errors in crop assessment. Sampling may fail to make representative collections in view of the spatial discontinuity of plankton. Phytoplankton is usually patchy, the patches varying greatly in dimensions in most seas (cf. Chapter 8). In addition, although the euphotic zone may be well mixed and the phytoplankton sometimes more or less evenly distributed vertically, stratification may occur, including microstratification, often characterized by alternating rich and poor layers of algae, which normal collecting fails to sample adequately. Sorokin (1960) has pointed to the importance of thermal stratification within the euphotic zone in relation to the phytoplankton; if thermal stratification is present within the euphotic zone the distribution of the phytoplankton may be greatly modified and production must be related amongst other factors to the distribution of the crop (cf. Sorokin, 1960 and Fig. 5.1). While the significance of the vertical stratification of phytoplankton has been well recognized in temperate and warmer oceans, pronounced differences in vertical distribution may occur also in colder waters (cf., for example, El-Sayed, 1970a and b for Antarctic waters, and cf. Chapter 6).

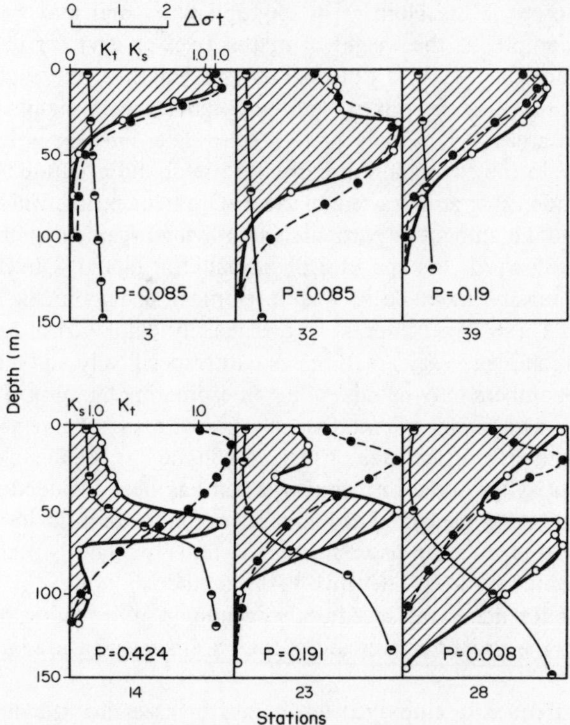

Fig. 5.1. The vertical distribution of the rate of primary production, K_s, and the dependence of the rate of photosynthesis on light, K_t, as influenced by density stratification (σ_t)

——○—— $= K_s$
- - -●- - - $= K_t$
——◒—— $= \sigma_t$

P gives the primary organic production in C g/m²/24 hr (Sorokin, 1960).

Apart from the discontinuous distribution of phytoplankton horizontally and vertically, the problem of ensuring the complete collection of all the autotrophic organisms from a known volume of water presents great difficulties. The size spectrum typical of the phytoplankton poses problems for quantitative sampling. It is now widely recognized that nanoplankton organisms make a substantial contribution to the total density of phytoplankton (cf. pp. 249 et seq). Sometimes in inshore areas and frequently in warmer oceanic seas, the nanoplankton may predominate in biomass, but there is great variability spatially and temporally in the ratio of nanoplankton to net phytoplankton (cf. Sournia, 1968; Tundisi, 1971; Malone, 1971, and Table 5.6).

The proper enumeration of nanoplankton organisms is exceedingly difficult; complete enumeration is probably still not possible. Centrifugation has been used but some of the more delicate cells are destroyed or are not brought down even with prolonged centrifuging. Filtration employing hardened filter papers, glass fibre, membrane or sintered glass filters is frequently used, and some membrane filters can be rendered translucent for subsequent cell counts. More usually, chlorophyll extraction is carried out on the phytoplankton filtered. To quantify the contributions of net and nanoplankton to the total phytoplankton (as chlorophyll), some type of fractional filtration may be attempted. Frequently a separate net phytoplankton sample is taken simultaneously with the total phytoplankton sample, obtained as a water-bottle collection (*vide infra*); the chlorophyll content is determined for each and the nanoplankton contribution estimated by difference. For a floristic assessment of the nanoplankton, counting is necessary; small samples of seawater collected by water bottle are allowed to sediment in long vertical columns, after fixation. The settled algae are counted by inverse microscope. Sedimentation methods are slow and are subject to the disadvantage that some of the most delicate cells are destroyed even by the best fixatives. Dilute Lugol's iodine is often used. The technique is useless in the presence of considerable amounts of detritus, often abundant in eutrophic waters; sedimented detritus obscures the nanoplankton. Another problem is that, with such exceedingly small organisms, it is not easy to decide whether the settled cells contain chlorophyll and are autotrophic, though certain stains and fluorescence microscopy have proved useful in identifying chlorophyll. Counting of unconcentrated seawater samples has been attempted but is not very accurate. The serial dilution technique has been occasionally used but it is very tedious and somewhat inaccurate. The Coulter Counter, an apparatus which estimates the size fractions of particulate matter, has been used to determine the size spectra of phytoplankton. The seawater sample containing the phytoplankton to be estimated is passed through the counter which measures the amount of electrolyte displaced as a particle moves through a small electric field. The measurement is, therefore, of particle volume, but this may be related to the diameter of a sphere of equivalent volume. Using suitable orifice sizes for the Coulter Counter, a spectrum of the relative abundance of different sizes of phytoplankton particles, ranging from *ca.* 1–100 μm (diameter), may be obtained. Cell chains cannot be distinguished from single cells. The method is rapid and gives useful comparative data in surveys of specific areas. Clearly, however, no taxonomic information is obtained; particles of detrital material and small, non-photosynthetic organisms will also be counted. Peaks in the particle spectrum may be identified with major algal species if a cursory microscopical examination of a similar water sample is made simultaneously.

The nanoplankton was crudely defined as that part of the phytoplankton (some authorities include heterotrophic organisms) not retained by the finest plankton net, the finest silk net then being the No. 25 (200 mesh per inch), average aperture size *ca.* 60 μm. (Some workers suggested an aperture size of *ca.* 40×50 μm for a wet net.) There can be no precision in catching power with a mesh dimension, however; age and method of manufacture of net, speed of towing, and especially the diversity of size and density of the phytoplankton, all may modify the size of organism retained. By definition, nanoplankton cannot be sampled by net hauls, though many plankton organisms far smaller than 40 μm may be retained to a variable extent, largely dependent on the degree of clogging.

Apart from the uselessness of nets for nanoplankton, net hauls for the quantitative sampling of the larger so-called net plankton can be subject to large errors which usually cannot be estimated. Once more, all the factors modifying effective mesh size, the variable dimensions of the phytoplankton species with chain-forming species, and above all the richness of the suspended particulate matter, both phytoplankton and detritus, will affect the catching power to an incalculable degree. Net hauls cannot, therefore, yield a quantitative analysis of phytoplankton crop. Instead, the investigation of total crop is now usually achieved by sampling known quantities, usually many litres, of seawater throughout the euphotic zone with large water samplers (e.g. Van Dorn bottles or Nansen bottles). Part of the sample is subjected to a cell count to assess the floristic composition; a sedimentation technique is often employed. Even with relatively large water samples, however, the density of the larger more widely distributed species is often misrepresented by cell counts (cf. Paasche, 1960). Usually another quantitative subsample of the seawater collected is used for evaluation of the chlorophyll *a* concentration. Filtration of the seawater is usually through a filter of nominal porosity 0.45 μm to attempt to retain the smallest nanoplankton, but the porosity employed and the quantity of water filtered may need to be adjusted if the filter clogs rapidly, as in some eutrophic waters with very high crop densities. The chlorophyll in the phytoplankton retained on the filter is subsequently extracted, normally with 90% acetone; magnesium carbonate is frequently used as a pad in filtration to prevent acid decomposition of chlorophyll. Grinding or sonification is sometimes employed to break down the phytoplankton cells. The extracted chlorophyll may be estimated spectrophotometrically against a standard. In many investigations only chlorophyll *a* is determined, at 665 nm. (A correction must be applied for absorption due to turbidity.) Since the extraction technique removes other chlorophylls and carotenoids, however, measurements of absorbance at different wavelengths can give values for chlorophyll *a*, *b* and *c* and for total carotenoids (cf. Richards with Thompson, 1952). A detailed account of the procedure may be obtained from Parsons (1966) and the SCOR handbook.

In oligotrophic areas where the chlorophyll concentration may be so low that accurate measurement by spectrophotometry is not possible, fluorometric measurement can give accurate estimates about an order of magnitude lower, but total chlorophyll is then measured. Fluorometry can yield information on the quantity of degraded chlorophyll as well as the amount of active chlorophyll. An outline of the method is given by Yentsch and Menzel (1963). The determination of the proportion of active and degraded chlorophyll is of great significance in primary production.

Whatever refinement of method is used to determine chlorophyll, the values represent comparative data indicating relative amounts of crop. Any assessment of biomass of phytoplankton organic matter must assume a fixed relationship between cell carbon, organic matter and chlorophyll. While the ratio of chlorophyll to carbon is not constant (Steele and Baird, 1962, 1965; Strickland and Parsons, 1965; Bunt, 1967; Zeitzschel, 1970), approximate conversion factors are available (Cushing, 1968; cf. Chapter 3). Chlorophyll determinations thus represent a rapid and extremely valuable method for assessing standing crop (cf. Chapter 9) but the limitations of the method, due especially to variable carbon/chlorophyll ratios, must be appreciated (cf. Bunt, 1967).

Regional Differences in Crop

Marked differences exist between the phytoplankton biomass of various regions. Spatial comparisons must be examined cautiously, however, if samplings have been made along a line of stations at fairly long intervals, and sampling has not been repeated to assess temporal variations and the effect of lateral advection of water. Generally, oceanic regions in temperate and moderately high latitudes have a fairly high standing crop whereas low latitudes, and especially the more central areas of oceanic gyres, are marked by low crops. This overall pattern is complicated by the effect of ocean currents, bottom topography and islands. Divergences between ocean currents with upwelling of deeper nutrient-rich water, and areas near islands, especially at some distance in the lee of oceanic islands, are characterized by high crops. This "island effect" (Doty and Oguri, 1956; Jones, 1962), which usually involves both crop and primary production, while doubtless partly attributable to increased nutrient concentration, may be experienced far from the island itself. Jones, for example, described an increased crop at 200 miles from the Marquesas Islands. While discontinuous accumulations of plankton consequent on hydrological movements may be partly responsible, other factors such as increased concentrations of trace metals and organic growth-promoting agents may be involved. Probably "land mass" and "island mass" effects are the outcome of many, not necessarily related, factors. Some examples of "land mass" effect, especially on production, are given in Chapter 9.

Along coastal regions, offshore winds may produce upwelling of rich water leading to high crops. On a major scale, longshore winds along continental coasts can cause water movements, which turn offshore under influence of Coriolis force, producing upwelling on a grand scale; such regions have very large phytoplankton crops. In addition, in most regions of the world, whatever the latitude, the plankton near the coast is normally considerably richer than in offshore and oceanic areas. The extra contribution of nutrients consequent on land run-off is partly associated with the higher crops, but stratification and stabilization of the near-surface layers with lowered salinity water from river outflows are among other contributory factors.

Many investigations of spatial differences in phytoplankton consider standing crop and primary production; the two are clearly linked and in order to save needless repetition both aspects of phytoplankton growth are sometimes discussed in the following pages. Chapter 9, dealing with Primary Production, also includes some examples where crop has been considered with production.

Hentschel and Wattenburg (1930), investigating the density of plankton, essentially phytoplankton, for the upper 50 m in the South Atlantic, demonstrated the comparative poverty of the central gyre of the ocean and of the lower latitudes generally (Fig. 5.2). Apart from rich crops in the region of marked upwelling off north-west Africa (Mauritania), and an upwelling area in equatorial regions off West Africa, a zone of relatively high crop extended far across the Atlantic in near equatorial latitudes. This zone is probably associated with sub-surface richer waters upwelling in relation to the divergence between the South Equatorial Current and the Equatorial Counter Current. The distribution of nanoplankton crop was similar to that of the net plankton. Zernova (1974) has confirmed the variation in phytoplankton biomass in different parts of the tropical Atlantic. Richer zones were characteristic of coastal upwelling areas off Africa, with a biomass exceeding 500 mg/m^3 in several places. One area receiving a branch of the rich Benguela Current, with very high phosphate concentration (2.39 μg-at P/l), had the highest biomass of 1300 mg/m³. High crops were also found in the mixed waters in the western Atlantic near the outflow of the River Amazon, and in a coastal shelf area of the Gulf of Mexico. Oceanic waters were low in nutrients and had a much reduced biomass, but a rich near-equatorial belt occurred in the open Atlantic with phytoplankton ranging from 10 to 100 mg/m³. North and south of latitude 10° the biomass decreased to <1 mg/m³. Zernova claims that the biomass in the tropical Atlantic is similar to that of the tropical Pacific. Seasonal variation in crop is apparent where there are changes in the hydrological regime. This is confirmed by Mahnken's (1969) studies of primary production in the tropical Atlantic (cf. Chapter 9). Zernova found that diatoms were dominant in coastal, especially eutrophic waters, and in parts of the equatorial zone; other algae (dinoflagellates, cyanophyceans, coccolithophorids) could be dominant elsewhere.

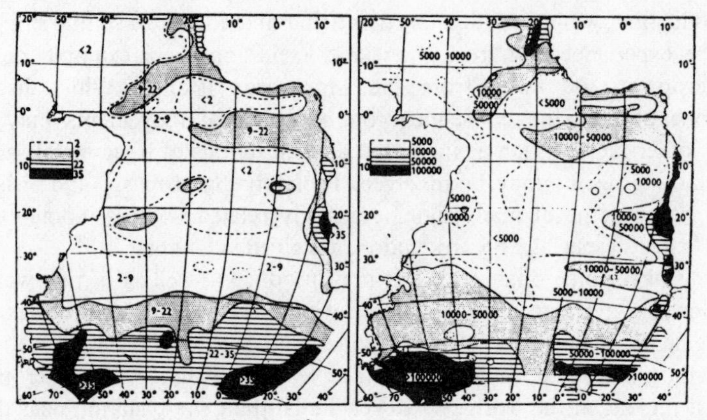

Fig. 5.2. Distribution of phosphate (left) in μgP/l and the density of plankton organisms (number per litre) (right) in the upper 50 m layer in the South Atlantic (after Hentschel and Wattenberg, 1930; Harvey, 1955).

Upwelling occurs on a seasonal basis along several regions of the west coast of Africa, though some extensive areas remain stratified throughout the year. Where seasonal upwelling occurs, it is usually followed by a considerable increase in phytoplankton production and crop. Off Morocco, Furnestin (1957) describes pronounced upwelling beginning in spring and reaching a maximum in summer. A very marked growth of phytoplankton occurs over the summer, sometimes, according to Furnestin, approaching "red tide" proportions. Seguin (1966), noting the extraordinary richness of phytoplankton which can be found off Mauritania and Senegal with upwelling, has followed the cycle of events off Dakar. Two outbursts of phytoplankton occurred, both during the main upwelling colder period, which lasts approximately from December to May. The first was during March to June and the second in December/January, both characterized by few algal species in great abundance. Diatoms dominated the catches, and there was a succession of dominant species including *Chaetoceros densus*, *C. pseudocurvisetus*, *Stephanopyxis palmeriana* and *C. socialis*. Upwelling off the Ivory Coast, from about June to October, is also accompanied by increased phytoplankton (cf. Seguin, 1970). Off Lagos, on the Nigerian coast, upwelling, while not quite regular, is generally experienced over August and September and, as a consequence, a phytoplankton maximum occurs (cf. Longhurst and Bainbridge, 1964).

Hentschel and Wattenburg (1930) also found generally dense crops of phytoplankton off south-west Africa in the marked upwelling region of the Benguela Current. South of the Sub-Tropical Convergence there is an obvious increase in crop characteristic of the nutrient-rich sub-Antarctic and Antarctic waters. The generally much greater density of phytoplankton at high southern latitudes is demonstrated by early data from Hart (1934) which show a remarkable fall in crop, as cell counts, on proceeding to lower latitudes. The data (Table 5.1) do not illustrate the full magnitude of the change, since the main bloom at the extreme southern stations had already passed when sampling occurred.

Table 5.1. Variation in phytoplankton crop with latitude (from Hart, 1934)

Stat. no.	Latitude	Diatomales		Dinoflagellata		Schizophyceae	
		Total	%	Total	%	Total	%
661	57° 36′ S	11,046,400	99.88	12,800	0.12	—	—
663	53° 34½′ S	6,249,200	100.00	—	—	—	—
666	49° 58¾′ S	354,600	98.42	5700	1.58	—	—
670	44° 52′ S	14,700	89.09	1800	10.91	—	—
671	43° 08′ S	52,200	76.99	15,600	23.01	—	—
673	38° 10½′ S	1200	26.67	2400	53.33	900	20.00
675	34° 08′ S	—	—	1800	100.00	—	—
677	31° 16¼′ S	4200	35.00	7800	65.00	—	—
679	26° 06½′ S	22,500	81.53	5100	18.47	—	—
681	21° 13′ S	2000	20.83	7600	79.17	—	—
684	15° 37′ S	400	8.69	3600	78.26	600	13.05
687	09° 47′ S	2600	26.00	7400	74.00	—	—
690	03° 17¾′ S	800	2.98	6000	22.39	20,000^{+}	74.63
693	02° 59′ N	5400	7.11	4600	6.05	66,000^{+}	86.84
699	14° 27¼′ N	1600	12.31	9000	69.23	2400	18.46

(From Hart, 1934).

A fairly similar distribution pattern of phytoplankton crop holds for the South Pacific; denser crops of phytoplankton are typical of higher southern latitudes, associated with raised nutrient concentrations. Phytoplankton biomass is by no means uniform, however (cf. El-Sayed, 1966, 1970). The variability of crop in the Antarctic is now recognized; in general the Pacific sector seems to be considerably less rich than the Atlantic (cf. Chapter 9). In the Pacific, outside high southern latitudes, the central areas of the sub-tropical and tropical ocean north of the Sub-Tropical Convergence have low levels of crop related to the nutrient-poor waters. The now well-known Cromwell Counter Current with the obvious divergence in near-equatorial regions causes upwelling of nutrient-rich water, responsible for increased phytoplankton crop in that area (Austin and Brock, 1959). Bogorov (1959) also noted large crops of phytoplankton in regions of upwelling near the equator in the Pacific. In the eastern equatorial Pacific, in oceanic waters off central America, some areas exhibit comparatively rich crops, apparently related to a complex current system with some upwelling in the Costa Rica Dome region. As in the South Atlantic, a zone of very active coastal upwelling is found along the eastern border of the South Pacific; the coasts of Chile and Peru are marked by very large phytoplankton crops, and a rich tongue extends towards the Galapagos.

Over much of the tropical and sub-tropical areas of the Indian Ocean, especially in the southern sub-tropical gyre, the standing stock of phytoplankton is poor. Much denser crops are characteristic of the northern Arabian Sea (Fig. 5.3). Denser phytoplankton is also typical of the sub-Antarctic and Antarctic waters, but large areas of the Indian sector of the Antarctic are poorly studied. The area around the Kerguelen Islands has been investigated; Philippon (1972) describes rich crops but these are close to the Islands. El-Sayed and Jitts (1973) also reported very high chlorophyll concentrations off the Kerguelen and Heard Islands, and in more oceanic waters near the Antarctic Convergence.

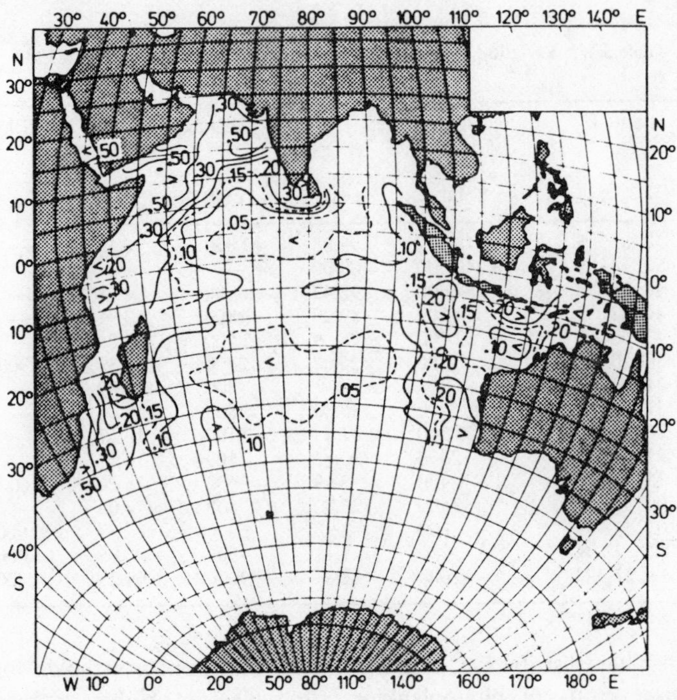

Fig. 5.3. Average concentration of chlorophyll a (mg/m^3) in the surface layer of 0–50 m during June–Sept. (Krey, 1973).

The Indian Ocean north of the equator is unlike the North Pacific and North Atlantic, not only in its limited extent, but by its division into the Bay of Bengal and Arabian Sea, with the consequent modifications of ocean circulation. Moreover, the Bay of Bengal especially receives the outflow of very large river systems, so that many northern areas of the Bay have somewhat lower salinity and may benefit from additional nutrients. These effects are greatly accentuated by the monsoonal climate. Both coasts of the Indian sub-continent experience vast contributions from rivers during the monsoonal seasons. Krey (1973) comments on rich nutrients supplied by the River Indus. It is not surprising that run-off, in addition to some coastal upwelling, for example off south-west India, leads to a general richness of phytoplankton crop, not only along the coast but more widely over much of the northern regions of the Indian Ocean.

Monsoonal effects are not limited to waters immediately adjacent to the Indian sub-continent. Very conspicuous is a region of intense upwelling with rich crop which develops seasonally with the south-west monsoon off east Africa (Somalia) and the south Arabian coast (Fig. 5.3). Rich phytoplankton is also characteristic of some more minor areas where complex upwelling occurs, as off the Andaman Island, Sumatra and north-west Australia (cf. Cushing, 1975).

A seasonal monsoonal gyre is a feature peculiar to the northern Indian Ocean; south of the equator is the sub-tropical gyre with generally nutrient-poor water as in the other oceans. The two systems meet in a front, more pronounced at sub-surface levels, at approximately latitude 10°S, but the change from the north-east to the south-west monsoon will modify the front (cf. Wyrtki, 1973). As in the other two oceans, an equatorial counter-current is present, and divergences near the equator can increase crop somewhat. Krey (1973), describing phytoplankton-geographical regions of the Indian Ocean, points out that the rich regions (Antarctic, sub-Antarctic, West Wind Drift and major upwelling areas) are dominated by diatoms and may have concentrations of 10^4 to 10^5 cells/l; less rich areas (Arabian Sea, equatorial current areas) have a great proportion of dinoflagellates, often with coccolithophorids with concentrations of *ca.* 10^4 cells/l; the poorest areas include the south sub-tropical gyre and central part of the Bay of Bengal, with mostly dinoflagellates and coccolithophorids amounting to $< 0.5 \times 10^4$ cells/l. A marked difference, according to Krey, in the Indian Ocean is the abundance of Cyanophyceae with massive blooms of *Trichodesmium*. Figure 5.3 illustrating average chlorophyll *a* concentrations from June to September, shows the richness of the upwelling regions and the area off south-east Africa, the moderate crops off the Madagascar Coast, west of Australia and south of the East Indies, and the poor crops north and south of the equator. In the December/March period there is considerable reduction in the extent of some of the richer areas; a crop of <0.05 mg chlorophyll a/m^3 over much of the central Indian Ocean underlines the poverty of the large oceanic region (Fig. 5.4).

It is curious that a markedly increased density of phytoplankton has not been observed off south-western Australia where it might have been expected from upwelling similar to that off the coasts of Peru and Chile in the Pacific and off south-west Africa in the Atlantic (cf. Smith, 1967). The lack of intense upwelling in this area is not fully understood. (Some further discussion of crop and production in the Indian Ocean is included in Chapter 9.)

In the North Atlantic and North Pacific the overall picture is of high standing crops of phytoplankton in temperate and boreal areas. South of the Kuroshio Current system in the Pacific, and south of the North Atlantic Drift in the Atlantic Ocean, crops tend to be lower, with especially low biomass in the central regions of gyres, particularly in the Sargasso Sea. Apart from the usual effect of increased vertical mixing, and hence enrichment of nutrients, with greater crop densities in areas of submarine banks, ridges and plateaux, the eastern edges of the northern oceans are characterized by upwelling on a very large scale, associated with increased phytoplankton crops off California and off north-west Africa, respectively. The more general richness of the boreal areas is exemplified by Bogorov's statement (Bogorov, 1958a) that the boreal Pacific is on the whole some ten times richer than tropical areas of that ocean (cf. also Holmes, 1958). He suggested phytoplankton crops exceeding 500 mg/m^3 south of the Kurile Islands, with less than 100 mg/m^3, often below 20 mg/m^3, in areas in the warm Kuroshio waters. Motoda, Taniguchi and Ikeda (1974) report crops of about 10^4 to 10^6 cells/l in sub-Arctic waters of the western North Pacific. A different flora occupies a transition zone with 10^3 to 10^4 cells/l between the Oyashio and Kuroshio Currents. To the south from the Kuroshio to the equator the plankton flora is comparatively uniform and less rich, viz. 10^2 to 10^3 cells/l.

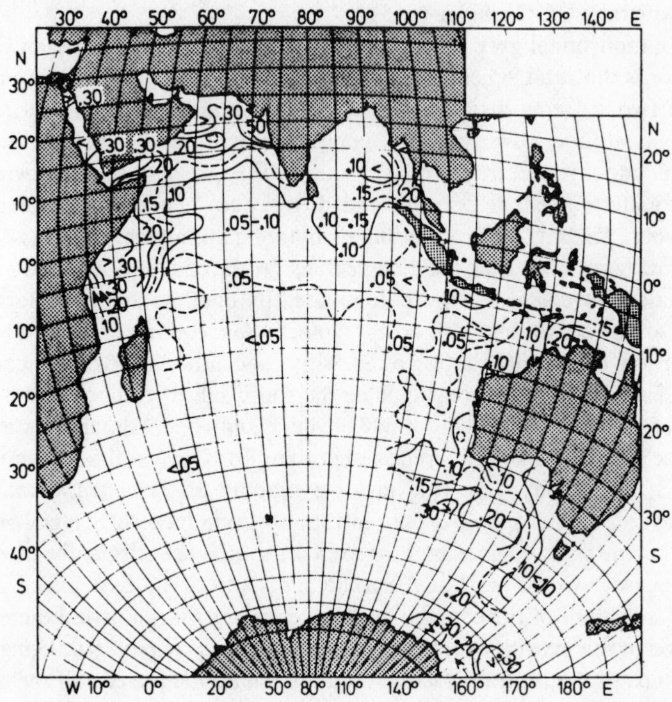

Fig. 5.4. Average concentration of chlorophyll *a* (mg/m³) in the surface layer of 0–50 m during Dec.–Mar. (Krey, 1973).

Temporal Variations in Crop

It is important that such comparisons of crop be viewed against the different degrees of temporal change in boreal and in tropical and sub-tropical areas. Whereas in lower latitudes the crop over the year shows relatively small fluctuations, at high latitudes there is a very marked seasonal difference with high crops over the productive part of the year, particularly in the spring, contrasting sharply with low biomass over winter.

Variations in the amplitude and the duration of phytoplankton (and zooplankton) seasonal cycles have been summarized by Cushing (1975), who points to the greater amplitude both in numbers and volume of plankton algal changes at higher latitudes. While the amplitude of change may vary between ten and a thousand times, the higher values are typical of cold temperate waters. Cushing finds that for crop expressed as chlorophyll the amplitude of seasonal change is usually below that of cell numbers, but it may be very large at high latitudes if only net phytoplankton is measured. For example, Digby's estimate of Greenland coastal plankton showed an amplitude of 9000. Cushing suggests that the duration of the production cycle may vary even as much as from 1 to 12 months in different regions, but more usually averages about 3 months. In Antarctic waters a calculation by Currie from Hart's data indicates an amplitude of forty times; the main production lasts only about 3 months, though the whole cycle extends to about 8 months. Hart described a compression of the cycle at higher southern latitudes (*vide infra*). By comparison, the algal stock in the high Arctic is low; peak production is limited to the order of 1 month, and Cushing describes the whole cycle as occupying only some 3 months, the amplitude being as little as six times. Zenkevitch (1963) also points to the substantially reduced crop at the highest latitudes. Although high crops of phytoplankton can occasionally be found in very cold Arctic waters, the flowering period is of very short duration. In the Arctic Basin the biomass may be only one-twentieth that of the northern Barents Sea. Low surface salinity may reduce productivity in the Arctic Ocean, but the thick ice cover which permits phytoplankton growth only for a very short period is mainly responsible. The northern Barents Sea is in turn far less productive on an annual basis than a warmer temperate area.

Significant differences may be observed between the crop of phytoplankton in coastal and more offshore areas. Robinson (1970), summarizing results for the Continuous Plankton Recorder samplings, while recognizing the limitations of the method particularly for phytoplankton in that only a small portion of the phytoplankton is retained and the assessment of chlorophyll is approximate, has drawn important general conclusions regarding phytoplankton abundance. There were differences in the type of plankton cycle in areas of the North Atlantic (Fig. 5.5). Surveys demonstrated that shallower waters were associated with high standing crop, with an early spring outbreak and with a relatively long productive season (Figs. 5.5 and 5.6). These three characteristics were not always associated. For example, an area off West Greenland had a very high standing stock but a relatively late outbreak and short season; coastal waters off Norway, Canada and the Bay of Biscay had low crops, although the season was early and long-lasting. Figure 5.6b shows the relative abundance of crop in various regions of the North Atlantic; Fig. 5.6a and c illustrate the date and length of season. Generally, deeper oceanic waters had a smaller phytoplankton biomass but over most of the region studied, bounded roughly by the North Atlantic Drift, the colder boreal areas had higher crops than warmer temperate areas, especially those off south-west Europe.

In the northern Pacific Anderson (1964) studied the phytoplankton crop, as chlorophyll *a*, off the Oregon and Washington coasts and found as in many other regions that the crop was richer in coastal areas than further offshore. Beyond the coastal areas algal density was higher in winter than in summer. Surface waters near the Columbia River outflow showed a greater biomass than open deeper waters, but the densest and most variable crops occurred off the Oregon coast during the summer when coastal upwelling was active. The vertical distribution of the phytoplankton as chlorophyll also showed some significant differences. Although sometimes more or less uniformly distributed, chlorophyll was massed mostly at the surface or a little deeper in the river plume, whereas, more offshore, in oceanic areas off Oregon during the summer period, Anderson (1969) demonstrated a sub-surface maximum of chlorophyll. This sub-surface layer extended beyond the edge of the continental shelf, roughly at a depth of between 50 and 75 m. The concentration was between three and ten times the surface chlorophyll and was associated with discontinuity layers, due to a permanent halocline and a shallower seasonal thermocline. The phytoplankton sub-surface maximum may be widely distributed in the North Pacific, at least between latitudes 35–48°N (cf. Chapter 6).

Currie (1958) also found a far richer phytoplankton crop near the Portuguese coast than further offshore, and a higher rate of photosynthesis at the inshore station, but offshore the euphotic column was much deeper as a consequence of the greater depth of light penetration. The length of the euphotic column in relation to crop is examined by Riley, Stommel and Bumpus (1949). In offshore waters of the western North Atlantic, at lower latitudes, the phytoplankton crop per unit area may not be on average very much smaller than a rich boreal area such as the Gulf of Maine, provided that the very rich spring flowering in boreal waters is excluded. Riley *et al.* point out that although surface crops show a seasonal range of 2500 to 45,000 plant pigment units/m^3 for Georges Bank, as against 1000 to 23,000 units/m^3 for slope waters, and 140 to 1900 units/m^3 for the Sargasso Sea, in view of the different depths of the euphotic column, if the spring peak density in the Georges Bank area is excluded, the annual plant crop per unit area (per m^2) is about the same in slope water as on Georges Bank. But in the Sargasso Sea, despite the considerably greater depth of the euphotic layer, the crop is

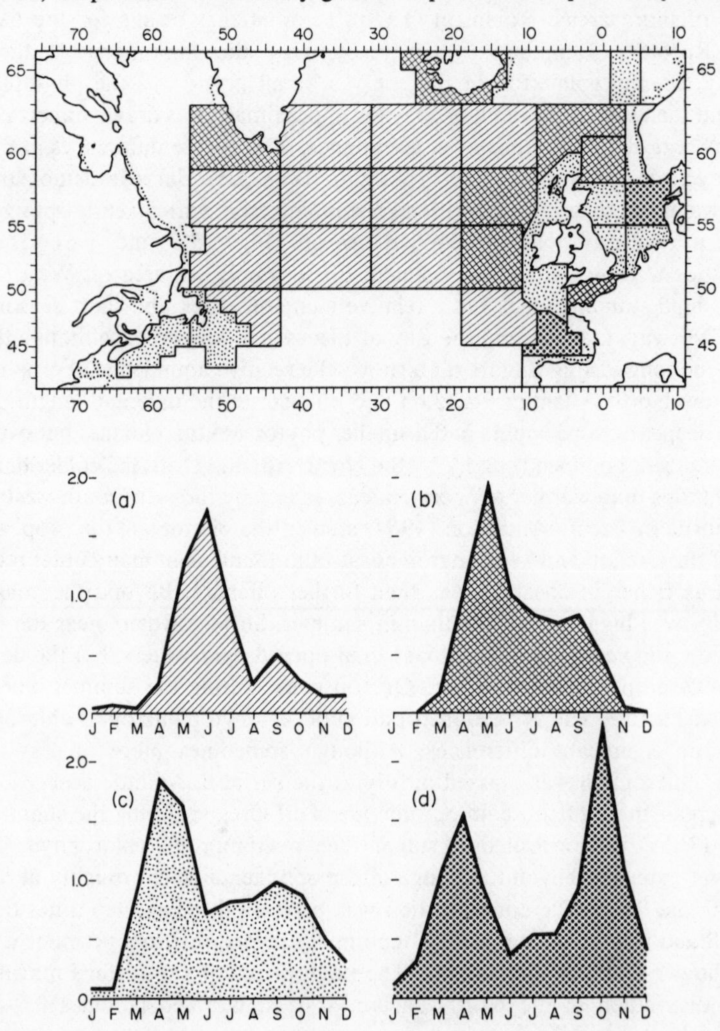

Fig. 5.5. Graphs showing four types of mean seasonal cycles of phytoplankton ((a), (b), (c), (d)) typical of different geographical areas, illustrated in upper map. (Robinson, 1970.)

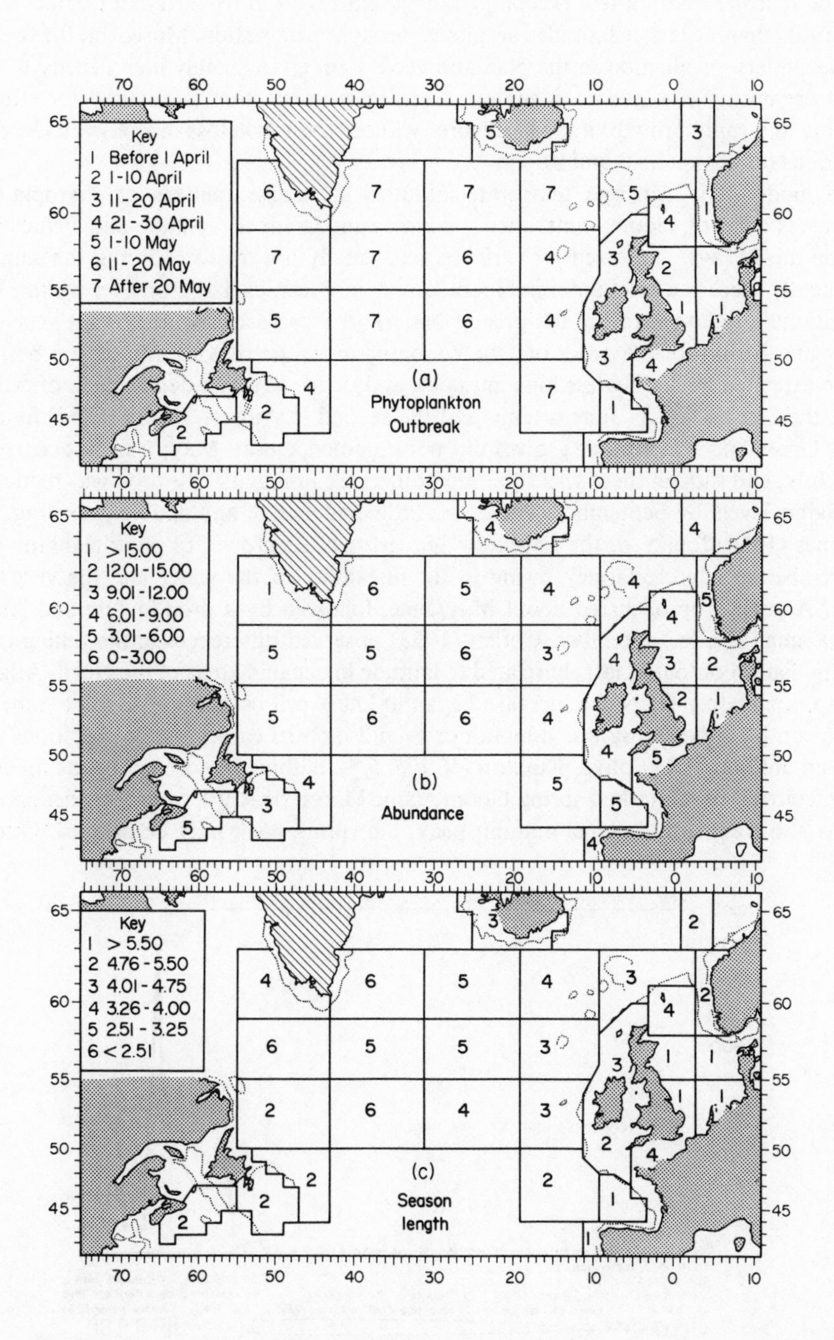

Fig. 5.6. Charts giving (a) the average time of the spring outbreak of phytoplankton; (b) the average abundance of phytoplankton; (c) the average season length in months of phytoplankton production in the standard areas of the Recorder Survey. The key to the numbers is given at the top left-hand corner of each chart. (Robinson, 1970).

distinctly lower. With a few exceptions, the biomass of sub-tropical and tropical waters is far less than at higher latitudes despite better light penetration. Moreover, for successful secondary production in the plankton ecosystem a reasonably high density of algal food organisms confined to a comparatively narrow stratum is required for effective feeding and rapid growth of the herbivores without too much loss of energy in searching for foods (cf. Chapters 8 and 9).

In moderately high and temperate latitudes the classic pattern of phytoplankton change is a spring bloom, consequent on increasing incident radiation and a shallowing of the mixed layer, followed by variable, but usually lower crops during the summer; frequently there is a brief autumn bloom, and a very low level of crop over winter. Wide variations exist, however, in the precise pattern in a sea area. Apart from a very short blooming period, of the order of 1 or 2 months in Arctic regions, the season becomes more extended at slightly less high latitudes, and there may be a double peak of production, the first as late as June or July and the second in August. Digby (1953) found in East Greenland waters that growth did not commence until May; a peak occurred in mid-July, and though there was some slight increase in August, the crop was reduced to negligible levels by September. There was little evidence of any double peak (Fig. 5.7). Holmes (1956) found for the Labrador Sea virtually no growth of phytoplankton from November to March, largely owing to the instability of the water and the very poor light. A peak crop occurred about May/June, followed by a drop in July and August and a small rise in September. Corlett (1953) observed differences in the timing of the spring diatom outburst, largely related to latitude in oceanic areas of the North Atlantic. At approximately 60°N the increase began in late April or May; at a point some 370 miles south the flowering was a month or even 2 months earlier. At both stations there was an autumn rise in phytoplankton (cf. Fig. 5.8). Although more temperate areas are characterized by a marked spring bloom about March or April, rather smaller summer crops and frequently a small autumn peak, the spring peak may be delayed. Often in

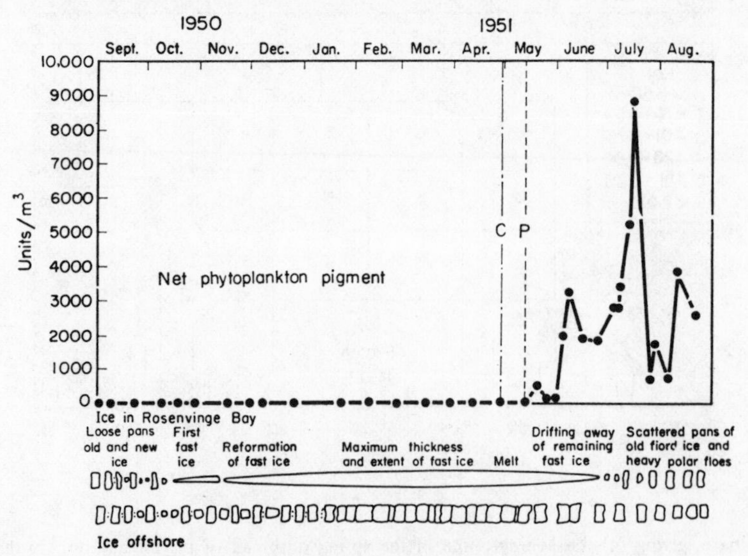

Fig. 5.7. Seasonal changes in growth of phytoplankton and in ice conditions in Scoresby Sound, Greenland. C = date of onset of phytoplankton growth as seen by cell counts. P = date by which phytoplankton growth was indicated by net pigment (after Digby, 1953).

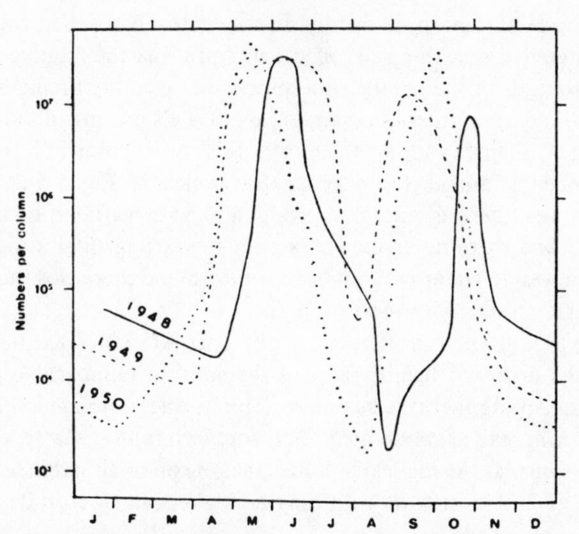

Fig. 5.8. Seasonal cycles of diatoms over 3 years from samplings from Ocean Weather Ship at Station "I" (from Corlett, 1953).

inshore, particularly estuarine environments, the maximum crop may not appear until about June with a decline or irregular fluctuations over the rest of the summer. The delay may be partly associated with increased turbidity and consequent poorer light penetration inshore (cf. p. 239).

The variations described by Robinson (1965, 1970) in the onset of the spring increase and duration of the flowering period for the North Atlantic have already been noted. Robinson's four main patterns for the seasonal cycle in the North Atlantic fairly reflect broad geographical distributions (cf. Fig. 5.5). The late spring maximum (June) with a very small second increase in September, typical of deep oceanic waters (type (a), Fig. 5.5), contrasts with the maximum occurring as early as March/April and a substantial autumn crop (type (c)), even exceeding the spring peak in type (d), both latter types being characteristic of many shallow areas such as much of the North Sea and western coastal regions of the British Isles. Over certain shallow waters the season of abundance may apparently be prolonged without pronounced spring and autumn peaks (cf. Colebrook and Robinson, 1961). Cushing (1975), examining some of the data, suggests that the variations in timing, intensity and duration of the phytoplankton cycles are mostly related to changes in the production ratio (compensation depth/depth of mixing) which is in turn largely dependent on seasonal variations in light and wind strength. The production ratio will increase as spring advances in temperate waters, since incident radiation will increase and wind strength will tend to decrease, reducing mixing. Sunshine and wind force, however, may vary over quite short distances. Thus, in the north-east Atlantic, oceanic regions beyond the continental shelf are subjected to comparatively deep mixing with the long fetch of the wind. The depth of mixing decreases only slowly during the spring and the phytoplankton peak is delayed till May or June in mid-ocean northern areas, but the productive season lasts for five months. In the shallower shelf areas off south-west England and western Scotland, some lessening of the fetch induces an earlier start to production with the peak at the end of April, but somewhat decreased light penetration offshore may become an increasingly important factor. A low amplitude autumn peak occurs. In the central North Sea with the shallowness and more sheltered water, the peak occurs in April, but the relative importance of turbidity and the strength of the summer thermocline affects the timing and size of the summer crop; a marked autumn crop occurs of about the same magnitude as the spring peak. The English Channel and north North Sea are somewhat intermediate in character between oceanic and shelf phytoplankton cycles (cf. Fig. 5.5). In the Southern Bight of the North Sea there is an early production with variable crop lasting from March to October. The extreme shallowness with the strong tidal streams mixes the algae throughout the whole water column to the bottom and increases the turbidity, but the degree of mixing is modified by wind strength.

Variations in the timing and duration of the phytoplankton increase have so far been related mainly to the northern hemisphere. In the south a comparable pattern exists, with the timing of the spring increase becoming later as one proceeds to higher southern latitudes. Similarly, whereas at moderately high southern latitudes a typical spring and autumn peak may occur, at the highest latitudes the season of abundance is compressed into a short period of 2 or 3 months with only a single peak of crop density (cf. Hart, 1942).

Heinrich (1962) emphasizes that spatial and temporal variations in crop must include consideration of the timing and intensity of grazing. Relating changes in phytoplankton density in parts of the North Pacific to the life histories and grazing activities of dominant zooplankton species, she points out that in oceanic areas of the Bering Sea the zooplankton broods maintain the spring phytoplankton crop approximately in balance. Differences in amplitude of seasonal change exist between the North Atlantic and North Pacific. Although both oceans show a seasonal effect at high latitudes, the dense crop in spring is usually more obvious in oceanic areas of the Atlantic. This has in part been attributed to more constant grazing in boreal and sub-Arctic areas of the Pacific Ocean. Examples of the effects on crop of different grazing patterns in other seas are considered more fully in Chapter 8. Cushing (1975) claims that the autumn phytoplankton outburst in temperate waters is more associated with a contemporaneous decline in the herbivorous zooplankton, especially associated with their migration to deeper layers, than with the disruption of the seasonal thermocline.

In contrast to the very marked changes in crop abundance between the rich productive season and the period of winter scarcity typical of higher latitudes, in warm oceanic regions where there is little change in water temperature, incident light or salinity throughout the year, the crop, though small, may suffer little annual variation. Such a condition assumes that the intensity of grazing is also more or less uniform (cf. Cushing, 1959a and b; Steemann Nielsen, 1958a, 1962) and (cf. Chapter 8). The quasi-equilibrium described for some areas of open ocean in the sub-Arctic Pacific resembles to some extent this near balance achieved in warm oceanic areas, but in warm seas the crop maintained at more or less steady level by constant grazing tends to be much smaller. Determinations of standing crop made merely on a few occasions during a year may possibly then serve as a reasonable basis for annual estimates.

In contrast to the warm open oceans, many neritic tropical areas experience large fluctuations in phytoplankton crop, due partly to local climatic conditions, partly to a different grazing intensity. Subramanyan (1959), for example, has suggested a maximum amplitude of algal crop of 150 times for some India waters. Malone (1971) states that in the eastern tropical Pacific the average grazing pressure in oceanic areas is some three times that of inshore waters, so that the crop inshore is more variable than in the open ocean.

In the open tropical eastern Indian Ocean, Tranter (1973) observed seasonal changes in the phytoplankton and the zooplankton but these were usually low in amplitude, except that the algal and zooplankton cycles could become out of phase. Zernova's (1974) observations on phytoplankton standing crop in the tropical Atlantic have already been noted. Seasonal changes could be identified in those areas where seasonal variations occurred in the hydrological regime, in particular, changes in the positioning and intensity of upwelling.

Blackburn, Laurs, Owen and Zeitzschel (1970) studied differences in standing stocks of phytoplankton, zooplankton and of two groups of micro-nekton in the tropical eastern Pacific (approximately 16°N to 15°S). The phytoplankton in the more western areas showed a simple seasonal effect with one maximum and one minimum and with an amplitude of less than twice (values about 15–25 mg chlorophyll a/m^2). The maximum could occur at any time between April and September, the minimum between October and January (cf. Fig. 5.9 and 5.10). In more eastern regions of the area under study, sampling in February/March and in August/September showed a small seasonal change in some areas, with the phytoplankton maximum in the summer, but other areas exhibited no significant change. Fluctuations in zooplankton density indicated a slightly higher population in the winter period (Fig. 5.10). No clear reason may be put forward for the slightly increased populations of phytoplankton during the April to September period. Nearer the equator, as has frequently been observed, there was a higher standing stock, presumably related to divergent upwelling. This is one of the few direct investigations confirming the suggestion that in open tropical oceans fluctuations in crop during a year are frequently insignificant.

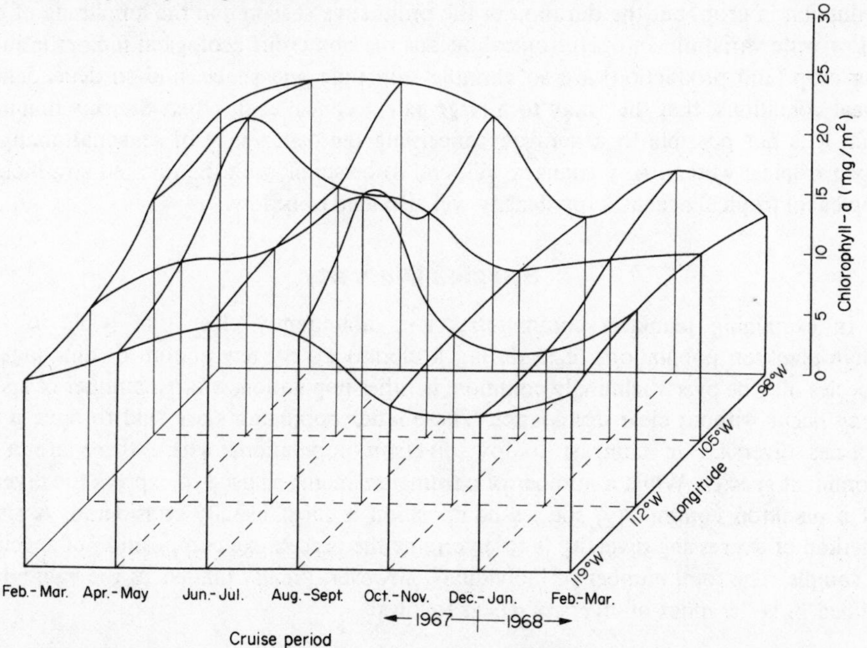

Fig. 5.9. Variation in chlorophyll a with cruise period and longitude in the enlarged western area. Data are for all latitudes combined. The longitudes shown are approximate (Blackburn *et al.*, 1970).

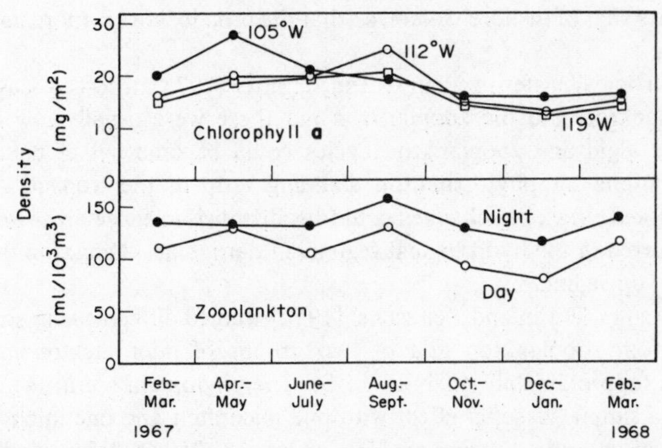

Fig. 5.10. Seasonal variations in standing stocks of phytoplankton and zooplankton in the western area of the tropical eastern Pacific (Blackburn *et al.*, 1970).

Sournia (1969) has given a description of seasonal changes in the phytoplankton crop of the world's oceans which summarizes our discussions. In coastal regions or in the vicinity of islands, in both high and low latitudes, a seasonal cycle in crop (and in primary production) is usual, though in lower latitudes the cycle is dependent almost entirely on the local climatic and hydrological conditions. In open oceans in temperate and high latitudes the biological seasons are pronounced, and of a more or less similar pattern. During winter, phytoplankton is extremely poor due to mixing and poor light conditions. A burst of growth occurs in spring, with stratification and the attainment of critical illumination; during summer impoverishment of nutrients may cause some reduction in crop, but the duration of the productive season and the amplitude of crop show wide variation. In open tropical oceans the numerous ecological factors influencing crop (and production) are so variable with time and space, and so dependent on local conditions, that they may to a large extent cancel each other. Sournia maintains that it is not possible to generalize concerning the occurrence of seasonal change in open tropical waters. Any changes will tend to be small; with the marked stratification typical of tropical oceans, crop density will also tend to be low.

Species Diversity

In examining plankton communities it is abundantly clear that while in some phytoplankton populations (e.g., during a bloom) a very few diatom or dinoflagellate species may be overwhelmingly common, in other populations a large number of species may occur without clear dominance. These latter populations are said to have a high species diversity, in contrast to low diversity populations where there are a few dominant species. While a number of parameters might be used to express the diversity of a plankton community, species composition is most usually considered. A simple method of expressing diversity is to determine the percentage composition of species in a sample. The total number of individuals, however, greatly influences the value determined. A better index of diversity d is as follows:

$$d = \frac{S}{\log_{10} N}$$

where S = number of species in the population,
N = number of individuals in the population.

An improved index is:

$$d = \frac{S-1}{\log_{10} N}$$

since theoretically if all the individuals in a population, or in a sample, belonged to one species, then d becomes zero.

Hulburt (1963) described some phytoplankton floras exhibiting marked differences in diversity. He first used as an index of dominance the ratio:

$$\frac{\text{Combined concentration of the two most abundant species}}{\text{Concentration of all the species}}$$

During spring and summer near Bermuda, Hulburt found that the poor nutrient conditions limited growth so that species which at another season could be dominant formed only a moderate proportion of the cells counted in samples. The relatively high salinity and considerable depth of the euphotic zone permitted the occurrence of numerous species, so that species diversity was high. During winter (December–March), with better nutrient conditions, the phytoplankton populations were larger and there was greater dominance, the major organism being *Coccolithus huxleyi*. This type of phytoplankton flora Hulburt compared with that of estuaries, shallow inshore bays and ponds, where conditions for active cell multiplication are very good but the physical environment is specialized. Phytoplankton populations there were very much larger and dominated by exceedingly few species, often of minute diatoms (*Thalassiosira nana*, *Chaetoceros simplex*) or of dinoflagellates or flagellates (e.g. *Nannochloris*, *Prorocentrum*, *Prymnesium*). Hulburt also examined the diversity index (α) of the various populations according to the formula:

$$S = \alpha \ln \left(\frac{N}{\alpha} + 1 \right)$$

where S = number of species, N = number of individuals. The higher values of the diversity index in open ocean and coastal populations as compared with estuarine populations and the effect of population size is illustrated in Fig. 5.11.

The high species diversity values for oligotrophic regions, typically open tropical oceans, as contrasted with the low diversity of eutrophic areas leads to the question whether the differences in diversity are in part a reflection of nutrient limitation on species (cf. Yentsch, 1974). The relationship between rate of uptake by algae and nutrient concentration is frequently best described by the Michaelis-Menten equation, and interspecific differences in the half-saturation constants (K_s) for ammonium and nitrate have been found to be especially significant for many phytoplankton species (cf. Chapter 7). Such varying abilities to utilize nutrients effectively at low concentrations, including nutrients other than inorganic nitrogen compounds, might possibly be linked to diversity. Yentsch points out that an important consideration in future investigations is whether the specific half-saturation constant itself is subject to change.

More sophisticated estimates of diversity have been developed, particularly by Margalef and his colleagues (e.g. Margalef, 1967). A brief review is given by Parsons and Takahashi (1973a). Margalef (1967) includes another expression for the diversity D of an assemblage of individuals (or sample):

$$D = \frac{1}{N} \log_2 \frac{N!}{Na!\, Nb! \dots Ns!}$$

where Na, Nb ... Ns are the numbers of individuals of species $a, b, \dots s$ and N is the total individuals. He points out that this latter formula is particularly useful for following changes in diversity through the dynamics of a mixed population.

It is necessary to discuss briefly what is meant by equitability in plankton samples, in relation to diversity. There must be a maximum value for the number of individuals, given the total number of species in a population, if all were equally represented. This would be a measure of equitability – roughly the opposite of dominance. Theoretically, the same degree of equitability may exist, whether a large or a small number of species is present in a population. In a population with only a few species (i.e. low diversity) each species might be represented by more or less equal numbers of individuals (high equitability). More usually, as in an inshore bloom of phytoplankton dominated by a few species, diversity and equitability are both low. Frequently, in open tropical oceanic environments, the population has a large number of species with relatively little dominance; diversity and equitability are both high (cf. Thorrington-Smith, 1970c, for Indian Ocean phytoplankton). Thorrington-Smith (1970c), also pointing to the two factors influencing species diversity, the number of species and their evenness (equitability) of distribution, emphasizes that it is sometimes important to separate the two rather than combining them as overall diversity. In the open Indian Ocean (i.e. apart from rich upwelling areas) regions with some rise in nutrient level tended to exhibit increased diversity. This was largely due to an increase in the species component,

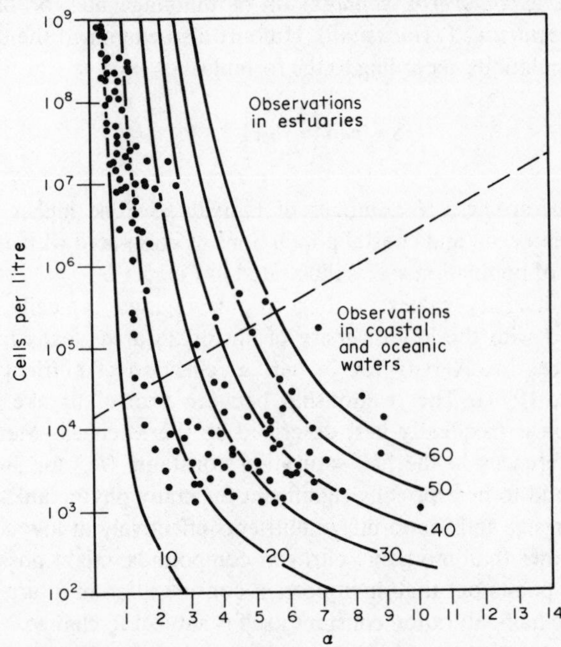

Fig. 5.11. The relation of population size to diversity. Curved lines indicate 10, 20, 30, etc., species per litre derived for various values of cell number and α (Hulbert, 1963).

although the more fluctuating conditions reduced equitability. The population appeared to be a relatively stable community.

An index of diversity is valuable *inter alia* in characterizing the flora of different regions and in examining species succession (*vide infra*). Margalef (1958) suggested three stages in the changing diversity of a plankton community:

(1) Turbulent waters with few species surviving; occasional blooms; low diversity.
(2) Inflowing water bringing more species, passively increasing diversity.
(3) Highly stratified water with a mature community; low crop with high diversity.

Margalef (1967) points out that while the mixing of two water masses may result in a passive increase in diversity, more generally the meeting of water masses causes stress (e.g. at thermoclines, marine fronts, etc.), though there is an enhancement of production. Frequently, better nutrient balance and other conditions in mixed waters lead to active growth of the population, but this is accompanied by a lowering of diversity – only some species react immediately to the improved conditions by rapid growth. Some species depend on the appearance of others so that there is a time lag in multiplication. Thus diversity falls in such *active* contacts of water masses, though the biomass increases.

Succession of species (*vide infra*) is related not only to interaction between a plant cell and its environment, but to interactions between species. Primary production per unit biomass falls as succession proceeds; average cell size increases, relative pigment content falls and there is an increase in motile types of phytoplankton. A general increase in diversity also occurs, sometimes with a decline in diversity towards the end of succession. These changes are interrupted if there is an increase in mixing processes. For a normal progression of succession, a general increase in stability is necessary. In general, succession, studied mainly in temperate regions, begins with mixing or upwelling fertilizing the euphotic zone; small celled diatoms and flagellates are then dominant. These are succeeded by *Chaetoceros*, followed by *Rhizosolenia*, coccolithophorids and then dinoflagellates. The cycle may be repeated and proceeds further in regions of greater stability, especially in oceanic warmer seas.

Communities and Species Succession

Areal differences in phytoplankton apply not only to differences in density, but to distinctions in floristic composition. As much of the earlier study of phytoplankton involved cell counts, the species composition of phytoplankton populations has been fairly well documented. The concept of communities of phytoplankton was early established, though some problems in typifying communities remain outstanding. Margalef (1967) discusses the practice of using groups of recurrent species to classify plankton communities; factor analysis may be employed to complete the analysis but it is sometimes difficult to understand the significance of the principal components. The groups are frequently related to selected environmental factors, usually temperature and salinity, since these, though not necessarily the factors controlling the distributions, are important in defining water masses. Thorrington-Smith (1971) prepared species lists of phytoplankton from stations from the Indian Ocean. After separating initially pre- and post-monsoon samples, stations were grouped according to the major hydrological distribution. A coefficient of dissimilarity was then calculated for each pair of samples. These derived values were subjected to a cluster analysis, and a dendrogram constructed with the various cluster groups appearing at different levels according to their degree of similarity. In this manner a phyto-hydrological classification of the major communities was achieved.

Species composition of the phytoplankton is of real significance, since though the size of crop is of paramount importance in the marine ecosytem, and though nearly all phytoplankton species are utilized as food by some particle feeder, certain zooplankton herbivores exhibit considerable selectivity in their diet. A suitable density of food organism of the right type at the appropriate time may be all important for the survival of a brood of zooplankton. Further, assuming that the timing of production of a new stock of fish larvae is more or less fixed, its measure of success as a year brood is largely dependent on a favourable food supply, usually of particular zooplankton species. Thus Cushing (1975) argues that the phytoplankton cycle is governed in its timing by the date at which the critical depth exceeds the depth of mixing (see p. 278) and the rate at which the production ratio develops. But the appropriate type of zooplankton food for the fish larvae must have developed from a suitable algal crop in sufficient quantity by the critical time.

In some regions the species composition of the phytoplankton is relatively stable. Some algae are fairly sharply restricted to particular areas of ocean, the distribution being limited most obviously by temperature and salinity characteristics (cf. Margalef, 1967). The geographical distribution of temperature and salinity in the world oceans has, however, a complex pattern, due mainly to ocean currents, which transport cells far from their normal endemic area. Active multiplication of the algae may persist, provided the quality of water in the current changes only slowly. Some phytoplankton species may, therefore, act as indicators of a particular water mass, or of the flow of a current. A mixture of species, typically found in different geographical areas, may indicate a mixture of water masses.

This use of plankton species as indicators is well known for the zooplankton (cf. Volume 2). Algal species have been used far less widely, partly because plant cells are small so that rapid precise recognition is difficult without detailed microscopic examination; another important limitation is that many species appear to have the capacity to survive over a considerable range of environmental conditions. Adaptation of plant plankton is a well-known phenomenon; the species are more widespread and possibly less environmentally specialized. Hart and Currie (1960) point to this important distinction, often not recognized by earlier investigators. While a few species are clearly neritic, or obviously polar or tropical stenothermal forms, many are not limited to particular water masses in a comparable manner to zooplankton. The great number of phytoplankton species common to the cold temperate areas of both North Atlantic and Pacific is notable; *Denticula*, with one species dominant in the sub-arctic Pacific, is exceptional in being restricted to that ocean. In the warm oceans, phytoplankton species are usually circum-global in equatorial regions, and often spread north or south to sub-tropical areas, sometimes being common to all three regions. Many diatoms are more or less cosmopolitan, appearing almost anywhere in the world's oceans and becoming locally important, even dominant, when conditions are suitable. Although some zooplankton species tend to be cosmopolitan, the majority exhibit fairly clearly defined distribution patterns. Given good conditions, an expatriate phytoplankton species with its comparatively rapid rate of division may quickly build up a recognizable population. Genetic isolating mechanisms do not appear to be strong in phytoplankton; vegetative reproduction may be a significant factor. Zooplankton expatriates, by contrast, are usually characterized by much slower reproductive rates, and isolating mechanisms appear to evolve more readily.

Despite these limitations of phytoplankton indicator species, considerable knowledge has accrued concerning the geographical distribution of planktonic algae. Early observations such as those of Gran distinguished Arctic, Antarctic, boreal, temperate and tropical species, in addition to neritic and oceanic forms. Temperature and salinity were early regarded as environmental factors limiting distribution. But the quality of the water including the concentration of trace metals and other inorganic and organic constituents, the biological history of the water with its bacterial population, exudates from phytoplankton and other cells, and products of land drainage, can exert more subtle influences on distribution. Water depth itself confers certain conditions, such as the ability of some inshore phytoplankton species to use resting spores. Light quality and intensity and nutrient concentration may also affect floristic composition.

A clear demonstration of distribution of phytoplankton species in relation to many of these environmental factors is often difficult. Certain species, however, show fairly clearly defined environmental limits associated with temperature. Gessner (1970), analysing the distribution of dinoflagellates in relation to temperature, listed only two definitely cold-water species, *Dinophysis arctica* and *Ceratium arcticum*, but temperature could be a significant factor with other dinoflagellate species. Thus, in the South Atlantic Ocean, the 16°C isotherm served as a border, especially for the genera *Amphidinium* and *Ornithocercus*. Of a total of ninety dinoflagellate species, sixty-eight occurred north of the 16° isotherm (i.e. these are warm-water species) and seventeen species occurred in areas south of the isotherm; only five species were common to both areas. For the Caribbean, Gessner lists only 16 out of 116 species as strictly tropical dinoflagellates. They include *Ceratium humile*, *C. lunula*, *Peridinium grande* and *Pyrodinium bahamense.*

Williams (1971), reviewing studies on the distribution of dinoflagellates, dealt particularly with investigations on *Ceratium*. Although in the South Atlantic, where natural regions were suggested, temperature was a factor distinguishing the major geographical divisions, the influence of environmental temperature between 15°C and 27°C was not so obvious. A single species, *Ceratium pentagonum robustum*, was found in the cold southern boundary region, whereas thirty-three species occurred in warm and twenty-one species in warm-to-cool waters. Greater species diversity occurred in waters of high salinity and low phosphate. Other studies in the North Atlantic and Pacific oceans relating distribution of *Ceratium* species to temperature, salinity and nutrient levels, however, suggested no obvious correlation with phosphate concentration.

Smayda (1958), discussing the biogeographical distribution of phytoplankton, particularly diatoms, describes *Thalassiosira antarctica* as a characteristic and important circumpolar species, limited to Antarctic waters. It is a stenothermal species with a temperature range from only −1.8° to +3.5°C and is found within a fairly narrow range of salinity. Other diatom species typical of the Antarctic include *Chaetoceros criophilus*, *C. neglectus*, *Eucampia balaustium*, *Fragilaria curta* and *Synedra reinboldii*, together with the dinoflagellate *Peridinium applanatum*.

Thalassiosira hyalina is an Arctic diatom, but its distribution is extended somewhat from the Arctic by cold currents (cf. Fig. 5.12). Like *T. antarctica*, *T. hyalina* is stenothermal; although it shows a temperature range from −1.35° to +9.0°C it apparently does not grow actively at temperatures exceeding 5°C. The species is also stenohaline. Other Arctic species listed by Smayda (1958) are *Achnanthes taeniata*, *Bacteriosira fragilis*, *Chaetoceros furcellatus*, *Fragilaria oceanica*, *Navicula vanhoeffeni* and the dinoflagellate *Ceratium arcticum*.

Planktoniella sol is often regarded as a typical warm-water diatom. Its distribution is circum-tropical and it is found in all the major tropical current systems, but its very wide distribution includes at least some temperate areas. Although its occasional appearance in relatively cold currents might be disregarded, since it may have been dying, and the species probably does not grow actively at temperatures much below 10–15°C, its range in distribution including waters of 30°C, suggests that it is really more eurythermal. *P. sol* appears to be confined to waters of high salinity; probably, therefore, its usual sub-tropical and tropical distribution is largely determined by its

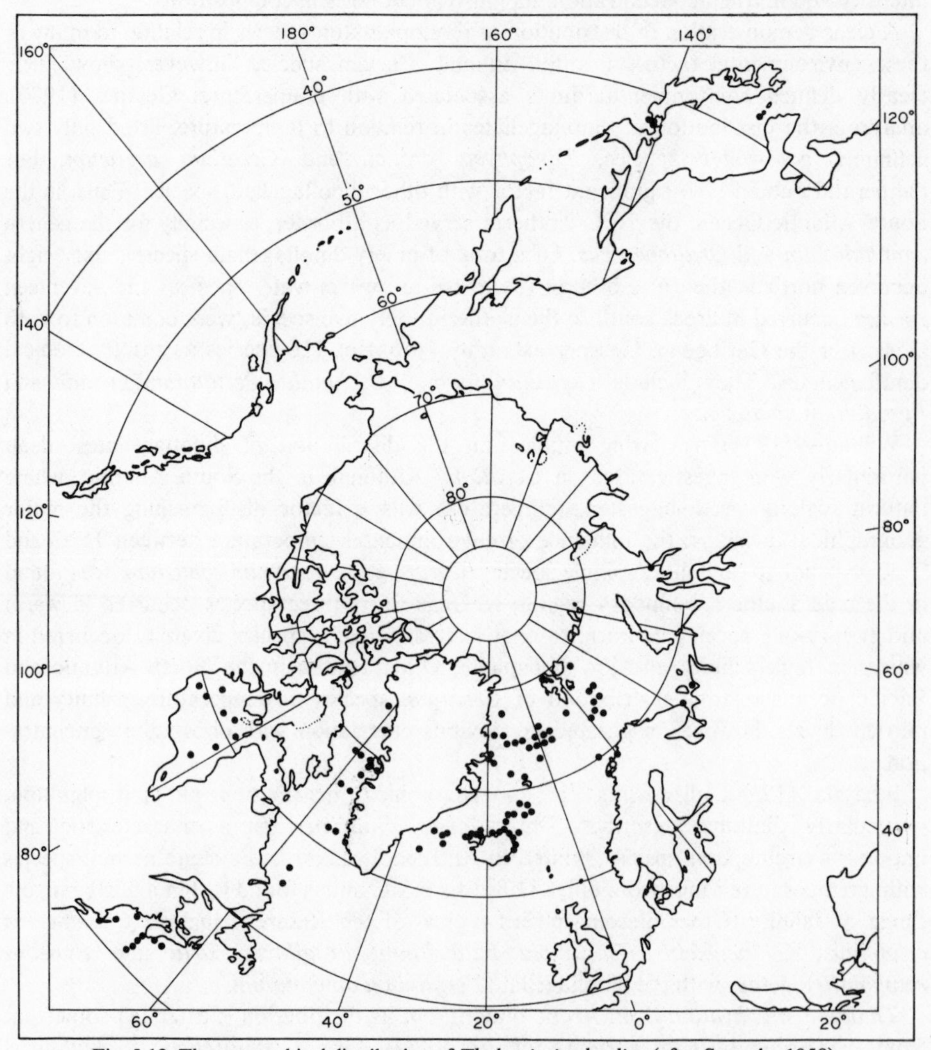

Fig. 5.12. The geographical distribution of *Thalassiosira hyalina* (after Smayda, 1958).

toleration of high temperatures and need for relatively high salinity. Tropical diatoms include *Chaetoceros laevis*, *Hemiaulus membranaceus* and *Rhizosolenia robusta*. Other tropical phytoplankton species are the dinoflagellate *Peridinium elegans* and the coccolithophorid *Coccolithus sibogae*. Smayda suggests that although there are relatively few truly tropical diatoms, a surprisingly large number of temperate diatom species can thrive in tropical waters.

An example of a cosmopolitan phytoplankton species is *Thalassionema nitzschioides*, circum-global in distribution, though found mainly in neritic areas. *T. nitzschioides* is extremely eurythermal; it occurs in the high Arctic at lowest sea temperatures, though surprisingly it is absent from Antarctic waters. At the opposite temperature extreme it is known from waters exceeding 30°C. The species is also extremely euryhaline, with a salinity range from about 4‰ to above 34‰. Other cosmopolitan diatoms are *Chaetoceros compressus*, *Leptocylindricus danicus*, *Skeletonema costatum*, with representatives of other algal classes such as *Exuviaella baltica*, *Prorocentrum micans*, *Ceratium furca* and *Coccolithus huxleyi*. A few species have been described as bipolar, since they are known from high northern and southern latitudes (e.g. *Thalassiothrix longissima*, *Rhizosolenia hebetata* f. *semispina*). Bipolarity, if defined as the near-surface occurrence of a species at very high northern and southern latitudes and its continuity at depth but absence from the surface at lower latitudes, cannot be true for phytoplankton, since they are unable to maintain themselves on submergence to avoid the higher temperatures of lower latitudes. The so-called bipolar diatoms have a more or less continuous distribution and are well known in near-surface warmer waters (cf. Smayda, 1958).

In temperate latitudes, where there may be considerable variation in temperature, light and nutrient concentration throughout the year, there is frequently a marked change in species composition of the phytoplankton, accompanying fluctuations in crop. This seasonal succession is also well known at high latitudes. Even in low latitudes where usually environmental variations are smaller, the species composition of the phytoplankton often shows what might be described as a seasonal succession. The floristic changes may sometimes be recognized as associated with changing environmental conditions; for example, a succession may be very obvious in areas subject to monsoons. Succession may also be related to periods of upwelling in certain regions.

In some cases, particular environmental factors governing the geographical distribution of species can be recognized as influencing also their temporal succession. For instance, in sub-Arctic areas temperature may be the chief factor determining the time of appearance of the Arctic forms, *Thalassiosira hyalina* and *Bacteriosira fragilis* (Smayda, 1963). The various factors influencing distribution and presumably, therefore, succession interact, however. Some species have fairly specific requirements for temperature, light intensity, salinity and nutrients; nutrients may include as well as inorganic compounds of nitrogen and phosphorus, silicon, trace metals and a number of organic factors. The effects of a variety of metabolic products which are liberated by algae and bacteria are difficult to assess (cf. Johnston, 1955, 1963 a and b, 1964; Aubert, Pesando and Gauthier, 1970; Aubert and Pesando, 1971; Aubert and Gambarotta, 1972). While some of these "ectocrines" are recognized as inhibitory, stimulatory or even toxic (cf. Chapter 3), many can have more subtle effects. Moreover, their concentration will be greatly affected by decomposition, absorption and other factors. It seems likely that the succession of species is a result of the simultaneous, interacting action of a number of these factors rather than the effect of one parameter.

Sverdrup, Johnson and Fleming (1946) described seasonal succession in the Gulf of Maine and quoted the following series of species:

April	3°C	*Thalassiosira nordenskioldii*
		Porosira glacialis
		Chaetoceros diadema
May	6°C	*Chaetoceros debilis*
June	9°C	*Chaetoceros compressus*
August	12°C	*Chaetoceros constrictus*
		Chaetoceros cinctus
		Skeletonema costatum

Sverdrup *et al.* suggested that surface temperature was to some extent responsible for the change in phytoplankton composition. Barker (1935) found that dinoflagellates often became more abundant in temperate areas later in the summer; he associated the floristic change with nutrient requirements. Gran and Braarud (1935) suggested that not only higher temperature but lower nutrient levels might be reponsible for the growth of *Ceratium*, *Pontosphaera* and the diatom *Rhizosolenia alata* during the later summer in the Gulf of Maine. Not all dinoflagellates have low nutrient requirements, however (cf. Ryther, 1954).

Lillick (1940), describing species succession in the Gulf of Maine, found the winter flora usually dominated by *Coscinodiscus* together with peridinians, particularly *Ceratium*. The spring outburst of diatoms was mainly due to *Thalassiosira*, with species of *Chaetoceros* following soon afterwards. Over the summer a mixture of diatoms occurred, but the plankton consisted largely of peridinians with coccolithophorids. In the later summer there was another flowering of diatoms, the major forms being *Rhizosolenia*, *Guinardia*, and often *Skeletonema*; peridinians and coccolithophorids were reasonably abundant. Even different areas of the Gulf of Maine, however, showed considerable variations in the species succession (cf. Figs. 5.13 and 5.14). Conover (1956), in a rather similar study in Long Island Sound, described the early spring flowering as dominated by *Thalassiosira nordenskioldii* and *Skeletonema costatum*, with *Thalassiosira* being somewhat more abundant at the low light intensity and lower temperature of the earliest part of the season. Other species including *Chaetoceros* spp., *Leptocylindricus danicus*, *Peridinium trochoideum*, *Asterionella japonica*, *Lauderia borealis* and *Schroederella delicatula*, appeared a little later. Over the summer months dinoflagellates and small flagellates were more abundant than diatoms, but in the autumn *Chaetoceros* became abundant. Conover considers conditioning of the water as one of the environmental factors important in species succession.

Brockman, Eberlein, Hosumber, Trageser, Maier-Reimer, Schone and Junge (1977) comment on the stabilizing effect of species succession on the ecosystem in temperate waters. They point out that although a few observations have been made on phytoplankton succession in large tank or column experiments, those experiments involved the addition of nutrients which could have influenced the natural succession. A commonly observed result of enclosing phytoplankton populations is a lowering of species diversity (cf. Thomas and Seibert, 1977). The larger-celled diatoms also usually show a rapid decline (cf. Chapter 8). Brockman *et al.* carried out small-scale experi-

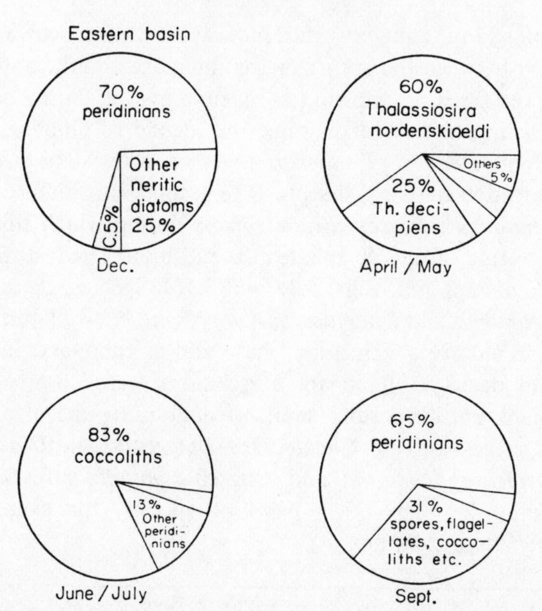

Fig. 5.13. Percentage composition of the phytoplankton in the Eastern Basin, Gulf of Maine, over different months (modified from Lillick, 1940).

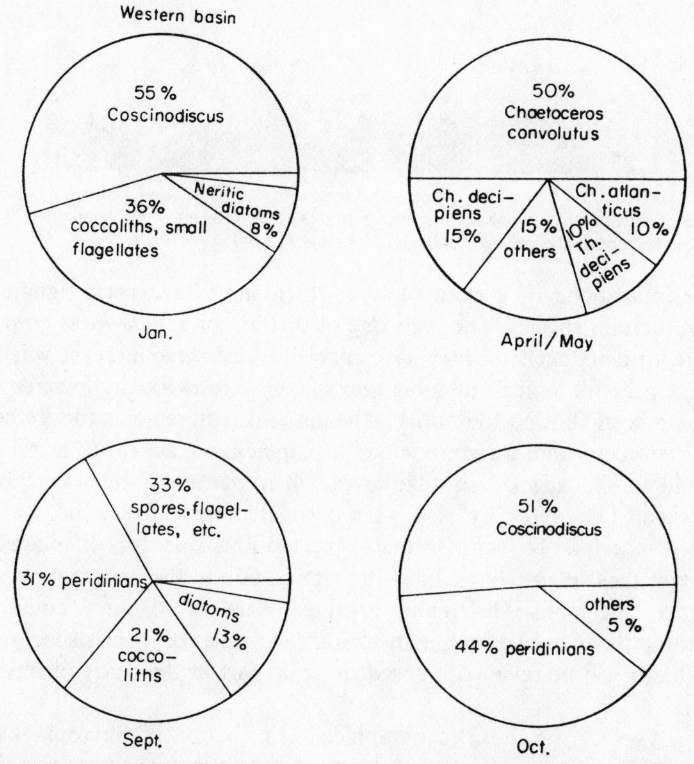

Fig. 5.14. Percentage composition of the phytoplankton in the Western Basin, Gulf of Maine, over different months (modified from Lillick, 1940).

ments in plastic tanks (3 m^3 capacity) with natural North Sea seawater over the period of nutrient depletion between the spring diatom increase and the autumn dinoflagellate bloom. They observed three phytoplankton maxima over a 28-day period, the first and third maxima consisting mainly of diatoms, the second of dinoflagellates. During the first maximum, dominated by *Lauderia*, together with some *Ceratium* spp., the ammonium concentration fell very sharply. The second maximum consisted mainly of *Dinophysis acuminata* and *Prorocentrum micans* during which time nitrate declined and there was a lowering of the phosphate concentration. The last maximum consisted mainly of *Rhizosolenia* spp. (cf. Figs. 5.15 and 5.16). Species, especially diatoms such as *Lauderia*, *Skeletonema* and *Chaetoceros compressus*, showed much the same type of growth pattern as in cultures, with a log phase and a stationary phase, followed by a phase of decline in density. Significant correlations were observed between certain species of phytoplankton, suggesting similar reactions by the algae to those factors influencing species succession (e.g. *Chaetoceros compressus* and *Dinophysis acuminata*; and again, *Rhizosolenia delicatula* and *Nitzschia longissima*). Significant negative correlations appeared between other pairs of species, for example, *Skeletonema costatum* and *Diplopeltopsis minor*.

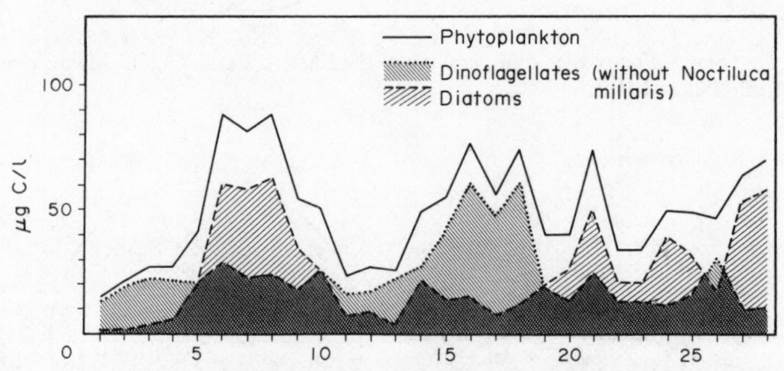

Fig. 5.15. Biomass of total phytoplankton and contributions of diatoms and dinoflagellates. Carbon content is calculated from cell count and cell size (after Brockman *et al.*, 1977).

There are indications of certain factors which may have been significant in the dominance of certain species. The high rate of division of *Lauderia* as compared with the lower rate for dinoflagellates may have enabled *Lauderia* to increase when abundant nutrients were present. When ammonia and silicon were markedly lowered dinoflagellates may have been allowed to flourish. The nutrient requirements for *Rhizosolenia*, if the half-saturation constant for ammonium is an indication, appear to be relatively low; this might allow the species to take over when nutrients were sharply reduced. Temperature and light intensity also were correlated with the abundance of certain phytoplankton species. To some extent light requirements for dinoflagellates were modified in that these algae showed a definite migration to the upper layers even in the limited depth of the tanks. Differential grazing by the zooplankton could have been another factor in the pattern of dominance. Some examples of species succession in high and low latitudes will be reviewed subsequently as part of the study of phytoplankton communities.

Earlier analyses of the areal (geographical) distribution of phytoplankton species such as those of Gran emphasized temperature as an important factor (cf. p. 225). Gran also distinguished neritic species from oceanic forms found generally far offshore.

Braarud, Gaarder and Grontvedt (1953) indicated, however, that these terms were often not precise; a number of phytoplankton species overlapped both regions. For example, while *Corethron hystrix* appeared to be truly oceanic, and *Asterionella japonica* and *Thalassiosira gravida* strictly neritic, *Coccolithus huxleyi* while mainly oceanic, may thrive in coastal waters. Conversely, *T. decipiens* is really neritic but can thrive in oceanic water. Smayda (1958) has also criticized the terms neritic and oceanic, partly

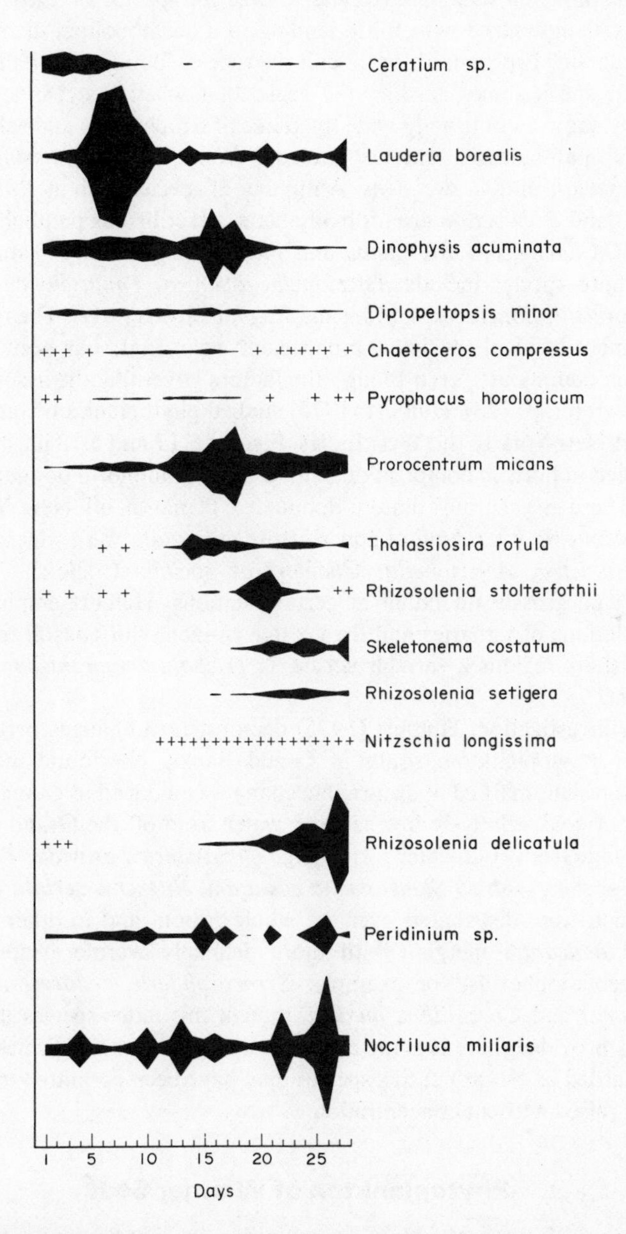

Fig. 5.16. Temporal distribution and biomass of the most important phytoplankton species. + = low biomass concentrations (after Brockman *et al.*, 1977).

on the grounds that coastal conditions are so very diverse that further sub-division is required.

Hart and Currie (1960), dealing with the diatoms of the Benguela Current, pointed to the difficulty of describing species as neritic or oceanic. Some oceanic diatoms, abundant at all times in the open sea, occurred sometimes plentifully near coasts. Neritic species could equally form a large part of the phytoplankton well beyond the limits of the continental shelf. They prefer to describe species as restricted to particular water masses as compared with those tending to a cosmopolitan distribution, and the term "panthalassic" is proposed for ubiquitous species found both offshore and inshore. Cosmopolitan species may exist in the sense that whether oceanic, neritic or panthalassic, they show an unusually wide tolerance of temperature and salinity conditions. Examples of panthalassic species include *Nitzschia delicatissima*, *N. seriata*, *Chaetoceros sociale* and *C. decipiens*. A number of species such as *Rhizosolenia alata*, *R. styliformis* and *R. hebetata* are probably better described as panthalassic rather than cosmopolitan. *Chaetoceros atlanticum* and *Planktoniella sol* are examples of offshore diatoms; inshore species include *Asterionella japonica*, *Thalassionema nitzschioides*, *Fragilaria karsteni*, *Chaetoceros affine* and *Stephanopyxis turris*. The repeated association of a number of algal species in a particular water mass has been recognized as a phytoplankton community, even though the factors governing the association are often not fully known (cf. p. 223). Hulburt (1970) studied phytoplankton communities at stations from off New York to the West Indies. Figures 5.17 and 5.18 illustrate some of the differences seen in floristic composition, with a coccolithophorid dominated plankton in the Sargasso and a generally diatom-dominated plankton off New York. In coastal areas *Thalassionema nitzschioides* and *Rhizosolenia alata* were widespread, but a few forms (*Skeletonema*, *Asterionella*, *Chaetoceros socialis* (*sociale*), *Leptocylindricus danicus*) became grossly abundant at certain stations. Hulburt emphasizes the high-density populations of estuaries and the greater variability of coastal regions generally. In tropical inshore regions a variable species is *Trichodesmium thiebautii* which can be very abundant.

In another investigation, Hulburt (1975) demonstrated changes between colder and warmer-water phytoplankton, south of Grand Banks, Newfoundland. If competing species were present, marked hydrographic changes could lead to changes in flora. With the intrusion of cold, relatively low salinity, water from off the Grand Banks, the flora showed a change from cold-water forms (e.g. *Thalassiosira gravida*, *T. nordenskioldii*, *Lauderia*) to species such as *Skeletonema costatum*, *Nitzschia seriata* and *Coccolithus pelagicus* which were distributed over the whole region, and to other species such as *Rhizosolenia delicatula* merging with more definitely warmer water forms. These included coccolithophorids, for example, *Syracosphaera mediterranea*, *Cyclococcolithus leptoporus* and *Coccolithus huxleyi*, though this latter species may be found in colder waters provided there is some enrichment. An "unusual" species was *Eutreptia*, normally regarded as estuarine; this species may have been permitted to colonize under conditions of raised nutrient concentration.

Phytoplankton of Warmer Seas

Investigations of phytoplankton communities in warmer seas have frequently included observations of species succession. Riley (1957), examining the phytoplankton of the North Central Sargasso, classified the diatoms in the flora as winter, spring and summer/autumn species. Temperature was not the sole factor in the species succession since some species showed bursts of abundance at different seasons. The summer/autumn density of diatoms was much lower than that of dinoflagellates; nanoplankton flagellates also showed changes in density through the year. Hulburt, Ryther and Guillard (1960) observed a marked seasonal change in phytoplankton

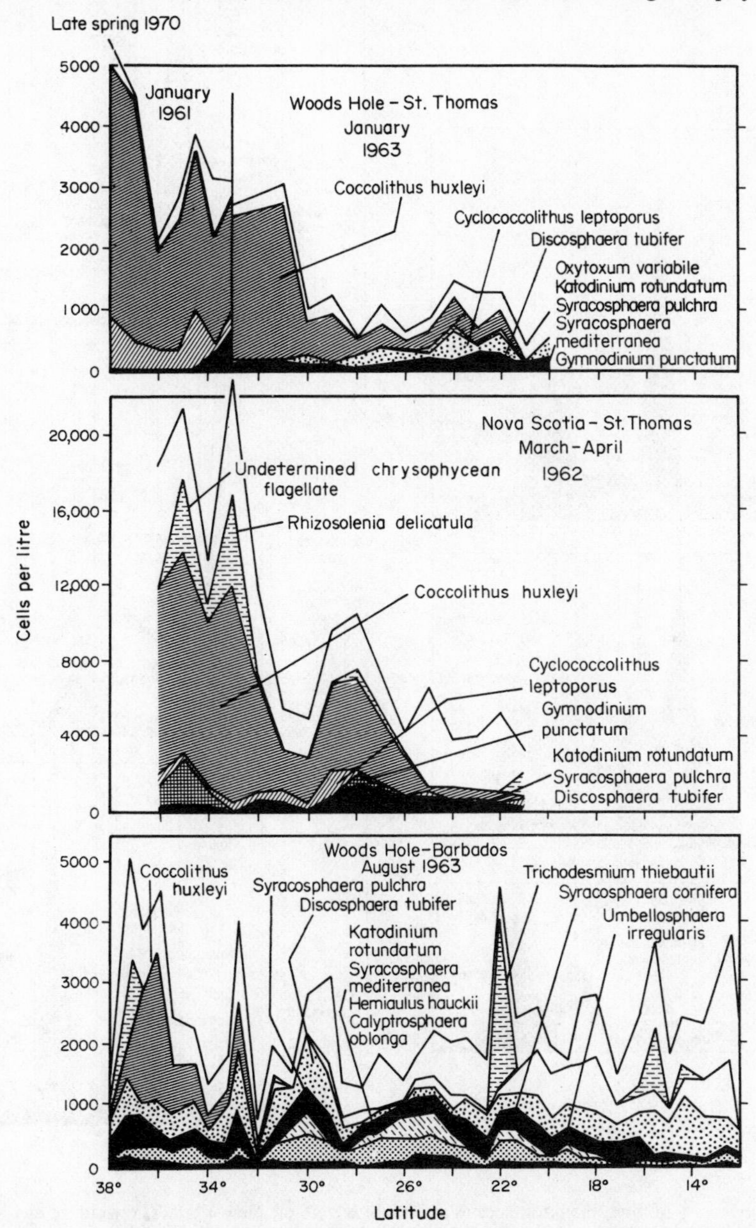

Fig. 5.17. The density of phytoplankton species from north to south in the western North Atlantic Ocean. The number of cells of *Trichodesmium* is about 66 times the number of filaments shown. The width of any shaded area gives the density of the indicated species (after Hulburt, 1970).

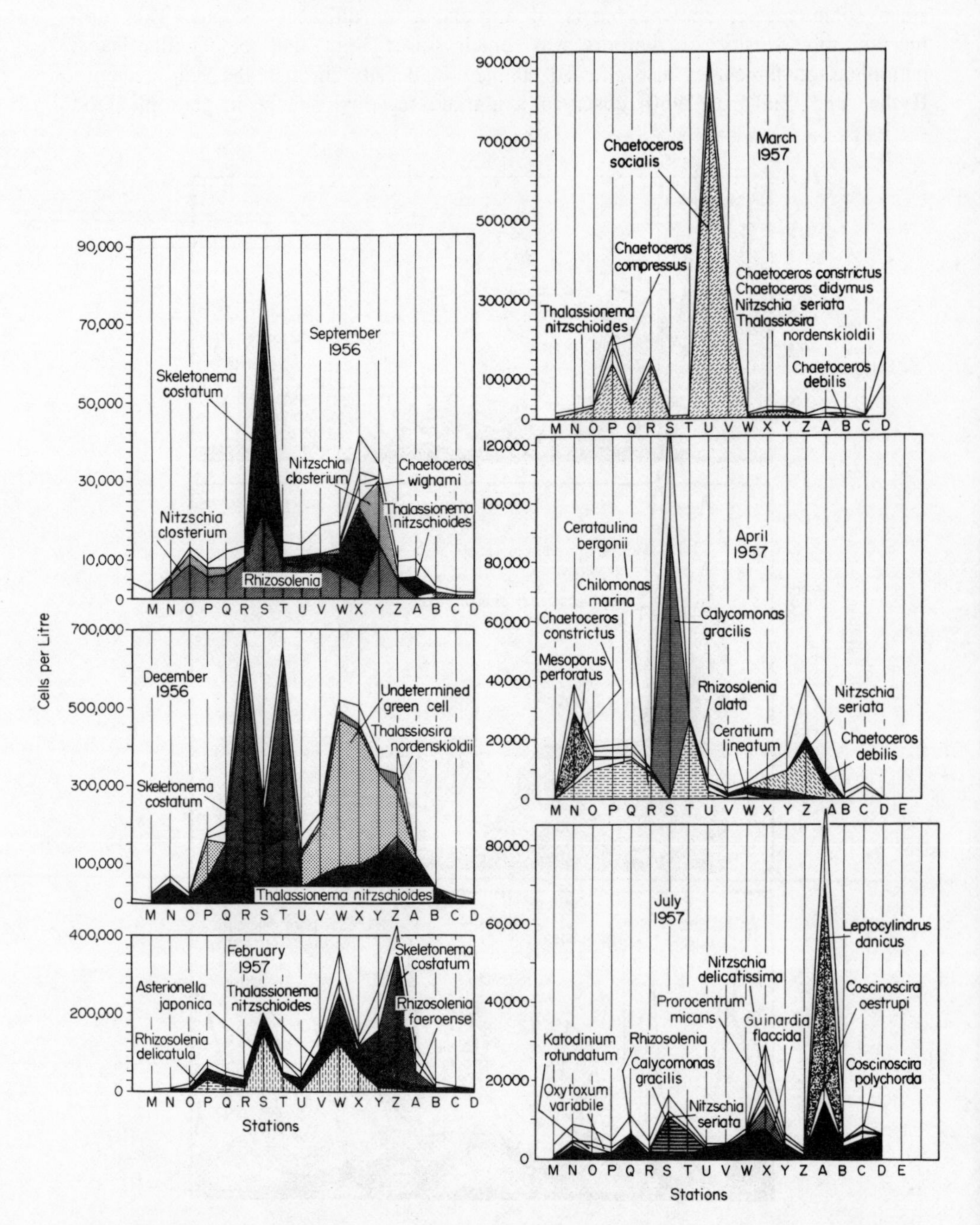

Fig. 5.18. The density of phytoplankton species in coastal waters off New York. The width of any shaded area gives the density of the indicated species (after Hulburt, 1970).

density from about 1000 to more than 200,000 cells per litre, with the lowest population in summer (May to October) in warm oceanic waters off Bermuda. Diatoms were generally insignificant, except during a spring bloom about April, when approximately ten species (*Bacteriastrum delicatulum*, *Chaetoceros* spp., *Leptocylindricus danicus*, *Rhizosolenia stolterfothii*, *Eucampia zoodiacus*, *Nitzschia delicatissima*, *Thalassiosira rotula*) were more or less equally abundant, Coccolithophorids dominated the phytoplankton over most of the year, with *Coccolithus huxleyi* dominant over the winter. Revelante and Gilmartin (1976) found that while nanoplankton dominated the phytoplankton of the northern Adriatic, blooms were due to the larger phytoplankton, mainly diatoms, and a species succession of diatoms occurred. In "winter" (October--January) the major species were *Rhizosolenia stolterfothii*, *Leptocylindricus danicus*, *Nitzschia longissima* and *Lauderia borealis*; in "late winter" (February–April) *Skeletonema*, *Chaetoceros* spp. and *Bacteriastrum delicatulum* were characteristic. "spring" (May–July), with increased stability of the water column, saw mainly dinoflagellates, especially *Prorocentrum*; in "summer" (August–September) *Nitzschia seriata* was typical.

In a tropical but more inshore area Smayda (1963, 1966) demonstrated that the phytoplankton was remarkably constant in composition over the Gulf of Panama and was dominated by diatoms. A species succession was evident, with *Skeletonema costatum* and *Chaetoceros compressus* as the main species during the period of poorer crop over the rainy season, and at the beginning of the increasing crop during the seasonal upwelling commencing about December/January. With rising crop, these species were followed by *Rhizosolenia delicatula*, *Nitzschia delicatissima*, and later, about April, by *Rhizosolenia stolterfothii*. Steven (1966) describes the phytoplankton flora of Kingston Harbour, Jamaica. There was probably a considerable variety of species, since he lists thirty-two diatom species, seven dinoflagellates, six Cyanophyceae and four Chlorophyceae, but remarks that the list was far from complete. Diatoms, especially *Rhizosolenia stolterfothii*, *Chaetoceros* spp., *Nitzschia seriata* and *Skeletonema costatum*, were unusually dominant, with some blooms attaining concentrations of $2–5 \times 10^6$ cells/l; dinoflagellates, mainly *Ceratium tripos*, *C. furca* and *Dinophysis diegenesis*, were usually in lower densities. A single bloom of *Exuviaella*, reaching a maximum density of 29×10^6 cells/l, was also recorded. Both nutrient concentration, especially nitrate, and also freshwater run-off following heavy rainfall, were particularly high just before the bloom. Steven points to the marked influence of land, especially the contribution of organic nutrients, on the crop and production of phytoplankton in near-shore tropical environments.

The marked seasonal variations in phytoplankton which may occur in tropical areas related to monsoonal changes are exemplified by the investigations of Subramanyan and Sarma (1965) off Calicut, south-west India. Standing crop, estimated by fractional filtration followed by pigment measurement, confirmed the very large size of crop, especially during the south-west monsoon period. Although on some occasions (June 1959, August 1960) nanoplankton comprised more than 80% of the crop, on average it constituted only some 40%. The results thus contrast with most investigations in warm oceans, for example, those of Hulburt *et al.* (1960) who concluded that nanoplankton might be a hundred times more numerous than net plankton and the biomass roughly equal. Even so, in terms of numbers, the nanoplankton crop off the west coast of India must have been enormous. A comparison of the productivity of the nanoplankton and net plankton would be of considerable interest (cf. Malone, 1971). Seasonal fluctuation of the nanoplankton off the west Indian coast was not so pronounced as that of the net plankton. Small diatoms and dinoflagellates predominated in the nanoplankton; few microflagellates were recovered. Large crops of diatoms would seem typical of much of the phytoplankton off the Indian coast (cf. the bloom of *Asterionella* reported by Subba Rao (1969) off the Bay of Bengal). Krishnamurthy and Purushothaman (1972) also recorded high and variable phytoplankton crops (as plant pigment) in the Vellar Estuary (south-east India). Describing the flora of three stations in the estuary where the composition was fairly similar, they found diatoms to be important. Seven genera were common; *Pyrocystis*, *Peridinium* and *Ceratium* were abundant amongst dinoflagellates. Although Krishnamurthy and Purushothaman do not discuss diversity of the phytoplankton, it would appear to be rather low as in temperate estuarine plankton.

Gopinathan (1972) carried out a quantitative and qualitative study of the phytoplankton of the Cochin backwaters (south-west India). He confirmed maximal abundance of the crop during the monsoon, roughly May to July/August, with a second smaller peak in October or November. The species appear to be adapted to the greatly lowered salinity and somewhat decreased temperature over the monsoon period, and so can benefit from the abundant nutrients. Nanoplankton was believed to form the major part of the phytoplankton crop (approximately 70%) in the Cochin backwaters, but in net hauls diatoms preponderated, forming about 97%. *Skeletonema* was important, blooms occurring on occasions. Other diatoms found through the year included three species of *Coscinodiscus*, *Planktoniella sol*, *Ditylum*, *Triceratium*, *Fragilaria oceanica*, *Biddulphia mobiliensis* and *B. sinensis*. Three species of dinoflagellates, *Peridinium depressum*, *Ceratium furca*, *Diplopsalis lenticula*, were also important. Altogether seventy-four species of diatom were relatively common through the year. This diversity is certainly greater than that typical of temperate estuaries, but according to Gopinathan is considerably less than in oceanic tropical areas.

Among the complex factors influencing species succession, specific nutrients may be significant at least in some warmer seas. Menzel, Hulburt and Ryther (1962) carried out enrichment experiments on Sargasso Sea waters. Although they found, as had other investigators, that diatoms were not usually predominant in the plankton of the Sargasso, with enrichment of silicate, diatoms tended to dominate the phytoplankton, with *Chaetoceros* spp., *Nitzschia closterium* and *Skeletonema costatum* among the more abundant species. Without addition of silicon, flagellates predominated, though in some experiments a small growth of *Skeletonema* occurred initially. With iron in addition to nitrogen and phosphorus, *Coccolithus huxleyi*, normally the most abundant alga, as well as flagellates, showed strong growth. Enrichment with silicon was always associated with growth of diatoms, but the major species varied with the mix of other nutrients and the duration of the experiment. For example, marked growth of *Chaetoceros simplex* was typical of a relatively rapid flowering upon enrichment with nitrogen, phosphorus and iron. With addition of nitrogen and phosphorus only, and with a longer period (7 days) for development of maximum population, *Skeletonema* dominated, with *Chaetoceros curvisetus*, *C. decipiens* and especially *C. didymus*. Another rather remarkable observation was that surface water enriched with nitrogen, phosphorus and iron showed an overwhelming abundance of *C. simplex*, whereas surface water mixed with deeper (700 m) water and enriched also with nitrogen, phosphorus and iron showed a clear dominance of *Nitzschia closterium*.

Many phytoplankton species have vitamin requirements, usually for B_1, biotin and B_{12}, but specific requirements differ (cf. Provasoli, 1963), so that species succession may be related to vitamin content of the seawater. Menzel and Spaeth's (1962b) investigations in the Sargasso indicate that vitamin B_{12} concentration may control species composition to some degree. *Coccolithus huxleyi*, the normally dominant species, has no B_{12} requirement. A small increase in the diatom flora of the Sargasso was associated with the temporary enrichment of vitamin B_{12}; all the commoner diatom species identified are known to require B_{12}. The effect was on species composition; Menzel and Spaeth found no effect on primary production with enrichment with vitamin B_{12}. On the other hand, iron would appear to limit production at certain times in the Sargasso, but iron also exerts a marked effect on species composition (cf. Provasoli, 1963).

Enrichment experiments by Barber and Ryther (1969) with natural phytoplankton samples from the equatorial eastern Pacific region indicated that recently upwelled water from the Cromwell Current did not support a high rate of phytoplankton growth, despite its high content of inorganic nutrients (nitrate, phosphate and silicon), even when trace elements and vitamins were added to the seawater. Only the addition of the chelating agent EDTA or an undefined zooplankton extraction improved the water quality. North and south of the equator the somewhat older upwelled water, rich in nutrients, gave good growth; this is believed to result from organic substances released in older water acting as natural chelators. Natural chelators might act either as facilitating uptake of trace metals or by reducing the toxicity of some heavy metal in unusually high concentration in upwelling water (cf. Steemann Nielsen and Wium-Andersen, 1970, on the effect of copper on photosynthesis). Although the experiments of Barber and Ryther dealt with production, it seems inevitable that species succession must also be influenced by the age and history of upwelled water.

That certain conditions causing stress might be a factor in species succession as well as in crop fluctuations is indicated from the results of experiments by Vaccaro, Azam and Hodson (1977) using CEPEX columns (cf. Chapter 8). When phytoplankton was exposed to small additions of copper, algal excretion and death resulted in considerable enrichment of organic material which stimulated heterotrophic bacterial activity. This increased activity led to the release of abundant nutrients which in turn produced substantial phytoplankton crops, but only those species which were more resistant to the stress conditions could benefit. Other stress factors might conceivably play a part in limiting species diversity and influencing succession, without necessarily reducing crop. The situation in some upwelling areas where only a few species flourish, but at high density, may to some extent simulate such a situation.

A study by Smayda (1964, 1971) also demonstrated remarkable differences in the quality of oceanic water. Using *Cyclotella nana* as an assay diatom, Smayda found that water from 0, 50, and 100 m depth respectively, from stations between Puerto Rico and Bermuda, could affect growth very differently. Thus water from 100 m caused some mortality of the algal inoculum, but addition of nutrients and iron complexed with EDTA removed this inimical effect. Addition of vitamins further stimulated growth but trace metal addition was ineffective. When the inorganic nutrients were added to surface waters, growth was often inhibited by vitamins which had a stimulating effect. The factors responsible for the different effects of surface and deeper water are not known, but it is probable that such differences in quality could be significant in species succession.

Estuarine Phytoplankton of Temperate Regions

The variable and usually large algal crop and the generally lower species diversity of estuarine phytoplankton in tropical areas has already been noted. In temperate latitudes the estuarine flora may be exceedingly abundant at certain times of the year. Massive blooms may be related in some estuaries to eutrophication, with extensive urban development and pollution. Lackey (1967) summarizes the type of flora in temperate estuaries as consisting of a few chlorophycean genera (*Carteria*, *Dunaliella*, *Nannochloris*), euglenoids such as *Eutreptia*, a few widely found cryptomonads, dinoflagellates usually of small size (*Exuviaella*, *Prorocentrum*, *Peridinium*) and diatoms which at least at times greatly outnumber the rest of the population. With the restriction in the total number of species, estuarine and near-shore plankton may be very monotonous, with a diatom such as *Skeletonema* ubiquitous and overwhelmingly dominant. In some inshore environments, phytoplankton other than diatoms may be dominant (cf. Ryther, 1954; Hulburt, 1970). Smayda (1957) describes the extraordinary abundance of crop in the inshore area, Narragansett Bay, and he includes comparative values for other sheltered areas such as Loch Striven and Kiel Bay, as well as for the exposed coast of California (Table 5.2). In Narragansett Bay, four flagellate groups (*Gymnodinium* spp., a gymnodinoid, *Prorocentrum* and unidentified microflagellates) could be significant elements in the flora, but nine species of diatoms, in addition to these flagellates, accounted for 94% of the total phytoplankton. The low diversity of the phytoplankton is clear. *Skeletonema* was overwhelmingly dominant (forming 80% of the total) so that equitability was probably low. Figure 5.19 illustrates the major species and gives some indication of species succession. Marumo, Sano and Murano (1974) also noted that certain neritic diatoms (e.g. *Chaetoceros debilis*, *Eucampia zoodiacus*, *Cerataulina bergonii*, *Nitzschia seriata*, *Skeletonema costatum*) were particularly important in the phytoplankton of Tokyo Bay. The dinoflagellates *Prorocentrum* and *Exuviaella* were also important. In part of the area *Skeletonema costatum* formed more than 90% of the total diatom population, but seasonal succession, particularly with reference to *Skeletonema*, varied considerably over the years of study, emphasizing the complicated interaction of the factors determining succession. In Narragansett Bay the major bloom of phytoplankton occurred in winter, with a smaller autumn peak; this appeared to be correlated with an unusual phosphate cycle with a winter minimum and summer regeneration. Probably the shallowness and special features of the Bay allowed

Table 5.2. Phytoplankton densities in several temperate inshore, mostly sheltered areas (data from Smayda, 1957)

	Mean crop cells × 10^3/l
Block Island Sound	471
Gulf of Maine	256
Long Island Sound	2500
Lower Narragansett Bay	6700
Vineyard Sound	23
Kiel Bay	627
Loch Striven	1000
Californian coast	50

reasonable growth of phytoplankton in winter. In some other near-shore areas also, apparently only the very beginning of seasonal light increase is needed for the phytoplankton flowering. Riley (1967) suggests, for instance, late winter/early spring as the usual time for Long Island Sound; the species succession may be compared with that observed by Smayda.

In other estuarine areas, however, there is such instability and poor light penetration due to very low transparency of the water associated with heavy silt load that a peak crop does not occur until much later in the spring. Bakker and de Pauw (1974) investigated the phytoplankton of a brackish stagnant lake (L. Veere) and of the estuarine Westerschelde, Holland, and confirmed the great reduction in light with depth. In the tidal Westerschelde this was largely owing to sediment load; maximum stocks of

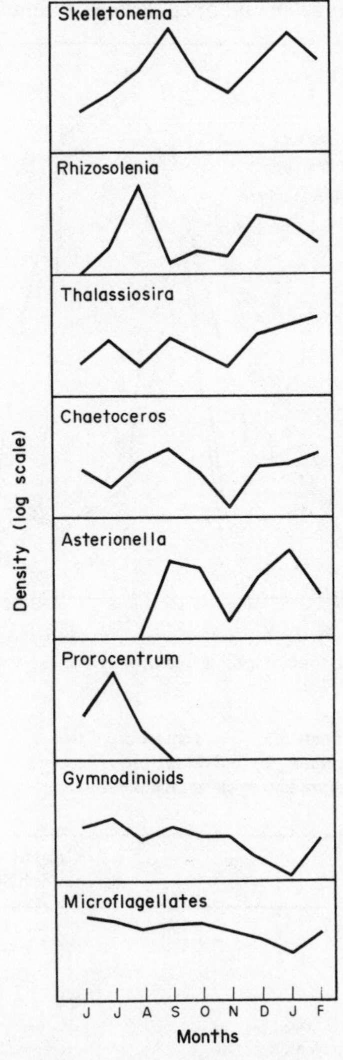

Fig. 5.19. Seasonal changes in abundance of the major species of phytoplankton in Narragansett Bay. Density as log scale (modified and redrawn after Smayda, 1957).

phytoplankton were not encountered until summer and constant turbulence was a major factor reducing effective light penetration. In the relatively stagnant L. Veere the phytoplankton maximum came earlier with stabilization being early established, and light reduction was due mainly to intense blooming of nanoplankton. In both areas nutrient level was high, as in most estuaries. The Westerschelde phytoplankton was dominated chiefly by diatoms, mostly of marine origin. In L. Veere the nanoplankton was of great importance, but some diatoms, including *Skeletonema costatum*, *Coscinodiscus commutatus*, *Chaetoceros mulleri*, *Actinocyclus ehrenbergii* and *Biddulphia sinensis*, made substantial contributions.

In Southampton Water the maximum density of algae usually does not occur until early summer (Fig. 5.20); there can be very large blooms, but great variation in crop is experienced, largely related to local weather conditions. Some comparison of the richness of crop in inshore, estuarine and open-sea environments is given in Table 5.3.

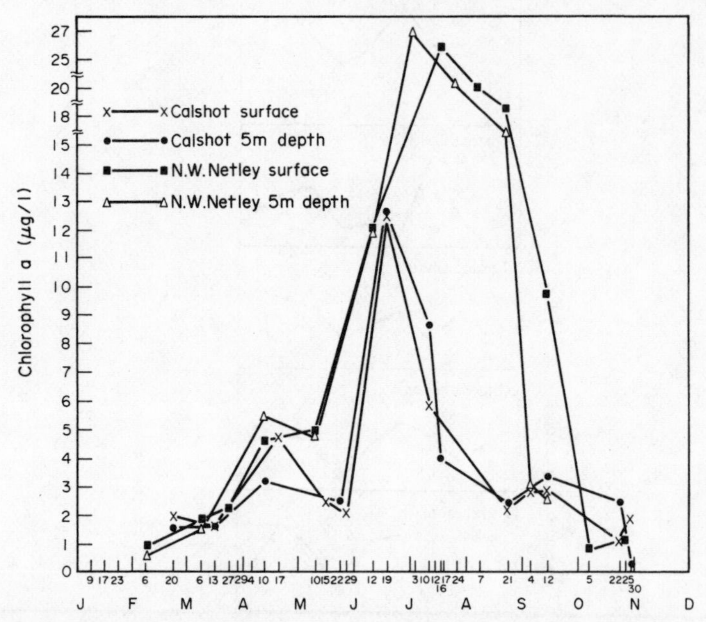

Fig. 5.20. Seasonal changes in phytoplankton, as chlorophyll *a*, at two stations and two depths in Southampton Water (from Diwan, 1978).

Table 5.3. A comparison of the maximum phytoplankton crops in some inshore and open sea temperate areas

Area	Crop as mg chlorophyll/m^3
North Sea	2–9
Forge River	20– <50
Aberdeen	3–20
Plymouth	0.2–1.6
Long Island Sound	2.5–5
Southampton Water	<10– >35
Open Sea	~0.5–5

Despite the usually very high levels of nutrients in estuarine waters (cf. Jeffries, 1962; Ketchum, 1967) the huge flowering of algae may greatly reduce nutrient concentrations, especially if during the summer, in addition to a halocline, a thermocline has developed so that there is insufficient mixing with deeper water. Though the rate of regeneration of nutrients, especially the release of ammonium within the shallow euphotic layer, may be very high, with the surface seaward flow of water in addition to the high utilization by algae, nutrient levels, especially of nitrogen, may be insufficient. Many investigations of near-shore waters show a reduction to less than 1 μg–atom N/l during high summer. The crop may be substantially smaller at least over part of the summer; though primary producton continues at a high rate, regenerated nutrients being used immediately, intensive grazing by the rich zooplankton can effectively reduce the crop.

In some estuaries the spring diatom outburst is succeeded by a summer flowering mainly of dinoflagellates. The lower nutrients and higher temperatures typical of summer conditions were considered to favour dinoflagellates. Riley (1967) and Smayda (1957, 1966) find little support for this view and many exceptions to this type of succession may be quoted; some dinoflagellates have relatively high nutrient demands and can exist over a considerable range of temperature.

An example of the frequently observed greater richness, particularly of diatom crop, in inshore as compared with offshore waters in temperate latitudes is described by Hulburt and Corwin (1970). Casco Bay, Gulf of Maine, is characterized by strong mixing and generally high nutrient levels. Diatoms dominated the phytoplankton through the growth season, eight major species forming more than 80% of the population, and the crop density was relatively high, varying from 100 to 5000 cells/ml. By contrast, outside the Bay, in the Gulf of Maine waters, there was considerable nutrient depletion in the euphotic zone during summer. The coccolithophorid, *Coccolithus huxleyi*, then dominated the population but the crop was much reduced.

The classic diatom-to-dinoflagellate succession is described by Mandelli, Burkholder, Doheny and Brody (1970) for a region of Long Island Sound. During late winter and spring Mandelli *et al.* observed a burst of diatoms (*Skeletonema*, either with *Thalassiosira* and *Chaetoceros* spp. or with *Thalassionema* and *Nitzschia* spp.) which was followed by an increasing population of dinoflagellates and small green flagellates from about April to August. The biomass of phytoplankton showed the very high and variable quantities typical of a temperate estuarine area; high chlorophyll (5–28 mg/m^3) during April to August; low values (1–3 mg/m^3) during autumn and winter. Slightly further to seaward the seasonal pattern was reversed with higher quantities of phytoplankton during autumn and winter (2.5–9 mg/m^3) and lower biomass (1.5–3.5 mg/m^3) during spring/summer. The crop was less variable outside estuarine waters, and surface primary production decreased from the estuary seawards. Patten (1962) has also described a change from a diatom to a dinoflagellate flora for the Raritan Estuary on the east coast of North America. During spring the phytoplankton was dominated by *Skeletonema*; over the summer a dinoflagellate and flagellate flora occurred, dominated by *Nannochloris*. Peak populations were found in spring and late summer. The low diversity agrees with the findings of other investigators; diversity increased to seaward.

Gieskes and Kraay (1975) include a detailed analysis of species succession in Dutch coastal waters. From early spring to the beginning of April diatoms were dominant, with a variety of forms including species of *Coscinodiscus*, *Biddulphia*, *Thalassiosira*,

Rhizosolenia, etc. *Chaetoceros socialis* became very important in late March/early April. In April/May *Phaeocystis* was the major alga, to be succeeded in May/June by armoured dinoflagellates, particularly *Ceratium* spp. and *Dinophysis* spp.

Phytoplankton of High Latitudes

The importance of diatoms in phytoplankton communities of high latitudes has already been emphasized. Thus Zenkevitch (1963) investigating the phytoplankton of the seas of the USSR, found that whereas in the western Barents Sea diatom species formed some 50% of the crop with dinoflagellates amounting to nearly 40%, in the central Arctic Basin the total number of phytoplankton species was only about one-quarter that of the Barents Sea, but diatoms were overwhelmingly important, forming nearly 80% of the species. In biomass the difference was even more obvious; diatoms comprised 98% of the biomass in the central Arctic Ocean. *Phaeocystis* is one of the few non-diatom algae which may be exceedingly plentiful at very high latitudes. In the Barents Sea, apart from *Phaeocystis*, three species of *Ceratium* (*C. longipes*, *C. fusus*, *C. arcticum*), three species of *Peridinium* (*P. depressum*, *P. oratum*, *P. pallidum*) and *Halosphaera viridis*, Zenkevitch reported only six diatom species which were really important contributors to the considerable crop: *Chaetoceros diadema*, *Corethron criophilum*, *Skeletonema costatum*, *Coscinodiscus subbulliens*, *Rhizosolenia styliformis* and *R. semispina*.

In the Antarctic the dominance of diatoms, often of relatively few species, was early emphasized (e.g. Hart, 1942). Hart listed the major species according to their oceanic or neritic character; of phytoplankton other than diatoms, he included only *Phaeocystis* and one silicoflagellate, *Distephanus speculus*, as relatively common. Balech and El-Sayed (1965), while accepting that diatoms dominated Antarctic phytoplankton, considered that some dinoflagellates could make a significant contribution. Balech (1970) listed about fifty species of dinoflagellates for Antarctic waters; the genus *Ceratium* was poorly represented, though two species occurred near the Antarctic Convergence. A number of species of *Peridinium* were found in the Antarctic, about 80% of the total species being confined to the Antarctic waters. Radiolarians may be also abundant, and since some species possess zooxanthellae, they can contribute to primary production. Zernova (1970) is one of the very few investigators who examined Antarctic phytoplankton from Nansen bottle samplings (i.e. quantitative samples) and from net hauls (qualitative samples). In the Indian and Pacific sectors of the Antarctic, apart from the well-established occurrence of *Phaeocystis*, and of the silicoflagellate, *Distephanus*, more than 99% of the total cells were diatoms belonging to about eighty species. Seventeen species of dinoflagellates were also reported. Zernova's quantitative analysis is of particular significance since it has been suggested that the reported importance of diatoms in Antarctic seas was perhaps exaggerated by the use of nets in earlier collections, and that nanoplankton might make a significant contribution. The overwhelming importance of diatoms would appear to be genuine (cf. Fogg, 1975). Other authors according to Zernova have recorded eighty to ninety diatom species for Antarctic waters. The major diatoms include *Chaetoceros neglectus*, *C. dichaeta*, *C. atlanticus*, *Corethron criophilum*, *Nitzschia* spp. and *Fragilariopsis* spp.; these diatoms can occur in high concentrations exceeding 10^5 cells/l. Other species such as *Rhizosolenia alata* were only slightly less plentiful. Among species which are more typical of the ice edge are: *Biddulphia weissflogii*, *Nitzschia closterium*, *Chaetoceros neglectus* and *Eucampia balaustium*. A species succession was observed with ice-edge species tending to bloom first during the spring, but then declining as summer species take over; these summer forms include *Chaetoceros criophilum*, *Rhizosolenia alata*, *R. chunii*, *R. hebetata*, *Dactyliosolen antarctica* and *Corethron criophilum*.

Apart from the generally high standing crops of summer phytoplankton characteristic of certain regions of the Antarctic (e.g. Gerlache Strait, South Georgia, South Orkney Islands, cf. Chapter 9), huge blooms usually attributable to one or two species of diatoms have been described from a number of Antarctic areas. Among the species of diatom involved are: *Thalassiothrix antarctica*, *Chaetoceros sociale*, *C. criophilum*, *Corethron criophilum* and *Biddulphia weissflogii* (cf. El-Sayed, 1971). El-Sayed has also described a huge bloom of *Thalassiosira tumida* in the south-west Weddell Sea.

A study of cold-water plankton communities and of seasonal changes at slightly lower northern latitudes comes from Holmes' (1956) investigation in the Labrador Sea. Autochthonous species were *Corethron hystrix*, *Fragilaria nana* and *Peridinium oceanicum*, but immigrant species included *Chaetoceros atlanticus*, *C. borealis*, *Fragilaria oceanica*, *Rhizosolenia styliformis* and *Thalassiosira gravida*, brought by the West Greenland and Irminger Currents. During the long winter period, roughly November to April, the phytoplankton stock was very low. While *Fragilaria nana* was particularly prominent over the flowering period, April to June, other forms (*Chaetoceros* spp., *Thalassiosira*, *Rhizosolenia*, *Nitzschia* spp.) tended to bloom in late May or June. A fairly classic spring blooming occurred, with neritic species and also oceanic species of Arctic, boreal and temperate origin all contributing; the population could rise from about 5000 cells/l to 87,000 cells/l in a week. A rather irregular summer decline was followed by a very brief September peak which involved mainly boreal and temperate species. Three dinoflagellates, including the ubiquitous *Exuviaella*, were also important in a seasonal cycle generally resembling the diatoms. Halldal (1953) followed the yearly cycle in the Norwegian Sea; phytoplankton numbers increased somewhat in spring lasting until June, but the main increase and maximum populations were not achieved till summer (June to September). A succession of diatom, dinoflagellate and coccolithophorid species was observed. Many of the important species were common to those described by Holmes. *Fragilaria nana* was outstandingly important in summer.

Phytoplankton of Boreal Waters

The Continuous Plankton Recorder studies of Robinson and his colleagues (e.g. Robinson, 1961, 1965; Colebrook, Glover and Robinson, 1961), which are concerned with phytoplankton distribution generally over the boreal North Atlantic, have the great advantage of several years of investigation. Despite the limitations in capturing only the larger species, oceanic and neritic and colder and warmer water species have been recognized and their distributions mapped. Thus, certain species (*Ceratium carriense*, *C. azoricum*, *C. hexacanthum*, *C. arcticum*, *Rhizosolenia alata indica*) are oceanic, and in the order given illustrate a change from more southerly warmer water forms to more northerly cold species. Some species (*Ceratium lineatum*, *Dactyliosolen mediterraneus*, *Thalassiosira longissima*, *Nitzschia seriata* and *N. delicatissima*) are typical of mixed intermediate oceanic/neritic areas but tend to be more oceanic; others (*Ceratium macroceros*, *C. longipes*) are also intermediate but tend to be more neritic in distribution. *Biddulphia sinensis*, *B. aurita* and *Asterionella japonica* are examples of typically neritic species; others (*Ceratium tripos*, *C. fusus*, *C. furca*, *Thalassionema nitzschioides*, *Rhizosolenia styliformis*, *R. hebetata*) occurred in all areas of the survey. Some of these are described as panthalassic species by Hart; *Thalassionema nitzschioides* is a cosmopolitan form (Smayda, 1958). The species succession was similar in all areas of the Plankton Recorder survey, though the timing varied; the succession was more gradual in the North Sea. Over the whole area the spring flowering consisted of diatoms, particularly *Thalassionema nitzschioides*, *Chaetoceros* spp.,

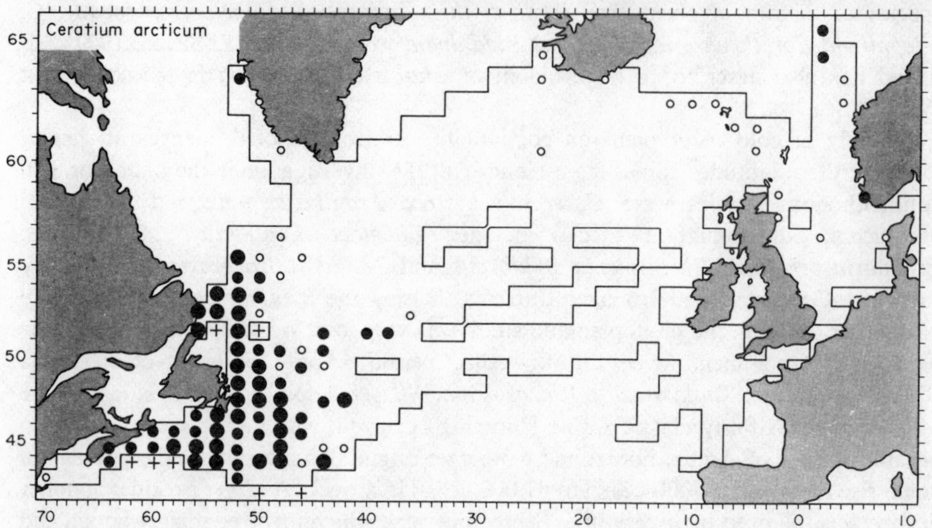

Fig. 5.21. Distribution of *Ceratium arcticum*. Small open circle, small solid circles and larger solid circle indicate three increasing density levels of the organism. A cross indicates its presence outside the regular sampling area. (*Plankton Atlas*, Edinburgh, 1973).

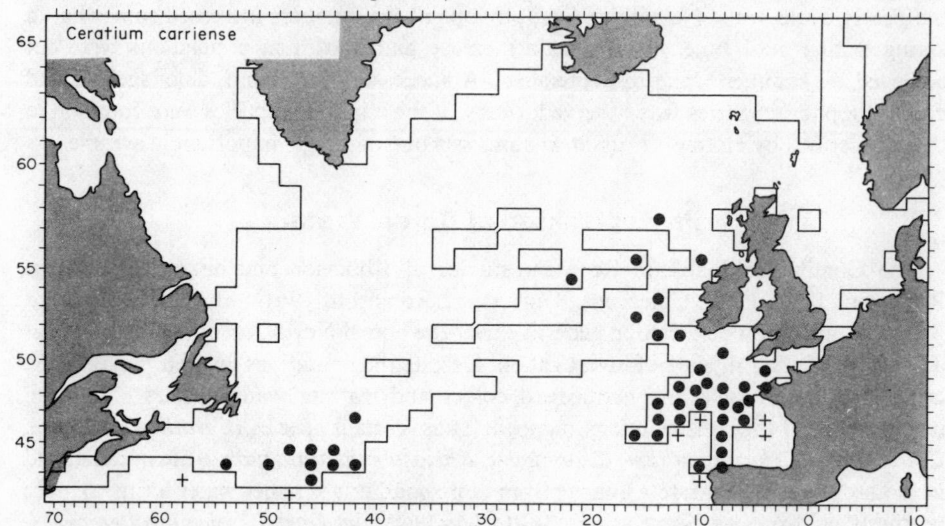

Fig. 5.22. Distribution of *Ceratium carriense*. Small open circle, small solid circles and larger solid circle indicate three increasing density levels of the organism. A cross indicates its presence outside the regular sampling area (*Plankton Atlas*, Edinburgh, 1973).

Thalassiosira spp. and *Rhizosolenia hebetata*, with *Thalassiosira longissima* and *Nitzschia* spp. important in the Atlantic. Summer species were mainly dinoflagellates, chiefly *Ceratium fusus*, *C. furca* and *C. tripos*, with *Rhizosolenia alata alata* in shallow areas. *Dactyliosolen* and *Rhizosolenia alata indica* were abundant in autumn, but almost entirely in more oceanic regions. *Biddulphia* and *Rhizosolenia styliformis* were abundant during autumn in the North Sea. Some of the more striking patterns of geographical distribution are included in the Continuous Plankton Records, *Plankton Atlas* (Edinburgh, 1973). For example, contrasting distributions of four species of *Ceratium* are illustrated in Figs. 5.21, 5.22, 5.23 and 5.24. Braarud, Gaarder and

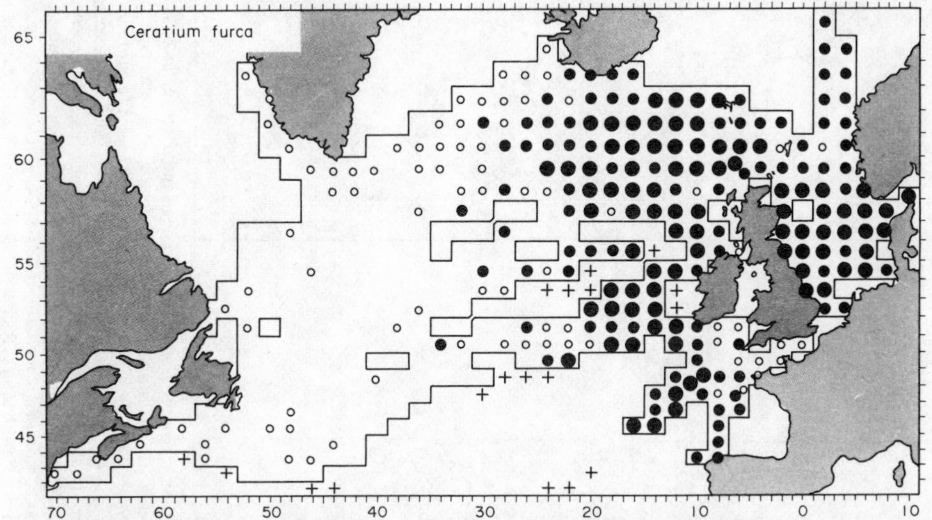

Fig. 5.23. Distribution of *Ceratium furca*. Small open circle, small solid circles and larger solid circle indicate three increasing density levels of the organism. A cross indicates its presence outside the regular sampling area (*Plankton Atlas*, Edinburgh, 1973).

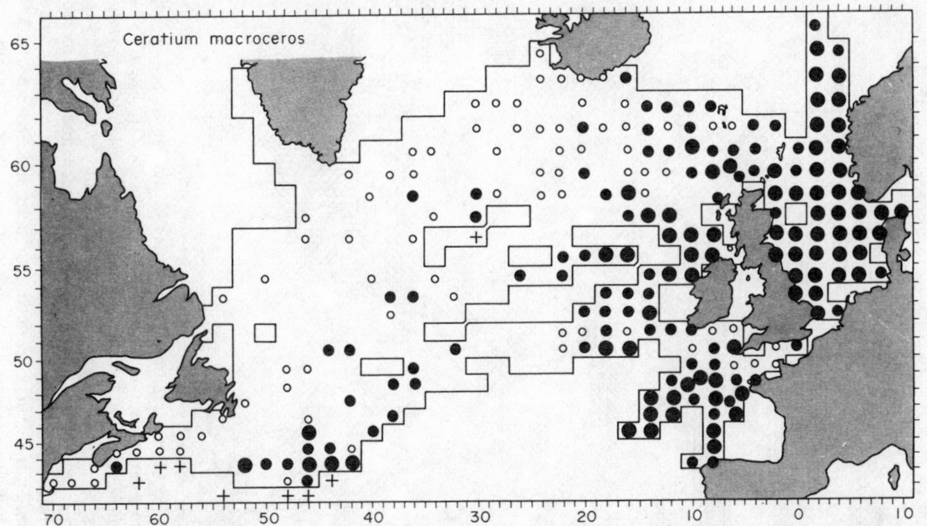

Fig. 5.24. Distribution of *Ceratium macroceros*. Small open circle, small solid circles and larger solid circle indicate three increasing density levels of the organism. A cross indicates its presence outside the regular sampling area. (*Plankton Atlas*, Edinburgh, 1973.)

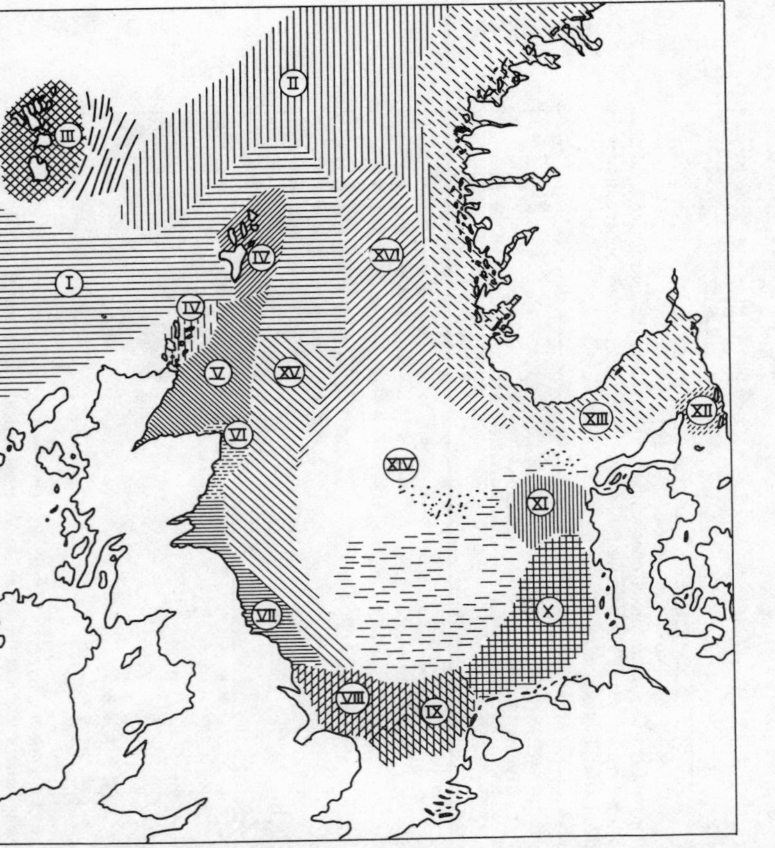

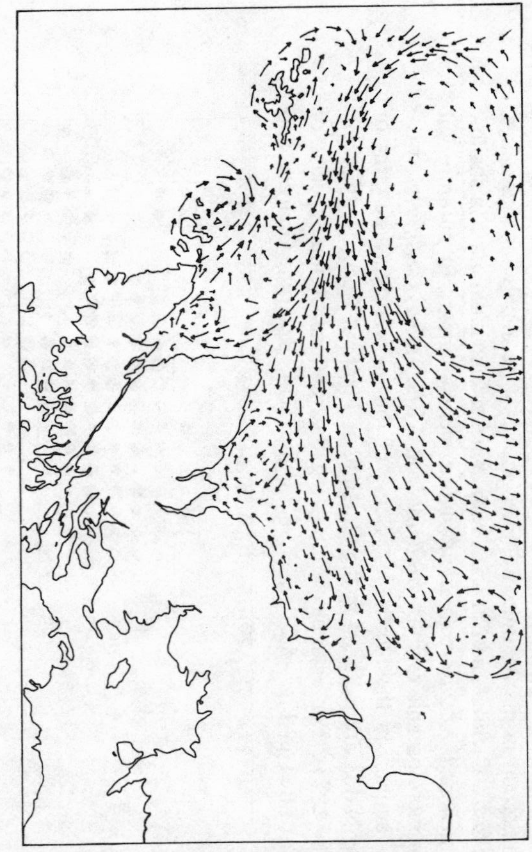

Fig. 5.25. Phytoplankton communities in the North Sea. I. Inflowing Atlantic water. *Coccolithus huxleyi* and *Exuviaella baltica*. II. Local Atlantic water. *Skeletonema costatum*, *Thalassiosira gravida*. III. Faero Isles. *T. gravida*, *S. costatum*, *Chaetoceros debilis* and other spp., *Nitzschia delicatissima* and *Coccolithus huxleyi*. IV. Shetland and Orkney Islands. *Asterionella japonica*, *C. debilis*, *C. decipiens*, *T. gravida* and *S. costatum*. V. *T. gravida* was the major species. VI. Off Rattray Head. *Rhizosolenia fragilissima*. VII. Coastal current. *A. japonica*, *Chaetoceros* spp. and *N. delicatissima*. VIII. Coastal water. *Nitzschia closterium* mixed with *A. japonica* and *Chaetoceros danicus*. IX. Southern Bight. Benthic species, e.g. *Bellerochea malleus*, *Biddulphia* spp., *Campylosira cymbelliformis*, *Cymatosira belgica*, *Navicula membranacea*. X, XI. German and Danish coasts. *A. japonica* and *C. debilis* and further off shore *Cerataulina bergoni*, *Eucampia zoodiacus* and *Phaeocystis*. XII. Kattegat. *Chaetoceros tortissimus*. XIII. Skagerak and Norwegian coastal current. Diatoms (*Chaetoceros compressus*, *N. delicatissima*, *Leptocylindrus danicus* and *R. fragilissima*). Richer in Dinoflagellates, *Ceratium* spp., *Exuviaella baltica* and *Peridinium trochoideum*. XIV. Central North Sea and Dogger Bank. *Rhizosolenia imbricata* v. *shrubsolei*. XV. South-flowering Atlantic water. *C. debilis*, *C.* [illegible] *Exuviaella baltica* and *Coccolithus huxleyi*. XVI. Mixed water of the north North Sea. *A.*

Grontved (1953) had earlier described phytoplankton communities in the North Sea and adjacent areas. The detailed distribution of a few species (e.g. *Asterionella japonica*, *Nitzschia delicatissima*, *Exuviaella baltica*) were described, and the floras of some areas related to major bodies of water (cf. Fig. 5.25). Distribution was interpreted in terms of surface currents and variation in timing of the succession of the spring/summer populations especially in relation to stability of the water. The dominance of such species as *Nitzschia closterium* was related to river outflow promoting an estuarine type of phytoplankton.

Biogeographical studies on phytoplankton communities in Pacific boreal waters, due to Marumo (1967), demonstrated that the Oyashio Current and Alaska Current formed a boundary demarcating a mainly sub-Arctic cold-water community to the north. Similarly the warm Kuroshio Current limited the southern phytoplankton distributions and acted as a barrier to mixing populations (cf. Figs. 5.26, 5.27). Marumo found diatom crops north of about 45°N were very much richer (10^3–10^6 cells/l) than crops to the south; the difference could be a hundred or even a thousand times (cf. p. 211). There were four major diatom communities. Sub-Arctic populations were typical of the Bering Sea, the Aleutian Islands and Oyashio areas, with *Chaetoceros* spp., *Rhizosolenia hebetata*, *Fragilaria* spp., *Thalassionema nitzschioides*, *Thallasiosira nordenskioldii*, *Corethron hystrix*, and *Leptocylindricus danicus*, amongst other species. Differences were apparent between coastal and oceanic areas, and between major spring and summer species. To the south were two fairly distinct warm-water communities: the first dominated by *Rhizosolenia*, especially *R. styliformis*, *R. calcar-avis* and *R. stolterfothii*,

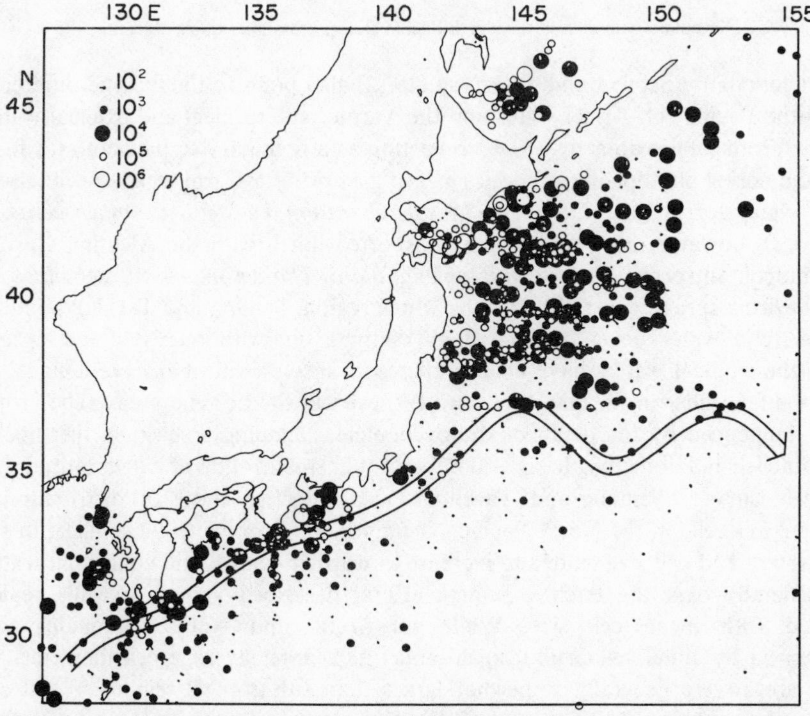

Fig. 5.26. Density of diatoms in seas off Japan during summer (numbers: cells/litre). Arrow indicates Kuroshio Current (after Marumo, 1967).

the other, a very monotonous community, consisting mainly of *Hemiaulus*, with some *R. styliformis*. Between the sub-tropical and sub-Arctic communities in a relatively narrow band, approximately 45–50°N latitude, was a very characteristic community dominated to a remarkable degree by the diatom *Nitzschia seriata*. More typically neritic species were found close to the Japanese coast, but *Eucampia zoodiacus* could spread out from the coast north of the Kuroshio Current.

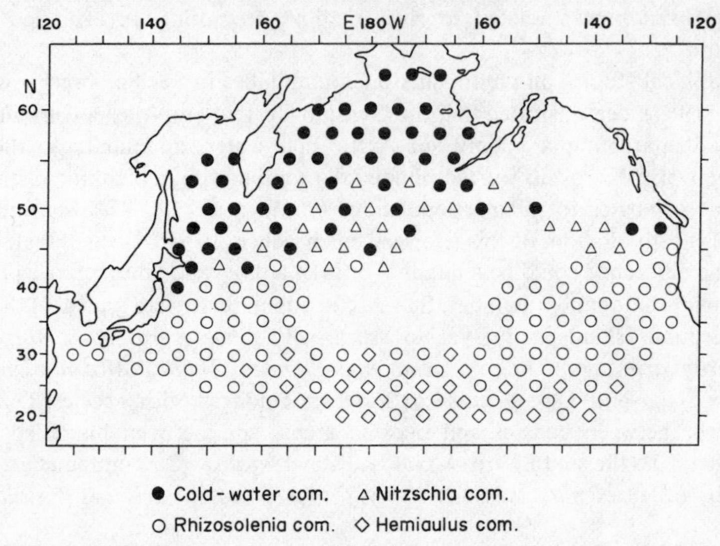

Fig. 5.27. Diatom communities in the North Pacific Ocean in summer (after Marumo, 1967).

Investigations by Semina and Tarkhova (1972) also point to the floristic differences between the Pacific sub-Arctic water and the warmer sub-tropical and tropical waters, the Polar Front (approximately 35–40°N) acting as a boundary to the south for many Arctic and boreal phytoplankton species and to the north for warm-water species. Some warmer water forms (e.g. *Planktoniella sol, Ceratium karsteni, C. macroceros, C. massiliense*), however, are successfully transported north with the Aleutian Current, and cosmopolitan species such as *Rhizosolenia alata, Thalassionema nitzschioides* and *Ceratium fusus* spread generally over the whole region. Semina and Tarkhova refer to the sub-Arctic water being dominated by diatoms but with relatively few species, whereas the tropical waters have an abundance of species with many peridinians, but the crop is far richer in the sub-Arctic waters, even in the oceanic areas. The crop is strongly influenced by the depth of the pycnocline, including a seasonal thermocline where established, but melting ice is also important in stabilization at higher latitudes.

In their survey, Semina and Tarkhova also refer to the size distribution of phytoplankton cells in the North Pacific. The algae were shown to be smallest in sub-Arctic waters and cell size tended to increase in warmer central and equatorial waters. More generally over the Pacific, Semina (1972) described various oceanic regions correlated with mean cell size. While sub-Arctic and Antarctic regions were characterized by small cells, sub-tropical areas had larger average cell diameter, and tropical areas were generally somewhat larger than sub-tropical regions in cell size. However, the equatorial belt and part of the tropical region in the southern hemisphere showed some decrease in cell size. Generally also the warm oceans showed a much greater range in cell size (cf. Table 5.4). Semina believes that this distribution can be referred to the oceanic circulation and that average cell size is directly related to the velocity and direction of the vertical water movement. Size is also related to the density gradient in the main pycnocline and to phosphate concentration, but all three factors interact. Parsons and Takahashi (1973b) have suggested that additional environmental factors (rate of uptake of inorganic nitrogen, light intensity and extinction coefficient, depth of the mixed layer) may affect cell size, and some discussion has followed, especially concerning size and rate of nitrate uptake (cf. Hecky and Kilhan, 1974). Size of phytoplankton is of considerable significance to the type of grazer.

Table 5.4. Size of phytoplankton cells in some Pacific regions (from Semina, 1972)

	Cell diameter (μm)		
	average	minimum	maximum
Arctic-boreal region	20 ± 2	10	35
Zone of mixing	36 ± 4	10	103
Tropical region	102 ± 2	10	305

Coccolithophorids and the Nanoplankton

Certain species of coccolithophorids have been included in descriptions of some phytoplankton communities. They can be particularly good indicators of water masses (cf. Gessner, 1970) and may be the dominant phytoplankton of some seas, but most of the species are small enough to be part of the nanoplankton. A more detailed treatment of these algae is, therefore, now included, with some account of the importance of nanoplankton.

Gaarder (1971) has emphasized the numerical abundance of coccolithophorids. Some species have a worldwide distribution while others are considerably restricted but it is often difficult to identify coccolithophorids to species in view of the minuteness of the cells. For example, the very common *Coccolithus huxleyi* may be confused with *Gephyrocapsa oceanica*, or with the small form of *Cyclococcolithus fragilis*. Coccolithophorids have been little studied in some areas such as the major part of the Indian Ocean.

Coccolithophorids are more generally typical of warmer seas; Lohmann (1908, 1911), for example, found significant populations only during August at Laboe, whereas in the Mediterranean coccolithophorids formed a major part of the phytoplankton over much of the year (cf. Bernard, 1967). Hasle (1959a) described coccolithophorids as more abundant than diatoms or dinoflagellates in the equatorial Pacific, especially in the upper 100 to 150 m. In the neritic waters of the Great Barrier Reef, diatoms were an important part of the phytoplankton, occasionally with *Trichodesmium*, but coccolithophorids became abundant for part of the year (August to November) (Marshall, 1933). We have already noted the importance particularly of *Coccolithus huxleyi* in the phytoplankton off Bermuda (Hulburt, Ryther and Guillard, 1960). Marshall (1968) found that whereas coccolithophorids could be abundant in the Sargasso in the upper part of the euphotic zone, they might also occur in considerable densities at deeper levels – 200, 400, 500 m, or even deeper. By contrast, diatoms and dinoflagellates were largely restricted to the upper 100-m layer. Coccolithophorids with dinoflagellates tended to be predominant in waters of high salinity, diatoms more plentiful in coastal areas; *C. huxleyi* was ubiquitous. The number of species of coccolithophorids increased with distance offshore with fewer dominant forms (i.e. greater diversity). A few species of coccolithophorids may be important on occasions in colder seas (cf. Lillick, 1940). Gran and Braarud (1935) considered that low nutrient levels during summer might favour coccolithophorids in the Gulf of Maine (cf. Hulburt and Corwin, 1970). In the North Sea, *C. huxleyi* can be relatively abundant in certain areas (cf. Fig. 5.25), particularly where Atlantic water intrudes around the north of Scotland. Halldal (1953) describes coccolithophorids becoming abundant in the Norwegian Sea during June (*C. pelagicus*), and again during early September (*C. huxleyi* and other species) (cf. Fig. 5.28).

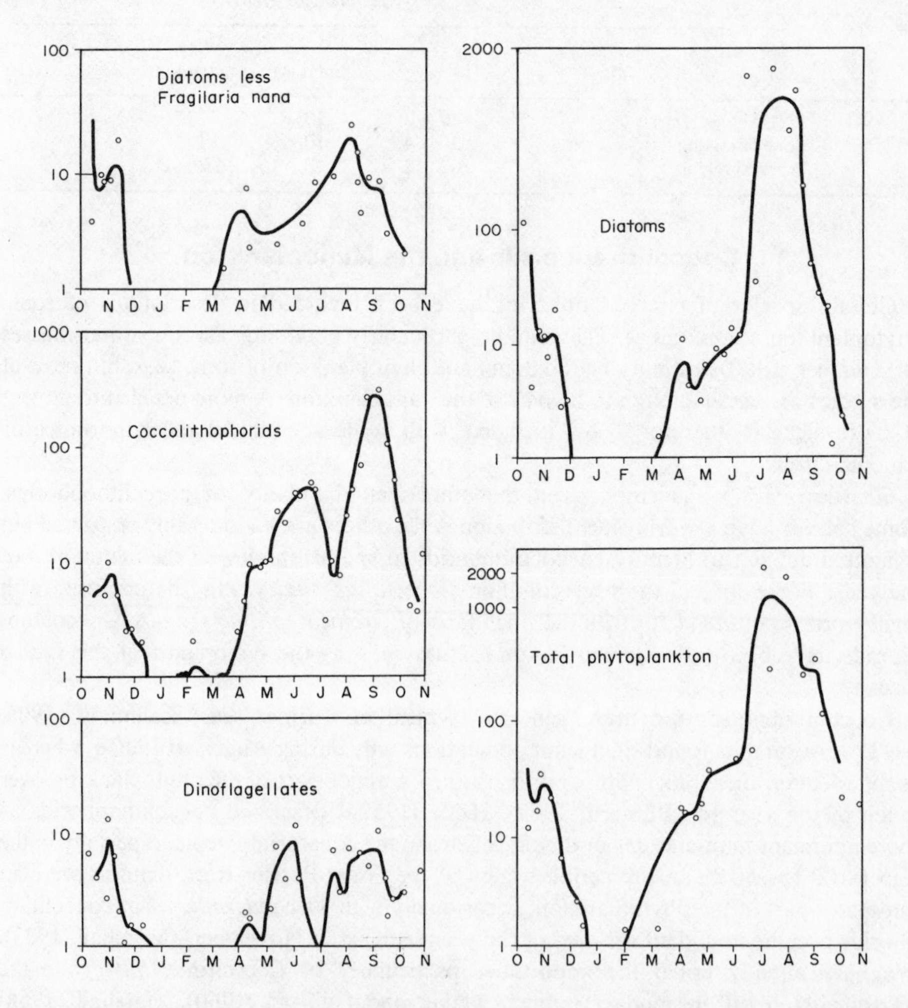

Fig. 5.28. The seasonal changes in diatoms, diatoms less *Fragilaria nana*, coccolithophorids, dinoflagellates and total population at Weather Station M from October 1948 to the end of October 1949. Logarithmic scale, numbers in thousands per litre (after Halldal, 1953).

Of the few species occurring in colder waters, *Coccolithus* (*Emiliana*) *huxleyi* is found in all except polar seas; it has an unusually wide temperature and salinity range. In the polluted Oslo Fjord the species occurs in water as low as 18‰ salinity (Braarud, 1962). It is sometimes described as cosmopolitan, being one of the most widely distributed of all coccolithophorids occurring in inshore and offshore waters, though possibly then as different strains. According to Paasche (1968c) in laboratory culture *C. huxleyi* has a very narrow temperature optimum of 20–23°; at 10°C growth is only 25% of that at optimal temperature. In the sea its greatest natural concentration occurs in waters at about 10°C, and there is some growth even at 2–3°C. It apparently disappears from tropical plankton at about 27°C. Berge (1962) found up to 1.15×10^8 *C. huxleyi* per litre in surface waters of coastal and fjord waters of western Norway; records in other years indicated similar coccolithophorid blooming. *C. huxleyi* is found even in sub-Antarctic waters; Hasle (1970) recorded up to 3×10^5 cells/l. Very high light intensities are tolerated since in Norwegian waters the richest concentrations were at the surface in May/June.

As regards warm waters, Bernard (1967) emphasized the importance of coccolithophorids, especially *Cyclococcolithus fragilis*, in the Mediterranean. This species comprised 50–90% of the mean total volume of euphotic phytoplankton, and 90–99% of the aphotic phytoplankton, everywhere in the southern Mediterranean. Coccolithophorids dominated the phytoplankton for about two-thirds of the year. An average doubling time for *C. fragilis* was about 10 days, and the average population about 270 cells/ml. Although naked flagellates, mainly Chrysophyceae, could be much more abundant (up to 500 cells/ml), they were so much smaller that their contribution to the phytoplankton volume was only some 3–15%. In western Algerian waters, diatoms (e.g. *Skeletonema*, *Leptocylindricus*, *Chaetoceros* spp.) could show outbursts from March to June, but their average annual contribution to the phytoplankton volume was only 10%; dinoflagellates averaged about 13%. The maximum density recorded in the euphotic zone for *Cyclococcolithus fragilis* was 1800 cells/ml; in deep outflowing Mediterranean water from Algiers to Gibraltar, occasional maxima of 3000–7000 cells/ml were observed. In the palmelloid stage *C. fragilis* may reach densities of 5000 cells/ml. Bernard observed red "plates" of *Cyclococcolithus*, visible to the naked eye, each containing some $1-18 \times 10^4$ cells (cf. Chapter 4). The outstanding significance of this coccolithophorid is clear but other species can be important in deeper ocean waters (Bernard, 1953).

Okada and Honjo (1973) examined the distribution of coccolithophorids in the north and central Pacific Ocean, from approximately 50°N to 10°S latitude. Dominant species are listed for the different zones from north to south (cf. Figs. 5.29 and 5.30). The variety of species in tropical waters contrasts with the almost exclusive occurrence of one species *Coccolithus (Emiliana) huxleyi* in sub-Arctic regions. The standing crop of coccolithophorids, however, was highest in the sub-Arctic region; it was moderately high in equatorial regions. In sub-Arctic waters the high density was very close to the surface and decreased rapidly with depth; an average density throughout the upper 30 m was of the order of 1.2×10^5 cells/l. Further south, the maximum density could be as deep as 50 or 100 m, with in some areas a more or less even distribution to about 200 m. With this extended distribution the different layers were inhabited by characteristic species (cf. Fig. 5.29). Temperature and light were believed to be the main factors limiting distribution.

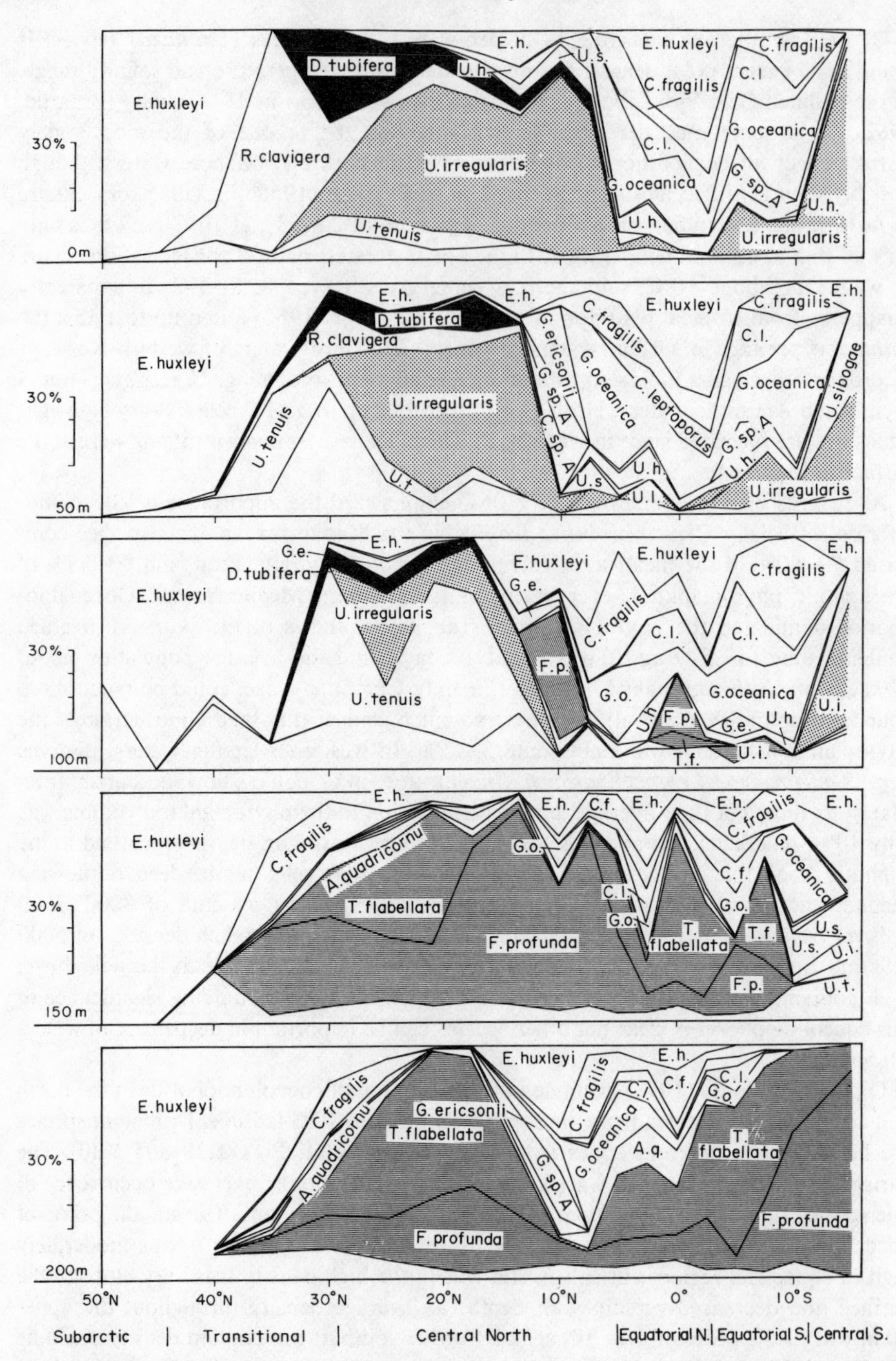

Fig. 5.29. Species assemblage of coccolithophorids at the surface and four sub-surface levels from 50 m to 200 m, along traverse 4 (155°W). Abbreviations of genera:

A = *Anthosphaera*
D = *Discosphaera*
F = *Florisphaera*
S = *Syracosphaera*

U.h = *Umbilicosphaera hulburtiana*
U.s = *Umbilicosphaera sibogae*
U.i = *Umbellosphaera irregularis*
U.t = *Umbellosphaera tenuis*

C = *Cyclococcolithina*
E = *Emiliana*
G = *Gephyrocapsa*
T = *Thorosphaera*
R = *Rhabdosphaera*

(after Okada and Honjo, 1973).

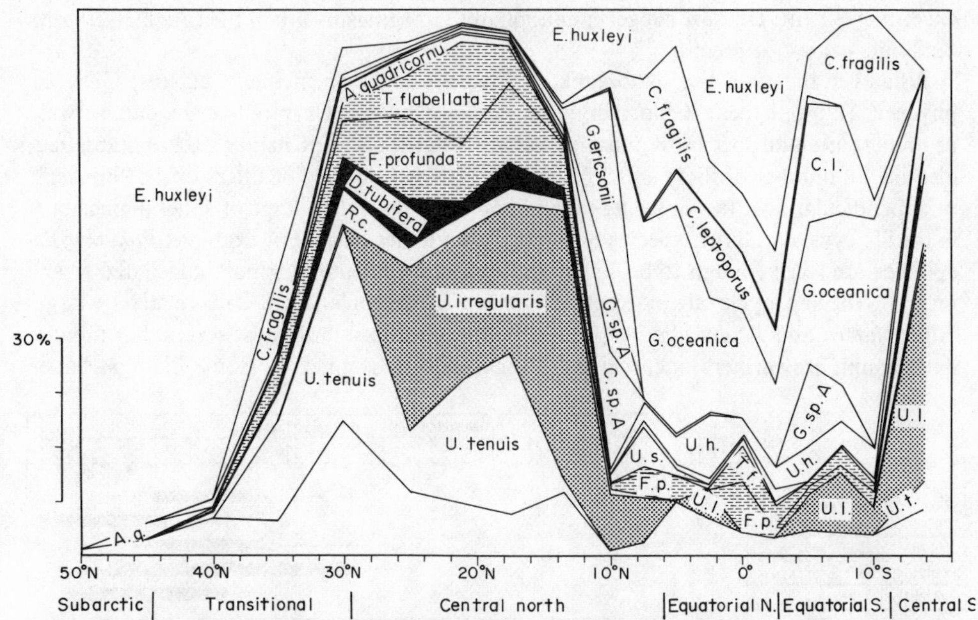

Fig. 5.30. Species assemblage obtained from the 200-m water column along traverse 4° (155°W). Coccolithophorids below the 200-m level are omitted. Abbreviations as Fig. 5.29 (after Okada and Honjo, 1973).

McIntyre and Bé (1967), describing the geographical distribution of coccolithophorids from surface Atlantic waters, also found that although the majority of species are warm water inhabitants, with diversity decreasing to high latitudes, the few species which flourish in boreal waters may be extremely abundant. At times, for example, *Coccolithus huxleyi*, attained 3.5×10^7 cells/l. They confirmed *C. huxleyi* as the widest ranging and most ubiquitous of species, often comprising 50% of the coccolithophorid population in tropical waters, and nearly 100% in sub-Arctic and sub-Antarctic waters. A summary of the distribution of species with temperature is given in Fig. 5.31. Seasonal changes in three dominant species in Bermuda waters are also illustrated (Fig. 5.32). A few coccolithophorids are stenothermal, e.g. *Umbellosphaera irregularis* – a tropical species confined to waters of 21–28°C.

The coccolithophorids frequently form a substantial part of the nanoplankton. The importance of the nanoplankton fraction in evaluating the total phytoplankton and some of the problems in devising methods for the quantitative assessment of the nanoplankton have already been outlined (cf. pp. 205–206). The contribution of the nanoplankton to different phytoplankton populations will now be examined.

Since the original definition of nanoplankton referred to those organisms normally not retained by the "finest" (No. 25) silk net, of nominal aperture size 60 μm, or, according to Harvey, 50×40 μm, the maximum diameter of nanoplankton cells would be of the order of 50 or 60 μm. Even though only the photosynthetic organisms are regarded as nanoplankton a very considerable size range will be covered in the nanoplankton, since many of the algal species are only 2–3 μm in diameter or even less. Today, nets of much smaller mesh dimensions (20–30 μm) are available, permitting further fractionation of the plankton. Some authorities suggest separating the very smallest cells (<5 μm) as "ultra plankton" and to limit the nanoplankton to algae not exceeding 30 μm. Dussart suggests nanoplankton organisms are in the range 2–20 μm, but limits cannot be precise.

Whatever the size, the nanoplankton includes algae from many classes, Chrysophyceae, Haptophyceae, Chlorophyceae, Prasinophyceae, Cryptophyceae, etc., as well as numerous small species of diatoms and dinoflagellates (cf. Chapter 4). Lohmann had identified a number of these small organisms retained on the fine filters of the "houses" of appendicularians. In some cases the nanoplankton cell is a stage of small dimensions in the life cycle of a larger species normally taken in net hauls. For example, *Phaeocystis* colonies are large enough to be visible to the naked eye but the minute flagellated swarmers in the life cycle are correctly regarded as nanoplankton. The relatively large *Halosphaera* and *Pachysphaera* (Prasinophyceae) are similarly cyst stages, but minute nanoplankton swarmers occur in the life cycles (cf. Chapter 4). Some dinoflagellates

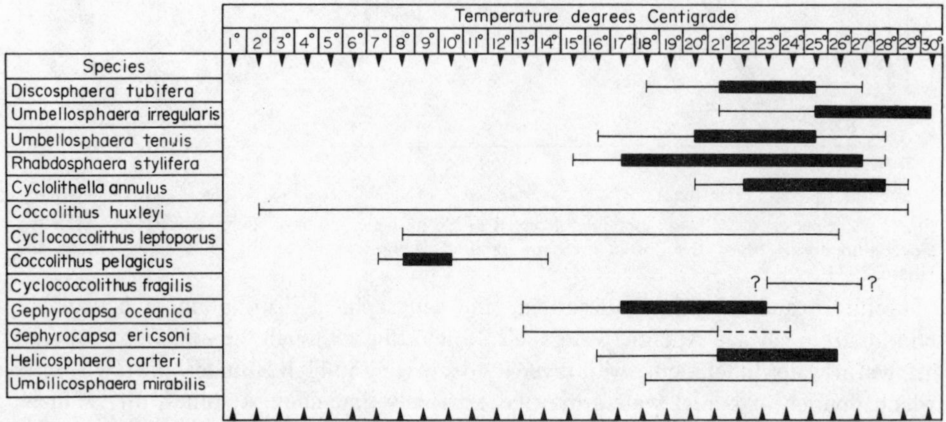

Fig. 5.31. Maximum and, where ascertained, optimum temperature ranges of thirteen dominant coccolithophorid species (McIntyre and Bé, 1967).

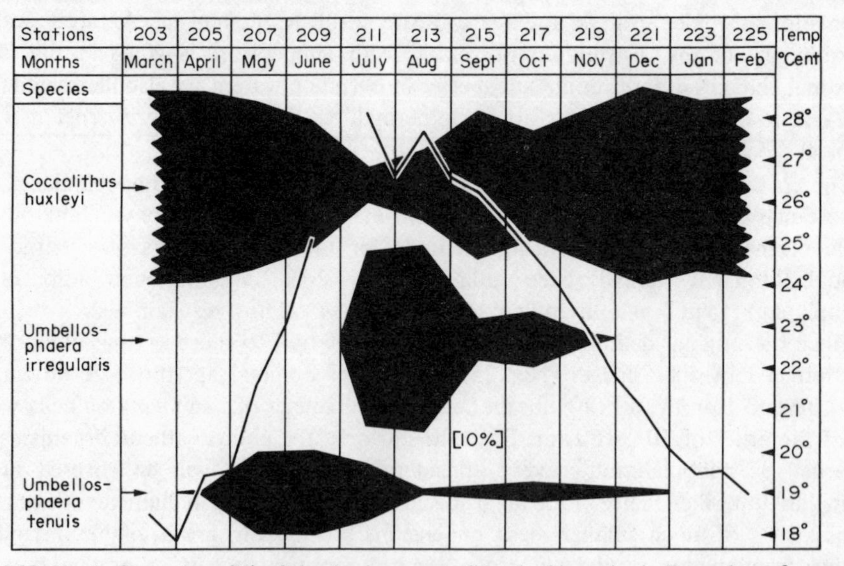

Fig. 5.32. Yearly fluctuations (%) of the dominant coccolithophorid species in Bermuda surface waters compared with temperature (McIntyre and Bé, 1967).

also have complex cycles with minute swarming stages. Aside from these species, each of the algal classes includes a number of now well-known genera of minute size, e.g. *Micromonas* (Prasinophyceae); *Chromulina* (Chrysophyceae); *Nannochloris*, *Carteria* (Chlorophyceae); *Chrysochromulina*, *Pavlova*, *Isochrysis*, *Dicrateria* (Haptophyceae); *Cryptomonas*, *Hemiselmis*, *Rhodomonas* (Cryptophyceae).

The silicoflagellates are an important group of Chrysophyceae and the coccolithophorids, which have already been discussed in some detail, well-known members of the Haptophyceae. Though the life histories and the detailed structure of many of these minute nanoplankton species must yet be elucidated, the painstaking and brilliant researches of such investigators as Parke, Manton, Leadbeater, Paasche, Dodge, Boney, Bernard and many other have now revealed a wealth of detail. Chapter 4 deals with some of this work in the various algal classes.

An important feature of many nanoplankton species is that their optimal rate of multiplication is generally higher than that of the larger net phytoplankton, following the very broad generalization that rate of uptake, growth and division is higher with reduced cell size. It is often suggested that oceanic nanoplankton may have an advantage over the larger-celled species in competition for nutrients in low concentration, since smaller-celled and usually faster growing oceanic species frequently have lower K_s values (cf. Friebele, Correll and Faust, 1978). The efficiency of utilization of light energy is one factor which may modify the effect of nutrient level. Parsons and Takahashi (1973b) computed growth rates for a comparatively large-celled (*Ditylum brightwellii*) and small-celled species (*Coccolithus huxleyi*), applying average values for incident radiation, average light intensity in the water column, mixed layer depth, nitrate concentration and upwelling intensity. The growth rate for *Coccolithus* was higher in regions of water stability; the larger-celled algae appeared to grow faster in upwelling and coastal areas. Some nanoplankton species may, given optimal conditions, divide rapidly, for example, every 5 or 6 hours (cf. Lanskaya, Parke, etc.). Dense populations may, therefore, be built up even more rapidly than with some of the larger diatoms and dinoflagellates. For temperate waters, it has been suggested that in many areas nanoplankton becomes important after the typical spring diatom outburst (cf. Hulburt and also Braarud on summer populations of coccolithophorids; also Holligan and Harbour, 1977). Nanoplankton has sometimes been claimed to be of greater significance in certain inshore waters. Though coccolithophorids are of great importance in some seas, minute diatoms may be extraordinarily abundant in the nanoplankton (cf. Collier and Murphy, 1962; Thorrington-Smith, 1970b). Other algae (Cryptophyceae, Prasinophyceae) may also be exceedingly abundant; the very minute prasinophyte, *Micromonas pusilla*, may be the commonest organism in inshore British waters (Knight Jones, 1951) (cf. Chapter 4). Harvey (1950), using filters, showed that the crop of phytoplankton was far greater than that estimated from net hauls in waters off Plymouth, though the nanoplankton species were not determined. A similar conclusion was reached by Qasim, Wellershaus, Bhattathiri and Abidi (1969) for eutrophic Indian waters.

Gross, Orr, Marshall and Raymont (1947) and Gross, Nutman, Gauld and Raymont (1950) found that nanoplankton formed a very appreciable part of the algal crop in shallow temperate waters. Yentsch and Ryther (1959) called attention to the importance of the nanoplankton fraction in waters off Woods Hole. Average net plankton was only 9% of total cell density: flagellates formed only a small portion of the nanoplankton as compared with minute diatom species. Small flagellates have been reported in considerable concentration in the North Sea (Grontved, 1952); in the Baltic and in northern seas by Steemann Nielsen (1951, 1935), Braarud (1935), Halldal (1953) and Bursa (1963). Studies by Savage (1969) in Southampton Water, using size fractionation techniques, demonstrated the outstanding importance of nanoplankton; the great majority of the crop as chlorophyll (*ca.* 90%) was attributed to the "small" phytoplankton (<55 μm in size), much of it less than 10 μm diameter. Similarly the greater part of ^{14}C assimilation (73%) was due to the nanoplankton. Burkill (1978) more recently confirmed, using size fractionation on samples taken in the same area, that 89% of the crop (as chlorophyll) was <50 μm in size. McAllister, Parsons and Strickland (1960) found at Station P in the North Pacific that the great majority of the phytoplankton passed through the finest nets and was mainly less than 10 μm. Further investigations by Parsons (1972) confirmed that nanoplankton (size range 8–16 μm) predominated in the sub-Arctic Pacific. Pingree, Holligan, Mardell and Head (1976) confirmed the importance of the nanoplankton in the Celtic Sea (south-west of the British Isles). In terms of carbon fixed, nanoplankton contributed $>70\%$ before and after the spring outburst and, although the relative contribution fell to less than 10% during the spring bloom, actual production of nanoplankton increased over that season.

Undoubtedly nanoplankton is of particular significance in warmer seas, especially in oligotrophic ocean regions. The investigations of Lohmann (1908, 1911) and of Bernard (1953) in the Mediterranean, of Riley (1957) for the Sargasso Sea, and of Bsharah (1957) for the Florida Current, call attention to the importance of nanoplankton in warmer waters. Hulburt, Ryther and Guillard (1960) report on the prevalence of coccolithophorids in waters off Bermuda; though the nanoplankton was not quantitatively measured, small coccolithophorids, flagellates and other nanoplankton forms made a considerable contribution to the total crop. Hulburt (1962), dealing with the tropical north Atlantic, finds that small flagellates and coccolithophorids are fairly regularly distributed in these warmer waters together with small diatoms and dinoflagellates. Wood (1963) considers that the nanoplankton generally becomes increasingly important in tropical seas, although there are exceptions, such as a considerable abundance of diatoms occurring off the Australian coast, and off the west coast of India. Revelante and Gilmartin (1976) found that nanoplankton, defined as cells of less than 20 μm diameter and consisting mainly of microflagellates, with some naviculoid diatoms and dinoflagellates, dominated the phytoplankton of the northern Adriatic, averaging 74–88% by cell number of the total phytoplankton. The larger phytoplankton, mainly diatoms, showed some seasonal increases.

A comparison between the nanoplankton and larger net phytoplankton by Teixeira (1963) and Teixeira and Kutner (1963) for Brazilian waters, demonstrated that in shallow lagoon waters the net plankton, consisting mainly of large diatoms, comprised only about 3% of the total which was dominated by the nanoplankton. Even in offshore waters up to 400 miles from the coast the nanoplankton was dominant, the net plankton contributing less than a fifth to the total crop. Net plankton contributed only 10% to the total photosynthetic activity in offshore waters. A few nanoplankton species seem characteristic of polluted waters with high organic content (e.g. Ryther, 1954), but many nanoplankton algae are particularly adapted to oligotrophic waters where they may contribute considerably to overall production. Tundisi (1971) has summarized work in the equatorial Atlantic indicating that 87–93% of the photosynthetic activity and

71–83% of the standing crop of phytoplankton is attributable to the nanoplankton fraction. He compares these results with those of Mullin for the Indian Ocean where the majority of particulate carbon was in the 1–10 μm size range and was believed to be mainly autotrophic nanoplankton, and with the findings of Saijo that a fair proportion of the phytoplankton was even below 5 μm in size in the Indian Ocean. In Brazilian mangrove swamp areas nanoplankton was the major fraction contributing to primary production and dominating the phytoplankton, but net plankton showed some increase over the summer.

Malone (1971) for the Caribbean and eastern tropical Pacific found that nanoplankton was the much more important part of the total phytoplankton, though a greater proportion of net plankton appeared in neritic waters. Chlorophyll *a* density and surface primary productivity for nanoplankton exceeded net plankton at every station (cf. Table 5.5.).

Table 5.5. Contributions of nanoplankton and net plankton to standing crop and primary production

	Neritic		Tropical surface		Caribbean	
	Nanoplankton	Net plankton	Nanoplankton	Net plankton	Nanoplankton	Net plankton
P.P.	1.54	0.74	0.67	0.115	0.30	0.03
S.C.	0.23	0.15	0.18	0.046	0.048	0.008

P.P. = Primary production (mg $C/m^3/hr$).
S.C. = Standing crop (mg chlor a/m^3).
Values are averages of all stations in the three areas.
Tropical surface waters do *not* include the Peruvian Current.
(Data from Malone, 1971).

Table 5.6. Standing stock of the net phytoplankton as a percentage of total phytoplankton at different latitudes (from Tundisi, 1971)

Locality	Author	Method	Approximate latitude	Net phytoplankton(%)
Scoresby Sound Greenland	Digby, 1953	Membrane filter 0.6 μm, pore size	70°N	66
English Channel	Harvey, 1950	Membrane filter	50°N	10–26
Long Island Sound	Riley, 1941	Membrane filter	41°N	9–56
Vineyard Sound (USA)	Yentsch and Ryther, 1959	Membrane filter	41°N	2–47
Tortugas	Riley and Gorgy, 1948	Membrane filter	24°N	1
Equatorial Atlantic	Teixeira, 1963	Millipore filter, H.A., 0.45 μm pore size	4°N	28
Equatorial Atlantic	Teixeira, 1963	Millipore filter, H.A., 0.45 μm pore size	0°	17
New South Wales	Wood and Davis, 1956	Centrifugation	33°S	3–4

The importance of diatoms at high latitudes has already been noted. Although nanoplankton algae are found in very cold waters, there is some evidence that the net plankton makes a greater contribution to total phytoplankton at higher latitudes (cf. Table 5.6). Since different methods of assessment of nanoplankton must introduce errors, the estimates in Table 5.6 should be regarded as only approximate. Considerable differences exist also between oceanic and coastal waters. Nevertheless, in broadest terms, the greater importance of the larger phytoplankton at high latitudes is confirmed.

Chapter 6
Factors Limiting Primary Production: Light and Temperature

Whatever the areal and temporal variations in productivity, primary production can extend over the whole ocean surface.

The outline of the mechanism of photosynthesis stressed the importance of light, and despite the modifications in cell physiology in different phytoplankton species described in Chapter 3 (variations in photosynthetic pigment, adaptations to light intensity, wavelength, temperature, nutrient concentration, etc.), light is the ultimate limiting factor, supplying the radiant energy for photosynthesis. The depth of the productive or euphotic zone and the intensity of photosynthesis must, therefore, vary in the ocean. Many factors including latitude, season, cloudiness and other weather conditions, wave action and the degree of turbidity, which itself will vary with factors such as currents, wave force, bottom topography and distance from land, affect photosynthetic activity and the depth of the euphotic zone.

Incident Light and its Transmission in Seawater

Although many species among phytoplankton populations are adapted to photosynthesize at relatively low light intensities, with the attenuation of light in the sea primary production almost everywhere is ultimately limited in depth. Only in very shallow and unusually clear waters does the euphotic zone reach the bottom. Of the incident radiation falling on the sea surface, relatively little is reflected unless the sun is at a low altitude; reflection is hardly significant at altitudes above 30°. At very low altitudes the surface loss may be very large indeed (e.g. 25% loss at 10° altitude) and this is a major factor at high latitudes during winter, where the incident light is from a low altitude for a significant part of the very short day. In addition, in polar regions during winter, with large areas of sea ice, a very large proportion of the small amount of incident light may be entirely reflected from the ice surface (cf. Bunt, 1966; El-Sayed, 1966).

The total radiation reaching the earth's surface normal to the sun's rays is nearly 2.0 g cal/cm^2/min (= ly/min)* (the solar constant). Since the earth is approximately spherical, the incident radiation spread over the surface throughout the 24-hour period would average about 0.5 ly/min but this is reduced by transmission through the atmosphere to 0.2 ly/min. Incident radiation will vary with time of day, latitude, cloudiness, etc., and is particularly subject to very marked seasonal fluctuations,

* One langley (ly) = 1 g cal/cm^2; as far as possible the langley is used as the unit of measurement for radiation throughout the text.

especially at very high latitudes. Thus instantaneous values for incident radiation near the equator can approach 2.0 ly/min. Some indication of the differences in the amount of total radiation reaching the sea surface in various months is given in Table 6.1 based on Kimball's original data. Although the variations between mid-summer and mid-winter at high latitudes are especially striking and there are marked differences between temperate latitudes and the tropics in certain months, differences in the *annual totals* of solar radiation received at high and low latitudes are not as great as sometimes supposed. Thus Steemann Nielsen (1974), quoting data from Kimball, points out that the total annual radiation at 65°N (Alaska) is about 50% of that at 26°N (Miami). It is the great seasonal variation which emphasizes the difference. The average daily insolation during mid-winter at 65°N is only 0.8% of that in mid-summer; at 26°N latitude it is 55%. Figure 6.1 illustrates the average integrated incident illumination for each month near Copenhagen, and also the change in radiation caused by exceptionally dark and bright weather. The maximum, in June, is about 144 to 180 ly/day; the minimum (December) would appear to approach only 3.6 ly/day. In more general terms Strickland (1965) estimated that whereas during a bright day in mid-summer temperate areas could receive a maximum of about 1.4 ly/min and only 0.5 ly/min at maximum during a mid-winter day, tropical areas receive about 1.6 ly/min with clearest skies during many months of the year.

Table 6.1. Average amount of radiation (ly/min) from sun and sky which reaches the sea surface in the stated localities (after Kimball)

Locality		Month						
Latitude	Longitude	Jan.	Feb.	Apr.	June	Aug.	Oct.	Dec.
60°N	7°E–56°W	.002	.053	.207	.292	.212	.074	0
52°N	10°W	.048	.089	.219	.267	.211	.104	.041
42°N	66–70°W	.094	.138	.272	.329	.267	.174	.086
30°N	65–77°W	.146	.165	.285	.310	.282	.188	.142
10°N	61–69°W	.254	.276	.305	.276	.292	.269	.239
0	48°W & 170°E	.261	.265	.297	.300	.340	.362	.278
10°S	72–171°E	.290	.308	.289	.253	.306	.313	.303
30°S	17 and 116°E	.452	.406	.254	.148	.214	.362	.430
52°S	58°W	.289	.237	.112	.039	.097	.222	.302
60°S	45°W	.213	.171	.056	0	.054	.156	.221

Water, in comparison with other liquids, is fairly transparent to solar radiation, but it is vastly less transparent than air. Of the radiation reaching the sea surface more than 50% is made up of wavelengths ranging from about 2500 to 730 nm (the infra-red or "heat rays") (Fig. 6.2). Water is particularly opaque to infra-red rays so that approximately 98% of the total infra-red radiation is absorbed in the upper 2 m of sea, whatever the value of the incident radiation. This relatively longer wave radiation is the source of the heat energy but is restricted to the surface layers (cf. Chapter 1). Ultra-violet radiation, approximately of wavelengths less than 380 nm, which forms only a small percentage of the incident radiation, is also rapidly absorbed by water, so that very little penetrates even in clear ocean waters. In turbid waters the ultra-violet radiation is almost confined to the surface. Strickland suggests that in the clearest waters ultra-violet radiation may penetrate to 10 m or more but in turbid seas penetration does not exceed 1–2 m depth. On a bright summer day in temperate regions in clear waters, therefore, the maximum solar energy available for photosynthesis cannot exceed 0.4 to 0.5 ly/min.

If consideration is now restricted to visible light (approximately 730 to 380 nm), which is about half the total incident radiation (Fig. 6.2), water is opaque as compared with air, even though visible light penetrates far more readily than infra-red rays. Absorption of light in water is approximately logarithmic: in pure (distilled) water about one-half the incident visible light energy remains at a depth of only 10 m, <5% at

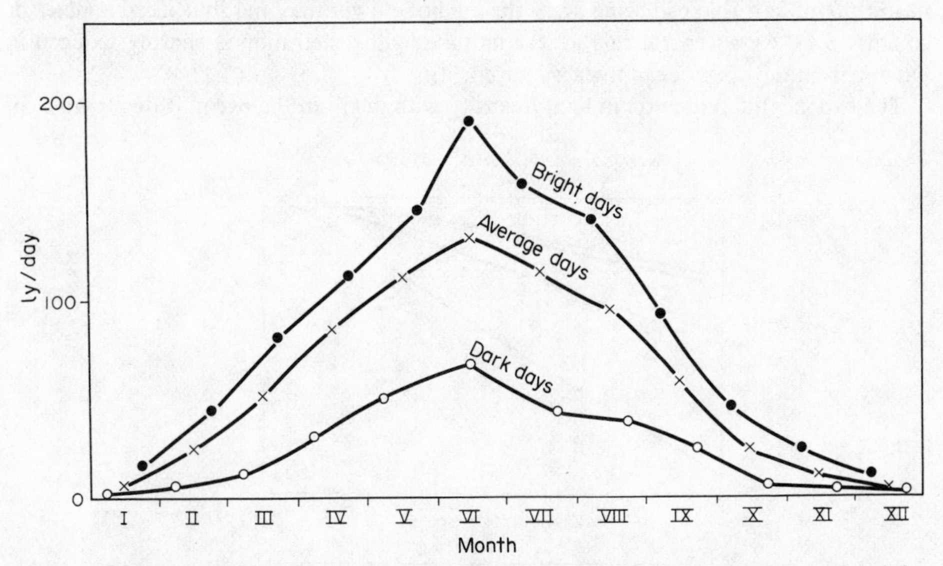

Fig. 6.1. The integrated daily illumination during the year at Copenhagen for bright, average and dark days (Steemann Nielsen, 1974).

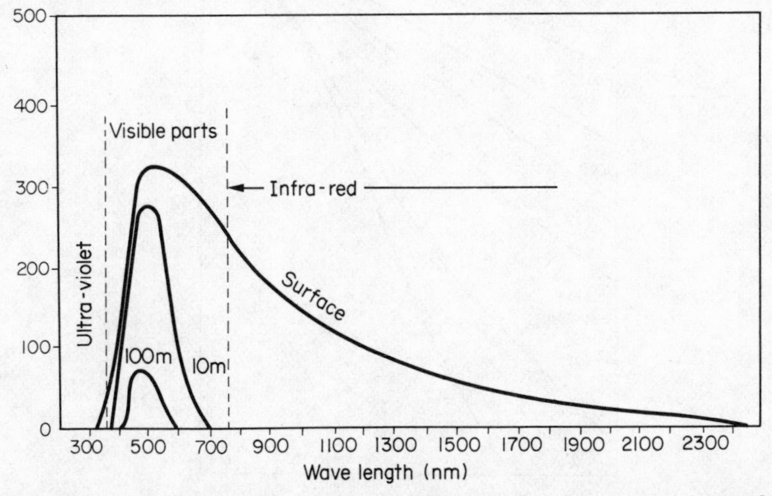

Fig. 6.2. Incident solar radiation falling on the sea surface, and the amounts remaining at 10 and 100 m. Visible wavelengths and especially infra-red rays are very rapidly absorbed. (Energy units along ordinate are arbitrary.) (Slightly altered from Sverdrup *et al.*, 1942.)

about 100 m, and a mere 1% at about 130 m. The transparency of the clearest sea water, typical of open tropical oceans, is only slightly less (Fig. 6.3), but the exponential rate of absorption of light with depth must severely restrict the thickness of the photosynthetic (euphotic) zone in any ocean.

Although the clearest open tropical waters approximate to distilled water in transparency, ocean waters at higher latitudes, and especially waters from inshore areas, show substantially reduced transparency because of suspended particles and organisms. Whereas in oceanic tropical areas the photosynthetic zone may certainly exceed 100 m with the maximum reaching 120 m, in temperate oceanic regions the maximum depth is nearer 50 m; in turbid estuarine areas the euphotic layer may hardly exceed a metre or so (Fig. 6.4). As a general rule at any latitude light penetration is sharply reduced in passing from the open ocean towards the coast.

The exponential reduction in light intensity with depth in the ocean is due to absorp-

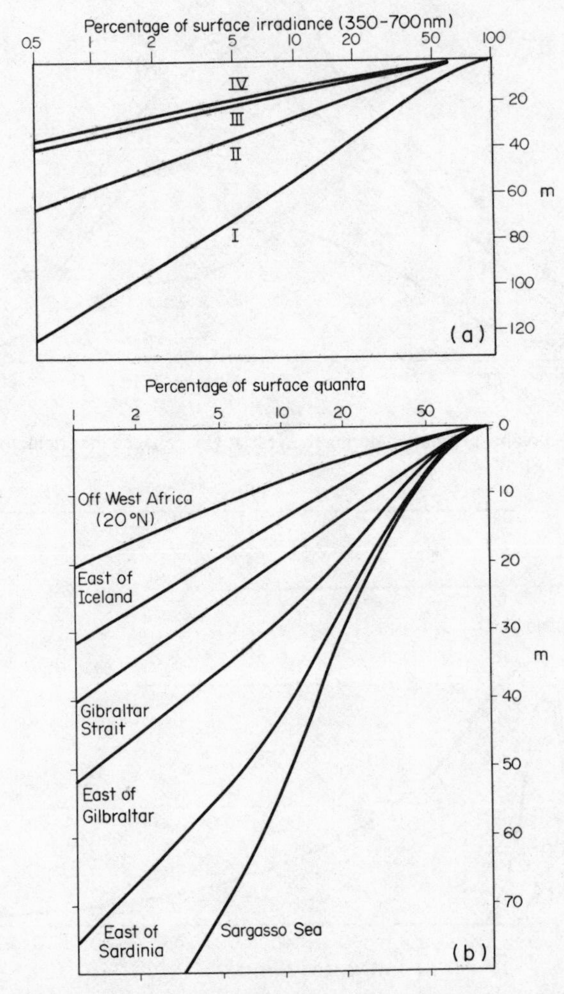

Fig. 6.3. The attenuation of visible light (350–700 nm) with depth in the sea. (a) Three types of ocean water (I, II and III) are represented, I being the clearest water. IV represents clear coastal water. (b) Attenuation of light in different regions (after Jerlov, 1968, 1976).

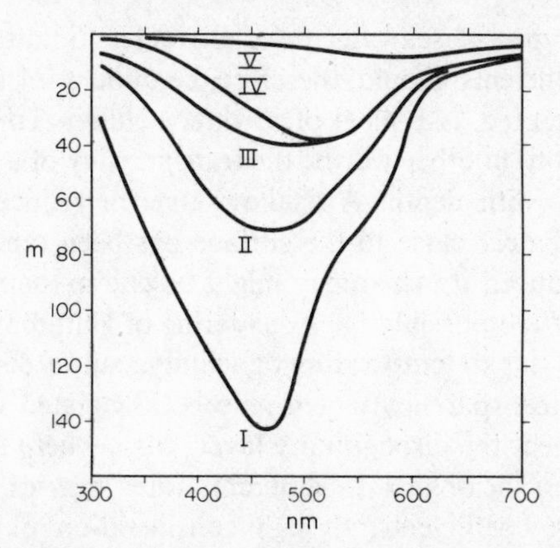

Fig. 6.4. Depths at which downward irradiance is 1% of the surface value for different types of seawater. I–IV are the different types of seawater listed in Fig. 6.3, V represents an extremely turbid coastal water. *Note*: the attenuation varies with the wavelength (from Jerlov, 1968).

tion and scattering by the water and suspended particles. It is better described as attenuation. The extinction or attenuation coefficient (k) is calculated:

$$k = \frac{2.3\,(\log I_1 - \log I_2)}{d_2 - d_1}$$

where I_1, I_2 represent light energy (irradiance) at two depths, d_1 and d_2, respectively.

The value of the extinction coefficient varies considerably from one sea area to another. Table 6.2 illustrates differences in the attenuation coefficient in waters ranging from very clear oceanic (Sargasso) to turbid coastal (Baltic) areas. It should be noted that the attenuation coefficient varies with wavelength within visible light for pure water,

Table 6.2. Observations of the attenuation coefficient relative to pure water (c—c_w) and of the scattering coefficient for particles (b_p) for two wavelengths

	$c - c_w/\mathrm{m}^{-1}$		b_p/m^{-1}	
Region	380 nm	655 nm	380 nm	655 nm
Sargasso Sea	0.05	0.04	0.03	0.02
	0.06	0.03	0.04	0.02
Caribbean Sea	0.11	0.06	0.06	0.06
Eq. Central Pacific	0.11		0.05	
Romanche Deep	0.14		0.07	
Mediterranean	0.11	0.04	0.04	0.03
	0.16	0.11	0.06	0.07
		0.15		0.11
Galapagos	0.25	0.13	0.09	0.08
Bermuda waters	0.20	0.10	0.10	0.11
	0.25	0.16	0.11	0.12
		0.32		0.23
Kattegat	(0.54)	0.23	0.16	0.15
Baltic Sea	(1.15)	0.27	0.21	0.20
Bothnian Sea	(1.72)	0.38	0.31	0.28

and that the different types of seawater have characteristic attenuation patterns (*vide infra*). Attenuation coefficients should, therefore, be quoted for particular wavelengths. Coefficients are often quoted as typical of a water column. To a large extent this is a reasonable approximation; in other words, the transparency of a single water mass does not change appreciably with depth. A shallow zone of reduced transparency in the uppermost layer immediately close to the surface has been reported for some waters. Suggestions that this reduced transparency might be due to foam near the surface have not been entirely supported; possibly back scattering of light may be involved. Where a pronounced pycnocline, due to temperature or salinity, causes discontinuity, there is frequently a reduction in transparency which may be associated with particulate matter tending to accumulate near the discontinuity layer. Also, where the vertical structure of the water includes layers belonging to different water masses, the layers may show characteristic extinction coefficients, though consideration of such differences with regard to photosynthetic activity must be confined to relatively shallow layers; the total thickness of the euphotic zone anywhere in the ocean is comparatively small.

Although the attenuation of light in natural waters is due partly to the absorption of light by water molecules, it is strongly affected by scattering due to suspended particulate matter; traces of pigment may also modify attenuation. The significant effect of particles in reducing light penetration, especially due to scattering, is indicated by the increase in the scattering coefficient (b_p) in turbid waters, calculated for two wavelengths, 380 nm (ultra-violet) and 655 nm (red) (cf. Table 6.2). The directional character of the light may be altered to some extent with depth. The total incident light from sun and sky appears as a cone inclined to the vertical at the surface. Despite the considerable scattering effect, the directional character of the light is more or less maintained, certainly within the euphotic zone. Because of the marked attenuation occurring at greater depths, the directional character may be largely lost, the cone becoming vertical (Fig. 6.5).

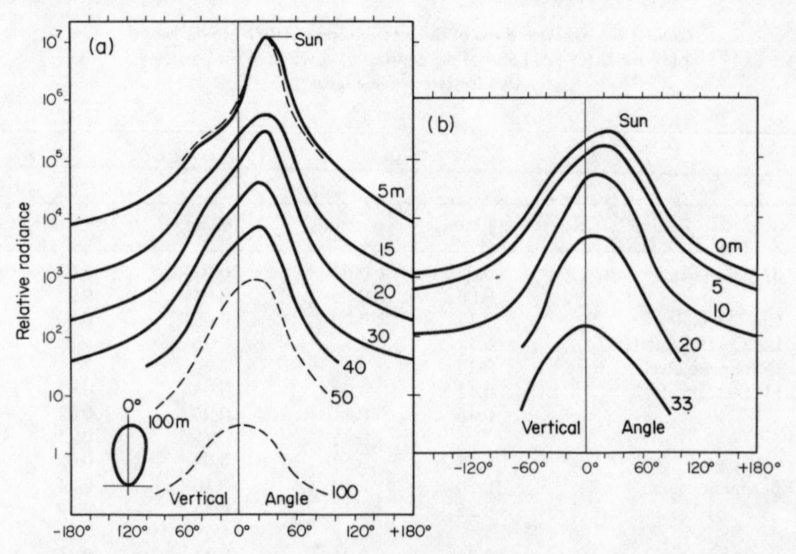

Fig. 6.5. Change, with increase of depth, of radiance in the vertical plane of the sun. (a): Baltic sea, green light. (The dashed and continuous lines represent two series of observations.) (b): Gullmar Fjord, blue light (Jerlov, 1968).

Methods for Measuring the Attenuation of Light and Variation in Transparency in the Sea

A widely used method for measuring the attenuation of light in the sea is to suspend a photometer enclosed in a watertight casing to various depths in the sea. The photocell is mounted behind a clear-glass window with a diffusing disc on the outside. The photocell is suspended by an electrical cable so that the current generated at various depths can be read on a sensitive micro-ammeter. The incident light intensity can, however, vary greatly, especially with clouds temporarily obscuring the sun. Therefore a matched photocell in a similar housing is mounted on the deck of the ship on which the measurements are being made, so that the incident solar radiation can be measured simultaneously against any sub-surface intensity. (Care must be taken in siting the photometer on the deck of the ship to avoid shadow.) Thus depth profiles of the irradiance can be obtained and the extinction coefficient calculated from the equation on p. 263.

Before the development of the submarine photometer by such workers as Atkins, Poole, Clarke and Jerlov, the use of the Secchi Disc was prevalent. This is a white disc of standard size which is lowered from the research vessel; the depth is noted at which the Disc just disappears from sight. Clearly such a method is open to error owing to differences in personal observation and to many other factors. Nevertheless it is a useful method for obtaining a rough estimate of the transparency of the water. Harvey demonstrated that an approximate value for the extinction coefficient can be calculated from Secchi Disc observations, provided these have been obtained by a well-trained observer. The formula is as follows:

$$k = \frac{1.7}{d},$$

where k = the extinction coefficient and
d = depth at which the Disc just disappears from sight.

Holmes (1970) proposes the equation:

$$k = \frac{1.44}{d}$$

as more appropriate for turbid waters. Although measurement of the attenuation of light in the sea is complicated by the directional pattern, from the point of view of primary production it is the total quantity of light received by an algal cell, whatever the direction, which is significant.

In place of the submersible photometer for measuring light, *in situ* transmittance or attenuance meters may be employed. In these meters a parallel beam of light emitted by a standard source passes through a column of the seawater whose characteristics are to be measured, and the light received is measured by a photocell (cf. Fig. 6.6).

Scattered light must be excluded as far as possible and baffles may be mounted in the apparatus to reduce "interfering light". The length of the light path is critical for effective attenuation and an increased path length can be obtained by using reflecting mirrors. A total path length of about 10 m can be employed without building an exceptionally large and cumbersome instrument. In measurements in connection with primary production data, the instrument may be first standardized using, for example, distilled water in the water path, and a series of measurements subsequently taken at various depths to the base of the euphotic zone (cf. Jerlov, 1968).

The concentration of suspended particles, which may be inorganic material and/or organic detritus, as well as living phytoplankton cells, strongly affects transparency. The marked turbidity of inshore and estuarine waters, often combined with a high density of living matter, thus contrasts in its excessively low transparency with the clear, relatively sterile, waters of open tropical oceans ("the blue waters"). There is a well-known empirical relationship between colour and chlorophyll concentration.

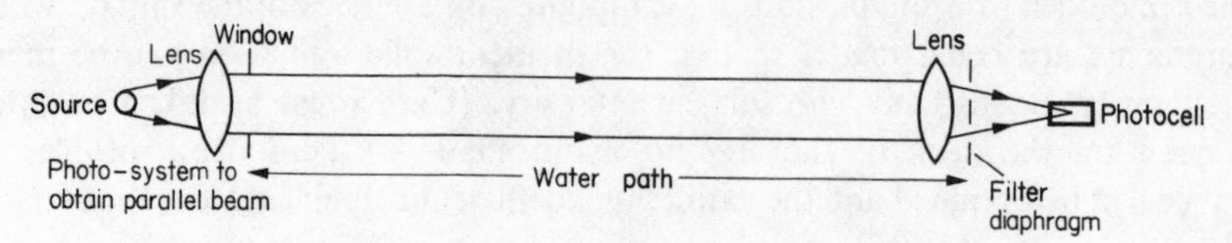

Fig. 6.6. Diagram of submersible transmittance meter.

Jerlov (1963) has classified natural marine waters into a series of categories, dependent upon their transparency, ranging from ocean waters of varied transmission to coastal waters; ocean currents and water masses may also be characterized by their attenuation coefficients. Jerlov suggests a broad correlation between particle content and biological activity for the upper layers of the ocean. Frequently a correlation exists between particle content and water flow, and with salinity, at least in coastal areas.

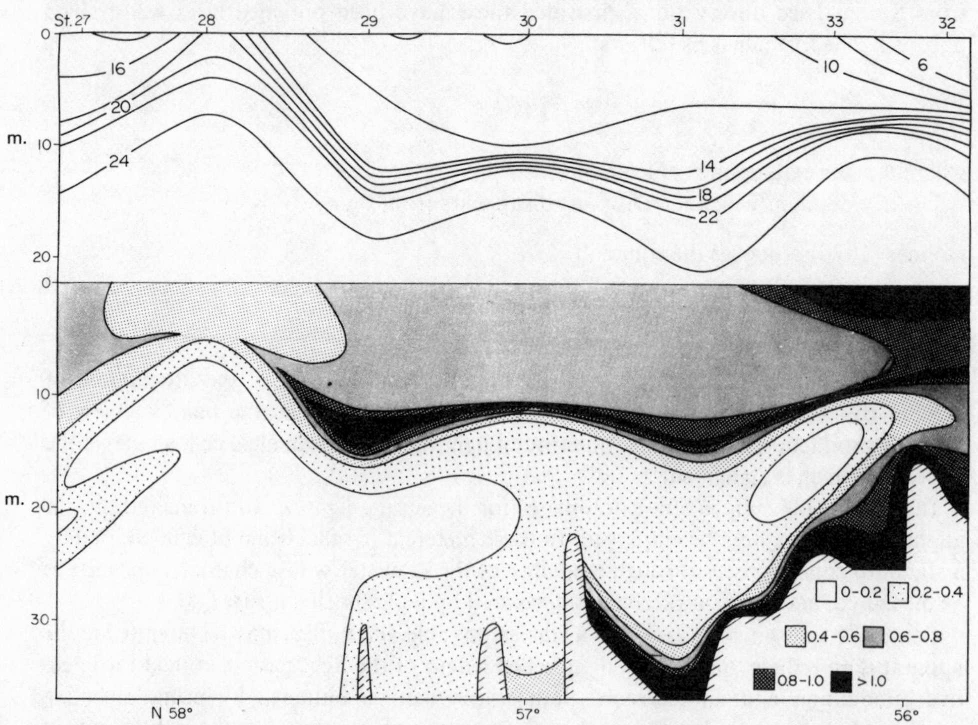

Fig. 6.7. Correlation between distribution of density (upper fig.) and beam attenuation coefficient (lower fig.) in the Kattegat and Sound. (Baltic Water enters at the surface to the right of diagram.) (Data of Joseph, from Jerlov, 1963.)

Particle accumulation in relation to a discontinuity layer, with the consequent sharp reduction in light energy transmission, may sometimes be traced across considerable areas of sea (cf. Jerlov, 1963, Fig. 6.7).

Quality of Transmitted Light in the Sea

Apart from the rapid absorption of infra-red and ultra-violet light in the sea, the various wavelengths of the visible spectrum are differentially transmitted. It is worth noting also that the spectral composition of the visible radiation as direct sunlight differs from that from the sky, and that there are considerable differences between when the sun is at its zenith and approaching the horizon. Measurement of the differential wavelength transmission in the ocean presents difficulties in selecting the appropriate photocell and filters for obtaining the necessary sensitivity over the required range of wavelengths. Strickland (1965) mentioned the possible use of a thermopile radio-meter but there are a number of practical problems with its use in the sea; response is apparently slow and sensitivity limited. The use of the Jerlov–Nygaard light meter, so that light measurements can be made in terms of energy units or quanta, is recommended by Steemann Nielsen and Willemoes (1971). The authors suggest the use of quanta for ecological studies (cf. Halldal, *vide infra*).

Figures 6.8 and 6.9 show that clear ocean water is most transparent to blue rays (about 470 to 490 nm), 1% reaching a depth approaching 140 m. Transparency is reduced for green and violet wavelengths, but the sharpest reduction in transmission is for red rays. For wavelengths of about 650 nm, 1% of the incident radiation reaches a depth of only some 10 m.

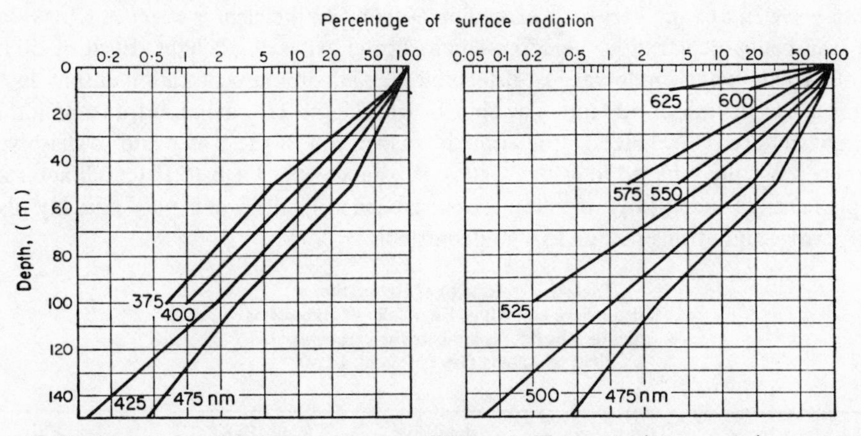

Fig. 6.8. Penetration of daylight in oceanic waters off Bermuda. The penetration expressed as percentage of surface radiation is shown for various wavelengths (after Jerlov, 1951).

In more coastal waters there is a shift in the pattern of differential transmission, the maximum changing toward the longer wavelengths. Green rays show maximum transmission in offshore waters such as the Gulf of Maine and English Channel, though the 1% level of incident light is reduced to a depth of some 40 m. In turbid inshore waters there is a further shift towards longer (yellow) rays, 1% of the incident, reaching only very shallow depths of 10 to 15 m or less. Inshore waters vary in their attenuation characteristics.

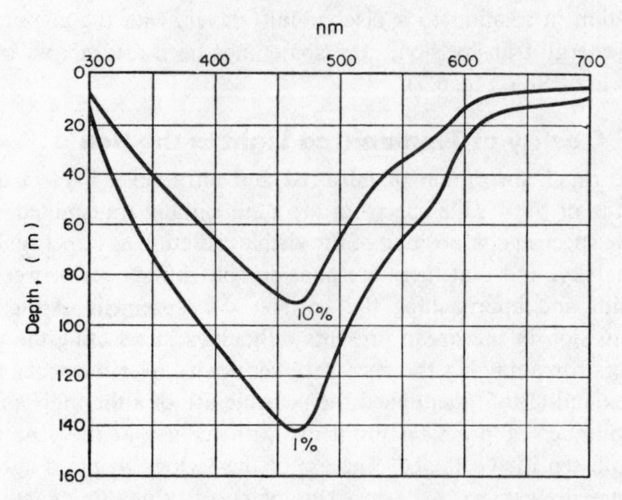

Fig. 6.9. Depths at which the percentage of the surface radiation reaches 10% and 1%, respectively, for various wavelengths in the clearest ocean waters (after Jerlov, 1951).

Halldal (1974) compared light penetration at two coastal areas, Kristineberg (Sweden) and Kings Bay (Spitzbergen), in terms of quanta since the values are more directly related to the potential for excitation of chlorophyll molecules in primary production. Incident radiation was similar at both localities. On the west coast of Sweden, not only was the attenuation very rapid (Table 6.3) but the best light penetration was for 570 nm wavelength. Below 30 m depth no light was present of less than 450 nm wavelength, and very little exceeding 600 nm. In the clearer water of Kings Bay there was better penetration including some shorter wavelength light. Even at 50 m, light of 410 nm wavelength was just detectable; the maximum penetration at that depth was for a wavelength of 500 nm. The shift in wavelength of maximum transmission in more coastal areas is related to materials in suspension in the water and to dissolved substances, usually referred to as the "yellow substance" or "Gelbstoff" (cf. Chapter 2). The pigments of living phytoplankton cells suspended in the water may possibly also modify wavelength transmission to a limited extent.

Table 6.3. Total light attenuation at Kristineberg and Kings Bay: values reckoned as quanta. The figures indicate the percentage of light remaining (after Halldal, 1974)

Depth (m)	Kristineberg	Kings Bay
5	10	16.4
10	1.27	6.18
25	0.24	0.58
50	0.0037	0.053

Jerlov (1963, 1968) speaks of the yellow substance as formed from disintegrating organic material, mainly carbohydrates and amino acids, which form carbohydrate-humic acid complexes or melanoidines. These generally yellow-coloured substances are fairly stable and may occur everywhere in the ocean, but are especially abundant in high-productivity areas. While chemically complex, their composition is almost certainly not uniform throughout the world's oceans. For example, some of the material is derived from the breakdown of organic matter, probably mainly plant detritus, from land, and is brought down by rivers entering the sea. The quantities of such materials may, therefore, be relatively high in areas like the Baltic. Jerlov states that the colloidal and high molecular fractions of yellow substance originate mainly in rivers, and a large proportion of this material precipitates in seawater with the increase in salinity. A negative correlation can be established between the quantity of yellow substance and salinity. The yellow substance is always accompanied, however, by soluble matter, believed to be non-terrestrial in origin, and derived from the breakdown of living materials, essentially plankton, formed in the seawater itself.

One of the most detailed studies of Gelbstoff is due to Kalle (1966). He distinguished between brown humus-like compounds, mainly in a colloidal state and derived chiefly from river water, and true marine Gelbstoff, a soluble selected fraction of the total materials which gives blue fluorescence when irradiated with longer wave ultra-violet light. Although the amount of soluble fluorescent Gelbstoff decreases with increased salinity, the proportion of the marine fluorescent matter to total Gelbstoff increases with salinity. For example, the ratio increases eight times from low salinity Baltic water to the more marine Kattegat. The soluble fluorescent fraction would, therefore, appear to originate in the sea. Humic materials originating from land consist essentially of lignin derivatives, whereas the melanoidines derived from carbohydrate and amino acids are formed in seawater and may be regarded as of oceanic origin.

According to Prakash (1971) coastal waters of eastern Canada, amongst other areas, receive very large amounts of humus-laden water from land. Only a limited amount of precipitation occurs on mixing with seawater, however; a considerable quantity of the organic material is represented by relatively low molecular-weight fractions which do not precipitate rapidly. Prakash (1971) claims remarkable physiological effects of these land-drainage materials in increasing coastal fertility; they can act as chelators, but can promote growth by enhancing permeability and uptake in plants by increasing enzyme activity, amongst other effects (cf. also Burkholder, Burkholder and Almodovar, 1967).

The maximum absorption of light energy by chlorophyll is at wavelengths of about 670 and 440 nm (cf. Chapter 3). The rapid loss of the longer wavelengths of radiant energy in the sea might be expected, therefore, to result in a lowering of photosynthetic efficiency. The accessory photosynthetic pigments of marine algae, however, permit the effective use of other wavelengths. Thus, fucoxanthin and peridinin utilize light from about 500 to 550 nm, and biliproteins make use of other, including shorter, wavelengths (cf. Chapter 3). Since in the deeper layers of the euphotic zone the major light available is of wavelengths from 450 to 560 nm, the significance of the accessory pigments in photosynthesis is obvious. It is the *total* light energy which is of real importance in primary production. Eppley, Reid and Strickland (1970) used the incident irradiance between 400 and 700 nm in calculating photosynthetic rate. They regard the rate as proportional to the total light energy for depths where the intensity varied between 1% and 20% of the surface illumination, but believe that the rate becomes constant at intensities exceeding 20% of the incident illumination. Platt, Denman and Jassby (1977) point out that for widely different types of water, maximal photosynthesis usually occurs where the available light is equivalent to 30–50% of the surface value.

A factor related to the quality of incident radiation which may have caused some errors in productivity estimates in the past concerns ultra-violet radiation. Lorenzen (1976) has confirmed suggestions by Steemann Nielsen that reduction of photosynthetic activity due to ultra-violet radiation can be important at the sea surface, and may extend to some depth in the euphotic zone in clear ocean water. The use of vessels in ^{14}C experiments which are opaque to ultra-violet has obscured this possible source of error. Lorenzen believes that carbon fixation estimates may have been over-estimated by as much as 25% in some investigations.

Light and Seasonal Effects

Although in tropical areas the average daily incident illumination shows little variation as compared with high latitudes, despite short-lived changes due, for example, to thunderstorms, marked climatic changes in some tropical countries can affect daily solar radiation substantially. Ryther and Menzel (1961) found a five-fold difference for the Sargasso Sea. Smayda's (1966) weekly incident radiation data appear to show an even greater variation in one year (1956) for a coastal region in the Gulf of Panama, with the change from the mainly dry to the rainy season. However, the average monthly incident radiation varied only about one and half times, in the Gulf area (cf. p. 235). Thus, the relative constancy of incident energy in tropical regions contrasts sharply with the great variability of incident energy at highest latitudes. Towards the poles incident radiation will be virtually zero over two to three winter months; such polar seas will receive abundant light only during 2 to 3 months round mid-summer (Table 6.1). El-Sayed and Mandelli (1965) report the average incident radiation during December and January as 171 ly/day for the north Weddell Sea (Antarctica). The light energy available for primary production in polar seas cannot, however, be quoted so broadly. Considerable variability occurs even during the short period available for primary production in the vast Arctic and Antarctic regions, both from day to day and with area, and there are consequently marked differences in level of production. For example, a maximum for solar radiation in the Weddell Sea area was recorded as 750 ly/day. This is only slightly less than the maximum radiation quoted for the tropical Gulf of Panama. Some of the variations in the incident illumination at high latitudes are related to local climatic conditions, for example, persistent heavy cloud cover in certain areas.

An indication of the considerable variation in incident illumination in some major regions of the Antarctic appears in investigations by El-Sayed and Mandelli (1965). Figure 6.10 shows a marked difference in total radiation on two succeeding days. Aside from the maximum of 750 ly/min on 27 December 1963, already quoted, the average total solar radiation varied greatly between north and south:

Northern Weddell Sea	171 ly/day
Eastern Weddell Sea	222 ly/day
Southern Weddell Sea	470 ly/day

A further comparison from Burkholder and Mandelli's (1965a) observations gives the following incident radiation:

Gerlache Strait on a sunny day	650 ly/day
Bellingshausen Sea area in overcast weather	172 ly/day

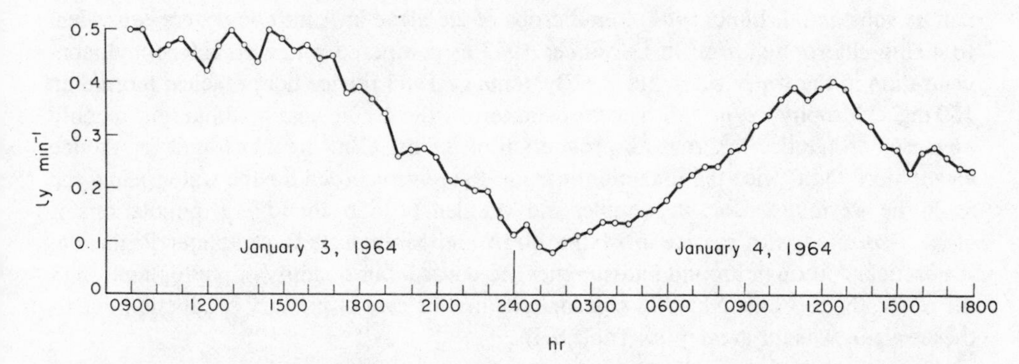

Fig. 6.10. Variations in the incident solar radiation during summer on two successive days in the southern Weddell Sea (from El-Sayed and Mandelli, 1965).

Primary production in the Antarctic will not only be affected by areal variation in incident illumination but also will be markedly influenced by the depth of the mixed layer, and the thickness and degree of permanence of the ice cover. Bunt (1966) discusses the break-up and thinning of ice necessary to allow production to proceed in Antarctic regions, and comments on the great variability from north to south over the enormous areas covered by Antarctic sea ice. The effect of ice on global production is emphasized by the estimate that about 12% of the world's ocean surface is covered by sea ice during the southern winter. Maps indicate that the pack ice in the Antarctic is reduced from 2.4×10^7 km^2 in winter (September) to 1.8×10^7 km^2 in summer (February) (El-Sayed, 1970b). The ice cover is subject to much less seasonal change in the Arctic. Such differences must be reflected in the total crop and production over the brief summer. Although data for ice cover over the whole Arctic are not available, Collin and Dunbar (1964) illustrate the great areas in the Canadian Arctic subject to ice. Lewis and Weeks (1970) suggest the following comparative data for sea ice: for the Antarctic 2.55×10^7 km^2; for the Arctic, 1.51×10^7 km^2. At maximum, therefore, the Antarctic cover is approximately 1.7 times that of the Arctic.

Apart from the cessation of primary production over the lightless winter months, rapid growth is impossible during those months of short day length, with substantial surface reflection of light off the ice, and poor light penetration through ice snow cover. Over the short summer, while growth may proceed rapidly in open water, primary production beneath ice will depend largely on the thickness of ice and snow.

Significance of Ice Flora

Investigations such as those of Bunt (e.g. 1964, 1967), Bunt and Lee (1970), Horner and Alexander (1972), Meguro (1962), Burkholder and Mandelli (1965) and Andersen (1977) have drawn attention to the importance of populations of algae associated with ice and snow in polar seas. A number of species of diatoms and a few other algae have been identified. In investigating both crop and production Bunt and Lee emphasized that, despite the remarkable adaptation to low light conditions by "ice algae", the amount of light available, especially in relation to the amount of snow cover, limited production at the prevailing low temperature. Nevertheless, the production of ice algae can be substantial. Bunt (1964) found crops of ice algae in Antarctic waters equivalent to 40 mg chlorophyll a/m^3 in December 1962 as compared with zero chlorophyll concentration in the water below the ice. By January 1963 the ice flora reached more than 150 mg chlorophyll a/m^3, with phytoplankton in the water below amounting to only 16 mg chlorophyll a/m^3. A peak production of 3.8 mg $C/m^3/hr$ was found for the ice algae, more than twice the maximum carbon fixation recorded for the water below the ice in the previous season. Burkholder and Mandelli (1965b) found large populations of algae associated with sea ice in Matha Strait and Marguerite Bay, Palmer Peninsula, Antarctica. Although around mid-summer there was a fair quantity of phytoplankton in the water, the crop, estimated as chlorophyll, and the rate of primary production due to the ice algae, was far greater (cf. Table 6.4).

Table 6.4. Contributions of the ice flora and the phytoplankton to standing crop and production in Antarctic waters (from Burkholder and Mandelli, 1965)

	Carbon fixed ($mg/m^3/hr$)	Chlorophyll *a* (mg/m^3)
Matha Strait (water sample)	6.24	2.56
Matha Strait (ice sample)	1080	407
Marguerite Bay (ice sample)	797	305

The ice flora consisted mainly of diatoms with a number of species common to the phytoplankton community, but there was a preponderance of very small species. Burkholder and Mandelli calculated that for the whole thickness of the ice flora layer, receiving between 25% and 1% of the incident light, production might amount to 0.19 g $C/m^2/day$. In the open-water column over summer in two Antarctic areas they estimated production as: Bellingshausen Sea 0.09 g $C/m^2/day$; Gerlache Strait 0.66 g $C/m^2/day$. Andersen's (1977) data for algae found in ice in Greenland are somewhat difficult to compare. In late March/early April the peak production in the ice appears to have been only 40 mg $C/m^3/$*day* and the crop of chlorophyll 7 mg chlorophyll/m^3. He comments that the *total* production in and adjacent to the ice (4.5 cm above to 4.5 cm below the interface) was only 0.15 g C/m^2, as compared with 90 g C/m^2 per year for the phytoplankton in the water; the crop in the ice was also small. Fogg (1977) quotes low values for primary production by Arctic ice algae, ranging from 7.7 mg $C/m^2/hr$ near the peak of an algal bloom to only 0.3 mg $C/m^2/hr$ before the ice thaw.

Presumably the ice and snow flora must be adapted to withstand considerable salinity changes with the thawing and freezing of pockets in sea ice. Andersen (1977) comments that although ice flora cells can withstand remarkably low salinities (cf. also Horner and Alexander, 1972), considerable destruction was observed of diatoms caught in sea ice. He attributes this to osmotic change rather than to temperature. The near-freezing temperatures in addition to lowered light intensities must, however, lower the efficiency of photosynthesis in the surviving algae.

Apart from the significant production by ice flora in polar seas, the trapped algal cells are of great importance when released from sea ice in providing "seed" for the great outburst of phytoplankton growth over the summer.

Meguro (1962) pointed to the brown coloration of much of the sea ice in the Antarctic pack, apparently due to rich populations of phytoplankton, mainly diatoms, with some Chrysophyta. This layer lay at sea level, beneath a layer of snow, and above a thick sea ice stratum, and is believed to have been formed from early fallen snow which had sunk to sea level and then was penetrated by seawater. Whatever the precise method of formation, this intermediate layer was exceedingly rich in phytoplankton. Burkholder and Mandelli (1965b) describe their ice flora as more similar to that identified by Meguro. They believe that there are at least two kinds of algal communities in sea ice. The algae found in ice in Matha Strait and Marguerite Bay had much higher light requirements (saturation at about 0.108 ly/min; inhibition at 0.173 ly/min) than the shade-adapted epontic species described by Bunt (saturation at about 0.006 ly/min), but they are capable of contributing considerably to production. Meguro (1962) also demonstrated that abundant nutrients were present in the ice habitat rich in algae (about 2.5–3.0 μg-at PO_4–P/l; 10–20 μg-at NO_3–N/l) and that 15–25% of the incident illumination (i.e. much more light than at the bottom of pack ice) was available to these algae. Considerable levels of chlorophyll were quoted (e.g. 0.67 mg chlor *a*/1) for the snow/ice layer. Thus a habitat existed where intense growth of phytoplankton could occur, though this growth is masked as the ice breaks up in summer. Such an ice flora may be widespread in Antarctic areas.

In Arctic waters (Point Barrow), Meguro, Ito and Fukushima (1966) identified a comparatively rich ice flora, consisting mostly of phytoplankton diatoms. The chlorophyll concentration was 100 times greater than in the surrounding seawater. The algae were actively dividing, despite the reduced salinity of the habitat, and nutrients were abundant. Meguro *et al.* suggest that such algae may contribute substantially to primary production in a region where production in the water is very low.

Light and Regional Differences in Crop and Production

Growth of algae, whether phytoplankton or sea ice flora, at high latitudes must be limited by the outstanding seasonal change in incident radiation, as well as varying during the growth season with area (cf. Zenkevitch, 1963). Ryther (1963) cites exceedingly low annual production in the true Arctic Ocean (1 g C/m^2), a value similar to Bunt's (1971) estimate for a shaded station in McMurdo Sound. The overall estimate of 100 g C/m^2/year for Antarctic production given by Ryther must presumably be subject to considerable areal variation.

At middle latitudes, incident illumination is also subject to very considerable and more or less regular changes over the year. Time of year, with latitude and with typical weather pattern, combine to give a characteristic yearly cycle. Harvey (1955) pointed out that at a latitude of 50°N the maximum incident illumination for the year, experienced during the May/June period, approximated each day to *ca.* 216 ly/day. This quantity is of the same order as the maximum quoted by Steemann Nielsen (1974) for Copenhagen. The average daily value must depend to a considerable degree, however, on weather conditions (cf. Fig. 6.1). Strickland (1965) mentioned that even a moderately overcast sky could reduce light intensity to one-third or less that found in clear weather. Pronounced differences in weather from one year to another may, therefore, affect the time of the spring outburst in a temperate area. Early observations by Marshall and Orr (1927, 1930) indicated that the timing of the spring outburst was more related to total incident light than to the amount of spring sunshine. In the locality of Southampton Water a very late phytoplankton growth in 1972 has been tentatively related to very poor light conditions in the spring of that year.

Hitchcock and Smayda (1977) associated a marked delay in 1973 in the onset of the winter-spring bloom in Narragansett Bay, normally beginning in December, with abnormally low incident light. The decrease in average daily incident illumination from early December to early January 1973 was followed by an increase in mid-January (Table 6.5), but illumination in the 8-m mixed water column exceeded an intensity of about 40 ly/day, estimated as the critical value for phytoplankton blooms in coastal estuaries, only during the third week of January. The February bloom included *Skeletonema*, *Detonula* and *Thalassiosira*. Neither temperature nor nutrients limited growth of *Skeletonema*, which was present over the winter, during December; grazing intensity, due mainly to *Acartia*, was also insufficient to prevent an algal bloom. The limiting

Table 6.5. Daily incident radiation (averaged for previous 7 days) and size of phytoplankton crop (as chlorophyll a) (from Hitchcock and Smayda, 1977)

Date	Incident light (ly/day)	Chlorophyll (μg/l)
4 Dec.	152.3	3.26
11	62.8	2.00
18	121.4	1.89
26	63.8	0.51
2 Jan.	98.8	1.18
8	125.3	1.02
15	200.6	0.83
23	185.9	2.50
29	136.2	4.13
5 Feb.	175.6	4.25
12	175.9	4.50
19	254.0	7.49
26	295.6	25.2

factor appeared to be low incident light. Incident radiation over several years during early December (i.e. preceding the phytoplankton bloom) ranged from 122 to 173 ly/day, according to Pratt (1965), and there was no clear relationship between incident illumination and the timing of the bloom. Provided the incident radiation exceeded about 120 ly/day, logarithmic growth of the algae was possible, but in December 1972 the mean illumination was only 91 ly/day. Smayda (1973) had earlier indicated that incident light between 0.05 and 0.10 ly/min, dependent on temperature, was sufficient for growth of *Skeletonema*, though later he suggested that the value should be increased two to three times to allow for sinking and other losses of production. The critical value for winter incident illumination, given by Pratt, of 120 ly/day would be equivalent to 0.2 ly/min, reckoning a 10-hour day.

While variations from year to year may be of significance in relation to the onset of phytoplankton growth, primary production in temperate latitudes is largely conditional on the more or less regular seasonal changes in incident illumination. At latitude 50°N, as compared with the maximum incident illumination of 216 ly/day, Harvey (1955) quotes a mean figure of only a ninth of the total incident illumination (*ca.* 24 ly/day) during the mid-winter period. The variation during mid-winter is also noteworthy. Harvey suggests at noon on a sunny day the intensity may reach 0.15 ly/min, as compared with a cloudy mid-winter day when the value would be nearer 0.03 ly/min. Reckoning on an 8-hour day in mid-December, the daily radiation would hardly exceed 10 ly/day in overcast weather. Estimates by Harvey may be compared with data from Parsons and Takahashi (1973a). For 50°N latitude, maximal and minimal radiation (June and December, respectively) was calculated as 769 and 131 ly/day. Records obtained in British Columbia in 1969 showed the actual maximum occurring in July (622 ly/day) and the minimum in December (44 ly/day). At a slightly lower latitude, off the Oregon coast, Curl and Small (1965) record summer total incident radiation sometimes exceeding 350 ly/day with low winter values of *ca.* 30–40 ly/day. Values are also quoted by Pingree, Holligan, Mardell and Head (1976) for monthly total solar radiation in south-west Britain. The maximum in June would approximate to 560 ly/day (cf. Table 6.8).

The greater constancy of incident radiation in the tropics, already quoted, is illustrated by Harvey's calculations that approximately 216 ly/day falls virtually every day on tropical regions. Harvey's estimate for the tropics appears to be distinctly lower than some recent data, but his values, both for the tropics and for temperate latitudes, refer to visible light energy, whereas most data quoted refer to total incident radiation, i.e. including the infra-red. For the Gulf of Panama, Smayda (1966) found that over 3 years the total incident mean monthly radiation showed a maximum approaching 600 ly/day (>600 in December, 1957); minimum quantities during the rainy season were between 350 and 400 ly/day. Only in June and July 1956 were unusually low values recorded of below 300 ly/day. Qasim, Wellershaus, Bhattathiri and Abidi (1969) give data for the Cochin area with a maximum incident radiation in February of 574 ly/day and a minimum in June of 264 ly/day, values which are in reasonable agreement with those of Smayda, bearing in mind that the influence of the monsoon in southern India reduces incident light more than the rainy season in Panama.

Although therefore marked changes in incident illumination may be seen in inshore areas of the tropics under the influence of local climatic conditions, overall incident radiation is very much greater than at middle and higher latitudes. Smayda (1966) makes the following striking comparison: for the Gulf of Panama (8°45′N latitude) waters, the light intensity during the period of poor illumination over the rainy season, at the 50% isolume level, exceeded the *incident* illumination at 40°, 60° and 80°N latitude for 4, 6 and 8 months, respectively. At the depth of 50% isolume, approximately 3 to 4 m, during the rainy season of 1956, the worst year climatically, the mean illumination was 184 ly/day. Smayda compares this with data from S. Conover that in Long Island Sound at 5 m the radiation *never* exceeded 60 ly/day.

The thickness of the euphotic zone, while varying between >100 m in oceanic tropical waters and a few metres in turbid waters at higher latitudes, is not a static entity. In the tropics, despite diel variations, the relative constancy of total light conditions, at least offshore, ensures that the average depth of the euphotic zone is fairly similar throughout the year; in middle and high latitudes the depth changes sharply between summer and winter. A thickness of some 50 m is a reasonable average for the euphotic layer in a moderately high latitude open ocean, but this is true only usually over spring and summer when incident light is optimal. During winter, decreased incident radiation, increased surface reflection, shorter daylight hours, and some increase in turbidity with winter storms, so reduce light penetration that the euphotic zone is a mere surface skin of a few metres. Marshall and Orr (1928, 1930) demonstrated for Scottish waters that the thickness of the euphotic layer, indicated by the compensation depth, was reduced from 20–30 m over summer to close to the surface over winter (cf. also Jenkin, 1937). Strickland (1965) points out that consideration of incident light alone would indicate that no mid-winter primary production would be possible beyond latitudes of about 60°, that only small seasonal variation should be experienced at latitudes less than about 30°, but that marked seasonal effects could occur between 45° and 55°. Stability, temperature, nutrients, nearness to shore and other factors may all modify the effects of incident light. An example of changes in the depth of the euphotic layer at low latitudes is taken from a tropical inshore area, the Gulf of Panama. Over most of one year the bottom of the euphotic zone was between 20 and 30 m, reaching >40 m (i.e. the sea floor) during September and October. During the following year, when incident radiation was generally higher, the euphotic zone extended to the sea floor throughout the year, except for brief periods in July and in October/November (Smayda, 1966).

Aside from the very marked shallowing of the photosynthetic layer over winter in middle and high latitudes, increased turbulence transports algal cells out of the euphotic zone over part of the day, reducing productivity. Except in particular areas where marked stratification may retain phytoplankton cells in the extreme upper layers permitting them to photosynthesize effectively over winter, production over middle and high latitudes during winter is negligible. The increase in incident light energy and in light penetration with the coming of spring produces conditions suitable for the phytoplankton spring outburst. The gradual increase in incident light over the 6 months, January to June, at latitude 52°N, and the steady thickening of the euphotic zone with increased light penetration, is illustrated by data from Cushing (1959b) (see Table 6.6).

Table 6.6. Average energy in ly/day in lat. 52°N, January to June, reaching the surface and reduced by the energy extinction coefficient of coastal water (type 1 of Jerlov, 1951) (after Cushing, 1959)

	Jan.	Feb.	Mar.	April	May	June
Surface	18.98	44.91	92.09	164.25	219.20	242.06
1 m	7.00	16.57	33.98	60.61	80.88	89.32
5 m	2.70	6.38	13.08	23.32	31.13	34.37
10 m	1.12	2.65	5.43	9.69	12.93	14.28
20 m			1.20	2.14	2.85	3.15
30 m					0.66	0.73

Stability and Temperature: Turbidity

Even with the increase in light during spring, the vernal blooming of phytoplankton, a characteristic feature in temperate and high latitudes, is dependent to a large extent on how far turbulence is active. Normally some degree of stratification of the water is essential to reduce the passage of algal cells from the relatively shallow photosynthetic zone. Riley (1946) stressed the importance of stratification in initiating the spring increase in temperate latitudes. For the Gulf of Maine, he demonstrated a direct correlation between the rate of increase of the phytoplankton and the stability of the water column (Fig. 6.11). This stability depends to a large extent on temperature in so far as warming of the surface layers will cause them to become less dense, restricting mixing with underlying layers (cf. Chapter 1). Stability therefore tends to increase in spring with rising temperature, and at higher and middle latitudes there is a correlation early in the spring between rate of primary production and temperature increase.

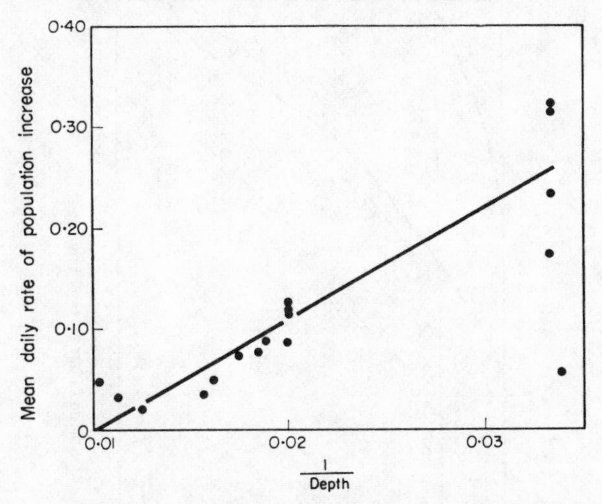

Fig. 6.11. Relation between the estimated daily rate of photosynthesis during March–April and the stability of the water column, expressed as the reciprocal of depth of the zone of vertical turbulence (from Riley, 1942).

Reduction in surface salinity may also play a part in stabilization; some of the earlier and more marked production of coastal shallow as compared with offshore waters can be attributed to this increased stability (cf. Braarud and Klem, 1931; Gross, Marshall, Orr and Raymont, 1947; Marshall and Orr, 1948). Under such conditions phytoplankton production may be possible during the winter, but must be restricted to an extremely shallow surface layer. In parts of the Baltic reduced salinity of the upper layers produces stability throughout the year so that production of phytoplankton continues over winter months (cf. Steemann Nielsen, 1935; 1940). A detailed comparison of two areas showed that the Kattegat had a remarkably stable rate of primary production for surface algae over much of the year. Only in December and January was incident light apparently limiting for surface algae; the efficient production was due to marked stratification. By contrast, the Great Belt with reduced stratification showed less constant production; illumination was limiting over much of the year for surface algae (cf. Chapter 9).

In areas very close inshore, though run-off, by assisting stabilization, can act as an important factor in the onset of blooms, the burst of algal growth may not be always linked to increased stability. Pratt (1965) found in Narragansett Bay that the incident light in most winters was sufficient for algal growth owing to the extreme shallowness of the Bay.

The importance of light penetration in relation to the depth of the mixed (i.e. unstratified) water layer was emphasized by Riley (1942–46). Sverdrup (1953) used the term "critical depth" (D_{cr}), which may be defined as that depth at which the total photosynthesis of the algae in the water column is equivalent to their total respiration. In a thoroughly mixed upper water layer plankton cells might be assumed to be practically evenly distributed; production would, therefore, decrease logarithmically with depth with the decrease in light, but respiration would be approximately constant with depth (cf. Fig. 6.12). For effective production the critical depth must exceed the thickness of the mixed layer.

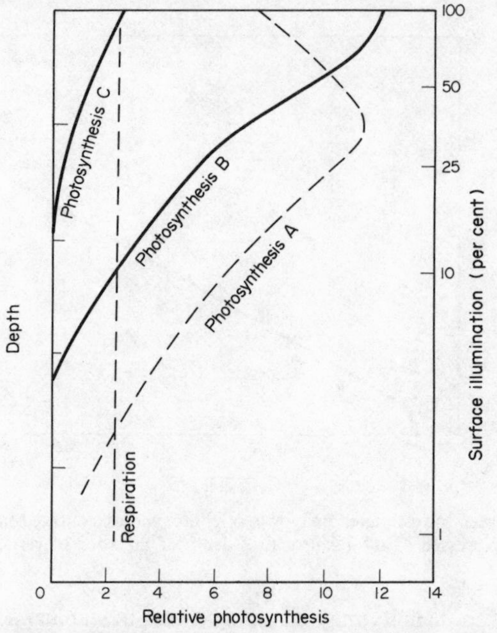

Fig. 6.12. The relation between respiration and integrated daily photosynthesis from surface to depths of penetration of 1% surface light. Three days are shown: (A) summer with bright sunshine, (B) summer with cloud, (C) winter (redrawn from Ryther, 1956).

Let D_c = compensation depth, i.e. that depth at which the photosynthetic production of a cell and its respiration are in balance, and I_c be the compensation light intensity. Sverdrup reduced the incident solar radiation (I_o) by a factor (0.2) to allow for the very rapid absorption of the ultra-violet and infra-red wavelengths which are not photosynthetically active. If k = average extinction coefficient for the remaining radiation (approximately 400 to 700 nm), then:

$$I_c = 0.2 \times I_0 \times e^{-kD_c}.$$

Algal cells move up and down across the compensation depth but they receive an average illumination ($\bar{I}$). The depth where the average light intensity for the water column equals the compensation light intensity is the critical depth (D_{cr}). This average compensation light intensity ($\bar{I}_c$) is:

$$\bar{I}_c = 0.2 \times I_0 \int_0^{D_{cr}} \frac{e^{-kD}}{D_{cr}} dD,$$

$$\bar{I}_c = \frac{0.2 I_0}{k D_{cr}} (1 - e^{-kD_{cr}}).$$

Thus the critical depth is related to incident radiation, average extinction coefficient and compensation intensity (cf. Fig. 6.13). The significance of critical depth is that if the depth of mixing is greater, nett production cannot occur. The model assumes such factors as an even distribution and constant respiration of algae with depth, an excess of nutrients, and production as proportional to radiation. Parsons and Takahashi (1973a) point out that Sverdrup's factor of 0.2 for reducing incident radiation is excessive for shallow water columns.

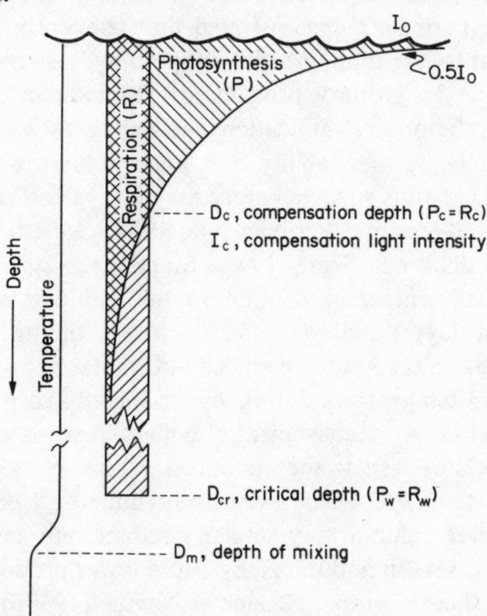

Fig. 6.13. The relationship between critical depth and compensation depth and depth of mixing (after Parsons and Takahashi, 1973).

In the Norwegian Sea, Sverdrup demonstrated that the time of the spring phytoplankton increase (early April) coincided with a time of increasing stability; the mixed layer was about 50 to 100 m depth, whereas the critical depth exceeded 100 m. Halldal (1953) found that a fairly early spring growth commenced with increased light penetration in the Norwegian Sea, but that the considerable mixing of the water layers prevented a real increase of phytoplankton until May/June when stability was established. The main growth (June/September) more or less coincided with the period of stability. Holmes (1956) found that in the Labrador Sea there was very little phytoplankton from November to April with the marked instability of the water column. Over the May/June period, increasing stability and increased radiation were mainly responsible for phytoplankton production. In Arctic waters off Bear Island, Marshall (1958) showed that the effective production of phytoplankton commenced during March/April when the mixed layer shallowed to less than the critical depth; in the warmer Atlantic water nearby, the mixed layer only became shallower than the critical depth by about May and June, and phytoplankton production was correspondingly delayed. At these high latitudes the stability of the water in late spring may be due in part to the melting of ice; the phytoplankton outburst in polar regions frequently seems to follow the melting ice edge (cf. Braarud, 1935; Hart, 1934). Zenkevitch (1963) also describes the algal bloom following the ice melt in Arctic seas; he emphasizes the importance of the critical depth exceeding the mixed turbulent layer.

Many species, particularly diatoms, live normally in very cold polar seas; a few species are confined to Arctic or Antarctic waters (cf. Chapter 5). Although few are apparently obligate cryophilic species, and for many species photosynthetic activity is not maximal at the very low temperature of high latitude seas, the algae are able to grow effectively. In any event, for many species, near maximal photosynthetic rates achieved at higher temperatures also require relatively high light intensities (cf. Chapter 3). An example of a reduced growth rate with lowered temperature is given by Smayda (1973); laboratory and field experiments demonstrated that temperatures below 3°C reduced growth of *Skeletonema* to less than one division per day. Clearly, lowered temperature of itself is not a barrier to primary production. The indirect effects of temperature, however, especially in relation to stratification, can be of great significance.

Another factor influencing the stability of a water column is wind strength. Strong winds will reduce stability; thus sheltered areas may bloom earlier than exposed regions at similar latitudes (e.g. Kreps and Verjbinskaya, 1932; Corlett, 1953; Conover, 1956; Fish, 1925; Bigelow, Lillick and Sears, 1940). Some discussion by Cushing (1975) of wind strength and shallowness, in relation to the production ratio (compensation depth/depth of mixed layer) and its effects on the timing and duration of the phytoplankton bloom has already been included in Chapter 5.

The continued rise in temperature during late spring and early summer in temperate regions typically results in the establishment of a marked seasonal thermocline over the summer. This thermocline restricts the continued supply of essential nutrients to the euphotic zone where they are being extensively utilized. Reduced surface salinity producing a stable water column may similarly reduce nutrient replenishment to the upper layers later in the season and thus may cause lower productivity when regeneration of nutrients is insufficiently rapid. Steemann Nielsen (1958b) quotes examples from coastal areas off Greenland (cf. Chapter 9).

Some field investigations already described such as those by Marshall and Orr, Cushing, Hitchcock and Smayda and others have demonstrated a relationship between the growth of phytoplankton at temperate latitudes, the light intensity and stabilization of the water column. Frequently, concomitant changes in other factors (nutrient concentration, grazing intensity) obscure a clear relationship. A few additional field studies emphasizing the role of light and/or stabilization may now be discussed.

Becacos-Kontos and Svansson (1969) found a good correlation between primary production and incident irradiant energy for a station in the Aegean Sea, for a period between the end of December and March. Table 6.7 indicates computed values for the total light energy at various depths and the measurements of primary production for certain dates. Maximum photosynthesis was generally at about 5 to 10 m depth, corresponding to 13 to 50 ly/half-day total energy. Good agreement was also obtained between observed primary production and estimates based on incident radiation and attenuation data according to a formula devised by Steele (cf. Chapter 9).

The importance of stability of the water column in relation to development of phytoplankton in the Celtic Sea (south-west of the British Isles) was studied by Pingree, Holligan, Mardell and Head (1976). During the winter they found the column well mixed to the bottom, due to tides near the bottom, and to wind and cooling at the surface; plant crop was minimal and nutrients maximal. Though increasing illumination during spring could cause some rise in phytoplankton, the outburst of spring growth was mainly dependent on stabilization of the water. With increase in temperature on the

Table 6.7. Total energy (*E*) as ly/half solar day, and primary production (*P*) as mg C/m³/day and as percentage of maximum value, in relation to depth for various dates. Only selection of data shown (from Becacos-Kontos and Svansson, 1969)

Water depth	23 Dec. 1963			Water depth	6 Feb. 1964			Water depth	30 Mar. 1964		
(m)	*E*	*P*	%	(m)	*E*	*P*	%	(m)	*E*	*P*	%
0	36.1	0.47	29.4	0	44.5	4.46	78.5	0	153.8	3.75	41.7
2.5	22.5	0.96	60.0	2.5	30.3	5.68	100	2.5	104.6	4.86	54.0
5	19.1	1.60	100	5	24.0	5.36	94.4	5	81.5	7.94	88.2
10	12.4	1.42	88.8	10	16.5	4.86	85.6	10	52.3	9.00	100
20	6.3	1.13	70.6	20	9.3	4.11	72.4	20	26.1	3.49	38.8
40	1.7	0.52	32.5	40	2.9	1.78	31.3	40	7.7	2.10	23.3
60	0.4	0.07	4.4	60	0.9	0.60	10.6	60	1.5	0.62	6.9
80	0	0.03	1.9	80	0.1	0.12	2.1	80	0.5	0.55	6.1
100	0	0	0	100	0	0.15	2.6	100	0	0.12	1.3

surface and lessened interchange between near surface and deep waters, algae in the upper 15 to 20 m multiplied with the increased light. A thermocline was early established only in certain areas of the Celtic Sea, corresponding to regions of least tidal streams, for example, in an area south of the Nymphe Bank (south of Ireland; 6–7°W). Temperature difference there between surface and bottom exceeded 0.6°C; chlorophyll *a* reached a concentration of 10 mg/m³ in the spring bloom, and nutrients were sharply reduced. At the surface nitrate was reduced by $>98\%$, silicate about 85% and phosphate approximately 70% (cf. Figs. 6.14 and 6.15). The reduction of nutrients did not occur in areas where the water column remained well mixed in the spring and the phytoplankton was low.

The spring bloom followed a gradual extension of the thermocline through the area, eventually reaching the mouth of the English Channel. Stabilization of the water column was the essential factor in the algal multiplication; the progress of the flowering could not be attributed to advection. A local contributing factor to stabilization in the Celtic Sea on occasions may be excessive precipitation and land run-off.

After the spring the plant crop fell sharply; summer growth depended apparently on the rate of supply of nutrients by vertical mixing from the bottom layers. Where tidal effects are strong, the column will be mixed, allowing transport of nutrients; production is then mainly dependent upon average light (transparency and depth – Table 6.8). Where there is moderate tidal influence, provided there is sufficient vertical transport, better light conditions will give increased production. Thus west of the Scilly Isles, relatively strong stratification gave poor production; close to Land's End and to the French coast considerable mixing, indicated by the lower surface temperature, was also associated with poor production. Between such areas, as in a front north-west of Ushant, there was high production (Fig. 6.16). The thermocline region itself tended to be rich in chlorophyll with the degree of stability and extra nutrients.

During autumn the thermocline retreated from its wide spread towards the Nymphe Bank. Although greater mixing allowed enrichment of surface layers, phytoplankton was generally low. Surface light was reduced; by October it was less than 50% of that in April, and the depth of the mixed layer was nearly equal to the critical depth. In autumn therefore phytoplankton blooms were possible only with unusually sunny and calm

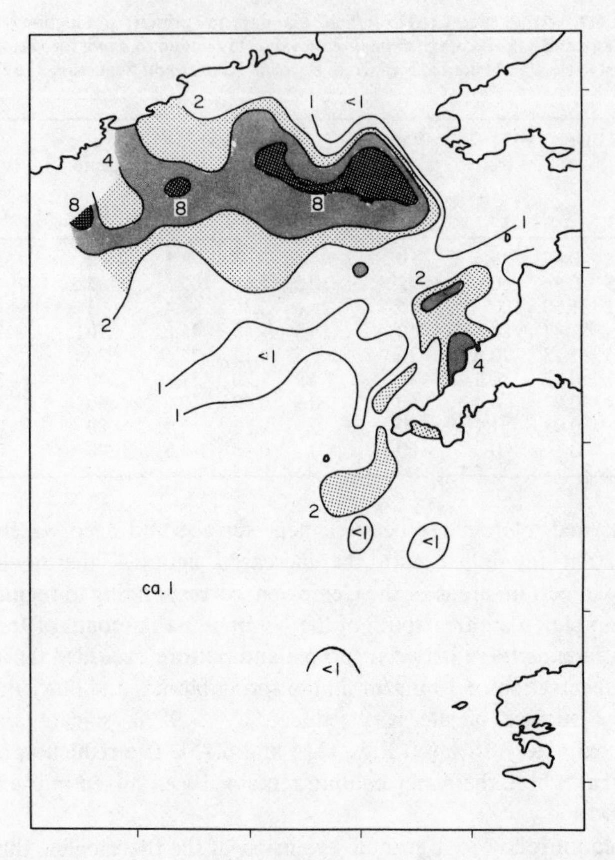

Fig. 6.14. Surface distribution of chlorophyll *a* (mg/m³) 9–21 April 1975, in the Celtic Sea (after Pingree *et al.*, 1976).

Table 6.8. Critical and compensation depths for water with an extinction coefficient, *K*, of 0.10/m

Month	Mean total solar radiation ly × 10³/month (Day, 1961)	Mean critical depth (m)	Mean compensation depth (m)
J	2.1	15	—
F	4.1	30	—
M	7.5	60	25
A	12.9	100	31
M	15.5	120	33
J	16.8	130	33
J	14.2	110	32
A	12.9	100	31
S	9.0	70	27
O	5.2	40	22
N	2.6	20	—
D	1.5	10	—

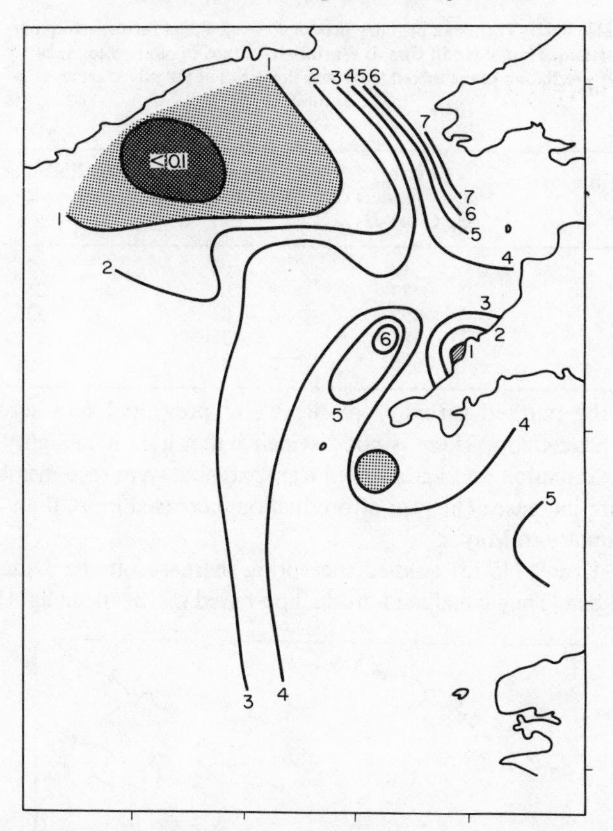

Fig. 6.15. Surface distribution of nitrate (μg-at N/l); date and location as on Fig. 6.14.

weather, and locally where the temperature gradient was still maintained. Thus the stability of the water column, related to thermocline formation and breakdown, and to tidal bottom movements, exerts a profound influence on algal growth in the Celtic Sea.

Barlow's (1958) studies on the changes in phytoplankton in a deep estuary on the Washington coast during spring are also of interest since a number of factors were relatively stable. A regular brackish surface layer was present during spring over the estuary; with the marked stratification there was very little effective vertical turbulent transport of phytoplankton below the surface layer. Variations in temperature and salinity were very small; nutrient concentration, monitored as phosphate, appeared high at all times so that nutrients were unlikely to have been limiting. Though precise information was not available on the zooplankton, substantial changes in grazing activity also appeared unlikely. Changes in radiation and transparency of the water have, therefore, been fairly clearly related to primary production.

The estuary was a region of relatively high turbidity and changes in turbidity occurred during the spring. The mean radiation through the mixed layer showed marked variations during March and April, reaching about 50 ly/day (cf. Table 6.9). Barlow claims that a value in excess of the order of 43 ly/day is necessary for the initiation of the spring bloom (cf. p. 274). Integrated primary production rates, though somewhat variable, remained low until March/April when there was a very large increase. Relative to the incident radiation such an increase might have been expected as early as

Table 6.9. The mean primary production calculated for five sampling stations in the Hood Canal, Washington State, in relation to mean radiation in the mixed layer and the depth of the mixed layer (after Barlow, 1958)

Month	Mean primary production (mg C/m²/day)	Mean depth of mixed layer (m)	Radiation (ly/day)
Jan.	12	9	3
Feb.	73	9	25
Mar.	452	8	52
April	686	7	51
May	2646	8	110

February, but the marked turbidity of the water prevented any substantial rise in photosynthetic production. There is good evidence that light availability, dependent on changes in solar radiation and variation in transparency, were responsible for the initiation of the spring increase. The rate of production increased more than 200 times over the 5 months January to May.

Gieskes and Kraay (1975) studied the spring increase off the Dutch coast in the southern North Sea. They calculated production based on the mean light intensity in the

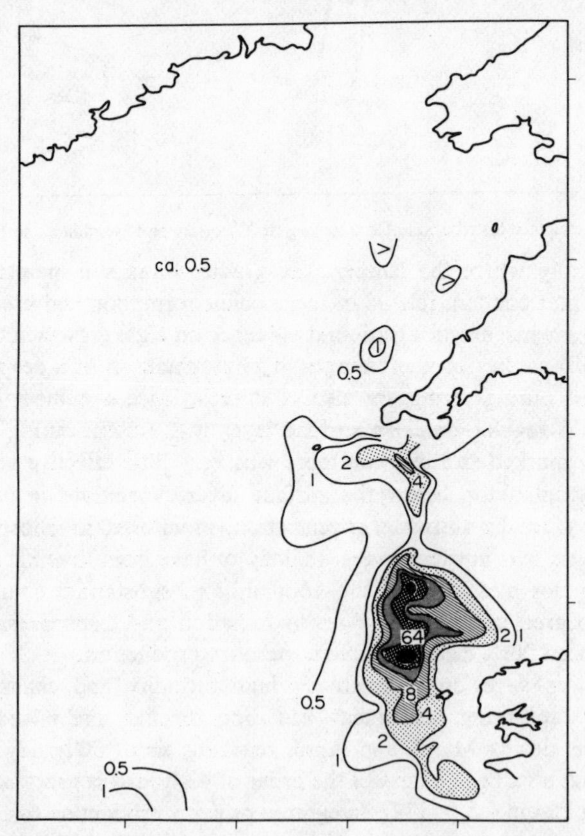

Fig. 6.16. Surface distribution of chlorophyll *a* (mg/m³) 16–31 July 1965 in the mouth of the English Channel (after Pingree *et al.*, 1976).

mixed water column, assuming that light saturation was unlikely during February-March at that latitude. They found that active production by the algae in the Southern Bight began when the average light intensity exceeded about 0.03 ly/min, the same value calculated earlier by Riley for the critical intensity off the east coast of the USA. The intensity was reached about mid-February offshore but only by March in inshore turbid areas. During February/March the authors observed that the critical depth was two to two and a half times the mixed layer offshore, whereas close to the coast the critical depth hardly exceeded the mixed layer depth.

They used a model to predict primary production:

$$P_{rel} = K_b \frac{I_e}{K_b + K_q} - 3.12(Z)$$

where P_{rel} = estimated production,
I_e = incident radiation (assumed at 250 ly/day), reduced for reflection, 20% available for photosynthesis,
K_b, K_q = light attenuation due to biological and physical processes, respectively,
Z = mixing depth,
3.12 = energy at compensation depth (ly/day).

Figure 6.17, showing P_{rel} plotted against K_b, indicates that production increased more rapidly in sub-areas V to VII than in sub-areas I to IV (more inshore). Sub-area III had very high turbidity. Thus offshore areas in the Southern Bight should bloom, in the absence of grazing, more rapidly, even if primary production in all areas started together. Observations confirmed the predictions; sub-area III did not develop substantial phytoplankton before mid-April; VI and VII developed in February; I and II were intermediate (March). Intensive grazing was not apparent during the earlier period of phytoplankton development.

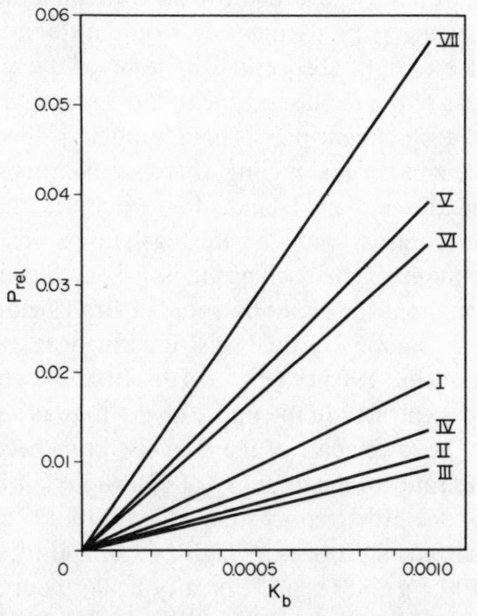

Fig. 6.17. Relation between relative primary production per m² and light attentuation due to biological processes in various sub-areas of the Southern Bight (after Gieskes and Kraay, 1975).

A factor in light penetration which may become of significance in areas of eutrophic productivity is attenuation due to phytoplankton cells. "Self-shading" has been long recognized as a factor limiting light penetration in laboratory cultures, and various methods (illumination from several aspects; central illumination with annular ring cultures; circulation of cell cultures past the light source) have been used to reduce its effect. In the field, in eutrophic lagoons and estuaries, the euphotic layer is often thin; the shallowness is in part due to the large quantities of suspended inorganic and dead organic material, but reduction in depth is partly attributable to the phytoplankton itself. Two examples from temperate areas where thinning of the euphotic zone is probably partly due to phytoplankton density are the Raritan River (Patten, 1962) and Moriches Bay (Ryther, 1954). Burkholder, Burkholder and Almodovar (1967) comment on the shallowness of the euphotic layer due to great concentrations of algal cells in tropical lagoons and mangrove swamps, despite the very high incident radiation (cf. also Teixeira, Tundisi and Santoro, 1969). Even in less eutrophic seas, part of the reduction in radiant energy must be due to the algae. Platt has distinguished between physical and biological light attenuation (cf. Chapter 9).

In some areas of the Antarctic, reduction of light by the phytoplankton is believed to be a factor in the thickness of the euphotic zone. El-Sayed (1966) reports variation in depth over Antarctic areas of about 7 m to more than 50 m; he attributes the greatly reduced depth of light penetration in the Gerlache and Bransfield Straits in part to the richness of the phytoplankton. Burkholder and Mandelli (1965a) relate the thinness of the euphotic layer in Gerlache Strait and more obviously in Deception Island (less than 10 m) to the more abundant plankton as compared with the Bellingshausen Sea.

Stratification and the Vertical Distribution of Phytoplankton

While density of algal crop may be a factor in light attenuation in the sea, the vertical distribution of the algae, which is partly dependent on the stability of the water column, can also be related to rate of light absorption. In some of the examples just quoted of differences in the thickness of the euphotic zone in the Antarctic, vertical distribution of the phytoplankton plays a significant part. Though much of the open Antarctic Ocean shows marked turbulence and vertical mixing, there are characteristic areal differences. In Gerlache Strait and Deception Island, for example, the surface waters are temperature stratified, in contrast with the Bellingshausen Sea and Bransfield Strait, which show little stabilization. In the Bellingshausen Sea chlorophyll *a* was uniformly low (*ca.* 0.2 mg/m^3) throughout the euphotic zone; in Bransfield Strait chlorophyll was higher, but at any one station vertical distribution was more or less uniform (cf. Burkholder and Mandelli, 1965a) (Fig. 6.18). In the two stratified waters the phytoplankton was so concentrated in the upper layers that as summer advanced much of the diatom crop near the lower part of the euphotic zone became cut off from light. For Deception Island the ratio of the quantity of chlorophyll *a* in the euphotic zone to that in the "hypophotic zone" (defined as from a depth of 1% surface light to a depth equivalent to 0.0001% surface light) was as high as 1.0, and total chlorophyll was very rich, averaging nearly 400 mg chlor *a*/m^2. For a typically poor station in the Bellingshausen Sea the ratio was 0.66 and the total chlorophyll only about 20 mg chlor *a*/m^2 (cf. Fig. 6.19) (cf. also Chapter 9). Anderson (1969) has described a somewhat similar

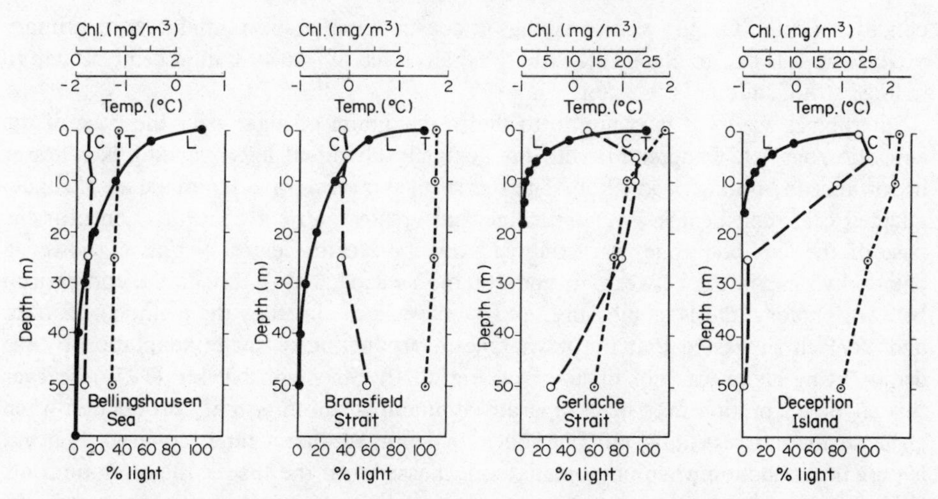

Fig. 6.18. Vertical distribution of temperature and chlorophyll and the attenuation of light with depth in various Antarctic seas (chlorophyll as mg/m^3; light as per cent attenuation). (After Burkholder and Mandelli, 1965a.)

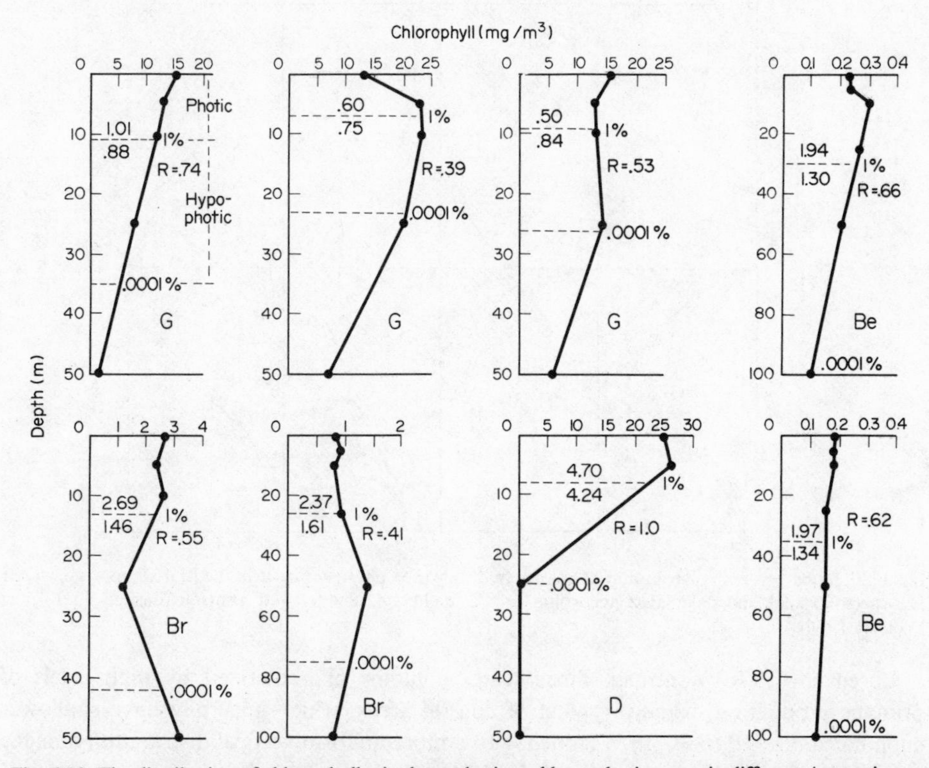

Fig. 6.19. The distribution of chlorophyll *a* in the euphotic and hypophotic zones in different Antarctic seas. Chlorophyll as mg/m^2 for each layer. The ratio of chlorophyll in the two layers is indicated by the numbers labelled R. Assimilation numbers of the algae are indicated just above and below the dotted lines separating the two layers. G = Gerlache Strait; Be = Bellingshausen Sea; Br = Bransfield Strait; D = Deception Island (after Burkholder and Mandelli, 1965a).

condition off the Oregon and Washington coasts where a substantial part of primary production was due to algae below the 1% light level, their distribution being related to marked stratification of the water.

In tropical waters it is common to find a maximum of algae near the base of the euphotic zone (cf. Chapter 3); with the generally abundant light, shading is of lesser importance in limiting production. The sub-surface maximum, often made up of shade-adapted cells, can benefit from the somewhat greater nutrient concentration near the base of the euphotic zone. An example from the North Central Sargasso showed a chlorophyll maximum between 75 and 100 m (Yentsch, 1974), though the comparison between photosynthesis as measured *in situ* and as calculated by the method of Ryther and Yentsch suggested that the advantage to production of shade adaptation by the deeper living algae was not obvious (cf. Fig. 6.20). Steemann Nielsen (1974) believes that shade adaptation may be comparatively unimportant in primary production when algae are more or less uniformly distributed in depth, but that it can become a significant feature in production when phytoplankton is massed near the base of the euphotic zone. Such a distribution is common in warm oceans but is not confined to those waters. In contrast, at fairly high latitudes production can continue in winter in highly stratified waters provided the algae are massed in a very thin surface layer.

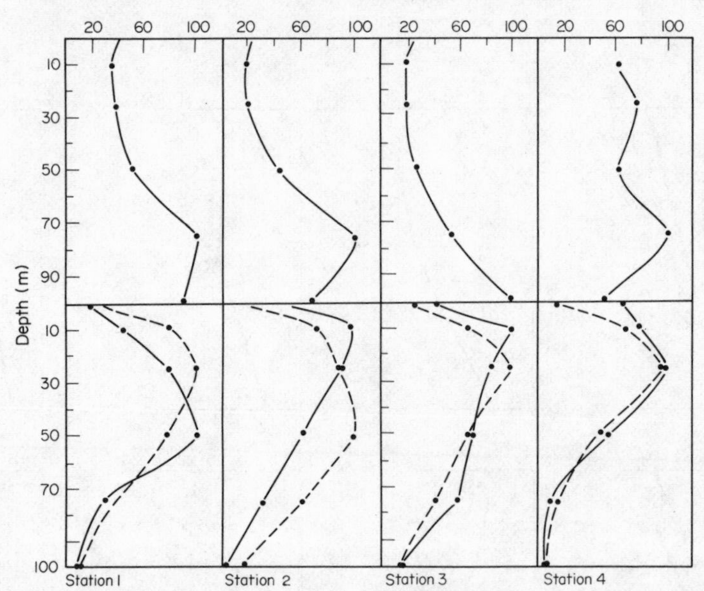

Fig. 6.20. Upper = depth distribution of chlorophyll. Lower = photosynthesis at the four stations measured *in situ* (solid line) and calculated according to the method of Ryther and Yentsch (dashed line) (after Yentsch, 1974).

Lorenzen (1976) contrasts those water columns characterized by high levels of primary production, usually typical of coastal areas, which generally have shallower euphotic zones (10 to 30 m), a tendency to a more uniform vertical distribution of algae and a rate of production exceeding 0.5 g C/m^2/day, with central regions of open oceans usually of low productivity, where the euphotic zone is more than 50 m and may exceed 100 m, the biomass of algae tends to concentrate near the base of the euphotic layer giving a chlorophyll maximum, and the rate of production is less than 0.5 g C/m^2/day.

The depth of maximal productivity is usually below the immediate surface in all water columns, but whereas in productive regions it is high in the euphotic zone, averaging about 2.5 m in a 10-m-thick euphotic layer, it is deeper in poorer regions, approaching 40 to 50 m in a euphotic zone of 100 m thickness. An analysis by Lorenzen also demonstrated that chlorophyll content (algal biomass) decreased logarithmically with increasing depth of the euphotic layer and that the productivity of the whole water column declined with increasing euphotic zone thickness. The absorption of the available light by the algae was much greater at lesser thicknesses of the euphotic layer (cf. Figs. 6.21 and 6.22). In a 10-m euphotic zone Lorenzen calculated that 56% of the light energy could be absorbed by algae as contrasted with only 1% in a

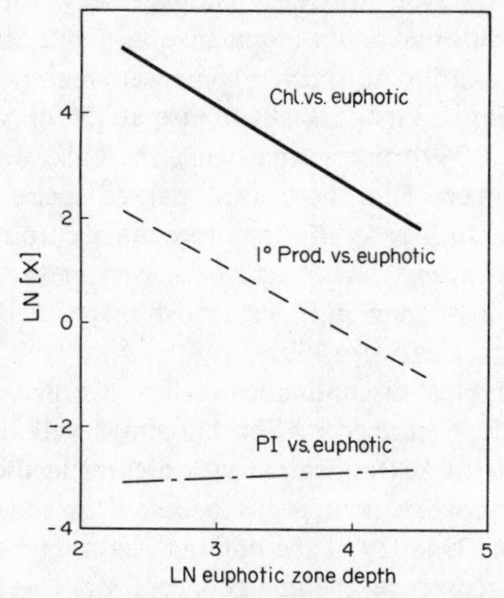

Fig. 6.21. Relationship between euphotic zone chlorophyll (mg/m^2), euphotic primary production (mg C/m^2/d), and Productivity Index (mg C/mg/d) as a function of total euphotic zone depth (after Lorenzen in Cushing, 1974).

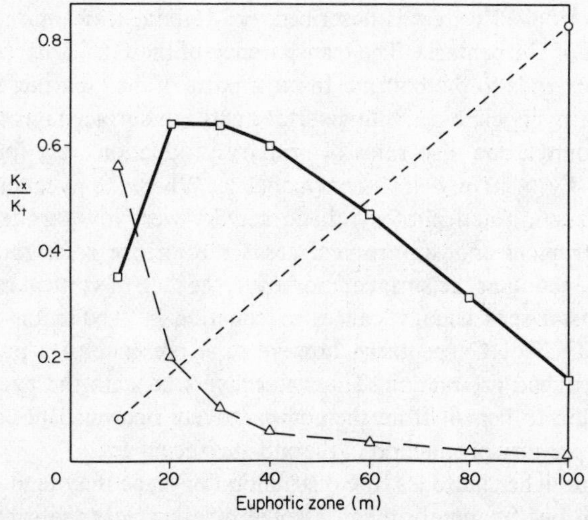

Fig. 6.22. Partitioning of total light attenuation within euphotic zone into fractions attributable to water (circles), plants (triangles), and other materials (squares) as a function of total euphotic zone depth (after Lorenzen in Cushing, 1974).

100 m euphotic zone. The productivity index, according to Lorenzen, did not change significantly with euphotic zone depth, and would therefore not appear to be sensitive to the generally lower nutrient content of greater euphotic zone depths, though chlorophyll cell content could modify this conclusion to some degree. The nutrient level, however, is presumably clearly related to the total biomass of algae.

Different species of phytoplankton may exhibit depth preferences within the euphotic zone, though with the reduced stratification more typical of temperate and high latitudes and the lesser thickness of the euphotic layer, such differences will tend to be more apparent in tropical seas. Hasle (1959a) found that while the bulk of the phytoplankton in the equatorial Pacific was present in the upper 100 m, diatoms were mainly in the uppermost 25-m or 50-m layer, whereas dinoflagellates, especially gymnodinians, occurred slightly deeper with maximum abundance at about 50 m. Coccolithophorids showed a greater range, according to species; some were mainly in the top 50 or 100 m, but *Thorosphaera flabellata* had its peak abundance at 150 m depth (cf. Chapter 5). In the Red Sea, Sukhanova (1969) reports that while the bulk of the phytoplankton was distributed through the upper 75 m, there were marked species differences. *Ceratium* spp. occupied the upper 50 m; *Pyrocystis* spp. were mainly from 25 to 50 m. The most abundant diatom, *Climacodium frauenfeldianum*, was in the uppermost 20 m, but *Planktoniella sol*, next most abundant, was most plentiful below 25 m and never occurred at the surface (but cf. Hasle, 1959a).

An example of a fairly typical discontinuous vertical distribution of phytoplankton in the western tropical Pacific, described by Sorokin and Tsvetkova (1972), showed that provided no significant thermal stratification was present in the euphotic layer to the depth of the regular thermocline, two peaks occurred in actively synthesizing algal (chlorophyll) concentration. One lay at the optimum light intensity depth (20 to 30 m) but was less constant in occurrence; the main concentration lay at the upper boundary of the thermocline – 50 to 70 m in the equatorial divergence region; 100 to 130 m in the convergence area.

One example of the effect of stratification and algal distribution on primary production in inshore sub-tropical waters is described by Motoda, Kawamura and Taniguchi (1978) for the Gulf of Carpentaria. The transparency of the Gulf waters is so great that the euphotic zone extends to the bottom. In most parts of the Gulf there is pronounced stratification with a pycnocline delimiting surface and sub-surface layers. Higher standing crops as chlorophyll and also rates of primary production were found in the sub-surface waters, mostly >30 m, below the pycnocline. Where the pycnocline was absent, chlorophyll concentration and photosynthetic activity were low throughout the water column. In most tropical and sub-tropical areas the intense solar radiation restricts phytoplankton production at the surface; moreover, the lack of vertical mixing generally in sub-tropical coastal seas usually causes a reduction in production due to nutrient restriction. In the Gulf of Carpentaria, however, the presence of a pycnocline is not accompanied by reduced production. The water layers beneath the pycnocline receive sufficient illumination to benefit from the comparatively rich nutrient concentration to give a high level of primary production in the sub-surface layers.

In temperate and higher latitudes, the distribution of algae may tend to a more even vertical distribution but frequently there is some massing near the surface and sometimes in deeper layers. Sinking may be sufficiently rapid to modify the vertical pattern. Some deeper chlorophyll maxima are thus due to deteriorating algal blooms. Ketchum and Corwin (1965) reported on the vertical distribution of chlorophyll in the upper 50 m (euphotic zone) in the Gulf of Maine in April during a 10-day bloom. Though the concentration at 1 m and 10 m showed a decline, there was a general large increase in crop amounting to 118 mg/m^2. Pigments at deeper levels also increased over the 10 days but were believed to be mainly decomposed chlorophyll. The studies of Anderson, Burkholder and Mandelli and others, however, indicate that actively photosynthesizing algae can occur at deeper levels in some areas.

The effect of stability in a temperate area of the Sea of Japan, as influenced by seasonal change in temperature, on the vertical distribution of phytoplankton is described by Ohwada (1972). Living and dead diatoms taken in collections were identified, mainly on the condition of the chromatophores so that the genuine distribution would not be obscured by dead sinking cells. In winter, live diatoms were found mostly near the surface at stations very close to the coast, but turbulence could carry them to the bottom. Live cells were found to about 150 m depth. Offshore, the diatoms could reach 300 m with the mixing of an almost homogeneous water column, species composition and overall density being fairly uniform to that depth. In summer, with the development of a strong pycnocline, the algae were limited to less than 100 m. In near-shore areas the phytoplankton was mostly close to the surface with those species predominating which preferred fairly high temperature and light intensity. Further offshore, two maxima occurred, the deeper at 50 to 75 m composed of species preferring colder water and lower light intensities.

In estuaries which are highly stratified the phytoplankton is frequently confined to a narrow vertical zone, below the surface, lower salinity, water. The rapid attenuation of light, especially with turbid conditions, markedly restricts the depth of the phytoplankton in the lower, more saline layer, but the comparatively rich nutrients can lead to massive algal concentrations, and self-shading by the algae may then be significant (cf. Riley 1967).

Sub-surface maxima of chlorophyll have been reported by a number of workers (*vide supra*). Steele and Yentsch (1960) discuss the problem that algal concentrations represented by chlorophyll maxima are usually found either close to the surface or near the base of the euphotic layer in temperate latitudes. The sinking of living algal cells at constant speed, however, is difficult to reconcile with the presence of a maximum crop sometimes close to or even below the compensation depth. With a decreasing sinking rate of the algae, on the other hand, a chlorophyll maximum could theoretically occur below the compensation depth of the community (i.e. the balance between photosynthesis and algal respiration plus grazing) (cf. Fig. 6.23). Steele and Yentsch demonstrated that differences in water density alone, as might occur with temperature stratification in the euphotic zone, could not satisfactorily account for variable sinking rates of the algae, but that changes in cell buoyancy, mainly due to light and nutrient concentration, could modify the rate (cf. Chapter 4). Older algal cells are believed to settle faster, but on sinking into the deeper euphotic layers they typically encounter richer nutrients as well as reduced light. Experiments indicated that older cells when enriched, especially in darkness, have a reduced settling rate. The algae, therefore, tend to accumulate near the base of the euphotic layer. By contrast, during early spring in middle or higher latitudes, algae grow rapidly with the abundance of nutrients and the young cells will tend to form a near-surface accumulation with their low sinking rate. Akinina (1969) has confirmed variation in the sinking rate of algae in relation to their physiological state. She refers to young, actively dividing diatoms sinking more slowly than older cells and quotes an extreme range of 0.02 to 20 m/day. The dinoflagellates *Prorocentrum micans* and *Gymnodinium kovalevskii* were found to sink most rapidly at night, when division rate is lowest, and most slowly during the morning when division rate is high. Aged cultures and cells deficient in nutrients (phosphorus) also sank more rapidly. Steele (1956) in discussing effects of changes in sinking rate gave a range of 1.4 to 4.0 m/day.

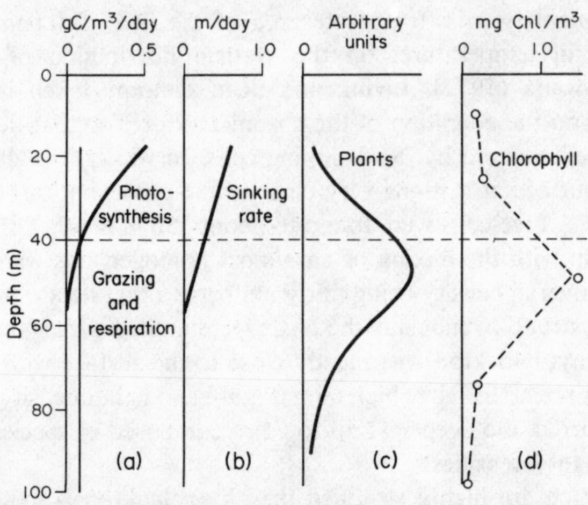

Fig. 6.23. Vertical distribution of phytoplankton. (a) indicates photosynthesis balancing grazing and respiration at 40 m; (b) the effect of linearly decreasing sinking rate reaching zero at 58 m. The resulting vertical distribution is shown in (c), with an actual profile for a station in waters off New York (d) (Steele and Yentsch, 1960).

A pronounced sub-surface chlorophyll maximum is described by Taniguchi and Kawamura (1972) as associated during summer with a seasonal thermocline in the Oyashio Current waters off Japan. During spring, with well-mixed waters, there was no sub-surface maximum and crop and primary production were high. During summer a maximum developed at the discontinuity, and though Taniguchi and Kawamura list a number of factors which may have contributed to the chlorophyll maximum, they consider sinking and accumulation of algae as the potent factors in Oyashio waters. Though crop and production throughout the whole euphotic zone were reduced in summer, the pycnocline limited the loss of algal cells and sufficient nutrients remained, despite the considerable reduction in concentration, to permit a reasonable production in the sub-surface algal layer (cf. Fig. 6.24).

Holligan and Harbour (1977) suggest that sub-surface chlorophyll maxima in inshore (continental shelf) temperate waters usually differ from those in oceanic situations in that there is a shallower discontinuity (a seasonal thermocline lying between 10 and 40 m), tidal influences on mixing bottom water are significant, and conditions generally are less stable. They believe, however, that the favourable light conditions and nutrient supply are sufficient to produce the dense sub-surface algal maxima.

Occasionally in sheltered temperate areas it has been possible to follow a gradual deepening of the productive algal layer, particularly in a bloom, with time. While this could reflect the gradual sinking of algal cells, the deepening might be associated with a depletion of nutrients from the surface downwards (cf. Marshall and Orr, 1927, 1930). Such a pattern could occur only under unusually stable and sheltered conditions. A contributing factor could be the increase in light intensity in spring, with the more light-sensitive species tending to flourish at sub-surface rather than surface depths. Parsons and Takahashi (1973a) describe the theoretical changes in the vertical distribution of algae in a stratified water column, following vertical mixing and a homogeneous distribution. A sub-surface maximum can develop with surface light inhibition and the normal attenuation of light with depth. The maximum will approach the surface as algal density increases due to self-shading. The compensation depth decreases, and a shallow bloom will occur, provided nutrients are sufficient. As nutrients are exhausted, the maximum algal crop is found in the deeper euphotic layer. The pattern will, of course, be greatly modified in nature, especially with grazing.

A few studies have been made on changes in the vertical distribution of phytoplankton with time. Holligan and Harbour (1977) studied the vertical distribution

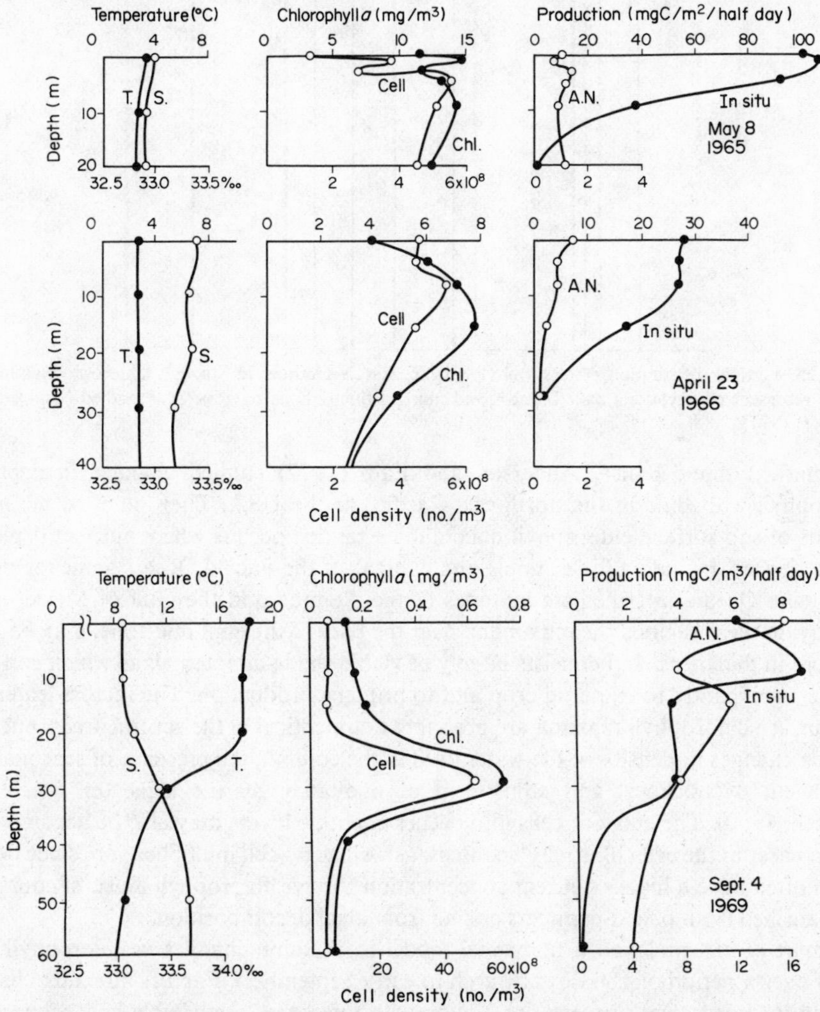

Fig. 6.24. Seasonal changes in the vertical distribution of phytoplankton crop (as cell number and as chlorophyll), in primary production and in assimilation number. Temperature and salinity profiles are indicated (Taniguchi and Kawamura, 1972).

of phytoplankton over the productive period of the year in the western English Channel (Station E1). The spring outburst was dominated by diatoms, first in the surface and throughout the column, and later below the thermocline. From June to August the flora was dominated by dinoflagellates, mainly in the thermocline region. During September diatoms were again abundant for a short period (cf. Fig. 6.25).

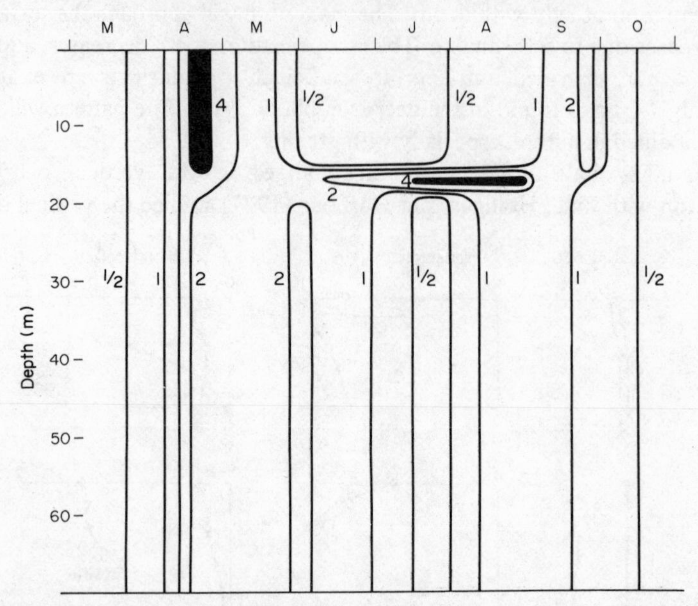

Fig. 6.25. Vertical distribution of chlorophyll *a* from March–October at Stn. E1. Chlorophyll as mg/m^3. Values represent observations over 2 years and marked fluctuations have been smoothed (Holligan and Harbour, 1977).

Jamart, Winter, Banse, Anderson and Lam (1977) studied changes in depth of chlorophyll with time in the north-east Pacific off the USA. They point to the many records of sub-surface chlorophyll maxima in stratified oceans where nutrient depletion is experienced, including for example, many areas of the Pacific Ocean, some regions of the Indian Ocean, intermediate latitudes of the Atlantic and the Gulf of Mexico. Frequently, in these regions the maximum, near the base of the euphotic zone, may be 10 m or more in thickness, and consists largely of viable shade-adapted algae which can contribute significantly to standing crop and to primary production. The factors leading to sub-surface chlorophyll maxima are probably not identical in the several areas, but may include changes in density of the water itself in the column, the presence of seasonal and permanent pycnoclines, and adjustment of buoyancy by the algae (cf. Steele and Yentsch, 1960). The entry of cells into deeper euphotic layers may also be accompanied by increases in the cell chlorophyll content, as well as by cell multiplication, since deeper layers often have a higher nutrient concentration. Active chlorophyll must, of course, be distinguished from phaeo-pigments arising from algal decomposition.

Jamart *et al.* employed a numerical model to examine changes in chlorophyll with depth over a period of 180 days (March to early September). Factors affecting changes in crop included algal primary production with reference to available light, concentration of nitrate and ammonia, algal respiration, variable sinking rate linked to nutrient concentration, turbulent vertical mixing and grazing. Until about Day 95 (i.e. early

April) the rise in chlorophyll was slow, reflecting the increase in incident light, but rapid stabilization of the water column then produced a bloom of phytoplankton. During the next month (May) the chlorophyll was mainly in the upper 20 m and both nitrate and ammonia were reduced in the upper layer. From early June there was a very marked depletion of nitrate and ammonia; the maximum algal crop deepened to >20 m and became deeper and more obvious from July to the end of the period under study (cf. Fig. 6.26). The development of the deeper chlorophyll maximum could thus be largely attributed to algal growth in the upper part of the deeper layer, with the substantial nutrients and with adequate light penetration, but attenuation of light prevented its continued deepening. Its contribution to daily production over the summer was considerable, though after mid-July primary production declined with time (Fig. 6.27). By September the lower incident radiation, with continued grazing pressure, led to the beginning of a decline in the sub-surface chlorophyll. Indeed, variation in grazing intensity can greatly modify the overall computed pattern, with its direct effect on crop and also indirectly through the supply of ammonia.

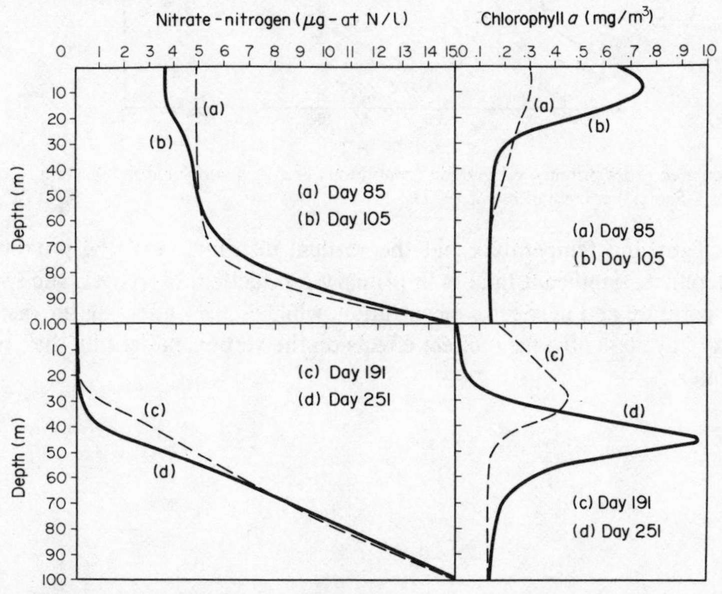

Fig. 6.26. Calculated changes in the vertical distribution of chlorophyll *a* and nitrate – N on 4 days during the period March–September (after Jamart *et al.*, 1977).

Towards the end of the productive season the presence of a a deeper algal layer could be of further significance in contributing viable cells to a possible autumn phytoplankton outburst when the seasonal thermocline was broken by turbulence and nutrients became available in the upper layers.

In considering vertical distribution in relation to primary production, the efficiency of carbon assimilation in stratified waters may also be important. Ignatiades (1973) studying the waters of Loch Etive, Scotland, found that stratification occurred over the summer. The crop of phytoplankton, estimated as pigment concentration and as cell counts, was highest at the surface, though species composition was similar at all depths. When samples of phytoplankton were taken from three depths, 1 m, 10 m and 45 m (the last below the euphotic layer), and all were exposed at 1 m depth, carbon fixation was highest in the 1 m sample. Fixation of ^{14}C in the sample from 10 m was only 64% of that from 1 m depth. This was in part due to lower cell concentration in the 10-m sample. The assimilation ratio was the same for the samples from the two depths, but almost twice the amount of assimilated carbon was excreted by the 10 m sample though the total carbon fixation was lower. Cells collected from the 45-m layer gave a poor assimilation rate when exposed at 1 m depth; carbon fixation was only 6% of that of the 1-m sample and excretion increased sharply (cf. Chapter 3).

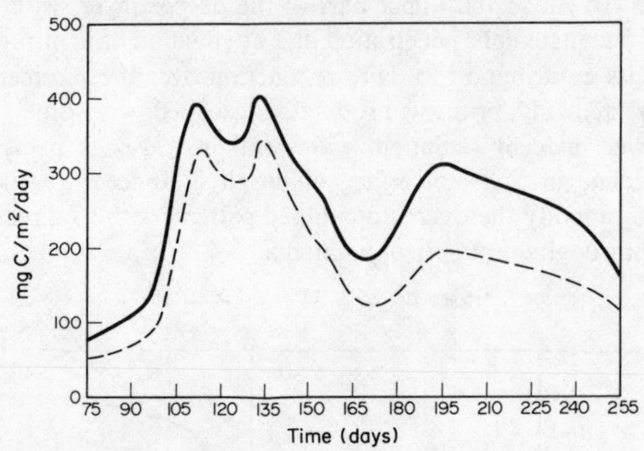

Fig. 6.27. Calculated gross primary production (continuous line) and zooplankton ingestion rate (hatched line) from March–September (Jamart *et al.*, 1977).

Not only light and temperature but the vertical distribution of the phytoplankton itself are, therefore, significant factors in primary production. Moreover, such variables as light, temperature and nutrient concentration, while directly affecting the distribution of algae, may have less obvious indirect effects on the vertical pattern by their influence on sinking rates.

Chapter 7
Factors Limiting Primary Production: Nutrients

Phytoplankton must obtain a range of substances from their environment in order to sustain growth and division. Some of the universally required constituents are available in abundance relative to what is needed, for example, CO_2 and ions such as Na^+, K^+, Mg^{2+}, Ca^{2+} and SO_4^{2-} (cf. Chapter 2). Certain other abundant constituents are known to be essential only for restricted groups of organisms. Thus, strontium is commonly taken up in association with calcium but so far as is known it is essential only for a group of zooplankton, the acantharians, which form skeletal celestite (strontium sulphate).

In addition to these abundant constituents, phytoplankton require a variety of elements present at lower concentrations. The most important of these are the micronutrients nitrogen, phosphorus and silicon and it is with these that this chapter is mainly concerned. Nitrogen and phosphorus are essential for all algae. The known requirement for silicon is limited to those groups which form siliceous skeletons although silicon is apparently required for purposes other than skeletal deposition. Silicon is also taken up by phytoplankton organisms which have no known requirement for the element.

Certain other trace elements are essential for phytoplankton growth, including the metals Fe, Mn, Zn, Cu and Co which are incorporated into essential organic molecules, particularly a variety of enzyme cofactors which enter into photosynthetic reactions. Phytoplankton accumulate, often to a high degree, many apparently non-essential elements and there is uncertainty with some elements as to whether or not their presence in cells serves an essential function. The requirements for trace metals are generally very low and despite the low concentrations of many of them in seawater they will normally be present in considerable excess of the amounts needed by the phytoplankton. Iron, however, is present largely in forms which are not readily available and with this element, and possibly some others, routine analytical measurements of concentrations may be uninformative in relation to biological processes. Little is known about the complex relationships between chemical speciation of metals and biological availability. It is possible that organic molecules which complex metals may influence availability of some elements. Certain organic compounds, including vitamin B_{12}, thiamine and biotin, are themselves essential for the growth of at least some algal species. A variety of inorganic and organic forms may thus play important roles in determining the relative growths of different algal species and may influence patterns of population composition and succession. There is little firm evidence, however, regarding the roles of trace constituents other than the principal micronutrients.

Early ideas about the role of nitrogen, silicon and phosphorus in the sea stemmed largely from the work of Brandt and of Raben. As discussed by various authors, including Cushing (1975), this early thinking was influenced by ideas from the field of agriculture, including the concept due to Liebig (e.g. Liebig, 1843) that yield is essentially determined by the amount of that nutrient which is in shortest supply. Progress in the study of distributions of nutrient elements and their relation to production in the sea depended largely upon the availability of suitable methods for measurement of the low concentrations of the principal micronutrients. The development of sensitive absorptiometric methods led to major advances beginning in the 1920s with the initiation of a long series of systematic measurements in the English Channel (see, for example, Harvey, 1926; Atkins, 1928; Cooper, 1933; Armstrong and Butler, 1968). This was followed by intensive investigations of distributions and seasonal variations in other coastal regions, notably the Gulf of Maine, Puget Sound, the North Sea and the Barents Sea, and by extensive surveys of oceanic distributions. Much of this work was summarized by Sverdrup, Johnson and Fleming (1942) and Barnes (1957). Accounts of the chemistry and distribution of the principal micronutrients which provide substantial coverage of the later literature include those by Armstrong (1965a, b), Vaccaro (1965) and Spencer (1975), while reviews focused on uptake by phytoplankton, and regeneration, have been given by Corner and Davies (1971) and Dugdale (1972, 1976).

The course of studies of these micronutrients was to some degree influenced unavoidably by analytical considerations. In the earlier years most information was obtained on phosphate, which could be measured much more readily than nitrate. It was in fact only with the introduction by Morris and Riley (1963) of a column-reductor

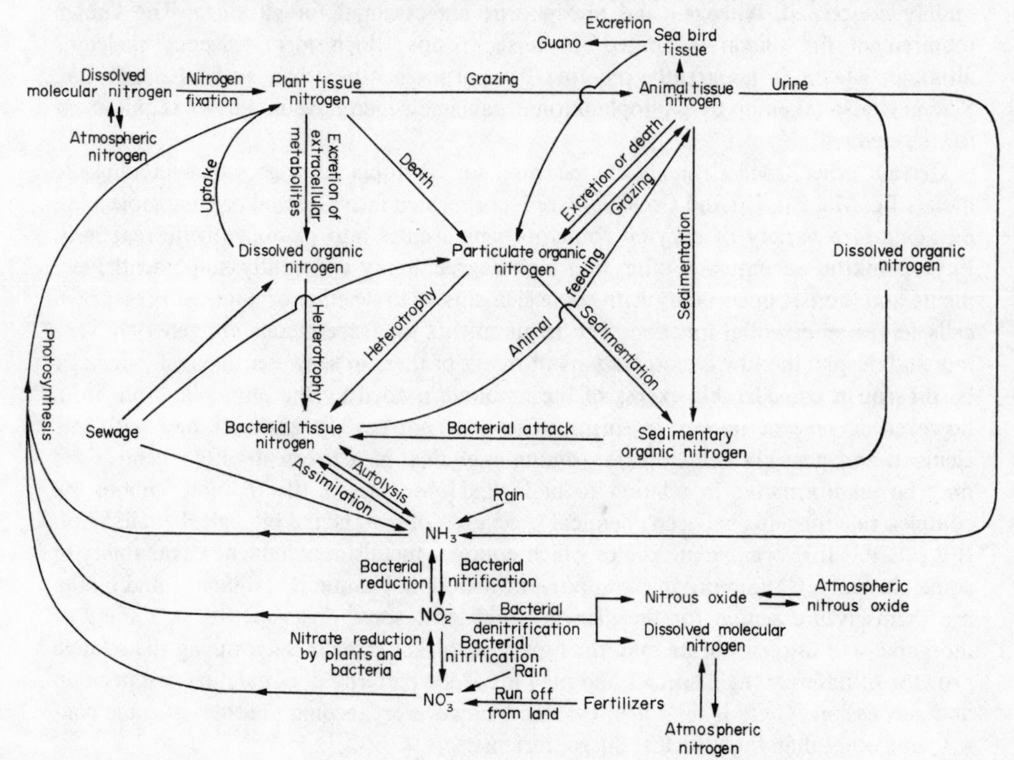

Fig. 7.1. The nitrogen cycle in the sea (from Riley and Chester, 1971).

method that a reliable, rapid and reasonably convenient shipboard technique for nitrate was available. Methods for other forms of combined nitrogen, notably ammonium and urea, have subsequently been greatly improved. During the period since the early 1960s, considerable emphasis has been placed on nitrogen, in keeping with the generally accepted view that "nitrogen is normally at least as limiting as any other component" (Steele, 1974). The availability of improved methods for analysis of transient components in the nutrient cycles and stable-isotopic labelling techniques using nitrogen-15 and, more recently, silicon-29, has led to a much greater knowledge of the complex processes of uptake and cycling of the various chemical species. Some recent workers have emphasized the importance of silicon, which had become relatively neglected in this context.

In this chapter the various chemical forms of the principal micronutrients and the processes of their interconversions are first outlined. This outline, which amplifies the more general information provided in Chapter 2, is followed by a discussion of some of the main features in the spatial distributions of these micronutrients and the marked seasonal variations in concentrations which occur in some regions. Studies of the uptake of the micronutrients by phytoplankton and the relationship between micronutrient concentrations and productivity in the ocean are then considered. The final section summarizes information on other trace constituents in relation to productivity.

Forms of the Principal Micronutrients and their Interconversions in Seawater

Nitrogen

Nitrogen is notable for the wide range of dissolved inorganic forms in which it can occur in seawater, reflecting the complexity of the chemistry of the element and of the nitrogen cycle in the sea (Figs. 7.1 and 7.2). The forms which have been detected are the dissolved gases, molecular nitrogen (N_2) and nitrous oxide (N_2O), and the ionic forms nitrate (NO_3^-), nitrite (NO_2^-) and ammonium (NH_4^+) which is in equilibrium with

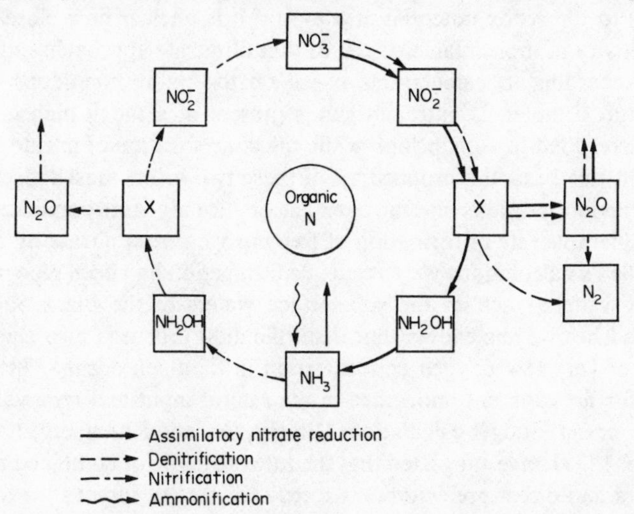

Fig. 7.2. Pathways of nitrate reduction, denitrification and nitrification (after Hahn and Junge, 1977).

a small fraction of unassociated ammonia (NH_3). Nitric oxide (NO) has been detected as a product in experimental studies of biological interconversions of nitrogen species and it is likely that other intermediates occur which have not yet been detected, such as hydroxylamine (NH_2OH). Fiadeiro, Solorzano and Strickland (1967) failed to detect hydroxylamine in seawater and found that it cannot accumulate to a significant level in water containing oxygen at a concentration even as low as 0.2 ml/l. Cline and Richards (1972) inferred that the concentration was below 0.3 μg-at N/l in sub-oxic waters where the oxygen concentrations were below 0.01 ml/l in some cases; a similar upper limit was indicated for sub-oxic waters in the work of Fiadeiro *et al.* (1967).

Urea and a range of organic nitrogen compounds also occur in solution in seawater. The latter include free and combined amino acids as well as biologically refractory material (cf. Chapter 2). Determinations of total dissolved organic nitrogen give little useful information in the context of productivity, since the labile material, which may be turned over quite rapidly, may be a relatively small fraction of the total. Particulate forms of nitrogen, dominantly organic, are also present.

As discussed in Chapter 2, oxic samples of seawater throughout the oceans contain dissolved molecular nitrogen at concentrations which are close to those corresponding to the 100% saturation values under surface conditions. Significant rates of fixation of molecular nitrogen are found in terrestrial and freshwater environments, nodulated leguminous plants being particularly important. It is now known that molecular nitrogen is utilized by some marine organisms, particularly by colonies of the blue-green alga, *Trichodesmium* (=*Oscillatoria*). The rate of utilization is so low, however, relative to replenishment by atmospheric exchange that the water concentrations of the dissolved gas are not significantly affected. Biological fixation may nevertheless be significant in relation to the oceanic budget of combined nitrogen (Delwiche, 1970). As discussed below, free nitrogen is produced by denitrification in anoxic waters, and, under some conditions, in sub-oxic waters, and this process may lead to concentrations significantly above the 100% saturation values in isolated water masses.

The dominant form of combined nitrogen in the ocean is the nitrate ion. Because of uncertainty as to the redox potential of seawater it is unclear how closely the average ratio of the activity of molecular nitrogen to that of nitrate approaches thermodynamic equilibrium. According to calculations based on the redox conditions postulated by Sillen (1961) (cf. Chapter 2), nitrogen gas is present at a much higher concentration than would correspond to equilibrium, while the concentration of nitrate is correspondingly lower. On this basis the proportions of these two forms must reflect either highly metastable behaviour of molecular nitrogen gas or a steady-state condition which would require a considerable rate of formation of free nitrogen from nitrate by denitrification. At the time Sillen's calculations were made, denitrification had been recognized to occur only in anoxic waters, such as the sub-surface waters of the Black Sea. Subsequent work, discussed below, has shown that denitrification proceeds to a significant extent also in zones of very low oxygen concentration in the open ocean. This process may also account for an apparent imbalance in the rate of input and removal of combined nitrogen in the ocean. Budget calculations (Emery, Orr and Rittenberg, 1955; Delwiche, 1970; Holland, 1973) have suggested that the rate of supply of combined nitrogen to the ocean by rivers and direct precipitation exceeds the rate of removal to sediments by a magnitude of 10^{13} to 10^{14} g/yr. The rates of denitrification recently estimated for sub-oxic oceanic waters suggest that this process may account for a substantial part of the apparent imbalance.

An alternative view to that of Sillen concerning the redox speciation of nitrogen in the ocean has been given by Breck (1974) who suggested that the redox potential of oxic seawater may be considerably lower than was assumed by Sillen (cf. Chapter 2). On this view the balance between free nitrogen and nitrate is closer to the equilibrium prediction.

Nitrate shows major variations in its concentration in the ocean. Thus, while it is the dominant form of combined nitrogen in the ocean considered as a whole, other forms are important in certain sectors. This is often the case in the euphotic zone where under some circumstances nitrate may be undetectable. These other forms include the reduced species which are the end products of biological activity, namely organic nitrogen compounds (including amino-acids), urea and ammonium. Ammonia is released in the decomposition of organic nitrogen compounds but is also a major excretory product of many invertebrates, including zooplankton. Nitrite is often detectable in the mixed layer and in certain sub-surface waters. It is formed as an intermediate in the inter-conversions of nitrate and ammonium. The roles of some of these forms have been investigated to a considerable extent in relation to productivity and the cycles of the nutrients. The following general summary of the inter-conversions between the forms is given as a basis for subsequent discussions of the distribution and utilization of the nutrients.

The reduction of nitrate to ammonium and the reverse oxidative reaction are of particular importance. These processes occur in two stages with nitrite and other forms as intermediates. The relevant reactions are shown in Fig. 7.2. The oxidative processes, described as nitrification, can be shown as:

$$NH_4^+ + OH^- + 1.5O_2 \rightleftharpoons H^+ + NO_2^- + 2H_2O \qquad (1)$$

and

$$NO_2^- + 0.5O_2 \rightleftharpoons NO_3^- \qquad (2)$$

Each reaction is exothermic, the free energy changes being −59.4 kcal and −18.0 kcal for (1) and (2) respectively. It has long been known that in terrestrial systems a major role in the regeneration of nitrate is played by two main groups of autotrophic nitrifying bacteria in soils, the ammonium oxidizers (e.g. *Nitrosomonas*) and the nitrite oxidizers (e.g. *Nitrobacter*). Such organisms utilize the energy released in the above reactions to fix carbon dioxide. Direct evidence for the occurrence of nitrification in seawater was only obtained more recently (Vargues and Brisou, 1963; Watson, 1965; Rheinheimer, 1967). The process requires the presence of free oxygen but very low concentrations suffice; oxygen-deficient conditions in fact favour higher rates of nitrification (Carlucci and McNally, 1969). Nitrifying bacteria have been isolated from seawater from various regions and it is apparent that distinct marine types are widely distributed in the sea. Generally they have been recognized by measurement of their activity following isolation, but Watson (1965) has described in detail the ammonium-oxidizing species *Nitrosocystis oceanus*, an obligate autotroph, and Watson and Waterbury (1971) have identified two nitrite-oxidizing bacteria. Studies on the marine nitrifying bacteria have indicated that with the bacterial numbers and concentrations of ammonium generally found in the sea, the production of nitrite by nitrifying bacteria is very slow (Watson,

1965; Carlucci and Strickland, 1968). Examination of seven oceanic bacteria for capacity to oxidize ammonia (Carlucci, Hartwig and Bowes, 1970) gave no evidence of heterotrophic nitrification and they concluded that in the sea this process is not important. It is possible that nitrification may be more significant in certain micro-environments where ammonia is being released in the decomposition of organic material. Carlucci *et al.* (1970) investigated this by measuring rates of oxidation of ammonium by *N. oceanus* in the presence of decomposing organic particles. They found that the oxidation of ammonium released under such conditions was more rapid than that for ammonium added to seawater but the rates appear to be inadequate to account for the levels of nitrite often found in surface seawaters. In this respect the role of release of nitrite by phytoplankton, as discussed below, appears most significant. The fact that rates of nitrification are increased under sub-oxic conditions has led to the suggestion that nitrification may contribute to the accumulation of nitrite in sub-oxic zones in the open sea (Carlucci and McNally, 1969). Such accelerated rates of nitrification remain inadequate, however, to explain the rate of formation of nitrate in the ocean.

The possible role of non-biological oxidation of ammonium has not been adequately ascertained. Vaccaro (1962) found that some oxidation of ammonium usually occurred within 48 hours when aerated seawater was kept in the dark. He considered that chemical mechanisms probably contribute to *in situ* oxidation, especially at the surface. Hamilton (1964) concluded from experiments using natural and artificial illumination that the role of photochemical oxidation of ammonia was unimportant in the marine nitrogen cycle.

While several mechanisms have been identified which bring about the regeneration of nitrate from reduced forms it remains unclear as to how the overall process is dominantly mediated in the ocean. The disparity between the rate of formation of nitrate which appears to occur and the likely rates of the biological processes which have been examined may imply an important role of other processes, possibly inorganic, or may reflect inadequate knowledge of the true environmental rates of nitrification.

Phytoplankton organisms are able to utilize various forms of combined nitrogen (cf. p. 324) so that considerable recycling may occur without the completion of the sequence of oxidative reactions which regenerate nitrate. When nitrate is utilized it must be reduced in the cell to ammonia for the synthesis of organic nitrogen compounds. This cellular reductive pathway to organic nitrogen (assimilatory nitrate reduction) involves the initial reduction of nitrate by nitrate reductase with cofactor NADPH or NADH as an electron donor. Further reduction to ammonia is carried out by nitrite reductase, reduced ferredoxin being the probable cofactor. As the reactions are endothermic, energy is required which is provided by solar radiation. There is some evidence that ammonium can be a preferred form of combined nitrogen in uptake by phytoplankton, which conforms with the energetics of the reductive processes.

Under certain circumstances a considerable leakage of nitrite from cells may occur. As discussed subsequently (p. 313), this appears to happen particularly at the base of the euphotic zone where concentrations of nitrate are increased relative to waters nearer the surface and where the light intensity is low. Kiefer, Olson and Holm-Hansen (1976) have proposed the model illustrated in Fig. 7.3 to explain this phenomenon. Nitrate and nitrite ions are actively transported into cells where reduction occurs. It is suggested that the leakage of nitrite occurs in the form of undissociated nitrous acid molecules which can pass through the cell membranes more readily than can nitrite ions. Rates of extracellular nitrite production increase with increase in the concentration gradient of nitrite across the cell membrane. Maximum release may therefore be related to a high rate of intracellular nitrate reduction associated with relatively high intracellular concentrations of nitrate. This can explain the environmental observations of an association of increased concentrations of extracellular nitrite with the region in the euphotic zone where the concentration of nitrate is beginning to increase (cf. p. 313). At a slightly greater depth the process is inhibited because the radiant energy is insufficient to sustain cellular reductive processes.

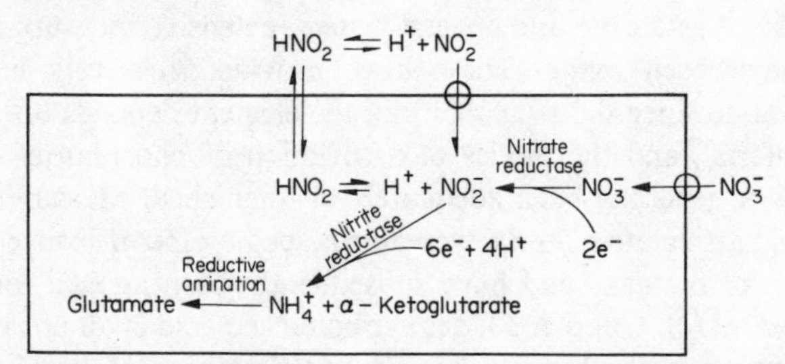

Fig. 7.3. Model for production of extracellular nitrite by phytoplankton (from Kiefer *et al.*, 1976).

Extracellular reductive processes are also important in the marine transformations of nitrogen species. Hamilton (1964) showed that ultra-violet radiation causes photochemical reduction of nitrate to nitrite and purely physicochemical processes of this kind may be significant in the oceanic surface microlayer. Of more general importance are reactions in which bacteria exploit nitrate and nitrite is terminal electron acceptors in the oxidation of organic material from which energy is obtained. These heterotrophic processes may involve nitrate reduction (i.e. reduction to nitrite) and denitrification, which is the term used to describe the dissimilatory reduction of nitrate to nitric and nitrous oxides and molecular nitrogen. These forms of respiration occur only when molecular oxygen is not available as the terminal electron acceptor. They are carried out by many facultative anaerobic bacteria but in most environments bacteria which can reduce nitrate to nitrite are present in greater numbers than are the denitrifiers (Rheinheimer, 1974). In marine environments these processes were initially recognized in anoxic regimes in sediments and in isolated bottom waters such as those in the Black Sea where they form an intermediate stage, thermodynamically, between the use of free oxygen and sulphate for the oxidation of organic matter (cf. Chapter 2). They occur also, as described subsequently (p. 315), in certain comparatively extensive areas of the open ocean below the thermocline, where oxygen concentrations have become exceptionally low as a result of particular conditions of circulation.

Several investigations have been made recently of nitrous oxide in the ocean. The gas is present at concentrations which range from below the 100% saturation values to substantially supersaturated levels, particularly in tropical regions (Seiler and Schmidt, 1974). Supersaturation levels are evidently associated with biological production of the gas. There has been uncertainty, however, as to the relative roles of denitrification and nitrification in its formation, since it is formed in both processes (see Fig. 7.2). Cohen (1978) has shown that in Saanich Inlet, surface oxic waters contain nitrous oxide but that the concentrations decrease to undetectable levels in the anoxic zone, presumably reflecting the total utilization of the gas as a terminal electron acceptor under the conditions of extreme denitrification. Cohen and Gordon (1978) found that there is nett consumption of nitrous oxide in sub-oxic waters in the eastern tropical North Pacific, where partial denitrification occurs. These workers therefore concluded that nitrification is the important mechanism for the production of nitrous oxide (see also Hahn and Junge, 1977).

Phosphorus

The chemistry of phosphorus in seawater is considerably simpler than that of nitrogen, as is indicated by the cycle of the element shown in Fig. 7.4. In some investigations, especially in estuarine and coastal waters, extensive measurements of particulate phosphorus have been made. These have provided relatively little information of interpretative value since the availability of the element depends upon the nature of the particulate material, and the modes of association of phosphorus with different components have not generally been adequately distinguished. Measurements of adenosine triphosphate in particulate material were introduced by Holm-Hansen and Booth (1966) as an indicator of biomass and have subsequently been applied for this purpose to a considerable extent (cf. Chapter 5); deoxyribonucleic acid is an unsatisfactory indicator in this respect as it is present in non-living organic remains (Holm-Hansen, 1969).

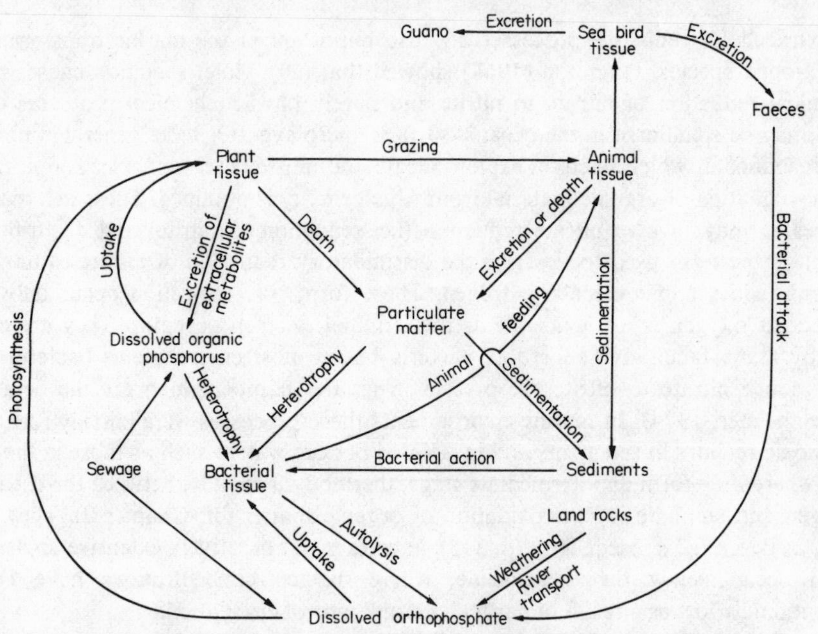

Fig. 7.4. The phosphorus cycle in the sea (from Riley and Chester, 1971).

The principal form of dissolved phosphorus in the ocean is inorganic orthophosphate, usually designated simply as phosphate. In the pH range characteristic of seawater, orthophosphate ions (PO_4^{3-}) are largely associated as HPO_4^{2-} (Kester and Pytkowicz, 1967). Under most circumstances the analytically defined fraction usually described as reactive phosphate probably approximates closely to the concentration of the orthophosphate ion and its inorganic associations. Organic compounds of phosphorus, produced by biological processes, form a significant, and in some cases a dominant, part of the total dissolved phosphorus under certain conditions, notably in water bodies sustaining a high level of productivity. There has been little progress towards the characterization of these organic compounds but they can be expected to include a number of substances which contain orthophosphate groupings, such as phospholipids, sugar esters and the polynucleotide chains of nucleic acid, and polyphosphate groupings as in pyridine nucleotides. Since aminophosphonic acids, containing carbon-phosphorus bonds, have been found to occur in some marine organisms, including plankton (Kittredge, Horiguchi and Williams, 1969), they may also be expected to contribute to the pool of dissolved organic phosphorus. The relative importance of these various compounds has not been established. Strickland and Solorzano (1966) were unable to detect esterase hydrolyzable phosphate at the level of 0.03 μg-at P/l in open oceanic waters. In inshore waters the maximum value found was 0.5 μg-at P/l in samples influenced by a decaying red-tide bloom, and values generally amounted to less than 30% of the inorganic phosphate concentrations. Concentrations of inorganic condensed phosphate appear to amount to less than 0.05 μg-at P/l in unpolluted seawater (Armstrong and Tibbetts, 1968; Solorzano and Strickland, 1968).

There is little detailed information on the rates of conversion of dissolved organic phosphorus to inorganic phosphate. The low concentrations of the organic fraction during the winter in waters which show a marked seasonal variation in phytoplankton production suggest that mineralization is quite rapid but there have been observations of substantial amounts of dissolved organic phosphorus in some deep-water samples. Comprehensive measurements using modern analytical approaches are required to establish the distribution of the fraction more definitively.

As with the organic nitrogen compounds it must be borne in mind that the analytically determined fraction of dissolved organic phosphorus is not a single entity from either the chemical or metabolic standpoints and that it contains compounds which are turned over at different rates. There is, however, no evidence that significant amounts of organic phosphorus compounds occur in the long-term refractory pool of dissolved organic material.

Silicon

In the cycle of nutrient utilization and regeneration the behaviour of silicon is still simpler than that of phosphorus since, whereas the latter involves the production and subsequent mineralization of organic forms, the major cycle of silicon involves only inorganic forms. It essentially amounts to the production of opaline silica, by utilization of dissolved silicon, and its dissolution following the death of organisms. The requirement of silicon by diatoms is not, however, entirely limited to skeletal formation. Darley and Volcani (1969) showed that the element is needed for DNA synthesis in the diatom *Cylindrotheca fusiformis*.

Substantial dissolution of biogenous silica takes place in the upper part of the water column (Hurd, 1972; Wollast, 1974) and this can be very significant in terms of the recycling of silicon in depleted surface waters (cf. p. 340).

Silicic acid (H_4SiO_4) remains largely undissociated under the conditions normally encountered in seawater, and use of the term dissolved silicon is thus preferable to that of dissolved silicate which has, however, been widely used in the literature. The analytical methods used for the determination of dissolved silicon measure monomeric silicic acid and polymers of short chain length which are reactive under the conditions employed. Burton, Leatherland and Liss (1970) showed that polymeric forms of silicic acid are rapidly depolymerized when added to seawater and they were unable to detect any unreactive forms of dissolved silicon in ocean waters.

The Distribution of the Principal Micronutrients and the Processes by which they arise

The concentrations and some general features of the processes affecting the distributions of the principal micronutrient elements were discussed in Chapter 2.* The major large-scale feature in their distributions is the marked depletion in the mixed layer and pronounced increase in concentration in intermediate and deep waters. These features are illustrated by the vertical profiles for major oceanic regions in Fig. 7.5 (see also Fig. 2.5, p. 56). The most rapid changes in concentration with depth are associated with the region of the permanent thermocline and vertical distributions are more uniform in regions where this feature is absent. The distributions of phosphate and nitrate are rather similar since, when considered on a sufficiently extensive spatial and temporal basis, nitrogen and phosphorus are utilized and regenerated in a relatively constant ratio (see p. 326). In vertical profiles of phosphate in the Atlantic Ocean a maximum is apparent at intermediate depths, somewhat below the minimum in the vertical distribution of dissolved oxygen. This feature is observable in southern parts of the Indian Ocean but is less well defined or absent in other regions. The vertical distributions of nitrate show less clear maxima and where they occur they tend to be slightly deeper than for phosphate. Vertical profiles of dissolved silicon, whilst showing broadly similar trends, show differences in detail from those of phosphate and nitrate. In particular,

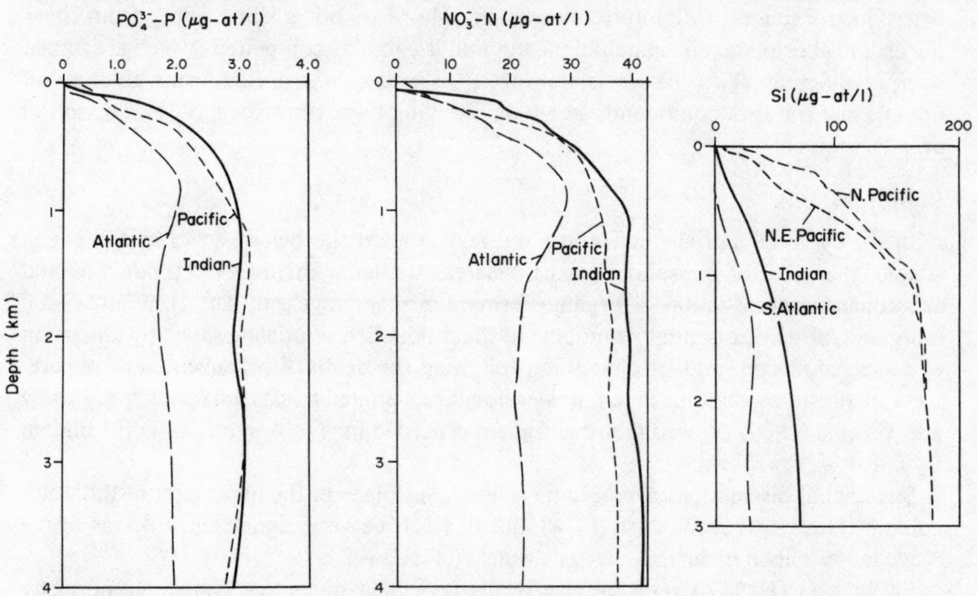

Fig. 7.5. Vertical distributions of the principal micronutrients in the major oceanic regions (from Sverdrup *et al.*, 1942).

* As discussed in Chapter 2, concentrations of the principal micronutrients are frequently given in μg-at/l, rather than μg/l, and these units are used here for the various forms of nitrogen, phosphorus and silicon. Conversions are as follows: 1 μg-at N/l = 14.008 μg N/l; 1 μg-at P/l = 30.975 μg P/l; 1 μg-at Si/l = 28.09 μg Si/l.

they generally show an essentially monotonic increase in concentration with depth over the whole profile. The profiles displayed in Fig. 7.5 should be considered only as illustrative of the general trends in major regions. Depending on localized factors, and particularly upon the water masses present at various depths, variations in detail occur within relatively small areas. These are exemplified by the comparison shown in Fig. 7.6 of profiles of dissolved silicon in the Antilles Arc region.

The deep waters of the various major oceanic regions show distinctive differences in their concentrations of the trace nutrients. The large vertical fluxes of the elements generated by biological uptake in surface waters and downwards transport in association with sinking particles lead to an increase in concentration along the direction of the major circulation of deep water from the North Atlantic Ocean through the South Atlantic, Indian and South Pacific Oceans to the North Pacific. Despite the complicating influence of other features in the deep-water circulation, the variations in deep-water concentrations broadly reflect the combined effects of the above processes (Broecker,

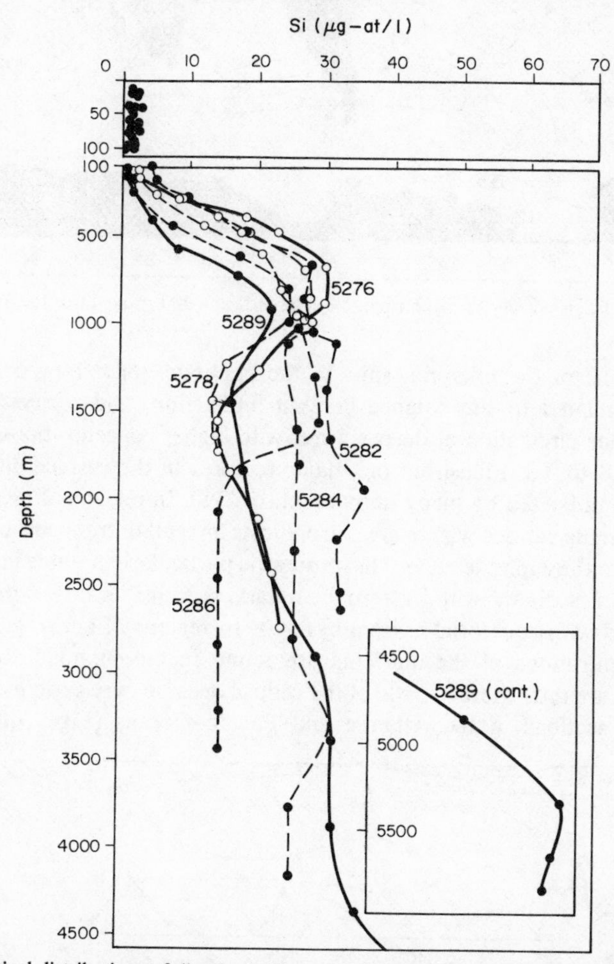

Fig. 7.6. Vertical distributions of dissolved silicon in the Antilles Arc region. Location of stations: 5276/8, Atlantic Ocean, east of Windward Islands; 5282/4, Venezuelan Basin; 5286, Jungfern Passage region; 5289, Puerto Rico Trench (after Richards, 1958).

1974), the deep Pacific Ocean waters containing about twice as much phosphate and nitrate, and five times as much dissolved silicon, as the deep Atlantic Ocean waters. The distribution of phosphate in the World Ocean at 2000 m, shown in Fig. 7.7, illustrates the changes in concentration.

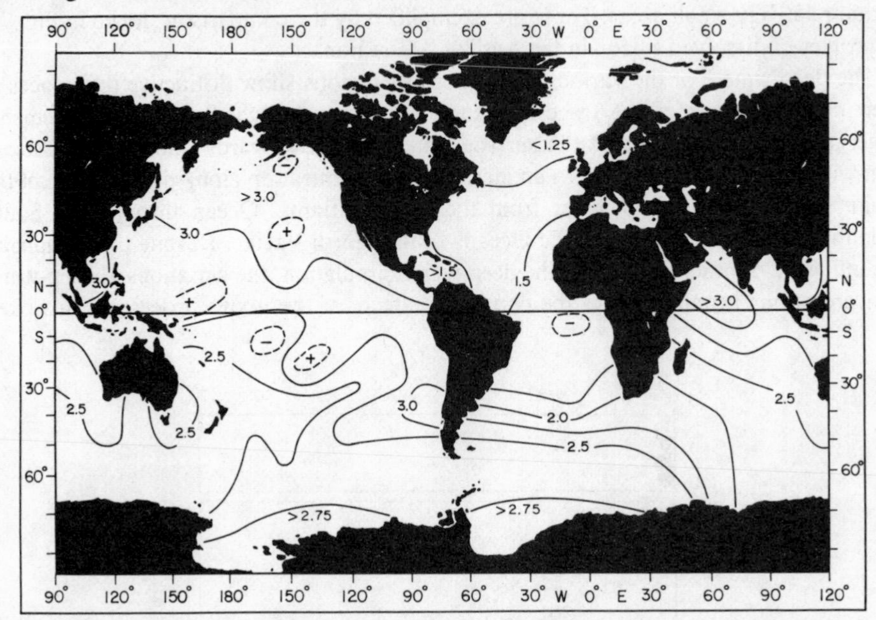

Fig. 7.7. Distribution of concentrations of phosphate (μg-at P/l) at 2000 m depth in the World Ocean (from Redfield, 1958).

The concentrations of the micronutrients in the euphotic zone show considerable spatial variations, related to the balance between utilization, and replenishment by upwelling and surface circulation of deeper water with higher concentrations. Data for phosphate (Figs. 7.8 to 7.11) illustrate the major features in the surface distributions which have been summarized by many authors (cf. p. 298). In regions of divergence in the water circulation the surface waters are eutrophic as a result of transport of nutrient-rich deeper waters to the euphotic zone. This process is particularly notable in the major areas of upwelling associated with eastern boundary currents, such as that off the Peruvian coast, and with equatorial current systems. In regions of convergence rather uniform low concentrations of the nutrients are found in the euphotic zone. These oligotrophic waters are thus characteristic of the central oceanic gyre systems.

The meridional sections in the Atlantic and Pacific Oceans (Figs. 7.8 and 7.9,

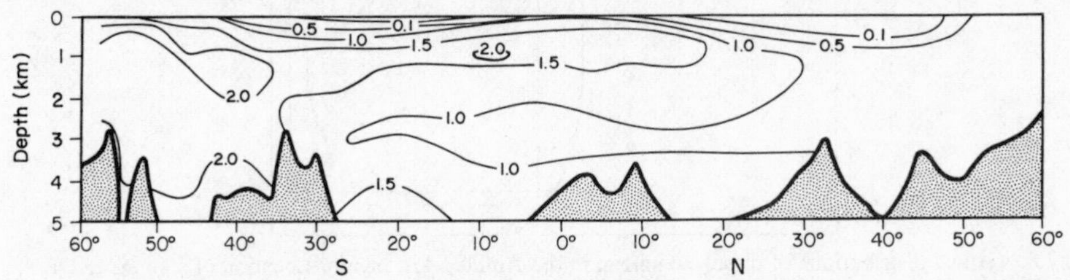

Fig. 7.8. Variations in concentration of phosphate (μg-at P/l) with latitude and depth in a longitudinal section of the central Atlantic Ocean (from Sverdrup *et al.*, 1942).

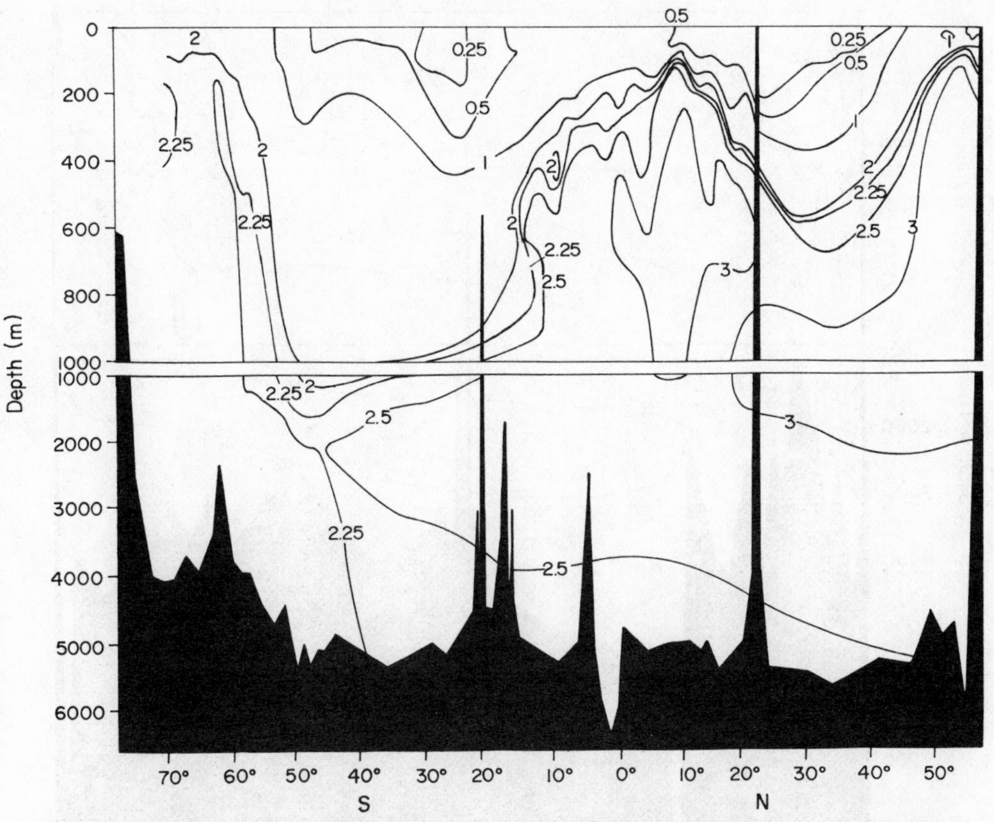

Fig. 7.9. Variations in concentration of phosphate (μg-at P/l) with latitude and depth in a longitudinal section (*ca.* 160°W) of the Pacific Ocean (from Reid, 1965).

respectively) show clearly the high concentrations of phosphate in surface waters at high latitudes, the oligotrophic tropical regions, and the increased concentrations in the equatorial regions. The extent of the oligotrophic regions is emphasized by the latitudinal section in the Pacific Ocean shown in Fig. 7.10, which also illustrates the pronounced effects of upwelling in enriching the surface waters in the eastern margin, off Peru. These figures demonstrate the variable depth of the nutrient-poor surface layer, which is thickest in the mid-latitudes in each hemisphere. Where the surface layer is thin the gradient in concentration below it tends to be steepest.

The influence of circulation upon surface nutrient concentrations is illustrated also by the data given by McGill (1973) for the western Indian Ocean (see Fig. 7.11). These data were obtained during the period of the south-west Monsoon. At this time the surface water is moved away from the African coast by the strong monsoon winds and is replaced by water from greater depths with higher concentrations of nutrients, leading to pronounced surface enrichment off the coast around 10°N. A marked contrast is evident between the eutrophic sector in the northern waters of the area surveyed and the oligotrophic waters which dominate most of the region.

The pronounced oligotrophic character of the surface waters of the Mediterranean Sea is shown by the data given in Fig. 7.12. They show also the low concentrations in the deeper waters, reflecting the restricted exchange of the landlocked water with the

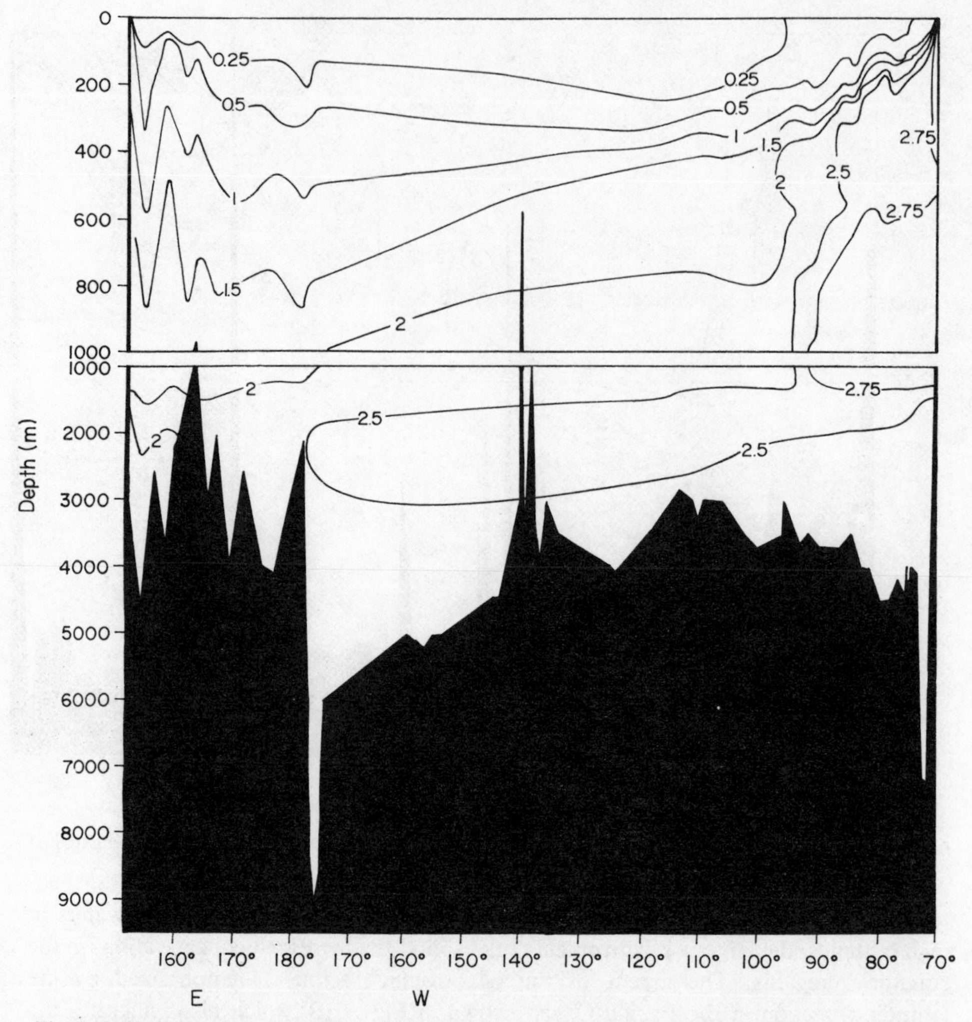

Fig. 7.10. Variations in concentration of phosphate (μg-at P/l) with longitude and depth in a latitudinal section (*ca*. 27°S) of the South Pacific Ocean (from Reid, 1965).

waters of the adjacent Atlantic Ocean. In these circumstances exchanges of surface water with deeper water have comparatively little effect upon the fertilization of the euphotic zone.

In waters of the shelf areas the concentrations of nutrients are influenced by those in the bodies of seawater entering the region and the local effects of discharges from the continents. Such a situation is illustrated by the data for the North Sea and adjacent regions, given in Fig. 7.13. The effects of major river inputs are seen clearly. In this region anthropogenic influences on certain river inputs of nitrogen and phosphorus are considerable. The southern North Sea receives amounts in river discharge which represent some 50% of the total phosphorus input to the area from all sources and some 60% of the total input of combined nitrogen (Postma, 1973).

Irrespective of man's influence, the concentrations of phosphate and nitrate, and also that of dissolved silicon, are characteristically higher in river waters than in surface seawater of the shelf regions and this leads to a general enhancement of nutrients in estuaries and in coastal waters influenced by land drainage. Such effects are not, however, invariable. In some cases the concentration of a nutrient in the river input may be similar to or lower than that in the local seawater and there is some evidence (Liss, 1976) that under certain circumstances interactions in estuarine waters may "buffer" the concentrations of phosphate at a rather constant level. Nevertheless, local nutrient enrichments of the kind illustrated in Fig. 7.13 are important in many regions and not only in those where man's influence is manifested. Except where there is large-scale

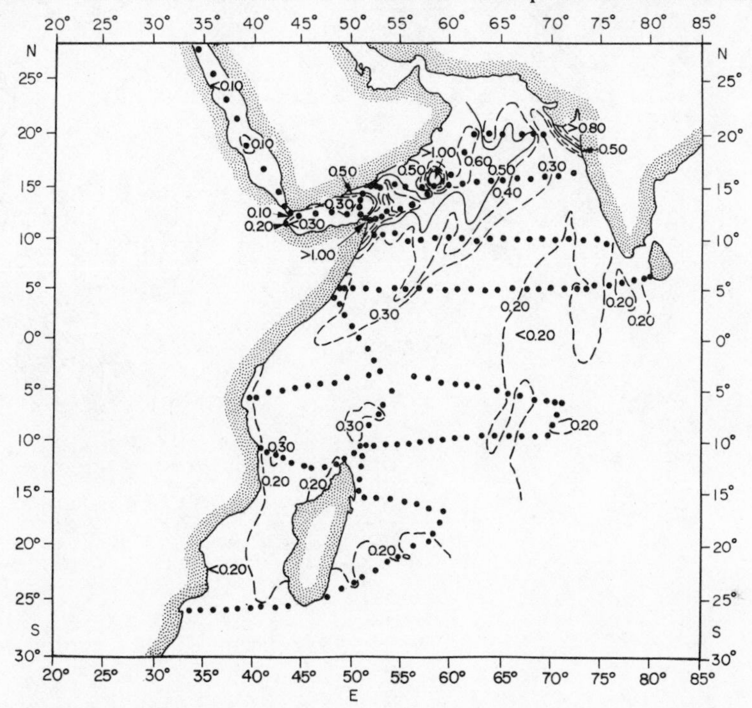

Fig. 7.11. Surface distribution of phosphate (μg-at P/l) in the western Indian Ocean during the period of the south-west Monsoon (from McGill, 1973).

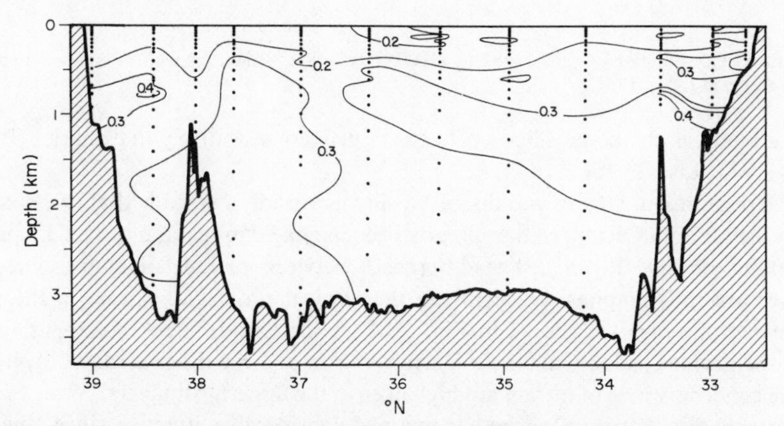

Fig. 7.12. Variation in concentration of phosphate (μg-at P/l) with latitude and depth in a longitudinal section (*ca.* 20°E) of the Central Mediterranean Sea (from McGill, 1970).

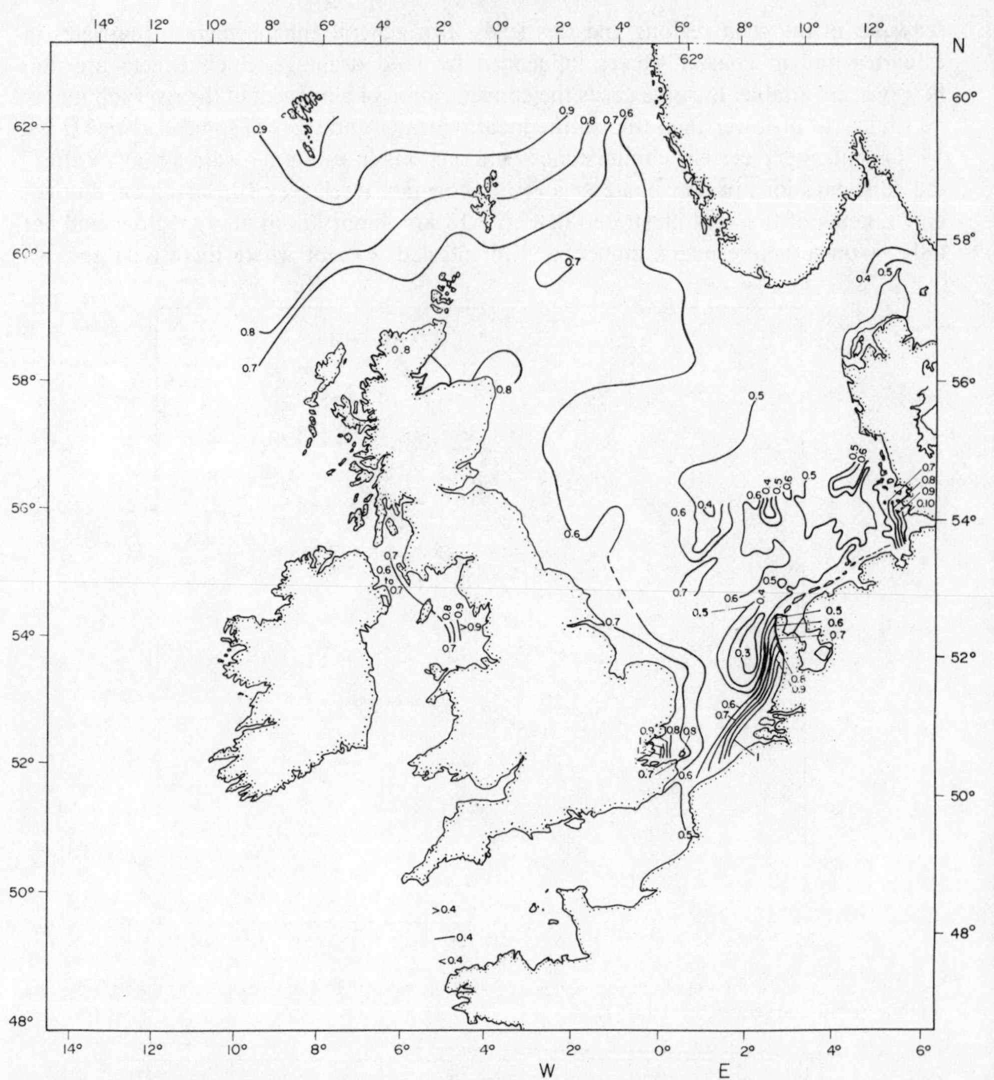

Fig. 7.13. Surface distribution of phosphate (μg-at P/l) in the North Sea during winter (January–February) (from Johnston and Jones, 1965).

biological removal, dissolved silicon behaves largely conservatively in estuaries (Burton and Liss, 1973; Liss, 1976).

The distributions of nitrate and dissolved silicon broadly resemble that of phosphate although they differ in detail, as has already been indicated in relation to vertical profiles in the major oceanic regions. The differences between regions of convergence and divergence are thus emphasized again by the vertical profiles of nitrate in the upper 180 m of various regions, shown in Fig. 7.14. A marked difference is apparent also between the profile for the California Current region and the other areas of divergence where the concentrations of nitrate are high even in the uppermost layers.

The oceanic distribution of nitrite has received considerable attention since, due to its role as an intermediate in the inter-conversion of nitrate and ammonium, its presence in

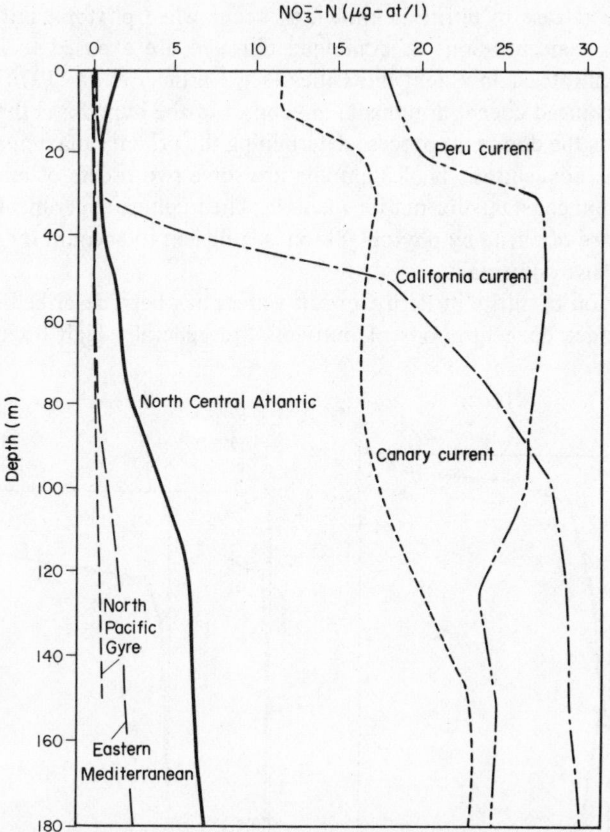

Fig. 7.14. Vertical distributions of nitrate (μg-at N/l) in various regions (data of Walsh, after Dugdale, 1976).

significant amounts provides an indication of biological processes involving such interconversion. A widely observed feature in oceanic waters is a significant accumulation of nitrite in a narrow depth range around the base of the euphotic zone, with concentrations ranging from 0.1 to 2.5 μg-at N/l. This primary nitrite maximum has been observed in the Sargasso Sea (Vaccaro and Ryther, 1960) and in a number of areas in the Pacific Ocean (Brandhorst, 1958, 1959; Wooster, Chow and Barrett, 1965; Kiefer *et al.*, 1976); examples are given in Fig. 7.15. It occurs at a similar depth to the chlorophyll maximum. The observations of Kiefer *et al.* (1976) in the central North Pacific show clearly a maximum located in the zone where the nitrate concentration is beginning to increase with depth (see Fig. 7.16). The high concentrations of nitrite in the primary maxima are found in association with variable concentrations of dissolved oxygen, ranging up to at least 5 ml/l, and this establishes that they cannot be the result of extracellular reduction of nitrate.

Brandhorst (1958, 1959) suggested that the peak in nitrite was due to bacterial oxidation of ammonia. Experimental studies in which an increased production of nitrite has been observed when samples containing added ammonium are maintained in the dark have been interpreted as supporting this hypothesis (Wada and Hattori, 1971; Hattori and Wada, 1972; Miyazaki, Wada and Hattori, 1973). Vaccaro and Ryther (1960) suggested that the maximum resulted from the release of nitrite by phytoplankton (cf. p.

302). Increased release of nitrite is known to occur when phytoplankton which have adapted to low concentrations of combined nitrogen are exposed to increased concentrations of nitrate at low light intensities (see Carlucci *et al.*, 1970). Kiefer *et al.* (1976) have advanced cogent arguments in support of the hypothesis that release from phytoplankton is the dominant process determining the primary maximum, particularly the fact that the concentrations of ammonia are some two orders of magnitude below the half-saturation constants for marine bacteria. Their culture experiments indicate that the rate of release of nitrite by phytoplankton is sufficient to account for the concentrations observed environmentally.

The distribution of nitrite in Peru Current waters has been described by Wooster *et al.* (1965). Surface concentrations of nutrients are generally high because of coastal

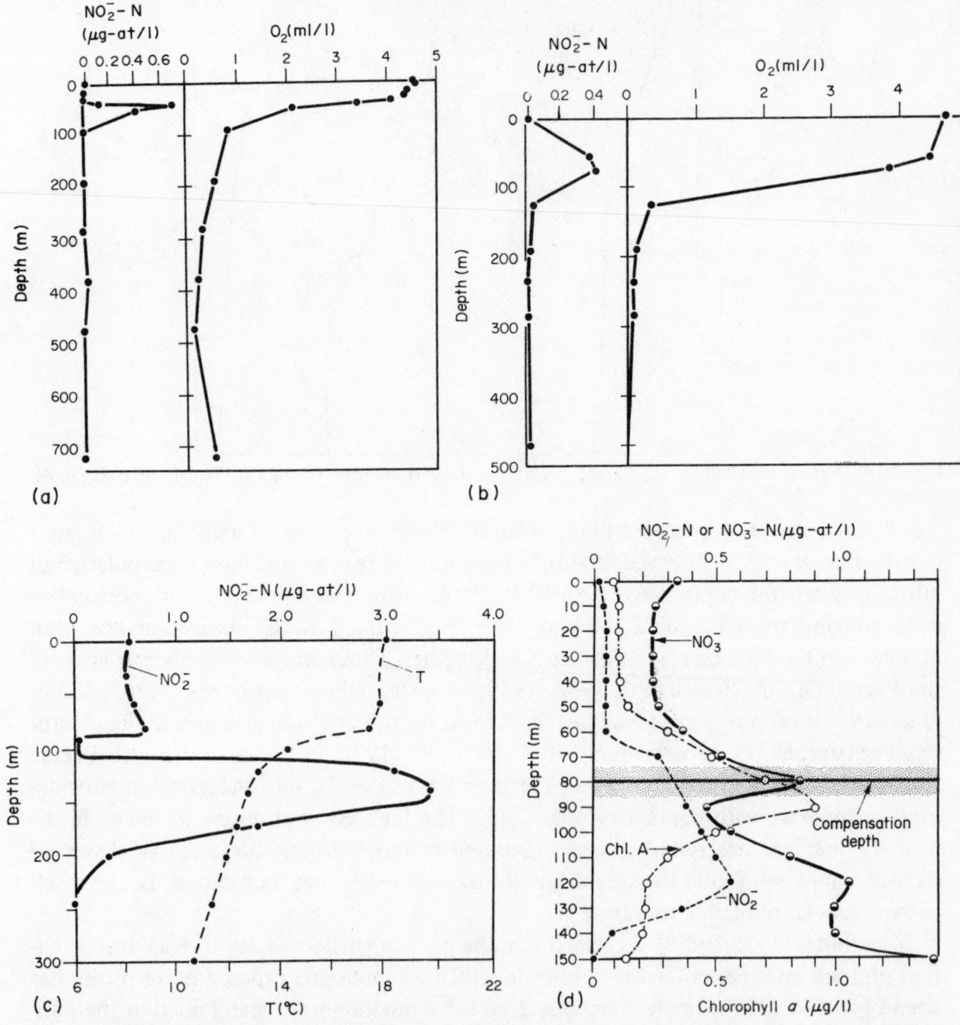

Fig. 7.15. Primary maxima of nitrite at various stations and their relationships to other oceanographic features: (a) East Pacific, 6°N, 80°W; (b) East Pacific 23°N, 113°W; (c) East Pacific, 12°S, 82°W; (d) Sargasso Sea, *ca.* 25 km north of Bermuda. ((a), (b) from Brandhorst, 1959; (c) from Wooster *et al.*, 1965; (d) data of Menzel, from Vaccaro and Ryther, 1960.)

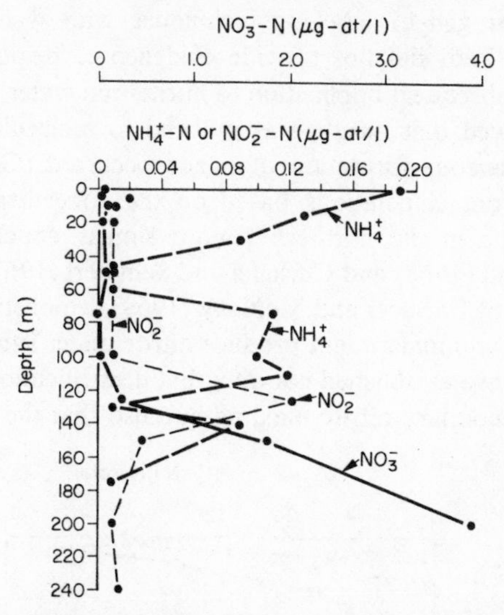

Fig. 7.16. Primary maximum of nitrite at 28°N, 155°W and its relationship to the vertical distributions of other forms of combined inorganic nitrogen (from Kiefer *et al.*, 1976).

upwelling and those of nitrite exceed 0.5 μg-at N/l in places. At some stations relatively high and uniform concentrations occur throughout the near-surface waters. The horizontal distribution of values in the primary maxima broadly resembles that for the surface values except that there is an offshore region with exceptionally high concentrations, up to 2.5 μg-at N/l, in the primary maxima.

A secondary nitrite maximum at greater depth has been observed in certain regions. It is restricted to, but not a necessary feature of, water bodies with very low concentrations of oxygen (<0.1 ml/l). This phenomenon was first described in detail by Brandhorst (1958, 1959) for the eastern tropical Pacific Ocean, having been recognized earlier in the northern Arabian Sea by Gilson (1937).

In the eastern tropical Pacific Ocean there is an extensive oxygen minimum layer on both sides of the equator in the depth range of 100 to 1000 m. This layer occurs in coastal waters between about 30°N and 25°S with tongue-like seaward extensions, that in the north hemisphere reaching the mid-ocean area. In these waters the concentrations of oxygen fall below 0.01 ml/l in places (Cline and Richards, 1972). A secondary nitrite maximum below the thermocline in the north-eastern tropical Pacific extends from 8°N to 21°N, with a maximum westward extension to 118°W. In the region of the Peru Current, secondary maxima, sometimes two in a single profile, have been observed south of 10°S (Wooster *et al.*, 1965). They occur in the upper half only of the rather thick oxygen-poor layer of this region which itself has a northern limit of 2–3°S in the area studied. The high values of nitrite are associated with the salinity maximum of the Peru–Chile undercurrent. Circulation is important also in determining the occurrence of secondary nitrite maxima, containing up to 3.9 μg-at N/l, in the Arabian Sea region (Deuser, Ross and Mlodzinska, 1978). Examples of secondary nitrite maxima are given in Fig. 7.17. The distribution of nitrite in a section in the Pacific Ocean off the coast of South America is given in Fig. 7.18; the primary and secondary maxima are apparent.

Brandhorst (1959) suggested that the secondary nitrite maximum arose by bacterial reduction of nitrate under the conditions of very low oxygen concentration. Initial experiments using nitrogen-15 labelling techniques with water from the secondary maximum region off Peru did not provide evidence of denitrification (Goering and Dugdale, 1966) but subsequent application to nitrite-rich water from below the thermocline off Mexico showed that nitrate was reduced to molecular nitrogen and that in some samples simultaneous formation of nitrite occurred (Goering, 1968). Thomas (1966a) concluded from calculations based on the concentrations of nutrients that denitrification occurred in the northern region; similar conclusions were drawn by Fiadeiro and Strickland (1968) and Carlucci and Schubert (1969) regarding the area off Peru. While the work of Carlucci and McNally (1969) demonstrated the possibility that bacterial oxidation of ammonia might produce nitrite under sub-oxic conditions, subsequent investigations have established not only that denitrification plays a dominant role in the formation of secondary nitrite maxima but also that the process is significant in

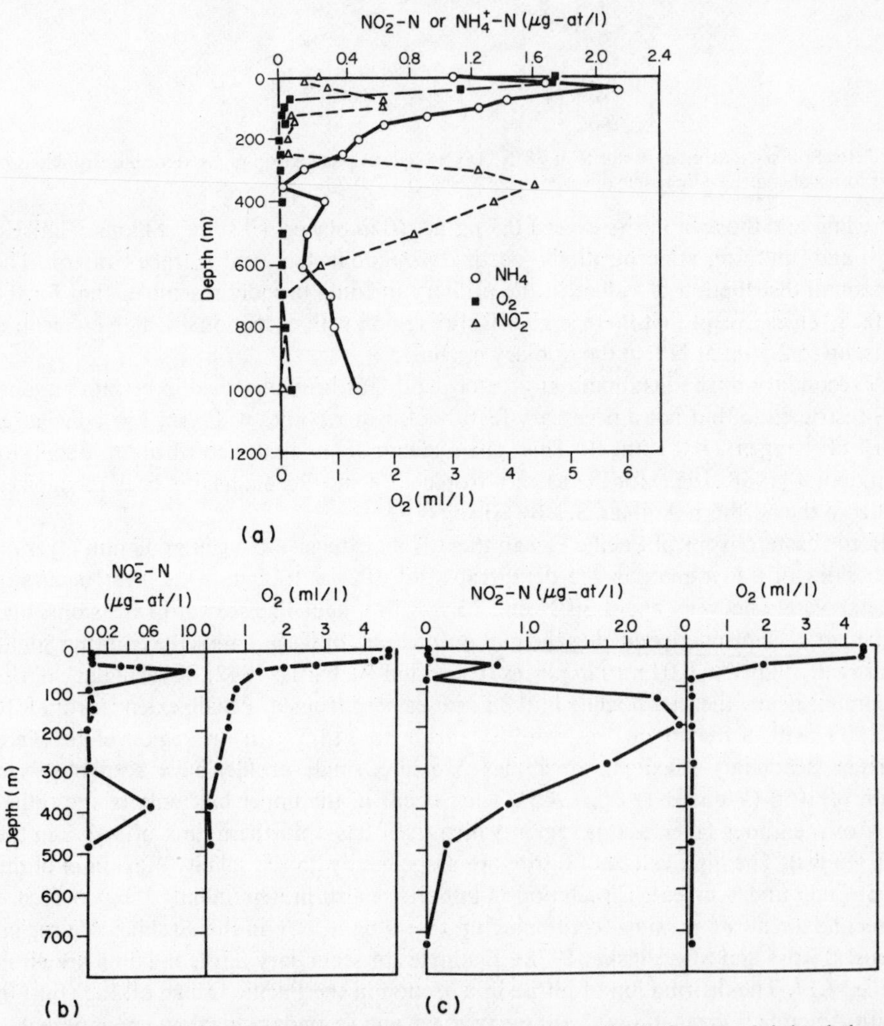

Fig. 7.17. Secondary maxima of nitrite at various station in the East Pacific Ocean and their relationships to other oceanographic features: (a) 13°N, 92°W; (b) 9°N, 86°W; (c) 20°N 106°W. ((a) from Cline and Richards, 1972; (b), (c) from Brandhorst, 1959.)

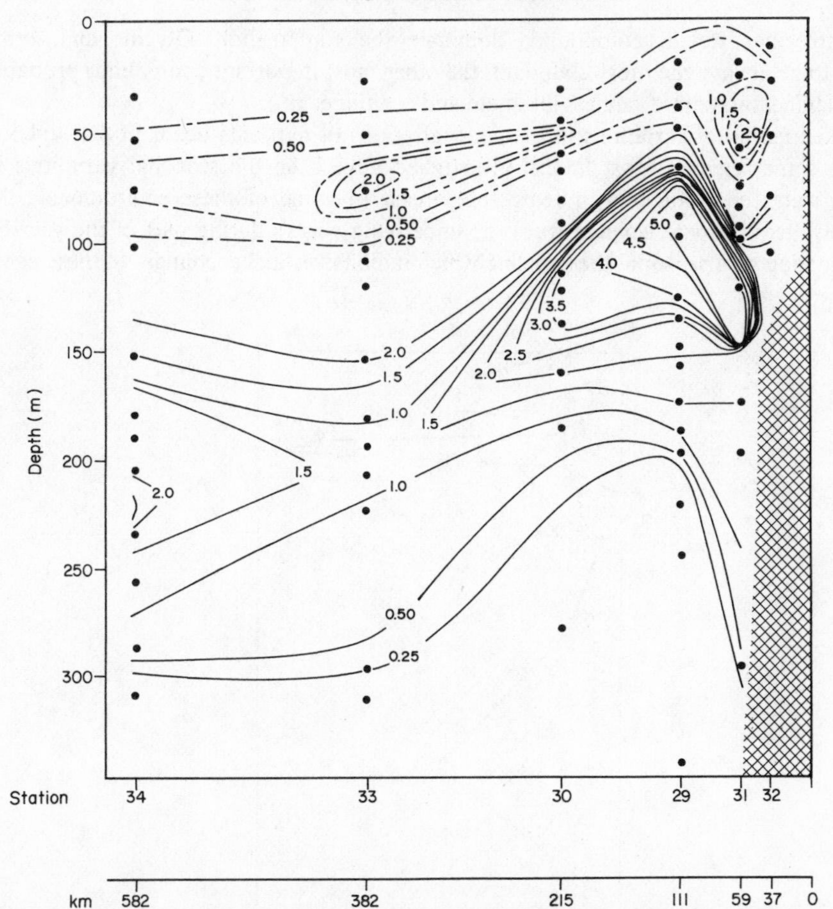

Fig. 7.18. Section off the Peruvian coast, showing both primary (slotted line) and secondary (continuous line) nitrite (μg-at N/l) layers (from Fiadeiro and Strickland, 1968).

global terms in converting nitrogen to molecular nitrogen (Cline and Richards, 1972; Cline and Kaplan, 1975; Codispoti and Richards, 1976; Deuser *et al.*, 1978).

Ammonium often constitutes an important part of the total resources of combined nitrogen in the euphotic zone in oligotrophic regions and during periods of seasonal depletion of nutrients in temperate waters. Vertical distributions of ammonium in relation to either nitrate or nitrite are shown for the north-western Sargasso Sea (Fig. 7.19) and for New England coastal waters (Fig. 7.20). Similar data for Peruvian coastal waters are shown in Figs. 7.16 and 7.17(a). In many of these profiles, maxima in the concentration of ammonium are apparent.

Information has been obtained only relatively recently on the concentrations of urea and organic nitrogen compounds. Remsen (1971) reported values for various American coastal waters to be generally in the range of 0.25–5 μg-at N/l. Some of his data for vertical profiles are given in Fig. 7.21. Measurements by McCarthy (1970) gave values for surface waters off California and Peru which ranged from the detection limit to 1 μg-at N/l.

The concentration of total free amino-acids in seawater is usually about 0.2 to 1 μg-at N/l. Variations with depth and location seem to be relatively small (see Williams, 1975).

Low molecular weight compounds dominate the composition. Glycine and serine appear to be always the most abundant, the other most important compounds probably being alanine, threonine, valine, glutamate and ornithine.

Marked temporal variations in the concentrations of nutrients occur in the euphotic zone in some regions. Most intensively studied have been the seasonal variations in shallow-water environments in temperate areas. In some of these environments the surface waters become isolated from the underlying waters during part of the year by the formation of a seasonal thermocline which inhibits vertical exchange. In these cases

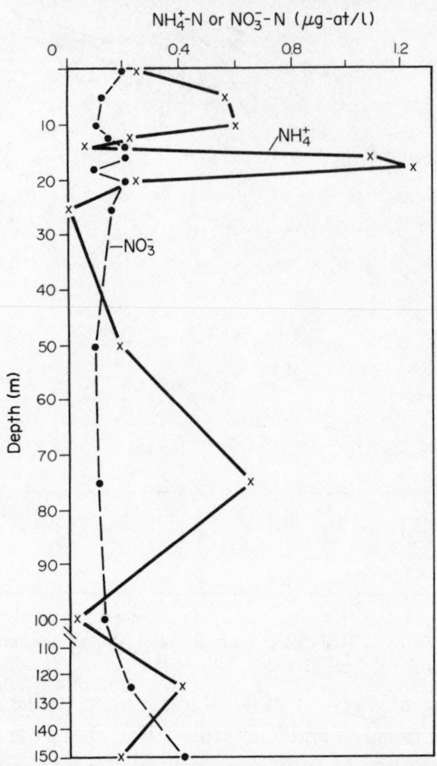

Fig. 7.19. Vertical distribution of ammonium and nitrate at a station in the Sargasso Sea (from Menzel and Spaeth, 1962a).

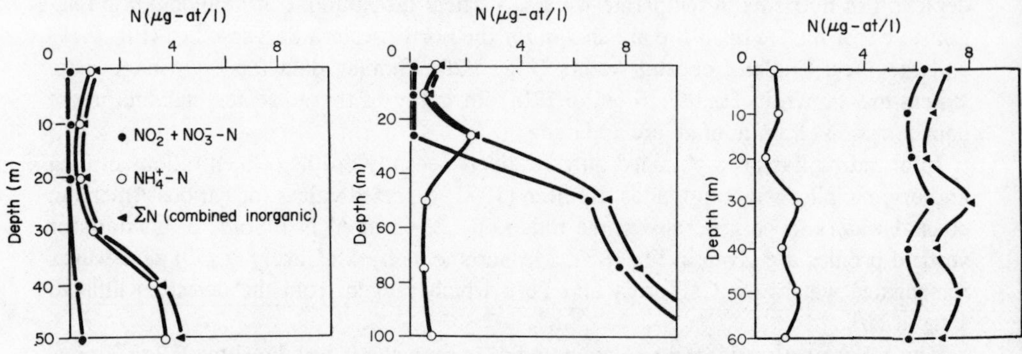

Fig. 7.20. Vertical distributions of ammonium and other nitrogen species at stations in New England coastal waters (from Vaccaro, 1963).

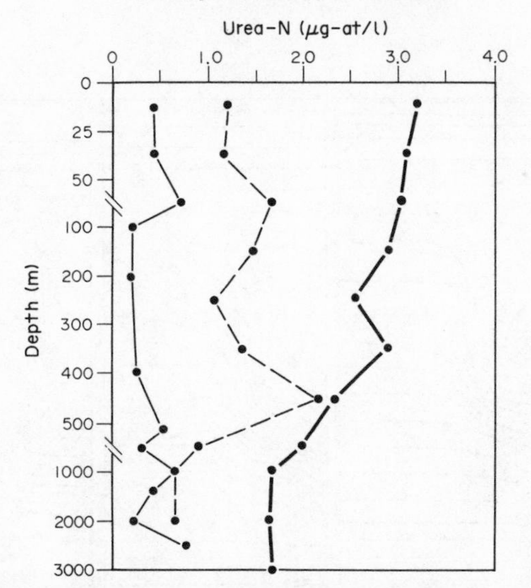

Fig. 7.21. Vertical distributions of urea in three regions: (a) Sargasso Sea, 32°N, 64°W (unbroken line); (b) between Cape Cod and Cape May (slotted line); (c) Peruvian coast (heavy line) (from Remsen, 1971).

the effects of seasonal variations in utilization of nutrients do not influence the whole water column, but in some areas a seasonal thermocline does not develop and concentrations are relatively uniform over the vertical profile.

Seasonal variations influenced also by the development of a thermocline are well illustrated by data for the English Channel. In Fig. 7.22(a) the concentrations of phosphate at the International Hydrographic Station E1 are shown, based on approximately monthly sampling in the years 1962–5 (Armstrong and Butler, 1968). They show the essentially uniform vertical concentrations during the winter period when the water column is generally well mixed, and the marked depletion in the surface concentrations following intensive utilization by phytoplankton in the spring and summer. In these years the thermocline was established in the May–June period (see Fig. 7.22(b)). Detailed patterns of development varied but in the first of the 3 years shown the thermocline developed initially at 15–20 m, deepened during the summer and disappeared in the October–November period. The maximum concentrations of phosphate at this station have varied considerably over a prolonged series of measurements. In the years illustrated they were lower than average and the winter maximum value of 0.41 µg-at P/l for the water column in 1965 was the third lowest since the measurements of phosphate were begun in 1922. Changes in dissolved silicon show a pattern which is broadly similar but with more erratic variations, probably reflecting variable coastal influences.

Essentially similar seasonal trends have been observed in other temperate shelf seas. Many studies have been made in waters along the eastern coast of the United States. An example is given in Fig. 7.23, which shows seasonal changes in the stratification of the water column and in the concentration of nitrate for a section off the coast of New England. In the winter, concentrations are high and relatively uniform in the shelf waters reflecting the lack of stratification; this is attributable to cooling of the surface water and the effects of increased mixing through wind stress. Nitrate concentrations are relatively

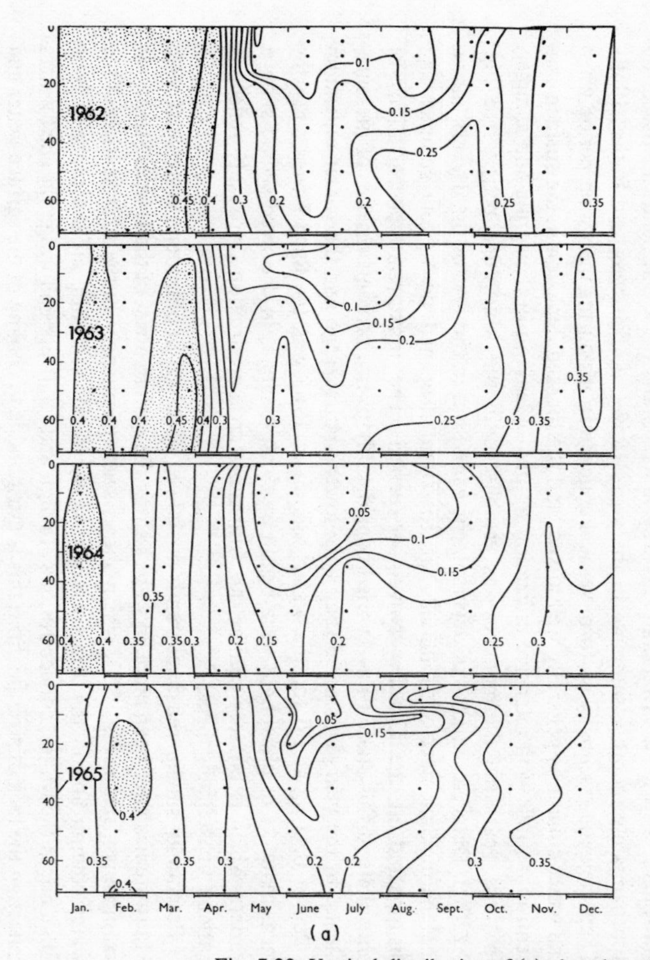

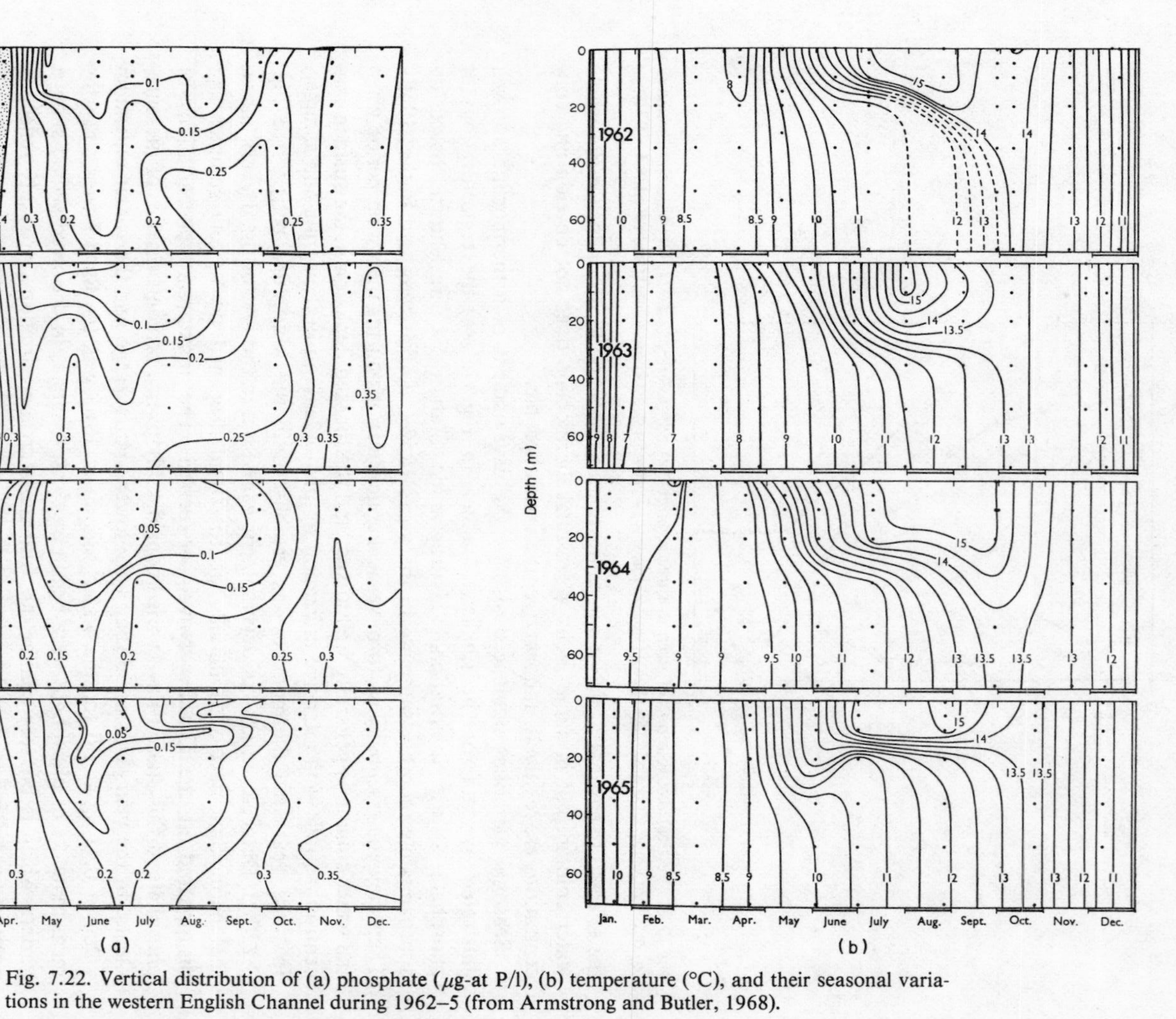

Fig. 7.22. Vertical distribution of (a) phosphate (μg-at P/l), (b) temperature (°C), and their seasonal variations in the western English Channel during 1962–5 (from Armstrong and Butler, 1968).

uniform to depths of some 100 m. Nutrient utilization occurs with the spring bloom of phytoplankton and by the early summer marked stratification has developed with a warm mixed layer of about 25 m depth, which is eventually depleted of nitrate. Phosphorus follows similar trends (Ketchum, Vaccaro and Corwin, 1958) but with a difference in the timing as it is not depleted as early in the summer as is nitrate. Other studies in this area include those of Riley and Conover (1956), Harris (1959) and Vaccaro (1963).

Information on seasonal cycles in tropical waters, much of which has been summarized by Armstrong (1965a) and Corner and Davies (1971), presents a very variable picture. In some environments, concentrations of nutrients show little change seasonally. The large variations found in certain areas are related primarily to drastic climatic changes, particularly monsoon conditions and storms. In near-shore waters, events such as massive run-off from the adjacent land, and the stirring up of bottom material may influence concentrations. In areas with a shallow permanent thermocline high concentrations of nutrients can exist not far below the euphotic zone so that increased turbulence can lead to abrupt changes in the concentrations in the mixed layer. Seasonal variations in the north-western Sargasso Sea were studied by Menzel and Ryther (1960). Concentrations of nitrate and phosphate in the euphotic zone were low throughout the year. The maximum values of *ca.* 2 μg-at N/l and 0.2 μg-at P/l occurred in the winter and were associated with the breakdown of the seasonal thermocline which allows mixing with water down to the permanent thermocline. This water, while richer than that present in the euphotic zone at other seasons, is still poor in nutrients in absolute terms. These data are illustrated and discussed in Chapter 9. Near-shore waters in the same region were investigated by Beers and Herman (1969). Their results, shown in Fig. 7.24, show a somewhat similar pattern for nitrate to that in the more open waters but the phosphate maxima appear to be out of phase with those for nitrate.

Seasonal variations in the concentrations of nitrate and phosphate are accompanied by large changes in the relative importance of other forms of the elements concerned. These are well illustrated by the data of Vaccaro (1963) shown in Fig. 7.20. The concentrations of ammonium in the waters off New England remain fairly constant throughout the year at about 1 μg-at N/l. In August this represents the most abundant supply of combined nitrogen in the upper layers whereas in the winter it constitutes less than 20% of the total combined nitrogen. In a parallel manner organic phosphorus can be an important part of the total dissolved phosphorus under conditions where the inorganic phosphate is depleted.

In the Sargasso Sea region, ammonium shows variations in concentration on several time-scales. Seasonal variations were examined by Menzel and Spaeth (1962a) and Beers and Herman (1969). The findings of the latter workers, for Bermuda coastal waters, are shown in Fig. 7.25; the periods of higher concentrations tended to coincide with those when phosphate was also high. Diurnal variability in waters off Bermuda was studied by Beers and Kelly (1965). The average concentrations in the euphotic zone were high during the mid and late morning and low around midnight. From 200 to 500 m, the values were maximal after noon and minimal in the early morning. The variations were probably explicable largely in terms of the patterns of diurnal vertical migration of zooplankton which excrete ammonium, and assimilation of this source by phytoplankton.

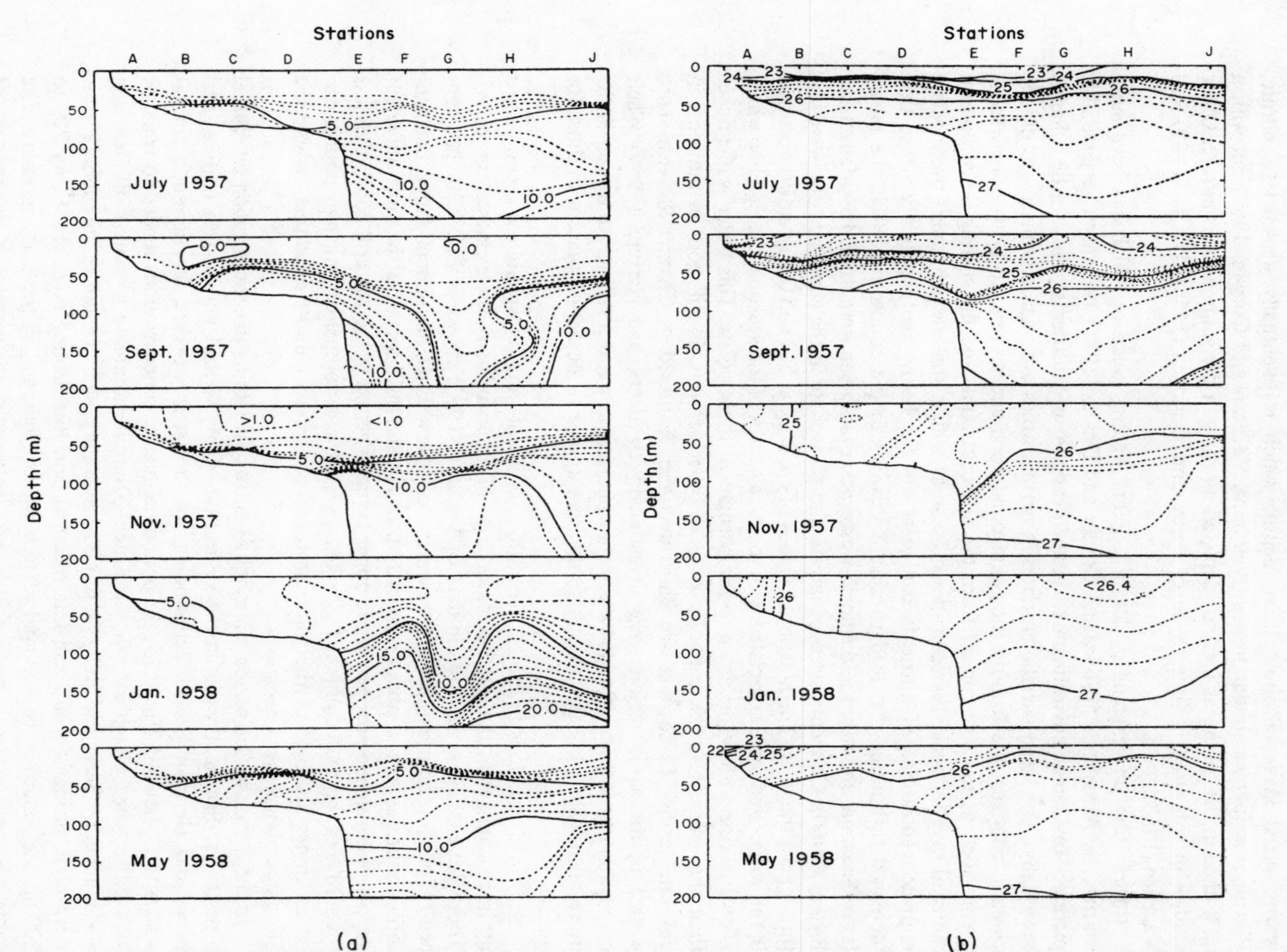

Fig. 7.23. Distribution of (a) nitrate (μg-at N/l), (b) sigma-t, and their seasonal variations, in a section off the coast of New England (from Ketchum *et al.*, 1958).

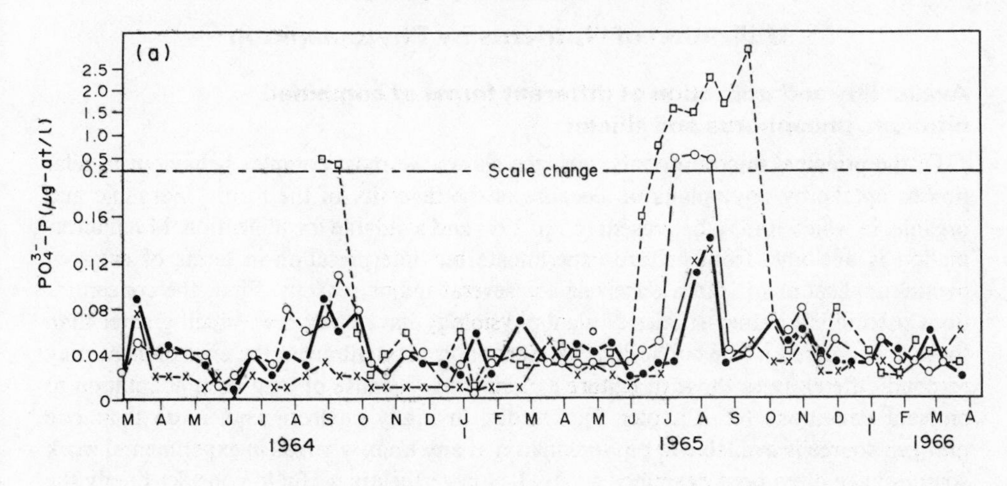

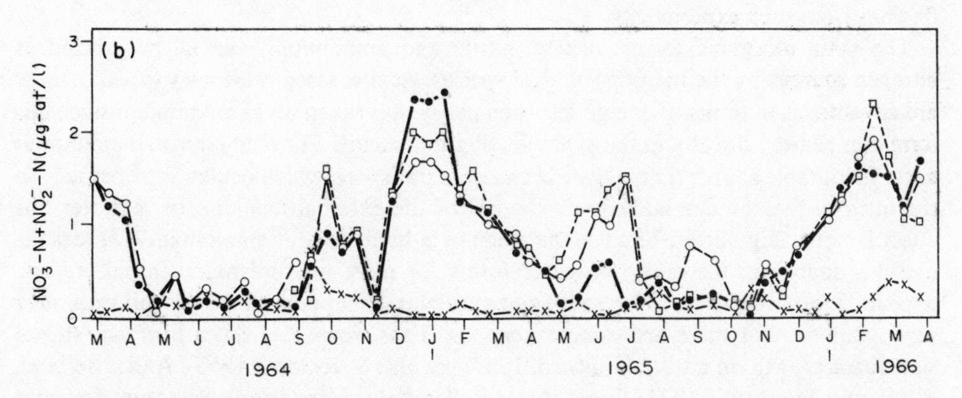

Fig. 7.24. Seasonal variations in (a) phosphate (μg-at P/l), (b) nitrate (μg-at N/l), in Bermudan coastal waters. (Harrington Sound; ●, 1 to 14 m; ○, 21 m; □, 25 m; North Lagoon: ×, 1 to 12 m) (from Beers and Herman, 1969).

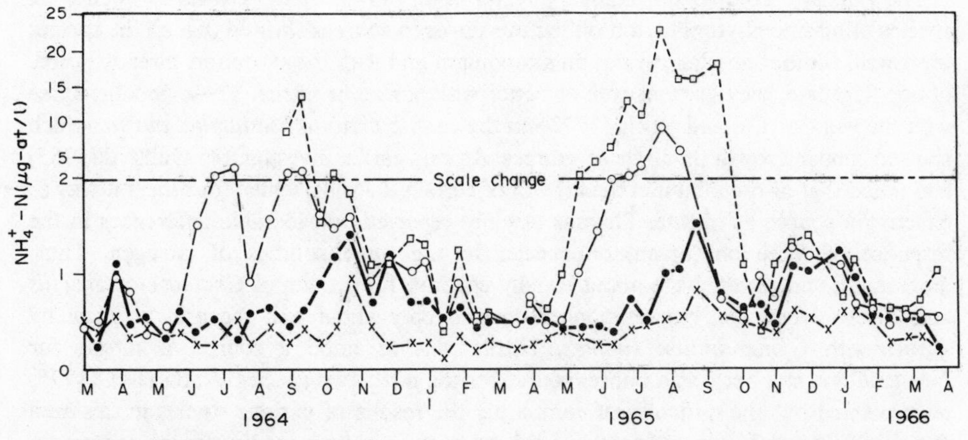

Fig. 7.25. Seasonal variation in ammonium (μg-at N/l) in Bermuda coastal waters. For key see Fig. 7.24 (from Beers and Herman, 1969).

Utilization of Nutrients by Phytoplankton

Availability and utilization of different forms of combined nitrogen, phosphorus and silicon

Of the principal micronutrients, nitrogen shows the most complex behaviour in relation to uptake by phytoplankton because of the diversity of the forms, inorganic and organic, in which it may be present (cf. p. 299) and available for utilization. Much information is available from culture experiments but interpretation in terms of environmental implications is often uncertain for several major reasons. First, the concentrations used in laboratory studies of algal physiology have often been much greater than those encountered in the euphotic zone. Secondly, populations in the environment may respond differently to those in culture experiments, because of long-term adaptation to ambient conditions (cf. Chapter 3). Thirdly, in many environments more than one nitrogen source is available to phytoplankton at any time, whereas in experimental work sources have often been examined singly. It is nevertheless useful to consider briefly the findings from such experiments.

The main inorganic forms, nitrate, nitrite and ammonium, can all be utilized as nitrogen sources by the majority of algal species. As discussed previously (p. 302), there are advantages in terms of energetics when nitrogen is taken up as ammonium since this form can be used directly in the synthesis of amino-acids. For land plants ammonium is a less favourable source than nitrate because its transport, which occurs preferentially in the form of free ammonia, reduces the pH of the external medium. In seawater this effect is not so significant because the medium is buffered and ammonium is at least as useful a source as the other inorganic forms for most phytoplankton in culture and appears, moreover, to be universally utilizable. Nitrate supports good growth with most phytoplankton but there are some exceptions. Thus, some flagellates have not shown significant growth on nitrate (Guillard, 1963; see also Strickland, 1965). Antia, Berland, Bonin and Maestrini (1975) found that one flagellate, *Hemiselmis virescens*, among a number examined, was unable to utilize nitrate and that this organism could also not grow on nitrite. Ammonium and to a lesser extent nitrate can be taken up in the dark at low rates by cells deficient in nitrogen (see, for example, Dugdale and Goering (1967)).

The extensive comparison made by Antia *et al.* (1975) of the growth of twenty-six species of marine phytoplankton on various nitrogen sources showed that all the species grew well, without any lag period, on ammonium and with the exception, already noted, of one flagellate, they grew as well or better with nitrate or nitrite. These findings agree with the work of Lui and Roels (1972) on the centric diatom, *Biddulphia aurita*, which showed similar growth on all three sources. Among earlier investigators Guillard (1963) had noted that nitrite inhibited the growth of certain diatoms, whilst for others it was as effective a source as nitrate. Thomas (1966b) reported considerable differences in the response of three phytoplankton species to the three sources of nitrogen. Thus, ammonium and nitrate were about equally effective for growth of *Chaetoceros gracilis* and *Nannochloris* sp., but ammonium yielded only about half the growth given by nitrate with *Gymnodinium simplex*. Nitrite was as good a source as nitrate for *Nannochloris* sp., but was a poorer source for the other two species. Antia *et al.* (1975) have pointed out the difficulty of comparing the results of various studies in this area due to differences in experimental conditions, such as culture media, source concentrations and criteria of growth. The problem of comparability is emphasized by the fact that the measurements by Antia *et al.* were made on cells which were not starved of nitrogen, whereas Thomas (1966b) worked with nitrogen-deficient cells. It is apparent, however, that most species can utilize the three main inorganic forms, although with some species the forms may not be equally effective. Failure to grow on nitrate or nitrite may be related to the absence of the cellular reductase systems for reduction to ammonia (Antia *et al.*, 1975).

Studies have also been made of the interactions between the inorganic sources in relation to phytoplankton growth. Guillard (1963) observed that growth on nitrate was reduced by the presence of ammonium, an effect shown also in other culture experiments (Morris and Syrett, 1963; Grant, Madgwick and Dal Pont, 1967; see also McCarthy and Eppley, 1972). This effect is probably explained by the repression of the nitrate reductase system when high concentrations of ammonium are present. The experiments of Eppley, Coatsworth and Solorzano (1969) indicate that this repressive effect is unlikely to be very significant at concentrations of ammonium up to about 1 μg-at N/l and hence its general environmental significance is uncertain. MacIsaac and Dugdale (1972) give a somewhat lower value of about 0.5 μg-at N/l as the lower limit for significant inhibitory effects of ammonium. With estuarine phytoplankton, Bates (1976) found that nitrate was taken up, although at a reduced rate, at concentrations of ammonium above 1 μg-at N/l. The effects of light conditioning probably had an important influence on the extent to which nitrate uptake was inhibited. Concomitant uptake of nitrate and nitrite or ammonium has been shown to occur in culture experiments with some planktonic diatoms (Eppley and Rogers, 1970; Eppley and Renger, 1974). In the absence of inhibitory effects, as in the studies with single sources, culture experiments indicate that there is generally little difference in growth regardless of the nature of the nitrogen sources. Caution is needed, however, in extrapolating these findings to natural environments with sub-inhibitory concentrations of ammonium. In the culture experiments abundant light energy was available. The thermodynamic advantage of ammonium utilization may be more relevant under environmental conditions.

Considerable information has also been obtained from culture experiments concerning the ability of phytoplankton to grow on urea and a variety of organic nitrogen compounds; much of this work has been summarized by Antia *et al.* (1975). The responses of phytoplankton to these substances vary much more than with the inorganic sources. Urea is at least as effective as ammonium for the growth of many species (see, for example, Thomas, 1966b and Antia *et al.*, 1975) although some species appear unable to utilize it. There is evidence that urea uptake is suppressed by high concentrations of ammonium (McCarthy and Eppley, 1972; McCarthy, 1972). Wheeler, North and Stephens (1974) found that most of the twenty-five algal species which they examined in axenic cultures could utilize several of the nine amino-acids tested at a concentration of 1 mg-at N/l, although growth was generally somewhat below that with an equivalent concentration of nitrate. Two species of *Platymonas* grew well on glycine, serine, alanine, aspartate, threonine, valine, glutamate and ornithine (a group which includes the most abundant amino-acids in natural seawater) but not on lysine (which is one of the less abundant amino-acids in the sea). *Stephanopyxis* and *Cyclotella* could not utilize any of these compounds. Between these extremes there was a spectrum of versatility in utilization but nine of the species could use at least half the number of compounds examined. There was no evidence to suggest that the effects of individual compounds were significantly modified by the presence of other amino-acids. It was established that a species that could grow on the high concentration of an amino-acid which was supplied could generally also take up the compound at low concentration (*ca.* 1 μg-at N/l). Antia *et al.* (1975) found that about half of the twenty-six species which they examined grew on glycine, growth being comparable with that on ammonium for eight species. The purine, hypoxanthine, was a more versatile nitrogen source and gave growth comparable to that with ammonium for some species. Glucosamine was an ineffective source although it has been reported to be utilized by certain diatoms which produce chitan, a material similar to chitin (McLachlan and Craigie, 1966). North (1975) studied the uptake of primary amino-nitrogen in California near-shore waters using *Platymonas*. This alga is able to utilize a variety of amino-acids at rates which are markedly increased under conditions of nitrogen limitation (North and Stephens, 1971). About 50% of the primary amino-nitrogen was utilizable; it was suggested that the remaining material may have been peptides.

It thus seems probable that many algae are able to use a variety of nitrogen sources at environmental concentrations. Variations in the relative proportions of different forms of nitrogen between environments, or at different times in a particular environment, may be more important as an influence upon the species composition of a population and species succession, than in determining the total productivity.

Assimilation of phosphorus by phytoplankton is generally accounted for by uptake of orthophosphate but it appears that some organic phosphorus compounds can be used (see, for example, Harvey, 1953 and Provasoli and McLaughlin, 1963). The work of Kuenzler and Perras (1965) suggests that organic phosphate esters are utilized by enzymatic hydrolysis at the cell surface with uptake of the orthophosphate thus released. The presence of orthophosphate in the culture medium inhibits the cellular production of alkaline phosphatase but it is produced under conditions of phosphate deficiency. Some algae can utilize polyphosphate in the presence of excess nitrate (Parsons, Takahashi and Hargrave, 1977), perhaps also by extracellular hydrolysis.

It appears that the dissolved silicon in seawater is wholly present in chemically reactive forms (cf. p. 305) and there is no reason to suppose that any of it is unavailable for utilization. There have, however, been observations in culture experiments that nett uptake of dissolved silicon can cease before the medium is entirely depleted of the element (cf. p. 333). The assimilation of silicon is unusual by comparison with that of other nutrients in that it occurs almost wholly during a specific part of the cycle of cell division, and coupling between silicification and cell division has been shown in culture experiments using marine diatoms on light/dark cycles (Chisholm, Azam and Eppley, 1978). Uptake of silica continues at significant rates over 24-hour cycles (Goering, Nelson and Carter, 1973; Azam and Chisholm, 1976; Nelson and Goering, 1978), contrasting in this respect with carbon fixation and nitrogen assimilation. The requirement of silicon for DNA synthesis in a diatom species has already been noted (cf. p. 305). Uptake of silicon occurs in *Platymonas* sp. and some other algae which have no known requirement for the element (Fuhrman, Chisholm and Guillard, 1978).

Relative utilization of nitrogen, phosphorus and silicon

It was recognized relatively early in the study of nutrients that the ratio of nitrogen to phosphorus (and also to carbon) is generally similar in plankton collected in various areas and at various times and, moreover, that the N/P ratio is similar to that of combined nitrogen to phosphorus in seawater (Redfield, 1934; Cooper, 1937). Data given by Redfield, Ketchum and Richards (1963) indicate an average atomic ratio of nitrogen to phosphorus of 16, with a value for phytoplankton alone of 15.5. These values closely resemble the slope of the line fitted to observations of the relationship between the concentrations of nitrate-nitrogen and phosphate-phosphorus in samples from the western Atlantic Ocean (Fig. 7.26). The usefulness of these general relationships is greatest when considering the overall cycle of nutrients, i.e. in relation to large-scale utilization and the relationship between the regeneration of nutrients and the consumption of dissolved oxygen (cf. Chapter 2). From the previous discussions it is clear that very wide variations can occur in N/P ratios in the euphotic zone and the concept of a quasi-stoicheiometric composition for plankton is valid only on a wide-scale average basis. These facts were emphasized by Ryther and Dunstan (1971) in drawing attention to what they regard as the very limited relevance of the approach when considering the behaviour of nutrients in the euphotic zone. Despite these reservations, it is worthwhile to consider briefly information on the proportions in which nutrients may be taken up from seawater.

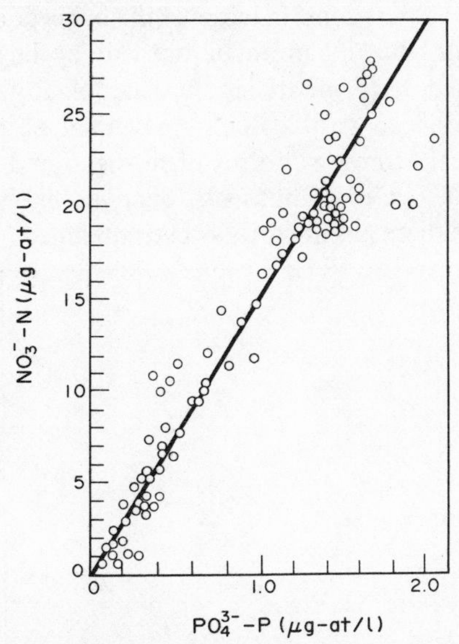

Fig. 7.26. Relationship between concentrations of nitrate (μg-at N/l) and phosphate (μg-at P/l) in waters of the western Atlantic Ocean (from Redfield, 1934).

Very wide variations in the N/P atomic ratio can be brought about in culture experiments by variation of the composition of the medium. Phytoplankton exhibit the phenomenon of so-called luxury consumption (cf. p. 334) when a nutrient is present in excess in the medium, and under conditions of nutrient-deficiency cells with a much reduced proportion of the deficient nutrient can be produced. Thus Redfield *et al.* (1963) cite data obtained with the freshwater alga, *Chlorella pyrenoidosa*, which demonstrate a thirteen-fold variation in the N/P ratio in cells, between extreme limits of phosphorus deficiency and nitrogen deficiency.

In surface waters generally, and coastal areas particularly, there may also be N/P ratios in seawater which are very different from the characteristic values. Thus, Pratt (1965) reported ratios as low as unity for Narragansett Bay. In shallow coastal environments it is possible in some circumstances that phosphate may be released from suspended material or sediments when the water concentrations become low, so that phosphorus is replenished without a concomitant supply of nitrogen; a review of evidence on this "buffering" effect is given by Liss (1976).

Information on the quasi-stoicheiometry of changes brought about by biological utilization in natural waters has been summarized by Redfield *et al.* (1963) and Corner and Davies (1971). They point out that even in environments with atypical N/P ratios the relative changes due to biological activity may still reflect utilization within the range of normal proportions. From regression analysis of water concentrations the relative change $\Delta N/\Delta P$, described as the assimilation ratio, can be derived. If the assimilation ratio remains constant, then in environments with low N/P ratios phosphorus will still be available when nitrogen has been exhausted. This situation is illustrated by the data for coastal waters south of Long Island, shown in Fig. 7.27. Under the conditions of depleted nitrogen concentration, production of phytoplankton can occur with lower assimilation ratios (McAllister, Parsons and Strickland, 1960; Stefansson and Richards, 1963). It is, however, difficult to interpret adequately data from environments in which nutrients are at very low concentrations, in terms of the concept of an assimilation ratio, without full information on the dynamics of nutrient cycling. Phytoplankton may utilize any input of a deficient nutrient, arising, for example, by mixing of richer water, as rapidly as it can be provided so that the supply may be greater than would be indicated by water analyses. Moreover, forms of nitrogen and phosphorus which are intermediates in the overall cycles of nutrients, such as ammonia, may be directly utilized, and information on these has not always been obtained.

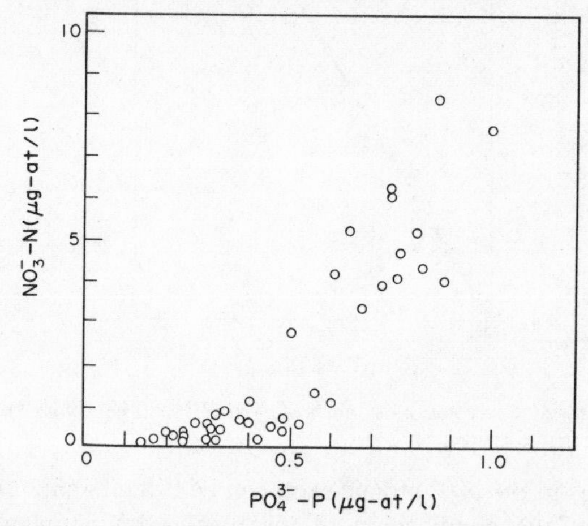

Fig. 7.27. Relationship between concentrations of nitrate (µg-at N/l) and phosphate (µg-at P/l) in coastal waters south of Long Island (from Ketchum *et al.*, 1958).

Although the underlying processes may be considerably more complex than is immediately evident from the apparently simple concept of the assimilation ratio, it does appear that the ratio of nitrogen to phosphorus in phytoplankton growing in an environment subject to considerable seasonal changes in composition sometimes remains surprisingly close to the characteristic value of 16. For example, Harris and Riley (1956) found ratios varying from 12.6 to 19.8 in phytoplankton from Long Island Sound over a period of 12 months. A very similar range has been observed during the course of an algal bloom in a large volume of water enclosed in a plastic sphere (Antia, McAllister, Parsons, Stephens and Strickland, 1963). These ranges are much smaller than might be expected on the basis of laboratory experiments with cultures. Attention has been drawn by Banse (1974) to the difficulties in deriving elemental ratios in plankton from the results of either large-scale culture experiments on enclosed water bodies or from field samples, particularly those arising through the presence of other organic material.

As discussed in Chapter 2 there is a quasi-stoicheiometric relationship not only between nitrogen and phosphorus in phytoplankton and zooplankton but also between these elements and carbon, the atomic proportions C:N:P being 106:16:1. Silicon is not an essential requirement of all organisms. It behaves in some respects independently in the cycle of biological material even though the general pattern of utilization and regeneration is similar. Assimilation ratios for silicon are of limited value but it is useful to have an approximate indication of its relative involvement in the biological cycle. Richards (1958) found that changes in the concentrations of silicon and phosphorus in western Atlantic Ocean waters correspond to an average ratio ($\Delta Si/\Delta P$) of 15, silicon thus being apparently assimilated in a very similar proportion as nitrogen. The corresponding atomic ratio Si/C of 0.14 is lower than that of 0.34 suggested by Strickland (1965) as typical of marine phytoplankton populations.

Kinetics of uptake of nutrients by phytoplankton

The detailed study of the kinetics of uptake of micronutrients by phytoplankton and the relationship between nutrient uptake and growth of phytoplankton has received considerable impetus in the last decade with the development of approaches which have allowed interpretation on a more quantitative basis. It is convenient to consider first the uptake of micronutrients.

The work of Ketchum (1939) on utilization of nitrate by *Nitzschia closterium* (= *Phaeodactylum tricornutum*) demonstrated several important features. First, the rate of uptake in the light measured over a short period depended upon the concentration of nitrate in the medium up to a certain level above which it was constant; similar features were found with phosphate. Secondly, while phosphate could be taken up when no nitrate was detectable in the medium, the uptake was increased when more nitrate was supplied.

The results of Ketchum, and also those of Harvey (1957), indicated a hyperbolic relationship between the rate of uptake of nitrate or phosphate and the concentration of the ion in the medium. Dugdale (1967) and Caperon (1967) suggested, respectively, that the rate of uptake of a limiting nutrient, and the specific growth rate, are related to nutrient concentration in the medium, by an equation of the form which describes the Langmuir adsorption isotherm and which corresponds also to the Michaelis-Menten equation used to describe enzyme kinetics. This equation was already known to describe the rate of growth as a function of the concentration of a limiting nutrient in continuous cultures of bacteria (Monod, 1942) and had been shown to apply to the uptake of substrates in some heterotrophic processes in seawater (Parsons and Strickland, 1962; Vaccaro and Jannasch, 1966). Such relationships are to be expected on the basis that uptake of nutrients involves a number of enzymatic processes. On this basis a similar relationship may be expected between nutrient concentrations in the medium and cell growth parameters and this question is considered in the following section (p. 333).

The Michaelis-Menten equation derives from the kinetics of enzyme reactions in which a substrate combines reversibly with an enzyme giving a complex which yields a product according to the schematic equation

$$S+E \rightleftharpoons C \rightarrow E+P. \tag{3}$$

The present use of the Michaelis-Menten equation is to relate the specific rate of nutrient uptake, V, which is identical to the rate of removal of nutrient from the medium, to its concentration (S) in the medium

$$V = -\frac{dS}{dt} = \frac{V_m.S}{K_s+S} \tag{4}$$

where V_m is the maximum specific rate of nutrient uptake for the particular conditions and K_s is the substrate concentration at which $V = V_m/2$. The values of V_m and K_s (usually termed either the half-saturation constant or the Michaelis constant) are constants for a particular system. The half-saturation constant, which is temperature dependent, gives the value of S which corresponds to an uptake rate of 50% of the maximum rate. It is an indication of the capacity of a particular species to take up a given nutrient at low concentrations and has come to be treated as determining the limiting concentration for the particular system. The relationship is shown schematically in Fig. 7.28(a). For the derivation of the constants it is convenient to use a linear transformation, one form of which, as shown in Fig. 7.28(b), is:

$$S = V_m\left(\frac{S}{V}\right) - K_s. \tag{5}$$

Thus, in Fig. 7.28(b) the intercept on the abscissa corresponds to $-K_s$ and the slope of the line is $1/V_m$. The use of the linear form to estimate K_s and V_m has been criticized by Caperon and Meyer (1972b) on the basis that it involves weighting the original data; they recommend estimation by fitting to the hyperbolic form.

Dugdale (1967) emphasized the importance of determining the kinetic parameters, V_m and K_s, for phytoplankton characteristic of different regimes of productivity in order to

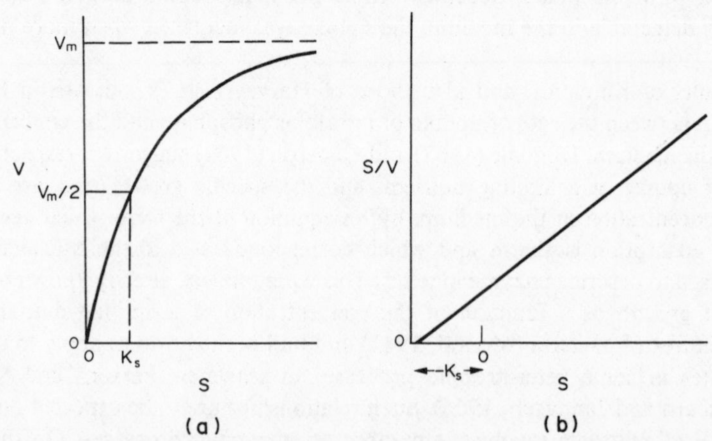

Fig. 7.28. (a) Schematic representation of the uptake of a nutrient as a function of its concentration in the external medium, according to the Michaelis-Menten equation. (b) A linear transformation of the same equation. For symbols see text.

clarify the significance of nutrient limitation especially with respect to the adaptation of populations to oligotrophic conditions, and responses to increases in nutrient concentrations, including anthropogenic inputs, in coastal waters. For example, the success of a particular species in competing for a nutrient depends upon both V_m and K_s, since, up to a certain nutrient concentration, a low value of K_s can compensate for a low value of V_m (see p. 335). Considerable effort has thus been devoted to their measurement using single organisms in culture, and natural populations.

Caperon and Meyer (1972b) have pointed out that significant differences in K_s between species imply differences in the mechanism of uptake. The constant K_s reflects the rate constants for the reactions involved in the process of uptake (equation (3)) and at a given temperature should always have the same value for a given mechanism. Thus, the application of the equation to a mixed population is valid only if the individual species have the same values of K_s and an assumption to this effect is implicit in use of the Michaelis-Menten equation with natural populations. Williams (1973) has investigated this question using a mathematical model. He showed that the simplification involved in the use of equation (4) for a heterogeneous population leads to under-estimation of uptake rates at low concentrations, the discrepancy becoming greater with increased diversity of the population. Where the kinetics of uptake by a mixed population show conformity to equation (4), however, the use of a single constant, whilst not rigorous, does give an empirical approach which is useful for comparative ecological purposes. Values of V_m, being related to the number of uptake sites per unit population, can be altered for a given uptake mechanism and population by adaptation to particular regimes of nutrient concentration, through the induction of enzymes involved in uptake.

Measurements of uptake of nitrogenous forms by natural populations have been possible through the development of sensitive techniques using nitrogen sources labelled by enrichment with the stable isotope ^{15}N. Samples are incubated in natural light with different concentrations of the labelled source and the uptake determined by mass-spectrometric measurements on the particulate material. An account of the technique and its limitations is given by MacIsaac and Dugdale (1969). The presence of nitrogenous detritus leads to under-estimation of V_m by this method. Problems arise if, within the range below V_m, uptake becomes limited by a nutrient other than that under investigation. This raises the fundamental question as to the nature of the interactive effects of two potentially limiting nutrients (cf. p. 335). It is not possible to measure K_s if the water sample has a concentration exceeding V_m for the nutrient under study. Eppley, Renger, Venrick and Mullin (1973) found it necessary to increase the base-line concentration of the nitrogen sources in oligotrophic seawater in order to obtain sufficient sensitivity by this method. Allowance can be made for this in the estimation of assimilation rates but the additions may lead to over-estimation because they may cause a rapid transient stimulation of uptake.

Initial studies employing the nitrogen-15 method with natural populations, using nitrate and ammonium, were made by MacIsaac and Dugdale (1969). The validity of the Michaelis-Menten equation was confirmed and it was found that the values of K_s varied with the nutrient and the productivity regime. Eppley and his co-workers (Eppley and Coatsworth, 1968; Eppley and Thomas, 1969; Eppley, Rogers and McCarthy, 1969) examined uptake of these nutrients in culture experiments by measuring changes in concentration in the culture medium; these also conformed to the Michaelis-Menten equation. Subsequent studies of these sources have been made on natural populations

(MacIsaac and Dugdale, 1972; Eppley *et al.*, 1973) and cultures (e.g. Carpenter and Guillard, 1971; Caperon and Meyer, 1972b; Eppley and Renger, 1974). It has been established that the kinetics of uptake of nitrite and nitrate by *Skeletonema costatum* are identical (Conway, cited by Dugdale, 1976). Uptake of urea by this alga also follows Michaelis-Menten kinetics (Carpenter, Remsen and Watson, 1972).

Investigations in this field have been reviewed by Dugdale (1976) whose summary of information on natural populations is given in Table 7.1. Some data obtained with single species are summarized in Table 7.2. These data show a striking difference between the

Table 7.1. Values of K_s and V_m for uptake of nitrate and ammonium observed for natural populations in oligotrophic and eutrophic areas[(a)]

Region		Nitrate-N		Ammonium-N	
		K_s (μg-at/l)	V_m (hr^{-1})	K_s (μg-at/l)	V_m (hr^{-1})
Oligotrophic					
Tropical Pacific	(1)	0.07	0.003	0.42	0.007
Pacific central gyre	(2)	—	0.003	—	0.005
North Pacific central gyre	(3)	—	—	0.15	—
Mediterranean	(2)	0.1–0.3	0.002	<0.1	0.004
Eutrophic					
Tropical Pacific	(1)	0.98	0.036	1.3	0.036
Sub-Arctic Pacific	(1)	4.2	0.016	1.3	0.036
Peruvian coast	(2)		0.024	1.1	0.012

[(a)] After Dugdale (1976). Original sources, indicated by numbers in parentheses in the table, are as follows: (1) MacIsaac and Dugdale (1969); (2) MacIsaac and Dugdale (1972); (3) Eppley *et al.* (1973). Data for V_m should be taken as illustrative of the differences between the areas as the original values were under-estimated by probably at least 30%.

Table 7.2. Values of K_s for uptake of nitrate and ammonium by phytoplankton in culture[(a)]

Organisms	Nitrate-N	Ammonium-N
	K_s (μg-at/l)	K_s (μg-at/l)
Oceanic species	0.1–0.7	0.1–0.5
Neritic diatoms	0.4–5.1	0.5–9.3
Neritic or littoral flagellates	0.1–10	0.2–2.4

[(a)] Range of values is for different organisms (or isolates). Data from Eppley *et al.* (1969).

values of K_s and V_m for populations in oligotrophic and eutrophic areas; this difference applies to both nitrate and ammonium. They correspond broadly with the results for individual organisms in that K_s tends to be low in small-celled oceanic species. The values of K_s for oligotrophic and eutrophic regions are similar, as would be expected, to the concentrations of nutrients characteristic of the areas concerned.

The kinetics of uptake of phosphate have been examined less extensively. Dugdale (1967) demonstrated that the data of Harvey (1957) on the response of *Phaeodactylum tricornutum* to added phosphate conform to the Michaelis-Menten relationship with $K_s = 0.33$ μg-at/l. Values for K_s of 0.58 μg-at/l and 1.71 μg-at/l have been obtained for *Thalassiosira pseudonana* and *T. fluviatilis*, respectively, in nutrient-limited culture experiments (Fuhs, Demmerle, Canelli and Chen, 1972).

Data for uptake of silicon by diatoms can also be interpreted in terms of the Michaelis-Menten equation. Values for K_s, summarized by Dugdale (1976), range from 0.2 to 3.4 μg-at/l. Paasche (1973), in experiments on uptake of silicon by five diatom species, observed that nett uptake of silicon ceases while significant concentrations remain in the medium, i.e. there is a threshold concentration for uptake. In equation (4) the term S then represents the difference between the medium concentration and the threshold concentration. In Paasche's experiments, threshold concentrations varied with species from 0.3 to 1.3 μg-at Si/l. A small threshold effect has also been detected in some studies on nitrate (Caperon and Meyer, 1972b). The results of initial studies on uptake of silicon by phytoplankton in the Peru upwelling system showed conformity to Michaelis-Menten kinetics; K_s and V_m were 2.93 μg-at Si/l and 0.075 hr^{-1}, respectively. Germanic acid labelled with radioactive ^{68}Ge has also been used as a tracer for the uptake of silicon by natural phytoplankton populations. The method was used by Azam and Chisholm (1976) with populations from the Gulf of California. Values of K_s for two stations were 1.59 and 2.53 μg-at Si/l.

The uptake of nutrients in the environment is affected not only by their concentration but also by interactions with other variables, particularly temperature and light. For example, uptake of either nitrate or ammonium under conditions where the concentration is not limiting shows an approximately hyperbolic increase with light intensity (MacIsaac and Dugdale, 1972).

Nutrient uptake and phytoplankton growth

The rate of growth of phytoplankton under conditions of nutrient limitation can be described by Michaelis-Menten kinetics but it has become recognized that the rate is only exceptionally determined directly by the concentration of nutrient in the medium. When this is the case we have the relationship:

$$\mu = \frac{\mu_m S}{K_{s(g)} + S} \qquad (6)$$

where μ = specific growth rate,
μ_m = maximum specific growth rate,
$K_{s(g)}$ = half saturation constant for growth.

Because values for the half-saturation constants for uptake can be made relatively easily the use of these constants in relation to growth would be attractive if it were valid. Eppley and Strickland (1968) pointed out that cells acquire differing concentrations of nitrogen and phosphorus at different stages of growth in cultures where nutrients are not limiting. Cellular concentrations are high during rapid growth and fall as growth decreases. This is demonstrated, for example, in the work of Kuenzler and Ketchum (1962). The initial high uptake in terms of cellular content has been termed luxury consumption, but such variations may be related to growth requirements in different phases. Under conditions of nutrient depletion in the medium, growth can occur at a lower rate with a reduction in the cellular concentrations of phosphorus and nitrogen. There is, however, a minimum cellular concentration below which cell division does not continue. Eppley and Strickland suggested that cellular concentrations of nutrients, rather than concentrations in the medium, might determine growth rates. As pointed out by Droop (1974), this seems reasonable also from the standpoint that cell surface enzyme systems are involved in uptake but the enzymatic reactions involved in growth occur within the cell.

Under steady-state conditions, the rate of nutrient uptake is proportional to the growth rate. The ratio V/μ gives the amount of nutrient per unit of population (i.e. the cellular nutrient content) and is termed the cell quota (Q), equivalent also to the demand coefficient if excretion is neglected. Using this relationship and equations (4) and (6), Eppley and Thomas (1969) showed that the specific growth rate is independent of the cell quota only when the half-saturation constants for uptake and growth are equal.

A large body of evidence has been obtained from continuous culture experiments, using chemostats, showing that Q varies with steady-state growth rate and that the relationships between Q and growth rate fit rectangular hyperbolas as with Michaelis-Menten kinetics. This evidence, substantiating the view that growth rates under limitation depend upon the cellular nutrient content, includes studies on vitamin B_{12} (Droop, 1968, 1970, 1973, 1974), nitrate and ammonium (Caperon, 1968; Caperon and Meyer, 1972a), phosphorus (Fuhs, 1969) and iron (Davies, 1970; Droop, 1973).

The work of Droop (1968) has led to the use in many subsequent studies of the following equation to express the relationship between growth and cellular nutrient concentration:

$$\mu = \mu_m \left(\frac{Q - k_q}{Q} \right) \tag{7}$$

where k_q is the minimum concentration of limiting nutrient per cell required for growth to occur.

While the importance of cellular nutrient content in determining growth rates has become fully recognized, environmental growth rates must be ultimately controlled by nutrient concentrations in the medium. Caperon and Meyer (1972b) developed a treatment of the relationship between steady-state growth rate and nutrient concentration in the medium, which takes into account the above considerations. Their approach involves the use of four simultaneous equations which define the relationship. These are:

(1) the two hyperbolic equations relating growth rate to intracellular nutrient concentration

$$\mu = \frac{\mu_m (Q - Q_0)}{K_q + (Q - Q_0)} \tag{8}$$

and nutrient uptake rate to nutrient concentration in the medium

$$V = \frac{V_m (S - S_0)}{K_s + (S - S_0)}. \tag{9}$$

These equations follow the previous nomenclature, the terms $(Q-Q_0)$ and $(S-S_0)$ allowing for the threshold effect (e.g. S_0 is the finite nutrient concentration in the medium at which the uptake rate is zero).

(2) The proportionality equations

$$V=\mu Q \tag{10}$$

and

$$V_m=b\mu. \tag{11}$$

The relationship between growth rate and nutrient concentration in the medium is approximately hyperbolic and was reasonably well fitted, with the data of Caperon and Meyer (1972b), by an equation, derived from the above, with some simplification:

$$\mu=\mu_m\frac{(S-S_0)-Q_0K_s/b}{(S-S_0)}. \tag{12}$$

In relation to growth measurements, the value of S_0 was important in these experiments since it was similar to the half-saturation constant for growth. When the half-saturation constants for uptake and growth were both expressed in terms of the concentration of nitrate in the medium, the latter value was the greater by a factor of 3.

Nutrient limitation and competition between species

The mathematical treatment of phytoplankton growth in terms of Michaelis-Menten kinetics gives a basis for the modelling of the competition between species for nutrient sources, which is important in relation to species diversity and succession. Dugdale (1967) demonstrated this for the competition of two algal species for nitrate, using hypothetical values for V_m and K_s (see Fig. 7.29). At high nitrate concentrations the species with high V_m is able to take up nitrate at a higher rate. When the concentration is depleted below a certain level, indicated by the crossover of the curves in Fig. 7.29, a species with a lower V_m can attain a higher uptake rate through a low value of K_s. Since species with low values of V_m show low values also of K_s, they are able to dominate in nutrient-poor regimes. In circumstances where nutrients are abundant, however, organisms with high values of V_m must dominate.

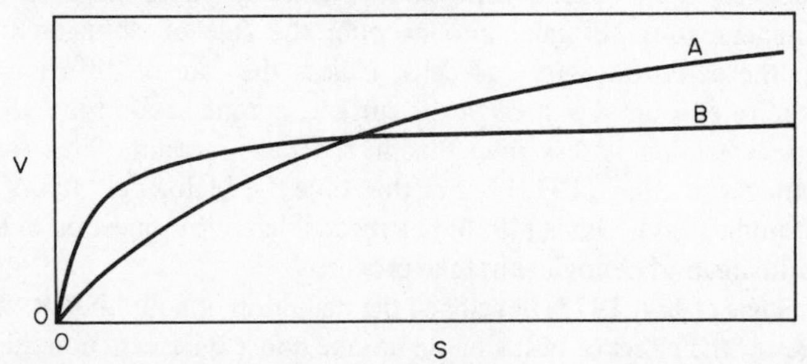

Fig. 7.29. Schematic representation of the uptake of a nutrient by two algal species with contrasting values of K_s and V_m. Species A, $V_m=0.068$/hr, $K_s=2.5$ μg-at/l; Species B, $V_m=0.034$/hr, $K_s=0.25$ μg-at/l (following Dugdale, 1967).

The question of competition between a number of species for several nutrients (or other substrates) has been examined theoretically, for steady-state continuous culture conditions, by Taylor and Williams (1975), assuming Michaelis-Menten growth kinetics. The question arises in such modelling as to whether, in a situation where a species competes for several nutrients, the effects on growth are multiplicative, when all nutrients have some controlling effect on growth, or whether growth is limited by a single nutrient at a time. Droop (1974) developed the equations necessary to test these hypotheses by chemostat experiments, and investigated the effects of two nutrients simultaneously on growth, using phosphorus and vitamin B_{12}, with *Monochrysis lutheri*. He found that growth followed more closely the threshold model in which a single nutrient controls growth. Both models were considered by Taylor and Williams (1975). Each leads to a similar qualitative conclusion for a multiple system of substrates and species under steady-state continuous growth conditions, namely that, within the constraints of the growth kinetics assumed, stable populations of several competing species can exist only if the number of growth-limiting substrates equals or exceeds the number of species. This suggests that in natural systems in which a single or a few nutrients are highly deficient, species diversity will be low. The model shows, however, that with many populations unsuccessful species will be eliminated slowly. In a chemostat the process of elimination could go to completion but in a natural environment changing conditions may impose fluctuating patterns of relative growth. Competition in natural populations is probably influenced also by the direct effects of one species on the growth of another through production of growth inhibitors or promoters.

A further aspect of competition concerns the ways in which one species may restrict the growth of another by reducing the supply of a nutrient required by the second species. Droop (1968, 1974) has discussed the binding of vitamin B_{12} by excreted proteins as a mechanism by which a species may restrict supply of the nutrient to its competitors. Fuhrman *et al.* (1978) have pointed out that the uptake of silicon by a species which has no requirement for the element may have an advantage to that species through reduction of the supply for other algae which do have such a requirement.

Diffusion transport of nutrients in relation to phytoplankton growth

The above discussion of uptake rates of nutrients by phytoplankton in terms of biological parameters does not take into account the role of diffusion transport of nutrient ions to the external surface of cells. Unless the rate of diffusion to the cell surface equals the rate of uptake through the surface, a zone around the cell will occur in which the concentration is less than that in the bulk medium. This question was addressed by Munk and Riley (1952) but at that time the biological factors controlling uptake were not understood. Gavis (1976) has reconsidered this question in terms of the interaction of diffusion and biological uptake processes.

Pasciak and Gavis (1974, 1975) developed the definition of a number P which gives a quantitative index of the effect of diffusion on uptake under quiescent conditions:

$$P = \frac{14.4\pi R\alpha D K_s}{V_m} \tag{13}$$

where R (cm) is a linear dimension expressing the distance from the cell surface to its centre, α is a factor varying with the shape of the organism, and D (cm^2/sec) is the diffusion coefficient. Gavis (1976) has shown mathematically that diffusion transport can influence uptake rates when both P and the ratio of the medium concentration to the half-saturation constant for uptake are small. Relative motion of the organism and medium increases the uptake rate, as realized by Munk and Riley. Decrease in cell size leads to an increased value of P, although not by the simple proportionality suggested by equation (13), and thus to a reduced effect of diffusion transport. It is suggested by Gavis that this may be a factor in the predominance of small-celled species in oligotrophic waters. It is evident from equation (13) that if the ratio of K_s to V_m decreases, diffusion transport assumes greater importance in relation to uptake. It has been noted earlier (cf. p. 332) that organisms showing low values of K_s also tend to show low values for V_m.

Environmental Relationships between Concentrations of Nutrients and Primary Productivity

A general positive correlation between the concentrations of the principal micronutrients and primary productivity is apparent from an examination of large-scale distributions of nutrients and indices of productivity. As noted already, the oligotrophic waters of the central oceanic gyres are regions of low productivity, while enhanced productivity is a feature of upwelling regions in which nutrient-rich waters are supplied to the surface layers. The supply of nutrients to the euphotic zone has long been recognized as a major factor in the differences in productivity between various areas (see, for example, Sverdrup *et al.*, 1942). The relationship between these processes is not, however, simple. There are major interactions between concentrations of nutrients and other factors influencing productivity, particularly light (see Chapters 6 and 9). Turbulent mixing with deeper water is essential for the enrichment of the surface waters but stability of the water column may have to be re-established if a large phytoplankton population is to develop. For example, in the region of coastal upwelling off north-west Africa, Huntsman and Barber (1977) noted that at times when there was mixing by wind action, supplying nutrients to the euphotic zone, productivity was reduced by comparison with stratified conditions. Furthermore, the complexity of nutrient cycles, with intermediates in the regenerative cycle having an important influence on the pattern of phytoplankton growth under certain circumstances, implies considerable variations between areas in the relationships between individual nutrients and productivity.

For the last two decades it has been widely accepted that nitrogen is generally the most important element in relation to nutrient limitation of productivity in marine environments. The concept of a single limiting nutrient is an over-simplification, but when the overall population of primary producers is considered, substantial progress in modelling processes can be made on this basis. Evidence for the particular significance of nitrogen has come from nutrient-enrichment experiments with nitrate, using coastal water samples during periods of nutrient depletion (Conover, 1956), and nutrient-poor oceanic waters (Thomas, 1969), with ammonium, using eutrophic coastal waters (Ryther and Dunstan, 1971), and with various nitrogen sources, using coastal surface waters (Eppley, Carlucci, Holm-Hansen, Kiefer, McCarthy, Venrick and Williams, 1971). Further evidence is provided by observations that low concentrations of phosphate can usually be detected in surface waters which have become greatly depleted in nitrogenous species. Observations of such an effect were initially based on measurements of nitrate alone (see, for example, Ketchum *et al.*, 1958, and Fig. 7.28) but some effect remains even when other nitrogen sources are included (Thomas, 1966b; see also Ryther and Dunstan, 1971). Dugdale and Goering (1967) emphasized the importance also of nitrogen fluxes in productivity studies in that the assimilation of this element should give a satisfactory measure of population growth in marine ecosystems, because it is used primarily in formation of structural organic material and not involved to the extent of phosphorus and carbon in energetic processes leading to rapid turnover of material. Recent work has given particular attention to nitrogen in the two contrasting environments of the oligotrophic mid-oceanic regions and the upwelling areas exemplified by Peruvian coastal waters.

In the former areas the euphotic zone is isolated from rapid exchange with the deeper waters by a thermocline. In an area such as the Sargasso Sea the permanent thermocline is at a considerably greater depth than the base of the euphotic zone. The stratification at some 100 m is there a seasonal feature, the breakdown of which allows the mixing of somewhat richer waters into the euphotic zone. In areas with marked stratification near the base of the euphotic zone, two sectors of the water column may be defined (Dugdale, 1967). These are an upper sector where the light intensity is adequate and where nutrient concentrations are limiting (the nutrient-limited sector) and a lower (light-limited) sector where nutrient concentrations are increasing with depth around the thermocline, but light intensities become inadequate.

In the nutrient-limited sector recycling of nitrogenous material plays an important part in phytoplankton production. A simplified model, shown in Fig. 7.30, of the cycle of inorganic nitrogen in the euphotic zone was used by Dugdale and Goering (1967) to clarify the important distinction between primary productivity sustained by regenerated forms of nitrogen, internally cycled in the reservoir, and that sustained by newly available nitrogen. The latter includes nitrate entering by turbulent diffusion from below the mixed layer and any molecular nitrogen fixed by organisms. The regenerated nitrogen is represented in the model by ammonium, but urea and organic nitrogen are also in this category. This distinction is fundamentally important in relation to the transfer of material to higher trophic levels since, as is evident from Fig. 7.30, only the fraction produced from new nutrients represents an increase in phytoplankton population and is potentially available to support increased production at secondary and higher levels. The use of ^{15}N labelling of different nitrogen sources to distinguish these fractions was introduced by Dugdale and Goering (1967) as a powerful tool for quantitative studies.

At a steady state the removal of combined nitrogen from the mixed layer by sinking of particles and excretion by migrating zooplankton must be balanced by turbulent diffusive transport into the layer and any other inputs of combined nitrogen entering with rainfall or by fixation of molecular nitrogen. In the central gyre of the North Pacific Ocean, Eppley *et al.* (1973) found that inputs of ammonium in rain and rates of

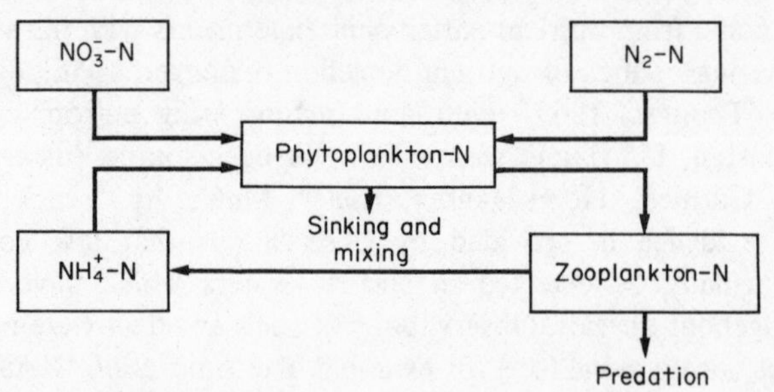

Fig. 7.30. Pathways of transference of inorganic nitrogen species between pools in the euphotic zone (following Dugdale, 1967).

nitrogen fixation were negligible relative to the excretion of nitrogenous material by zooplankton. The results of Eppley *et al.* (1973) suggest that urea, as well as ammonium, is a major excretory product of zooplankton. The recycling of excreted material is of dominant importance in this sector. Dugdale and Goering (1967) found that in subtropical waters the uptake of nitrate was low relative to that of ammonium. The measurements by Eppley *et al.* (1973) for a series of stations in the North Pacific central gyre region showed a generally insignificant role of nitrate, relative to ammonium and urea, to depths of 120 m or more; the concentration of nitrate was <1 μg-at N/l, too low for assimilation.

In the light-limited sector, below the mixed layer, nitrate was found by Goering, Wallen and Naumann (1970) to assume greater importance as a nitrogen source. In the area studied by Eppley *et al.* (1973) the assimilation of nitrate was low in the light-limited sector despite the adequate concentration of the ion. This was attributed to inhibition of assimilation at the low light intensities; under these conditions, as discussed above (pp. 303, 314), nitrite is commonly excreted by algal cells.

The low proportion of nitrate uptake relative to that of ammonium, found by Eppley *et al.* (1973) in the central North Pacific Ocean, represents an extreme of oligotrophic conditions. Nitrate uptake accounts for less than 5% of the total uptake of nitrogen as nitrate, ammonium and urea. For most regions measurements of uptake of urea have not been made but a marked contrast between oligotrophic and eutrophic waters is apparent on the basis of the measurements of uptake of nitrate and ammonium. The earliest such data were those of Dugdale and Goering (1967). In the Sargasso Sea, the uptake of nitrate as a fraction of the uptake of nitrate and ammonium varied between 0.5 and 19%, the high values being associated with the replenishment of nitrate in the euphotic zone which occurs when the seasonal thermocline breaks down. The mean values in the Sargasso Sea and the North Pacific Ocean were similar at about 8 to 9%. Coastal waters from several regions showed corresponding values of 20 to 40%. MacIsaac and Dugdale (1972) also found differences in relative uptake between various eutrophic and oligotrophic waters, nitrate uptake reaching 68% of the combined uptake of the two forms in the Peruvian upwelling area; their results suggest that in eutrophic waters less than 50% of the nitrogen assimilated is taken up as ammonium. The increases in the relative importance of nitrate are correlated approximately with increases in the absolute uptake rate for nitrate.

The extreme oligotrophic regions are thus characterized by the recycling of nutrient resources with a low contribution of new production. It appears that nitrate especially is utilized considerably less effectively in oligotrophic waters than in eutrophic waters (MacIsaac and Dugdale, 1972). The reasons for the incapacity of populations which are not normally exposed to considerable concentrations of nitrate to utilize the ion effectively probably lies in a lack of nitrate reductase in their cells.

Eppley *et al.* (1973) have emphasized the close coupling in such an environment between primary production and grazing and excretion by zooplankton. They draw an analogy with the steady-state condition of nutrient-limited continuous culture in a chemostat. Under this condition, with a constant algal standing stock, the rate of input of the limiting nutrient is compensated by the rate of removal by assimilation and outflow from the culture vessel. The cell concentration is diluted by the continuous flow of medium and the rate of dilution is equal to the rate of growth. In the central oceanic gyre the dilution occurs through grazing and sinking. The analogy with the chemostat is not exact, but can be developed further in relation to a range of environmental conditions (cf. Dugdale, 1976).

In the extreme oligotrophic regions the assimilation of nitrogen is greatly limited by the low concentrations of nitrogen sources in the waters. There are indications, however, of considerable variability in the responses of populations to enrichment (MacIsaac and Dugdale, 1972) and this suggests the possible importance of other limiting factors in some waters (see also p. 342).

In eutrophic waters with nitrate at high concentrations, uptake of the ion may be controlled by light but with a reduction in concentration the uptake becomes limited by the nitrate concentration (MacIsaac and Dugdale, 1972). The greater importance of nitrate relative to ammonium in these waters has been noted above but the regenerative processes are still significant in the supply of nitrogenous material. From an examination of the nitrogen budget of part of the Peruvian upwelling system, using a simulation model, Walsh and Dugdale (1971) showed that recycled nitrogen was important in phytoplankton growth. In this area excretion by the anchovy is a major source of reduced nitrogen (see Dugdale, 1976). Release of ammonium by zooplankton has been found to play a significant part in meeting the nitrogen requirements of phytoplankton in the upwelling system of north-west Africa (Smith and Whitledge, 1977).

In temperate coastal waters, marked seasonal variations occur in the concentrations of nutrients in the euphotic zone, which are related to seasonal changes in productivity and stratification of the water column. These variations were described in an earlier section (p. 318). At the beginning of the spring bloom nitrate is readily available, but later in the year other forms account for a substantial part of the total combined nitrogen available to phytoplankton. This was illustrated by the data for ammonium in Fig. 7.20.

The importance of silicon in relation to nutrient-limitation of productivity has recently received greater attention. MacIsaac and Dugdale (1972) found that with waters of the Saronikos Gulf (eastern Mediterranean) addition of silicon stimulated the uptake of ammonium, suggesting that the natural silicon concentrations might limit nitrogen assimilation. Enrichment with silicon was also found by Azam and Chisholm (1976) to stimulate photosynthesis in samples taken from the Gulf of California.

Dissolved silicon has particular importance potentially in coastal upwelling regions where diatoms form a dominant part of the phytoplankton. The Peruvian upwelling system has been considered from this standpoint by Dugdale (1972) and Walsh (1975). Silicon is there depleted to a potentially limiting concentration more rapidly than other nutrients as the upwelled water moves offshore. Simulation model studies (Walsh, 1975) suggested that phytoplankton growth is limited by nitrogen in the upstream sector of the system with silicon becoming limiting downstream. The regeneration of silicon by dissolution of skeletal material in the euphotic zone seems to play little part in the cycle of the element in this region.

In the upwelling region off north-west Africa, Nelson and Goering (1977) examined rates of dissolution of silica, using the stable isotopic labelling technique. In contrast with the Peruvian region, silicon dissolution in the upper 50 m was adequate to supply the total uptake by phytoplankton. Dissolution of silica appears to be very important also in the nutrient regime of the Baja California upwelling system (Nelson and Goering, 1978). Nitrogen and silicon are supplied to the euphotic zone there at similar rates with silicon being utilized at the higher rate. Primary production is not, however, silicon-limited, implying a more rapid regeneration rate from particulate silicon than from the nitrogenous material.

Trace Substances other than Nitrogen, Phosphorus and Silicon

A number of substances, other than nitrogen, phosphorus and silicon, may affect productivity. They include vitamins, organic chelators, trace metals and organic toxins. The ways in which interactions between the effects of such substances can arise are complex and interpretation is made difficult by several factors:

(1) There are uncertainties concerning the forms, and in some cases even the concentrations, of some potentially important trace elements in seawater (see Chapter 2).
(2) The roles of inorganic and organic species can be inter-related in a complex way because of chemical interactions between them.
(3) Some metals, such as copper and zinc, which are essential for growth have toxic effects at relatively low concentrations and thus can possibly inhibit productivity. Effects at such low concentrations are, however, difficult to establish and physico-chemical limitations make environmental interpretation difficult (Burton, 1979).
(4) Knowledge is lacking as to the forms in which many trace materials are taken up by plant cells and the detailed mechanisms of uptake. Thus there is no firm theoretical basis on which to predict the ways in which changes in chemical speciation in seawater may influence the stimulation or inhibition of growth.

Some trace metals are universally essential for the growth of phytoplankton (e.g. Mn, Fe, Co, Cu, Zn) and others may be necessary for certain organisms. Generally, the total concentrations of these elements are in considerable excess of likely requirements. In some cases, however, the availability of the element may be restricted by the nature of the chemical forms in which it is present. Such effects have been discussed extensively concerning the behaviour of iron in relation to uptake by cells and effects on phytoplankton growth. The work on this element illustrates a number of aspects of the general problems indicated above. It must be stressed, however, that the behaviour of iron differs considerably from that of other metals known to be essential, because of its very limited solubility. Solid $Fe(OH)_3$ is the overwhelmingly dominant form of Fe(III) in seawater and most of the reported dissolved iron is probably colloidal material.

There is a marked difference in availability between the iron present in surface oceanic waters and freshly introduced forms of the element. This has been strikingly shown by the much more rapid uptake in the food chain of radioactive ^{55}Fe, introduced in fallout, relative to that of stable iron already present (Jenkins, 1969). Such an effect may be related in part to the phenomenon of ageing of hydrolysed Fe(III) into progressively less reactive colloidal forms. Davies (1967) showed that the reactivity of hydrous ferric hydroxide, in terms of its adhesion to the surface of *Phaeodactylum tricornutum*, was reduced on ageing. Most of the iron remained on the cell surfaces, however, and the implications in relation to uptake into cells is uncertain. Additions of Fe(III) to culture media have been observed to increase diatom growth even though the added material precipitated (Harvey, 1937; Goldberg, 1952). This has been interpreted to indicate the use of colloidal material as an iron source but Jackson and Morgan (1978) point out that the results could be explained also by the scavenging, by the precipitate, of copper, which can inhibit growth at low concentrations (see below).

Several studies have shown that the addition of the strong chelator ethylenediaminetetra-acetic acid (EDTA) to seawater affects its capacity to support developing phytoplankton populations. Johnston (1964) showed by culture assay experiments that the supply of EDTA or other chelators played an important role in phytoplankton growth. Barber and Ryther (1969) observed that freshly upwelled nutrient-rich water in the Cromwell Current region supported phytoplankton growth less effectively than water which had been longer at the surface. The productivity was not improved by addition of various nutrients, trace metals, vitamins or amino-acids but was enhanced by the addition of EDTA or the dissolved material in a zooplankton homogenate. It appeared that the surface waters in this region become conditioned by the influence of biological activity following upwelling. These and related findings are discussed further in Chapter 9. Like those from Johnston's experiments they were interpreted in terms of an increased availability of trace metals by chelation. In view of its chemical behaviour, iron is a particularly likely element to be influenced in this way. Lewin and Chen (1971) found that addition of EDTA could make otherwise unavailable particulate forms of iron, produced in the preparation of a culture medium, available to phytoplankton; this was not the case, however, if the seawater used in the medium was stored before use.

The presence of EDTA at a micromolar concentration increases the equilibrium concentration of Fe(III) in the solution phase by two orders of magnitude (Jackson and Morgan, 1978). Three mechanisms by which additions of EDTA to seawater might increase the transport of Fe(III) to cells have been discussed by Jackson and Morgan (1978). The direct passage of the complex is unlikely in view of evidence that EDTA molecules are not transported. The chelator could serve either to carry iron in solution to the cell membrane where the iron might be transferred to a ligand acting analogously to an iron-transfer protein or to increase the supply of the element to the cell surface, relative to the diffusion of uncomplexed iron, by dissociation of the complex at the surface. Detailed calculations by Jackson and Morgan suggest that neither process could account for observed effects of EDTA on growth rates.

Some observations on natural waters have been considered indicative of a possible role of iron limitation of primary production. Experiments on the effects of added substances on photosynthesis in surface seawater from the Sargasso Sea were carried out by Ryther and Guillard (1959). Despite the low concentrations of nitrate and phosphate in the natural seawater, additions of these ions did not increase photosynthetic carbon fixation but a mixture of trace metals did have this effect, a phenomenon which was later shown (Menzel and Ryther, 1961b) to occur with Fe(III) alone, either as the chloride or the EDTA complex. It was suggested from these studies that iron was the most critical limiting factor in this region and Tranter and Newell (1963) drew similar conclusions from experiments on the eastern Indian Ocean. Later studies on the Sargasso Sea by Menzel, Hulburt and Ryther (1963) indicated that if the experimental period was lengthened, addition of phosphate and nitrate stimulated growth effectively without the addition of iron. Moreover, in some experiments aluminium, which has no known direct biological role, had similar effects to iron. This suggests the possibility of indirect effects of these elements by scavenging other metals, such as copper, which can inhibit growth (Jackson and Morgan, 1978). The role of iron in relation to productivity remains uncertain (see also Chapter 9).

Studies of the effects of copper on phytoplankton growth illustrate a number of additional considerations. Steemann Nielsen and Wium-Andersen (1970) showed effects of low concentrations of copper on the growth and photosynthetic rate of freshwater algae. They suggested that the findings of Barber and Ryther (1969), referred to above (p. 342), on the stimulation of phytoplankton growth in recently upwelled seawater by addition of chelators, could be explained in terms of detoxification of copper. This view was based on their observations of effects at concentrations as low as about 1 μg/l, then thought to be below the average concentration in seawater although later work (see Chapter 2) indicates that this is probably about 0.1 to 0.2 μg/l. Despite the erroneous basis for the original suggestion, the general point that factors affecting chemical speciation influence the biological effects of copper may remain valid. Manahan and Smith (1973) found in experiments with copper-deficient cultures of freshwater algae that there was a relationship between growth and the activity of cupric ions. Experiments with estuarine and marine species have shown certain inhibitory effects of free copper on growth rates. The data from several studies have been examined by Jackson and Morgan (1978) in the form given in Fig. 7.31. The results indicate an increasing toxic effect of free cupric ions over a range of concentrations. Sunda and Guillard (1976) also report a dependence of growth-rate inhibition, in culture experiments, on the activity of cupric ions. The data in Fig. 7.31, are particularly interesting as they extend into a sufficiently low range of concentrations to suggest possible effects in seawater. They also show convincingly the broadly similar effects of a given concentration of free ions produced by very different total copper concentrations adjusted to give the same concentration of free ions by varying the extent of chelation.

A great deal remains to be learned as to the effects of metals on photosynthesis and growth over the characteristic ranges of environmental concentrations. The mechanisms by which chelation affects responses of phytoplankton to metals is largely uncertain,

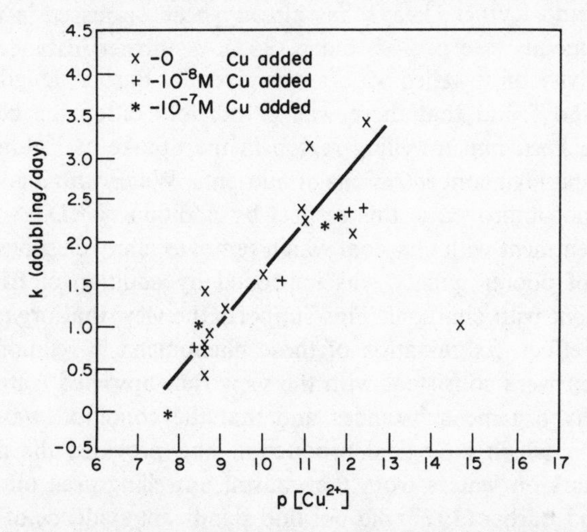

Fig. 7.31. Growth rate of *Chaetoceros socialis* as a function of the estimated concentration of free cupric ion in upwelled seawater. The parameter p[Cu^{2+}] is analogous to pH, i.e. it is the negative logarithm of the concentration of cupric ion. Changes in p[Cu^{2+}] were produced by various additions of copper and chelators (from Jackson and Morgan, 1978).

particularly as little is known of the extent and nature of organic associations of metals in oceanic waters. It seems, however, from the limited evidence, that organic associations are more likely to influence the system by binding toxic metals in relatively unavailable forms than by increasing the uptake of essential metals present at low available concentrations.

Some organic substances in low concentrations also play an important part in the promotion and inhibition of phytoplankton growth. In the stimulation of growth the role of organic substances may be either indirect or direct.

Indirect effects could arise through the chelation of trace metals as discussed above. Prakash (1971) has summarized information on the effects of humic substances on phytoplankton growth in culture experiments and coastal waters, the latter in relation especially to dinoflagellate blooms. He suggests that the lower molecular-weight fractions are particularly effective in stimulating growth. These substances might act as chelators although it seems that their chelating capacity might be saturated at equilibrium in seawater by the abundant cations, calcium and magnesium (Mantoura, Dickson and Riley, 1978). A variety of organic substances, other than humic materials, may combine with trace metals in seawater but the concentrations of these compounds (e.g. amino-acids), are too low to be significant in terms of chelation of metals.

Although the evidence is not unequivocal there are indications of significant organic associations of some metals in estuarine and coastal waters (see Chapter 2). It has also been established that substantial fractions of copper, zinc and some other trace metals in coastal waters are electrochemically inactive in anodic stripping voltammetry and that such waters have a capacity to complex added trace metals, up to a certain concentration. The evidence on these topics has been summarized by Whitfield (1975c). It is unclear whether the associations which produce these effects involve organic complexation or binding to colloidal material but they clearly represent a mechanism by which toxic effects of metals may be reduced.

The phenomenon of apparent organic conditioning of upwelled surface water, described by Barber and Ryther (1969), has already been discussed in relation to the availability of trace metals (see p. 342 and p. 343). A further instance of a marked difference in productivity of upwelled waters was given by Barber, Dugdale, MacIsaac and Smith (1971) who found that there was a four-fold difference between certain surface waters in the Peruvian upwelling region in the uptake of ^{15}N-labelled nitrate, despite their similar and high concentrations of nutrients. Water with a good capacity to sustain growth was not improved in this respect by addition of EDTA but its quality was diminished by treatment with charcoal which removes many dissolved organic substances. The water of poorer quality was improved by addition of EDTA and was unaffected by treatment with charcoal. This supports the view that organic substances have a conditioning effect. Examination of these phenomena in relation to the water circulation of the area were consistent with the view that upwelled water initially was poorly conditioned by organic substances and that the condition was improved by mixing with waters in which biological production had provided the necessary substances. In experiments on waters from the coastal upwelling area off north-western Africa, Huntsman and Barber (1977) did not find significant evidence of stimulation of growth by addition of EDTA.

Experiments have been undertaken by Barber (1973) in which photo-oxidation with ultra-violet radiation was used to decompose organic material in natural seawater and the consequent effects on phytoplankton growth examined. The findings varied with two samples of seawater used but with one sample the decomposition of organic material reduced the capacity of the seawater to support phytoplankton growth. This capacity could be restored by addition of EDTA suggesting that the critical organic factor in the original sample was a chelator acting on growth by its interaction with other components rather than a substance promoting growth directly.

The direct effects of organic substances include both the stimulation and inhibition of growth. The role of organic compounds of nitrogen and phosphorus in relation to growth of phytoplankton has been discussed in some detail earlier in this chapter. Apart from these substances, most attention has been given to vitamin B_{12} (cyanocobalamin), thiamine and biotin. Vitamin B_{12} has been investigated to a greater extent because of its recognized importance to the growth of phytoplankton organisms, many of which are unable to synthesize it. It has been suggested that bacteria are major producers of vitamin B_{12} in the ocean (Provasoli, 1963) although planktonic algae can release vitamins (see, for example, Carlucci and Bowes, 1970). Riley and Chester (1971) summarized information on the occurrence of vitamin B_{12}. The data suggest that the concentrations in more productive oceanic regions are in the range of 0.2 to 2 ng/l, whereas in some oligotrophic regions, such as the Sargasso Sea, concentrations are lower, ranging below 0.1 ng/l. Despite these low concentrations it seems unlikely that vitamins normally have an overall limiting effect on production (Vishniac and Riley, 1961; Menzel and Spaeth, 1962b; Carlucci, 1970) but possible that they affect species composition.

Certain organic compounds exert toxic effects upon phytoplankton and other marine organisms (cf. Chapter 3). Such effects are demonstrated in an extreme way by the lethal effects upon fish populations of toxic materials produced by red tide blooms of dinoflagellates such as *Gymnodinium breve*. The characterization of materials released at very low concentrations presents difficult problems but progress has been made as to the composition of toxins which are concentrated by shellfish and may thus lead to poisoning in man (Faulkner and Andersen, 1974).

A wide range of organic substances released by algae, together with those released by bacteria, may play significant but complex roles in the regulation of growth of particular species. Much of the work in this field relates to cultures and is difficult to interpret in relation to environmental effects; useful reviews include those by Lucas (1961), Provasoli (1963), Fogg (1966) and Sieburth (1968). As an example of the complex interactions which may occur Droop (1968) suggested that organisms which produce extracellular material capable of binding vitamin B_{12} might monopolize the supply of this vitamin and thereby exert a competitive advantage over other algae which require the same growth factor.

Given the very wide range of trace metals and organic substances which may influence the relative growth of individual phytoplankton species and the complexity of the interactions between these substances, it is clear that generalizations as to their effects cannot be made. It does seem likely, however, that the effects of these substances are more likely to be apparent in the species composition of populations and the phenomena of species succession than in the total productivity of a particular body of water, although recently upwelled deep waters may in some cases be an exception. The information is so sparse, however, and the possibilities so complex, that this conclusion may well need to be revised in the light of further evidence.

Chapter 8
Factors Limiting Primary Production: Grazing

Zooplankton Grazing

While recognizing that phytoplankton crop and primary production are closely related, we have so far largely considered factors affecting, on the one hand, photosynthetic rate (incident light, length of day, transparency of the water, stabilization, nutrient concentration and temperature) and, on the other hand, factors concerning size of standing crop (cell number, cell volume, surface area, floristic composition, chlorophyll concentration). The size of standing crop may, however, be misleading since the removal of algal cells has not so far been fully considered. Algal cells may disappear by sinking beneath the euphotic zone (cf. Chapter 6), but in most marine regions the great majority is consumed by the grazing of the herbivorous zooplankton. Steele (1956) ascribed a steady deepening which he observed in the level of the maximum phytoplankton density from April to September on the Fladen Ground (North Sea) both to grazing and to sinking of the algae. Apart from variations in algal sinking rate (cf. Chapter 6), the grazing pressure varies not only with zooplankton abundance but with its age composition. Young stages, for instance, may feed more intensively per unit weight, and older copepodites may feed over a wider depth range.

Problems of grazing are to some extent made more complex by the well-recognized patchiness (spatial heterogeneity) of plankton populations. Some of the problems are discussed by Cushing (1962). This patchiness applies both to the phytoplankton and zooplankton and exists on a wide range of scales. Patches may occur varying in dimensions from a few metres to many kilometres. Steele (1976) describes long narrow patches of phytoplankton, a few metres wide and some hundreds of metres in length. Some of this smaller scale patchiness is undoubtedly related to surface-water movements, particularly to Langmuir circulation and vortices (cf. Fig. 8.1). Larger scale patches of algae might result from variations in the distribution of nutrients promoting different densities of phytoplankton growth. In investigations by McAllister, Parsons and Strickland (1960) in the North Pacific, however, where considerable patchiness in phytoplankton density was experienced, there was no clear relationship between patchiness and variations in nutrient levels. The whole biological history of different areas of water could be in part responsible for patchiness (cf. succession – Chapter 5).

The larger scale patches of plankton are frequently roughly elliptical with a diameter estimated in terms of kilometres, and length of the order of tens of kilometres. Such patchiness occurs usually without any obvious physical boundary. The effect of diffusion on a patch has been considered by Cushing and Vucetic (1963), Steele (1974, 1976) and others. Cushing (1964) quotes one very high rate for losses due to

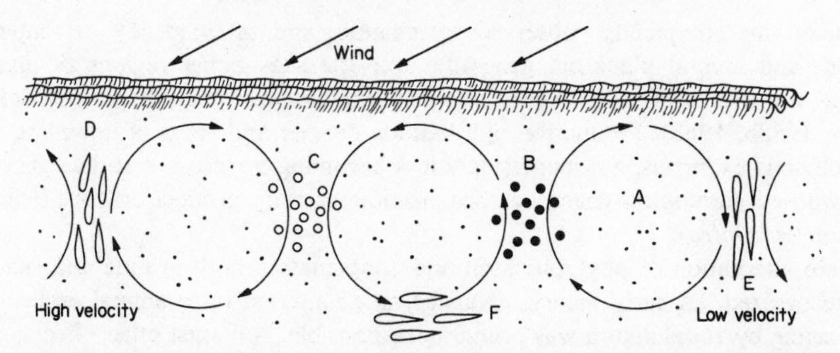

Fig. 8.1. Langmuir vortices and plankton distributions redrawn from Stavn (1971). A: Neutrally buoyant particles randomly distributed. B: Particles tending to sink, aggregated in upwellings. C: Particles tending to float, aggregated in downwellings. D: Organisms aggregated in high-velocity upwelling, swimming down. E: Organisms aggregated in low velocity downwelling, swimming up. F: Organisms aggregated between downwellings and upwellings where there is less relative current velocity than within the vortices (Parsons and Takahashi, 1973).

diffusion and sinking from an algal patch amounting to 15%. In view of the vertical distribution of plankton and zooplankton vertical migrations, horizontal diffusion should probably be considered as affected by vertical diffusion gradients as well as horizontal shear velocity. Steele (1974) points out that while a small plankton patch might tend to be smoothed out by turbulence, once a patch of sufficient size has developed, algal growth can more than compensate for diffusion losses and lead to the large-scale spatial heterogeneity. Steele (1976) indicates a critical diameter for a patch:

$$l_c = 4.8\sqrt{(k/a)}$$

where l_c = critical diameter, k = diffusion coefficient, a = algal growth rate. If the algal division rate is greater than the rate of dispersion, the algal density of the patch will increase with time. Critical size is obviously partly dependent on the algal division rate, but apart from this factor, the local concentration of algae will be controlled essentially by turbulence. Platt, Denman and Jassby (1977) suggest that while patchiness in phytoplankton for length scales less than about 100 m is controlled primarily by turbulence, for scales exceeding 100 m spatial variability parameters such as assimilation number and the community structure of the phytoplankton may play a part.

This preliminary analysis of patchiness has taken no account so far of interaction between algae and zooplankton. Frequently in the ocean, denser patches of phytoplankton and zooplankton tend to alternate with each other. Hardy (in Hardy and Gunther, 1935) suggested that very dense phytoplankton patches tended to exclude animals. The exclusion might operate particuarly by dense phytoplankton delaying the upward movement of zooplankton in their diurnal vertical migrations, resulting in differential lateral movement with the speed of flow of different water layers. These movements, in addition to a possible inimical effect of phytoplankton at very high densities, might result in an alternation of high density areas of phytoplankton and zooplankton. The lack of very positive evidence of any inimical effect, especially with the phytoplankton concentrations normally experienced in the oceans, has not given substantial support to the theory of animal exclusion. There is also some evidence for believing that where dense patches of phytoplankton occur, the zooplankton tends to feed actively round the edges of the patch (cf. Beklemishev, 1957).

Despite the frequently observed patchiness and a tendency to alternation of plant and animal plankton, generally over the seas richer regions of algal production are also regions of higher zooplankton biomass (cf. Chapter 5, Steemann Nielsen 1958b, 1963). Productive high latitude regions and areas of upwelling are the most obvious examples, but increasingly it is becoming apparent that divergences and areas whose hydrological characteristics promote primary production have richer zooplankton (*vide infra*).

Where alternation of phytoplankton and zooplankton, both in time and space, has been recognized, an early theory advanced as an alternative to animal exclusion was that grazing by zooplankton was primarily responsible. Amongst other effects, grazing could increase the critical size necessary for a persistent algal patch by reducing the effective algal growth rate. The whole spatial and temporal variations in the phytoplankton and zooplankton, including such factors as whether zooplankton reproduction is continuous or phased in cohorts, and whether developmental stages graze more intensively, can have obvious repercussions on patchiness.

The likely significance of grazing pressure in phytoplankton production appears from several earlier calculations, e.g. that *Eurytemora* might consume 40% of its body weight per day, that in the Plymouth region the herbivorous zooplankton might eat 50% of their body weight daily, and Riley's (1947) observation that the spring zooplankton might have a daily requirement of 30% of the body weight. Cushing (1963, 1964) also concluded that some 38 to 45% of body weight must be taken every day. Both Harvey (1950) and Riley (1947) found that the daily demand by copepods for maintenance and growth was less, 10% and 11 to 14%, respectively. A full discussion of more recent investigations in relation to daily demand is given in Volume 2. It is sufficient here, however, to draw attention to the high estimates for grazing, at least during the spring outburst, and to note the suggestion, attributed particularly to Beklemishev (1954, 1957 and 1962), that herbivorous zooplankton may indulge in excessive ("superfluous") feeding. While this may not have been clearly substantiated, the probable heavy demand of grazers on the phytoplankton is clear. Cushing, for example, states that *Calanus* might consume >300% of its body weight daily (cf. p. 359).

Among earlier field studies which stressed the importance of zooplankton grazing, Harvey, Cooper, Lebour and Russell's (1935) investigations drew attention to the remarkable reduction in phytoplankton crop immediately following the spring outburst in English Channel waters and noted that this was accompanied by marked grazing activity of the increasing zooplankton population. Over the summer, Harvey *et al.* concluded that although the restricted renewal of nutrients from below the thermocline limited production, and despite the importance of regeneration of nutrients within the euphotic zone, the generally rather low, but fluctuating, phytoplankton crop was largely determined by the rate of grazing of the zooplankton. Clarke (1939) also believed that over the productive period of phytoplankton growth in temperate latitudes, the zooplankton was largely responsible for regulating the extent of plant population. The importance of grazing was recognized in a number of other temperate and high latitude areas, for example, in Arctic waters by Braarud (1935); by Bigelow, Lillick and Sears (1940) for the Gulf of Maine; Vinogradov (1973) and fellow workers for the sub-Arctic Pacific; Wimpenny (1936, 1938), Cushing (1959a, 1963) and Steele (1956, 1974) for the North Sea; Riley (1946, 1947), Pratt (1965), etc., for shallow temperate north-western Atlantic waters; Parsons and colleagues for temperate Pacific areas; Holmes

(1956) for the Labrador Sea, and Hart (1942), Philippon (1972), Foxton (1966) and others for the Antarctic.

Grazing intensity of course may vary with season, especially at high latitudes. Halldal (1953) considers that in the Norwegian Sea it was much more intensive in spring than later in the year. Vinogradov and Arashkevich (1969) comment on the marked effect on the trophic structure of the plankton community in the north-west Pacific of seasonal migration by the zooplankton. They point to the clear predominance of filter feeders in the upper 200-m layer of the ocean, but emphasize the remarkable grazing effect of the interzonal zooplankton which migrates seasonally to the near-surface layers to feed on the spring and summer phytoplankton crop. The majority of migrating copepods, for example, are herbivores. This trophic pattern is believed to be true of most polar and boreal areas of the Pacific and Atlantic Oceans, as well as of the Antarctic. The upper interzonal species reinforced the actively reproducing surface-living zooplankton over the spring, the bloom of phytoplankton being essential for the increased intensity of grazing. According to Vinogradov (1973), with the extension of the depth of the phytoplankton crop over summer, due to the increased light penetration, the crop though reduced is often spread out vertically to some extent among the dominant herbivorous species.

The investigations of Voronina and Sukhanova (1977) in Antarctic waters confirmed the increase in grazing pressure with seasonal vertical migrations of zooplankton. Intensive grazing by four dominant copepods (*Calanoides acutus*, *Calanus propinquus*, *Rhincalanus gigas* and *Metridia gerlachei*) is described. When the copepodite stages of these herbivores arrive near the surface in their seasonal ascent, they feed on whatever diatoms are blooming over the brief summer period. The arrivals of the copepods are not simultaneous. Voronina and Sukhanova reported little difference in the diet of the copepod species, provided the flora composition was similar. Different copepodite stages also showed little or no dietary distinction. All diatoms ranging in size from 5 to 300 μm were consumed, although there was a suggestion that the larger spiny *Chaetoceros* spp. were not so actively grazed.

To the intensive grazing by copepods in Antarctic waters must be added the massive feeding of the herbivorous euphausiids (*Euphausia* spp., *Thysanoessa*) and of salps (cf. Hart, 1942; Hardy and Gunther, 1935; Foxton, 1966).

The grazing by copepods on the larger colonial diatoms has been examined by Peruyeva (1977) working in Arctic waters, since a suggestion has been advanced that colonial algae were less readily consumed. Preferential grazing with respect to size of diatoms was considered as possibly influencing their size range by Wimpenny (1973). Experiments showed an increase in size in grazed populations. Experiments by Peruyeva with the colonial diatom, *Chaetoceros crinitus*, dominant in an area of the White Sea over summer, showed that it was readily grazed by copepodites (Cop. IV) of *Calanus glacialis*. The amount of food increased with cell concentration to a maximum of about 13,000 cells/animal over 8 hours, the likely period of normal active grazing; this maximum ration was equivalent to about 13% of the body weight. There appeared to be no selection of diatom size in relation to the number of cells in the colony chain, and observations showed that the diatom was aligned by the feeding mouth parts of the copepod so that the whole colony could be ingested. Martin (1970) considered that longer chains of *Skeletonema* might be broken during the grazing process.

The recognized vast importance of grazing at high and temperate latitudes is paralleled in other oceanic regions. Areas of pronounced upwelling, which often varies in intensity with season, are characterized by heavy phytoplankton crops and large zooplankton densities with high grazing pressure. The structure of the plankton community is similar in many ways to that of high latitudes. For example, Longhurst (1967), studying the region of the Californian Current off southern California where upwelling was marked, found that the biomass of both chlorophyll and zooplankton was large. On the other hand, species diversity in plankton was low. *Calanus pacificus* and sometimes *Acartia tonsa* were the dominant copepod species but herbivorous copepods were generally overwhelmingly important and grazed heavily on the phytoplankton, mainly *Coscinodiscus* and *Chaetoceros*. A special characteristic feature of the particular region was marked grazing due to the pelagic crab, *Pleuroncodes planipes*. The carnivorous zooplankton was low in the upwelling areas amounting to only 5 to 10% in abundance of the total zooplankton. On the other hand, herbivores constituted more than 90% in abundance.

In contrast to the upwelling waters, the tropical waters off southern California were low in phytoplankton. For example, the algal biomass in non-upwelling areas was approximately 100 times less than that of a station in upwelling water. The surface zooplankton of tropical waters was characterized by high species diversity, low dominance and low biomass, with a substantial proportion of macrophagous species. Herbivores, for example, formed only from 30 to 50% of the total zooplankton abundance, although of those herbivores which were present, a large proportion were of small size and had the ability to feed on fine particles.

Areas at some distance from upwelling, but where reduced effects of eutrophication were experienced, had a greater biomass of phytoplankton and zooplankton than in warm tropical waters. There was also a considerable herbivorous zooplankton population but comparatively high species diversity which was partly due to advection causing a mixing of originally separate zooplankton populations. The proportion of carnivorous zooplankton in these mixed waters was intermediate between the upwelling and oligotrophic waters and amounted to 25 to 35% of the zooplankton. Longhurst points out that off southern California, where upwelling is seasonal and fairly shortlived, the dominance of a very few herbivores and the small numbers of carnivores, a trophic structure resembling high latitudes, provide an ecosystem which is well adapted for intensive grazing on sudden blooms of algae.

Vinogradov and Voronina (1962) also describe a larger biomass of zooplankton and an increased proportion of herbivores at divergences in the tropical Pacific, with carnivores displaced at some distance from the divergences. Intensive grazing by the zooplankton is not only true, however, of divergences and upwelling regions but is characteristic of the warm oceans of the world, though there it is usually on a relatively small algal crop. The biomass of zooplankton, especially of herbivores, is low but grazing pressure is severe. Grazing is indeed recognized as worldwide in its significance. Vinogradov (1966) suggests, as a generalization for open oceans, that 80% of the primary production is consumed by zooplankton, with only 10% sedimented in the bottom deposit. The trophic structure of the tropical plankton community, however, differs from that of high latitudes.

Vinogradov and Arashkevich (1969) and Vinogradov (1973) suggest that in tropical open oceans, the relatively small phytoplankton crop, showing only minor seasonal changes, is spread over a considerably greater thickness of euphotic zone. The epiplanktonic zooplankton, while usually not right at the surface, may show richer layers with different species being more abundant at various subsurface depths (25 to 50 m; 50 to 100 m, etc.). The chief distinction between the tropical and high latitude trophic pattern is, however, the lack of seasonal vertical migration of interzonal zooplankton. The tropical populations are comparatively stable through the year (*vide infra*).

Where vertical movement of tropical water, as at divergences, enriches surface waters leading to increased plant production, the trophic structure becomes more similar to the higher latitude pattern. Timonin (1969) describes such changes from investigations in the tropical Indian Ocean, in a large area of the South Equatorial Current and Equatorial Counter Current, where zones of marked upwelling and weaker upwelling, due to divergences, contrasted with the stable structure of the main body of the South Equatorial Current. Although he observed that the great majority of the zooplankton species were distributed throughout the whole area, a typical zoogeographical tropical region, substantial differences in biomass and dominance were apparent. Areas of strong upwelling had rich phytoplankton, an increased number and especially biomass of zooplankton, and were characterized by a comparatively few dominant zooplankton species, with an increase especially in the proportion of herbivorous copepods. Thus while Calanoida were generally numerically most abundant over the whole area, followed by cyclopoids, chaetognaths, ostracods and euphausiids, in upwelling zones 75% of the population were calanoids and of these about 75% was made up by only three species. In terms of biomass, whereas in stable stratified waters of the South Equatorial Current herbivorous calanoids accounted for only 28% of the comparatively low total biomass and predators 57%, in areas of strong upwelling herbivorous calanoids increased to 50% of a greater total biomass and predators (predatory copepods and chaetognaths) made up 28%.

In the typical oligotrophic stratified warm waters, while grazing is intense and continuous, there are doubts as to the crop being always sufficient to supply even the comparatively small stock of herbivores. It must be accepted, however, that zooplankton density may be more related to primary production than to standing stock. Petipa, Pavlova and Sorokin (1973) carried out experiments on feeding rates for a variety of tropical Pacific zooplankton which indicated that a constant feeding rate could be maintained only when maximal phytoplankton densities occurred in the waters. Herbivorous copepods were judged to be unable to satisfy their energy requirements with the usual level of phytoplankton production in the tropical Pacific without supplementing their nutrition with animal food. Mixed feeding was found to be common over a wide range of zooplankton species. Heavy grazing pressure is also regarded as typical of the warm eastern Indian Ocean by Tranter (1973). Some enrichment of the tropical surface waters occurs in winter, due to strengthening of the South Equatorial Current; this leads to a rapid increase in algal production which is immediately grazed by the herbivores. With the increased speed of mineralization in warm waters, the intense grazing releases nutrients which cause a more continuous algal multiplication. The drift of impoverished surface waters to strongly stratified subtropical regions, however, leads to reduced phytoplankton production in those areas. This combined with heavy grazing causes what Tranter describes as a "seasonal catastrophe" over the summer. The zooplankton starve and are reduced by predation. With apparently slower mineralization, the phytoplankton and zooplankton are more out of phase. Tranter calculates that in tropical regions the zooplankton content is 137% of the phytoplankton, as compared with 185% in subtropical areas of the Indian Ocean. On the assumption that zooplankton require one-twentieth of their body carbon as daily ration, the animals would consume 46% of the daily algal production in the tropics and 71% in the subtropics. But the value for algal production included detrital material; only about 30% of production is claimed to have consisted of living algal cells. The zooplankton would, therefore, appear to use detritus to supplement their diet, and despite heavy grazing and rapid re-cycling, starvation may occur at times in some areas.

Sorokin and Tsvetkova (1972) commented on the overall low primary production in the tropical western Pacific, though the level in equatorial divergent regions might be three to five times richer than in areas of convergence. Their observations indicated that grazing, which occurred mainly at night in the layer of maximum phytoplankton density, accounted for a very large fraction (approximately 70%) of the daily algal production, but even so, was insufficient to supply the required daily ration (cf. Petipa *et al.*, 1973).

The evidence for the enormous effect of grazing in reducing algal crop throughout the world's oceans is clear. Only in somewhat peculiar marine environments, often in very shallow waters, where phytoplankton crop can be excessively rich, and where the zooplankton population, due frequently to unusual unfavourable hydrographic factors, is low, does grazing appear to exert little influence on algal density (cf. Riley, 1952, 1956, 1959, 1967; Deevey, 1948; Gross, Marshall, Orr and Raymont, 1947; Teixeira, Tundisi and Santoro, 1969). Some of these examples will be discussed later but one unusual observation by Bogorov (1967) will be briefly described here as an illustration. During the 1930s a drop in level of the Caspian Sea caused an increased nutrient concentration in the water, especially in northern areas. This led to the proliferation of vast quantities of the diatom, *Rhizosolenia calcar-avis* in the plankton, but the indigenous Caspian zooplankton could not feed successfully on this comparatively large diatom. Bogorov states that not only was the zooplankton unable to exploit this massive crop; the zooplankton declined in density and this fall was followed by a drop in fish larvae, larvae of benthos, and fishes.

Harvey *et al.* (1935) obtained for an area of the western English Channel a minimal estimate of an overall production of phytoplankton calculated from the reduction in nutrient level over the period of the spring increase. For 1933 density was estimated at 85,000 plant pigment units/m^3, arbitrary pigment units being used as a measurement of chlorophyll, whereas the actual value of the standing crop was only 2500 units/m^3. For the following year, Harvey found that the standing crop was only 2–3% of the estimated total production. Hart (1942), following Harvey's calculations, suggested that the standing crop of phytoplankton in the highly productive South Georgia region of the Antarctic was only 2% of the calculated production; for oceanic Antarctic areas grazing pressure resulted in a standing crop only 0.5% of the calculated production.

Although a small standing crop of phytoplankton may, therefore, be observed when zooplankton density is high, leading to often recognized alternation of algae and herbivores, the time relation between phytoplankton and zooplankton is an important factor. A zooplankton population can graze down plant cells in a matter of a few days, but animals grow and reproduce relatively slowly. On the other hand, a small algal population can reproduce exceedingly quickly, and with good growth conditions and in absence of grazers can become a very dense population in a few days. The importance of this time factor in phytoplankton/zooplankton relationships has been well emphasized by both Steemann Nielsen (1937, 1963) and Clarke (1939). Parsons (1976) also comments on the difficulties with modelling in the marine plankton owing to time/scale relations.

Grazing in Mathematical Simulation

The importance of grazing as a factor in phytoplankton production has led to a number of mathematical treatments. Fleming (1939) expressed the difference between an initial population of phytoplankton and the population at some later time as the "increment". Total production can be equivalent to increment only if no death of cells or removal by grazing follows. In most areas little natural mortality of algal phytoplankton normally occurs, but a considerable proportion of the phytoplankton may be grazed. Fleming proposed the term "yield" for the difference between the total production and the increment, assuming that the difference was due to the removal of algae by zooplankton alone.

Fleming proposed an equation:

$$\frac{dP}{dt} = P(a-b-ct).$$

This expresses the rate of change of a phytoplankton population (P), where a = rate of division of the phytoplankton cells, assumed constant over a period; b = initial grazing rate; c = the increase of grazing rate which is assumed to be linear. If the grazing rate is assumed to be constant and if $(a-b)$ is positive (i.e. if the rate of division of the phytoplankton exceeds the constant grazing rate), the phytoplankton population must continue to increase. However, the rate of increase in density of the algae is slower than the true rate of division owing to grazing. The population increase may be so slow that the phytoplankton appears hardly to change at all although the actual fraction per day removed by grazers may be very considerable. If the rate of grazing exceeds the algal reproduction rate the population will decline; very intense grazing may lead to the typical rapid reduction following the spring bloom.

Using values for phytoplankton and zooplankton coefficients suggested by Harvey, Fleming obtained a symmetrical curve for the algal population change which agreed fairly well with field observations. He also computed the total production of algae, assuming a constant cell-division rate of one division in 36 hours over the period under investigation. The peak in standing crop was achieved in 37 days, and the total production was calculated over double this period, by which time the crop had declined to a very low value. Fleming obtained a calculated total production for the whole period of > 80,000 pigment units – a value which agreed remarkably well with Harvey's estimate from nutrient depletion. Thus in the second month of the spring increase the standing crop was a mere fraction of the total population and the yield increased enormously (cf. Fig. 8.2).

For Georges Bank (Gulf of Maine) Riley and Bumpus (1946) observed a rise in phytoplankton and a slower increase in zooplankton during late winter and spring. There was a change, however, from a positive correlation between the two populations in early spring to a negative correlation over May. Riley and Bumpus concluded that while the early correlation was due to some beneficial factor such as increasing

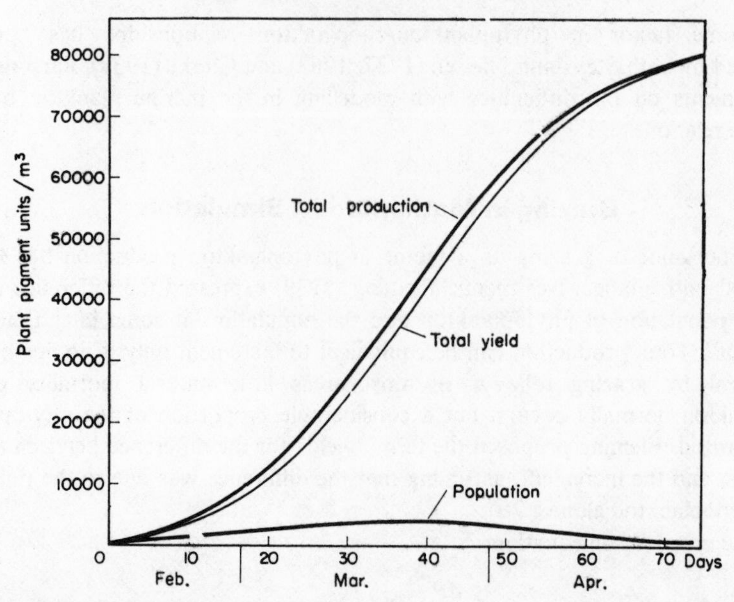

Fig. 8.2. Calculated total production, total yield, and population based on observations of Harvey *et al.* Total yield represents total amount removed by grazing (assumed rate of division once in 36 hours) (from Fleming, 1939).

temperature common to both phytoplankton and zooplankton, the sharp change to an inverse relationship in May was due to the grazing activity of the zooplankton. The date of the switch to the negative correlation varied with individual zooplankton species, depending upon when each species reached its seasonal peak population. The authors calculated that while the percentage of the phytoplankton production consumed by the zooplankton was relatively small (less than 10%) until April, it rose sharply in May to over 40%.

Riley (1946, 1963) developed a mathematical model following that proposed by Fleming for the rate of increase in a phytoplankton population, introducing loss due to algal respiration as affected by temperature, the effect of incident illumination and reduction of light energy with depth, and limitation due to nutrient depletion. He had estimated that 60–80% of the phytoplankton variation on Georges Bank could be accounted for in terms of depth and illumination, temperature, nutrients (phosphate and nitrate) and zooplankton density. The initial relationship may be expressed:

$$\frac{dP}{dt} = P(P_h - R - G)$$

where P is the total phytoplankton population, expressed per unit area of sea surface; P_h = photosynthetic coefficient; R = coefficient of phytoplankton respiration; G = zooplankton grazing coefficient. Riley states that these must be regarded as ecological variables rather than constants. Photosynthesis was found to vary more or less linearly with incident illumination during winter and early spring. It may thus be assumed to vary similarly with depth. By experiment, a photosynthetic constant (p) was established; its value, if light intensity is measured as ly/min, was 2.5. Thus:

$$P_h = pI \text{ (where } p \text{ is } ca. \text{ 2.5).}$$

At depth z the illumination I_z is:

$$I_z = I_0 e^{-kz}$$

(where I_0 is the incident illumination and k the extinction coefficient) (cf. Chapter 6). Riley assumed a limiting depth for photosynthesis where the light intensity was 0.0015 ly/min (i.e. this is the compensation intensity). If the depth at this intensity is z (i.e. the depth of the euphotic zone), then an average photosynthetic rate may be computed for the whole euphotic zone by integrating the illumination from the compensation depth to the surface and dividing by the depth. Thus:

$$P_h = \frac{pI_0}{kz}(1 - e^{-kz})$$

The effect of nutrient depletion on photosynthetic rate, designated as N, was calculated in terms of phosphate concentration, rate being held to be limited below a maximum of 0.55 μg-at P/l, i.e.

$$N = \frac{\mu\text{g--at P/l.}}{0.55}$$

If the thickness of the mixed layer is greater than the euphotic zone, turbulence (V) will also reduce the photosynthetic rate. Riley suggested that the depth of the mixed layer might be somewhat arbitrarily defined as the maximum depth at which the density does not exceed the surface density by more than 0.02 σt units. Thus:

$$V = \frac{\textit{depth of euphotic zone}}{\textit{depth of mixed layer}}$$

(provided the mixed layer is the greater). Thus:

$$P_h = \frac{pI_0}{kz}(1 - e^{-kz} NV).$$

Algal respiration (R) would reduce photosynthetic production and would be temperature dependent.
Thus:

$$R_T = R_0 e^{rT}$$

(where R_T = respiration at temperature T; R_o at temperature 0°; and r is a constant for the rate of change in respiration with temperature, experimentally determined as 0.069 for a 10°C increase).

Grazing rate G may be expressed as:

$$G = gZ$$

where g is the rate of reduction of phytoplankton per unit of animals; Z is the quantity (weight) of herbivorous zooplankton. It is likely that g will vary with temperature, though perhaps not linearly; different broods of herbivores and, of course, specific differences will affect the grazing rate. An approximation can be arrived at from the minimum daily food requirements of *Calanus* as deduced from the respiratory rate.

Riley supposed a daily food intake of from 1.2% (winter) to 7.1% (summer) of the body weight, and calculated an average value for g from these data. Finally:

$$\frac{dP}{dt}=P\left[\frac{pI_0}{kz}(1-e^{-kz}NV)-R_0e^{rT}-gZ\right].$$

The rate of change of phytoplankton population was thus expressed in terms of six ecological parameters: solar radiation, nutrients, transparency of water, depth of mixed layer, temperature and zooplankton quantity.

Riley obtained average numerical values from seasonal field data from Georges Bank. Over short periods of time approximate integration was possible, assuming a constant mean value for the variable over that period, and a relative curve of seasonal change was deduced. By statistically determining the best fit of the curve for all the phytoplankton cruise data, a seasonal cycle in absolute terms was obtained. Figure 8.3 shows the good measure of agreement between the theoretical curve and the actual cycle. Cushing (1975) has pointed out that a daily ration for herbivores of 7% body weight was too low and that the algal reproductive rate was under-estimated, though the ratio (R/G) would be approximately correct. More recent information would also indicate that production would be limited by nutrient depletion only at concentrations much lower than that proposed by Riley.

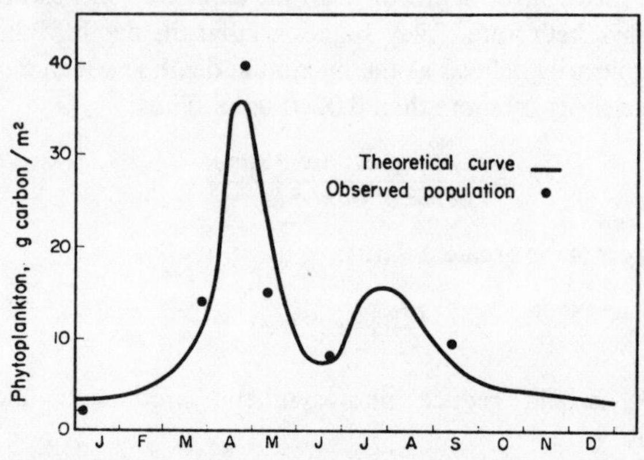

Fig. 8.3. The seasonal cycle of phytoplankton on Georges Bank calculated by approximate integration of the equation for the rate of change of the population. For comparison, observed quantities of phytoplankton are shown as dots (from Riley, 1946).

The phytoplankton cycle, deduced by the same method, for the coastal waters of New England off Woods Hole is considerably different from that of Georges Bank owing to different values of some parameters. A similar calculation for data obtained by Kokubo from Husan (Korea) (Riley, 1963) showed that changes in the phytoplankton in those waters differed quite appreciably over the two years, but the theoretical and field data agreed very well. The autumn of 1932 was more favourable to phytoplankton growth owing to better light and nutrients and sparser zooplankton (see Fig. 8.4).

Riley (1963) discusses certain refinements in modelling. A correction may be introduced, for example, for the loss of algal crop due to sinking and death, apart from grazing activity. The phosphate distribution, as one of the major limiting nutrients in the upper layers, may be computed by an assessment of the eddy diffusion in the water layers and the concentration of phosphate in deeper water. An equation for

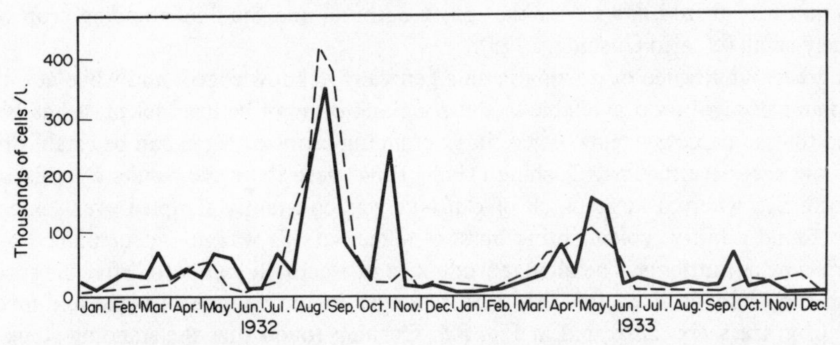

Fig. 8.4. Comparison of observed seasonal cycles of phytoplankton (solid lines) with theoretical cycles (dotted lines) at Husan (from Riley, 1963).

phytoplankton p at depth z, introducing these terms including algal loss, is as follows:

$$\frac{\mathrm{d}p}{\mathrm{d}t}=p(p_h-r_p-gh)+\frac{\delta}{\delta z}\frac{A}{\rho}\frac{\delta p}{\delta z}-\mathrm{v}\frac{\delta p}{\delta z},$$

where v = sinking rate, ρ = density of water, A = coefficient of vertical eddy diffusion.

Other computations introduced by Riley included estimating the vertical transfer of nutrient with the contribution from zooplankton excretion, and the consumption of grazers by carnivores. Production was estimated for a succession of depth intervals, from the measurements of parameters at each depth, from the surface to below the euphotic zone, assuming a steady state. Total production may thus be assessed.

Steele (1956) employed a formula similar to Riley's later equation to calculate primary production on the Fladen Ground, but he first applied equations to two depth intervals to estimate algal sinking and grazing rates, since these could not be assumed from existing data. Since Steele was concerned with seasonal changes and a steady state could not be assumed, he calculated the phytoplankton populations at the beginning and at the end of a time interval for a given depth stratum, knowing gradients in phytoplankton density and temperature at that depth interval and the zooplankton concentrations. The mean phytoplankton density was subsequently used to calculate population changes for successive depths. Steele compared the production deduced theoretically, with ^{14}C measurements, and obtained a good measure of agreement. In another model, simulating algal production in the North Sea, Steele (1958) considered the area as consisting of a two-layered system, with the thermocline fixed at 40 m, and the layers above and below homogeneous for the variable. Sinking of algae was set at 3 m/day; grazing at 3.4 m^3 swept clear/g zooplankton carbon. It was assumed that the rate of mixing of the two layers was constant and that nutrient concentration limited algal growth below 0.4 μg-at PO_4-P/l. No regeneration of phosphate was considered.

Cushing (1959a) developed a mathematical model for studying an area in the North Sea but no factor for nutrient depletion was included since he claimed that production was not dependent on nutrient concentration during the spring increase, regeneration being sufficient to maintain nutrient above a limiting level (*vide infra*). Thus the rate of change of the phytoplankton population was dependent on the algal rate of division related to light intensity, the production ratio (D_c/D_m), and the rate of change of the zooplankton population determining the grazing intensity. Grazing in that area of the North Sea was so intense, according to Cushing, that practically the whole of the production was grazed down over the spring outburst and the final standing crop was extremely small (cf. also Cushing, 1975).

The great significance of grazing is thus generally acknowledged, and while at times the amount of algal food available to the zooplankton may be insufficient, it has been claimed that some ecosystems, where large standing crops of algae can be established, may show excessive feeding. Cushing (1963, 1964) describing the results of cruises in the North Sea where a large patch of *Calanus* was constantly sampled over some $2\frac{1}{2}$ months found a fairly typical spring burst of algae but this was not accompanied by a sharp decline in nutrients. The algal reproductive rate actually *increased* after the spring peak in crop had declined (cf. Fig. 8.5). Changes in the volume of algae and in the weight of grazers are illustrated in Fig. 8.6. Cushing found that the standing stock of algae increased nearly fifty times from 0.02 to >0.90 mm^3/l, as estimated from counts, in 30 days. The apparent mean daily rate of production was thus 0.03 mm^3/l, but this was greatly reduced by grazing by the zooplankton population, which increased some thirty-fold over a 50-day period. The actual rate of algal production was deduced from estimating the decrease in cell size (transapical width) as the algae divided. Cushing estimated that at the peak of the spring bloom the diatoms were dividing at the high rate of 0.5 to 1.0 divisions per day (cf. Table 8.1). A second method for estimating algal production was based on the potential rate of reproduction to light intensity, the potential rate being raised by the production ratio (D_c/D_m). Table 8.2 indicates the quantity of phytoplankton produced and the amount grazed during cruises. The maximum production was about nine times the maximum standing stock and the amplitude of algal production about 400-fold (Table 8.1).

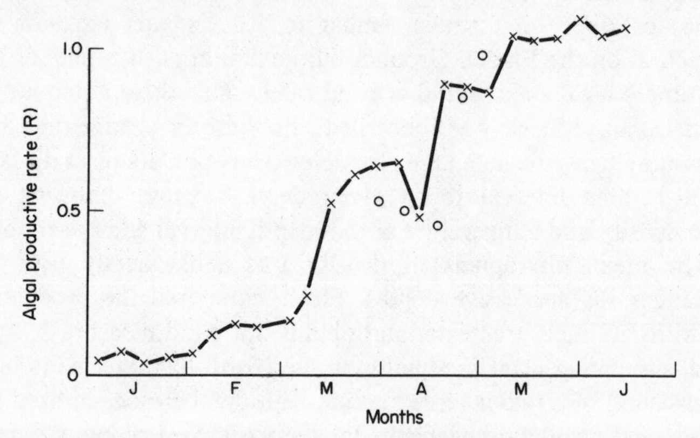

Fig. 8.5. Algal reproduction rates in the North Sea over 6 months, calculated from light data (×) and as observed from decrease in diatom cell width (o) (from Cushing, 1963).

Cushing calculated total mortality of the phytoplankton (Z) from the equation:

$$Z = R\frac{1}{t} \times \log_e P_1/P_0$$

where $R =$ daily algal reproductive rate, P_0 is the algal quantity in one cruise and P_1 the quantity in a subsequent cruise, with the time interval (t) in days. Assuming grazing was directly related to the weight of herbivores (cf. Volume 2), algal mortality due to grazing far exceeded death from any other cause (sinking, diffusion). Using estimates of the daily ration required by *Calanus* of 37.6% body weight (45% according to another

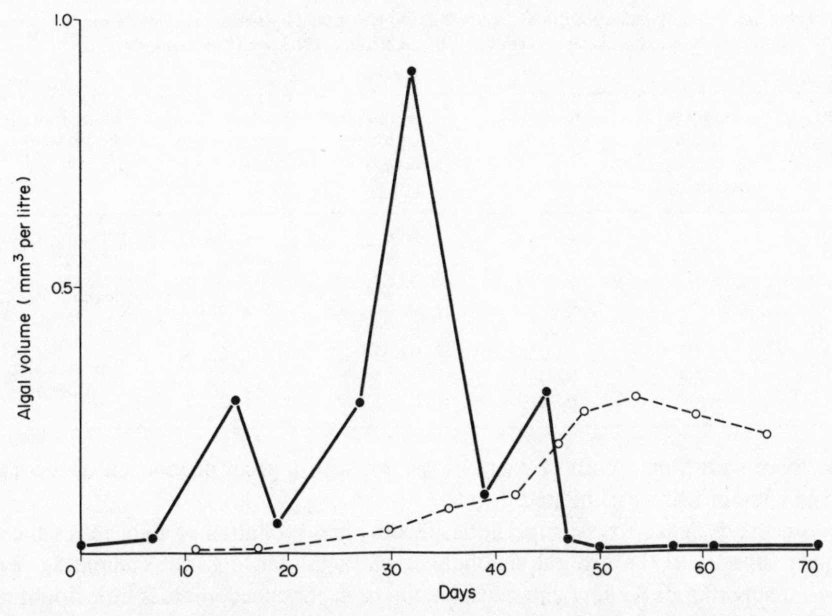

Fig. 8.6. The volume of algae (continuous line) and the volume of herbivores (broken line) as standing stock from mid-March to the beginning of June 1954. The quantity of animals is given in mg on the same scale as algal volume (from Cushing and Vucetic, 1963).

Table 8.1. Estimated phytoplankton production in the North Sea from mid-March to June 1954 (from Cushing, 1963)

Production sub-cruise	Standing stock (mm^3)	t (days)	Algal productive rate (R)	Production (mm^3)
2	0.025			
		7	0.53	2.80
3	0.286			
		7	0.55	0.33
4	0.055			
		3	0.48	0.60
5	0.279			
		8	0.88	7.87
6	0.898			
		5	0.88	1.11
7	0.107			
		6	0.86	5.06
8	0.298			
		6	0.88	0.30
9	0.022			
		3	1.04	0.31
10	0.001			
		3	1.04	0.02
11	0.005			
		6	1.04	0.03
12	0.004			
		4	1.04	0.03
13	0.005			
				Σ18.46 mm^3

calculation), Cushing suggested that superfluous feeding occurred on occasions. Thus during mid-April to the beginning of May the *Calanus* were grazing to excess. Cushing and Vucetic (1963) give daily rations equivalent to 350 and 390% body weight; for the first 10 days of May feeding was approximately just adequate (cf. Table 8.2). From that time to the beginning of June the grazers appeared to be short of food, though Cushing and Vucetic point out that if nanoplankton flagellates (not estimated) then formed an important part of the population, they might supply the necessary ration. Cushing believes that *Calanus* can filter its food requirements with very little swimming activity,

Table 8.2. Estimated amounts of phytoplankton consumed by herbivores in relation to crop and production in the North Sea (from Cushing, 1963 and 1964 – modified)

Cruise	Average standing stock (mm³/l)	Production (mm³/l)	Daily quantity eaten by one *Calanus* V (mg)	Quantity eaten as percentage body weight	Daily ration/ body weight
5–6	0.588	3.736	5.307	349.9	Superfluous feeding (370%)
6–7	0.501	1.603	3.714	390.3	
7–8	0.201	0.391	0.661	26.8	Adequate feeding (26%)
8–9	0.160	0.558	0.463	46.3	
9–10	0.011	0.022	0.036	4.3	
10–11	0.003	0.003	0.0034	0.05	Low feeding (0.8%)
11–12	0.004	0.003	0.015	1.15	
12–13	0.004	0.0019	0.012	1.23	

and therefore with a minimum of energy loss, but with a great proportion of the algae destroyed without being assimilated.

Experiments designed to test superfluous feeding and spoilation of algal cells have not lent much support to the general significance of excess feeding (cf. Volume 2). Even, however, if superfluous feeding can occasionally be experienced, there is little doubt that generally in the marine environment the planktonic herbivores are resource limited, i.e. that their food, the phytoplankton of the open sea, is eaten out almost as fast as it is formed. Nearly all the plant production passes through the planktonic herbivores. There is not much direct support for effective utilization of faecal pellet material from planktonic herbivores by other planktonic animals; in any case, faecal pellets seem to sink relatively rapidly to the bottom of the sea where they undoubtedly assist in the nutrition of the benthos. Some workers believe, nevertheless, that detrital material forms a considerable proportion of the food of some zooplankton, especially those of sub-tropical seas (*vide supra*) and of the deeper sea zooplankton. Vinogradov (e.g. 1973) and other Russian scientists in particular have stressed the importance of diurnal vertical migration in that food material originally arising from phytoplankton eaten in the euphotic zone is transported to deeper levels by the migrating zooplankton. The animal itself or the detritus is there consumed by mid-water pelagic animals which have also migrated vertically. They in turn carry food material to greater depth for the nutrition of the very deep-living zooplankton. A ladder of nutrition is thus established based on a chain of vertical migration of different depths and of different ranges (cf. Fig. 8.7).

Time Delays in Phytoplankton and Zooplankton Production: Seasonal Variations

While grazing is a major cause of reduction of phytoplankton crop, variations in the time lag between phytoplankton production and zooplankton increase can lead to remarkable differences in phytoplankton concentration. In open-ocean environments, particularly in the tropics, the lag may be very little; in some sub-tropical areas the lag may be somewhat greater. Where upwelling occurs close to the coast, a marked lag may occur between the very high primary production and the onset of grazing. Much of the plant production may then sink to the bottom. Some bottom areas of marked decomposition with an abundance of diatomaceous material which are found in a high-

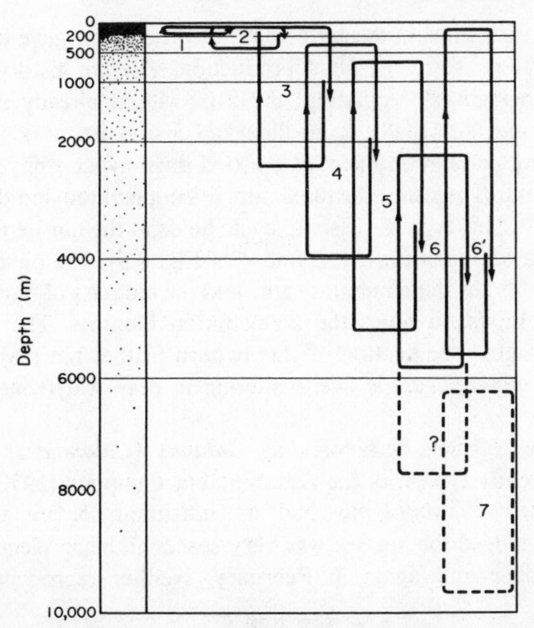

Fig. 8.7. A schematic representation of the different extent of vertical migrations of zooplankton. The density of points on the left of the diagram illustrates the variations in plankton abundance with depth. 1. Migrations of the surface species. 2. Migrations extending over the surface and transition zone. 3. Migrations extending over the surface, transition and upper layers' deep-sea zones. 4. Migrations extending over transition and part of the deep-sea zone. 5. Migrations within the whole deep-sea zone. 6. Irregular migrations of some species extending through the whole water column. 7. Range of distribution of ultra-abyssal animals (after Vinogradov, 1962).

production zone like Walvis Bay may result in part from a lag between production and grazing. On the other hand, Menzel (1967) found little sinking of algae in the rich Peruvian waters; grazing must have been very effective and presumably there was little lag in that area. Heron (1972) describes extraordinary rates of growth and reproduction of salp populations partly related to very effective utilization of blooms of phytoplankton with very little time lag (cf. Volume 2).

Cushing (1975) contrasts the comparatively low algal crop and the little delay in onset of grazing in warm oceans with the situation at high latitudes. In the Sargasso, for example, the compensation depth is about 120 m depth in summer; in winter, which is the more productive period, the compensation depth may shallow to about 60 m, largely as a result of the attenuation of light by the algae. The production ratio, as defined by Cushing (D_c/D_m – p. 217), nevertheless, varies little over the year. Primary production is more or less continuous, but seasonal amplitude is low and the small standing stock is grazed with little or no delay. By contrast, high latitude open-oceans, subject to great seasonal light changes and to deep mixing of ocean waters, are characterized by a slow start to production; delay in grazing is often marked with the slowness of zooplankton reproduction, and amplitude of phytoplankton change may be extremely large. Upwelling areas usually simulate high latitudes. At high latitudes, seasonal migration and the reproductive cycles of zooplankton can, however, modify the overall pattern.

Heinrich (1962) examined the patterns of seasonal changes in phytoplankton populations, especially in relation to the grazing of zooplankton and lag effect. Relatively rapid development of the zooplankton, and the coincidence of a dominant population of older stages (copepodites) with maximum phytoplankton production gave a more balanced seasonal cycle and fuller use of the plant production. The life history of herbivorous zooplankton is all-important in "regulating" the crop; with an already established brood of zooplankton, grazing is virtually immediate and a massive peak in spring is prevented. When breeding and development of a brood must await a phytoplankton crop, the peak of phytoplankton precedes the maximum in zooplankton and there is a marked tendency to instability. This type of seasonal cycle holds in the northern Atlantic, north Siberian seas and Sea of Japan (figs. 1, 2 and 4 of Fig. 8.8). The phytoplankton peak may reach a thousand times the minimum value and the biomass of phytoplankton may amount to ten to a thousand times the zooplankton biomass. The Norwegian Sea possibly presents a slight modification of this pattern in that the majority of grazers (especially *Calanus finmarchicus*) is over-wintering in deep water but migrates up to feed on the algal crop.

Similar cycles due to such herbivores as *Calanus* (*Calanoides*) *acutus* and *C. propinquus* are apparently typical of the Antarctic but Philippon (1972) found that off the Kerguelen Islands *C. simillimus* had its maximum *before* the pronounced phytoplankton peak; indeed the species was very seasonal, being plentiful only during September and October and again in February. Neither reproduction nor brood

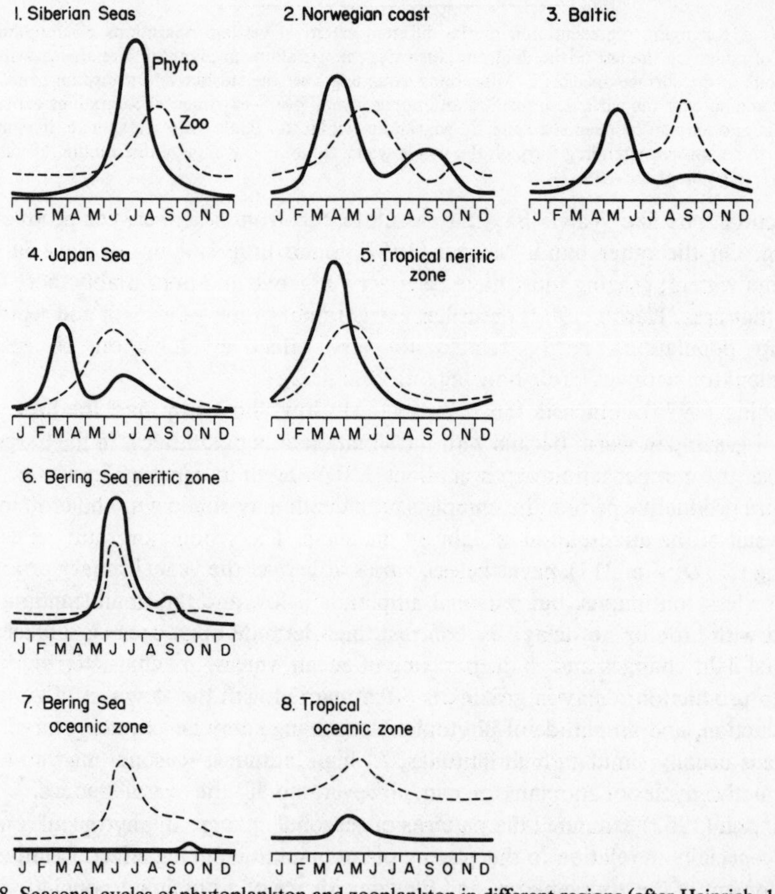

Fig. 8.8. Seasonal cycles of phytoplankton and zooplankton in different regions (after Heinrich, 1962).

development appeared to depend on a phytoplankton bloom. By contrast, the dominant copepods off Kerguelen, *Drepanopus forcipatus*, *D. pectinatus* and *Oithona helgolandica*, appeared to be entirely dependent on the phytoplankton outburst; though reproduction could apparently continue through the year, the major broods followed the main algal burst. The grazing activity of the herbivores in reducing the phytoplankton crop was very clear.

Colebrook and Robinson (1961) observed a difference in the cycles of phytoplankton and of copepods, as the major herbivores, between the fairly shallow waters off British coasts and the deep open North Atlantic. While in the shallower seas an early spring phytoplankton increase was followed by a rise in copepods in March/April to a maximum in May or June, in the deep Atlantic waters the numbers of copepods remained low during spring, despite a marked phytoplankton peak in April, and while some increase in copepod numbers followed in May or June, peak numbers were delayed till August or September (Fig. 8.9). The lag in grazing affects the cycle of algae. It is not obvious why the delay in zooplankton increase occurred over deeper waters; Colebrook and Robinson suggest that species composition might be significant. The main grazing species (*Calanus finmarchicus*, *Pseudocalanus* and *Paracalanus*), well represented in coastal waters, were not abundant in spring over the deeper waters.

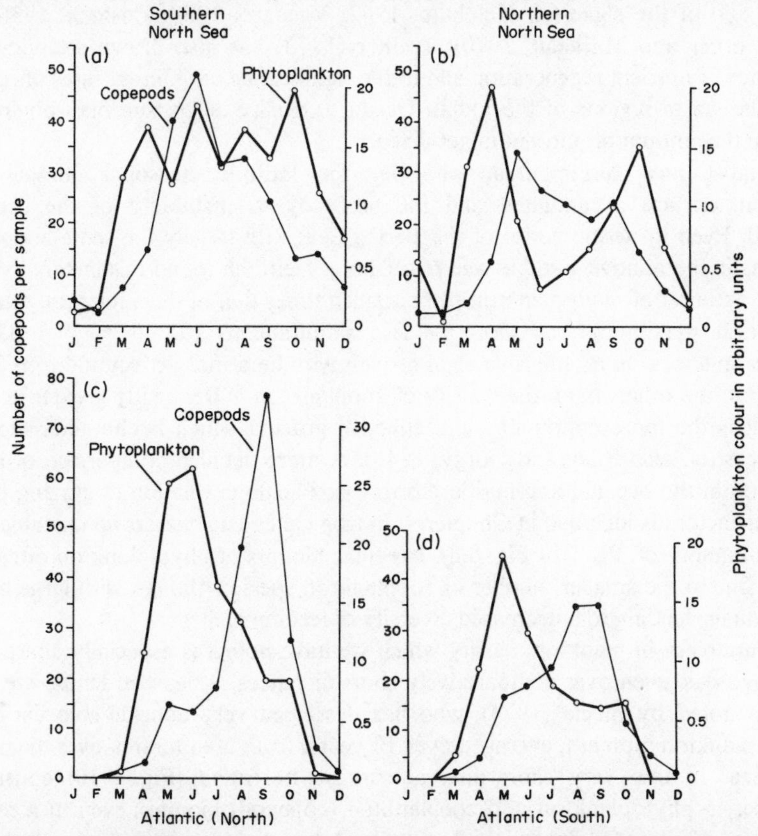

Fig. 8.9. Graphs showing the seasonal variations (at a depth of 10 m) of the abundance of copepods and the amount of green (phytoplankton) colour on the Continuous Plankton Recorder silks in four areas. (a and b), shallow areas <200 m; (c and d) deeper seas >200 m. The graphs were obtained by combining the results collected during the 9 years 1948–56 (after Colebrook and Robinson, 1961).

Heinrich contrasts the instability of the phytoplankton in the North Atlantic, Sea of Japan and the Antarctic, with other high latitude seas like the oceanic western Bering Sea and northern North Pacific. Although the bulk of the phytoplankton crop is produced fairly early (about June) in these latter areas, the extensive population of copepodites already present ensures that there is little delay in the onset of grazing. The system is well balanced and phytoplankton well utilized. According to Heinrich, the seasonal fluctuations in phytoplankton are very small, even by as little as a factor of 2; no spring peak in crop appears but a very small autumn peak is observed. The biomass of zooplankton exceeds the phytoplankton all the year. The cycle of events is more similar to that of the tropical open ocean, already described, though the crop of phytoplankton in tropical seas is usually much less (cf. figs. 7 and 8 of Fig. 8.8). In warm oceans the generally small sized, rapidly developing, herbivorous zooplankton assist in maintaining this balance and seasonal fluctuations in phytoplankton are slight (cf. Chapter 5).

In the more or less steady state established between phytoplankton production and zooplankton grazing typical of some temperate and most warm oceans, a critical factor associated with grazing is the rate of excretion of nutrients (e.g. ammonia, phosphate) by the zooplankton, permitting regeneration of limiting ions and allowing some reproduction of the algae (cf. Ketchum, 1962; Vaccaro, 1963; Cushing, 1958, 1975; Butler, Corner and Marshall, 1970). Tranter (1973) has also drawn attention to the importance of nutrient regeneration and to the significance of different rates of regeneration in the warm regions of the Indian Ocean. Intensity of grazing may obviously be related to the amount of nutrient mineralized.

In almost any neritic area, whatever the latitude, seasonal fluctuations of phytoplankton are accentuated and the tendency to instability of the ecosystem increased. Even in neritic zones of the Bering Sea, with largely the same zooplankton species as in the relatively stable waters offshore, Heinrich found this instability, with a seasonal outburst of phytoplankton some sixteen times that of the minimum, though the zooplankton maximum was more or less simultaneous (fig. 6 of Fig. 8.8). She attributes this to a more intensive algal growth with land mass effect and to fewer herbivores. On the other hand, the density of zooplankton is frequently greater in inshore waters; it is the more continuous and effective grazing which is characteristic of offshore areas (cf. also Riley and Gorgy, 1948). A more detailed comparison of primary production in the oceanic and neritic North-east Pacific in relation to grazing pressure and other factors is included in Chapter 9. In tropical coastal waters an unstable system is very common (cf. fig. 5 of Fig. 8.8). Irregular blooms of phytoplankton often occur, possibly due to the smaller number of zooplankton species (though with large numbers of individuals) leading to a decreased diversity of feeding niches.

The variability in plankton density which we have noted as especially characteristic of shallow seas, even over comparatively short distances, at least in temperate waters, has been noted by Steele (1974), who has described very considerable variation in summer plankton biomass, averaged over 10 years, from area to area over the northern North Sea. In two very short interval surveys he found (Fig. 8.10) considerable differences in phytoplankton and zooplankton (copepod) biomass even in a relatively small area, 40 by 60 kilometres. Apart from the obvious patchiness, which would appear to be real, there was some suggestion of an inverse relationship between copepod and phytoplankton density. From an examination of the typical seasonal plankton cycle in the northern North Sea based on long-term averages by Colebrook and Robinson, Steele (1974) suggested that the sharp late spring reduction in phytoplankton crop was the result of grazing (cf. Fig. 8.9(b)). Earlier estimates suggested that some 90 g C/m^2/year were available in the northern North Sea as phytoplankton production, equivalent to 900 kcal/m^2/year. Of the zooplankton, the major herbivore is *Calanus*. In broad terms, this copepod has three broods per annum in the North Sea (cf. Volume 2). From the average intervals between the various growth stages and the average weights of the stages, the production of zooplankton was estimated as equivalent to 175 kcal/m^2/year. Thus the yield from algal primary production to the grazer (herbivore) trophic level approached 19%.

Details of experimental work on food selection and on rates of feeding by zooplankton will be left until Volume 2, but models attempting to simulate and quantify the effects of grazing on plant production obviously depend to a considerable extent on the results of such laboratory experiments. Frequently it is difficult to relate experimental data to the complexities of the marine environment. If we may anticipate, however, some of the experimental work, we may look at a preliminary model proposed by Steele for the northern North Sea in so far as it examines cycles of phytoplankton subjected to grazing.

Steele proposed a minimum threshold value for grazing (but cf. Volume 2). With the relatively small range in temperature, he suggests that the metabolism of the

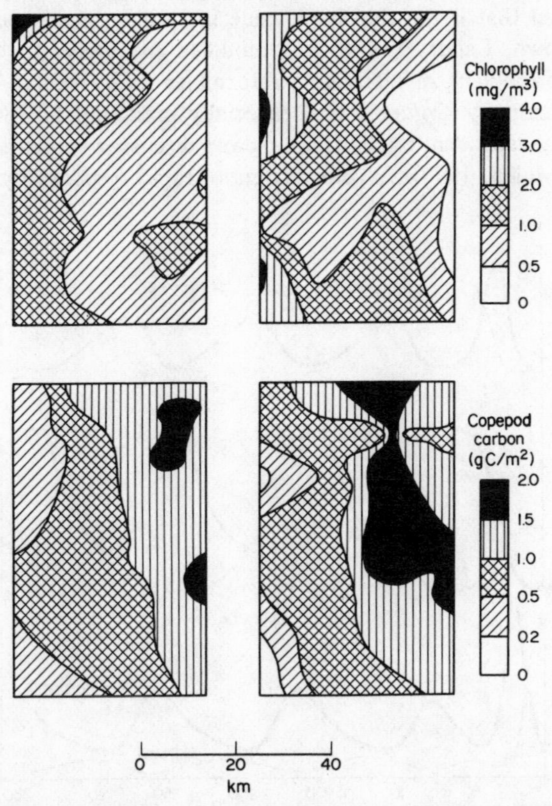

Fig. 8.10. Distribution of phytoplankton (as chlorophyll *a*) and of copepods (as carbon) in the North Sea on two successive plankton surveys (after Steele, 1974).

zooplankton might be regarded as independent of temperature over that range but must be related to feeding rate. The fluctuations in plant crop (as carbon) were then examined, assuming a single cohort of zooplankton grazing on the phytoplankton, and, from earlier estimates, a plant production over a period of approximately 200 days (the main production season for the North Sea) amounting to about 86 g C/m². Calculations allowed for the effect of some nutrient mixing; an average figure ($V = 0.01$) was assumed between the low rate of mixing typical of areas such as the Fladen Ground, and higher mixing rates experienced in shallow inshore waters of the North Sea. The excretion of ammonia by the zooplankton must be included since nitrogen supplies to the algae are thereby supplemented. Figure 8.11 shows the very large variations expected in plant crop. With a single cohort of zooplankton, a considerable peak of phytoplankton would occur in the spring, followed by a series of fluctuations. These fluctuations appear to be of the right order of magnitude, since the actual density of algae in the North Sea over the summer ranged from 100–400 g C/m². The major spring peak, which exceeds 500 mg C/m³, is mainly a result of a lag in effective grazing before the zooplankton population increased sufficiently. The small subsidiary peaks are presumably due to slight lags in grazing later during the season. With more than one cohort of zooplankton, the peaks would tend to be smoothed out. The total weight of zooplankton achieved at the peak (some 7 g C/m²) by the model is approximately correct, since Steele found 4–12 g C/m² as zooplankton over different years. Steele, however, observed that if the threshold value for grazing were reduced to zero, the model breaks down. Landry (1976) has modified the model to include the effects of such factors as zooplankton respiratory rate influenced by feeding, continuous (multi-cohort) reproduction by *Calanus*, and especially predation on the herbivore population as a density-dependent relationship, factors discussed subsequently by Steele (cf. also Steele and Mullin, 1977). A generally more rapid response by the herbivores to

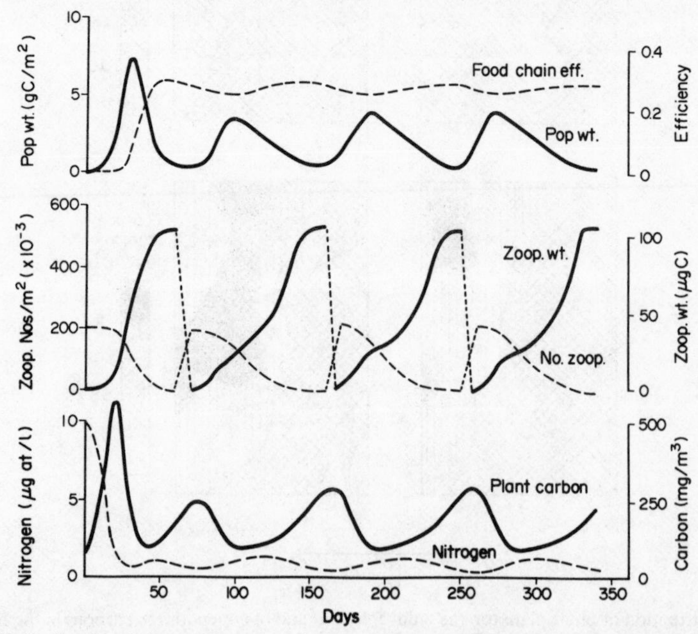

Fig. 8.11. Simulated changes in crops of phytoplankton and zooplankton (after Steele, 1974).

change in phytoplankton density is evident, giving a more stable system. Landry suggests that predation on the younger *Calanus* stages produces a greater degree of stability, rendering the assumption of a grazing threshold for the herbivores unnecessary, but proof of the presence or absence of a grazing threshold must await unequivocal laboratory results (cf. Volume 2).

Undoubtedly some of the obvious patchiness in the distribution of phytoplankton in the oceans is in part due to intensive grazing. In other words, the distribution of the herbivores causes the phytoplankton concentrations rather than the phytoplankton patches influencing the distribution of the grazers. Some impression of the complexity of the rates of change of phytoplankton and zooplankton populations, including the effects of different periods and varying intensities of grazing over a 24-hour period, variations in algal growth dependent on day length, changes in metabolic rate, etc., have been investigated, using a computer simulation model, by McAllister (1970) (cf. Volume 2).

Since phytoplankton production is limited to the euphotic zone whereas zooplankton grazing may occur throughout the water column, the standing crop of zooplankton beneath a unit area may be several times greater than the crop of phytoplankton. This apparent inversion of the typical trophic pyramid appears to be a reasonably common feature of phytoplankton/zooplankton relationships in the marine environment. Philippon (1972) points out, for example, that it is true of the older investigations carried out by Harvey *et al.* (1935) for the English Channel and by Riley (1956) and his colleagues for Long Island Sound. However, even in such areas, peak crops of phytoplankton may exceed the biomass of zooplankton on occasions. Although the abundance of copepods and phytoplankton in Colebrook and Robinson's (1961) data for various areas of the North Atlantic and North Sea is expressed in numbers and in arbitrary units, respectively, it would appear that the ratio of biomass of phytoplankton to zooplankton changes significantly during the productive season. For example, in the northern North Sea and the open Atlantic, while phytoplankton may exceed copepod abundance very considerably in spring, the reverse appears to be true in mid-summer (northern North Sea) or late summer (Atlantic). On the other hand, in the southern North Sea, using the same relative scales, copepods rarely seem to exceed phytoplankton (cf. Fig. 8.9). Steele (1956) compares the *maximum* crop of algae and of zooplankton for several of these seas (cf. Table 9.2, Chapter 9); in all cases the maximum biomass of zooplankton, at least at some period of the year, greatly exceeds the phytoplankton biomass. In part a greater zooplankton crop is due to the relatively slower turnover and slower reproductive rate of the zooplankton as compared with the phytoplankton. Philippon found the same inverse relationship in an investigation of an inshore area off the sub-Antarctic Kerguelen Islands. The zooplankton was dominated by very few species of copepods and amphipods, with some decapod larvae. Figure 8.12 illustrates the phytoplankton/zooplankton relationships over a 12-month period. When the dry weights of the phytoplankton and zooplankton were compared (Table 8.3) the biomass of zooplankton is almost always far greater (means: 14 mg/m^3 phytoplankton; 136 mg/m^3 zooplankton).

While suppression of a spring peak of phytoplankton is sometimes attributable to grazing (cf. Heinrich, etc., *vide supra*), the commencement of a bloom may be associated with a reduction in grazing. Even more characteristic, in poor "blue water" environments even a brief reduction in grazing can produce a minor bloom. Pratt (1965) investigated over several years the onset of the winter-spring phytoplankton bloom, which occurs about mid-December, in Narragansett Bay (cf. Chapter 6). Initiation was not apparently related to nutrient concentration, stabilization of the water, nor temperature, nor was there any clear correlation between incident illumination and the timing of the bloom in such very shallow waters (but cf. Hitchcock and Smayda, 1977, p. 274, Chapter 6). The flowering period ended in May/June; nutrient concentrations fell sharply at the peak blooming, nitrate being usually exhausted and silicon markedly reduced and apparently limiting growth. The major factor permitting the start of the bloom appears, however, to be a reduction in grazing pressure, especially of the two major copepods, *Acartia* and *Oithona.* In October, when the diatom population is approximately minimal, the zooplankton, though declining, is of sufficient density to exert considerable grazing pressure, but consistently over the years of study, the zooplankton reached a minimal density about November, just before the beginning of the phytoplankton growth. *Skeletonema,* which dominates the phytoplankton, was used

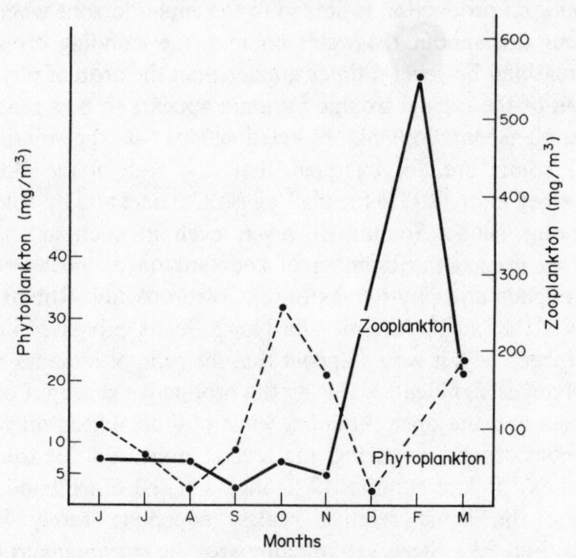

Fig. 8.12. Seasonal variations in dry weight of phytoplankton and zooplankton off Kerguelen (after Philippon, 1972).

Table 8.3. Mean monthly dry weights of phytoplankton and zooplankton (from Philippon, 1972)

Month	Phytoplankton mg/m³	Zooplankton mg/m³
May	16.4	
June	12.9	59.9
July	7.9	
August	3.3	54.1
September	9.1	22.5
October	31.6	54.7
November	21.2	39.1
December	2.7	
February	12.8	557.7
March	23.4	167.2

extensively by the two major copepods and Pratt believes that the minimum grazing about November/December allows the algae to begin multiplication. The date of onset of the phytoplankton increase, and the timing of the peak population tend to change in moving seaward from the Bay, and the size of the maximum population is greater close inside the Bay, probably reflecting the greater nutrient concentrations.

Martin (1970), analysing the effects of varying intensities of a number of factors on algal growth (nutrients, light, grazing) in Narragansett Bay, also demonstrated the great importance of grazing. During spring, a large grazing zooplankton population was present but reproduction of the algae was high enough to maintain the crop. In June, despite a tenfold increase in grazing rate partly in relation to temperature, a high level of crop was maintained, mainly by the growth of *Skeletonema*, until the end of the month. Subsequently, in July, a rapid fall in the grazing population was observed, due to the predation of ctenophores on the herbivorous copepod population, but only a small standing crop of algae was maintained, since the algae were then limited by factors other than grazing, possibly senescence. During the autumn, only a low algal crop was present owing chiefly to poor light conditions, but grazing activity was renewed though not at the highest level; as a result the crop of algae declined even further. By November, despite a fall in zooplankton population and very little grazing, the lowest crop of algae was recorded, consequent upon the very poor conditions for algal growth. An early winter/spring burst of phytoplankton growth in December/January led to a considerable increase in diatom crop, uninhibited by the continued lack of grazing.

Attempts to quantify the effects of various factors and their changes with time on phytoplankton/zooplankton relationships using simulation models are of great value, but are subject to certain limitations. Some of the difficulties stem from our lack of knowledge of the precise physiological effects of external factors on the various taxa of phytoplankton and on particular groups of zooplankton, and of the effects of adaptation by these organisms. Although computers permit the simulation of more and more complex systems of plankton production, the basic processes involved cannot yet be deduced theoretically, and the acknowledged variability of the plankton makes the task of calculating coefficients and parameters from field and experimental data for use in constructing models and for monitoring their results very difficult.

Platt, Denman and Jassby's (1977) review makes the important point that modelling of plankton production is mainly concerned with the prediction of varying quantities in the marine environment. Since the phytoplankton living in the mixed (euphotic) layer is characterized by variability in space and time the monitoring of any model by direct measurements must be related to an appropriate time/space domain. Recognizing the fundamental significance of light, they point out that primary production (P^B), normalized to chlorophyll, can be readily related to light energy (I), the two important quantities being (P_m^B) (mg C/mg chl *a*/hr) at light saturation (i.e. the assimilation number) and (α), the initial slope of the P^B vs. I curve which measures the maximal photosynthetic efficiency per unit of chlorophyll. The photosynthesis/light relationship is well represented by these two quantities, possibly with a parameter for photoinhibition. Since the assimilation number is affected by such factors as temperature, preadaptation to light and nutrient level, and (α), though less variable, appears to vary with season amongst other factors (both vary also with species), Platt *et al.* believe that, having determined the photosynthesis/light relationship, modelling should concentrate on the effect of environmental variables on the values of (α) and (P_m^B). They have themselves developed a general model for predicting phytoplankton biomass. After determining the basic photosynthesis/light model as modified by nutrient effects, other factors (e.g. sinking of algae, turbulent horizontal and vertical diffusion, upwelling, grazing) are considered, for which coefficients have been calculated, many from data obtained from literature. Since a very wide range of marine environments from estuarine to open ocean were drawn upon, the coefficients show a wide range, in most cases over several orders of magnitude. The range would be considerably narrowed if the data were restricted to a more limited marine environment, but any one of the processes appears to be capable of dominating the equation for production, given the appropriate conditions. Though as Platt *et al.* remark, "... almost nowhere in the modelling of the productivity of marine phytoplankton are we much beyond the semi-empirical stage ...", such models point the way for future studies on plankton prediction and emphasize the need for experimental and theoretical analyses of the effects of environmental variables, apart from light, on assimilation number ($P_m{}^B$) and (α).

Steele and Mullin (1977) also emphasize the necessity for a detailed analysis of the zooplankton in modelling plankton production. They point out that whereas earlier models of plant production often used a single parameter to represent grazing, the complexities of variability in the zooplankton (e.g. horizontal and vertical distribution, vertical migration, reproduction, growth rate, predation) must be represented separately in model simulations. Rate of change of herbivores (copepods) is considered with reference to ingestion, egestion, metabolism and predation. An expression is developed for the change in weight of a copepod in terms of losses and gains of energy, and for a cohort as a function of numbers assuming that all the animals are finally consumed. Experimental data are used as far as possible to include expressions for ingestion, assimilation efficiences, respiration, reproduction, growth and losses due to predation.

A model simulation studied the changes in phytoplankton (as C/m^3), nutrients, and zooplankton biomass (as g C/m^2), assuming a twelve-layer water column with vertical exchange between the layers, a single copepod grazer similar in weight, growth and reproduction to *Calanus finmarchicus*, and six cohorts of the copepod, separated by recruitment after a fixed period of 20 days. The depth distribution for maximal photosynthesis was related to nutrient depletion, and sinking rates were computed. Figure 8.13 illustrates the changes in zooplankton biomass and in phytoplankton assuming suitable values for ingestion, assimilation, growth, respiration and predation. The initial large peaks in phytoplankton and zooplankton are followed by a succession of smaller fluctuations. Figure 8.14 shows the differences introduced by varying maximum ingestion and lowering the rate of increase of ingestion ($I_m = 2.5$ instead of 2.0; $\delta = 100$ instead of 200). The greater variability is clear; moreover, successive cohorts have fast and slow growth, and over a period of 40 days (two life-cycles of the copepod) there is a high biomass with low phytoplankton crop. Periodicities in copepod recruitment thus can affect the phytoplankton as much as physical factors. The problem of whether a threshold food concentration exists in grazing (P_o) is important in the simulation. With a twelve-layer model in contrast to the single-layer model earlier studied by Steele (p. 366), an output in terms of phytoplankton and zooplankton biomass is possible, although there is greater variability. Predation on herbivores can also be varied in intensity but the effect is much affected by the grazing threshold assumed for the herbivores. This threshold, according to Steele and Mullin, allows variations in the other parameters to be absorbed to a considerable degree. A much greater range in maximum phytoplankton crop is experienced when vertical migration of the copepods is introduced, especially if the animals spend more time in one stratum.

Steele and Mullin point out that spatial and temporal variations in phytoplankton and zooplankton are typical of the natural environment and the model indicates that this is due as much to the population structure, behaviour and physiology of the zooplankton as to the physical environment. However, our knowledge of the experimental basis for predictions of zooplankton production is very limited; in particular our field data hardly distinguish between the effects of different species. Future work must attempt to distinguish between different groups of zooplankton, even on a size basis, rather than on an "average" herbivore or "average" predator.

To some extent the modelling proposed by Vinogradov and Menshutkin (1977) for a pelagic tropical community takes account of some of these different trophic groups in the zooplankton. They first comment on the apparent discrepancy between the biomass of zooplankton and the small average crop of phytoplankton in tropical oligotrophic

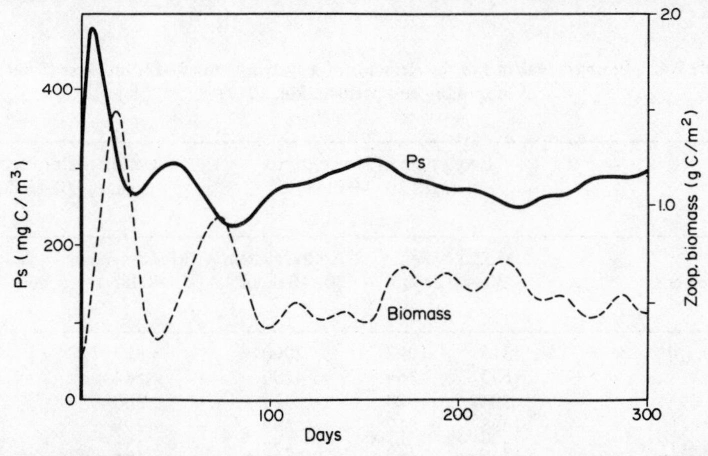

Fig. 8.13. Simulated phytoplankton (P_s) and zooplankton (biomass) variations with time, assuming maximum ingestion (I_m) = 2; rate of increase of ingestion (δ) = 200; threshold ingestion concentration (P_0) = 50 (after Steele and Mullin, 1977).

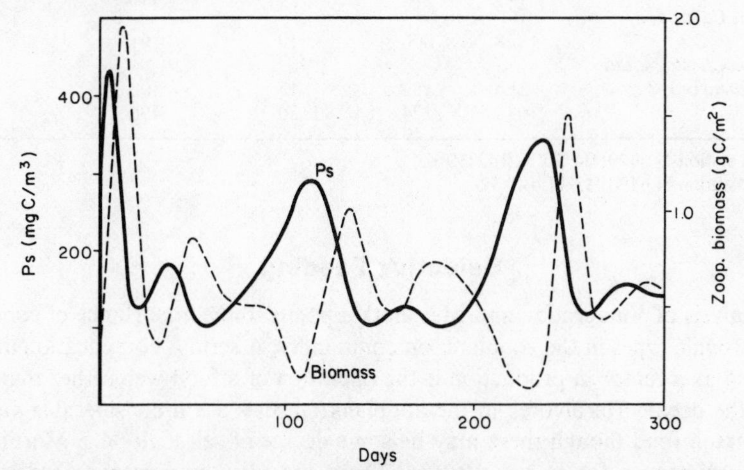

Fig. 8.14. Simulated phytoplankton (P_s) and zooplankton (biomass) variations, with same parameters as Fig. 8.13, except I_m = 2.5; δ = 100 (after Steele and Mullin, 1977).

areas, but explain this as due to the alternation of richer layers of algae and bacterioplankton with poor strata. Their model describes energy flow through a tropical open-ocean community, beginning with upwelling of nutrient-rich water as in a near-equatorial divergence, and using values for food consumption with selection for different trophic types, and for respiration, assimilation and growth efficiency. Daily production for zooplankton is based on respiration and growth efficiency and is related to ^{14}C measurements for phytoplankton and counts of bacterial crop. A model output illustrates the time scale and biomass for the development of various elements of the ecosystem, commencing with such an upwelling. Thus, while a phytoplankton peak is estimated to develop in 5 to 7 days, small herbivores reach a maximum in about 25 days and various carnivorous groups from 35–50 days. Table 8.4 compares the biomass of various elements estimated by the model with observations made in tropical waters. The time delays in the development of the herbivores, and various carnivorous groups, may be compared with their observed distributions from upwelling regions (cf. Volume 2).

Table 8.4. Biomass (cal/m^2) of the elements of a system in the 0–150-m layer (from Vinogradov and Menshutkin, 1977)

	Ecosystem of low maturity (ages 30–40 days)			A more mature ecosystem (ages 60–80 days)	
Elements	Model system 30–40 days		Actual system 30–40 days[(a)]	Model system 80 days	Actual system 60–80 days[(a)]
Phytoplankton (p)	1319	1092	2000	827	900
Bacteria (b)	1673	864	4100	564	2180
Nauplii (f_2)	394	303	321	300	?
Small herbivores (f_3)	1338	612	525	290	74
Large herbivores (f_4)	1416	726	420	252	164
Σf	3148	1641	1266	842	
Carnivorous *Cyclopoida* (s_1)	624	491	495	203	236
Carnivorous *Calanidae* (s_2)	288	600	610	191	175
Carnivorous *Chaetognatha* + *Polychaeta* (s_3)	184	183	15	102	51
Σs	1096	1274	1120	496	462

(a) "Vitjaz" station N 6429: 04°30′N, 142°30′E.
(b) "Vitjaz" station N 6493: 13°N, 140°E.

Selective Feeding

The analysis of Vinogradov and Menshutkin points to the importance of considering different trophic types in the zooplankton community. A serious complication in assessing grazing as a factor in production is the operation of a food web rather than a food chain in the ocean. Herbivores in the zooplankton may use a considerable variety of phytoplankton food though there may be some degree of selectivity e.g. Martin (1970) reported preferences for *Acartia tonsa*. There are also omnivorous feeders in the plankton. Predation by carnivorous members of the zooplankton introduces an even greater measure of complexity; while carnivorous forms may be primary, secondary or even tertiary carnivores, many of the animals can change their dietary habits depending upon the particular exigencies of the environment (cf. Volume 2). Even if it is possible to make a crude adjustment to the system under analysis for predation, the role of microzooplankton in the food web is a complex factor which is very difficult to quantify. Parsons and LeBrasseur (1970) described two basic types of food chains:

(a) nanoplankton → microzooplankton → macrozooplankton
(b) net phytoplankton → macrozooplankton,

mainly from studies in the North Pacific. They showed marked differences in the proportions of net and nanoplankton in inshore and offshore regions. More recently, Parsons (1976) has commented that the food web in the marine plankton ecosystem, with alternative pathways for energy flow, is one of the chief features of uncertainty in attempting modelling of food chain production. An example quoted by Parsons relates to two studies at different periods in the Peruvian upwelling region (cf. Chapter 9).

The generalization is frequently made that, when food resources are limited, animals cannot afford to feed selectively. Undoubtedly amongst the herbivorous zooplankton many larger copepods such as *Calanus* tend to select larger particles within the size range which they can consume when food is fairly abundant. Experimental evidence dealing with these aspects of feeding is referred to in Volume 2. Discussing filter feeding, Conover (1976) quotes workers such as Poulet (1973) who suggests that the copepod *Pseudocalanus* covers a very wide range of particle size in feeding but is also opportunistic within limits in its feeding habits, concentrating on whatever size of phytoplankton particle is most abundant. This conclusion is supported to some extent by the investigations of Parsons, LeBrasseur, Fulton and Kennedy (1969) in which a combination of field and experimental data demonstrated extensive grazing by selected zooplankton species on natural phytoplankton populations in the Strait of Georgia (British Columbia). The diatoms *Skeletonema costatum*, *Thalassiosira nordenskioldii* and *T. rotula* were heavily grazed, especially by *Calanus plumchrus* and *C. pacificus*. The daily ration generally exceeded 10% and even reached 60% of the body weight. Nanoplankton flagellates were sometimes effectively grazed, usually as a supplementary food, but these smaller phytoplankton particles, especially when present at lowered concentrations, proved to be inadequate and a poor growth of zooplankton resulted from the diet. A growth rate, as percentage weight increase per day, was calculated from the intervals between major stages and their weights. For *C. plumchrus* the growth rate was 14% from Nauplus I to Nauplus VI; 3.5% from copepodite I to copepodite III, and 8.8% from copepodite III to copepodite V. From these average growth rates the standing stock of *C. plumchrus* was estimated from the stock existing in February to that expected in May; field observed and computed stocks showed good agreement (cf. Parsons *et al.*, 1969, Fig. 8.15). For an effective conversion of primary to secondary production, availability of plentiful phytoplankton and a suitable size of phytoplankton particle was a requirement. Parsons *et al.* consider that a minimum concentration of food particles is necessary to stimulate feeding activity; cessation of grazing at very low phytoplankton concentrations would give some opportunity for the phytoplankton population to recover (cf. Adams and Steele, 1966; Steele, 1974). Computations of grazing activity by Cushing (1959b, 1968), however, assume a filtering rate in water devoid of food particles. Further discussion of this problem is included in

Volume 2. Some zooplankton appear to feed normally in the euphotic zone only during the night-time, after ascent to higher levels. When at deeper levels below the euphotic zone, the zooplankton may more or less cease feeding, due to the very much decreased concentration of food particles. Such a situation may be especially typical of temperate and boreal waters during the productive period when the stock of phytoplankton in the euphotic zone will be relatively high.

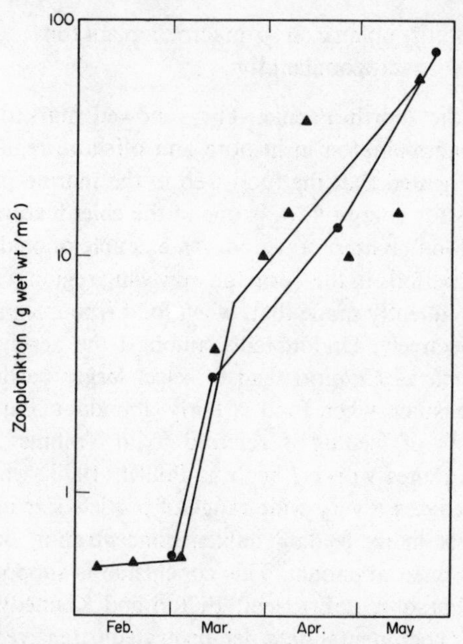

Fig. 8.15. The *in situ* (▲) and predicted (●) standing stock of *Calanus* (from Parsons *et al.*, 1969).

The degree of selectively by grazers could have an effect on the size distribution of the phytoplankton. Wimpenny's (1973) observations on the differences in size spectrum of grazed and ungrazed diatom populations are relevant. If, however, larger particles are selected by some herbivores only when food is abundant, and if many grazers follow the habit of *Pseudocalanus* in feeding chiefly on size ranges of particles in maximal abundance, grazing might tend to smooth out variations in algal size abundance in the field. Malone (1971) questioned whether differences in the proportions of nanoplankton and net plankton might be reflected in the distribution and abundance of herbivores. In regions of the eastern Pacific he found that grazing pressure was much higher in oceanic as compared with neritic areas. While nanoplankton was the more important producer in all areas, net plankton was significantly higher in neritic regions and showed higher growth rates and there was selective grazing against the large-celled forms. Parsons (1976) points out that larger zooplankton can feed at lower concentrations of algae than the smaller-sized grazers, but with their higher reproductive rates and ability to graze on finer particles, small zooplankton can compete in oceanic regions. They will tend to release more nutrients with the increased excretion also, thus tending to stabilize the ecosystem. The metabolic advantage of larger size is noteworthy, however (cf. Table 8.5), even though efficiency may decline with age.

On the sparser phytoplankton stocks typical of warmer waters, especially with the low densities typical of "blue water" environments, grazing pressure will generally tend to remain steady and phytoplankton and zooplankton stocks be maintained at a more nearly constant level. Sheldon, Sutcliffe and Prakash (1973) point out that with open-ocean low-density phytoplankton populations, reduction in grazing pressure will lead to rapid phytoplankton proliferation. They found that the standing stock of particulate matter, both living and detrital, at a number of oceanic stations tended to be fairly stable, suggesting that production and consumption maintained an approximate equilibrium. Cultivation of water samples as small isolated volumes enabled some assessment to be made of the growth of these particle populations in the absence of zooplankton grazers. For Sargasso Sea water, very considerable crop increases were obtained; particles of mean size 20 μm had a doubling time of 2.5 hours, equivalent to a production of 28 μg C/l/day. This is the order of magnitude of blooms of phytoplankton found previously in Sargasso Sea investigations. Menzel and Ryther's estimate of production during summer (July) in the area was only 1–3 μg C/l/day. Clearly, any reduction in grazing pressure in a "blue water" environment like the Sargasso can lead to blooms; indeed, Sheldon *et al.* calculate that a decrease of only about 10% in grazing would cause the type of phytoplankton crop increases reported by Menzel and Ryther for the Sargasso.

Table 8.5. Food requirements for boreal species of different size assuming growth (7% per day) and assimilation efficiency (80%) (after Parsons, 1976, based on Ikeda's data)

Animal wet wt. (mg)	Metabolism	Growth	Total	Food intake	Growth efficiency (K_1%)
	(per cent body weight per day)				
0.005	27	7	34	42	17
0.05	13	7	20	25	28
0.5	6.4	7	13.4	16.8	42
5.0	2.8	7	9.8	12.3	57

The investigations by Menzel and Ryther (1961a) dealt with the annual cycle of zooplankton between the surface and 500 m in the Sargasso Sea off Bermuda. The zooplankton crop showed a small spring increase but little variation through the remainder of the year; the biomass was comparatively small, apart from an apparent large increase in the spring of 1958, this value being suspect. The metabolic requirements of the grazing zooplankton population were calculated, first for the zooplankton population to 500 m depth, and secondly for the total zooplankton population to 2000 m. Over most of the 3 years (1958–60) very good agreement was found between nett primary production as measured by the ^{14}C method and the metabolic requirements of the zooplankton (cf. Fig. 8.16). The zooplankton crop would appear to be controlled over most of the time by the amount of food available; similarly, reduced grazing would presumably result in a small but rapid rise in algal crop. In general the steady grazing expected in such an area appears to be confirmed; any increase in primary production as, for example, a small spring flowering during 1959 and 1960 was accompanied by, or closely followed by, an increase in the zooplankton. Very little, if any, lag occurred between extra food becoming available and increased grazing effect. Grazing was very intense and there was little food wastage or food surplus in such an ecosystem, but any temporary decline in grazing intensity could produce a small peak in algal production (cf. Sheldon *et al.*, 1973).

Riley and Gorgy (1942) suggested that coastal zooplankton consumed a far greater amount of the phytoplankton crop than oceanic plankton, but the *proportion* of crop grazed was larger in poor waters such as the Sargasso. Later, Riley (1956) estimated nett production in Long Island Sound against consumption due to the zooplankton, the microzooplankton with bacteria, and to the benthos. Of some 205 g C/m^2 nett production, only about 25% appeared to be utilized by the larger members of the zooplankton, but some 69% was used by the zooplankton, microzooplankton and bacteria together, the microzooplankton forming an important fraction of the grazing population. The remainder of the plant material sank to the bottom where it was used by benthos. During the spring in one year, however, the zooplankton did not increase so rapidly. Only about a tenth of the phytoplankton flowering was consumed over that spring; the remainder settled fairly rapidly to the bottom. Despite a comparatively large phytoplankton production in Long Island Sound, utilization by the larger zooplankton appears to be relatively inefficient. Riley suggests that this might be associated with copepods of the genus *Acartia*, believed to be inefficient grazers, being dominant members of the zooplankton there. Less efficient grazing by the zooplankton may be particularly true when the crop consists largely of the smaller nanoplankton. Smayda (1976) concluded that the zooplankton population increases in importance in its role as a grazer in depths beyond about 100 m, but becomes less significant with depth in waters shallower than 100 m.

The importance of grazing as opposed to nutrient lack in limiting algal crop, provided light is adequate, has been repeatedly stressed by Cushing. While this may hold for some areas such as parts of the North Sea over certain periods of the plankton cycle,

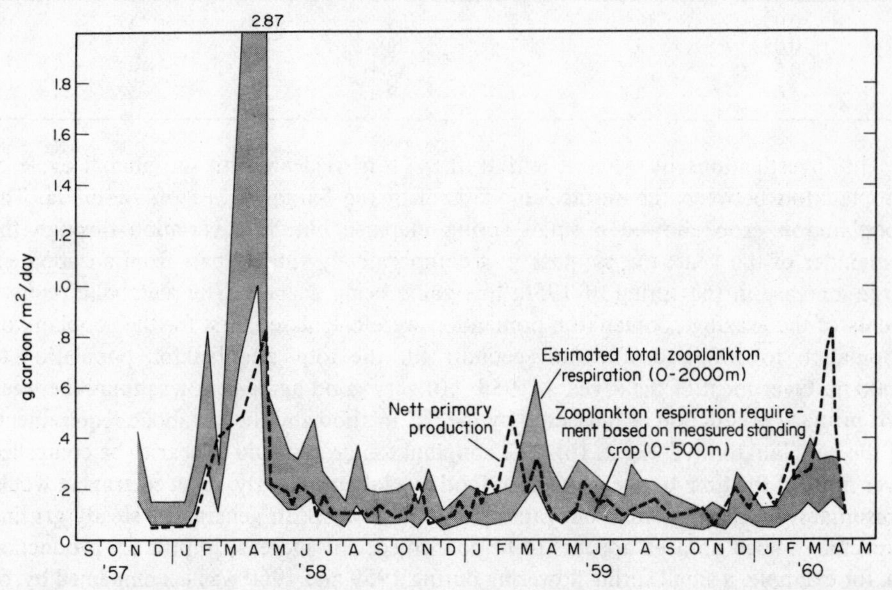

Fig. 8.16. The relation between nett primary production (broken line) and the metabolic requirements of the zooplankton represented as the stippled envelope. The lower limit based on measured zooplankton populations 0–500 m and the upper limit based on the estimated total zooplankton population 0–2000 m (from Menzel and Ryther, 1961a).

the situation can change with time. Holligan and Harbour (1977) found that while off Plymouth the role of grazing is of great importance, lack of nutrients may limit phytoplankton growth at times during the typical seasonal cycle, for instance, after a spring diatom bloom and during a mid-summer growth of dinoflagellates. Although dinoflagellate growth may be limited at that period, autotrophic nanoplankton can still grow effectively; the nanoplankton population is then mainly controlled by grazing. The rate of development of a phytoplankton crop may play a significant part in determining how far nutrient reduction and grazing are the controlling factors.

Lorenzen (1976) distinguishes between the rate of supply of nutrients which is more closely related to the rate of production and the total quantities of the nutrients which limit the absolute size of standing crop. As Cushing (1975) argues, the total amount of nitrogen and phosphorus should really be related to the total biomass of plankton, and higher links in the food chain since, with the slower turnover of the zooplankton, nutrient is temporarily stored as animal tissue. The rate of supply of nutrients must take into account regeneration due to zooplankton grazing and any changes in concentration due to advection. Lorenzen (1976) has used phaeo-pigment analysis as an index of grazing and regeneration. Grazing results in the formation of phaeophorbide with traces of phaeophytin. On a molar basis, passage of chlorophyll through the gut of a herbivorous copepod can occur with 100% transfer, so that 1 mg of ingested chlorophyll would be equivalent to 0.66 mg phaeophorbide in faecal pellets. But phaeophorbide is destroyed by light, about 99% reduction occurring at the sea surface and 30% at the 10% incident light depth. The level of phaeophorbide may thus be used as an indicator of grazing activity, less loss of pigment due to photo-oxidation, and *in situ* nutrient regeneration. Lorenzen suggests the time scale is less than a week.

Taniguchi and Kawamura (1972), studying grazing in an area in the Oyashio Current off Japan, found that grazing pressure was more related to production than to the standing crop of phytoplankton. Although crop fell substantially from 172.5 mg/m² in spring to 14.6 mg/m² in late summer, primary productivity declined only to about one-third of the spring maximum. Grazing accounted for about the same fraction of primary production (22 and 28%, respectively) over the two seasons (cf. Table 8.6).

Table 8.6. Daily primary production, standing crop of phytoplankton in the euphotic zone, zooplankton biomass in the upper 150-m water column and calculated food requirement of zooplankton (Taniguchi and Kawamura, 1972)

						Mean	
	May 1965	Apr. 1966	Apr. 1967	Sept. 1969		Spring	Summer
Daily production (g C/m²/day)	1.97	0.99	1.57	0.42	0.47	1.51	0.45
Phytoplankton crop							
Chlorophyll *a* (mg/m²)	243.3	182.1	92.0	14.9	14.2	172.5	14.6
Zooplankton biomass (g wet-wt/m²)	79.2	33.0	145.4	32.9	30.0	85.9	31.5
Food requirement of zooplankton (g C/m²/day)	0.31	0.13	0.57	0.13	0.12	0.34	0.13

We have already noted that variations in the onset and intensity of grazing, in addition to other factors, may greatly influence the pattern of seasonal phytoplankton change in a restricted geographical area, especially in coastal waters (cf. Pratt's conclusions, p. 369). Riley (1947) found an earlier phytoplankton flowering off Woods Hole, probably associated with increased stability of the water, but the extent and timing of grazing by herbivorous zooplankton also differed between Woods Hole and Georges Bank (Fig. 8.17). There are other differences between the two areas; Georges Bank has greater availability of nutrients but the shallowness at Woods Hole permits small quantities of nutrients to reach the euphotic zone at irregular intervals, giving short bursts of algal growth. Observed and calculated changes in phytoplankton density off Woods Hole (Fig. 8.18) showed fair agreement and emphasized the significance of grazing, in addition to other factors, in determining crop changes. Table 8.7, giving data for Georges Bank, illustrates the changing and, at times, very large proportion of the crop which is consumed.

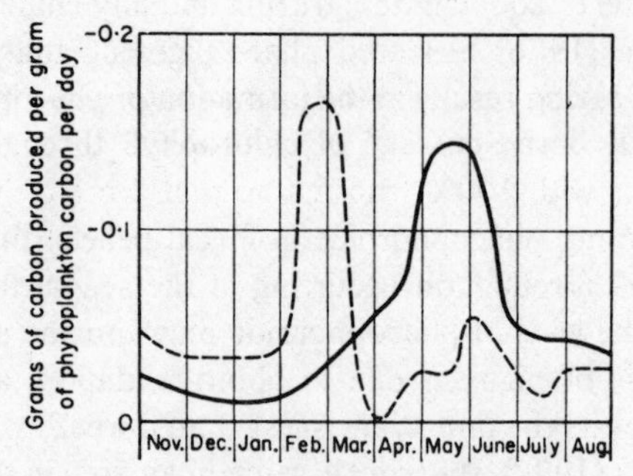

Fig. 8.17. Estimated zooplankton grazing rates for Georges Bank (continuous line) and for Woods Hole (broken line) (after Riley, 1947).

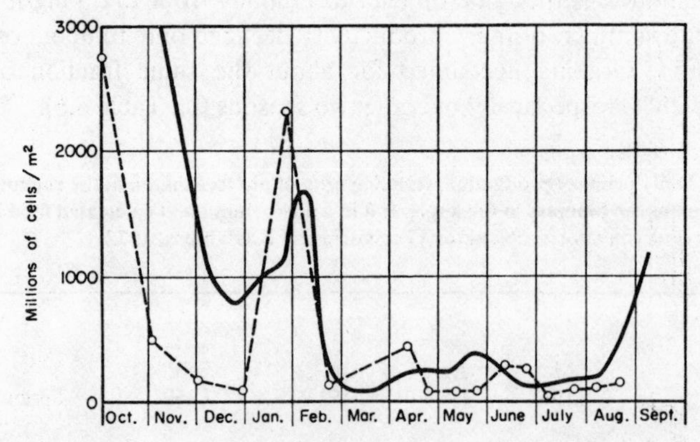

Fig. 8.18. Calculated (solid line) and observed (dotted line) seasonal changes in the density of phytoplankton off Woods Hole (after Riley, 1947).

A comparison of Long Island Sound, which has been shown to develop a large phytoplankton crop and to have a high level of primary production, with another northern temperate inshore area, the English Channel, illustrates the greater richness of algae in Long Island Sound, but indicates that the zooplankton is only slightly greater and hence grazing is less severe (Table 8.8). Winter, Banse and Anderson (1975) have described by a numerical model successive blooms of phytoplankton over spring in the

Table 8.7. Quantitative comparisons of phytoplankton and zooplankton with estimates of grazing, on Georges Bank (modified from Riley and Bumpus, 1946)

Month	Mean plant pigments (10^3 Harvey units/m^2)	Dry weight of phyto-plankton (g/m^2)	Mean no. of animals/m^2	Estimated total con-sumption (10^3 Harvey units/m^2)	% of phyto-plankton crop consumed	Food require-ments, % of zooplankton weight
Sept.	560	19.6	135,000	35	6	13
Jan.	120	4.2	14,000	−7	−5	−7
March	830	29.0	24,000	15	2	3
Apr.	2300	80.5	32,000	170	7	30
May	870	30.4	106,000	371	43	34
June	480	16.8	103,000	56	12	13

Table 8.8. Comparison of mean annual standing crops and organic production (g organic matter/m^2) in the English Channel (EC) and Central Long Island Sound (LIS) (modified from Riley, 1956)

	Standing crop		Daily production	
	EC	LIS	EC	LIS
Phytoplankton	4	16	—	3.2[a]
		—	0.4–0.5[b]	1.07[b]
Zooplankton	1.5	2	0.15	—
Pelagic fish	1.8	—	0.0016	
Demersal fish	1–1.25	—	0.001	—
Epi- and in-fauna	17	9	0.03	—
Bacteria	0.1	—	—	—

[a] Photosynthetic glucose production.
[b] Phytoplankton production in excess of respiratory requirement.

nutrient-rich upwelling waters of Puget Sound (cf. Chapter 9). Differences in computed algal crop between the normal level of grazing and in the absence of grazing did not appear to be very great, an unusual condition for a temperate area. The natural stock of grazers appears to be low; both copepod and euphausiid densities were below those of the Strait of Georgia, although primary production was higher. Possibly microzooplankton adds substantially to the grazing pressure in Puget Sound (cf. Fig. 8.19).

Less intensive grazing may be seen in some other environments, especially in relatively enclosed shallow marine areas where conditions other than food supply may not favour a rich zooplankton (cf. p. 352). In Moriches Bay (Ryther, 1954) grazing appears to have little effect on the very rich crop of flagellates. Deevey (1948) found that changes in algal crop in Tisbury Great Pond were not correlated with grazing intensity. In estuarine plankton communities the varying conditions often do not permit the establishment of a diverse grazing population, though rich algal blooms are usually characteristic. Riley (1947), Jeffries (1962, 1964, 1967), Cronin, Daiber and Hulburt (1962) and others comment on the low diversity of the zooplankton in estuaries. A few copepods, *Oithona*, *Pseudocalanus*, *Paracalanus*, *Acartia* and *Eurytemora*, are the major genera; various Acartiidae are found in estuaries all over the world (cf. Yamazi 1956, 1959; Tranter and Abraham, 1971). Cladocerans and a few mysids are amongst the other herbivorous estuarine zooplankton (cf. Bosch and Taylor, 1973), together with some meroplanktonic herbivorous larvae, especially cirripede nauplii (cf. Bousfield, 1955). Jeffries has pointed to the remarkable dominance of *Acartia* and *Eurytemora* in some estuaries, particularly in the eastern United States. Although the density of such a restricted zooplankton may at times be exceedingly high, the low diversity, the variable conditions of salinity, temperature, pH, and oxygen concentration and, on occasions, possible inimical effects of the dense phytoplankton itself, may reduce the grazing intensity of the zooplankton. Gross *et al.* (1947, 1950) commented on the very low zooplankton diversity and grazing pressure during fertilization experiments in a brackish sea loch (Loch Craiglin), although the phytoplankton could be excessively high (cf. Marshall, 1947; Marshall and Orr, 1948). Bakker and dePauw (1975) found a very restricted zooplankton in estuarine waters in Holland. Though blooms of phytoplankton occurred, grazing by the zooplankton did not appear to be a major factor controlling the algal crop. Bakker and de Pauw found that a large proportion of the phytoplankton consisted of small-sized particles; there was less food of suitable size for the larger zooplankton. The microzooplankton fraction was of considerable importance, and probably the abundant detritus was a significant food source.

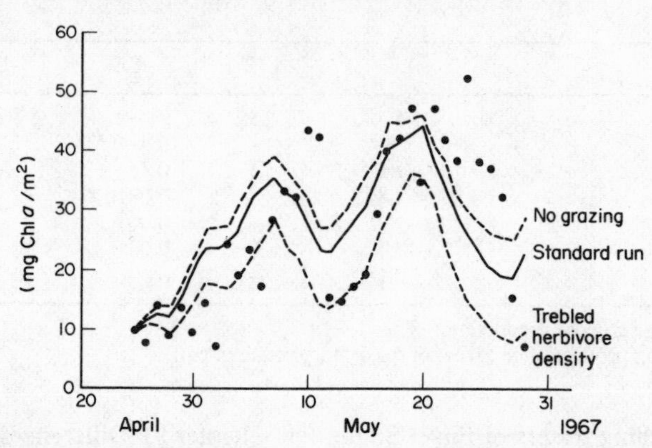

Fig. 8.19. Simulated study of the effect of different densities of zooplankton herbivores on standing stock of algae (from Winter *et al.*, 1975).

The poverty of species and at times low density of total zooplankton in estuarine situations may be true of tropical as well as temperate environments. Qasim, Wellershaus, Bhattathiri and Abidi (1969) comment on the relative paucity of grazers in the Cochin Backwaters. Only about 25% of the primary production was consumed by the zooplankton and much of the unconsumed food fell below the euphotic layer (cf. Fig. 8.20). Teixeira, Tundisi and Santoro (1969) comment on the poor zooplankton populations of mangrove swamp regions in Brazil and on the low grazing intensity. They quote similar observations by Bainbridge from estuarine regions of Sierra Leone. A high productivity is characteristic of these specialized environments but much of the crop may be "wasted" by the zooplankton, at least over certain periods, and descends to the bottom where it is used by benthic organisms, including rich bacterial populations.

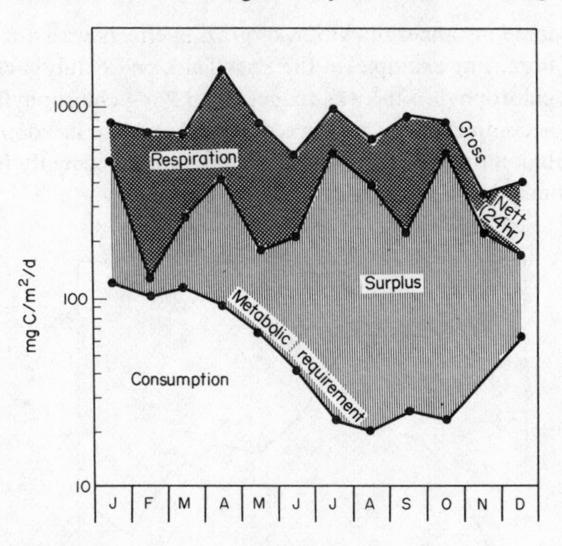

Fig. 8.20. Relation between metabolic requirement of zooplankton and gross and nett (24 hr) primary production. The lower portion shows the approximate consumption of primary production by zooplankton herbivores; the middle portion shows the approximate surplus of basic food and the upper portion shows the average respiration (24 hr) (from Qasim *et al.*, 1969).

Artificial Plankton Ecosystems

Reduction in grazing pressure may be encountered in large-scale experiments where the production of natural populations of plankton has been stimulated. The restricted zooplankton with the low grazing pressure in the fertilization experiment in Loch Sween has already been mentioned. Gross and Clarke (unpublished) stimulated phytoplankton production by the addition of nutrients to natural seawater populations. Although high crops of zooplankton, mainly *Oithona*, *Tisbe* and *Pseudocalanus*, were obtained in some cultures giving high grazing pressure, on many occasions, despite the rich phytoplankton, zooplankton remained low. Raymont and Miller (1962), employing tanks of 20 m^3 capacity at Woods Hole, obtained very rich phytoplankton crops with the addition of fertilizers; over several weeks densities exceeded 40 or even 50 g chlorophyll/m^3. Dense populations of some copepod grazers (*Eurytemora*, *Oithona*, *Paracalanus*, and especially *Acartia tonsa*) were observed on a number of occasions, reaching densities exceeding 100 or even several hundred copepods (all stages) per litre, but over substantial periods of the experiment, zooplankton populations were relatively low. Fluctuating environmental conditions (pH, temperature, salinity) were partly responsible for the restricted zooplankton, but the low diversity of phytoplankton may have been a contributory factor. During the periods of very low grazing intensities, very large algal crops were built up. For example, over about 3 weeks one tank had a population dominated by *Nannochloris* of $1\text{–}2 \times 10^6$ cells/ml; the other tank had $2.9\text{–}4.9 \times 10^5$ cells/ml, composed overwhelmingly of *Nitzschia closterium*. In these tank experiments there was generally a lack of correlation between phytoplankton and zooplankton densities; certainly at very high phytoplankton densities zooplankton was exercising no control. There were some instances, however, of grazing effects with a more mixed and moderate density of flora. For example, in the East Tank on 26 July a crop amounting to more than 20 mg chlorophyll a/m^3 was reduced to 3.7 mg chlorophyll a/m^3 within a week, and this was accompanied by an exceedingly sharp rise in zooplankton which reached a peak population on 2 August. The zooplankton subsequently fell sharply and the phytoplankton climbed to a new peak value (Fig. 8.21).

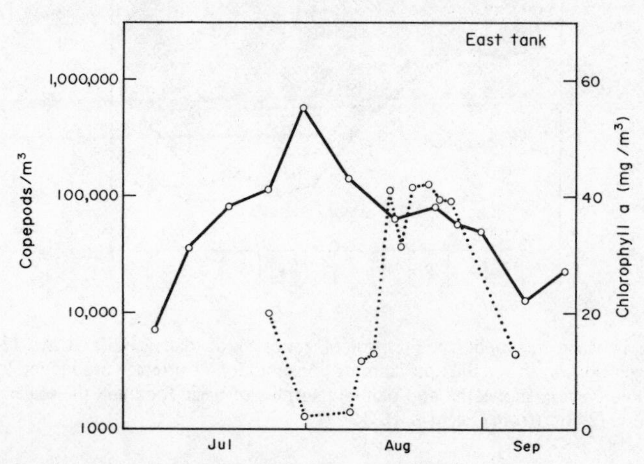

Fig. 8.21. Changes in phytoplankton abundance (as chlorophyll *a*) (continuous line) and in density of copepods (dotted line) in tank experiment (from Raymont and Miller, 1962).

More recently Parsons, Brockel, Koeller, Takahashi, Reeve and Holm-Hansen (1977) conducted large-scale experiments with enclosed seawater columns of 62 m^3 capacity to investigate the effects of the addition of nutrients (eutrophication) on phytoplankton and zooplankton production. Nitrogen, phosphorus and silicon were added in the proportion of 10:1:10 at approximately 3-day intervals over one month (August). Nutrients (nitrate) remained more or less constant in the unenriched control column (J); chlorophyll, indicating phytoplankton crop, declined somewhat initially, but then increased. The enriched column (M) showed a rise in nitrate and the phytoplankton crop showed a general increase (cf. Figs. 8.22 and 8.23). Primary production was quite high even in the control column – 127 mg carbon/m^2/day – but showed an increase in the other columns which received different amounts of fertilizer, reaching a maximum of 384 mg carbon/m^2/day in column (M) receiving most nutrient ($\equiv$17.6 g N *in toto*). Considerable recycling of nitrogen must have taken place in the control column (J) to maintain production; ammonia increased in (J) to 1.5 μg-at NH_4-N/l during the first 10 days. The data suggest that the equivalent of more than 1.5 mg-at NO_3-N/m^2 must have been recycled per day. If this recycling lasted for the total productive period in the same area (approximately 200 days), this would be equivalent to the formation of 300 mg-at NO_3-N, which exceeds the estimated amount for the upper 10 m (250 mg-at/m^2) found in earlier field observations in the area.

As regards secondary production in the experimental columns, following the initial dying off of a population of the copepod *Pseudocalanus minutus*, the growth of a cohort of *Paracalanus parvus* was observed together with the growth of ctenophores, representing tertiary production (cf. Fig. 8.22). Secondary production by the

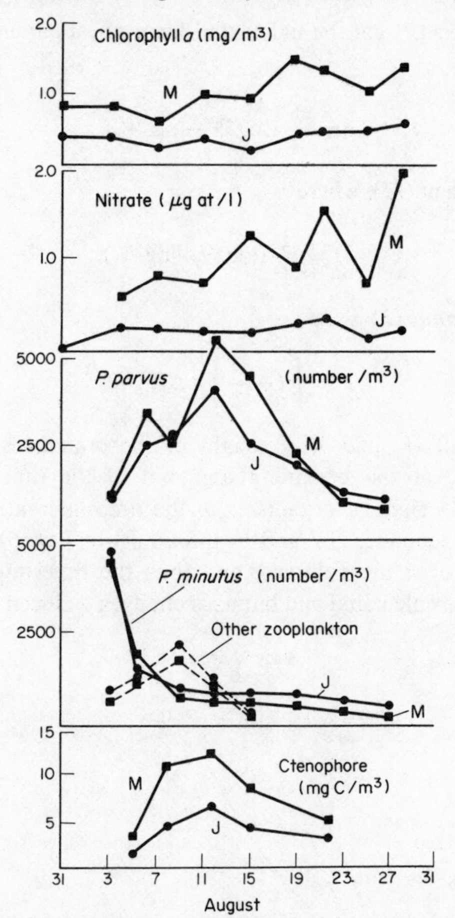

Fig. 8.22. Changes in nutrient concentration, chlorophyll and zooplankton in control column (J) and enriched column (M) over one month (from Parsons *et al.*, 1977a).

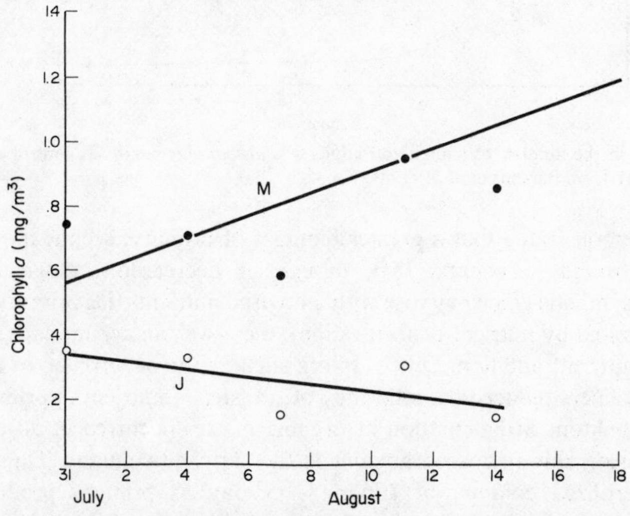

Fig. 8.23. Changes in the standing stock of chlorophyll *a* (0–10 m) in experimental columns (J) and (M) during the first part of the experiment (from Parsons *et al.*, 1977a).

Paracalanus population (*P*) can be calculated from an apparent growth constant (*g*), where:

$$g = \frac{1}{t}(\log W_t - \log W_0)$$

and a mortality coefficient (*M*), where

$$M = \frac{1}{t}(\log N_t - \log N_0).$$

Production of *Paracalanus* (*P*) is:

$$P = \left(1 + \frac{M}{g}\right) \times \Delta W_t$$

where t = time interval; W_0 and W_t = weight of the organisms and N_0 and N_t = the number of organisms, at the beginning and end of the time period, respectively; ΔW_t = the increase in weight of organisms in the time interval $t_1 - t_0$. Changes in the numbers of nauplii, copepodites IV and V and adults of *Paracalanus* were monitored over the month in two of the columns, and thus the time intervals and mortalities between stages could be calculated and biomass changes deduced (cf. Fig. 8.24).

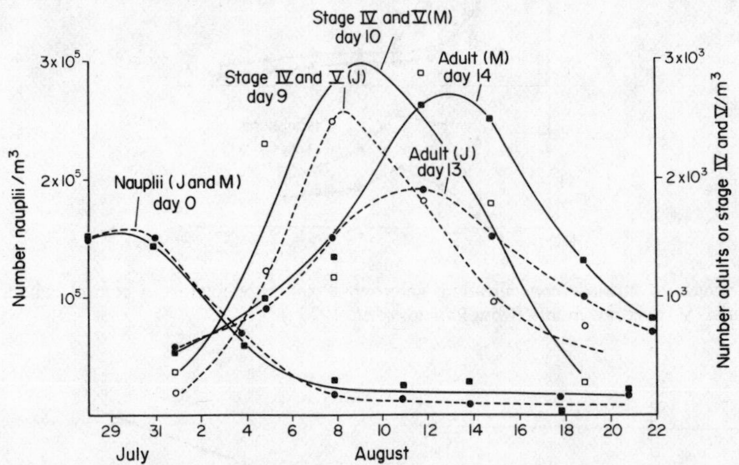

Fig. 8.24. Changes in the number of nauplii and adults of *Paracalanus parvus* in control column (J) and enriched column (M) (from Parsons *et al.*, 1977a).

The results demonstrated that a greater biomass of primary, secondary and tertiary producers was found in column (M), though at decreased ecological efficiency. Although photosynthetic efficiency rose with increased nutrients (i.e. primary production appears to be limited by nutrient concentration), there was an accumulation of nutrient as nitrate with nutrient addition. Other factors such as the occurrence of algae on the walls of the containers reduced the efficiency of transfer of nutrient to primary production by phytoplankton. Sedimentation of organic matter occurred in all columns but there was a considerably increased amount in the enriched columns. Thus in (M), the most heavily fertilized column, of 306 mg C/m²/day as primary production, only 244 mg C/m²/day was available for secondary production. A substantial amount of the phytoplankton in column (M) went to increasing the algal standing crop, and an appreciable amount sedimented, in addition to the settling of faeces from grazing. Grazing would appear to have been less efficient in the enriched column, perhaps partly due to the unsuitable phytoplankton species becoming dominant. In the control column (J), grazing was more efficient, producing a *decrease* in the phytoplankton standing crop (cf. Fig. 8.23); the sediment in (J), assuming a 70% assimilation efficiency for *Paracalanus*, was likely to have been almost entirely faecal pellets.

Differences in ecological efficiency are discussed in an account of another column experiment by Parsons, Thomas, Seibert, Beers, Gillespie and Bawden (1977). The addition of nutrients in the same proportions and at 3-day intervals over July and August, resulted in a marked difference in primary production between unenriched and enriched columns. Table 8.9 gives a summary for the period 28 July to 13 August.

Table 8.9. Primary, secondary and tertiary production (mg C/m²/day) in unenriched (J) and enriched (M) experimental columns (Parsons *et al.*, 1977b)

		Unenriched (J) column	Enriched (M) column
Primary productivity		109	306
Paracalanus parvus		46	50
Ctenophores		6.3	10.8
Sedimented material		41	105
Chlorophyll *a*	initial	0.20	0.58
(mg/m³)	final	0.46	1.22

Parsons *et al.* point out that the difference between enriched and unenriched columns in productivity is mainly at the level of primary production and of sedimented organic carbon, and amounts to nearly three times. There is less difference for the standing stock of phytoplankton and for secondary and tertiary production, emphasizing a loss of efficiency at the latter two levels. There is some suggestion that with heavy phytoplankton production there may be a loss in grazing efficiency, in part possibly attributable to the type of phytoplankton. Phytoplankton diversity declined with the enclosure of water in all experimental columns, but there were some differences between enriched and unenriched columns. For example, *Chaetoceros* sp. was the dominant species in both columns initially; relatively larger-celled phytoplankton persisted in the eutrophicated column, presumably in response to the silicon addition, though at reduced densities; nanoplankton flagellates increased in both columns, though they became more dominant in the unenriched column (J).

Some examples of variations in algal crop and grazing intensity may also be drawn from a number of CEPEX experiments. Beers, Stewart and Hoskins (1977), reporting on the microzooplankton in experimental columns, observed that on the whole the crop of phytoplankton was high in relation to the microzooplankton biomass, but this was much more obvious in those columns subjected to sub-lethal copper addition. The microzooplankton which consisted mainly of ciliates, especially tintinnids and oligotrichs among the protozoans, became substantially reduced by the addition of copper. The larger zooplankton decreased very rapidly in both controls and copper-treated columns, so that any grazing on the microzooplankton was slight. The combined low density of larger zooplankton and of microzooplankton gave an overall low grazing pressure on phytoplankton. Thus, the ratio of the biomass of ciliates (the major fraction of the microzooplankton) to the biomass of algal microflagellates (the major constituent of the phytoplankton) amounted to 22% and 44% respectively, in two control columns. In both copper-treated experimental columns the ratio fell to only 8%. Algal growth thus appeared to outstrip the zooplankton, especially in the copper-treated columns. To some extent lessened grazing may be related to reduced species diversity. Thomas and Siebert (1977) showed that species diversity declined in enclosed columns, but more obviously in those treated with copper, with microflagellates (*Ochromonas*, *Chrysochromulina* and *Dicrateria*) and two pennate diatoms becoming dominant; chain diatoms (*Chaetoceros* spp.) declined in importance.

Sedimentation of organic matter, as already described by Parsons *et al.* (*vide supra*), was observed in a number of CEPEX experiments. Sedimentation, according to Harrison, Eppley and Renger (1977), was considerable but consisted almost entirely of intact phytoplankton or of broken fragments rather than faecal pellet material in control and copper-treated columns, when both had received additional nutrients. With increased primary production consequent on enrichment, the algal crop largely settled, apparently due to insufficient grazing pressure. Herbivorous zooplankton was present but at too low a density. In the columns treated with copper the reduction of zooplankton produced a particularly marked decrease in grazing pressure. Reeve, Gamble and Walter (1977) describe differences in the standing crop of phytoplankton between controls and copper-treated CEPEX columns. While the algal crop was low initially in all columns, it remained comparatively low in the control column, but with the very marked reduction in zooplankton grazing in the copper-treated column, the phytoplankton later bloomed. In experiments carried out in columns in Loch Ewe, Gamble Davies and Steele (1977) also noted settlement of organic material, though this was largely as faecal pellets, which suggested pronounced grazing before the addition of sublethal copper concentrations. After the addition of copper to one column, there was a relative increase in sedimentation in this column as compared with the control, though it is not obvious whether this was a negative effect on the buoyancy of the phytoplankton, or whether there was an increase in the proportion of whole algal cells, an indication of reduced grazing. Gamble *et al.*, however, single out changes in the density of the dinoflagellate, *Peridinium depressum*, which may indicate grazing pressure changes. The dinoflagellates are favoured as a food by cladocerans, dominated by the species *Evadne nordmanni* in the Loch Ewe experiments. In one experimental column to which copper was added, the cladocerans were reduced by the toxicity as compared with a control column. *Peridinium* was much denser in the copper-treated column, presumably a direct effect of reduced grazing by the cladocerans. Such studies indicate that owing to factors other than food quantity, the zooplankton may vary greatly in density and may not act as a continuous brake on phytoplankton crop which can then achieve remarkably high values, but such investigations serve only to highlight the normally marked controlling influence of grazing in most natural environments except for certain highly specialized areas.

Passing reference has been made to grazing by the microzooplankton (cf. Riley, 1956). Bakker and de Pauw (1975) also mentioned the use of detritus by microzooplankton. The possible use of organic detritus by larger zooplankton herbivores should be included in calculations of secondary production, though Paffenhofer and

Strickland (1970) obtained no evidence for filtration of natural detritus by female *Calanus*. Artificial detritus prepared from dead *Ditylum* and *Skeletonema* cultures and also faeces from actively feeding *Calanus* were filtered to some extent, and some faecal pellets produced. Normal algal cultures were, however, filtered at a much higher rate and there was no indication that any of the detrital material consumed was efficiently assimilated. On the other hand, Gerber and Marshall (1974) claimed that differences in the carbon/nitrogen ratio inside and outside the lagoon in Eniwetok Atoll supported the occurrence of detrital feeding by zooplankton, and that direct evidence from gut contents for *Undinula vulgaris* and *Oikopleura longicaudata* suggested that detritus was an important food source inside the lagoon; fluorescence data indicated that about 90% of the gut contents was detritus. Similar tests of the gut contents of the temperate copepod, *Acartia tonsa*, taken from Narragansett Bay, suggested that only 34% of gut contents was detrital material.

Roman (1977) found that the same species, *Acartia tonsa*, would filter organic detritus derived from decomposed *Fucus* at about the same rate as the diatom *Nitzschia closterium*, and that the incorporation of material into the copepod body, as determined by labelled isotope studies, was approximately the same for both food materials. No clear growth of the copepod occurred on the detrital diet, however, and the copepod populations declined. Heinle, Harris, Ustach and Flemer (1977) reported growth and egg production by the two copepods, *Eurytemora* and *Scottolana*, on a detrital diet, but the diet was not as suitable as normal algal cultures. The detrital diet was substantially improved when abundant bacteria and protozoans were present. Ciliates were especially valuable as a food source. Algae and detritus mixed together formed an adequate diet and the authors suggest that the copepods may obtain a substantial quantity of their energy requirements from detritus naturally occurring in the estuarine environment, though supplementation of the diet by algae appears to be essential. Lenz (1977) also believes that organic particulate detritus is a useful supplementary food to planktonic filter feeders; dissolved organic matter may be absorbed onto detritus, providing a suitable substrate for bacteria. Investigations in a eutrophic semi-enclosed anoxic estuary in Sweden by Rosenberg, Olssen and Olundh (1977) indicated that although primary production was high, the zooplankton, dominated by the copepods *Oithona nana* and *O. similis*, consumed mainly bacteria, including possibly chemoautotrophic forms, as well as detritus, but relatively little phytoplankton.

Among the few investigations yielding clear evidence for detrital feeding by zooplankton is that of Poulet (1976). He found that non-living carbon, estimated as the difference between total carbon and living carbon determined from ATP measurements, was the dominant fraction throughout the year of the total particulate carbon in Bedford Basin, Nova Scotia. The non-living fraction was dominant, even during the spring. Non-living carbon also formed the major food of *Pseudocalanus minutus* over the year, though rare peaks in the proportion of living carbon food were occasionally seen in summer (cf. Fig. 8.25). The average proportions of living and non-living particulate carbon present in the water over the year were 21% and 79% respectively; the amounts taken as food by *Pseudocalanus* averaged 29% and 71%. Total carbon, non-living carbon and nitrogen intake were maximal in spring and minimal during winter. There was no significant selection of living particles but grazing was related to the particle size peaks of the total particulate matter available. Poulet claims that indirect evidence from body size and egg production by the copepods strongly suggests that detrital organic matter is truly assimilated, that detritus is not to be regarded as merely a supplementary food over periods of winter scarcity, and that the ability of at least some of the smaller copepod species to feed partly on detritus of suitable size and composition may be a significant factor in the succession of copepod species and their generation times.

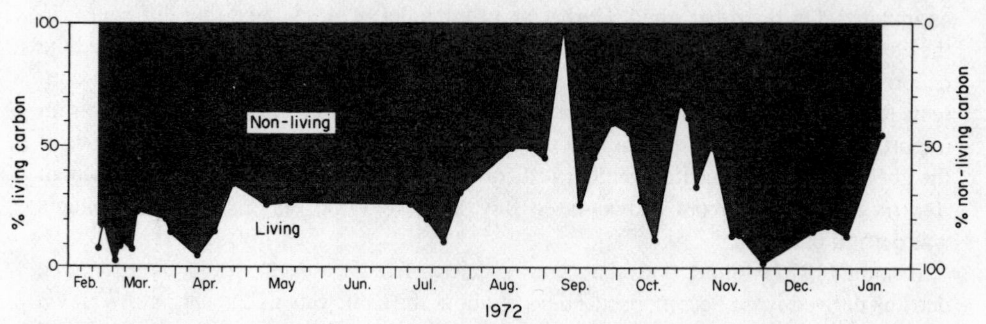

Fig. 8.25. Percentages of non-living and living carbon in diet of *Pseudocalanus minutus* (from Poulet, 1976).

Despite the paucity of direct evidence for detrital feeding by planktonic filterers, there are numerous references to its likely occurrence, more especially in warmer seas (cf. Tranter, 1973). Heinle *et al.* (1977) quote findings by Petipa, Monakov, Pablyutin and Sorokin (1974), for example, that fresh detritus, though not humus, is a suitable food for the warm water copepod, *Undinula darwini.* Detritus may also be of particular significance as a food source in the deeper waters of open oceans (cf. Raymont, 1971). Reference has earlier been made to Vinogradov's "ladder of nutrition". The accelerated descent of faecal pellets in the oceans has frequently been cited as an important factor contributing to detritus in the deeper layers, and Turner's (1977) recent measurements suggest that copepod pellets may sink at a rate of two to four orders of magnitude faster than the corresponding algal cells. But rates are highly variable and there is still no firm evidence for the importance of faecal pellets in the diet of deeper-living zooplankton. It is also especially difficult to assess the relative importance of detritus itself, as a food source, since microzooplankton, both bacteria and protozoans, are frequently associated with organic detrital matter.

The importance of microzooplankton is also stressed by Sorokin (1977). In late summer at a station in the Sea of Japan, when primary production was reduced, a high biomass of bacteria, heterotrophic flagellates, essentially a single species of *Bodo*, and ciliates was found, especially in the upper region of the thermocline. Ciliates, mainly represented by tintinnids (*Parafavella* spp., *Undella* sp. and *Acanthostomella* sp.) reached 1.5×10^4/l; flagellates even 3–5×10^4/l. While much of the microzooplankton fed on detritus, usually via bacteria, some protozoans grazed directly on the phytoplankton. Sorokin emphasizes that the higher turnover rate of bacteria and protozoans makes them an important part of the ecosystem. He had earlier claimed (Sorokin, 1971) that bacteria could be an important food supply of zooplankton in tropical waters. Comparatively high densities of heterotrophic bacteria were recorded for the upper layers of open-ocean waters of the tropical Pacific, but the density varied greatly with depth. For example, maximal concentrations were recorded in strata at 70–90 m and 400–600 m. Sorokin argued that the heterotrophic bacterial production could rival phytoplankton primary production, and this might account for the considerable biomass of zooplankton often encountered in tropical regions despite the usually low algal production. Steemann Nielsen (1972) has criticized this view, arguing that the level of primary production is sufficient to satisfy the requirements of the average zooplankton biomass in the tropical Pacific. Whatever the precise density of the bacterial population, the heterotrophs must depend on organic matter for their nutrition. Sorokin believes that a comparatively rich supply is derived from high latitudes, being transported equatorwards by deep ocean currents. It is then carried upwards where higher near-surface temperatures will encourage bacterial multiplication. There has been, however, no direct observation of this ascent of richer waters in open tropical areas.

While protozoans may feed to some extent on detritus and certainly on bacteria, in the upper layers of the ocean they may represent an important grazing element directly on the phytoplankton. In one area in Peruvian waters ciliates were believed to be the chief grazers on nanoplankton (cf. Chapter 9). Beers and Stewart (1970) reported on the microzooplankton samples retained on 35-μm-mesh filters, the water having previously been screened through 200-μm-mesh to exclude the larger zooplankton. More than 95% by number of the microzooplankton captured by this technique consisted of protozoans of which ciliates were easily the dominant group. Some of the smaller mostly non-tintinnid ciliates passed through the 35-μm net and were not quantitatively sampled. The number of microzooplankton decreased with distance offshore, average populations being $2700 \times 10^3/m^3$, $1100 \times 10^3/m^3$, and $850 \times 10^3/m^3$, respectively, for three stations extending seawards from the coast. Copepod nauplii made up the major group of metazoans among the microzooplankton, and though inferior in number to the protozoans, made a very considerable contribution to the biomass. The protozoans formed between 23% and 32% of the total biomass, depending on the station (cf. Table 8.10).

Table 8.10. Average numerical abundance and percentage contribution of different groups of the microzooplankton in the pigment layer of Stations 1, 2 and 3 at weekly intervals (26 April – 13 September 1967) (from Beers and Stewart, 1970)

	Average numerical abundance (no/m^3)			Average organic carbon as a percentage of total microzooplankton		
	Station 1	Station 2	Station 3	Station 1	Station 2	Station 3
Protozoa						
Foraminifera	5000	9600	9800	0.4	1.5	1.9
Radiolaria	18,000	11,000	6900	0.9	1.1	0.9
Tintinnida	1,200,000	390,000	260,000	23.0	12.0	12.0
Ciliata other than Tintinnida	1,400,000	650,000	550,000	8.0	10.0	9.0
Total Protozoa	2,600,000	1,100,000	820,000	32.0	24.0	23.0
Metazoa						
Copepoda, naupliar	63,000	33,000	26,000	44.0	45.0	45.0
Copepoda, post-naupliar	7200	4900	4800	18.0	24.0	30.0
Other Metazoa	7500	2500	400	6.3	6.1	1.7
Total Metazoa	78,000	41,000	32,000	68.0	76.0	77.0
Total microzooplankton	2,700,000	1,100,000	850,000			

Within the depth where phytoplankton was reasonably abundant, tintinnid ciliates made a considerable contribution to the biomass; for example, on one day they accounted for more than 90% of the protozoan biomass and more than 50% of the total. The standing crop of tintinnids, however, showed considerable fluctuations which in general accompanied changes in the stock and production of phytoplankton. Below the phytoplankton layer, the tintinnid population was very much smaller. Tintinnids appeared to exert a heavy grazing effect on the phytoplankton; marked grazing activity must also be attributed to the copepod nauplii. Beers and Stewart emphasize that trophic relationships among the microzooplankton are little known. They believe that each microzooplankton group is at least partly herbivorous, detritus with associated bacteria being also consumed; presumably some microzooplankton may be carnivorous. They have estimated the grazing activities of the major groups of microzooplankton, however, assuming that they are entirely herbivorous. The comparatively rapid growth and reproductive rates for protozoans will greatly influence grazing intensities. A division rate for ciliates, other than tintinnids, is suggested as once in 24 hours, and for tintinnids, once in 48 hours; copepod life histories by contrast amount to several weeks. Thus ciliates might consume three times the body weight of food per day, copepod nauplii an amount equivalent to the body weight daily and copepod post-nauplii 0.3 times the body weight per day. On this basis, calculations indicated that microzooplankton played an important part in grazing activity, perhaps amounting to 16 to 23% of the phytoplankton production, depending on the station sampled. The small ciliates may be particularly significant in grazing on the very small phytoplankton particles which are often not such a useful food for the larger zooplankton. For example, at an inshore station, these small ciliates were estimated to consume on average 54% of the small phytoplankton particles less than 10 μm in size, and some 23% of all phytoplankton particles less than 20 μm in size. Though this estimate may be slightly exaggerated since some of the food may be other than phytoplankton, the number of protozoans estimated was probably very conservative; those ciliates below the phytoplankton layers were not included. Nor was any estimate made of feeding activity by flagellates since many of these protozoans feed largely on dissolved organic matter. The total grazing activity of the zooplankton in the area studied must have been enormous since the fraction removed by the larger zooplankton ($>200\ \mu$m) was not included in the calculations. The crop of larger zooplankton was certainly considerable because the microzooplankton, so far discussed, formed only about 20% of the total.

Mullin and Brooks (1970) examined the grazing activity of the larger zooplankton component in the same area. An important constituent of the herbivore population was *Calanus pacificus*. From mortality rates for nauplii and for young and old copepodites and from data on ingestion and growth rates, Mullin and Brooks calculated the total crop (per m^2) of *C. pacificus*, in the field, the total biomass of zooplankton, as well as determining the phytoplankton crop for various stations. Table 8.11 shows that the biomass of total zooplankton sometimes exceeded the biomass of phytoplankton (cf. p. 367). Average crop and average primary production for two stations in the area are given in Table 8.12. According to these data, if *Calanus* used only phytoplankton as food it would consume at Station 2 about 26% of the daily nett primary production, and at Station 3 about 14% of primary production, or an average of about 20% of the nett daily production. (The calculations assume that the mean food concentration in the field did not limit ingestion rates). Only nauplii and immature copepodite stages of *C.*

pacificus were included in calculations. But late copepodite stages dominated the population so that grazing should allow for some adult female copepods. The total grazing activity of the whole zooplankton including the microzooplankton must, therefore, have been very considerable. Mullin and Brooks believe from the estimated concentration of phytoplankton in the area that more food must have been available to give the observed growth and reproduction of the zooplankton. The larger, normally herbivorous zooplankton might have fed partly on microzooplankton, or grazing might have been concentrated on relatively rich microlayers of algae. Horizontal and vertical patchiness must indeed be regarded as the characteristic pattern of plankton distribution. Not only may it be essential for the adequate nutrition of herbivores; carnivorous zooplankton such as fish larvae must probably concentrate their feeding on rich plankton patches. Parsons (1976) has commented on the apparently high food requirements of zooplankton as indicated by experiments and from successful laboratory culture techniques. Although the requirements in the field may not be quite so large, the necessary ration is substantial, and probably aggregations of food in the natural environment must be encountered for survival and successful reproduction.

As grazing is one of the major factors concerned in plankton production, accurate assessments of grazing pressure are essential. These must depend in part on precise experimental analyses and in part upon strict and detailed quantitative investigations in the field of phytoplankton and zooplankton crop and production. Despite the lack of accurate data in many regions there is clear evidence of the vast significance of grazing in determining algal production in all parts of the world oceans except perhaps for a few very specialized marine environments.

Table 8.11. Mean standing crop of phytoplankton, *Calanus* and total zooplankton, over 4 months at two stations. Values are mg carbon/m^2 (from Mullin and Brooks, 1970)

Period	Station	Phytoplankton	Calanus	Total zooplankton
18.4–19.5	2	5900	908	1800
	3	1400	790	1300
20.5–30.6	2	1400	1390	3000
	3	1600	439	1700
1.7–31.7	2	1400	424	2500
	3	950	160	1500
1.8–18.8	2	1100	86	1900
	3	440	25	830

Table 8.12. Primary production and phytoplankton crop for two stations off California. The biomass of *Calanus* and total zooplankton is shown (data from Mullin and Brooks, 1970)

	Crop as g C/m^2			Primary production
	Phytoplankton	*Calanus*	Total zooplankton	
Station 2	2.6	0.82	2.4	1.1 g C/m^2/day
Station 3	1.4	0.35	1.4	0.8 g C/m^2/day

Chapter 9
Primary Production: Global Considerations

Whatever the regional differences in phytoplankton biomass and floristic composition, it is the regional variations and temporal changes in the rate of primary production which are of paramount importance in the transfer of matter and energy through the marine ecosystem. While, as we have mentioned in earlier chapters, regions of high standing crop of phytoplankton tend to be areas of high productivity (cf. Steemann Nielsen, 1958a, 1963; Bogorov, 1958; Holmes, 1958; Koblentz-Mishke, Volkovinsky and Kabanova, 1970), and indeed it is not easy to divorce considerations of crop from productivity, the rate of primary production does not always reflect size of standing crop (cf. Chapter 8). Variations in primary production must now therefore be considered, together with an overall view of the controlling factors.

A broad picture of primary production over most regions of the world's oceans is now available, due largely to the widespread use of the radiocarbon method for estimating production, first introduced by Steemann Nielsen (cf. Steemann Nielsen, 1952).

Methods for Measuring Primary Production in the Oceans

In the radiocarbon method, the amount of photosynthesis over a time period is estimated from the uptake of the radioactive isotope, ^{14}C. While such measurements have contributed to a remarkable extent to our knowledge of plankton productivity, the limitations of the method must not be overlooked. One difficulty which has been mentioned previously (Chapter 3) is that the method is usually supposed to estimate nett rather than gross primary production, or some intermediate value. As Steemann Nielsen and Aabye Jensen (1957) and Steemann Nielsen (e.g. 1958c, 1964b) have also repeatedly emphasized, the method must be followed precisely to avoid serious errors. Details are given by Strickland and Parsons (1965) and Strickland (1960) (cf. also Steemann Nielsen and Willemoes, 1971).

An outline of the method follows. The isotope, ^{14}C, is added as bicarbonate to a sample bottle of seawater, the productivity of which is to be measured. The total carbon dioxide content of the water must be known and, assuming that the labelled carbon has been assimilated by the algae, the total amount of carbon photosynthesized may be calculated by determining the amount of ^{14}C present in the plankton. The plankton is removed by filtration (usual pore width maximum 0.4 μm) and the amount of ^{14}C measured as beta-radiation from the plankton retained on the filter. Counting is usually carried out with an end window gasflow counter. This method is usually sufficiently sensitive but greater sensitivity can be obtained by using a scintillation counter. Among other assumptions, it is necessary to accept that ^{14}C and stable C are absorbed in a constant proportion by the algae. The method of filtering (e.g. the type of micro-filter and the pressure used) may be critical. The enclosure of a photosynthetic population in a static volume of seawater may not reflect conditions operating in the natural environment; a "bottle effect" correction may be made. To measure production through the euphotic zone, samples may be taken at a series of depths, usually corresponding to 100%, 50%, 25%, 10% and 1%, respectively, of the incident illumination, the extinction coefficient for the particular water mass being previously known or measured at the time of the experiment. (The 1% illumination depth is an approximation for the base of the euphotic zone – cf. Chapter 3.) Since in most ^{14}C experiments today the test bottles are not lowered into the sea and maintained there over the period of experiment (the *in situ* method), but a simulated incubator technique is used on board ship while the vessel is in progress, one of the major problems is to ensure that the phytoplankton in each bottle is illuminated with light of the correct intensity and quality. Possible errors due to ultra-violet radiation as investigated by Lorenzen (1976) have already been mentioned. Problems concerned with simulating the spectral composition of light at different depths have already been discussed in Chapter 6 (cf. also Steemann Nielsen and Willemoes, 1971). The temperature of the incubator should match that of the natural environment, and care should be exercised in avoiding thermal and other shock effects to the algae during collection and transfer to the incubator. Errors due to uptake of ^{14}C by coccolithophorids or other organisms with calcium carbonate inclusions can be largely corrected by the treatment of filters with fumes of hydrochloric acid. Dark fixation by the algae and by non-photosynthetic organisms can be partly overcome by the use of a dark bottle carbon assimilation correction. The uptake of ^{14}C is measured simultaneously with the normal light experimental bottles by a sample of the plankton population enclosed in a light-tight bottle. There is, however, considerable argument about how far this correction should be used (cf. Morris, Yentsch and Yentsch, 1971a; Yentsch, 1974). Morris *et al.* (1971b), for example, demonstrated that dark ^{14}C fixation varied with distance offshore in Florida waters; the light/dark fixation ratio changed with cell density. With cultures, nitrogen deficiency also appeared to increase dark CO_2 fixation, but it was not obvious that natural populations in oligotrophic waters behaved similarly. Time of day for incubation in ^{14}C experiments and the duration of the incubation may also introduce difficulties in determining primary production in view of diurnal variations in photosynthetic rate.

Apart from these complexities, it is possible to devise a formula for calculating primary production from the light energy available over a time period, relating incident illumination at the sun's highest altitude at noon to time of day, day length, water depth and the extinction coefficient. Formulae have been produced by Vollenweider and by Ikushima (cf. Parsons and Takahashi, 1973a). Formulae assume clear skies; a correction must be applied for overcast conditions.

As regards other possible errors in the ^{14}C method, Williams, Berman and Holm-Hansen (1972) demonstrated filtration errors and also labelled organic impurities in the bicarbonate solution. There has been considerable discussion over errors introduced by poor filtration techniques. The relevance of precise filtration technique to the amount of ^{14}C apparently fixed and excreted as extracellular material by algae has been noted in Chapter 3. One of the problems in ^{14}C assimilation experiments is to assess the amount of genuine algae excretion (cf. Chapter 3). Perhaps the main argument that remains in interpreting ^{14}C-fixation results, however, concerns whether the method measures nett production or some value intermediate between nett and gross productivity.

Various empirical methods have been proposed for calculating total production throughout the euphotic zone from ^{14}C measurements (cf. Koblentz-Mishke *et al.*, 1970). Steemann Nielsen and Aabye Jensen (1957) suggested determining the incubator productivity for surface illumination and for the 10% and 1% illumination levels. The formula, as slightly modified by Strickland (1965), is:

$$P_{\text{euphotic}} = \frac{1}{5}(2P_{100}+2P_{10}+P_1)\times\frac{D}{2}NK$$

where P_{euphotic} is the total production for the column; P_{100}, P_{10}, and P_1 are productivity at the three light levels, respectively; N = day length; D = depth of 1% illumination; K is a constant which must be determined for the individual incubator with light intensity employed and sea location. The intensity need not be I_{max}. For tropical phytoplankton, Steemann Nielsen found that K was about unity for fluorescent lighting of 0.108 ly/min. Obviously, high production strata of algae at intermediate depths could introduce errors.

A number of refinements and precautions in employing the ^{14}C technique are described by Steemann Nielsen (1964a); for example, the importance of time of day, suitable illumination and duration of incubation, and the necessity to take samples of plankton in good physiological condition.

Of other methods for estimating primary productivity, the oxygen-bottle technique, already described, is very useful, though it is perhaps of less value for experiments of long duration when a bottle effect may be too large. The standard oxygen method is much less sensitive (cf. Qasim, Wellershaus, Bhattathiri and Abidi, 1959), though Bryan, Riley and Williams (1976) have recently developed a very sensitive micro-method which can rival the radiocarbon technique for precision.

Riley (1957) used oceanographic data to calculate an approximate level of nett primary production. Estimates involved the determination of oxygen changes in the upper water column due to biological processes, after calculating changes due to vertical diffusion, and subsequently converting the oxygen data to carbon production and consumption. For the Sargasso, a mean value of 0.13 g C/m^2/day was found for production above the compensation depth, but allowing for consumption by zooplankton and bacteria this value might be raised to about 0.15 g C/m^2/day, equivalent to 50 g C/m^2/year (cf. 150 g C/m^2/year for Georges Bank and 205 g C/m^2/year for Long Island Sound).

Some investigators have made direct estimates of the amount of carbon produced, carbon being measured usually as CO_2. Measurements of a photosynthate (e.g. carbohydrate) may be made, though estimates of total production are then difficult. Comparisons have been made of production using various techniques. McAllister, Parsons, Stephens and Strickland (1961) enclosed a natural population of coastal phytoplankton in a large plastic sphere sunk just beneath the sea surface. Primary production rates estimated by the oxygen bottle and ^{14}C methods showed rather wide differences, though some of the discrepancies were reduced by assuming that the ^{14}C method measured nett photosynthesis and, in the oxygen experiments, that the photosynthetic quotient was fairly high (1.3). Nett production of organic particulate carbon was estimated over 3 weeks by five methods: O_2 production, ^{14}C, changes in CO_2 calculated from changes in pH, cell counts, and oxidizable particulate carbon produced. There was a reasonable measure of agreement, especially with three of the techniques (cf. Fig. 9.1). In a later study by Antia, McAllister, Parsons, Stephens and Strickland (1963) using the plastic sphere, the ^{14}C method gave good agreement with production estimated as particulate phosphorus. Excellent agreement was obtained by Barber, Dugdale, MacIsaac and Smith (1971) in estimating production by phytoplankton by long-term ^{14}C-uptake and by increase in particulate carbon. A direct estimate of primary production as change in cell concentration over some time period is a convenient method for uni-algal cultures, but when applied to natural phytoplankton populations suffers from the difficulty of making complete cell counts of all the flora, and from problems in relating disparate cell size (volumes) to cell carbon (cf. p. 402). Changes in the concentration of chlorophyll over a time interval have been used in place of cell concentrations, but the estimations must then assume a constant carbon/chlorophyll ratio, usually 35 or 40. Variations in the ratio, which can be large, are discussed in Chapter 3. Another method for calculating primary production, due to Cushing (1959a, 1963, 1975), based on reproductive rate of alga related to cell width, has already been reviewed in relation to grazing estimations (cf. Chapter 8).

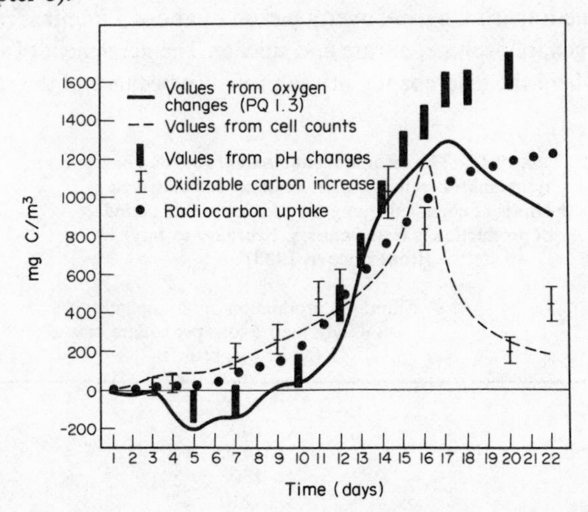

Fig. 9.1. The nett production of carbon in plastic sphere experiments as measured by five independent methods (from McAllister *et al.*, 1961).

Comparisons of production have been made using tank mass-culture techniques by Ansell, Raymont and Lander (1963) and Ansell, Coughlan, Lander and Loosmore (1964). Estimates based on O_2 bottle experiments, on changes in pH and in phosphate concentration, and on packed cell volume after harvesting the algae were found to show fair agreement. Ryther and Menzel (1965) have compared the direct weighing of carbon assimilated, with the ^{14}C method for estimating production.

Changes in the quantities of nutrients such as nitrate and phosphate, essential for the upbuilding of cell tissue, have been used as a measure of production, not only in cultures but in sea areas. Unless a steady state in water movement can be assumed, any lateral exchange of water through the area may introduce major errors in calculation. Vertical exchange of water must also be considered. In a deep oceanic area the extent of vertical mixing may be assessed so that the amount of nutrient brought into the photosynthetic zone from deeper water may be quantified. In shallow water it may be possible to integrate the total changes in nutrient from surface to bottom; leaching of nutrients from the bottom deposits may then be significant. Perhaps the greatest difficulty with using nutrient changes as a measure of production turns on the problem of regeneration. The recycling of inorganic nutrients from organic matter (mineralization) can be rapid under certain conditions (cf. Chapter 7). Harris (1959) suggested that a portion of the phosphorus and some nitrogen, particularly ammonia, may be fairly rapidly recycled. Ketchum (1962) found that phosphorus could be cycled some six or seven times a year in continental shelf waters off Woods Hole; similar determinations for the North Sea over the 7 months of main phytoplankton growth suggested that phosphorus might go through six regeneration cycles. Vaccaro (1963) pointed to the importance of the rapid regeneration of ammonia in the upper layers maintaining the cycle of phytoplankton growth, more or less through the year in warm seas, and over the period of the summer thermocline in temperate areas. If considerable regeneration of nutrient occurs, estimates of primary production calculated from changes in nutrient level may give only minimal values. These errors were recognized in the early work of Atkins, Harvey and Cooper for the English Channel; Cooper (1933) estimated minimal production of phytoplankton for the English Channel over a period of about 5 months from changes in carbon dioxide, oxygen, phosphate, nitrate and silicate. The agreement (Table 9.1) is fair except for silicate where the discrepancy arises probably because of the rapid recycling of this element.

Table 9.1. The theoretical minimal production of phytoplankton in the English Channel calculated on the basis of chemical changes in the water (the period of production is from January/February to July) (from Cooper, 1933)

Basis	Minimum production of phytoplankton wet weight metric tons per square km (g wet weight/m^2)
CO_2	1600
O_2	1000
Phosphate	1400
Nitrate	1500
Silicate	110

Steele (1956, 1958) used changes in phosphate concentration to calculate production of phytoplankton on the Fladen Ground (North Sea), where apparently little lateral transport of water occurs. Temperature profiles were used to estimate vertical diffusion; the effects of vertical mixing on the vertical distribution of phosphate were then subtracted to estimate changes in phosphate due to biological activities which could be used as an estimate of primary production. Possible errors include regeneration of phosphate, and the occurrence of phosphorus-deficient algal cells, since a direct relationship must be assumed between phosphate removed from the euphotic zone and carbon produced. Calculations for different areas of the North Sea indicated that mixing due to wind and also to tidal effects further inshore are important factors in primary production in the North Sea. A comparison of production on several inshore grounds is included in Table 9.2.

Table 9.2. Comparison of Northern North Sea, English Channel and Georges Bank (from Steele, 1956)

	Winter phosphate (μg-at/l)	Yearly production (g carbon/m^2)	Maximum plant population (mg chlorophyll/m^2)	Maximum zooplankton (g carbon/m^2)
Fladen } North Sea	0.7	54–82	100	5
Inshore } North Sea	0.6	127	175	5
English Channel	0.35	55–91	210	5
Georges Bank	1.1	120–300	660	30

Steele also developed a mathematical model, as described in Chapter 8, which included algal optimal production rate, effects of nutrient deficiency, algal sinking, zooplankton grazing and vertical mixing. One of the most important results from the model operation was the demonstration that mixing rate markedly affected primary production. Not only could very low mixing decrease production; excessively high mixing rates could also lower production, causing a slower outburst of growth in spring and smaller maximal production. With an appropriate mixing rate for the Fladen Ground, Steele obtained reasonable agreement over 4 months between the theoretical curve for production, photosynthesis as measured by the ^{14}C method, and production calculated from phosphate data (cf. Fig. 9.2).

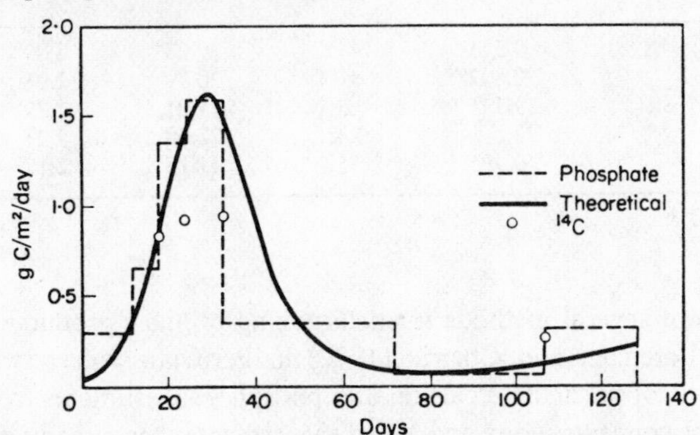

Fig. 9.2. Estimates of production for Fladen (North Sea) based on phosphate, ^{14}C, and theoretical model (from Steele, 1958).

Changes in O_2-content of the water over a selected sea area have been used to compute production. Apart from the problem of advection of water, any exchange of oxygen across the water/air interface might introduce serious errors; in a bloom, for example, substantial quantities of oxygen may be lost from supersaturated surface waters. If changes in concentration of oxygen (or nutrients) are to be interpreted in terms of organic production, a stoicheiometric relation must be assumed. The following atomic ratios can be employed:

$$\Delta O : \Delta C : \Delta N : \Delta P = -276 : 106 : 16 : 1$$

(cf. Corner and Davies, 1971, and Chapter 2).

Ketchum and Corwin (1965) used a drogue to maintain sampling of the same water mass over 10 days of a spring phytoplankton bloom in the Gulf of Maine (cf. Chapter 3). During the investigation a loss of oxygen to the atmosphere occurred owing to supersaturation; the O_2 gain in the upper 50 m (the euphotic zone) due to photosynthesis was therefore corrected for the loss, though there was some doubt about the precise rate. Organic production was then estimated from the O_2-change. Simultaneously, Ketchum and Corwin measured the changes in dissolved inorganic, dissolved organic and particulate phosphorus over the same area. Though a considerable amount of particulate phosphorus sank below the euphotic zone, there was no change in any fraction below about 136 m. Primary production was calculated on the basis of the phosphorus changes, regeneration being assumed to be of little significance over such a short period. Carbon production was also calculated from chlorophyll data, which increased over the 10 days by 118 mg/m², assuming a carbon/chlorophyll ratio of 55. The comparison of production showed a substantial measure of agreement (Table 9.3).

Table 9.3. Carbon production computed from changes in phosphorus, oxygen and chlorophyll *a* (from Ketchum and Corwin, 1965)

Variable measured	Change (g/m²)	Conversion factor by weight	Carbon production		
			g C/m²	g C/m²/day	Percentage of maximum
P removed	0.511[a]	41:1	20.95	2.09	95
Organic P produced	0.507[a]	41:1	20.79	2.08	94
Oxygen	59.7[b]	0.3:1	17.91	1.79	81
Oxygen corrected	73.6[b]	0.3:1	22.08	2.21	100
Chlorophyll *a*	0.385	55:1	21.17	2.17	98

[a] mg-at P/m² × 31 × 10^{-3}.
[b] litres/m² × 1.4285.

The comparison of several methods for determining primary production by Ryther, Menzel, Hulburt, Lorenzen and Corwin (1970) in Peruvian waters, which included direct determinations of particulate carbon and phosphorus, estimates from changes in oxygen and nutrient concentrations and ^{14}C measurements, has already been discussed (Chapter 3).

Teal and Kanwisher (1966) have advocated using changes in carbon dioxide content of seawater to estimate primary production, since they claim that variations in CO_2 concentration due to biological activity are relatively persistent, and also since the infra-red analyser technique is a sensitive method for the measurement of CO_2. In all areas, the horizontal distribution of CO_2 was patchy; for the Gulf of Maine the concentration varied during the spring bloom between 240 and 300 ppm, with patches measuring between 2 and 20 km in length. Lowered CO_2 concentration was correlated with high phytoplankton density during the spring bloom. A formula based on CO_2 changes and the depth of the mixed layer was used to estimate primary production. For Vineyard Sound, primary production was calculated also from chlorophyll concentration and solar radiation, according to the method of Ryther and Yentsch (*vide infra*). Production began early in Vineyard Sound (about December), with a winter bloom, and continued until about March or April. Data for Wilkinson Basin and Georges Bank, though less complete, indicated that the spring bloom was delayed in those areas until March, when stability was established. Nett production calculated from CO_2 data, i.e. photosynthesis less respiration of the phytoplankton and zooplankton, for a 6-month period from the beginning of the winter/spring bloom, showed considerable differences over the region: 17 g C/m^2 for Vineyard Sound and Wilkinson Basin; 70 g C/m^2 for Nantucket Sound and >100 g C/m^2 for Georges Bank (cf. Riley's estimate for Georges Bank, Chapter 8).

The direct dependence of photosynthetic rate on light intensity at levels not exceeding light saturation (cf. Chapter 3) and the logarithmic attenuation of light with depth in the ocean have led to mathematical expressions for calculating primary production, essentially in terms of incident light and depth. Some of these mathematical models have already been examined earlier in relation to evaluating factors affecting production, for instance that of Steemann Nielsen (p. 394), Riley's and Steele's various models described in Chapter 8 and the model of Gieskes and Kraay (1975) in Chapter 6. Some further examples will now be mentioned.

Steele (1962) calculated primary production from photosynthesis/light intensity relationships, using data first obtained by Ryther and Menzel for phytoplankton from three depths in the Sargasso, since these data covered a wide range in light values. The stability of the water allowed a fairly clear separation of "sun" and "shade" phytoplankton. Production was related to light and chlorophyll concentration, taking into consideration adaptation of algae to light intensity, decrease in production due to light inhibition especially by the deeper algae, and changes in the carbon/chlorophyll ratio with incident light and nutrient concentration. Both factors were important, nutrient lack being especially significant in affecting the carbon/chlorophyll ratio while having comparatively little direct influence on the photosynthetic rate per chlorophyll unit. Steele deduced from light and nutrient data a theoretical seasonal pattern of variation in the ratio chlorophyll/carbon which strongly resembled the typical seasonal pattern of production in temperate latitudes, with spring and autumn peaks in chlorophyll/carbon.

Strickland (1965) describes a method by Sorokin for assessing the level of primary production throughout the euphotic column, based also on photosynthesis/light intensity relationships. Sorokin used the well-known expression, relative photosynthesis (f), which is simply defined as:

$$f = \frac{P}{P_{max}}$$

where P_{max} is the photosynthetic rate at saturation light intensity (I_{max}). Relative photosynthesis may then be plotted against light intensity (f versus I). In Sorokin's method, samples of phytoplankton were taken from the surface and from various depths in the euphotic zone to take uneven vertical distribution into account, and the relative productivity of samples was measured with the samples illuminated in a deck incubator at an intensity below I_{max}. A sample was also exposed on deck, or at the sea surface, to determine the absolute photosynthetic rate with the algae exposed to incident solar radiation. A composite sample of phytoplankton from the whole column was also taken, and aliquots exposed at a few depths at determined light intensities. This enabled the f versus I relationship (relative photosynthesis versus intensity) to be established from which a factor (K_t) can be deduced relating photosynthetic rate at surface light intensity to the rate at the known illumination at the depth of each sample. This factor is multiplied by another factor (K_p) which allows for any unequal distribution of the algae.

The factor (K_p) is a ratio:

$$K_p = \frac{\text{photosynthetic rate in incubator for each depth sample}}{\text{photosynthetic rate in incubator for the surface sample}}$$

With the surface illumination known, production for each depth may then be calculated.

Strickland points out that the method cannot be accurate if "sun" and "shade" algal productions share the euphotic zone, or if populations and water characteristics can change rapidly.

Ryther and Yentsch (1957) developed a widely used method for estimating gross primary production based on the relationship between light intensity and photosynthesis. In an earlier study Ryther (1956a) suggested a formula, based on a general f versus I curve for a natural phytoplankton profile:

$$P_{\text{euphotic}} = \frac{R}{k} \times P_{\text{max}}$$

where P_{euphotic} = total production as mg carbon/m²/day,
R = relative photosynthesis for that population at the incident radiation,
P_{max} = maximal production, determined at I_{max} in an incubator,
k = extinction coefficient.

The method assumes a uniformly distributed population and that the f versus I curve is applicable for a given area over some considerable time. In their later study, Ryther and Yentsch proposed a new formula:

$$P = \frac{R}{k} \times C \times 3.7$$

where P = photosynthesis as g C/m²/day,
R is relative photosynthetic rate depending on surface light intensity,
C = g chlorophyll/m³ in a homogeneous column,
k is the extinction coefficient.

If the vertical distribution of chlorophyll is not homogeneous, separate calculations must be made for primary production at each depth interval depending on the chlorophyll concentration and light available. The constant (3.7) is an average figure for the amount of carbon assimilated/hr/unit of chlorophyll at light saturation and was obtained after a series of measurements on plankton populations and cultures. It is thus equivalent to the assimilation number (cf. Chapter 3), but this value may be subject to some change. Curl and Small (1965) suggest that several factors may affect the assimilation number and that it cannot be regarded as constant. They found a range from 6 to 21 (mean 8.6) and believe that variation is especially related to nutrient concentration (cf. also Platt, Denman and Jassby, 1977). Fogg (1975) states that the assimilation number varies in laboratory cultures and there is considerable evidence for variation in the field. Strickland (1965) suggests a variation of 1 to 10, but more usually from 2 to 6. (All measurements must obviously exclude any inactive chlorophyll.) Reports of effects of different factors on the assimilation number are conflicting. Strickland's view that nutrient deficiency reduced the value of the assimilation number is not supported by Steele and Baird's (1961) findings. El-Sayed, Mandelli and Sugimura (1964) found pronounced variations in the assimilation number (index) from 2.61 to

10.47 over late summer in Drake Passage and Bransfield Strait; a seasonal effect may also be evident. Steemann Nielsen and Park (1964) found a change with the light intensity to which cells were adapted. Burkholder and Mandelli (1965a) described variations in photosynthetic efficiency in natural phytoplankton populations in the Antarctic (e.g. Gerlache Strait, Bransfield Strait, Bellingshausen Sea, Deception Island) (cf. Fig. 6.19). Species composition differences were apparently insufficient to account for the variations. According to Burkholder and Mandelli, adaptations to light intensity and other environmental factors and post-bloom effects could modify the assimilation number. Fogg suggests that with the involvement of a whole complex of enzymes in photosynthesis, the assimilation number can hardly be expected to be fixed. Precise environmental conditions and the exact state of the algae, and in experimental work, even slight variations in technique, could modify its value. A better relationship may be with total plant pigment rather than with chlorophyll.

Platt (1969) argued that, since primary production in different areas and at different times is complicated by variations in incident radiation, temperature, salinity, inorganic nutrients, crop and species composition, to obviate chance variations in these parameters, a dimensionless coefficient be established (K_b), being the attenuation of visible radiation in the sea due to biological processes *only*. The total optical extinction coefficient (α) is thus made up of (K_p) (attenuation due to physical processes) and (K_b) (the fraction of the available light absorbed in photosynthesis):

$$\alpha = K_b + K_p.$$

Chlorophyll concentration is probably the most important factor related to (K_b); Platt found a marked correlation between the two for data from four depths. In an area off eastern Canada (α) varied between 0.10 and 0.15 but (K_b) was only about 1%; in open waters the biological component would presumably be larger. Primary production of a water mass can be calculated, after determining (K_b), provided illiumination (and K_p) and chlorophyll data are available. Reference has already been made in Chapter 6 to the work of Gieskes and Kraay (1975) using this approach. In view of possible effects of a number of factors on the coefficient, Platt suggests that (K_b) should be determined for each water mass.

In determinations of primary production the concentration of chlorophyll is one of the parameters in several methods of estimation. Some assume carbon production as directly proportional to chlorophyll; in others a carbon/chlorophyll ratio is employed. Data on carbon/chlorophyll ratios are thus essential (cf. Chapter 3). Estimates of phytoplankton crop other than chlorophyll are also widely used (cf. Chapter 5) and it is sometimes necessary to convert data on abundance to chlorophyll. A number of conversion formulae have been proposed (cf. Cushing, 1958, Table 9.4). Krey (1958) gave a wide variation in chlorophyll content for calculated cell volumes of various species

Table 9.4. Carbon and chlorophyll equivalents in phytoplankton (from Cushing, 1958)

1 μg chlorophyll	=	13.6 to 17.3 μg C
1 mm^3 algal volume	=	0.10 to 0.125 mg C
1 μg chlorophyll	=	0.139 mm^2 algal volume algal
1 μg chlorophyll	=	34.8 μg dry/organic matter

(Table 9.5). A critical review of some of the problems in determining crbon/chlorophyll ratios and other conversion factors is that of Banse (1977). Eppley, Reid and Strickland (1970) used equations devised by Mullin and his colleagues for converting measured cell volume (V ml) to cell carbon (g C):

For diatoms	$\log C = 0.76, \log V - 0.29$
For all other algae	$\log C = 0.94, \log V - 0.60$

They also computed carbon/chlorophyll ratios by determining C-assimilation at intervals and calculating specific growth rates, then back calculating to the initial C content and determining chlorophyll concentrations directly by fluorescence. They obtained ratios ranging from 33 for nitrate concentrations $>1\ \mu$M to 98 for nitrate below that concentration.

Table 9.5. Volumes of phytoplankton containing 1 mg chlorophyll (from Krey, 1958)

Volume	Organism
50 mm^3	*Chlorella*
86 mm^3	*Chaetoceros gracilis*
560 mm^3	*Thalassiosira gravida*
59 mm^3	*Nitzschia closterium*
97 mm^3	*Gymnodinium* sp.
260 mm^3	*Hemiselmis rufescens*
181 mm^3	mixed phytoplankton (Gillbricht)
279 mm^3	dinoflagellates (Gillbricht)
139 mm^3	diatoms (Gillbricht)
47 mm^3	phytoplankton (Riley)

Criteria for assessing phytoplankton crop in relation to estimates of primary production have been reviewed by Paasche (1960) who considered measurement of cell number, cell volume and cell surface area. Paasche pointed out that Lohmann as early as 1908 had advocated plasma volume as a more suitable characteristic. One problem with cell number concerns the great inter-specific variation in the size of phytoplankton (cf. Table 9.6). Intra-specific variation can also be large. Smayda (1966) measured more than 100 species of tropical plankton. He found cell volume ranged from 51 μm^3 (*Nitzschia delicatissima*) to $1.2 \times 10^7\ \mu m^3$ (*Rhizosolenia acuminata*); thus one *Rhizosolenia* cell would be equivalent to 2.3×10^5 *Nitzschia* cells.

Parsons (1976) summarizes some considerations of factors which affect the size of phytoplankton cells. Apart from vertical movement and density gradients in the water column, light and nutrient concentration are regarded as significant factors. Parsons claims from earlier experimental data that the growth rate of a large phytoplankton organism (*Ditylum*) only exceeded that of the small celled *Coccolithus huxleyi* in the presence of higher nutrient concentrations (nitrate) and higher illumination. An equation, introducing also sinking and upwelling rates and the depth of the mixed layer, allowed calculations indicating that larger celled phytoplankton would occur in temperate coastal areas during spring, in tropical upwelling regions throughout the year, and in the Antarctic during summer where upwelling occurred. Field observations confirmed these predictions and that only small-celled forms would be dominant in the subarctic oceanic North Pacific and in the Sargasso.

Paasche (1960) believed that surface area gave a much truer indication of biomass than volume or cell count and obtained a fair correlation between surface area and chlorophyll (cf. Table 9.7). A relatively large proportion of cell volume is occupied by the cell vacuole which is to a considerable degree not active photosynthetically. Plasma volume might be a better criterion of cell activity, but it is somewhat difficult to compute, and since the chloroplasts line the vacuole in the plasma layer, cell surface area is likely to be a reasonable measure for estimating photosynthetic activity. Smayda

Table 9.6. Average cell surface areas and cell volumes of some plankton algae (from Paasche, 1960)

	Number of cells measured	Surface area (μm^2)	Volume (μm^3)
Diatoms			
Chaetoceros borealis	42	9000	19,000
Chaetoceros debilis	84	730	1400
Chaetoceros decipiens	54	2600	9700
Chaetoceros densus	58	5000	7600
Coscinodiscus centralis	25	103,000	2,400,000
Coscinodiscus eccentricus	12	8600	60,000
Fragilariopsis atlantica	14	650	700
Fragilariopsis nana	20	25	15
Nitzschia delicatissima	50	180	70
Nitzschia seriata	40	2000	2500
Rhizosolenia alata	13	14,000	37,000
Rhizosolenia hebetata f. *semispina*	40	10,000	18,000
Rhizosolenia styliformis	60	33,700	210,000
Thalassiosira bioculata var. *raripora*	50	400	600
Thalassiosira gravida	43	1900	5500
Thalassiothrix longissima	23	17,400	87,500
Others			
Coccolithus huxleyi	31	125	130
Exuviaella baltica	100	450	700
Gyrodinium grenlandicum	20	700	1000
Phaeocystis pouchetii	10	25	10

Table 9.7. Number of cells, cell surface area, and cell volume per unit of chlorophyll in some phytoplankton species (from Paasche, 1960)

Species	Number of cells per μg chlorophyll	Specific cell surface area μm^2	Specific cell volume μm^3	Cell surface area (mm^2) per chlorophyll (μg)	Cell volume ($\mu^3 \times 10^6$) per chlorophyll (μg)
Diatoms					
Biddulphia regia	745	107,000	2.2 mill.	80	1640
Rhizosolenia alata	23,500	14,000	37,000	330	870
Thalassiosira gravida	123,000	1600	4500	195	550
Chaetoceros vanheurckii	170,000	1800	5430	305	920
Chaetoceros gracilis	286,000	220	230	63	66
"*Nitzschia closterium* f. *minutissima*"	743,000	35	18	26	13
Others					
Gymnodinium sp.	58,000	700	1700	40	100
Dicrateria inornata	563,000	50	33	28	19

(1965) also found plasma volume and surface area satisfactory for assessing crop; inter-specific differences in cell size were greatly reduced with the use of either parameter, and intra-specific variation in size was reduced by employing the ratio volume:area. Cell number used to estimate chlorophyll or biomass can be misleading, apart from the direct size effect, since generally the smaller celled species show greater activity, absorb nutrients, and respire and divide faster than larger celled species. During the spring increase in the field, the large-celled species are usually present in lower density than many small-celled species. If cell volume is the basis for calculation, a few very large-celled algae can grossly overshadow the true value since the vacuole is largely inactive. Smayda found, as expected, that ^{14}C uptake in 24-hour experiments varied directly with standing-crop size, especially with surface area and plasma volume, and, to a considerable degree, irrespective of floristic or environmental changes. However, senescent and chlorotic cells, often found in tropical phytoplankton communities, could modify the relationship.

Regional Differences in Primary Production

A world map of primary production, based mainly on ^{14}C determinations, such as that prepared by Koblentz-Mishke, Volkovinsky and Kabanova (1970), with slight modifications by Parsons and Takahashi (1973a), is remarkably similar to maps of standing phytoplankton crop (cf. Fig. 9.3). In general, productivity in warm oceans, especially in the centre of major gyres, is low, apart from areas of upwelling and

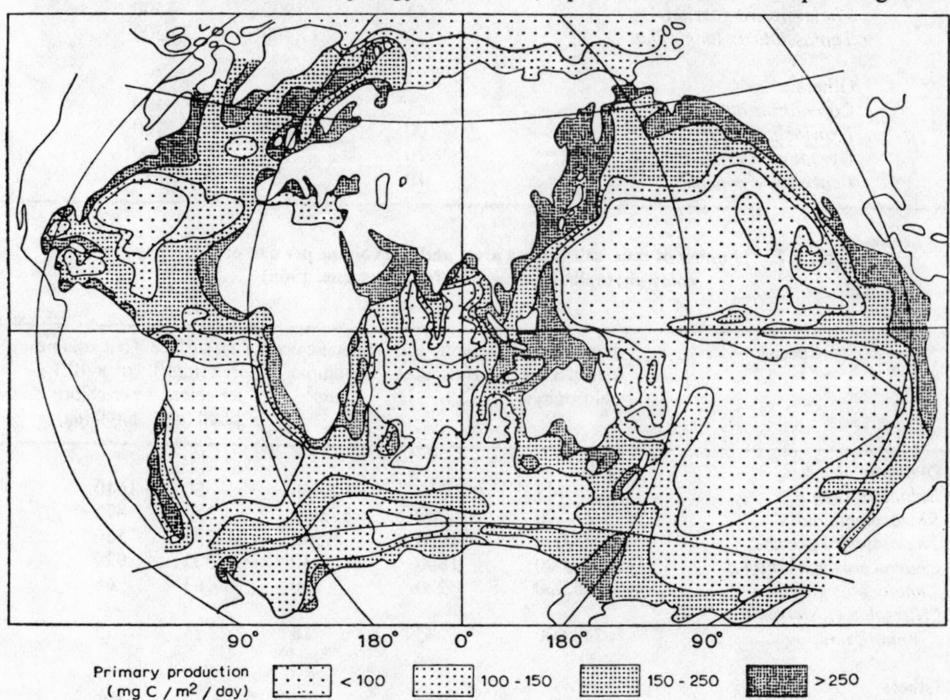

Fig. 9.3. Distribution of primary production in the World Ocean (after Parsons and Takahashi, 1973, redrawn from Koblentz-Mishke *et al.*, 1970).

divergences, and is subject to relatively small and often irregular fluctuations over the year. By contrast, seas at moderately high latitudes may show a much higher level of primary production, though this is strongly influenced by pronounced seasonal fluctuation. Areas near coasts also are marked by increased productivity (cf. Fig. 9.3). A more detailed analysis of spatial and temporal variations in primary production is now required.

1. Tropical and sub-tropical seas

Smayda (1963, 1965, 1966) has carried out one of the few detailed investigations of primary production over a considerable period of time in a tropical area. Working in the Gulf of Panama, he found seasonal and monthly variations in carbon assimilation (see Fig. 9.4) which generally paralleled standing-crop fluctuations. The area is characterized by considerable seasonal change; off-shore winds cause marked upwelling approximately over the period January to April. During this time, colder saline and nutrient-rich water comes towards the surface. By contrast, during the summer and autumn months a rainy period intervenes with considerable run-off from the land. Smayda showed that the assimilation rate varied from about 30 mg C/m^3/day during the upwelling period to only 12 mg C/m^3/day in the rainy season, rates being calculated as mean values at 10 m depth. Annual production was about 180 g C/m^2, of which about half was produced during the upwelling season. The total incident radiation varied from 103 to 692 ly/day and saturation intensity for the plankton measured about 70 ly/day, so that some surface inhibition might have been expected. Rather remarkably, maximum carbon assimilation seemed to be at the surface despite the high light intensities. The great run-off in the rainy season produced marked turbidity which, with the lower incident radiation due to the poor weather and with increased stabilization of the water, reduced both crop and production (cf. Chapter 6). The accretion of nutrients due to run-off was too poor to stimulate production; in fact, rainfall and run-off were accompanied by a marked *reduction* in nutrients. Thus reduction in crop and production was probably due mainly to lack of nutrients.

River outflows in tropical and sub-tropical countries are believed to have a beneficial effect on primary production owing to the contribution of nutrients. Smayda's work

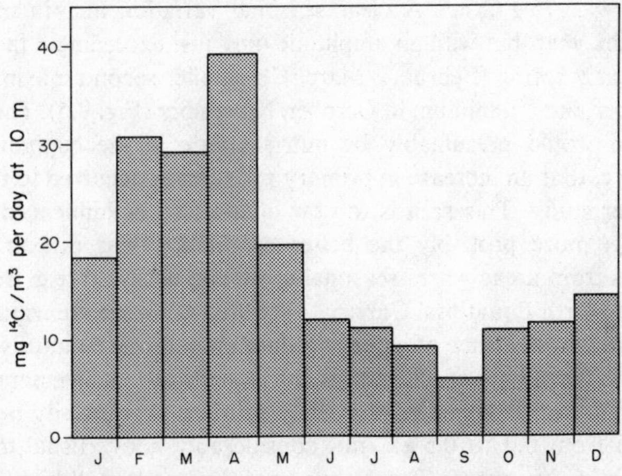

Fig. 9.4. Mean monthly ^{14}C assimilation at 10 m at the hydro-biological station from November 1954 to June 1959, based on seventy-nine experiments (modified from Forsbergh, 1963) (from Smayda, 1966).

indicates that run-off from the land is not always sufficiently rich in nutrients to have this effect. Ryther, Menzel and Corwin (1967), studying the outflow of the River Amazon, also found that nutrients and phytoplankton appeared to be generally lower in the lower salinity water extending even to hundreds of miles from the River Amazon mouth. Possibly, a number of rivers in tropical areas are nutrient-poor and do not contribute substantially to neritic production. In the description by McGill (1973) of nutrient concentrations over the Indian Ocean, the great influence of the monsoons and their effects on current flows and upwelling makes it difficult to assess separately the effect of river outflows. One interesting comment due to Banse was that off Madras, river discharge during the south-west monsoon did not contribute substantially to phosphate concentration in the sea. Ryther *et al.* (1967) found considerable enrichment at the edges of the outflow from the Amazon, possibly due to some upwelling. In contrast, the Mississippi outflow exerts a clear fertilizing effect over a considerable area of the Gulf of Mexico; there is also a most pronounced productive effect of the outflow of the Nile (cf. Halim, 1960).

Earlier observations in the Great Barrier Reef area of the Pacific by Marshall (1933) indicated that there was little, or no, seasonal fluctuation in a phytoplankton which was dominated by diatoms. Possibly, however, the nanoplankton, which was not sampled in this earlier work, might have shown seasonal effects.

The importance of the nanoplankton in contributing to primary production in tropical environments has already been mentioned (Chapter 5). Malone (1971), for the tropical Pacific and Caribbean, showed that surface productivity measured by the ^{14}C technique, was greater for the nanoplankton as compared with the net plankton for all stations and all regions, both oceanic and neritic. Though higher levels of production in neritic areas were due primarily to the higher levels of net plankton, the mean surface productivity due to nanoplankton in neritic areas was approximately twice that due to net plankton; for oceanic regions the difference was much greater (cf. Table 5.5).

One of the very few observations on seasonal fluctuations in productivity in tropical oceanic waters is that of Owen and Zeitzschel (1970) who examined ^{14}C uptake at two-monthly intervals over a year in the eastern Pacific. Mean production varied from 127 to 318 mg C/day for the depth integrated throughout the euphotic zone. The average yearly production was 75 g C/m^2. A clear seasonal variation in primary production occurred through the year, but with an amplitude only just exceeding a factor of 2. The maximum was in early spring (February/March), a smaller second maximum occurred in August/September and a minimum in October/November (Fig. 9.5). The main limitation on production would presumably be nutrient lack in the euphotic zone. It is interesting, therefore, that an increase in primary production occurred to the east of the Pacific region under study. This area is too far offshore to be influenced by the usual coastal enrichment; more probably the better production was due to transport of nutrient-rich waters from areas where seasonal upwelling occurred (e.g. from the Costa Rica Dome) by the North Equatorial Current, possibly with some contribution from the Peru Current. There was evidence of higher production near the equator where oceanic divergence increased the supply of nutrients in the euphotic zone. Steemann Nielsen and Aabye Jensen (1957) earlier suggested that daily primary productivity near equatorial divergences ranged from 0.2 to 0.5 g C/m^2, considerably above usual tropical ocean values (cf. Chapter 5 on crop). The richness of the now well-known regions of divergence near the equatorial belt in the Pacific Ocean was confirmed by Austin and

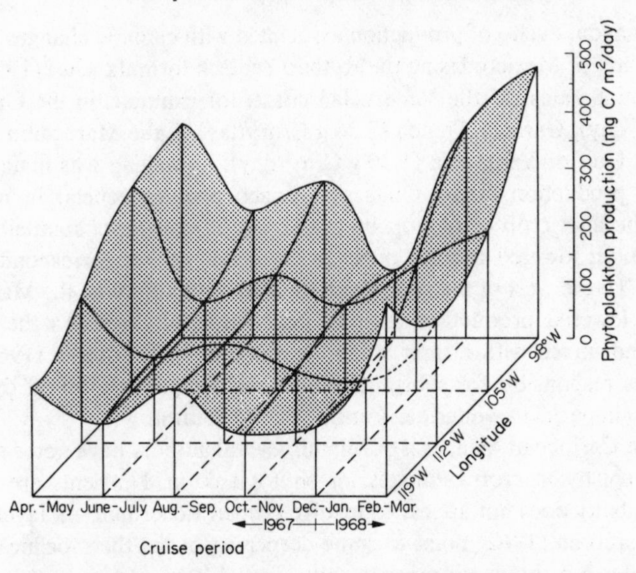

Fig. 9.5. Surface phytoplankton production (mg C/m^2/day) in the eastern tropical Pacific in relation to time and longitude. Shading indicates production levels specified on the figure itself; dashed lines demarcate grid limits (after Owen and Zeitzschel, 1970).

Brock (1959) who demonstrated high primary productivity and large crops of plankton in the nutrient-enriched water near the divergence.

In more coastal tropical areas, some degree of seasonal change is fairly well substantiated and Sournia (1969), reviewing both crop and primary production, lists a number of examples (cf. Chapter 5). In the tropical Atlantic a productive season was associated with a period of winter colder seasonal upwelling (December to April) off the west coast of Africa. Although in many cases only crop fluctuations were studied, in some areas, as off Dakar and Abidjan, a fairly clear seasonal maximum and minimum was found in rate of production and crop. The seasonal effect could be experienced at a considerable distance from the coast. Mahnken (1969) reported four zones in the tropical Atlantic showing raised primary production. Two of the areas (Senegal to Liberia and Gabon to the Congo) are fairly near the coast and show marked seasonal movement of a temperature front. Movement to the north of the front is at its maximum in July/August and the southern limit is reached in February/March. These movements cause seasonal cycles of enrichment, with high primary production off the Senegal-Guinea coast in February/March, reaching 1.61 g C/m^2/day. In July/August production is lower, though it is comparatively high for a tropical area. The second coastal region between Gabon and the Congo shows its maximum enrichment in August off Gabon, with carbon fixation reaching 1.68 g C/m^2/day. Whereas upwelling appears to be of little importance in the Cameroon-Gabon area, another region along the coast of Ghana and the Ivory Coast shows a high rate of carbon fixation, coinciding with a period of maximal seasonal upwelling. A fourth area of enrichment approximately follows the equatorial belt from the African coast to near South America. Seasonal changes in hydrographic structure cause surface enrichment, leading to changes in primary productivity, with highest values during July/August.

In the Caribbean, apart from the demonstration of a seasonal change in Panama waters by Smayda, cycles of production associated with climatic changes are known for parts of the Gulf of Mexico. Using the Ryther/Yentsch formula, Curl (1960) found high gross production rates off the Venezuelan coast; for example, in the Gulf of Cariaco (3.75 g C/m^2/day), Cariaco Trench (2.30 g C/m^2/day), Lake Maracaibo (1.23 g C/m^2/day) and the Gulf of Venezuela (1.20 g C/m^2/day). Upwelling was mainly responsible for the high production in the Gulfs of Cariaco and Venezuela. In northern Lake Maracaibo the high crop of phytoplankton (1.5×10^6 cells/l) substantially reduced the light penetration; lowered production in parts of this region corresponded with more turbid water. To the west of the region investigated, the outflow of the Magdalena River substantially lowered production, owing to the large silt load. As the water flowed north-east, and mixed with clearer water, production rose. Sournia (1969) states that upwelling was responsible for substantial seasonal changes in crop off the Venezuelan coast, with main peaks in winter/early spring and in autumn.

Outside the Caribbean area, east of Miami, several authors have noted some seasonal fluctuations, chiefly in crop estimates, of phytoplankton. Nutrients are very low and seasonal instability does not appear to involve deeper, richer nutrient layers. Alexander, Steele and Corcoran (1962) point to some deepening of the thermocline in the Florida Straits in winter but the depth reaches only about 120 m and since the euphotic zone is approximately 100 m, there is no effective seasonal replenishment of nutrients by vertical mixing. Nevertheless, some small seasonal variation is believed to occur, though the controlling factors do not seem apparent nor is the timing agreed. Thus, according to Vargo (1968), the spring is the most likely time for increased production, but four or five peaks occurred. Variable degrees of vertical mixing took place but the speed of current made correlation between enrichment and production very difficult. Corcoran and Alexander (1963) suggest little seasonal change in Florida Current waters, though they recorded occasional higher levels of production and crop.

A study dealing specifically with primary production is that of Beers, Steven and Lewis (1968) off Jamaica and off Barbados. Though sampling was some distance offshore, some island influence must be presumed. Near Jamaica, 24-hour ^{14}C incubation gave a nett production showing short-lived peaks in late spring and in autumn, alternating with lower productivity in summer and winter. The amplitude between maximum and minimum was about six-fold. Annual production was low; gross production averaged 0.18 g C/m^2/day, equivalent to 66 g C/m^2/year; nett production gave a mean of 0.11 g C/m^2/day, equivalent to 40 g C/m^2/year. Off Barbados production was higher; gross production averaged 0.38 g C/m^2/day, equivalent to 139 g C/m^2/year. Productivity tended to rise in the spring and was relatively high in summer, but fell in autumn to lower values over much of the winter. The amplitude in productivity was only about three-fold. In both Jamaica and Barbados, productivity is certainly limited by generally low nutrient concentrations. Throughout the year nutrients are low in the euphotic zone and the deeper, nutrient-richer water is largely prevented from mixing into the upper productive 100 m. Nutrient concentrations at the surface were 0.12 μg-at PO_4-P/l (Barbados) and 0.03 μg-at PO_4-P/l (Jamaica). Nitrate and nitrite together amounted to 0.25 μg-at N/l (Barbados) and 0.18 μg-at N/l (Jamaica). There were minor irregular fluctuations in nutrient concentration but they did not correspond with the peaks in production. Later investigations inshore and offshore near Barbados (Sander and Steven, 1973) confirmed that while small fluctuations in temperature and salinity occurred through the year, partly, it is believed, related to currents receiving water from the Orinoco and Amazon River outflows, these could not be associated with any significant seasonal variations in primary production or crop.

Results of the very few studies in the Sargasso (e.g. Riley, Stommel and Bumpus, 1949) and the Mediterranean (e.g. Bernard, 1939, 1957) indicated a small seasonal fluctuation in phytoplankton with a small productive winter season. Cushing (1959 a and b) indicated that closer to the equator, apart from areas of divergence, the fluctuation might be even less. For the Mediterranean, Becacos-Kontos (1968) found in Saronicos Gulf (Aegean) a clear maximum in gross primary production in early spring (334 mg C/m^2/day), nearly five times the minimum (70 mg C/m^2/day), but small fluctuations were typical over much of the year. Annual production amounted to 64 g C/m^2. The Gulf was oligotrophic, with low concentrations of both phosphate and nitrate, which were believed to act as limiting factors in turn. With the high light intensity, production was maximal at about 20 m depth. A later investigation in Petalion Gulf in the Aegean (Becacos-Kontos, 1977) gave rather similar results, with primary production ranging seasonally from 40 to 200 mg C/m^2/day. The maximum, somewhat surprisingly however, occurred in summer. Annual gross production in Petalion Gulf was low (33 g C/m^2), near the minimum recorded for the Mediterranean (cf. Table 9.8). Becacos-Kontos points to the very low annual productivity as approaching the minimum values (about 30 g C/m^2) suggested by earlier investigators for open warm oceans ("blue-water" areas).

Table 9.8. Gross primary production in g of carbon/m^2/yr in areas of the Mediterranean Sea (from Becacos-Kontos, 1968)

Sea area	No. of measurements	g carbon/m^2/year
Villefranche	12	64
Villefranche	8	83
Monaco	10	60
Cap Martin	5	40
Costa Catalana	15	75
Costa Catalana	13	85
Haifa (off Tira)	15	31
Haifa (off Tira)	7	39
Haifa (off Tira)	7	62
Haifa (off Tira)	11	65
Saronicos Gulf	20	64

Ryther and Menzel (1961a) reported on primary production in the open Sargasso Sea, south from Bermuda towards the Antilles. The mean rate of production was 0.05 g C/m^2/day by the ^{14}C method (presumably nett production) and 0.09 g C/m^2/day by the chlorophyll/incident radiation method (i.e. gross production). There was a considerable change in incident illumination with weather conditions from 70 to 380 ly/day, but this five-fold variation had little or no effect on primary production. Light would not be expected to be a limiting factor in such warm ocean conditions, but surface nutrient impoverishment is characteristic of the Sargasso. Ryther and Menzel noted that somewhat higher production values were observed near Bermuda but production fell towards the south, presumably with more intense stratification of the water. Menzel and Ryther (1960, 1961) state that generally for the Sargasso a winter bloom of phytoplankton is possible at any time between November and April. Normally there is little replenishment of nutrient-rich water into the euphotic zone through the year in the warm oceans, but if local cooling in the winter season is sufficient, especially if accompanied by strong winds, mixing may occur to a depth permitting a small flowering in the cooler season. Vertical mixing to nearly 400 m near the permanent thermocline would permit winter blooming. Thus areas of the Sargasso near Bermuda might show a bloom, depending on the degree of vertical mixing, but over the summer months, when a seasonal thermocline was evident at about 100 m, production is reduced. Further south, the stratification is so permanent that normally mixing cannot occur appreciably at any time and phytoplankton production varies little. Gross production off Bermuda averaged 0.44 g C/m^2/day, equivalent to 160 g C/m^2/year; nett production averaged 0.20 g C/m^2/day, equivalent to 72 g C/m^2/year. The nutrient level was always low; maximum 1.8 μg-at N/l and 0.16 μg-at P/l, with little seasonal variation (cf. Fig. 9.6).

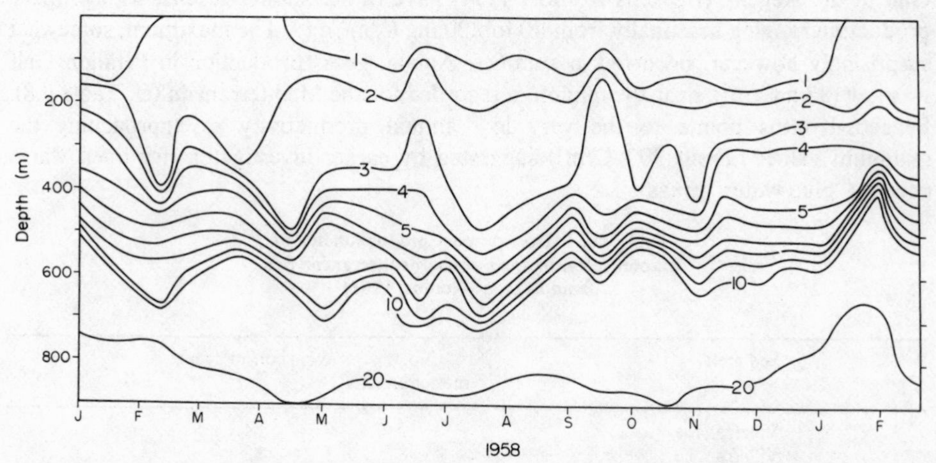

Fig. 9.6. The seasonal depth profile of nitrite and nitrate-nitrogen (μg-at/l) off Bermuda (after Menzel and Ryther, 1960).

Though the seasonal thermocline could be temporary near Bermuda, passage of nutrients through the permanent thermocline was negligible. However, the clear water and high incident radiation in winter, the moderately shallow mixed layer in winter and the rapid recycling of nutrients could permit a small winter production. Ryther and Menzel's values for production may be compared with earlier data from Riley's investigations. Production was computed from changes in oxygen profiles for the Sargasso. Nett production approximated to 0.15 g C/m^2/day (gross production about 0.27 g C/m^2/day), equivalent to an annual nett production of 50 g C/m^2 (Riley, 1957). Steemann Nielsen's rate from ^{14}C experiments, 0.05 g C/m^2/day, should probably be regarded as nett production during a period of generally lower rates, and this would agree both with Riley's (June) low estimate and Ryther and Menzel's lowest value. Variations in gross and nett production over the year (Menzel and Ryther, 1960) indicated an amplitude of five-fold to six-fold. Beers, Steven and Lewis (1968) also suggested that seasonal variations in production in continental slope waters south of New York and in waters off Bermuda were greater than those near Barbados or Jamaica, the amplitude near Bermuda being about ten-fold, as compared with three-fold off Barbados and six-fold near Jamaica (Fig. 9.7). Total annual production calculated

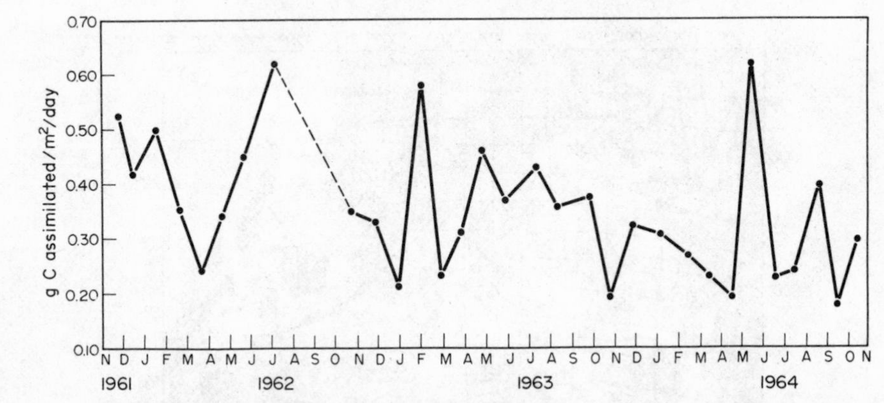

Fig. 9.7. Gross primary production off Barbados (after Beers *et al.*, 1968).

by Menzel and Ryther for Bermuda waters approached that of an open temperate sea area.

In the Indian Ocean the topography and the monsoonal climate have very conspicuous effects on plankton production in many regions of the Ocean. Some of the more obvious patterns in standing crop and their relation to the monsoons have been considered in Chapter 5. One of the most striking findings relates to the influence of the south-west monsoon which causes very intense seasonal upwelling along the coast of Somalia with the Somali Current, and along southern Arabia. Off south-east Arabia, for example, during a *Discovery* cruise in July 1963, pronounced upwelling was observed with high concentrations of phosphate extending to the surface. During the north-east monsoon, by contrast, upwelling was absent. Rich phytoplankton was observed in the coastal area, during the upwelling season, the richness extending to about 20 to 30 miles offshore. Currie, Fisher and Hargreaves (1973), point out that cell concentrations inshore were usually *ca.* 10^5 cells/l. During a subsequent cruise in August/September, whereas in the open waters of the western Indian Ocean phytoplankton was not abundant except in the region of the equatorial undercurrent, near the Somali coast nutrient levels were again high close inshore and phytoplankton was rich. Surface waters were characterized by comparatively low temperatures (<18°C) indicating vigorous upwelling, and phytoplankton was especially abundant where the cold upwelled water was mixed with warmer surface streams (Royal Society, 1965).

Although these observations are related to algal crop, the review of production in the Indian Ocean by Krey (1973) points out that over the summer (south-west) monsoon period, productivity estimates agreed well with crop evaluations. Maps of nutrient distribution published by Krey (1976) confirm the richness of both phosphate and nitrate off Somalia and southern Arabia and in other upwelling areas, in contrast to the general paucity of surface nutrients over large areas of the open tropical and subtropical Ocean. A marked correlation exists between the pattern of nutrient concentration and the distribution of phytoplankton.

Although far fewer data on primary production are available for many parts of the Indian Ocean, and Krey (1973) warns against the interpretation of some simulated *in situ* experiments on carbon-fixation, over the summer monsoon period high productivity is characteristic of the south-west Indian Ocean, the northern Arabian Sea, including the Somalia and Arabian coast, and the shelf area off north-west Australia (cf. Fig. 9.8).

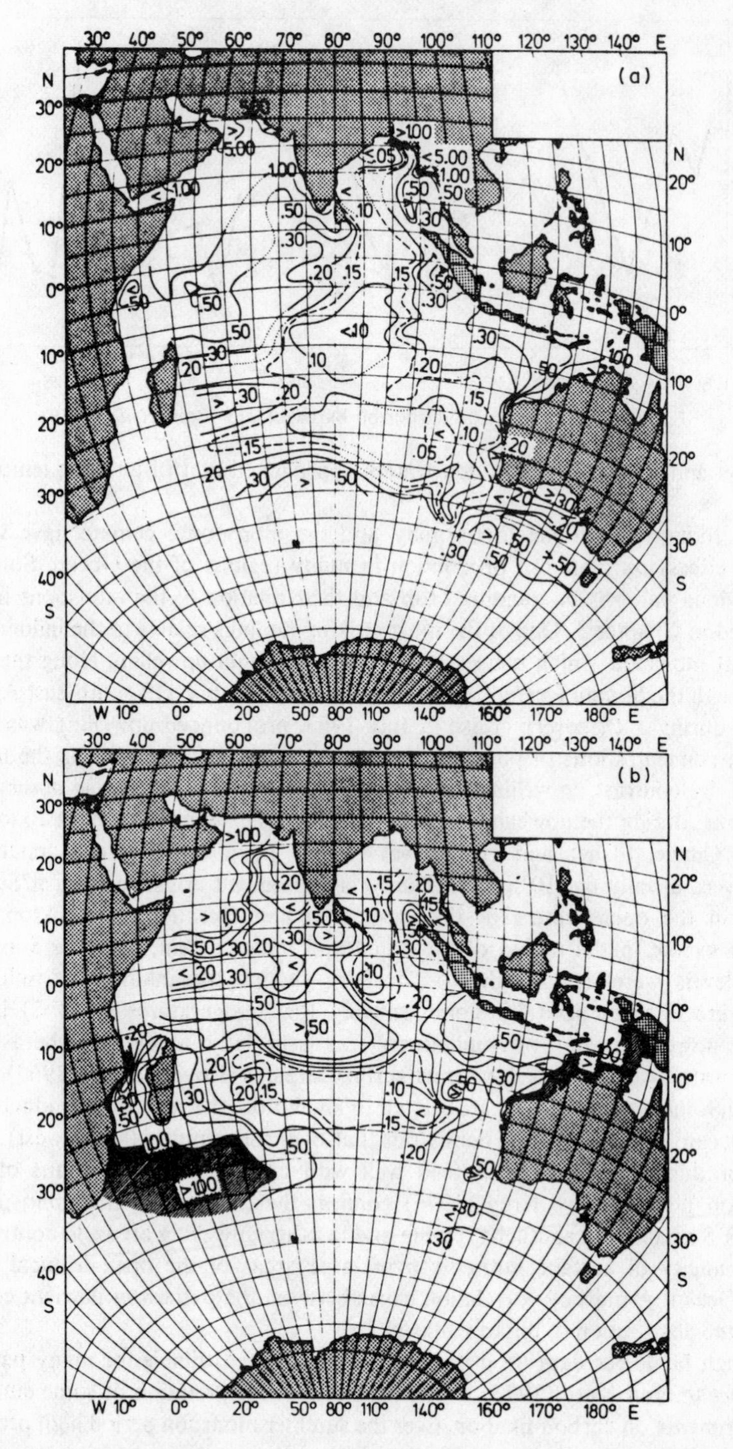

Fig. 9.8. Potential primary production in the Indian Ocean; averages for surface layer (0–50 m) in mg C/m³/hr. (a) December–March (b) June–September (Krey, 1973).

At higher southerly latitudes the production is high, perhaps varying between 0.50 and 1.50 mg C/m^3/hr. The region of very low productivity (0.1 to 0.2 mg C/m^3/hr) lie north and south of the equator, the productivity level of the near equatorial region being intermediate, presumably due to divergences (*vide infra*). During the weaker winter north-east monsoon, while the Arabian Sea exhibits a generally high productivity ($>$0.5 mg C/m^3/hr, with some areas exceeding 1.0 mg C/m^3/hr), and the higher latitudes south of 40°S show a high rate, the poor areas north and south of the equator have tended to enlarge and even spread across the equatorial region (Fig. 9.8). There is a general tendency, as in other warm oceans, for a decline in crop and primary productivity in proceeding from off the coast to the open ocean. In passing from near the coast off the tip of southern India, even though waters very close to the coast were not sampled, a decrease in phytoplankton production occurred. Potential primary productivity (^{14}C-uptake at illumination of about 0.066 ly/min) also fell from $>$0.50 mg C/m^3/hr to $<$0.05 mg C/m^3/hr (Fig. 9.9).

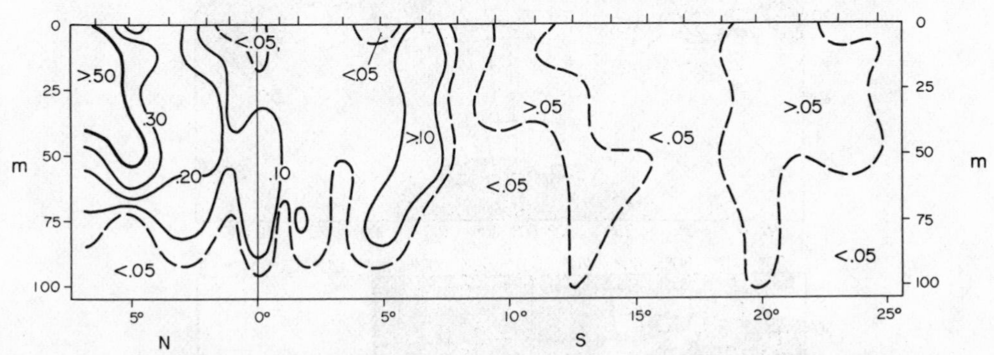

Fig. 9.9. Potential primary productivity (mg C/m^3/hr) from offshore to open ocean south of India (longitude approx. 77°E) (Krey and Babenerd, 1976).

Data published for primary production through the euphotic column also indicate high values ($>$500mg C/m^2/day) for the major areas of upwelling listed above during the south-west monsoon. These areas contrast with the central oceanic regions, especially south of the equator, but areas between the equator and about 10°S have more substantial levels of production. During the winter (north-east) monsoon, while production is relatively high over the whole Arabian Sea and at high latitudes, elsewhere it tends to decrease (cf. Fig. 9.10).

Aruga (1973), describing production in the eastern Indian Ocean, includes some data from several investigations permitting an indication of seasonal changes. For the south-east Ocean, primary production for the euphotic zone was high from about May to August (45 to 50 mg C/m^2/hr), attained a maximum of 62 mg C/m^2/hr in October, and then declined to a minimum in January (4 mg C/m^2/hr), thereafter rising only slowly to 25 mg C/m^2/hr in April. Mean productivity was 37 mg C/m^2/hr. Whereas the low January level and the rate of productivity in April was fairly uniform with latitude, over the period May to October there were distinctive latitudinal zones with seasonal changes where upwelling occurred, especially from 9° to 15°S. Aruga also gives data for the same region based on simulated *in situ* carbon-fixation experiments. From October to early May, daily rates were comparatively low (0.08 g C/m^2/day), but rose sharply to 0.18 g C/m^2 in late May and to 0.27 g C/m^2 in August. The south-west monsoon period

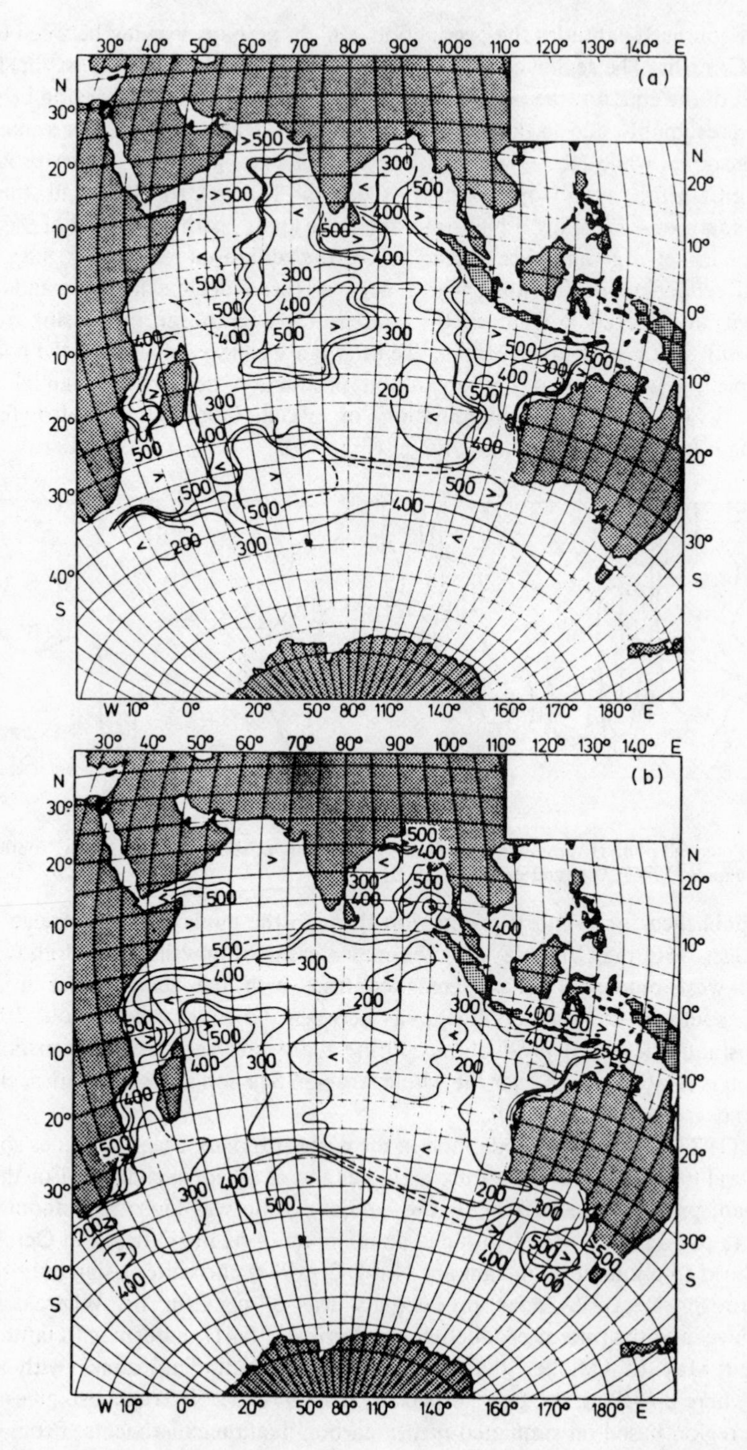

Fig. 9.10. Primary production in the Indian Ocean for the euphotic column in mg C/m²/day. (a) South-west monsoon season. (b) North-east monsoon season (Krey, 1973).

is generally believed to be more productive than the winter monsoon and Aruga ascribes much of raised production in tropical and sub-tropical regions to increased surface nutrient concentration. The tendency to increased crop and production in the near equatorial belt and to about 10°S has already been noted. Thorrington-Smith (1970c) commented on areas of higher phytoplankton crop at about 10°S in the Indian Ocean, in the area of the shear zone, and at the boundary of the equatorial under-current, both regions of somewhat elevated nutrient concentration. Ryther, Hall, Pease, Bakun and Jones (1966) found that whereas primary production could vary from <0.10 to >1.00 g C/m^2/day over the Indian Ocean, with low levels in the eastern region and north and south of the equator, moderate levels were encountered near the equator, possibly due to divergences between the South Equatorial Current and the Counter Current.

Throughout the Indian Ocean, monsoonal effects, particularly on upwelling, may have marked repercussions on production. As McGill (1973) has emphasized, however, there is apparently great diversity of water movements, including upwelling, with the monsoons, in different areas of the Indian Ocean. Apart from the major upwellings (Somalia, southern Arabia, north-west Australia, off the East Indies), many other areas have less conspicuous and less known surface enrichment. Off the east coast of India, coastal upwelling occurs off Waltair, though not off Madras. Along the south-west Indian continent, upwelling can occur from August to October, apparently spreading from south to north. Sournia (1969) reports observations by Subramanyan and his colleagues that the most obvious phytoplankton increase occurs during the south-west monsoon period. In addition to seasonal upwelling, it is suggested that the violent winds cause stirring of the offshore waters, liberating trace metals and other nutrients. Sournia (1968, 1969, 1972) observed that at Nosy Bé, Madagascar, the phytoplankton crop and primary production were much higher in neritic areas than in the open warm ocean. Production was, however, subject to considerable variation according to two seasons, a warm wet period with the north-east monsoon and a dry season with the south-west monsoon. The waters more inshore were thus warmer but more diluted from about December to May; cooler and more saline from June to November. Sournia describes two bursts of phytoplankton, one in winter and one in summer. Minimal values hold between September and January but in other months the values, though high, are very variable. To some extent, the lack of observations on nanoplankton may have obscured the complete cycle. Sournia attributes the seasonal variation inshore to a complex of factors – run-off from the land during the rainy season and some mixing of water, an influx of water from offshore in the dry season and perhaps some contributions from deeper inflows. At some distance offshore from Madagascar production was observed to increase during the rainy season, consequent on the north-east monsoon. Ryther *et al.* (1966) point out that monsoons play a part in local upwelling and assist in the general circulation of the Indian Ocean.

In the western Arabian Sea, Ryther and Menzel (1965) noted marked stratification and a very sharp gradient in nutrients, so that fertile water could be present even in the upper 100 m. Extremely high potential productivity is characteristic. With the fairly constant high illumination, any small alteration in stratification due to local winds, as well as larger scale monsoon effects, may lead to phytoplankton blooming. A marked patchiness of the plankton may be related to this capacity for sudden but restricted blooming. Generally the Arabian Sea gave a high level of production (1.5 to

1.8 g $C/m^2/day$), sudden outbursts even reaching 6 g $C/m^2/day$. If grazing is not immediate, heavy formation of detritus is likely to follow which might lead to lowered oxygen concentrations. The high production areas, however, contribute to nutrient enrichment in the Indian Ocean, since some of the organic material, after sinking and decomposition, is returned by major deeper water transport. El-Sayed and Jitts (1973) reported on *in situ* and simulated *in situ* measurements of ^{14}C-uptake for the south-east Indian Ocean, south of Australia. The survey included measurements in sub-Antarctic and Antarctic as well as sub-tropical waters. Primary production did not reach the very high levels in the major upwelling areas of the ocean off Arabia and Pakistan, but was similar to general oceanic regions of the north-west Indian Ocean. The mean value was 0.13 g $C/m^2/day$ and is comparable to earlier data obtained for lower latitudes (10–12°S) in the eastern Ocean. Much increased productivity was found near islands (Kerguelen and Heard Islands) and close to the Antarctic Convergence.

Production in very shallow warm seas

At any latitude phytoplankon production is enhanced in inshore waters (cf. Koblentz-Mishke, Volkovinsky and Kabanova, 1970 – Fig. 9.3), increased nutrient availability due to run-off and regeneration in shallow water, stabilization and other factors being involved (*vide infra*). An example of increased production in inshore waters of a warm ocean is illustrated by the observations of Burkholder, Burkholder and Almodovar (1967) who contrasted the poor production and crop in offshore Caribbean waters with the great richness of phosphorescent blooms of dinoflagellates in Phosphorescent Bay, Puerto Rico. The flagellates are notable in being able to bloom at very high light intensities, even full noon tropical sunlight. Primary production could be very high, ranging from 0.8 to 8.3 g $C/m^2/day$, in these dense inshore dinoflagellate blooms; the mean rate was about nine times that in offshore waters.

Oyster Bay, Jamaica, was shown by Taylor, Seliger, Fastie and McElroy (1966) to have enormous populations of dinoflagellates, overwhelmingly dominated by the bioluminescent species, *Pyrodinium bahamense*; peak densities reached 2×10^5 cells/l. The Bay receives abundant nutrients (phosphate and iron) as well as organic nutrients from land, with drainage from mangrove swamp. Photosynthetic rates, did not, however appear to be remarkably high. Sander and Steven (1973), on the other hand, have demonstrated a clear "island mass" effect off Barbados and report on a similar earlier finding for Jamaica. Primary production, measured as ^{14}C-fixation, increased progressively on proceeding inshore at Barbados; the difference from the farthest off-shore to the nearest inshore station was nearly five-fold, a value similar to that for Jamaica. Phytoplankton crop as cell number showed a similar gradient, though it was less obvious for chlorophyll. The zooplankton was also richer, an observation confirmed for both Jamaica and Barbados by Moore and Sander (1976, 1977), with copepods dominating the inshore zooplankton population. This dominance may be related to the extra primary production which is effectively used by herbivorous species; major variations in algal crop are therefore absent, with the intense grazing. Though some increase in nitrogen concentration is seen inshore, the difference is not great since any nutrient is almost immediately absorbed by the phytoplankton.

The "land mass" effect reflected in the abundance and higher production of plankton in coastal waters of tropical regions is partly associated with the input of nutrients and trace metals, as well as of organic factors. Prakash (1971) emphasized that coastal fertility is generally many times higher than oceanic waters in respect of phytoplankton production, and while increased inorganic nutrient concentration may be partly responsible, it cannot always account for the difference. Organic compounds undoubtedly play an important role and some of these are of terrigenous origin. Humus and humic-like substances are among the contributions, including soil and marsh sediment leachings, decomposition products of a wide variety of land, river and marsh plants, both macro- and micro-species, mangrove exudates and bacterial products. The nature and subsequent fate of these diverse humic-like substances is discussed in Chapter 2. Prakash cites evidence indicating that they are only partly precipitated on coming into contact with seawater, so that their effects may be comparatively long-term in coastal environments (cf. Chapter 6). He believes that whether the main function of humic substances is as a metal chelator or as a source of vitamin-like materials, humus acts as a potent growth stimulator (cf. use of soil extract in cultures). Addition of humic and fulvic acids in small amounts stimulated dinoflagellate growth, and in particular, enrichment by low molecular weight fractions of humic material led to high growth response of diatoms and dinoflagellates. Not only growth but yield and ^{14}C assimilation all showed a positive response. In field experiments Prakash found that growth of the large populations of *Pyrodinium bahamense* in Oyster Bay, Jamaica, was stimulated by leachings of mangrove roots.

Mangrove swamps may indeed show very high productivity, but as late as 1969, Qasim, Wellershaus, Bhattathiri and Abidi commented that almost nothing was known on the real effects of environmental factors on primary production in tropical estuaries, lagoons and mangrove swamps, though physical, chemical and biological factors are subject to extreme variations in shallow tropical waters. Teixeira, Tundisi and Santoro (1969) found considerable variation in daily primary production in a mangrove swamp area in Brazil; production could be very high – maximum (gross) 640 mg C/m^3/day in August and 580 mg C/m^3/day in March. Production was generally lower from April to July. There was marked variability of environmental factors. The main factor limiting photosynthesis was poor light penetration, mainly due to strong colouring matter ("yellow substance") exuded from the soil and from the roots of mangroves (cf. Chapter 6). The authors consider, on the other hand, that the leaching of humic materials to the water was partly responsible, in addition to the great shallowness of the area, for the very high productivity recorded on occasions, thus supporting Prakash's conclusions. Nanoplankton was responsible for the major part of the carbon assimilation. Teixeira (1973) reports that primary production in a tropical bay is lower than in an adjacent mangrove swamp area. For a station in the shallowest part of the bay he quotes 338 mg C/m^3 for surface daily assimilation during summer (February). This declines with distance offshore to 118 mg C/m^3 for a depth of about 20 m. A seasonal pattern is evident. During winter (July) production is approximately halved: 162 mg C/m^3 at the shallowest station and 56 mg C/m^3 at the deeper station. Teixeira quotes nutrient concentration as the major factor in production but increased stability of the water column promotes production in summer and permits inflow of deep nutrient-richer water. Land drainage does not apparently contribute much to fertility.

In the tropical estuary of Cochin Backwater, south-western India, Qasim and Gopinathan (1969) determined daily primary production using both oxygen-bottle and ^{14}C techniques. Primary production for an experiment in March amounted to 1500 mg C/m^3/day at the surface and about 400 mg C/m^3/day at a metre depth. The sharp reduction at 1 metre, despite the high intensity incident illumination, is related to the very high turbidity in Cochin Backwater. Solar radiation, according to Qasim Bhattathiri and Abidi (1968), was far more influenced by cloudiness during the monsoon period than by summer/winter. The average incident radiation during November to February was twice that of June and July (cf. Chapter 6). The compensation depth ranged between 2.5 m and 5.0 m, being least during the monsoon months. In a further study of the same area, Qasim, Wellershaus, Bhattathiri and Abidi (1969) found significant differences in primary production before and after midday; short-term experiments demonstrated an increase up to 14.00 hours and a sharp decline afterwards. The changes were apparently entirely associated with illumination, an illumination peak being found under normal weather conditions between 12.00 and 14.00 hours. There was no evidence for any inherent diurnal rhythm or that cell chlorophyll was lost during daylight by photo-oxidation. No reduction in photosynthesis was experienced with peak illumination. Figure 9.11 shows typical curves for production under uniformly sunny and uniformly dull conditions; normally the production curve lay between these two extremes. Surface phytoplankton was apparently well adapted to tropical conditions ("sun" type), with light saturation at times approaching 0.3 ly/min, though the plankton flora varied considerably, even over short periods.

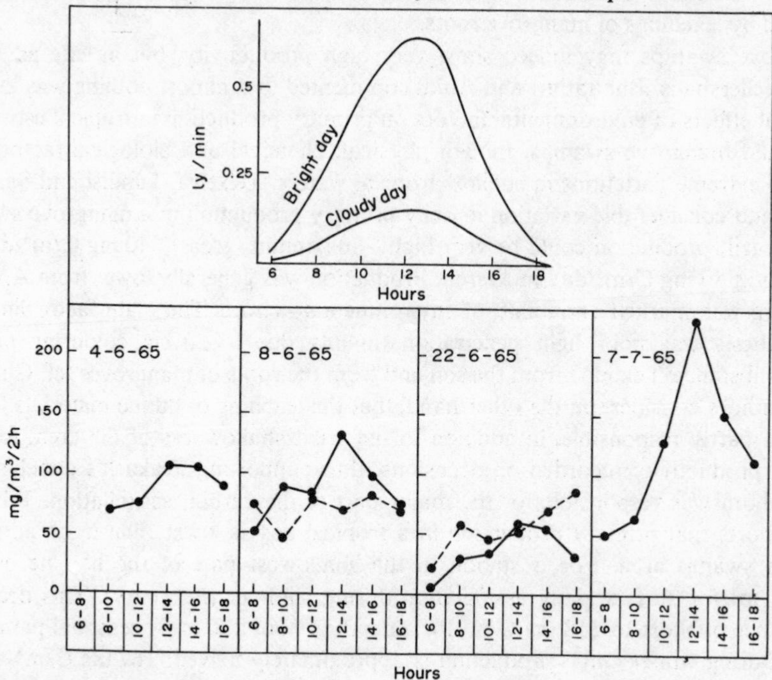

Fig. 9.11. Diurnal variation in photosynthetic activity in Cochin Backwater on four dates: (———) ^{14}C uptake; (– – – – –) nett O_2 production. Top figure indicates two extreme conditions of diurnal illumination (modified from Qasim *et al.*, 1969).

Seasonal changes in primary production were not very marked; the fluctuations were about three- to four-fold in amplitude, with higher productivity from April to August, and a lower period from September to March. Small peaks occurred in April, July and October but Qasim *et al.* believe that, as in other tropical inshore areas, these small peaks may be variable. The variations did not correspond with monthly changes in incident radiation. Over most of the year production was limited to a very shallow euphotic layer, about 4m deep; in May 1966 the compensation depth was < 1 m. Nutrient levels were generally high and greatly influenced by run-off, with nitrate low and the nitrogen/phosphorus ratio markedly reduced at times, but neither nutrients nor temperature limited production. The marked seasonal changes in salinity, due primarily to the monsoonal climate, appeared to be partly responsible for initiating brief plankton blooms.

The nanoplankton usually dominated the phytoplankton, though from November to March there was a larger contribution from the net plankton (cf. Chapter 5), but marked variations in light saturation and assimilation number values suggested that the floristic composition of the phytoplankton was very variable. Gross production was estimated at 281 g $C/m^2/yr$ (nett production 124 g $C/m^2/yr$), considerably less than that of shallow areas such as Long Island Sound (470 g $C/m^2/yr$) and a fjord in British Columbia (650 g $C/m^2/yr$), estimated by Gilmartin. The decreased production is accounted for largely by the very shallow euphotic layer. Even so, the annual production far exceeds estimates by Ryther for oceanic tropical areas; for example, the western Indian Ocean (127 g $C/m^2/yr$) and tropical oceans generally (70 to 120 g $C/m^2/yr$).

General conclusions regarding production in warm waters

It is difficult to describe a general cycle of primary production which would be true of all tropical and sub-tropical waters. Coastal influences even at some distance offshore as well as islands, local climate, run-off, local upwelling, and other factors can all influence production and give a simulated seasonal cycle. In open warm oceans, in the absence of major seasonal upwelling, fluctuations in productivity are probably comparatively small; irregular changes occur, though the factors responsible are often not evident (cf. Sournia, 1969).

It is the *rate* of supply of nutrients which is the most critical factor in the primary production in warm oceans, since light is available through the year. The presence of a permanent thermocline to a very large extent prevents vertical transport of nutrients to the euphotic zone, so that the supply of nutrients depends essentially on the rate of regeneration, chiefly influenced by grazing rate and by temperature. In the typical tropical open ocean, as Cushing and others have suggested (e.g. Cushing, 1975), the production cycle can be regarded as proceeding at a comparatively low but fairly steady level. Nutrients are almost continuously present but at very low concentrations; they are being continuously regenerated but are absorbed rapidly by a comparatively low crop of algae which is intensively grazed and is spread through a deep euphotic layer.

Ammonia, directly excreted by the zooplankton, is outstandingly important within the euphotic zone of warm oceans. Menzel and Spaeth (1962a) found its concentration was twice to four times that of nitrate. For the tropical Pacific, Thomas (1970) demonstrated that lack of nutrient in that region was essentially a lack of nitrogen, and that while nitrate and nitrite were almost undetectable, some ammonia was present. Phosphate was obviously available (see Table 9.9). Although the assimilation number of algae from nutrient-poor waters in the tropical Pacific was somewhat low, Thomas believes that the oligotrophic waters possess a phytoplankton which is maintained in reasonably healthy condition on low ammonia concentration. Some discussion of low half-saturation constants of oceanic oligotrophic algae is given in Chapter 7. Sudden increases in production and blooms are at best short-lived; they should probably be regarded, in the absence of recognized vertical mixing due to major upwelling or divergence, as more or less fortuitous.

Though the comparatively large depth of the euphotic zone may give a moderate level of productivity which is not much inferior to that of open temperate oceans (cf. Menzel and Ryther, 1960), mean production per unit volume is low (cf. Riley, Stommel and Bumpus, 1949).

The production ratio, D_c/D_m, is more or less constant in warm oceans, primary production is relatively continuous, the amplitude is low, and there is little lag in the onset of grazing.

Table 9.9. Nutrient concentrations (μg/l) in nitrate-poor tropical Pacific waters. Depth 10 m (from Thomas, 1970)

Position (104 to 105°W)	NO_3	NH_4	PO_4
13°N	0	6	7.5
8°N	0	6	7.5
4°N	0	5	7.5
7°N	0	6.5	8.5
14°N	0	8.5	8.0
Mean values for 14 stations	0	7.0	8.0

2. Temperate latitudes

In contrast to the lack of a general pattern in phytoplankton production in warm oceans, a cycle of production with some common characteristics is true for many temperate seas. During winter, despite the abundance of nutrients, production is exceedingly low, partly due to the presence of a relatively deep mixed layer which exceeds the critical depth, and to the low incident radiation. A conspicuous increase in primary production occurs in spring or early summer, the time largely dependent on latitude, with the increase in incident solar energy and the stabilization of the water column (cf. Chapter 6). Summer levels of primary production can be high with the increased illumination, but if stratification, usually due to a seasonal thermocline, is very pronounced, the vertical transfer of nutrients from the layers below the euphotic zone may be severely limited. As primary production depends on a constant nutrient supply, the maintenance of a high level of productivity is then conditional on the rate of regeneration of nutrients within the euphotic layer, a situation resembling the normal condition in tropical seas. The intensity of grazing and water temperature are among factors which influence nutrient regeneration rate. If primary production has been reduced over the summer due to severe restriction in nutrient supply, there is often a brief increase in the rate of primary production during early autumn. This is related to the disruption of the seasonal thermocline, due to changes in surface temperature and the effects of winds, allowing vertical transport of nutrients to the euphotic layer. Where there has been great impoverishment of nutrients the timing of this breakdown of the thermocline is very significant, since an increase in primary production can then occur only if the decreasing illumination is sufficient to stimulate production. Steele's (1956) model assessing seasonal variations in production in temperate waters indicated that an outburst of algal growth in autumn was largely dependent on whether a comparatively calm period followed the beginning of the breakdown of the seasonal thermocline. The depth of mixing was critical with the failing light. Incident radiation in the early autumn must be sufficient to raise the rate of gross photosynthesis to a level high enough to exceed the increased respiratory losses of the algae due to environmental temperature, which is higher than during spring time, and to offset losses due to grazing by the zooplankton which is usually still abundant in early autumn. The grazing activity is also temperature dependent (cf. Riley, 1959). S. M. Conover (1956) found marked variations in the intensity of autumn production in Long Island Sound which she believes were partly associated with different values of autumn incident radiation. In any event the decreasing incident light energy and lessening depth of light penetration during the autumn results in declining primary production, despite the re-establishment of the maximum nutrient level throughout the water column, leading to the typical winter minimum.

The majority of the earlier investigations on production cycles in temperate waters were concerned with estimates of standing crop, which can be greatly affected by grazing. The annual changes in phytoplankton density in the English Channel described by Harvey, Cooper, Lebour and Russell (1935), for example, show what has been regarded as the classic cycle with a large spring peak, variable but lower summer densities, a small autumn peak and a winter minimum. Concomitant changes in zooplankton population must also be studied, however (Fig. 9.12). When the tremendous grazing activity of the zooplankton and the significance of the lag effects as described in Chapter 8 are considered, the changes in phytoplankton crop can give at best only an indication of the likely course of the productive cycle. No accurate data for productivity rates can be deduced from the phytoplankton changes alone. The various models described earlier (cf. Chapter 8) permit estimates of the cycle of phytoplankton production in temperate areas in the absence of the grazing factor (cf., for instance, Riley's estimates for Georges Bank and Woods Hole phytoplankton – Fig. 9.13). However, while the series of small peaks over the summer seen, for instance, in Harvey's standing-crop analysis largely reflect grazing, some irregularities in primary production, especially over summer, are also likely in temperate waters. The fluctuations in zooplankton density and thus grazing activity will influence regeneration rates, affecting

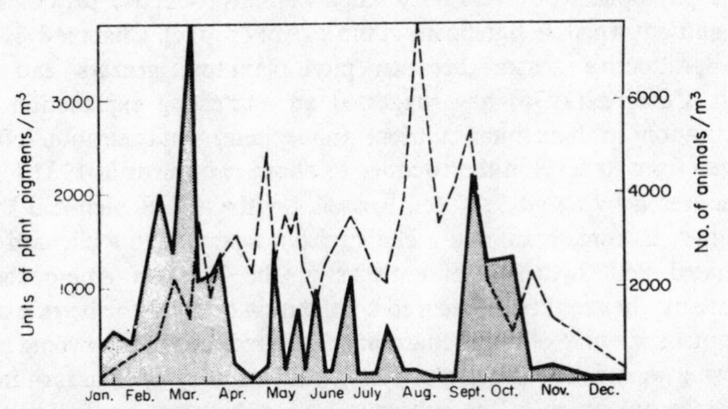

Fig. 9.12. Reciprocal relationship between the density of phytoplankton and zooplankton (continuous line – units of plant pigments per m³; dotted line – density of zooplankton) (from Harvey *et al.*, 1935).

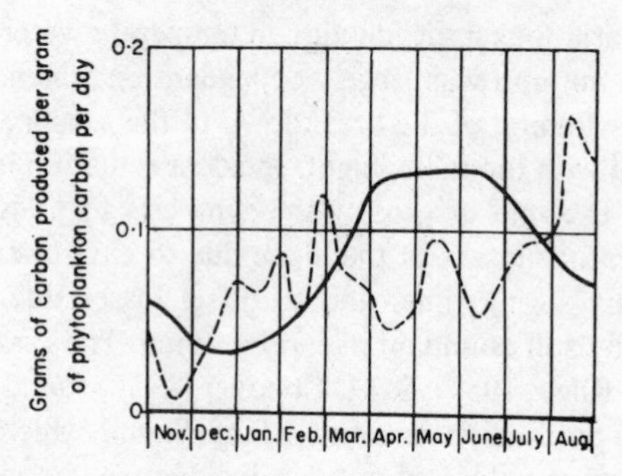

Fig. 9.13. Photosynthetic rates from Georges Bank (———) and for Woods Hole (– – – – – –) based on estimates of solar radiation, turbulence, transparency and nutrient depletion (from Riley, 1947).

the nutrient supply. Variations on the classic cycle in temperate waters may also occur with advection, and some degree of vertical mixing in less strongly stratified seas will affect nutrient concentration.

Earlier studies on changes in crop left little doubt, however, that nutrient limitation over summer restricted productivity in many areas. The work of Atkins, Cooper and Harvey for the western English Channel (cf. Harvey, 1950), of Marshall and Orr (e.g. 1927, 1930) in the Clyde Sea area and of Bigelow. Lillick and Sears (1940) for the Gulf of Maine, for example, all suggested that lack of nitrate or phosphate, whichever was in shorter supply, acted as a brake on production. Although lack of phosphate was formerly considered as the major limiting factor, Riley and his colleagues, working in Long Island Sound (e.g. Riley and Conover, 1956), suggested that limitation in production in the summer was mainly due to depletion of nitrogen. Strickland (1965) also emphasized that nitrate was the more important nutrient since it tends to be stripped first in most temperate areas.

The studies of Butler, Corner and Marshall (1969; 1970) illustrated variations in zooplankton excretory rate through the year in providing nutrients to algae, and the maintenance of an approximate nitrogen/phosphorus ratio. As Cushing has pointed out, decline in a nutrient is to some extent an indication of its storage in animal (zooplankton and/or fish) flesh. With normal grazing, fairly rapid regeneration of nutrient may maintain the phytoplankton cycle over much of the productive period. The absolute quantity of nutrient which is significant in the size of crop (cf. Chapter 8) is thus divided during the productive season between phytoplankton, grazers and carnivorous zooplankton. Cushing (1975) has suggested an interesting explanation of the well-known fluctuation in maximum nutrient (phosphate) concentration off Plymouth. Phosphate fell from its level in the twenties to about two-thirds in 1931 (cf. Fig. 9.14), and only recovered by about 1970 (cf. Russell, Southward, Boalch and Butler, 1971). Changes in the macroplankton and a change from a herring to a pilchard fishery have been associated with hydrological patterns in the Western Approaches. Cushing suggests that any changes first influenced competition between the herring and pilchard. A good summer spawning of the pilchard may have reduced the spawning of the herring in the following winter. The successful young pilchard may have in turn influenced the quantity of macroplankton in the subsequent year. Nutrient concentrations could be influenced in so far as a greater portion of the total quantity of phosphorus shared between water, phytoplankton, zooplankton and young fish was present as pilchard tissue.

Although primary production is directly influenced by the rate of absorption of a nutrient, algal growth is little affected, at least in some species, until nutrient levels are very low. The half-saturation constant (K_s) is considerably smaller for phytoplankton than was formerly believed (cf. Chapters 3, 7). Eppley, Rogers and McCarthy (1969) found that many oceanic plankton algae had particularly low half-saturation constants for nitrate and for ammonium. Values of K_s increased with cell size and probably in slower-growing species. The value of K_s for nitrate and ammonium was as little as 0.1 μmol/l, but could be fifty-fold or even more for some phytoplankton, especially neritic flagellates. Other data also suggested that natural oligotrophic plankton communities had much smaller half-saturation constants than eutrophic communities (cf. Chapter 7). Growth rate of phytoplankton species is, however, related to the level of illumination in addition to K_s values. Thus *Skeletonema* grows faster than *Coccolithus huxleyi* at relatively high nutrient and illumination levels (cf. p. 402). Cushing (1959a, 1963, 1975) has argued strongly that in temperate waters, nutrient depletion during the typical spring algal increase is not so great as to limit primary production significantly, especially since regeneration, due to grazing, maintains a considerable level of nutrient. The importance of ammonia especially, as a nitrogen source, is significant in the euphotic zone over summer in temperate latitudes. Ammonia is readily utilized in relation to other forms of nitrogen by most marine phytoplankton, and the marked buffering of seawater permits its continued absorption. Vaccaro (1963) demonstrated the importance of ammonia within the euphotic zone in maintaining production off the New England coast. While nitrate was totally depleted over part of the summer, ammonia was still present. Nitrate fluctuated twenty times between summer and winter

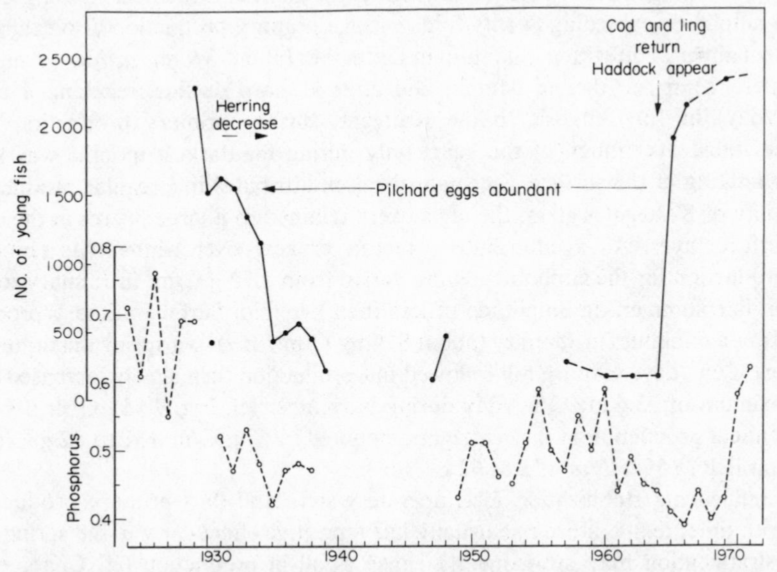

Fig. 9.14. Diagram to show (a) sums of average monthly catches of young fish (excluding clupeids) in standard hauls of the 2 m ring trawl off Plymouth (solid circle line), 1924–9 being shown as the average for those years; (b) winter maxima of inorganic phosphate (open circle line) as μg-at P/l. Changes in occurrence of herring, pilchard and certain gadoids are also given (from Russell, 1971).

at the surface; ammonia only two-fold. When nitrogen is reduced to trace concentrations algal division is halted, but Strickland (1965) points out that considerable biochemical adjustments, especially reduction of the ratio protein/lipid plus carbohydrate, are possible at low nitrogen concentrations to keep division continuing for some time. Phosphate, in contrast to nitrate, is temporarily stored in cells, often as polyphosphate. Exchange of phosphorus between cells and the medium can also be fairly rapid. These are possibly factors partly responsible for the less common restriction on production by phosphate as compared with nitrate (cf. Chapter 7). Anderson and Banse (1963), in emphasizing the importance of the replacement rate of nutrients in the euphotic zone over summer in stratified temperate waters, however, have pointed to several phosphorus budgets which suggested that the supply of phosphate due to vertical mixing was of great significance and a key factor in production.

Although Cushing has demonstrated an area of the North Sea where nutrient depletion did not appear to influence production significantly, at least earlier in the season, over the majority of temperate oceans, during part of the productive period, nutrient lack in the euphotic zone is a major factor limiting production. Some cycles of production in temperate and boreal waters have been referred to in earlier chapters. Some further examples will now be described.

The annual cycle of carbon assimilation estimated from ^{14}C measurements has been studied by Steemann Nielsen (1964a) in waters around Denmark. The differences in stability, due mainly to reduced salinity surface waters, between the Kattegat and Great Belt have already been described (Chapter 6). Great Belt waters showed the greater degree of mixing. A small production occurred in winter; there was a sharp spring increase but maximum production was found over 4 years during summer, possibly influenced by rapid regeneration of nutrients with higher temperature. For the euphotic column, monthly production varied from about 0.4 g C/m^2 in December to 9.2 g C/m^2 in July, an amplitude exceeding twenty-fold. Surface primary production also exhibited a large seasonal variation with a minimum in December (about 3.9 mg/m^3/day) rising to approximately 23 mg/m^3/day in March, and after a sharp decline, reaching a peak (29 mg/m^3/day) in July/August. In the Kattegat, surface primary production was remarkably stable over much of the year; only during the darkest months was light apparently limiting at the surface. Steemann Nielsen attributes this regular production to the stability of Kattegat waters; the algae were retained to a large degree in the very shallow surface layer. As a contributory factor, grazing over winter was reduced. Monthly production for the euphotic column varied from 1.75 g C/m^2 in January to *ca.* 8 g C/m^2 in late summer, an amplitude of less than five-fold. Surface primary production rose from a minimum in January (about 8.3 mg C/m^3/day) to a spring maximum of about 28 mg C/m^3/day. A sharp fall followed but production then slowly increased to a second maximum of 23.6 mg C/m^3/day during November (cf. Fig. 9.15). Over 6 years the mean annual production in the Kattegat amounted to 67 g C/m^2 (51 to 82 g C/m^2); for the Great Belt to 59 g C/m^2 (55 to 62 g C/m^2).

Factors influencing stabilization in temperate waters and thus primary production will vary with time; temperature rise usually has a positive effect early in the spring but increased stratification may subsequently cause a fall in production (cf. Chapter 6). Reduced surface salinity, while promoting production in Danish waters, may act as a barrier to the upward passage of nutrients. Off Greenland, Steemann Nielsen (1958) showed different levels of gross annual production in different coastal areas. Poor

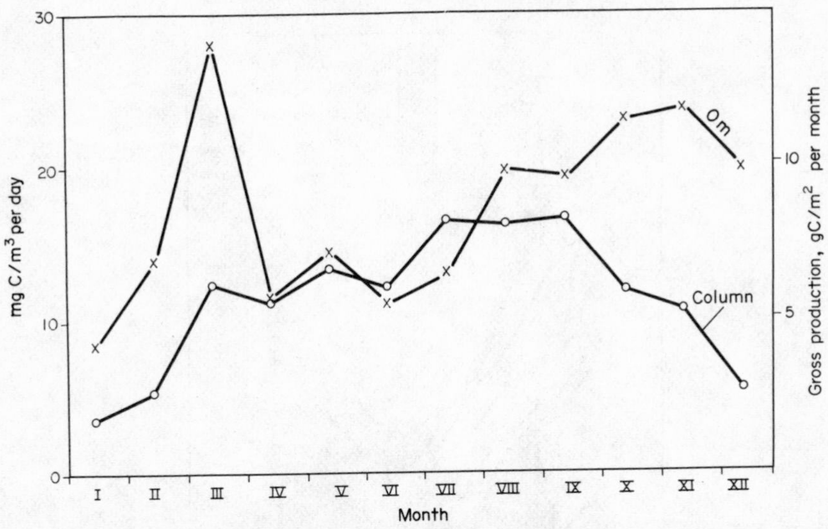

Fig. 9.15. Monthly rates of gross production (crossed line) at surface and (open circle line) for euphotic column at Anholt Nord, averaged 1954/60 (after Steemann Nielsen, 1964a).

production was associated with cold but less dense surface water (e.g. at Narssak), which prevented the nutrient-rich warmer oceanic water coming to the surface. On the other hand, where there was mixing as in Godthaab Fjord, a very high production occurred. Primary production rose from zero in mid-winter in Godthaab Fjord to > 1200 mg C/m²/day in late spring (cf. Fig. 9.16). Annual production at the mouth of Godthaab Fjord amounted to 98 g C/m²; at Narssak production was much lower (29 g C/m²/yr). More generally in waters round Greenland, where eddies brought nutrient-rich water to the surface, a high rate of production was found. Production during July/August ranged from < 100 to > 350 mg C/m²/day (Fig. 9.17).

In temperate as in other oceans, factors which increase mixing and reduce stratification will normally lead to higher production, but this will apply essentially to late spring and summer in temperate areas, and even then excessive turbulence may reduce production. Variations in the timing and extent of the spring increase, particularly in relation to mixing and critical depth have already been discussed (cf. Robinson, 1970; Cushing, 1975). Over the productive period, however, areas of mixing such as the Faroe-Shetland

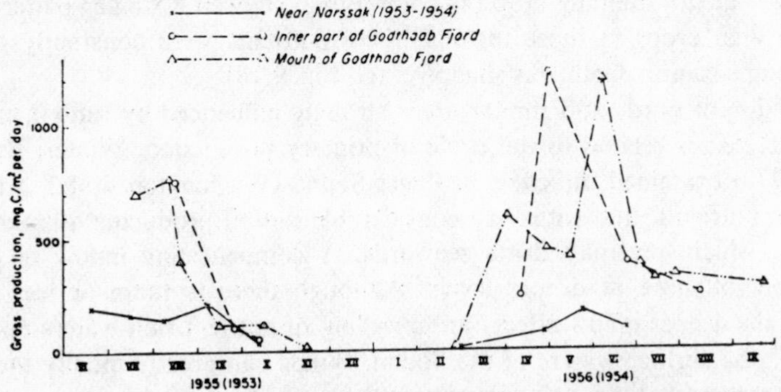

Fig. 9.16. Gross photosynthetic production at three coastal localities off Greenland: (a) = continuous line – Narssak; (b and c) = broken curves; two areas of Godthaab Fjord (after Steemann Nielsen, 1958).

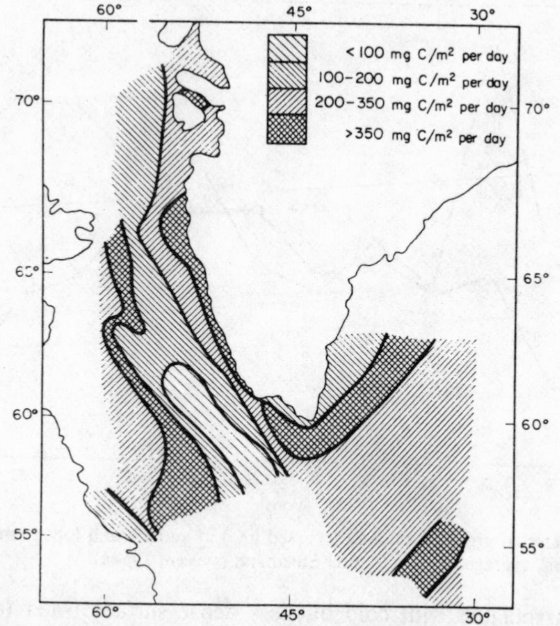

Fig. 9.17. Regions in the waters round Greenland showing marked differences in primary production (after Steemann Nielsen, 1958).

Ridge have long been recognized as showing higher surface nutrient concentrations and increased production.

Hansen (1961) found high production (500 to 750 mg C/m^2/day) over the growth period on either side of the Faroe-Shetland Channel; the water was not stabilized. Over the Faroe-Iceland Ridge, where there was constant upwelling, rates between 650 and 2700 mg C/m^2/day were recorded, even higher than in coastal Faroe water. In the East Polar Greenland Current, rates were low (130 to 230 mg C/m^2/day), and in the north-going Irminger Current, west of the Reykjanes Ridge, production was also fairly low (170 to 400 mg C/m^2/day) and was comparable to that of the main Atlantic water mass west of the Faroes (140 to 340 mg C/m^2/day). In the warmer Atlantic water between the North Atlantic Current and Ireland, nutrients were used up earlier in the year and by late summer production had fallen to 50 to 100 mg C/m^2/day. On the whole, variation in the size of the standing crop (as chlorophyll) showed a similar pattern to production, with high crops in those regions where nutrients were constantly available, though the compensation depth was shallower (cf. Fig. 9.18).

The circulation of fjords and similar areas strongly influenced by run-off may show special characteristics related to the cycle of primary production. Winter, Banse and Anderson (1975) examined the cycle in Puget Sound (Washington State), a fjord-like Sound, rich in nutrients, and with very considerable run-off producing a brackish near surface layer, which generally drifts seawards. A compensating inflow of seawater originates from offshore at deeper levels. Although there is more or less constant stratification, the deeper inflow effects an upwelling of nutrient-rich waters and carries algal cells into the surface waters of the Sound. Winds can greatly modify the circulation in the short-term, but the main influence is the longer term tidal cycle.

Winter *et al.* developed a model to study the effects of the hydrodynamic circulation

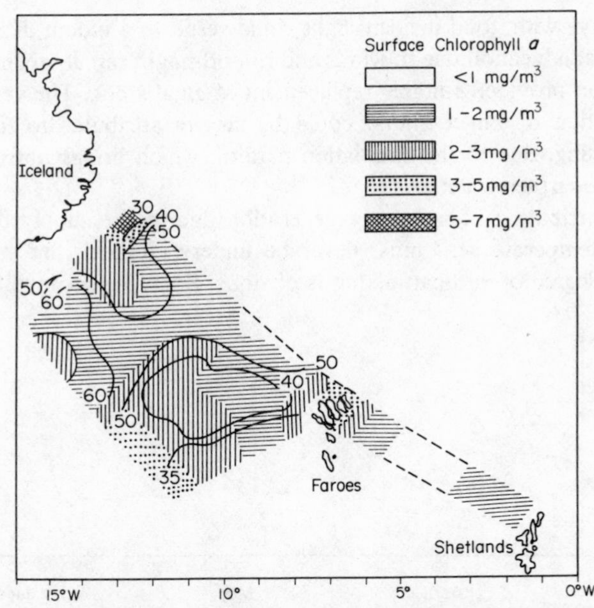

Fig. 9.18. Concentration of chlorophyll *a* at the surface between Iceland and Shetland Islands. Numbers indicate depths of the euphotic layer (June 1959) (Hansen, 1961).

on phytoplankton density and algal reproduction rate. Photosynthetic rate, respiration, sinking, vertical mixing and turbulence, horizontal advection, nutrient depletion and grazing were factors considered. Temperature changes were small and their effects on production rates were neglected.

Though incident illumination in Puget Sound from April to June 1966 was rarely less than 200 and occasionally exceeded 600 ly/day, growth of algae was light limited by the strong vertical mixing and turbulence, self-shading and suspended inorganic particles. Light inhibition at the surface occurred on a few occasions. In the absence of blooms large nitrate concentrations (15 to 20 μg-at/l) were present at the surface. Even when blooms reduced nitrate to almost undetectable levels, depletion usually lasted for only 1 to 2 days, persistent upwelling renewing nutrients at the surface.

Primary production at the northern end of Puget Sound was very high (460 to 470 g C/m^2/year); the phytoplankton exhibited a marked annual cycle with intense blooms between May and September. Simulated effects of run-off and of changing the tidal amplitude on the standing crop and on productivity are shown in Fig. 9.19. Though a considerable amount of chlorophyll was detected at deeper levels, this was attributed largely to viable cells brought in by the deeper inflow. Relatively little sinking of algal crop was believed to have taken place. The model demonstrated, however, that if the sinking rate, normally reckoned as 0.5 m/day except during nutrient depletion, increased to 3 m/day, the algal crop would have been severely reduced. There were also substantial differences in production according to whether algal respiration was calculated as a standard rate, or was corrected for differences between day and night and for depth.

Winter *et al.* found that the model adequately reproduced the main phytoplankton changes and could demonstrate a close relationship between the level of primary production and the circulation, the physical and chemical properties of the water, and light. Several days with good incident light could result in a bloom during spring, and though horizontal advection due to winds and run-off might rapidly reduce the crop, the deeper circulation provided suitable replacement of algal stock. The very high level of primary production of Puget Sound could be largely attributed to the marked and persistent upwelling, due to the circulation pattern, which brings nutrients as well as algal stock to the surface waters.

While the contribution of nutrient regeneration due to the supply of nitrogen and phosphorus in temperate seas must never be under-estimated, the importance over summer of the degree of vertical mixing is obvious. Some recent investigations which

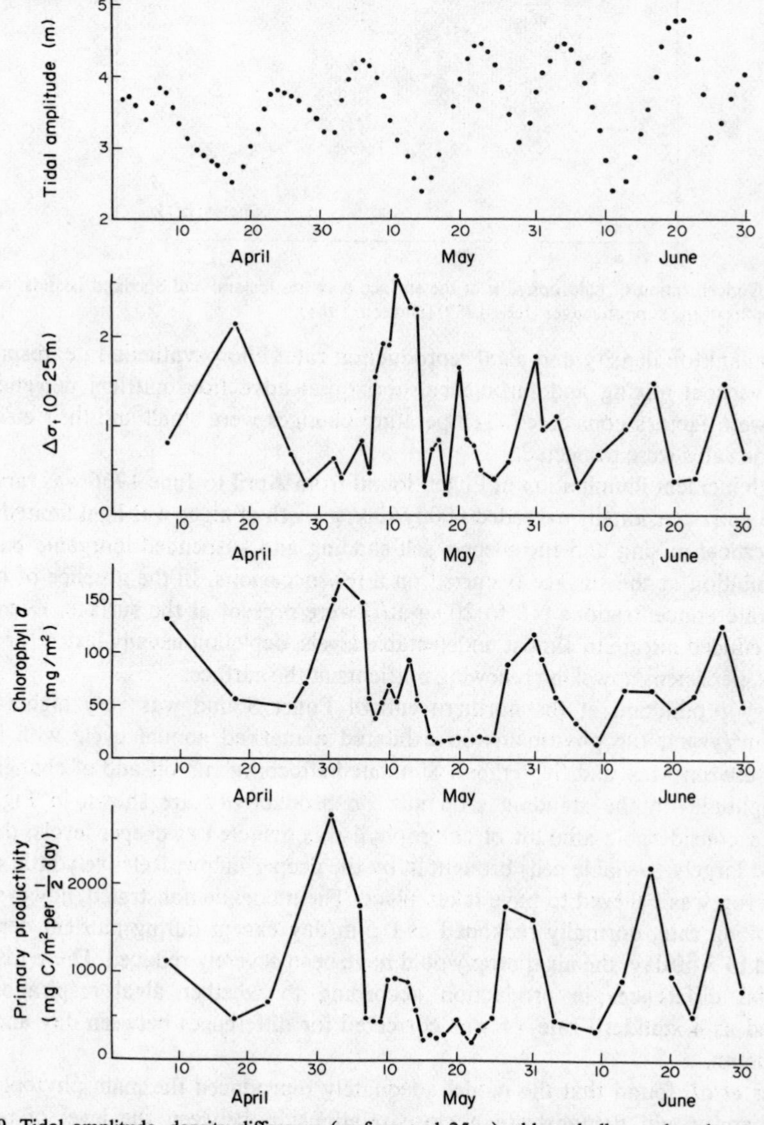

Fig. 9.19. Tidal amplitude, density difference (surface and 25 m), chlorophyll *a* and carbon uptake rate integrated from surface to 1% light depth (Winter *et al.*, 1975).

demonstrate this principle and to which reference has been made in earlier chapters include those of Pingree, Holligan, Mardell and Head (1976) and Jamart, Winter, Banse, Anderson and Lam (1977). In the temperate northern North Sea, Steele (1956) has described areas differing in the strength of the seasonal thermocline and degree of mixing. At an inshore group of stations there were substantial water movements, mainly due to tidal influence but also affected by wind; mixing was far greater than on the Fladen Ground. In a northern group of stations, mixing, though not as strong as inshore, was greater than on the Fladen Ground, and the thermocline broke down earlier. Primary production apparently reflected the degree of mixing. In 1953 production in the inshore area was nearly twice that on the Fladen Ground, with the northern group intermediate in value (cf. Table 9.10). The table also indicates considerable variation in production between years on the Fladen Ground. Steele believes the differences are mainly the result of different wind strengths, especially in early June and mid-September; stronger winds promote mixing and restrict the period of thermocline development. Table 9.2 compares production in the northern North Sea with two other temperate moderately shallow areas. Mixing in the English Channel station is similar to inshore stations in the North Sea and production is between that of Fladen and inshore stations, despite lower maximum phosphate. The considerable mixing and high maximum phosphate on Georges Bank may be linked with the greater annual production and especially with the high maximum crops of plankton (Table 9.2).

Table 9.10. Summary of the estimates of production (g C/m^2) (from Steele, 1956)

Group	1951	1952	1953				
			Station	1	2	3	4
Fladen	64.9	82.3		59.5	53.6	56.3	62.2
			Station	5	6	7	8
Northern	—	—		79.9	68.4	86.2	70.5
			Station	9	10	11	12
Inshore	—	—		127.2	103.9	108.1	108.4

In the north-east Pacific, Holmes (1958) obtained fair agreement between standing crop and rate of production of phytoplankton and the concentration of phosphate in the euphotic zone. Other factors being equal, productivity was closely related to the replenishment of nutrients in the upper layers. High values were typical of the water near the Aleutians, off the south Californian coast and off northern South America. Bogorov (1958a) similarly found a high rate of productivity for the boreal phytoplankton off the Kurile Islands and a productivity about an order of magnitude lower in the warmer Kuroshio waters (Fig. 9.20). Starodubtsev (1970) confirmed a higher rate of primary production in the boreal waters around the northern edge of the sub-Arctic Front in the Pacific (0.80 to 2.30 g C/m^2/day) than in the Kuroshio Current with low values of 0.15 g C/m^2/day. In waters lying between these two regions and subject to considerable mixing, production rates were intermediate.

McRoy, Goering and Shiels (1972) comment on the generally high primary production and standing stock of phytoplankton in the shallow eastern Bering Sea. They refer to earlier reports for the Aleutian region of fairly high productivity (e.g. 100 to 150 mg C/m^2/day for the column, 19 mg C/m^3/day for surface waters). In June/July

McRoy *et al.* found that with a mean standing crop of 1.88 mg chlor a/m^3 primary production amounted to 27.9 mg C/m^3/day in the eastern Aleutian area; integrated primary production was 243 mg C/m^2/day. Production tended to be higher in the lee of islands, but an especially high level was found for the Bering Strait (4100 mg C/m^2/day). The water was almost isothermal in the Strait and the standing stock of algae was so large that the euphotic depth was only 13 m. At other stations in the Aleutians the euphotic zone was much deeper. In winter in the Aleutian area primary production was reduced to only 1.2 mg C/m^3/day at the surface, and the crop of algae was also very low. On the underside of ice, however, there was a comparatively rich flora giving a primary production of 44.4 mg C/m^3/day. The authors believe that the annual cycle of production begins with a bloom of non-phytoplankton algae in the sea ice in early spring, and this is followed by a late spring burst of phytoplankton. Takahashi and Ichimura (1972) compared the high productivity of the Oyashio Current waters of the north-west Pacific with the very stable North Pacific central waters. Except for ammonium, nutrients were higher in Oyashio waters. Calculations of the daily photosynthetic rate per unit of chlorophyll and of the daily production (photosynthetic rate multiplied by chlorophyll concentration) suggested that production in the unstable "current" regions was distinctly higher than in the stable central waters (205 to 500 mg C/m^2/day as against 120–380 mg C/m^2/day). The higher production was partly due to the greater standing stock of algae in the unstable mixed waters but the amplitude of seasonal variation in production was much larger in the colder areas which are also richer.

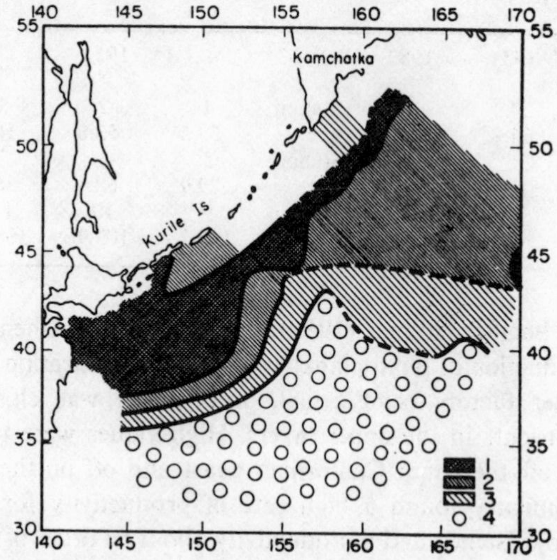

Fig. 9.20. Zones indicating different levels of primary production in the region south-west of the Kurile Islands, Pacific Ocean (after Bogorov, 1958).

Anderson (1964), investigating photosynthetic production off the west coast of the United States, showed that the area influenced by the surface outflow of the Columbia River had a high rate of photosynthesis. Production was seasonal, with a minimum in winter, a spring bloom (May), a summer minimum and a small autumn rise. The Columbia River outflow, however, mainly influenced the timing of the seasonal cycle, presumably by reducing the depth of the mixed layer, so that both inside and outside the plume area of the River production was similar (about 60 g C/m^2/year). Further south, off the Oregon coast, marked upwelling occurred during the summer; a very high production was evident with the abundant nutrients available. Anderson found a very variable assimilation number for the algae which may have been in part a reflection of nutrient deficiency over some periods of the year. With reference to a sub-surface chlorophyll maximum spreading beyond the continental shelf in the area (cf. Chapter 5), Anderson (1969) found considerably greater nutrient concentrations persisting at depth, although surface nutrients were markedly reduced during the seasonal bloom. The very low light intensities at the chlorophyll maximum were apparently utilized effectively in photosynthesis so that Anderson estimated that 15% of the total production was attributable to algae below the 1% surface light intensity level.

In middle and higher latitudes, the greater productivity near coasts as compared with that in the open ocean is usually attributed to nutrient concentration and greater stability. Strickland (1965) emphasized that though nutrient levels offshore may be lower, they are not always *demonstrably* limiting; the size of standing crop is an important factor in daily and annual production. The "land mass" effect may be of significance in so far as commonly the crop in open oceans is lower.

An example of land mass effect may be given by comparing data on production from an investigation by McAllister, Parsons and Strickland (1960) who measured primary productivity over 2 months (July/August) at Weather Station P in the north-east Pacific (latitude 50°N) with a study by Parsons, LeBrasseur and Barraclough (1970) for the Strait of Georgia about the same latitude. At Weather Station P, nutrients (nitrate, nitrite, ammonium, phosphate, silicate) never appeared to be limiting during the summer months, and this was abundantly confirmed by an experiment culturing natural phytoplankton populations on deck: satisfactory growth continued for 10 or 11 days until nitrate fell to <1 μg at/l and phosphate was still relatively abundant. Neither temperature nor incident illumination limited production. Production was not high – aproximately 7 mg C/m^3/day near the surface (200 mg C/m^2/day for the euphotic zone) and there was no bloom in production or crop. The productivity over the period varied only about three-fold, and even standing stock showed considerable constancy over the summer. McAllister (1972) states that gross production near Weather Station P ranged from about 45 to 72 g C/m^2/yr. A similar range is quoted by Parsons *et al.* (1970) for the sub-Arctic north-east Pacific (43 to 78 g C/m^2/yr). Nett production over a 6-year period at Station P varied from 34 to 51 g C/m^2/yr.

In contrast to this relatively stable primary production at Station P, Parsons *et al.* (1970) demonstrated that in the coastal region of the Strait of Georgia increasing solar radiation and shallowing of the mixed layer from January/February to April/May resulted in a large increase in primary production from 20 to 1200 mg C/m^2/day. The annual primary production was 120 g C/m^2, similar to that of other inshore areas at the same latitude, but considerably greater than at Weather Station P. The obvious spring bloom caused a fall in nutrient concentration in March (Fig. 9.21). Standing crop also increased to a maximum in June, but crop size varied over the area. The maximum density, near the River Fraser plume, even exceeded 5 mg/m^3 chlorophyll (cf. Fig. 9.22), i.e. at least ten to thirty times that at Weather Station P. There was also a difference in the type of phytoplankton stock. In the Strait of Georgia, diatoms dominated the spring bloom, with more dinoflagellates over the summer and nanoplankton mainly in winter. At Station P, coccolithophorids dominated the phytoplankton and net plankton formed a very small part of the crop.

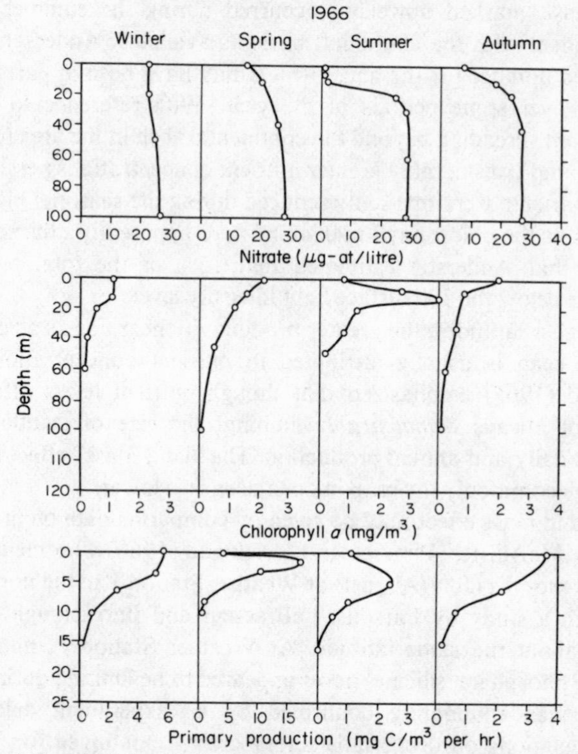

Fig. 9.21. Seasonal changes in the depth distribution of nitrate, chlorophyll *a*, and primary production in the Strait of Georgia (from Parsons *et al.*, 1970).

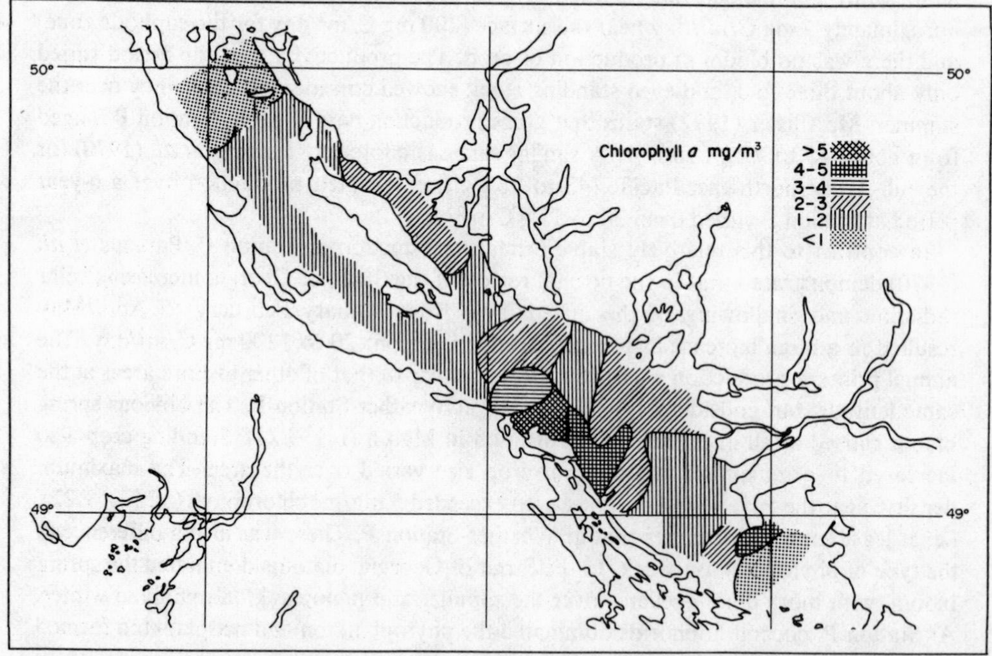

Fig. 9.22. Chlorophyll *a* distribution in the Strait of Georgia, 23–29 March 1966 (from Parsons *et al.*, 1970).

Strickland (1965) attributes lower production in the oceanic waters, despite good nutrient and light conditions which could have increased the crop at least fifty-fold, to more continuous and effective grazing offshore. The more stable situation is described as characteristic of a "fertile but overgrazed" area. The low standing crop typical of the open tropical ocean and, maintained as a quasi-equilibrium, is described as more similar to an "under-fertilized and spent" area. In the North Pacific the delay in the onset of grazing appears to be brief since herbivorous zooplankton (*Calanus cristatus*, *C. plumchrus* and *Eucalanus bungii*,) overwintering in deeper water, reproduce at least over part of the winter. In much of the boreal North Atlantic, where *Calanus finmarchicus* is the dominant herbivore, reproduction must await the spring growth of phytoplankton and bloom conditions are therefore more frequently encountered (cf. Chapter 8).

There are many examples, however, of increased nutrient concentration in inshore temperate waters enhancing production. With the stirring effect due to tides, wind and waves during summer, in addition to run-off from land, nutrient supply to the euphotic layer can be substantially greater than offshore. Inflow of deeper water in an estuarine type of circulation can also act as a nutriént trap.

Regeneration of nutrients from the bottom in shallow seas can make an additional contribution to the supply of nutrients, including trace metals, over the productive season. Steemann Nielsen (1964a) compared the fertility of waters of the Great Belt (Baltic), which receives supplies of salts from the bottom deposit, with the Kattegat, where the greater depth largely precludes influence of the bottom on nutrients. He also points to higher temperature in summer in shallow waters speeding up regeneration. Nutrient release from interstitial water flushed out from the bottom sediment was shown by Smetacek, Bodungen, Bröckel and Zeitzschel (1976) in plankton tower experiments in Kiel Bight to be important for phytoplankton production. Plankton blooms contributed substantially to sedimentation of organic matter on the bottom and subsequent blooms were usually preceded by a rise in nutrient content of bottom water. In open oceans, by contrast, regeneration can, in the short term, involve only the water column, and in very strongly stratified seas will concern very largely only the waters above the permanent thermocline.

Production in very shallow temperate seas

In temperate seas, primary production over winter is minimal; transient production in an exceedingly thin surface layer sometimes occurs. In very shallow areas such as estuaries and embayments, however, despite the strong mixing of the water, the depth is so limited that sufficient light may penetrate even in winter to permit growth. Such areas may be very productive. Riley (1952, 1956, 1967) found that a phytoplankton bloom could occur from late winter to early spring (end of January/early March) in the shallow waters of Block Island Sound and Long Island Sound, though there was considerable variation in the onset of primary production in Long Island Sound over a 6-year period. The level of incident radiation from mid-December onwards appeared to be the important factor, though low temperature, by reducing algal respiratory losses, had some small effect. Production could be related to incident radiation and temperature, with a minimal critical level of radiation of 40 ly/day. In some very shallow areas, incident radiation may not be limiting even in mid-winter. In lower Narragansett Bay (averaging *ca.* 10 m) the flowering may start in mid-December, whereas in Long Island

Sound (averaging 20 m depth) it is somewhat later (Smayda, 1976). Generally, Riley (1967) states that coastal waters of 20 to 50 m will bloom from mid-January to early March. Vertical stability is obviously a less important factor; the very shallow waters are more or less constantly mixed. In slightly deeper inshore waters the onset of spring production will depend on the establishment of some thermal stability in addition to the increasing incident light, and production is delayed until the end of March or later in the spring. Even in very shallow water, transparency is very important since turbidity will affect the start of production, but nutrient levels are not normally a factor. Only the duration and sometimes the magnitude of the algal bloom is conditional on nutrient level.

Cadée and Hegeman (1974) describe a spring peak and a broad summer peak in chlorophyll *a* and in primary production for the very shallow Dutch Wadden Sea. Production amounted to 100 to 120 g C/m^2/yr for western and eastern areas, but light was an important factor limiting primary production, due to the high turbidity, especially of the eastern Wadden Sea. The effect of silt load and constant turbulence in reducing light in the Westerschelde has already been noted (cf. Chapter 5). The turbidity of inshore Dutch waters was also found by Gieskes and Kraay (1975) to delay the spring plankton growth (cf. Chapter 6). Where a dense crop develops in a very limited depth as in estuarine waters, self-shading will also become a serious factor reducing light penetration. Though this may apply to estuaries in tropical regions of the world, it will be more critical with the lower incident radiation of temperature latitudes. Excessive mixing occurs in the Bay of Fundy (Nova Scotia) with constant tidal turbulence. Although the area is continuously rich in nutrient salts, the lack of stabilization prevents the phytoplankton being maintained in the euphotic zone, and with reduced light penetration due to the turbulence, the diatom crop in the Bay is much less than the value expected from the nutrient concentration.

Smayda (1976) discusses the decrease in primary production in proceeding offshore into deeper water. He quotes as illustration that along a transect between Long Island and the Sargasso, annual production decreases about ten-fold, from a maximum of *ca.* 500 g C/m^2 in waters shallower than 10 m. Similarly, from Narragansett Bay to Georges Bank, annual primary production declined from 310 g C/m^2 to 150 g C/m^2. The decline is not so dramatic in a fairly rich temperate region. Block Island and Long Island Sounds were intermediate in level. The start of the winter-spring phytoplankton production also becomes later in moving offshore. Smayda points out that this is also characteristic of the waters of the continental shelf off New York, beginning even in December at depths of less than 50 m, and being as late as April in the deeper waters.

3. Upwelling areas

Since production over much of the world ocean is limited by the supply of inorganic nitrogen and phosphorus to the algae, any upwelling of deeper nutrient-rich water will tend to stimulate production. This is especially evident in tropical and sub-tropical seas. Some examples have already been discussed of upwelling in middle latitudes. Major upwelling may also occur at high latitudes, as off Antarctica. The upward passage of water is associated with a drift of surface waters and the formation of deep bottom current, flowing away from the Antarctic Continent. It is especially characteristic of regions like the Weddell Sea. Primary production in the Antarctic will be reviewed later in the chapter. In warm oceans increased primary production has been noted in regions of equatorial divergences, the so-called dome areas occurring in eastern tropical oceans (e.g. the well-known Costa Rica dome) being sites of particularly active upwelling. Domes occur where an equatorial counter-current turns either north or south into an equatorial current flow, with a resulting complex circulation pattern. Upwelling may be found on a more restricted scale off the coasts of tropical and sub-tropical countries due to local climatic conditions, especially wind. Other upwellings are the result of current divergences and of currents impinging on submarine plateaux and ridges.

The rate of ascent of upwelled water is significant since too rapid vertical passage may lead to decreased production. Rates range from about 1 to 5 m/day. The speed of drift of water away from the upwelling zone will also in part determine the richness and the positioning of the productive belt. With seasonal variation in the intensity of upwelling, the timing of onset of grazing with its repercussions in regeneration is also important.

Strickland (1965) stated that production will tend to maximize if the rate of upwelling is approximately constant, and strikes a balance between nutrient replenishment and avoidance of the excessive transfer of algal cells out of the euphotic zone (cf. Guillen, Rojas de Mendiola and Izaguirre de Rondan, 1971). In an alternative pattern for achieving production, periods of excessive turbulence and hence nutrient replacement, due to local climatic conditions, may alternate with calm periods when, with stability established, a bloom of phytoplankton can follow. In oceanic regions experiencing upwelling where some degree of thermal stratification can be established, the fertility of upwelled water mixing with the usual oligotrophic warm surface water can be very great.

The enormous importance of upwelling, both on the large and on the more restricted scale in various regions of the Indian Ocean, mainly due to the monsoonal climate, has already been discussed. Well-known regions of pronounced upwelling in the other warm oceans of the world occur off the south-west coast of Africa, the west coast of South America, off north-west Africa, and off the coasts of California and Oregon. The upwelling is due to relatively constant winds parallel to the coastline which cause near-surface currents to move along the coast, gradually turning offshore under the influence of Coriolis force (cf. Chapter 1). These regions are particularly striking in their high levels of productivity which contrast with the generally oligotrophic oceanic waters offshore. The depth from which nutrient-rich colder water ascends is usually quite moderate (approximately 100 m and very rarely approaching 300 m).

Earlier studies demonstrated considerable upwelling off Chile and Peru. The Humboldt and Peru Coastal Current have fairly high nutrient concentrations but as the flow turns more offshore upwelling occurs. Gunther (1936) suggested that where upwelling was especially active high phosphate values, exceeding 1 μg-at PO_4-P/l, were found, and that phytoplankton density was especially rich in such areas. The rich zooplankton, enormous anchovy fishery and dense populations of fish-eating sea birds all reflect the very high productivity of the region. Posner (1957) recorded some areas with phosphate concentrations exceeding 2 μg-at PO_4-P/l. Nitrate was also high, though he believes it is more likely to be the limiting nutrient in production.

In a more recent study of Peruvian waters, Barber, Dugdale, MacIsaac and Smith (1971) found very high levels of nutrient in upwelled waters. At latitude about 15°S, phosphate usually exceeded 2 μg-at/l and nitrate ranged between 9 and >20 μg-at/l at different stations. Production reached 3500 mg C/m^2/day, with a dense crop of chlorophyll amounting to 26 mg/m^2. At one station, however, productivity was much lower (1075 mg C/m^2/day), despite the rich nutrient concentrations of the upwelled water and moderately dense phytoplankton crop. The salinity of the ascending water in this area was higher and the authors believe that it came from a different source. It is now widely recognized that some upwelled waters need a lag period before phytoplankton will flourish; the need for biological conditioning has been suggested. Chelation of trace metals is one possible requirement (cf. Chapter 7). Barber *et al.* carried out certain enrichment experiments which indicated that the addition of the chelating agent EDTA and/or trace metals enhanced ^{14}C uptake in water taken from the anomalous station, but had no effect on carbon-assimilation rates with waters from other areas. Though high productivity is generally true of the areas of intense vertical passage of water in upwelling regions, high nutrient level is obviously not the sole factor.

Smith (1968) reviewed the patterns of upwelling in different major regions and discussed some of the seasonal variations in intensity. He reported earlier data suggesting a productivity of > 1 g C/m^2/day and a standing crop of 2 mg/m^3 chlorophyll for a station in the Peru Current. Guillen and Izaguirre de Rondan (1968) demonstrated differences in primary production in different areas of Peruvian waters. Between 4° and 6°S latitude annual production was 97 g C/m^2; between 6 and 18°S latitude – 162 g C/m^2. The mean daily values were 0.27 and 0.45 g C/m^2, respectively. Maximum productivity was found for the Coastal Peru Current during the summer; the minimum occurred in winter, with an average for the year amounting to 190 g C/m^2 (i.e. 0.53 g C/m^2/day). Guillen compares this mean daily production with values of 5.7 to 6.4 g C/m^2/day for the Gulf of Oman – a very rich upwelling area. Guillen, Rojas de Mendiola and Izaguirre de Rondan (1971) confirmed that the high primary production off Peru, close to the coast south of 6°S, was associated with upwelling during summer. An area with very recently upwelled water might have few algae, but where some stability followed upwelling, there was high production. Areas of high productivity were associated with high densities of standing crop. During winter, despite intense upwelling, production was much lower. The authors attribute this decline to vertical turbulence, increased turbidity and reduced radiation, all lowering photosynthesis over winter. The difference between the mean productivity in summer (high) and winter (low) was between two and three times. The flora was dominated by five diatom species: *Eucampia zoodiacus*, *Rhizosolenia delicatula*, *Thalassiosira subtilis*, *Skeletonema costatum* and *Chaetoceros* sp. Over spring and autumn three diatoms made up 80% of the total phytoplankton. North of 6°S, in mainly equatorial waters, and far off the coast, production and crop were unusually low. The average daily primary production through the water column was between 300 and 400 mg C/m^2 for waters north of 6°S latitude, but exceeded 400 mg C/m^2 further south, with a maximum of *ca.* 1000 mg C/m^2 at 8 to 9°S latitude.

Guillen (1971) later quoted a very low production for the equatorial surface waters of < 0.05 g C/m^2/day as compared with a high production (0.80 g C/m^2/day, equivalent to 290 g C/m^2/yr) for an area of intense coastal upwelling. When El Niño caused equatorial surface water to push further south off the Peruvian coast, production was substantially reduced.

Another investigation of primary production in the Peru Coastal Current is that of Ryther, Menzel, Hulburt, Lorenzen and Corwin (1970). The area studied was some 10 miles offshore and there was an attempt to maintain observation over the same body of upwelled water by following a drogue over 5 days. Very large populations of phytoplankton (maximum 1.1×10^6 cells/l) dominated by diatoms, especially *Chaetoceros debile*, were observed. The low diversity of the phytoplankton is illustrated by the observation that eight species of diatom comprised 83% of the population. Very high nutrient levels were found in the euphotic zone but there was very rapid utilization. The algal population reached a peak density after 3 days and the subsequent decline was not due to nutrient lack, but was probably a result of intensive grazing. Estimates of productivity deduced from changes in nutrient levels and oxygen were of the order of 10 g C/m^2/day, and considerably exceeded measurements of ^{14}C uptake. The estimate is higher than data based on ^{14}C measurements from most other rich upwelling areas, though Fogg (1974) quotes a maximum rate of 9 g C/m^2/day found by Hobson and his colleagues for Peruvian waters. Cushing (1975) estimated that daily production in upwelling regions varies on average only about an order of magnitude between the different regions of the world (cf. Table 9.11), but differences in area and in the length of the upwelling productive season introduce a very much larger range in total production. He also claimed that the average production off the north-east coast of England during spring (1.25 g C/m^2/day) was similar to the general level for upwelling regions, but peaks in production in upwelling areas can certainly exceed this considerably.

Table 9.11. Some upwelling areas ranked in order of production intensity. The major upwelling areas were divided into two or three sub-areas (from Cushing, 1975)

g C/m^2 per day	Upwelling areas
1	Peru (1), Canary (1), Benguela (2), S. Arabia, Vietnam, Gulf of Thailand
0.3–1.0	Peru (2), Canary (3), Benguela (1), Somali, Guinea dome, Malagasy (Madagascar) wedge, Orissa coast, Java, Sri Lanka (Ceylon), Flores, Banda, E. Arafura
0.3	California, Costa Rica dome, Chile, Benguela (3), New Guinea, Andaman Islands, North-West Australia
0.1	Open ocean

For the Benguela Current, Hart and Currie (1960) confirmed earlier reports of rich nutrients; the upwelled coastal waters contrasted sharply with oceanic surface waters, with very high phosphate, at times exceeding 2 μg-at PO_4-P/l. Upwelling was not uniform over the area but was localized, consisting of a series of eddies with interlocking tongues of richer and sparser water (Fig. 9.23); the upwelling depth was about 200 to 300 m. The upwelling zone was extensive, some 80 miles wide and about 1000 miles long. Diatoms, especially species of *Chaetoceros*, dominated the very rich phytoplankton and the species diversity was not large as in other upwelling areas. A few dinoflagellates (*Ceratium* and *Peridinium*) and also *Trichodesmium* were important species. Some details of the flora are given in Chapter 5.

On the first survey the coastal populations were about four orders of magnitude, and on the second survey, three orders of magnitude greater than those in oceanic waters. Though a dense crop of zooplankton is typical of the Beneguela Current region, the zooplankton probably did not reach its maximum potential. Much of the algal primary

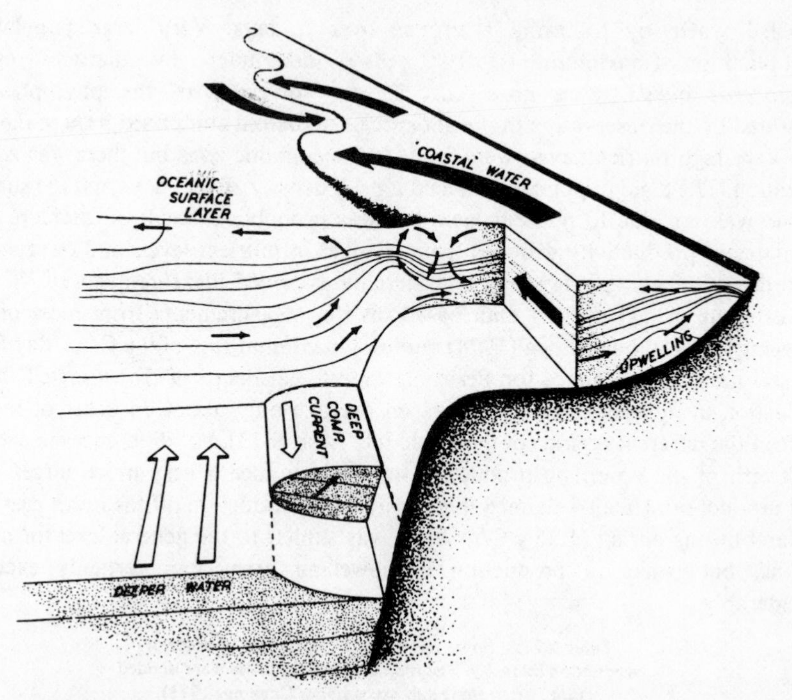

Fig. 9.23. A perspective diagram showing an idealized picture of the principal horizontal and vertical water-movements in the process of upwelling. The isosteres are represented by the thin lines on the "cut" faces of the water masses. The shaded sinuous line on the sea-surface represents the continuation of the boundary convection between the upwelled and oceanic surface-waters. The name of the deep compensation current has been abbreviated (from Hart and Currie, 1960).

production is lost to the upper layers by sinking and decay. The occasional marked mortalities among plankton and fish may be due to water movements over a bottom mud rendered anoxic by the rich deposit of organic matter.

Steemann Nielsen and Aabye Jensen (1957) comment on the enormously rich phytoplankton and the high concentrations of nutrient salts in the upwelling waters of the Benguela Current. Photosynthetic rate was high, the maximum at sub-surface level (cf. Fig. 9.24). Production varied between 0.46 and 2.5 g C/m^2/day for nine stations. Walvis Bay gave the remarkable level of 3.8 g C/m^2 for an extremely shallow layer. There appears to be no information on seasonal variation in rates of primary production. There were indications that in places very recently upwelled water did not give exceptionally high photosynthetic rates. The limitation is believed due to few algae being present in very newly ascended water.

This time factor is also considered of importance by Sverdrup and Allen (1939) for the upwelling region off southern California. There must be a period for development of the algae. On the other hand, they found that surface water, which on its characteristics of temperature and salinity appeared to have existed at the surface over a long period of time, was poor in diatoms, presumably owing to nutrients having been already depleted. In general, there was good correspondence between phytoplankton density and upwelling.

The earlier work by Sverdrup and Allen on upwelling off the south-west Pacific coast of North America has been supplemented by the investigations of Anderson already

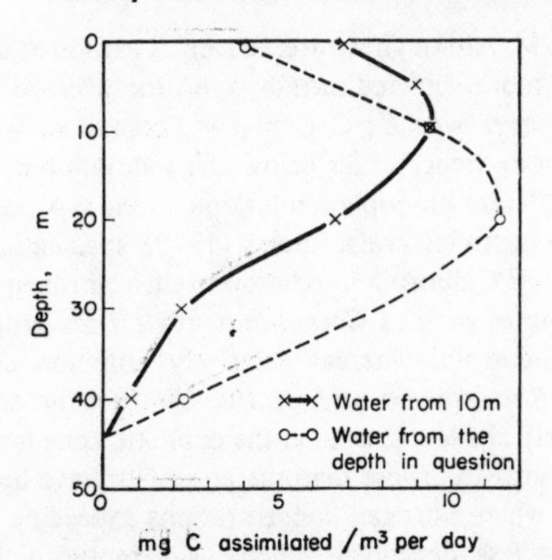

Fig. 9.24. Photosynthetic production with depth in Benguela Current, off Loanda, West Africa (from Steemann Nielsen, 1952).

described, and by the work of Strickland and his colleagues. Eppley, Reid and Strickland (1970), for example, found that off La Jolla, California, changes in production through the euphotic column were largely influenced by crop size. Both crop and primary production were highest during periods of upwelling and lower during extended periods of nitrate depletion in the mixed layer. The most seaward station (III) not surprisingly exhibited both crop and primary production at a lower level than stations I and II closer inshore (Fig. 9.25). The high value for primary production at station II in May may be associated with an especially high density of one species of diatom. Production compares very favourably with that in other upwelling regions. Different rates of carbon assimilation were found for phytoplankton above and below the depth where the nitrate

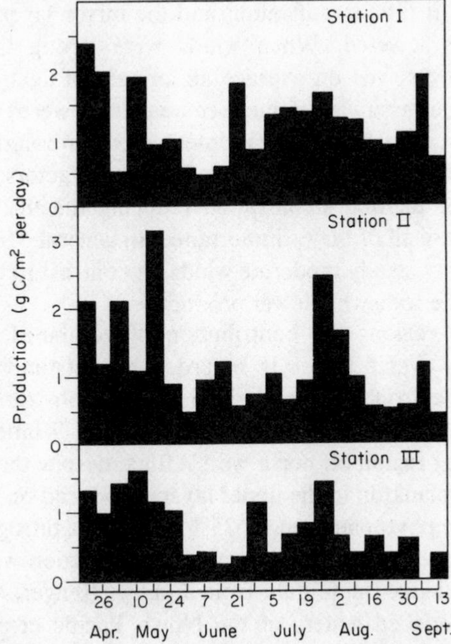

Fig. 9.25. Primary production at three stations off California (from Eppley *et al.*, 1970).

concentration reached 1 μM. Although a lowered rate of carbon assimilation would normally be expected in the more depleted nitrate layer, the average rate above the thermocline in the depleted layer was 6 g C/g chl *a*/hr, compared with a rate of about 3 g C/g chl *a*/hr in the nitrate-richer water below. This variation in primary production rate could be related to different phytoplankton populations associated with the average light intensities at the two depth intervals. Eppley (1970), speculating on the occurrence of phytoplankton species off California in relation to their nitrogen half-saturation constants, points out that species such as *Coccolithus huxleyi* are probably ubiquitous, in view of their low requirements, whereas relatively common diatom species (e.g. *Skeletonema costatum*, *Rhizosolenia stolterfothii*, *R. robusta* and *Leptocylindricus danicus*) must inhabit fairly shallow depths in the euphotic zone if they are to have sufficient nitrogen for rapid growth. These diatoms appear to have been most frequent as shallowly as 10 to 15 m where nitrogen concentrations exceeding 1 μM were present. *Gonyaulax polyedra* was less dependent since, as a motile dinoflagellate, it could migrate daily from a stratum with low light and high nitrate concentration to a shallower, high light, but low-nutrient layer.

In the other major upwelling region, off the north-west coast of Africa, recent studies by Hunstman and Barber (1977) suggested that although primary production is high, it does not on average achieve the level found, for example, in the Peruvian upwelling region. In general, with the offshore movement of the upwelled water, nutrient concentrations tended to decline from inshore outwards. Primary production, which averaged about 2 g C/m^2/day over spring, and chlorophyll concentration (averaging 68 g/m^2) also decreased offshore, but very close inshore both crop and production were lower. Huntsman and Barber attribute this reduction to heavy loads of suspended material inshore reducing the light available. They also point to the significance of wind strength over the whole region. Strong winds cause intense mixing so that, as in other upwelling regions, if there is little stratification and the mixed layer becomes too deep, photosynthetic activity is lowered. When winds were strong the majority of the phytoplankton population received on average an amount of light equivalent to about 10% of surface illumination; assimilation number was also lowered. On the other hand, when winds were very weak, upwelling was interrupted, and surface nutrients were lowered, thus also reducing primary production. The three factors, nutrient supply by upwelling, some degree of thermal stabilization reducing the thickness of the mixed layer, and light intensity, are all of major importance. In general, Huntsman and Barber believe that the lack of fairly steady moderate winds, in contrast to the Peruvian region, is mainly responsible for the somewhat lower production.

Although in upwelling regions the contribution of combined nitrogen (or other nutrient) from the deeper water layers is regarded as outstanding, the amount of ammonia and urea in the eutrophic zone, derived mainly from zooplankton excretion, may make a considerable contribution to fertility. Smith and Whitledge (1977) demonstrated that in the upwelling region off north-west Africa, despite the upward passage of nitrate-rich water, the zooplankton in the upper layers produced on average 44% of the ammonia demanded by the phytoplankton, or 25% of the total nitrogen requirements. In oligotrophic regions the contribution from zooplankton excretion would be expected to be larger. Smith and Whitledge quote data from Eppley, Renger, Venrick and Mullin (1973) that in the impoverished waters of the North Pacific central gyre, excretion supplied 54% and 40%, respectively, of the ammonia and total nitrogen demands of the phytoplankton. Generally in eutrophic areas, such as upwelling zones and inshore continental shelf regions, it appears that ammonia amounts to < 10% of the total nitrogen required; the north-west Africa upwelling region is exceptional in that the ammonia contribution is so much higher.

4. High latitudes

Production at very high latitudes is usually very low. As described in Chapter 6, the very poor illumination over much of the year with darkness over the extreme winter months, the loss of light due to reflection, and the greatly reduced penetration due to the thick ice cover, severely limit primary production. A very marked seasonal change may be evident, however, since over the very brief summer with the great day length, a bloom of phytoplankton, mainly diatoms, is possible. Seeding of diatoms is normally satisfactory at high latitudes since some species, especially resting stages, can survive in the ice. As soon as the snow cover disappears and the ice thins producing melt ponds, sufficient light can penetrate for growth, even at the very low temperature, since some indigenous species are adapted to photosynthesize at temperatures below 0°C, though growth may not be maximal (cf. Bunt, Chapter 3). Some degree of stabilization due to low salinity surface water may assist in maintaining a shallow mixed layer essential for effective photosynthesis.

Though some factors affecting primary production are therefore common to the Arctic and Antarctic, there are several characteristics which sharply separate the two great regions. The continental land mass of Antarctica at high latitudes is not paralleled in the northern hemisphere; the great ice-covered Arctic Ocean has no counterpart in the south. The Arctic Ocean is also more or less surrounded by great land masses; communication with the other oceans is practically restricted to the Atlantic, and even there, topography including bottom structure limits exchange of water. The great circumpolar east wind and west wind drifts in the far south are not reproduced at high northern latitudes and ocean currents are dissimilar in the far north and far south. A circumpolar belt of water instability, corresponding to the Antarctic Convergence, does not occur in the north, though more restricted polar fronts are known for North Pacific and North Atlantic Oceans. Though there is upwelling of deeper waters at high northern latitudes, this does not occur on the huge scale and over great areas of ocean as in the Antarctic, with its far-reaching repercussions on primary production. The great land masses of northern Canada and especially of Euro-Asia pour huge amounts of freshwater through their river systems into the Arctic Ocean. Extreme dilution has a direct effect on the marine phytoplankton and the river outflows increase turbidity.

Production in the two great regions must, therefore, be discussed separately. In the northern hemisphere some description of production at high latitudes, for example, off the Aleutians and off Greenland, has already been included. Further discussion regarding high northern latitudes will, therefore, be mainly confined to the Arctic Ocean itself.

In the Arctic Ocean, English (1959, 1961) found that light penetration through the snow-covered ice was so poor that production, in so far as chlorophyll density could be used as a criterion, increased only in late June to a single fairly small maximum in August. The productive cycle lasted only about 3 months and primary production was not exceptionally high even over that period, reaching 5 to 6 mg C/m^2/day for the euphotic column during the peak time. During winter, production was undetectable.

Strickland (1965) suggested that in mid-summer production in open Arctic waters could reach 1 g C/m^2/day but where ice cover was maintained, it could be as little as 0.03 g C/m^2/day. With the extremely brief summer, annual production must be extremely poor. Fogg (1977) quotes data for inshore waters over the period April to October; primary productivity reached a maximum of 30 mg C/m^3/day, but annual production was estimated at only 20 g C/m^2.

Although there are several studies of seasonal changes in phytoplankton density at slightly less high latitudes, there are few data on photosynthetic rates. Changes in standing crop may, however, give some indication of the production cycle. Zenkevitch (1963) pointed out that the phytoplankton bloom lasted barely more than a month in the Arctic Ocean and some 3 months in the Laptev Sea. Digby (1953) observed almost no growth of the phytoplankton off eastern Greenland before May; a peak occurred in July, with a smaller increase in August, and production ended in September. Kreps and Verjbinskaya (1930, 1932) and other investigators (e.g. Marshall, 1958) emphasize the great significance of stabilization in late spring and summer of the waters at high latitudes for the rich growth of phytoplankton. They demonstrated that with the establishment of a thermocline, rapid depletion of surface nutrients may follow so that growth may be slowed even before the shortening day length restricts phytoplankton development. Where the occurrence of submarine plateaux and ridges promotes vertical mixing, the greater abundance of nutrients can maintain phytoplankton production until the poor light and decreasing day length of the sub-Arctic latitudes bring about its cessation.

The considerably increased primary production which can occur over a limited summer period in somewhat lower latitudes of the Arctic is illustrated from studies of ^{14}C assimilation by Vedernikoy and Solovyeva (1972) at offshore stations near the Murman coast (Barents Sea). During the period June to August, surface primary production ranged from extreme values of 0.78 to 57.8 mg C/m^3/day, more usually between 5 to 25 mg C/m^3/day. Production was low at the beginning of June (45 mg C/m^2/day for the euphotic column), emphasizing the late beginning at these high latitudes; it increased sharply at the end of June or early in July (594 mg C/m^2/day at maximum) and fell somewhat in August. The depth of the euphotic zone was 35 to 50 m and maximum photosynthesis was usually a little below the surface, presumably reflecting the abundant light over the short summer.

The classic view of the Antarctic is a region of high primary production due mainly to the massive and extensive upwelling of nutrient-rich water. Regional variability in the crop of Antarctic phytoplankton was, however, recognized early, for instance by Hart (1934, 1942), who also called attention to differences in the duration of the flowering period in different Antarctic regions. Hart showed that the timing of the spring increase tended to become later as one proceeded south, in a somewhat similar pattern to that of the northern hemisphere, but more severe conditions are experienced at comparable latitudes in the southern hemisphere. At the highest southern latitudes Hart described a continuous burst of phytoplankton growth soon after the midsummer period; there was no autumn maximum. At a less extreme latitude off South Georgia, the phytoplankton increased during the southern spring, to a maximum in November; rather lower densities were found in December/January, with a slight increase in May indicative of a small autumn rise, before the decline to the very low winter densities (cf. Fig. 9.26). Hart considered that phytoplankton production at the highest Antarctic latitudes was slightly below that of somewhat less extreme latitudes. A more obvious difference in the richness of Antarctic phytoplankton was evident between oceanic and neritic regions. Hart suggested that northern, intermediate and southern regions of the Antarctic Ocean developed similar standing crops, but that their biomass approximated to only one-tenth that of the crop around South Georgia (Fig. 9.27). The so-called neritic area off South Georgia, however, extended for a great distance from the shore; its increased fertility might be associated with trace metals derived from land (*vide infra*). Hart pointed out that the biomass even of the Antarctic oceanic areas was comparable with that of a temperate neritic area such as the English Channel (Fig. 9.27).

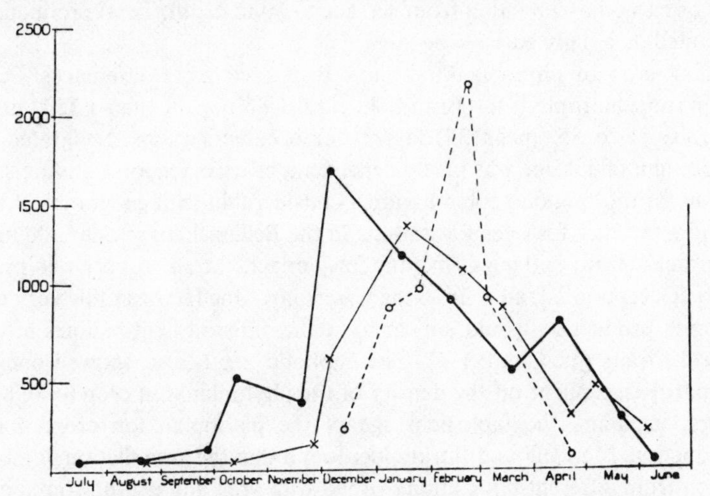

Fig. 9.26. Seasonal variation in phytoplankton, expressed as plant pigments per m^3, in three Antarctic oceanic regions. Thick line = Northern Region; thin line = Intermediate Region; pecked line = Southern Region (after Hart, 1942).

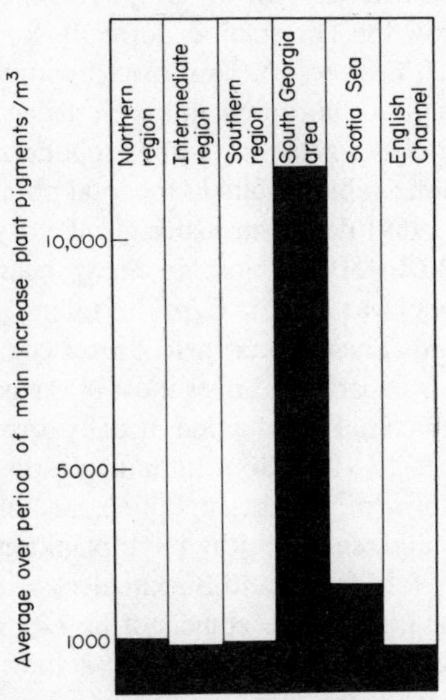

Fig. 9.27. Comparisons of the crop of phytoplankton, expressed as plant pigments per m^3, over the period of the main increase in different areas (from Hart, 1942).

Apart from differences between neritic and oceanic areas, variations in incident radiation due to differences in day length and local climatic conditions (cf. Chapter 6) and differences in the thickness of the mixed layer must affect the intensity of primary production in Antarctic waters. Burkholder and Mandelli (1965a) and Mandelli and Burkholder (1966) calculated, from estimates of the average incident illumination during the months of February/March for the Bellingshausen Sea, light curves for depths corresponding to a series of percentages of the incident light intensity. Using the ^{14}C technique, they determined photosynthetic rate of local phytoplankton against light intensity. Photosynthetic rate/light intensity curves were subsequently used to calculate hourly rates of carbon assimilation from surface to 50 m depth. Total production could then be estimated on a daily basis.

The standing crop of phytoplankton showed marked areal differences: 20 to 209 (mean 68) mg/m^2 chlorophyll for Bransfield Strait; 86 to 209 (mean 151) mg/m^2 for Gerlache Strait; 11 to 85 (mean 34) mg/m^2 for oceanic stations. Integrated primary production per unit of surface was partly dependent, as expected, on standing stock, but the production for the euphotic column varied considerably with geographical area, e.g. about 370 mg C/m^2/day for oceanic stations in the Bellingshausen Sea; 700 mg C/m^2/day for Bransfield Strait; 860 mg C/m^2/day for Gerlache Strait. A very rich production was found for Deception Island – 3620 mg C/m^2/day. One factor in this very considerable variation in production would appear to be the different light regimes of Gerlache and Bransfield Straits (cf. Chapter 6). The euphotic depth also showed considerable differences, partly dependent on the density of the phytoplankton crop (cf. Chapter 6). Other factors, including the biological age of the phytoplankton crop, may affect primary production. Mandelli and Burkholder found that the assimilation number of the phytoplankton from different populations varied with area and depth, from a minimum of 0.4 to a maximum exceeding 5.0. Low values were found in Gerlache Strait during the post-bloom period; high values with a taxonomically different population from Deception Island. A considerable portion of the phytoplankton population in some Antarctic areas occurred below the 1% isolume, normally considered as approximating to the compensation depth (cf. Chapter 3). This lower hypophotic zone contained 33% and 55%, respectively, of the total chlorophyll in the Gerlache and Bransfield Straits (cf. Fig. 6.19, Chapter 6). Some of the algae in the hypophotic zone proved to be photosynthetically active, contributing substantially to the total primary production.

Horne, Fogg and Eagle (1969) demonstrated high photosynthetic rates for inshore shallow Antarctic waters (20 to 50 m), close to Signy Island (latitude about 60°S). Maximum rate (early February) was 250 mg C/m^2/hr, falling to 75 mg C/m^2/hr towards the end of the season (March). These values held over a considerable part of the day. Marked surface inhibition was experienced over most of the growing season, due to the high light intensities, and maximal production usually occurred at a depth where incident radiation was reduced to 10 to 30%. Standing crop as chlorophyll fluctuated between 100 and 400 mg chlor a/m^2 for the euphotic zone, falling sharply to the end of the productive period. Much apparently healthy phytoplankton lay *beneath* the euphotic zone in the hypophotic zone (cf. Mandelli and Burkholder).

Daily variations in photosynthetic rate could not be estimated, but since El-Sayed and Mandelli found little variation for an Antarctic station at 77°S latitude, it was assumed that if the total 24-hour day were divided into five periods, only two periods would probably have a much lowered rate (about 5%). On this assumption, calculations showed a maximum production of 3.7 g C/m²/day in February. The seasonal production was calculated, assuming an ice-free season of roughly mid-December to April/May (approximately 130 to 140 days), as amounting to about 130 g C/m²/yr. This estimate is higher than that made by several earlier investigators (*vide infra*), but the shallow water column, abundant light and lack of stratification promote production. The close proximity to shore probably also precludes any serious nutrient depletion. The results support the view that primary production over the whole Antarctic must be subject to considerable variation, especially between oceanic and inshore areas.

El-Sayed and Mandelli (1965) observed regional differences in productivity in the Weddell Sea, with especially poor production in the east (average 0.18 g C/m²/day), compared with 0.53 g C/m²/day for northern and southern regions. The highest ^{14}C uptake (10 mg C/m³/hr) and the densest crops of phytoplankton (maximum chlorophyll 4.3 mg/m³) were found north-east of the South Orkney Islands and south of the South Sandwich Islands. The lesser stability of the waters in the eastern Weddell Sea may be one factor in the lower production. Overall production for the Weddell Sea, based on four summer months, was 84 g C/m². This may be compared with estimates of Antarctic annual production by other authors, reckoned on a growing season of about 100 days: e.g. 100 g C/m² (Ryther); 20 to 50 g C/m² (Anderson and Banse); 43 g C/m² (Currie); 100 g C/m² (Strickland). Near Antarctic islands production would be far higher.

Further studies (El-Sayed, 1967) demonstrated the low productivity of Drake Passage and the Bellingshausen Sea as contrasted with the high production off the Argentina Shelf and areas to the west of the Antarctic Peninsula. Large standing crops (as chlorophyll *a*) generally accompanied high production (cf. Fig. 9.28). The poor production of waters near the Antarctic Convergence probably reflects the lack of

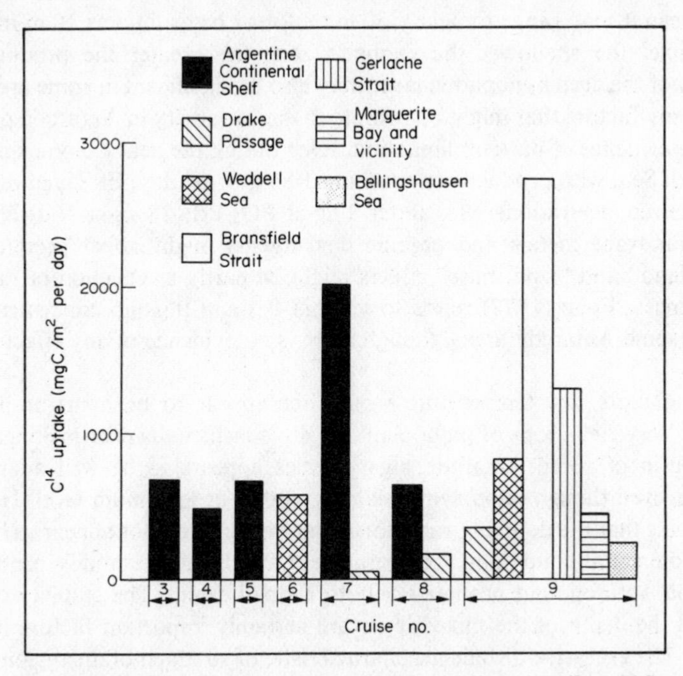

Fig. 9.28. Distribution of ^{14}C uptake (mg C/m²/day) in various Antarctic areas (from El-Sayed, 1967).

stabilization in that region. While some of the discrepancies in estimates of production may be attributable to variations in technique and conditions of experiment, undoubtedly much of the variation is real. In contrast to the poverty of the eastern and southern Weddell Sea, El-Sayed (1970) described in the southern region off the Filchner Ice Shelf, during February 1968, a huge phytoplankton bloom >6000 square miles in extent, and with very high photosynthetic activity (82 mg $C/m^3/hr$). The euphotic layer was shallow, however (10 m).

More broadly over the Antarctic, the richness of the crop around the Antarctic Peninsula contrasted with the general poverty of the Pacific Antarctic sector, though the Pacific stations were mostly oceanic. Uptake of ^{14}C for surface waters resembled the crop distribution, being low in the Pacific (< 5.0 mg $C/m^3/hr$) and higher but very variable in the Atlantic sector. The marked differences in both surface and total ^{14}C uptake in general agreed well with the chlorophyll distribution over the Atlantic and Pacific. A "land mass" effect is evident; neritic regions in the Atlantic are five times richer in crop and production than Atlantic oceanic areas. For the Indian Ocean sector, Fogg (1977) quotes data from several investigators indicating generally low rates of primary production, though higher rates are reported off the Crozet Islands and east of the Kerguelen-Heard plateau. El-Sayed (1977) has again pointed to the marked differences between the productive coastal and the poorer ocean waters. Primary production showed a good correlation with the distribution of standing crop, with high values near coasts and off Antarctic islands and also in the Scotia Sea, Bransfield Strait and Antarctic Peninsula, whereas oceanic waters, with Drake Passage, the Bellingshausen Sea and the Antarctic Convergence, were low. A *mean* productivity of only 0.134 g $C/m^2/day$ was calculated for Antarctic ocean waters.

The thickness of the euphotic zone is an important variable in Antarctic waters; in the open ocean it may range to nearly 80 m, but may be as little as 10 m in rich areas. On the whole, the shallower the euphotic zone the greater the productivity. The productivity of the deep hypophotic layer may also be significant in some areas.

Of the many factors that might contribute to the variability in Antarctic productivity, there is little evidence of nutrient limitation. Even during the heavy blooms in the southwest Weddell Sea, with productivity reaching 1.56 g $C/m^2/day$, El-Sayed reported very large nutrient concentrations, viz. about 2 μg-at PO_4-P/l; 25 μg-at NO_3-N/l; 68 μg-at Si/l. How far trace metals and organic constituents might affect production is not known. "Island" and "land mass" effects might be partly a reflection of the supply of such substances. Fogg (1977) refers to vitamin B_{12} and thiamin concentrations being very low in some Antarctic areas, though there is no evidence of any effect on primary production.

Of other factors low temperature would not appear to be a major influence on production. Very rich crops of phytoplankton are conspicuous near melting sea ice and the metabolism of some Antarctic algal species appears to be well adapted to low temperature, even though photosynthesis may not be at maximum level. However, El-Sayed suggests that the doubling rates for Antarctic algae do not appear to be very high under environmental conditions. Temperature may, therefore, modify production rate locally in the Antarctic but probably only to a small extent. The stability of the water column and the depth of the mixed layer are certainly important factors in Antarctic production. The excessive turbulence characteristic of so much of the region produces a great depth of the mixed layer and though the compensation depth for much of the phytoplankton is very low (cf. Chapter 3), there must be a considerable reduction in growth with algae carried to depth. Any stabilization such as may occur with a period of fine weather appears to promote production (cf. Fogg, 1977). Melting ice may assist in stabilization of the surface layers. Areas of more or less permanent intense mixing in the Antarctic (submarine ridges, the Antarctic Convergence) appear to have especially low production. Light is another factor in the variation in productivity over the Antarctic, due partly to the considerable range in incident illumination over the region, and partly to differences in the effectiveness of light penetration with varying ice cover and other factors. El-Sayed found a high degree of correlation between production rate and incident light, which could vary even during the summer in the Weddell Sea about five-fold between a bright sunny day and an overcast sky.

5. Summary of regional production

Despite examples of considerable variation in production between areas at approximately the same latitude, it is possible to attempt some generalizations concerning world primary production. At any latitude inshore waters tend to be richer than offshore. Ryther and Yentsch (1958), confirming this general view, pointed out that although offshore waters could occasionally show high rates of production comparable to inshore areas, these were due to temporary enrichment and were not sustained; production over a whole year was greater at inshore stations (cf. Fig. 9.29). Annual production from the nutrient-rich Long Island Sound to the impoverished Sargasso Sea waters was estimated:

		g C/m²/year
Long Island Sound	=	389
Shelf Waters	=	160–100 (depending largely on distance from coast)
North Sargasso Sea	=	78

Apart from the very shallow, rich inshore areas, production in continental shelf waters does not exceed that of offshore regions at the same latitude very greatly, though the

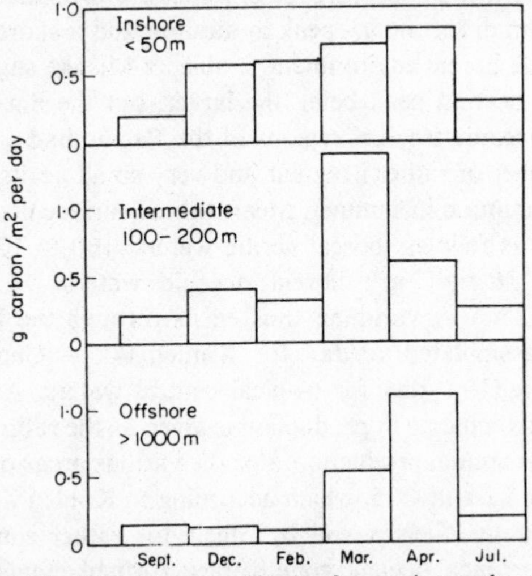

Fig. 9.29. Comparison of the daily primary production for certain months at shallow inshore, intermediate and deep offshore stations (from Ryther and Yentsch, 1958).

timing of the phytoplankton cycle may be strongly affected (cf. Smayda, 1976). This is especially obvious if, as Riley, Stommel and Bumpus (1949) had earlier suggested, the crop of phytoplankton is estimated per unit of sea surface. Differences between more inshore and continental slope areas of the western North Atlantic were not very marked, although the Sargasso was much poorer (cf. p. 214). Takahashi and Ichimura (1972) found that the euphotic column in the rich Oyashio Current waters was less than 25 m, as compared with a euphotic zone of about 100 m in central North Pacific waters. Although Oyashio waters were much richer, reckoned per unit of volume, the total crop was greater in central waters when estimated per unit area. Not only the depth of the euphotic column but the temporal variability in production is another factor to be considered in comparisons. Inshore environments at any latitude have a relatively shallower euphotic layer due to increased turbidity and greater algal concentration, but production tends to be more variable than in the open ocean, partly as a result of increased vertical mixing. Although, therefore, except for extremely shallow environments, production per unit of surface may not be very dissimilar inshore and offshore, production estimated per unit volume is normally far greater in shallower waters, and as we have already stressed, in terms of the ecosystem, the amount of organic matter per unit volume is of great significance. A grazing zooplankton population will expend far more energy feeding throughout an extended water column of low algal density than grazing on a more concentrated thin stratum.

Apart from the high productivity of upwelling areas, it is possible also to compare broadly production in different latitudes. Previous discussion in this chapter confirmed earlier views of high production in middle and moderately high latitudes and low levels in warm oceans, especially in regions of great stability. Production at extremely high latitudes may be low.

Latitudinal comparisons of primary production must be strongly influenced, however, by the sharp seasonal variations, due essentially to changes in incident light at high latitudes, in contrast to the near constancy of tropical regions. Thus Koblentz-Mishke (1965) discussing seasonal variation in primary production specifically in the North Pacific Ocean describes the occurrence of a massive production in coastal boreal areas in spring, with a small secondary peak in autumn and a short summer minimum (Fig. 9.30). In the oceanic boreal environment, Koblentz-Mishke suggests also a double peak in production, the second peak being the larger, but the maximum was far less than in coastal areas. Oceanic tropical regions of the Pacific had a rather low production overall, with a number of rather irregular and very small peaks, but the maximum was in winter and the minimum in summer. Mean annual surface production in terms of carbon assimilated was as follows: boreal neritic waters, 160 to 100 mg C/m^3; boreal more offshore waters, 16 mg C/m^3; boreal oceanic waters, 10 mg C/m^3; tropical oceanic waters, 2 mg C/m^3. A variation thus exists through the North Pacific from more than 100 mg C assimilated/m^3/day for Kamchatka – Canada neritic boreal waters to less than 2 mg C/m^3/day for tropical central waters. An indication of the different extent of seasonal change in production is given as the ratio between maximum and mean (not minimum) annual production. For the various areas of the boreal Pacific, the ratio ranges between 3.2 and 4.25, which according to Koblentz-Mishke agrees well with estimates by Steemann Nielsen and by Steele for rather similar regions of the Atlantic Ocean. For the tropical oceanic zone Koblentz-Mishke finds a value of 4.7.

Cushing (1975), reviewing production cycles in marine phytoplankton, points out

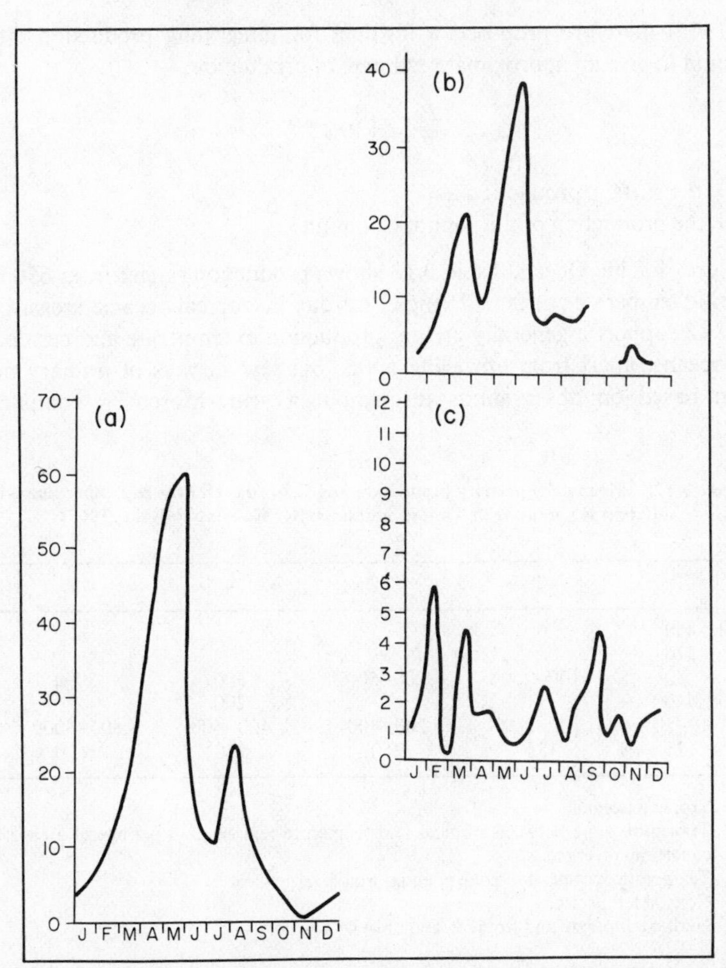

Fig. 9.30. Seasonal variation of primary production (averaged values) at the surface in various areas of the North Pacific: (a) coastal boreal region, (b) oceanic boreal region, (c) oceanic tropical region. X-axis – months; Y-axis – production (mg C/m^3) (after Koblentz-Mishke, 1965).

that estimates throughout a year based on ^{14}C uptake or oxygen production data are few. Certain features, however, are fairly clear. The duration of production cycles at higher latitudes may be as long as 8 months, whereas in warm waters (e.g. Sargasso, Mediterranean) the period for increased production is briefer, though a lower level of production persists for the remainder of the year. The amplitude of seasonal change (i.e. the ratio between maximum and minimum production) would appear to be about five-fold (cf. Menzel and Ryther, Becacos-Kontos), though occasional high very short-lived peaks and irregular fluctuations can occur. The amplitude of change at high latitudes is very considerable since production in mid-winter is extremely low (cf. Steemann Nielsen's data for the North Atlantic). A more detailed discussion of the limited data now available on seasonal amplitude in primary production is probably not warranted.

As we have emphasized, however, the strong seasonal influence on primary production at high latitudes could give a misleading impression of average daily production, especially with the marked difference in the thickness of the euphotic zone. Koblentz-

Mishke (1965) therefore proposed a formula for integrating production through the water column to give an approximate estimate of production:—

$$\frac{P_z}{P_0} = 1.9P_0^{-0.54}$$

where P_0 = the surface production and
P_z = the production of the euphotic column.

For the North Pacific Ocean, Table 9.12 shows production ranges from 650 mg C/m²/day in neritic temperate zones to 76 mg C/m²/day in tropical oceanic areas. Other data in Table 9.12 support a generally greater production in temperate and neritic seas than in warm oceans, apart from upwelling areas, but few surveys of primary production have been based on observations throughout a year. Moreover, comparisons are

Table 9.12. Mean daily primary production (mg C/m²/day) for the euphotic column in different sea areas, from various authors (after Koblentz-Mishke, 1965)

1	2	3	4	5
460				
270				
50	100–200	200–500	500	3000
440			200	
10–50	100–200	200–400	400–600	800–3000
76	135	250	280	650

1 Tropical oceanic.
2 Transition zone between tropical and temperate regions, and peripheral area of equatorial divergence.
3 Temperate oceanic, and zone of equatorial divergence.
4 Coastal temperate.
5 Neritic temperate and tropical, and upwelling waters.

difficult since more often neritic areas are surveyed and at any latitude production will then be higher. McAllister, Parsons and Strickland (1960) quote daily production of about 1 mg C/m³ for warm, near equatorial waters of the Pacific, 7 mg C/m³ for Weather Station P in the boreal waters, and 50 mg C/m³ for an area near the Aleutians. The earlier observations of Holmes (1958) and Bogorov (1958a and b) also testify to the low production of tropical and sub-tropical waters as compared with boreal areas of the Pacific. Bogorov suggests that productivity in warm waters, may be an order of magnitude lower. The estimates of primary production due to Steemann Nielsen and Aabye Jensen (1957) also contributed enormously to our knowledge of latitudinal variations, but data on seasonal change and detailed observations at some high latitudes are still wanting. Nevertheless, the world map (Fig. 9.3), to which we have already referred, gives a useful summary of present knowledge. Omitting from consideration very productive shallow inshore waters, a useful comparison of annual oceanic production is due to Ryther (1963):

(1) For tropical open oceans, 18 to 50 g C/m² (e.g. Sargasso); where some mixing occurs as off Bermuda the annual value may reach 70 g C/m².

(2) In temperate and sub-polar waters, 70 to 120 g C/m^2. At high latitudes very high rates may hold for short periods (even up to 5 g $C/m^2/day$) but the flowering period is short.
(3) Antarctic production – approximately 100 g $C/m^2/yr$ except for especially rich zones (e.g. South Georgia), but there is marked areal variation. Thus Fogg (1977) states that while data from El-Sayed might indicate Antarctic production as about twice the average world ocean productivity, estimates by Holm-Hansen suggest that Antarctic production approximates to one-half the average world ocean value.
(4) Arctic – very low production in the high Arctic – perhaps < 1 g $C/m^2/year$.

From such estimates and the world map of primary production, total world production by marine phytoplankton has been computed. The first estimate by Steemann Nielsen (1954) was 1.5×10^{10} metric tons of carbon per annum; Parsons and Takahashi (1973a) give a similar production (1.9×10^{10} tons carbon). Koblentz-Mishke *et al.* (1970) suggest that the highly productive inshore regions may have been under-estimated and that annual production might amount to 2.5 to 3.0×10^{10} tons carbon. The world production of fish has been variously estimated as 1.9×10^6 and 30.4×10^6 metric tons of carbon. Whatever the precise value, the enormous reduction in the transfer of material through perhaps five trophic levels is notable. One of the important factors is the complex food web in the marine environment. Every transfer in the web will result in some loss of energy and reduce the "ecological efficiency".

Transfer Efficiency

Ecological efficiency (E) has been defined (Parsons and Takahashi, 1973a) as:

$$E = \frac{\text{Amount of energy extracted from a trophic level}}{\text{Amount of energy supplied to a trophic level}}$$

If B = annual biomass at primary trophic level,
P = production at another level,
n = the number of trophic levels,
E = ecological efficiency, then:

$$P = BE^n.$$

An average value of 10% is often cited for ecological efficiency from Slobodkin's early studies, but transfer efficiency, at least at the herbivore grazing level in the marine plankton, is distinctly higher (cf. Volume 2). It probably varies with the trophic level, being lower at higher (secondary and tertiary carnivore) links in the food chain. Cushing (1975) suggests that Steele's analysis of transference of energy in the North Sea indicates an efficiency exceeding 10%. The higher efficiencies at lower trophic levels would appear to be consistent with higher-level predators being larger, longer lived, and storing considerable quantities of materials in their bodies. Moreover, not all the carbon material apparently lost in transfer is oxidized to CO_2; some is recycled in the complex marine food web. The number of trophic levels is clearly of the utmost significance to the amount of production at a relatively high level such as fish. In the equation the number of trophic levels (n) is an exponent. Reference has already been made (Chapter 8) to alternative food chains in the ocean and the use of microzooplankton in various marine environments. Ryther (1969) has indicated, however, that oceanic and coastal regions, probably with continental shelf areas, differ in their overall pattern of food chains. In oceanic regions nanoplankton is generally far more abundant than the larger net phytoplankton; the reverse is true for coastal areas. This difference in turn leads to a broad distinction in the food chain:

Oceanic

nanoplankton ⟶ microzooplankton ⟶ macrozooplankton
(e.g. protozoa) (e.g. herbivorous crustacea)

⟶megazooplankton ⟶ planktivorous ocean fish ⟶ tuna, etc.
(e.g. euphausiids) (e.g. myctophids)

Inshore

net phytoplankton ⟶ macrozooplankton ⟶ planktivorous herring, etc.
(herbivorous copepods, etc.)

In the oceanic environment, some shortening is possible, for example, in the type of food consumed by the final fish link; some zooplankton also exhibit a fairly wide range in diet. Nevertheless, an important distinction may be drawn between an average of perhaps five trophic levels in oceanic regions as against three in coastal and continental shelf areas. Table 9.13 indicates the difference in efficiency between the two and that coastal zones, despite their comparatively small area, contribute so substantially to world fish production. Even more striking is the great contribution from the major upwelling areas of the world, though they are yet again much smaller in total area. The shortening of the food chain in upwelling regions is remarkable. For example, in Peruvian waters anchovies appear to feed when adult mainly directly on the phytoplankton rather than on zooplankton, and some other planktivorous fish are also phytoplankton feeders (e.g. the Indian mackerel). Parsons (1976) cites two surveys of Peruvian waters (cf. p. 373) where strongly upwelled, less stable water produced mainly nanoplankton. They were fed on by ciliates and presumably four further trophic transfers would be needed to achieve a usable fish product. Assuming an ecological transfer efficiency of 20%, the unstable water would give a terminal production of 0.04 g C/m^2/day. In the second survey, marked by less actively upwelled water, diatoms preponderated in the crop which was directly grazed by anchovies. Terminal production was estimated as 1 g C/m^2/day (i.e. twenty-five times greater).

The ecological transfer coefficient not only varies with the trophic level; it may

Table 9.13. Comparison of primary production and the production of fish in different world areas (from Ryther, 1969)

	% of Ocean	Primary prod. (tons C × 10^9)	Trophic levels	Efficiency (%)	Fish prod. (tons)
Oceanic	9.0	16.3	5	10	16 × 10^5
Coastal Zone	9.9	3.6	3	15	12 × 10^7
Upwelling areas	0.1	0.1	1½	20	12 × 10^7
				Total	24 × 10^7

change with the amount of primary production. Cushing (1975) describes earlier studies in the Indian Ocean where extensive primary and secondary production estimates were made in upwelling regions of the Ocean and from central stable areas. A transfer coefficient was calculated simply as the ratio secondary production/primary production. Cushing demonstrated that the coefficient varied with the intensity of primary production. For medium levels of primary production, the transfer coefficient was 10%; with high primary production as in upwelling areas, the coefficient fell to 5%, and in low production (open ocean) areas it rose to 15%. The transfer of material from primary to secondary level is thus three times as efficient in the open ocean as in upwelling areas. This emphasizes once more that the near-equilibrium state between small crop and grazers typical of open warm oceans, with a near constant level of primary production, and with a high diversity of plankton, gives a steady transfer of material to the zooplankton with very little loss. Efficiency is thus high. By contrast, primary production and algal standing crop are high in upwelling areas; the diversity of the plankton is low and the transfer of material to the zooplankton is relatively inefficient. There may be delay before effective grazing commences. There is often a substantial loss of algae to the sea floor. To some extent simulation of the increased production of upwelling regions has been achieved in large-scale column and bag experiments with nutrient addition, as described in Chapter 8. Although several experiments demonstrated greatly increased primary production with fertilization, and in some there was also a considerable rise in secondary production (herbivorous zooplankton), a decline in ecological transfer efficiency was often observed due to such factors as excessive rises in algal crop, sometimes without a marked zooplankton increase, leading to algal blooms and excessive sedimentation ("wastage"). Tertiary production (e.g. ctenophore production), though showing a rise in several experiments, was marked by a lower transfer efficiency than secondary production.

In spite of the lower efficiency of the upwelling areas these support enormous fish stocks. The great concentrations of phytoplankton will support large stocks of herbivorous zooplankton, and dense aggregations of plankton-feeding fish can feed and grow rapidly without great expenditure of energy in food searching. Blackburn (1973) also concluded from studies in the eastern tropical Pacific that the relative amount of organic material transferred from one trophic level to another appeared to decrease where stock and production were high at lower trophic levels, i.e. transfer of material was apparently more efficient in oligotrophic than in eutrophic waters.

Amongst the difficulties in calculating transfer efficiencies are the collection of adequate field data, and the interpretation of field observations in the light of experimental results. McAllister (1972), however, has attempted an analysis of transfer coefficients over several years for Weather Station P (North Pacific). To compute secondary production, he first calculated instantaneous rates of phytoplankton growth from monthly mean photosynthetic rates and mean standing algal crops. After deducting estimated algal respiration, the monthly changes in standing crop were used to calculate grazing mortality rates (cf. Chapter 8). The food ration available to the zooplankton was estimated from the average (nocturnal) algal crop, assuming an assimilation efficiency of 65% and a daily respiration demand of 7% body weight, and zooplankton production calculated from the ration. The standing stock ratio (i.e. the ratio zooplankton/phytoplankton) was an important factor in transfer efficiencies. While the mean standing stock ratio was approximately unity, it varied greatly from year to year; in

1960, for instance, with a very high zooplankton density and low phytoplankton, it reached 2.1; in some years it was only 0.75. The transfer efficiency defined as the ratio secondary production/observed primary production, varied three- or four-fold over the years and generally was low over the years 1960 to 1962, when zooplankton crop and standing stock ratio was high. McAllister believes that the extensive zooplankton stock built up over those years was not reduced sufficiently by predators so that an excessive amount of primary production was used in maintenance requirements for the very large zooplankton stock. There was some indication also that transfer efficiency varied directly with primary production, but a clearer relationship was found between transfer efficiency and the ratio P/Z (mean annual phytoplankton stock/mean annual zooplankton stock).

While it is generally accepted that a shortening of the food chain, typically considerably extended in the deep open ocean, is accompanied by increased transfer efficiency, LeBrasseur (1972) has stressed the importance of possible switching to alternative diets in efficiency studies. He cites as an example the feeding of juvenile salmon in the North Pacific. To a considerable degree the young salmon feed on herbivorous zooplankton (copepods, *Limacina*, etc.) and so long as these are sufficiently dense for young salmon to feed abundantly, their growth is satisfactory. With a seasonal decline in the zooplankton herbivore population, however, juvenile salmon production is more efficient if they switch feeding to carnivorous zooplankton or to secondary carnivores (squid, myctophids). Both the primary and especially the secondary carnivores can act as a food store for the young salmon, in view of their longer life span, especially squid and myctophids. Despite, therefore, some lengthening of the food chain, this switching to different prey introduces a valuable food resource, helping to maintain the level of transfer efficiency.

Alternative food paths are also described by Parsons and LeBrasseur (1970) in relation to the feeding of euphausiids and *Calanus* in the North Pacific; the food chain may begin with nanoplankton or larger net plankton. A particular interesting example is *Euphausia pacifica*, since it may be herbivorous or carnivorous in the North Pacific. The choice of diet and the trophic level for animals in the plankton thus appear to depend largely on the expenditure of energy required to secure sufficient ration (cf. Parsons, 1976).

The relative proportions of nanoplankton and microzooplankton to net phytoplankton and larger zooplankton may well be one factor in the higher overall productivity at higher trophic levels of inshore as contrasted with oceanic regions. To some extent these considerations parallel the earlier comparison of upwelling and oceanic regions. In more inshore waters, concentration of the higher trophic level predators is also significant. A comparison of the importance of microzooplankton in inshore (Strait of Georgia) and oceanic (Weather Station P) regions has been made by LeBrasseur and Kennedy (1972). In brief, they found that although the annual mean density of zooplankton was similar in the two regions, the smaller zooplankton (<225 μm) was about three times more numerous at the oceanic station. They also believe that microzooplankton small enough to escape nets of 44 μm mesh, such as ciliates, may make an important contribution, particularly in open oceans. A more detailed discussion is given in Volume 2.

It would seem no chance happening, therefore, that the great fisheries of the world are in the upwelling regions and in the inshore and continental shelf areas. Though ecological transfer efficiency may be comparatively low in these seas, the great aggregations of fish are able to feed intensively on the dense plankton, relatively concentrated in depth. In many of the shelf areas the material which is apparently wasted may also promote the production of rich benthos which may in turn encourage a substantial demersal fishery.

References

Abbott, B. C. and D. Ballantine (1957) *J. Mar. Biol. Ass. U.K.* **36**, 169–189.

Abdullah, M. I., L. G. Royle and A. W. Morris (1972) *Nature, Lond.* **235**, 158–160.

Ackman, R. G., P. M. Jangaard, R. J. Hoyle and H. Brockerhoff (1964) *J. Fish. Res. Bd. Can.* **21**, 747–756.

Ackman, R. G., C. S. Tocher and J. McLachlan (1968) *J. Fish. Res. Bd. Can.* **25**, 1603–1620.

Adams, J. A. and J. H. Steele (1966) In H. Barnes (Ed.), *Some Contemporary Studies on Marine Science.* Allen & Unwin, London. pp. 19–35.

Ahrland, S. (1975) In E. D. Goldberg (Ed.), *The Nature of Seawater.* Dahlem Konferenzen, Berlin. pp. 219–244.

Akinina, D. K. (1966) *Oceanology* **6**, 702–708.

Akinina, D. K. (1969) *Oceanology* **9**, 248–251.

Alexander, J. E., J. H. Steele and E. F. Corcoran (1962) *Proc. Gulf Caribb. Fish. Invest.* **14**, 63–67.

Andersen, O. G. N. (1977) *Ophelia* **16**, 205–220.

Anderson, G. C. (1964) *Limnol. Oceanogr.* **9**, 284–302.

Anderson, G. C. (1969) *Limnol. Oceanogr.* **14**, 386–391.

Anderson, G. C. and K. Banse (1963) In M. S. Doty (Ed.), *Proc. Conf. on Primary Productivity Measurement, etc. Univ. Hawaii, 1961.* US Atomic Energy Commission. pp. 61–71.

Anderson, G. C. and R. P. Zeutschel (1970) *Limnol. Oceanogr.* **15**, 402–407.

Ansell, A. D., J. Coughlan, K. F. Lander and F. A. Loosmore (1964) *Limnol. Oceanogr.* **9**, 334–342.

Ansell, A. D., J. E. G. Raymont and K. F. Lander (1963a) *Limnol. Oceanogr.* **8**, 207–213.

Ansell, A. D., J. E. G. Raymont, K. F. Lander, E. Crowley and P. Shackley (1963b) *Limnol. Oceanogr.* **8**, 184–206.

Antia, N. J. and J. Y. Cheng (1970) *Phycologia* **9**, 179–183.

Anita, N. J., B. R. Berland, D. J. Bonin and S. Y. Maestrini (1975) *J. Mar. Biol. Ass. U.K.* **55**, 519–539.

Antia, N. J., C. D. McAllister, T. R. Parsons, K. Stephens and J. D. H. Strickland (1963) *Limnol. Oceanogr.* **8**, 166–183.

Apstein, C. (1908) In K. Brandt and C. Apstein (Eds.), *Nordisches Plankton. Botanischer Teil* **22**. Lipsius & Tischer, Kiel and Leipzig. pp. 1–5.

Armstrong, F. A. J. (1965a) In J. P. Riley and G. Skirrow (Eds.), *Chemical Oceanography*, Vol. 1, 1st ed. Academic Press, London. pp. 323–364.

Armstrong, F. A. J. (1965b) In J. P. Riley and G. Skirrow (Eds.), *Chemical Oceanography*, Vol. 1, 1st ed. Academic Press, London. pp. 409–432.

Armstrong, F. A. J. and E. I. Butler (1968) *J. Mar. Biol. Ass. U.K.* **48**, 153–160.

Armstrong, F. A. J. and S. Tibbitts (1968) *J. Mar. Biol. Ass. U.K.* **48**, 143–152.

Arthur C. R. and F. H. Rigler (1967) *Limnol. Oceanogr.* **12**, 121–124.

Aruga, Y. (1973) In B. Zeitzschel (Ed.), *The Biology of the Indian Ocean.* Chapman & Hall, London; Springer-Verlag, Berlin. pp. 127–130.

Atkins, W. R. G. (1928) *J. Mar. Biol. Ass. U.K.* **15**, 191–205.

Aubert, J. and J. P. Gambarotta (1972) *Rev. Int. Océanogr. Médicale* **25**, 39–47.

Aubert, M. and J. Aubert (1969) *Océanographie Médicale.* Gauthier–Villars, Paris. 298 pp.

Aubert, M. and D. Pesando (1971) *Rev. Int. Océanogr. Médicale* **21**, 17–22.

Aubert, M., D. Pesando, and M. Gauthier (1970) *Rev. Int. Océanogr. Médicale* **18–19**, 69–76.

Austin, T. S. and V. E. Brock (1959). *Int. Ocean. Congress Preprints.* A.A.A.S., Washington, D.C. pp. 130–131.

Azam, F. and S. W. Chisholm (1976) *Limnol. Oceanogr.* **21**, 427–435.

Bada, J. L. and C. Lee (1977) *Mar. Chem.* **5**, 523–534.

Bakker, C. and N. de Pauw (1974) *Hydrobiol. Bull.* **8**, 179–189.

Bakker, C. and N. de Pauw (1975) *Nethl. J. Sea Res.* **9**, 145–165.

Balech, E. (1970) In M. W. Holdgate (Ed.), *Antarctic Ecology*, Vol. 1. Academic Press, London. pp. 143–147.

Balech, E. and S. Z. El-Sayed (1965) *Biology of the Antarctic Seas.* II. *Ant. Res. Ser.* **5**, 107–124.

Banoub, M. W. and P. J. LeB. Williams (1973) *J. Mar. Biol. Ass. U.K.* **53**, 695–703.

Banse, K. (1974) *Limnol. Oceanogr.* **19**, 695–699.

Banse, K. (1977) *Mar. Biol.* **41**, 199–212.

BARBER, R. T. (1973) In P. C. SINGER (Ed.), *Trace Metals and Metal-Organic Interactions in Natural Waters*. Ann Arbor Science, Ann Arbor. pp. 321–338.
BARBER, R. T. and J. H. RYTHER (1969) *J. Exp. Mar. Biol. Ecol.* **3**, 191–199.
BARBER, R. T., R. C. DUGDALE, J. J. MACISAAC and R. L. SMITH (1971) *Inv. Pesq.* **35**, 171–193.
BARKER, H. A. (1935) *Arch. Mikrobiol.* **6**, 157–181.
BARLOW, J. P. (1958) *Biol. Bull.* **109**, 211–225.
BARNES, H. (1957) In J. W. HEDGPETH (Ed.) *Treatise on Marine Ecology and Palaeoecology*, Vol. 1. Geological Society of America, New York. pp. 297–343.
BARTH, T. F. W. (1952) *Theoretical Petrology*. Wiley, New York. 387 pp.
BATES, S. S. (1976) *Limnol. Oceanogr.* **21**, 212–218.
BECACOS-KONTOS, T. (1968) *Limnol. Oceanogr.* **13**, 485–489.
BECACOS-KONTOS, T. (1977) *Mar. Biol.* **42**, 93–98.
BECACOS-KONTOS, T. and A. SVANSSON (1969) *Mar. Biol.* **2**, 140–144.
BEERS, J. R. and A. C. KELLY (1965) *Deep-Sea Res.* **12**, 21–25.
BEERS, J. R. and S. S. HERMAN (1969) *Bull. Mar. Sci.* **19**, 253–278.
BEERS, J. R. and G. L. STEWART (1970) *Bull. Scripps Inst. Oceanogr.* **17**, 67–87.
BEERS, J. R., D. M. STEVEN and J. B. LEWIS (1968) *Bull. Mar. Sci.* **18**, 86–104.
BEERS, J. R., G. L. STEWART and K. D. HOSKINS (1977) *Bull. Mar. Sci.* **27**, 66–79.
BEKLEMISHEV, C. W. (1954) *Zool. J. Inst. Oceanól. Acad. Sci. U.S.S.R.* **33**, 1210–1229.
BEKLEMISHEV, C. W. (1957) *Trudy Inst. Okeanol.* **20**, 253–278.
BEKLEMISHEV, C. W. (1962) *Rapp. P-v. Réun. Cons. Perm. Int. Explor. Mer* **153**, 108–113.
BEKLEMISHEV, C. W., M. PETRIKOVA and G. I. SEMINA (1961) *Trudy Inst. Okeanol.* **51**, 31–36.
BENDER, M. L. and C. GAGNER (1976) *J. Mar. Res.* **34**, 327–339.
BENSON, B. B. and P. D. M. PARKER (1961) *Deep-Sea Res.* **7**, 237–253.
BERGE, G. (1962) *Sarsia* **6**, 27–40.
BERNARD, F. (1939) *J. Cons. Perm. Int. Explor. Mer* **14**, 228–241.
BERNARD, F. (1948) *J. Cons. Perm. Int. Explor. Mer* **15**, 177–188.
BERNARD, F. (1953) *Deep-Sea Res.* **1**, 34–46.
BERNARD, F. (1958) *Rapp. P-v. Réun. Cons. Perm. Int. Explor. Mer* **144**, 103–108.
BERNARD, F. (1963) *Pelagos, Bull. Inst. Ocean. d'Alger* **1**(2), 1–34.
BERNARD, F. (1967) *Oceanogr. Mar. Biol. Ann. Rev.* **5**, 205–229.
BERNARD, F. and B. ELKAIM (1962) *C. R. Hebd. Séanc. Acad. Sci. Paris* **254**, 4208–4210.
BERNER, R. (1971) *Principles of Chemical Sedimentology*. McGraw-Hill, New York. 240 pp.
BERNER, R. (1972) In D. DYRSSEN and D. JAGNER (Eds.), *The Changing Chemistry of the Oceans*. Wiley, New York; Almqvist & Wiksell, Stockholm. pp. 347–361.
BIGELOW, H. B., L. LILLICK and M. SEARS (1940) *Trans. Am. Phil. Soc.* **31**, 149–191.
BLACKBURN, M. (1973) *Limnol. Oceanogr.* **18**, 552–563.
BLACKBURN, M., R. M. LAURS, R. W. OWEN and B. ZEITZSCHEL (1970) *Mar. Biol.* **7**, 14–31.
BLACKWELDER, P. L., R. E. WEISS and K. M. WILBUR (1976) *Mar. Biol.* **34**, 11–16.
BLUMER, M., M. M. MULLIN and R. R. L. GUILLARD (1970) *Mar. Biol.* **6**, 226–235.
BLUMER, M., M. M. MULLIN and D. W. THOMAS (1964) *Helgolander Wiss. Meeresunters* **10**, 187–201.
BOALCH, G. T. and M. PARKE (1968) *Br. Phycol. Bull.* **3**, 600.
BOGORAD, L. (1962) In R. A. LEWIN (Ed.), *Physiology and Biochemistry of Algae*. Academic Press, New York. pp. 385–408.
BOGOROV, V. G. (1958a) *Rapp. P-v Réun. Cons. Perm. Int. Explor. Mer.* **144**, 117–121.
BOGOROV, V. G. (1958b) *Deep-Sea Res.* **5**, 149–161.
BOGOROV, V. G. (1959) *Int. Ocean. Congress Preprints*. A.A.A.S., Washington D.C. pp. 139–140.
BOGOROV, V. G. (1967) *Oceanology* **7**, 649–665.
BOHNECKE, G. (1922) *A. Geogr.-naturwiss.*, R.H., **10**, 1–34.
BOLEYN, B. J. (1972) *Int. Revue Ges. Hydrobiol.* **57**, 585–597.
BONEY, A. D. (1970) *Oceanogr. Mar. Biol. Ann. Rev.* **8**, 251–305.
BONEY, A. D. and A. BURROWS (1966) *J. Mar. Biol. Ass. U.K.* **46**, 295–319.
BONQUAHEUX, F. (1972) *Cah. Biol. Mar.* **13**, 1–8.
BOSCH, H. F. and W. R. TAYLOR (1973) *Mar. Biol.* **19**, 161–181.
BOUCAUD-CAMOU, E. (1966) *Bull. Soc. Linn. Normandie*, Ser. 10, **7**, 191–209.
BOUSFIELD, E. L. (1955) *Bull. Natn. Mus. Can.* **137**, 1–69.
BOUTRY, J. L., M. BARBIER and M. RICARD (1976) *J. Exp. Mar. Biol. Ecol.* **21**, 69–74.
BOYLE, E. and J. M. EDMOND (1975) *Nature, Lond.* **253**, 107–109.
BOYLE, E. A., J. M. EDMOND and E. R. SHOLKOVITZ (1977) *Geochim. Cosmochim. Acta* **41**, 1313–1324.
BOYLE, E. A., F. R. SCLATER and J. M. EDMOND (1976) *Nature, Lond.* **263**, 42–44.
BOYLE, E. A., F. R. SCLATER and J. M. EDMOND (1977) *Earth Planet. Sci. Lett.* **37**, 38–54.
BRAARUD, T. (1935) *Hvalrad Skr.* **10**, 1–173.

BRAARUD, T. (1961) In M. SEARS (Ed.), *Oceanography*. A.A.A.S., Washington, D.C. pp. 271–298.
BRAARUD, T. (1962) *J. Oceanogr. Soc. Japan*. 20th Ann. Vol., pp. 628–649.
BRAARUD, T. and A. KLEM (1931) *Hvalrad Skr.* **1**, 1–88.
BRAARUD, T., G. DEFLANDRE, P. HALLDAL and E. KAMPTNER (1955) *Micropalaeontology* **1**, 157–159.
BRAARUD, T., K. R. GAARDER and J. GRONTVED (1953) *Rapp. P-v. Réun. Cons. Perm. Int. Explor. Mer* **133**, 1–89.
BRAARUD, T., K. R. GAARDER, J. MARKALI and E. NORDLI (1953) *Nytt. Mag. Bot.* **1**, 129–134.
BRANDHORST, W. (1958) *Nature, Lond.* **182**, 679.
BRANDHORST, W. (1959) *J. Cons. Perm. Int. Explor. Mer* **25**, 3–20.
BRANDT, K. (1898) *Wiss. Meeresunters., Abt. Kiel, N.F.* **3**, 43–90.
BRECK, W. G. (1972) *J. Mar. Res.* **30**, 121–139.
BRECK, W. G. (1974) In E. D. GOLDBERG (Ed.), *The Sea*, Vol. 5, *Marine Chemistry*. Wiley-Interscience, New York. pp. 153–179.
BREWER, P. G. (1975) In J. P. RILEY and G. SKIRROW (Eds.), *Chemical Oceanography*, Vol. 1, 2nd ed. Academic Press, London. pp. 415–496.
BREWER, P. G. and A. BRADSHAW (1975) *J. Mar. Res.* **33**, 157–175.
BREWER, P. G., D. W. SPENCER and P. E. WILKNISS (1970) *Deep-Sea Res.* **17**, 1–7.
BROCKMAN, U. H., K. EBERLEIN, P. HOSUMBER, H. TRAGESER, E. MAIER-REIMER, K. SCHONE and H. D. JUNGE (1977) *Mar. Biol.* **43**, 1–17.
BROECKER, W. S. (1965) In T. ICHIYE (Ed.), *Symposium on Diffusion in Oceans and Fresh Waters*. Lamont Geological Observatory, Palisades, N.Y. pp. 116–144.
BROECKER, W. S. (1974) *Chemical Oceanography*. Harcourt, Brace, Jovanovich, New York. 214 pp.
BRONGERSMA-SANDERS, M. (1957) In J. W. HEDGPETH (Ed.), *Treatise on Marine Ecology and Palaeoecology*, Vol. 1. Geological Society of America, New York. pp. 941–1010.
BROWN, N. L., and B. V. HAMON (1961) *Deep-Sea Res.* **8**, 65–75.
BRULAND, K. W., G. A. KNAUER and J. H. MARTIN (1978a) *Limnol. Oceanogr.* **23**, 618–625.
BRULAND, K. W., G. A. KNAUER and J. H. MARTIN (1978b) *Nature, Lond.* **271**, 741–743.
BRYAN, J. R., J. P. RILEY and P. J. LEB. WILLIAMS (1976) *J. Exp. Mar. Biol. Ecol.* **21**, 191–197.
BSHARAH, L. (1957) *Bull. Mar. Sci. Gulf Caribb.* **7**, 201–251.
BUCHANAN, R. J. (1968) *J. Phycol.* **4**, 272–277.
BUNT, J. S. (1964) *Biology of the Antarctic Seas* I. *Ant. Res. Ser.* **1**, 13–31.
BUNT, J. S. (1966) *SCAR Symposium on Antarctic Oceanography*, pp. 198–218.
BUNT, J. S. (1967) *Biology of the Antarctic Seas* III. *Ant. Res. Ser.* **11**, 1–14.
BUNT, J. S. (1971) *Soc. Gen. Microbiol.* **21**, 333–354.
BUNT, J. S. and C. C. LEE (1970) *J. Mar. Res.* **28**, 304–320.
BUNT, J. S. and C. C. LEE (1972) *Limnol. Oceanogr.* **17**, 458–461.
BUNT, J. S., O. VAN H. OWENS and G. HOCH (1966) *J. Phycol.* **2**, 96–100.
BURKHOLDER, P. R. and E. F. MANDELLI (1965a) *Proc. Natn. Acad. Sci.* **54**, 437–444.
BURKHOLDER, P. R. and E. F. MANDELLI (1965b) *Science N.Y.* **149**, 872–874.
BURKHOLDER, P. R., L. M. BURKHOLDER and L. R. ALMODOVAR (1967) *Bull. Mar. Sci.* **17**, 1–15.
BURKILL, P. H. (1978) Ph.D. Thesis, University of Southampton.
BURRIS, J. E. (1977) *Mar. Biol.* **39**, 371–379.
BURSA, A. S. (1963) In C. H. OPPENHEIMER (Ed.), *Marine Microbiology*. Thomas, Springfield, Ill. pp. 625–628.
BURTON, J. D. (1979) *Phil. Trans. R. Soc. Lond.* B, **286**, 443–456.
BURTON, J. D. and P. S. LISS (1973) *Geochim. Cosmochim. Acta* **37**, 1761–1773.
BURTON, J. D. and P. S. LISS (1976) (Eds.) *Estuarine Chemistry*. Academic Press, London. 229 pp.
BURTON, J. D., T. M. LEATHERLAND and P. S. LISS (1970) *Limnol. Oceanogr.* **15**, 473–476.
BUTCHER, R. W. (1952) *J. Mar. Biol. Ass. U.K.* **31**, 175–191.
BUTCHER, R. W. (1959) *Fishery Invest., MAFF.* Ser. IV, **1**. 74 pp.
BUTCHER, R. W. (1961) *Fishery Invest. MAFF.* Ser. IV, **8**. 17 pp.
BUTCHER, R. W. (1967) *Fishery Invest. MAFF.* Ser. IV, **4**. 54 pp.
BUTLER, E. I., E. D. S. CORNER and S. M. MARSHALL (1969) *J. Mar. Biol. Ass. U.K.* **49**, 977–1001.
BUTLER, E. I., E. D. S. CORNER and S. M. MARSHALL (1970) *J. Mar. Biol. Ass. U.K.* **50**, 525–560.
CADEE, G. C. and J. HEGEMAN (1974) *Nethl. J. Sea Res.* **8**, 240–259.
CAPERON, J. (1967) *Ecology* **48**, 713–722.
CAPERON, J. (1968) *Ecology* **49**, 866–872.
CAPERON, J. and J. MEYER (1972a) *Deep-Sea Res.* **19**, 601–618.
CAPERON, J. and J. MEYER (1972b) *Deep-Sea Res.* **19**, 619–632.
CARLUCCI, A. F. (1970) *Bull. Scripps Inst. Oceanogr.* **17**, 23–31.
CARLUCCI, A. F. and P. M. BOWES (1970) *J. Phycol.* **6**, 351–357.
CARLUCCI, A. F. and P. M. MCNALLY (1969) *Limnol. Oceanogr.* **14**, 736–739.

Carlucci, A. F. and H. R. Schubert (1969) *Limnol. Oceanogr.* **14**, 187–193.
Carlucci, A. F. and J. D. H. Strickland (1968) *J. Exp. Mar. Biol. Ecol.* **2**, 156–166.
Carlucci, A. F., E. O. Hartwig and P. M. Bowes (1970) *Mar. Biol.* **7**, 161–166.
Carpenter, E. J. and R. R. L. Guillard (1971) *Ecology* **52**, 183–185.
Carpenter, E. J., C. C. Remsen and S. W. Watson (1972) *Limnol. Oceanogr.* **17**, 265–269.
Castenholz, R. W. (1964) *Physiologia Pl.* **17**, 951–963.
Chapman, V. J. and D. J. Chapman (1973) *The Algae*. Macmillan, London. 497 pp.
Chau, Y. K., L. Chuecas and J. P. Riley (1967) *J. Mar. Biol. Ass. U.K.* **47**, 543–554.
Cheng, J. Y. and N. J. Antia (1970) *J. Fish. Res. Bd. Can.* **27**, 335–346.
Chester, R. and J. H. Stoner (1972) *Nature, Lond.* **240**, 552–553.
Chisholm, S. W., F. Azam and R. W. Eppley (1978) *Limnol. Oceanogr.* **23**, 518–529.
Chong, B. J. (1970) M.Sc. Dissertation, University of Southampton.
Christensen, T. (1962) In T. W. Bocher, M. Lange and T. Sorensen (Eds.) *Alger Botanik*, 2nd ed. (Systematisk Botanik) Vol. 2. Munksgaard, Copenhagen. 178 pp.
Chuecas, L. and J. P. Riley (1969a) *J. Mar. Biol. Ass. U.K.* **49**, 97–116.
Chuecas, L. and J. P. Riley (1969b) *J. Mar. Biol. Ass. U.K.* **49**, 117–120.
Clark, R. C. and M. Blumer (1967) *Limnol. Oceanogr.* **12**, 79–87.
Clarke, G. L. (1939) *Q. Rev. Biol.* **14**, 60–64.
Cline, J. D. and I. R. Kaplan (1975) *Mar. Chem.* **3**, 271–299.
Cline, J. D. and F. A. Richards (1969) *Envir. Sci. Technol.* **3**, 838–843.
Cline, J. D. and F. A. Richards (1972) *Limnol. Oceanogr.* **17**, 885–900.
Codispoti, L. A. and F. A. Richards (1976) *Limnol. Oceanogr.* **21**, 379–388.
Cohen, Y. (1978) *Nature, Lond.* **272**, 235–237.
Cohen, Y. and L. I. Gordon (1978) *Deep-Sea Res.* **25**, 509–524.
Colebrook, J. M. and G. A. Robinson (1961) *J. Cons. Perm. Int. Explor. Mer* **26**, 156–165.
Colebrook, J. M., R. S. Glover and G. A. Robinson (1961) *Bull. Mar. Ecol.* **5**, 67–80.
Collier, A. (1953) *Trans. N. Am. Wildl. Conf.* 463–472.
Collier, A. (1958) *Limnol. Oceanogr.* **3**, 33–39.
Collier, A. and A. Murphy (1962) *Science N.Y.* **136**, 780–782.
Collin, A. E. and M. J. Dunbar (1964) *Oceanogr. Mar. Biol. Ann. Rev.* **2**, 45–75.
Collyer, D. M. and G. E. Fogg (1955) *J. Exp. Bot.* **6**, 256–275.
Conover, R. J. (1976) In O. Devik (Ed.), *Harvesting Polluted Waters*. Plenum Press, New York. pp. 67–85.
Conover, S. A. M. (1956) *Bull. Bingham Oceanogr. Coll.* **15**, 62–112.
Cooper, L. H. N. (1933) *J. Mar. Biol. Ass. U.K.* **18**, 677–728.
Cooper, L. H. N. (1937) *J. Mar. Biol. Ass. U.K.* **22**, 177–182.
Corcoran, E. F. and J. E. Alexander (1963) *Bull. Mar. Sci. Gulf Caribb.* **13**, 527–541.
Corlett, J. (1953) *J. Cons. Perm. Int. Explor. Mer* **19**, 178–190.
Corner, E. D. S. and C. B. Cowey (1968) *Biol. Rev.* **43**, 393–426.
Corner, E. D. S. and A. G. Davies (1971) *Adv. Mar. Biol.* **9**, 101–204.
Cowey, C. B. and E. D. S. Corner (1966) In H. Barnes (Ed.), *Some Contemporary Studies in Marine Science*. Allen & Unwin, London. pp. 225–231.
Cox, R. A., F. Culkin and J. P. Riley (1967) *Deep-Sea Res.* **14**, 203–220.
Craig, H. (1971) *J. Geophys. Res.* **76**, 5133–5139.
Cronin, L. E., J. C. Daiber and E. M. Hulburt (1962) *Chesapeake Sci.* **3**, 63–93.
Culkin, F. and R. A. Cox (1966) *Deep-Sea Res.* **13**, 789–804.
Curl, H. and G. C. McLeod (1961) *J. Mar. Res.* **19**, 70–88.
Curl, H. and L. F. Small (1965) *Limnol. Oceanogr., Suppl.* **10**, R67–R73.
Currie, R. I. (1958) *Rapp. P-v. Réun. Cons. Perm. Int. Explor. Mer* **144**, 96–102.
Currie, R. I., A. E. Fisher and P. M. Hargreaves (1973) In B. Zeitzschel (Ed.), *The Biology of the Indian Ocean*. Chapman & Hall, London; Springer-Verlag, Berlin. pp. 37–52.
Cushing, D. H. (1953) *J. Cons. Perm. Int. Explor. Mer* **19**, 3–22.
Cushing, D. H. (1958) *Rapp. P-v. Réun. Cons. Perm. Int. Explor. Mer* **144**, 32–33.
Cushing, D. H. (1959a) *Fishery Invest. (Lond.) Ser. II*, **22**(6). 40 pp.
Cushing, D. H. (1959b) *J. Cons. Perm. Int. Explor. Mer* **24**, 455–464.
Cushing, D. H. (1962) *Rapp. P-v. Réun Cons. Perm. Int. Explor. Mer* **153**, 198–199.
Cushing, D. H. (1963) *J. Mar. Biol. Ass. U.K.* **43**, 339–347.
Cushing, D. H. (1964) In D. J. Crisp (Ed.), *Grazing in Terrestrial and Marine Environments*. Blackwell, Oxford. pp. 207–225.
Cushing, D. H. (1968) *J. Cons. Perm. Int. Explor. Mer* **32**, 70–82.
Cushing, D. H. (1975) *Marine Ecology and Fisheries*. Cambridge University Press, London. 278 pp.
Cushing, D. H. and T. Vucetic (1963) *J. Mar. Biol. Ass. U.K.* **43**, 349–371.

CUSHING, D. H. and J. J. WALSH (1976) (Eds.) *The Ecology of the Seas*. Blackwell, Oxford. 467 pp.
DARLEY, W. M. and B. E. VOLCANI (1969) *Exp. Cell. Res.* **58**, 334–343.
DAVIES, A. G. (1967) In B. ABERG and F. P. HUNGATE (Eds.), *Radioecological Concentration Processes*. Pergamon Press, Oxford. pp. 983–991.
DAVIES, A. G. (1970) *J. Mar. Biol. Ass. U.K.* **50**, 65–86.
DAWSON, E. Y. (1966) *Marine Botany, An Introduction*. Holt, Rinehart & Winston, New York. 371 pp.
DEEVEY, G. B. (1948) *Bull. Bingham Oceanogr. Coll.* **12**, 1–44.
DEFANT, A. (1961) *Physical Oceanography*, Vol. 1, Pergamon Press, Oxford. 729 pp.
DEGENS, E. T. and D. A. ROSS (1969) (Eds.) *Hot Brines and Recent Heavy Metal Deposits in the Red Sea*. Springer-Verlag, New York. 600 pp.
DELWICHE, C. C. (1970) *Scient. Am.* **223**, 136–147.
DESIKACHARY, T. V. (1957) *J. R. Microsc. Soc.*, Ser. 3, **76**, 9–36.
DEUSER, W. G., E. H. ROSS and Z. J. MLODZINSKA (1978) *Deep-Sea Res.* **25**, 431–445.
DE SOUSA E. SILVA, E. (1962) *Notas Estud. Inst. Biol. Mar., Lisb.* **24**, 75–100.
DIGBY, P. S. B. (1953) *J. Anim. Ecol.* **22**, 289–322.
DIWAN, H. R. (1978) M.Phil. Thesis, University of Southampton.
DODGE, J. D. (1960) *Br. Phycol. Bull.* **2**, 14–15.
DODGE, J. D. (1963) *Br. Phycol. Bull.* **2**, 282–283.
DODGE, J. D. (1967) *Br. Phycol. Bull.* **3**, 327–337.
DODGE, J. D. (1973) *The Fine Structure of Algal Cells*. Academic Press, New York. 261 pp.
DOTY, M. S. and M. OGURI (1956) *J. Cons. Perm. Int. Explor. Mer* **22**, 33–37.
DOTY, M. S. and M. OGURI (1957) *Limnol. Oceanogr.* **2**, 37–40.
DREBES, G. (1966) *Helgolander Wiss. Meeresunters.* **13**, 101–114.
DREBES, G. (1967) *Mar. Biol.* **1**, 40–42.
DREBES, G. (1969) *Helgolander Wiss. Meeresunters.* **19**, 58–67.
DROOP, M. R. (1968) *J. Mar. Biol. Ass. U.K.* **48**, 689–733.
DROOP, M. R. (1970) *Helgolander Wiss. Meeresunters.* **20**, 629–636.
DROOP, M. R. (1973) *J. Phycol.* **9**, 264–272.
DROOP, M. R. (1974) *J. Mar. Biol. Ass. U.K.* **54**, 825–855.
DRUM, R. W. and J. T. HOPKINS (1966) *Protoplasma* **62**, 1–33.
DUCE, R. A. and E. K. DUURSMA (1977) *Mar. Chem.* **5**, 319–339.
DUGDALE, R. C. (1967) *Limnol. Oceanogr.* **12**, 685–695.
DUGDALE, R. C. (1972) *Geoforum* **11**, 47–61.
DUGDALE, R. C. (1976) In D. H. CUSHING and J. J. WALSH (Eds.), *The Ecology of the Seas*. Blackwell, Oxford. pp. 141–172.
DUGDALE, R. C. and J. J. GOERING (1967) *Limnol. Oceanogr.* **12**, 196–206.
DURBIN, E. G. (1978) *Mar. Biol.* **45**, 31–37.
DUURSMA, E. K. (1961) *Nethl. J. Sea Res.* **1**, 1–147.
DYER, K. R. (1973) *Estuaries: A Phyical Introduction*. John Wiley, Chichester. 140 pp.
DYRSSEN, D. and M. WEDBORG (1974) In E. D. GOLDBERG (Ed.), *The Sea*, Vol. 5: *Marine Chemistry*. Wiley-Interscience, New York. pp. 181–195.
ELDERFIELD, H. (1970) *Earth Planet. Sci. Lett.* **9**, 10–16.
EL-SAYED, S. Z. (1966) *SCAR Symposium on Antarctic Oceanography*, pp. 227–239.
EL-SAYED, S. Z. (1967) *Antarctic J. U.S.* **11**, 200–201.
EL-SAYED, S. Z. (1968) *Biology of the Antarctic Seas*. III. *Ant. Res. Ser.* **11**, 15–47.
EL-SAYED, S. Z. (1970a) In M. W. HOLDGATE (Ed.), *Antarctic Ecology*, Vol. 1. Academic Press, London. pp. 119–135.
EL-SAYED, S. Z. (1970b) *SCAR Symposium on Antarctic Ice and Water Masses*. Tokyo. pp. 35–54.
EL-SAYED, S. Z. (1971) *Biology of the Antarctic Seas*. IV. *Ant. Res. Ser.* **17**, 301–312.
EL-SAYED, S. Z. (1977) In M. A. MCWHINNIE (Ed.), *Proc. Symposium on Polar Research: To the Present and the Future*. A.A.A.S., Washington, D.C. pp. 1–16.
EL-SAYED, S. Z. and E. F. MANDELLI (1965) *Biology of the Antarctic Seas*. II. *Ant. Res. Ser.* **5**, 87–106.
& Hall, London. Springer-Verlag, Berlin. pp. 131–142.
EL-SAYED, S. Z. and E. F. MANDELLI (1965) *Biology of the Antarctic Seas II, Ant. Res. Ser.*, **5**, 87–106.
EL-SAYED, S. Z., E. F. MANDELLI and Y. SUGIMURA (1964) *Biology of the Antarctic Seas I. Ant. Res. Ser.* **1**, 1–11.
ELTRINGHAM, S. K. (1971) *Life in Mud and Sand*. English Universities Press, London. 218 pp.
EMERY, K. O., W. L. ORR and S. C. RITTENBERG (1955) In *Essays in Natural Science in Honor of Capt. A. Hancock*, University of South California Press, Los Angeles. pp. 299–309.
ENGLISH, T. S. (1959) *Int. Ocean. Congress Preprints*. A.A.A.S., Washington, D.C. pp. 838–839.
ENGLISH, T. S. (1961) *Sci. Repts. Arctic Inst. N. Am.* **15**, 1–79.

Eppley, R. W. (1970) *Bull. Scripps Inst. Oceanogr.* **17**, 43–49.
Eppley, R. W. and J. L. Coatsworth (1968) *J. Phycol.* **4**, 151–156.
Eppley, R. W. and E. M. Renger (1974) *J. Phycol.* **10**, 15–23.
Eppley, R. W. and J. N. Rogers (1970) *J. Phycol.* **6**, 344–351.
Eppley, R. W. and P. R. Sloan (1965) *J. Fish. Res. Bd. Can.* **22**, 1083–1097.
Eppley, R. W. and J. D. H. Strickland (1968) *Adv. Microbiol. Sea* **1**, 23–62.
Eppley, R. W. and W. H. Thomas (1969) *J. Phycol.* **5**, 375–379.
Eppley, R. W., J. L. Coatsworth and L. Solorzano (1969) *Limnol. Oceanogr.* **14**, 194–205.
Eppley, R. W., R. W. Holmes and E. Paasche (1967) *Arch. Mikrobiol.* **56**, 305–323.
Eppley, R. W., F. M. H. Reid and J. D. H. Strickland (1970) *Bull. Scripps Inst. Oceanogr.* **17**, 43–51.
Eppley, R. W., J. N. Rogers and J. J. McCarthy (1969) *Limnol. Oceanogr.* **14**, 912–920.
Eppley, R. W., E. H. Renger, E. L. Venrick and M. M. Mullin (1973) *Limnol. Oceanogr.* **18**, 534–551.
Eppley, R. W., A. F. Carlucci, O. Holm-Hansen, D. Kiefer, J. J. McCarthy, E. Venrick and P. M. Williams (1971) *Limnol. Oceanogr.* **16**, 741–751.
Faulkner, D. J. and R. J. Andersen (1974) In E. D. Goldberg (Ed.), *The Sea*, Vol. 5. *Marine Chemistry*. Wiley-Interscience, New York. pp. 679–714.
Ferguson, C. F. (1973) Ph.D. Thesis, University of Southampton.
Fiadeiro, M. and J. D. H. Strickland (1968) *J. Mar. Res.* **26**, 187–201.
Fiadeiro, M., L. Solorzano and J. D. H. Strickland (1967) *Limnol. Oceanogr.* **12**, 555–556.
Findlay, I. W. O. (1972) *Int. Revue Ges. Hydrobiol.* **57**, 523–533.
Fish, C. J. (1925) *Bull. U.S. Bur. Fish.* **41**, 91–179.
Fisher, F. H. (1972) *Geochim. Cosmochim. Acta* **36**, 99–101.
Fogg, G. E. (1963) *Br. Phycol. Bull.* **2**, 195–205.
Fogg, G. E. (1965) *Algal Cultures and Phytoplankton Ecology*. University Wisconsin Press, Madison. 126 pp.
Fogg, G. E. (1966) *Oceanogr. Mar. Biol. Ann. Rev.* **4**, 195–212.
Fogg, G. E. (1968) *Photosynthesis*. The English Universities Press Ltd., London. 116 pp.
Fogg, G. E. (1975) In J. P. Riley and G. Skirrow (Eds.), *Chemical Oceanography*, Vol. 2, 2nd ed. Academic Press, London. pp. 385–453.
Fogg, G. E. (1977) *Phil. Trans. R. Soc. Lond.* B **279**, 27–38.
Fogg, G. E., W. D. P. Stewart, P. Fay and A. E. Walsby (1973) *The Blue-Green Algae*. Academic Press, London. 459 pp.
Foster, P. and A. W. Morris (1971) *Deep-Sea Res.* **18**, 231–236.
Fournier, R. O. (1971) *Limnol. Oceanogr.* **16**, 952–961.
Foxton, P. (1966) *Discovery Rep.* **34**, 1–115.
Friebele, E. S., D. L. Correll and M. A. Faust (1978) *Mar. Biol.* **45**, 39–52.
Fuhs, G. W. (1969) *J. Phycol.* **5**, 312–321.
Fuhs, G. W., S. D. Demmerle, E. Canelli and M. Chen (1972) *Limnol. Oceanogr. Special Symp.* **1**, 113–133.
Fuhrman, J. A., S. W. Chisholm and R. R. L. Guillard (1978) *Nature, Lond.* **272**, 244–246.
Furnestin, M-L. (1957) *Rev. Trav. Inst. Pêches Marit.* **21**, 1–356.
Gaarder, K. R. (1954) *Rep. Scient. Results Michael Sars N. Atlant. Deep Sea Expd. 1910*, **2**, 1–20.
Gaarder, K. R. (1971) In B. M. Funnell and W. R. Riedel (Eds.), *The Micropalaeontology of Oceans*. Cambridge University Press, London. pp. 97–103.
Gaarder, T. and H. H. Gran (1927) *Rapp. P-v. Réun. Cons. Perm. Int. Explor. Mer* **42**, 1–48.
Gaines, A. G. and M. E. Q. Pilson (1972) *Limnol. Oceanogr.* **17**, 42–49.
Gamble, J. C., J. M. Davies and J. H. Steele (1977) *Bull. Mar. Sci.* **27**, 146–175.
Garrels, R. M. (1965) *Science, N.Y.* **148**, 69.
Garrels, R. M. and M. E. Thompson (1962) *Am. J. Sci.* **260**, 57–66.
Gavis, J. (1976) *J. Mar. Res.* **34**, 161–179.
Gerber, R. P. and N. Marshall (1974) *Limnol. Oceanogr.* **19**, 815–824.
Gessner, F. (1970) In O. Kinne (Ed.), *Marine Ecology I* (1). Wiley-Interscience, London. pp. 368–406.
Gibbs, M. (1962) In R. A. Lewin (Ed.), *Physiology and Biochemistry of Algae*. Academic Press, New York. pp. 61–90.
Gieskes, W. W. C. ad G. W. Kraay (1975) *Nethl. J. Sea Res.* **9**, 166–196.
Gilson, H. C. (1937) *Scient. Rep. John Murray Exped. 1933–1934*, **2**, 21–81.
Goering, J. J. (1968) *Deep-Sea Res.* **15**, 157–168.
Goering, J. J. and R. C. Dugdale (1966) *Science, N.Y.* **154**, 505–506.
Goering, J. J., D. M. Nelson and J. A. Carter (1973) *Deep-Sea Res.* **20**, 777–789.
Goering, J. J., D. D. Wallen and R. A. Naumann (1970) *Limnol. Oceanogr.* **15**, 789–796.
Goldberg, E. D. (1952) *Biol. Bull.* **102**, 243–248.
Goldberg, E. D. (1971) *Comments Earth Sci. Geophysics* **1**, 117–132.
Goldberg, E. D., and G. Arrhenius (1958) *Geochim. Cosmechim. Acta* **13**, 153–212.

GOPINATHAN, C. P. (1972) *J. Mar. Biol. Ass. India* **14**, 568–577.
GORDON, L. I., P. K. PARK, S. W. HAGER and T. R. PARSONS (1971) *J. Oceanogr. Soc. Japan* **27**, 81–90.
GRAN, H. H. and T. BRAARUD (1935) *J. Biol. Bd. Can.* **1**, 279–467.
GRANT, B. R., J. MADGWICK and G. DAL PONT (1967) *Aust. J. Mar. Freshwater Res.* **18**, 129–136.
GRASSHOFF, K. (1975) In J. P. RILEY and G. SKIRROW (Eds.), *Chemical Oceanography*, Vol. 2, 2nd ed. Academic Press, London. pp. 455–597.
GREEN, J. C. (1975) *J. Mar. Biol. Ass. U.K.* **55**, 785–793.
GREEN, J. C. (1976) *J. Mar. Biol. Ass. U.K.* **56**, 595–602.
GRIFFIN, J. J., H. WINDOM and E. D. GOLDBERG (1968) *Deep-Sea Res.* **15**, 433–459.
GRONTVED, J. (1952) *Meddr. Komm. Danm. Fisk. -og Havunders, Serie: Plankton* **5**, 1–49.
GRONTVED, J. (1960) *Meddr. Komm. Danm. Fisk -og Havunders, N.S.* **3**, 55–92.
GRONTVED, J. (1962) *Meddr. Komm. Danm. Fisk. -og Havunders, N.S.* **3**, 347–378.
GROSS, F. (1937) *Phil. Trans. R. Soc. Lond.* B **228**, 1–47.
GROSS, F. (1940) *J. Mar. Biol. Ass. U.K.* **24**, 381–415.
GROSS, F. and E. ZEUTHEN (1948) *Proc. R. Soc. Lond.* B **135**, 382–389.
GROSS, F., S. M. MARSHALL, A. P. ORR and J. E. G. RAYMONT (1947) *Proc. R. Soc. Edinb.* B **63**, 1–95.
GROSS, F., S. R. NUTMAN, D. T. GAULD and J. E. G. RAYMONT (1950) *Proc. R. Soc. Edinb.* B **64**, 1–135.
GUILLARD, R. R. L. (1963) In C. H. OPPENHEIMER (Ed.), *Symposium on Marine Microbiology*. Thomas, Springfield, Ill. pp. 93–104.
GUILLARD, R. R. L. and P. J. WANGERSKY (1958) *Limnol. Oceanogr.* **3**, 449–454.
GUILLEN, O. (1971) In J. D. COSTLOW (Ed.), *Fertility of the Sea*, Vol. 1. Gordon & Breach, New York. pp. 187–196.
GUILLEN, O. and R. IZAGUIRRE DE RONDAN (1968) *Biol. Inst. Mar. Peru* **1**, 349–376.
GUILLEN, O., B. ROJAS DE MENDIOLA and R. IZAGUIRRE DE RONDAN (1971) In J. D. COSTLOW (Ed.), *Fertility of the Sea*, Vol. 1. Gordon & Breach, New York. pp. 157–185.
GUNTER, G., R. H. WILLIAMS, C. C. DAVIS and F. G. WALTON SMITH (1948) *Ecol. Monogr.* **18**, 310–324.
GUNTHER, E. R. (1936) *Discovery Rep.* **13**, 109–276.
HAHN, J. and C. JUNGE (1977) *Z. Naturforsch.* **32A**, 190–214.
HALIM, Y. (1960) *J. Cons. Perm. Int. Explor. Mer* **26**, 57–67.
HALLDAL, P. (1953) *Hvalrad Skr.* **38**, 1–91.
HALLDAL, P. (1966) In C. H. OPPENHEIMER (Ed.), *Marine Biology* **2**. New York Academy Science, New York. pp. 37–83.
HALLDAL, P. (1974) In N. G. JERLOV and E. STEEMANN NIELSEN (Eds.), *Optical Aspects of Oceanography*. Academic Press, London. pp. 345–359.
HAMILTON, R. D. (1964) *Limnol. Oceanogr.* **9**, 107–111.
HANDA, N. (1969) *Mar. Biol.* **4**, 208–214.
HANDA, N. and H. TOMINAGA (1969) *Mar. Biol.* **2**, 228–235.
HANSEN, V. K. (1961) *Rapp. P-v. Réun. Cons. Perm. Int. Explor. Mer* **149**, 160–166.
HARDY, A. (1967) *Great Waters*. Collins, London. 542 pp.
HARDY, A. C. and E. R. GUNTHER (1935) *Discovery Rep.* **11**, 1–456.
HARRIS, E. (1959) *Bull. Bingham Oceanogr. Coll.* **17**, 31–65.
HARRIS, E. and G. A. RILEY (1956) *Bull. Bingham Oceanogr. Coll.* **15**, 315–323.
HARRISON, W. G., R. W. EPPLEY and E. H. RENGER (1977) *Bull. Mar. Sci.* **27**, 44–57.
HART, T. J. (1934) *Discovery Rep.* **8**, 1–268.
HART, T. J. (1942) *Discovery Rep.* **21**, 263–348.
HART, T. J. and R. I. CURRIE (1960) *Discovery Rep.* **31**, 123–298.
HARVEY, H. W. (1926) *J. Mar. Biol. Ass. U.K.* **14**, 71–88.
HARVEY, H. W. (1937) *J. Mar. Biol. Ass. U.K.* **22**, 205–219.
HARVEY, H. W. (1950) *J. Mar. Biol. Ass. U.K.* **29**, 97–136.
HARVEY, H. W. (1953) *J. Mar. Biol. Ass. U.K.* **31**, 475–476.
HARVEY, H. W. (1955) *The Chemistry and Fertility of Sea Waters*. Cambridge University Press, London. 224 pp.
HARVEY, H. W. (1957) *The Chemistry and Fertility of Sea Waters*, 2nd ed. Cambridge University Press, London. 240 pp.
HARVEY, H. W., L. H. N. COOPER, M. V. LEBOUR and F. S. RUSSELL (1935) *J. Mar. Biol. Ass. U.K.* **20**, 407–441.
HARVEY, J. G. (1976) *Atmosphere and Ocean*. Artemis Press, Horsham, Surrey. 143 pp.
HASLE, G. R. (1959a) *Deep-Sea Res.* **6**, 38–59.
HASLE, G. R. (1959b) *Int. Oceano. Congress Preprints*. A.A.A.S., Washington, D.C. pp. 156–157.
HASLE, G. R. (1970) In M. W. HOLDGATE (Ed.), *Antarctic Ecology*, Vol. 1. Academic Press, London. p. 187.
HATCHER, B. G., A. R. O. CHAPMAN and K. H. MANN (1977) *Mar. Biol.* **44**, 85–96.
HATTORI, A. and E. WADA (1972) In A. Y. TAKENOUTI (Ed.), *Biological Oceanography of the Northern North Pacific Ocean*. Idemitsu Shoten, Tokyo. pp. 279–287.
HAUG, H. and S. MYKLESTAD (1976) *Mar. Biol.* **34**, 217–222.

HAWLEY, J. and R. M. PYTKOWICZ (1969) *Geochim. Cosmochim. Acta* **33**, 1557–1561.
HAXO, F. T. (1960) In M. B. ALLEN (Ed.), *Comparative Biochemistry of Photoreactive Systems*. Academic Press, New York. pp. 339–360.
HEAD, P. C. (1976) In J. D. BURTON and P. S. LISS (Eds.), *Estuarine Chemistry*. Academic Press, London. pp. 53–91.
HECKY, R. E. and P. KILHAM (1974) *Limnol. Oceanogr.* **19**, 361–366.
HECKY, R. E., K. MOPPER, P. KILHAM and E. T. DEGENS (1973) *Mar. Biol.* **19**, 323–331.
HEINLE, D. R., R. P. HARRIS, J. F. USTACH and D. A. FLEMER (1977) *Mar. Biol.* **40**, 341–353.
HEINRICH, A. K. (1962) *J. Cons. Perm. Int. Explor. Mer* **27**, 15–24.
HELLEBUST, J. A. (1965) *Limnol. Oceanogr.* **10**, 192–206.
HELLEBUST, J. A. (1970) In O. KINNE (Ed.), *Marine Biology* **1**. Wiley-Interscience, London. pp. 125–158.
HELLEBUST, J. A. and J. TERBORGH (1967) *Limnol. Oceanogr.* **12**, 559–567.
HENDEY, N. I. (1954) *J. Mar. Biol. Ass. U.K.* **33**, 537–560.
HENDEY, N. I. (1964) *Fishery Invest.*, Ser. 4. 317 pp.
HENDEY, N. I. (1971) In B. M. FUNNELL and W. R. RIEDEL (Eds.), *The Micropalaeontology of Oceans*. Cambridge University Press, London. pp. 625–631.
HENDEY, N. I. (1974) *J. Mar. Biol. Ass. U.K.* **54**, 277–300.
HENTSCHEL, E. and H. WATTENBURG (1930) *Ann. Hydrogr. Berlin* **58**, 279.
HERON, A. C. (1972) *Oceanologia* **10**, 269–312.
HITCHCOCK, G. L. and T. J. SMAYDA (1977) *Limnol. Oceanogr.* **22**, 126–131.
HOBSON, L. A., D. W. MENZEL and R. T. BARBER (1973) *Mar. Biol.* **19**, 298–306.
HOLDSWORTH, E. S. and J. COLBECK (1976) *Mar. Biol.* **38**, 189–199.
HOLDWAY, P. A. (1976) Ph.D. Thesis, University of Southampton.
HOLEMAN, J. N. (1968) *Wat. Resources Res.* **4**, 737–747.
HOLLAND, H. D. (1973) In E. INGERSON (Ed.), *Proceedings of the Symposium on Hydrogeochemistry and Biogeochemistry*. Clarke, Washington, D.C. pp. 68–81.
HOLLIGAN, P. M. and D. S. HARBOUR (1977) *J. Mar. Biol. Ass. U.K.* **57**, 1075–1093.
HOLMES, R. W. (1956) *Bull. Bingham Oceanogr. Coll.* **16**, 1–74.
HOLMES, R. W. (1958) *Rapp. P-v. Réun. Cons. Perm. Int. Explor. Mer* **144**, 109–116.
HOLMES, R. W. (1970) *Limnol. Oceanogr.* **15**, 688–694.
HOLM-HANSEN, O. (1962) In R. A. LEWIN (Ed.), *Physiology and Biochemistry of Algae*. Academic Press, New York. pp. 25–45.
HOLM-HANSEN, O. (1969) *Limnol. Oceanogr.* **14**, 740–747.
HOLM-HANSEN, O. and C. R. BOOTH (1966) *Limnol. Oceanogr.* **11**, 510–519.
HOPKINS, J. T. and R. W. DRUM (1966) *Br. Phycol. Bull.* **3**, 63–67.
HORN, M. K. and J. A. S. ADAMS (1966) *Geochim. Cosmochim. Acta* **30**, 279–297.
HORNE, A. J., G. E. FOGG and D. J. EAGLE (1969) *J. Mar. Biol. Ass. U.K.* **49**, 393–405.
HORNER, R., and V. ALEXANDER (1972) *Limnol. Oceanogr.* **17**, 454–458.
HULBURT, E. M. (1962) *Limnol. Oceanogr.* **7**, 307–315.
HULBURT, E. M. (1963) *J. Mar. Res.* **21**, 81–93.
HULBURT, E. M. (1970) *Ecology* **51**, 475–484.
HULBURT, E. M. (1975) *Bull. Mar. Sci.* **25**, 1–8.
HULBURT, E. M., and N. CORWIN (1970) *J. Fish. Res. Bd. Can.* **27**, 2081–2090.
HULBURT, E. M., J. H. RYTHER and R. R. L. GUILLARD (1960) *J. Cons. Perm. Int. Explor. Mer* **25**, 115–128.
HUNTSMAN, S. A. and R. T. BARBER (1977) *Deep-Sea Res.* **24**, 25–33.
HURD, D. C. (1972) *Earth Planet. Sci. Lett.* **15**, 411–417.
ICES (1962) *Atlas of Mean Monthly Temperature and Salinity of the Surface Layers of the North Sea and adjacent Waters. 1905–1954.*
IGNATIADES, L. (1973) *J. Mar. Biol. Ass. U.K.* **53**, 923–935.
IGNATIADES, L. and G. E. FOGG (1973) *J. Mar. Biol. Ass. U.K.* **53**, 937–956.
IGNATIADES, L. and T. J. SMAYDA (1970a) *J. Physiol.* **6**, 332–339.
IGNATIADES, L. and T. J. SMAYDA (1970b) *J. Physiol.* **6**, 357–364.
JACKSON, A. H., (1976) In T. W. GOODWIN (Ed.), *Chemistry and Biochemistry of Plant Pigments*, 2nd ed., Vol. 1. Academic Press, London. pp. 1–63.
JACKSON, G. A. and J. J. MORGAN (1978) *Limnol. Oceanogr.* **23**, 268–282.
JACOBSEN, J. P. and M. KNUDSEN (1940) *Pubs. Scient. Ass. Oceanogr. Phys.* No. 7. 38 pp.
JAMART, B. M., D. F. WINTER, K. BANSE, G. C. ANDERSON and R. K. LAM (1977) *Deep-Sea Res.* **24**, 753–773.
JAROSCH, R. (1962) In R. A. LEWIN (Ed.), *Physiology and Biochemistry of Algae*. Academic Press, New York. pp. 573–581.
JEFFERIES, R. L. (1972) In R. S. K. BARNES and J. GREEN (Eds.), *The Estuarine Environment*. Applied Science Publishers, London. pp. 61–85.

JEFFREY, S. W. (1965) *Aust. J. Mar. Freshwat. Res.* **16**, 307–313.
JEFFREY, S. W. (1968) *Biochem. Biophys. Acta* **162**, 271–285.
JEFFREY, S. W. (1972) *Biochem. Biophys. Acta* **279**, 15–33.
JEFFREY, S. W. (1974) *Mar. Biol.* **26**, 101–110.
JEFFRIES, H. P. (1962) *Limnol. Oceanogr.* **7**, 354–364.
JEFFRIES, H. P. (1964) *Limnol. Oceanogr.* **9**, 348–358.
JEFFRIES, H. P. (1967) In G. H. LAUFF (Ed.), *Estuaries*. A.A.A.S., Washington, D.C. pp. 500–508.
JEFFRIES, H. P. (1970) *Limnol. Oceanogr.* **15**, 419–426.
JENKIN, P. M. (1937) *J. Mar. Biol. Ass. U.K.* **22**, 301–343.
JENKINS, C. E. (1969) *Hlth. Phys.* **17**, 507–512.
JERLOV, N. G. (1951) *Rep. Swed. Deep Sea Exped. 3, Physics and Chemistry, No. 1*, 1–59.
JERLOV, N. G. (1963) *Oceanogr. Mar. Biol. Ann. Rev.* **1**, 89–114.
JERLOV, N. G. (1968) *Optical Oceanography*, 1st ed. Elsevier, Amsterdam. 194 pp.
JERLOV, N. G. (1976) *Marine Optics*, 2nd ed. Elsevier, Amsterdam. 231 pp.
JITTS, H. R., C. D. MCALLISTER, K. STEPHENS and J. D. H. STRICKLAND (1964) *J. Fish. Res. Bd. Can.* **21**, 139–157.
JOHNSTON, R. (1955) *J. Mar. Biol. Ass. U.K.* **34**, 185–195.
JOHNSTON, R. (1963a) *J. Mar. Biol. Ass. U.K.* **43**, 427–456.
JOHNSTON, R. (1963b) *J. Mar. Biol. Ass. U.K.* **43**, 409–425.
JOHNSTON, R. (1964) *J. Mar. Biol. Ass. U.K.* **44**, 87–109.
JOHNSTON, R. and P. G. W. JONES (1965) *Inorganic Nutrients in the North Sea (Serial Atlas of the Marine Environment*, Folio 11). American Geographical Society.
JONES, E. C. (1962) *J. Cons. Perm. Int. Explor. Mer* **27**, 223–231.
JORGENSEN, E. G. (1966) *Physiol. Plant.* **19**, 789–799.
JORGENSEN, E. G. and E. STEEMANN NIELSEN (1961) *Physiol. Plant*, **14**, 896–908.
KAIN, J. M. (1977) *J. Mar. Biol. Ass. U.K.* **57**, 587–607.
KALLE, K. (1937) *Annln Hydrogr. Berl.* **65**, 276–282.
KALLE, K. (1966) *Oceanogr. Mar. Biol. Ann. Rev.* **4**, 91–104.
KANWISHER, J. W. (1966) In H. BARNES (Ed.) *Some Contemporary Studies in Marine Sciences*. Allen & Unwin, London. pp. 407–420.
KATES, M. and B. E. VOLCANI (1966) *Biochem. Biophys. Acta* **116**, 264–278.
KAYAMA, M., Y. TSUCHIYA and J. F. MEAD (1963) *Bull. Jap. Soc. Scient. Fish.* **29**, 452–458.
KEELING, C. D. (1973) In S. I. RASOOL (Ed.). *Chemistry of the Lower Atmosphere*. Plenum Press, New York. pp. 251–329.
KESTER, D. R. (1975) In J. P. RILEY and G. SKIRROW (Eds.), *Chemical Oceanography*, Vol. 1, 2nd ed. Academic Press, London. pp. 497–556.
KESTER, D. R. and R. M. PYTKOWICZ (1967) *Limnol. Oceanogr.* **12**, 243–252.
KESTER, D. R. and R. M. PYTKOWICZ (1969) *Limnol. Oceanogr.* **14**, 686–692.
KESTER, D. R. and R. M. PYTKOWICZ (1970) *Geochim. Cosmochim. Acta* **34**. 1039–1051.
KETCHUM, B. H. (1939) *Am. J. Bot.* **26**, 399–407.
KETCHUM, B. H. (1962) *Rapp. P-v. Réun. Cons. Perm. Int. Explor. Mer* **153**, 142–147.
KETCHUM, B. H. (1967) In G. H. LAUFF (Ed.), *Estuaries*. A.A.A.S., Washington, D.C. pp. 329–335.
KETCHUM, B. H. and N. CORWIN (1965) *Limnol. Oceanogr, Suppl.* **10**, R148–R161.
KETCHUM, B. H., and A. C. REDFIELD (1949) *J. Cell Comp. Physiol.* **33**, 281–299.
KETCHUM, B. H., R. F. VACCARO and N. CORWIN (1958) *J. Mar. Res.* **17**, 282–301.
KETCHUM, B. H., J. H. RYTHER, C. S. YENTSCH and N. CORWIN (1958) *Rapp. P-v. Réun. Cons. Perm. Int. Explor. Mer* **144**, 132–140.
KIEFER, D. A., R. J. OLSON and O. HOLM-HANSEN (1976) *Deep-Sea Res.* **23**, 1199–1208.
KIMOR, B. and B. GOLANDSKY (1977) *Mar. Biol.* **42**, 55–67.
KITTREDGE, J. S., M. HORIGUCHI and P. M. WILLIAMS (1969) *Comp. Biochem. Physiol.* **29**, 859–863.
KNIGHT-JONES, E. W. (1951) *J. Cons. Perm. Int. Explor. Mer* **17**, 140–155.
KOBLENTZ-MISHKE, O. J. (1965) *Oceanology* **5**, 104–116.
KOBLENTZ-MISHKE, O. J., V. V. VOLKOVINSKY and J. G. KABANOVA (1970) *Scientific Exploration of the South Pacific*. Nat. Acad. Sciences, Washington, D.C. pp. 183–193.
KORNMANN, P. (1955) *Helgolander Wiss. Meersunters.* **5**, 218–233.
KREPS, E. and N. VERJBINSKAYA (1930) *J. Cons. Perm. Int. Explor. Mer.* **5**, 327–346.
KREPS, E. and N. VERJBINSKAYA (1932) *J. Cons. Perm. Int. Explor. Mer.* **7**, 25–46.
KREY, J. (1958) *Rapp. P-v. Réun. Cons. Perm. Int. Explor. Mer* **144**, 20–27.
KREY, J. (1973) In B. ZEITZSCHEL (Ed.), *The Biology of the Indian Ocean*. Chapman & Hall, London. Springer-Verlag, Berlin. pp. 115–126.
KREY, J. and B. BABENERD (1976) *Atlas of the Indian Ocean Biological Expedition*. University Kiel – UNESCO.

KRISHNAMURTHY, K. and A. PURUSHOTHAMAN (1972) *J. Mar. Biol. Ass. India* **13**, 271–274.
KUENZLER, E. J. and B. H. KETCHUM (1962) *Biol. Bull.* **123**, 134–145.
KUENZLER, E. J. and J. P. PERRAS (1965) *Biol. Bull.* **128**, 271–284.
LACKEY, J. B. (1967) In G. H. LAUFF (Ed.), *Estuaries*. A.A.A.S., Washington, D.C. pp. 291–305.
LAL, D. (1977) *Science, N.Y.* **198**, 997–1009.
LANDRY, M. R. (1976) *Mar. Biol.* **35**, 1–7.
LANSKAYA, L. A. (1963) In C. H. OPPENHEIMER (Ed.), *Marine Microbiology*. Thomas, Springfield, Ill. pp. 127–132.
LEADBEATER, B. S. C. (1971) *J. Mar. Biol. Ass. U.K.* **51**, 207–217.
LEADBEATER, B. S. C. (1972) *J. Mar. Biol. Ass. U.K.* **52**, 67–79.
LEADBEATER, B. S. C. and J. D. DODGE (1966) *Br. Phycol. Bull.* **3**, 1–17.
LEADBEATER, B. S. C. and I. MANTON (1974) *J. Mar. Biol. Ass. U.K.* **54**, 269–276.
LEBRASSEUR, R. J. (1972) In A. Y. TAKENOUTI (Ed.), *Biological Oceanography of the Northern North Pacific Ocean*. Idemitsu Shoten, Tokyo. pp. 581–588.
LEBRASSEUR, R. J. and O. D. KENNEDY (1972) In A. Y. TAKENOUTI (Ed.), *Biological Oceanography of the Northern North Pacific Ocean*. Idemitsu Shoten, Tokyo. pp. 354–365.
LEBOUR, M. V. (1925) *The Dinoflagellates of Northern Seas*. Marine Biological Association, U.K. 250 pp.
LEBOUR, M. V. (1930) *The Planktonic Diatoms of Northern Seas*. Ray Society, London. 244 pp.
LEE, A. (1970) *Oceanogr. Mar. Biol. Ann. Rev.* **8**, 33–71.
LEE, R. F. and A. R. LOEBLICH (1971) *Phytochemistry* **10**, 593–602.
LEE, R. F., J. C. NEVENZEL and G. A. PAFFENHOFER (1971) *Mar. Biol.* **9**, 99–108.
LEE, R. F., J. C. NEVENZEL, G. A. PAFFENHOFER, A. A. BENSON, S. PATTON and T. E. KAVANAGH (1970) *Biochim. Biophys. Acta* **202**, 386–388.
LEEDALE, G. F. (1967) *Euglenoid Flagellates*. Prentice-Hall, Englewoods Cliffs, New Jersey. 242 pp.
LENZ, J. (1977) *Mar. Biol.* **41**, 39–48.
LEVRING, T., H. A. HOPPE and O. J. SCHMID (1969) *Marine Algae*. Cram, de Gruyter, Hamburg. 421 pp.
LEWIN, J. and C. H. CHEN (1971) *Limnol. Oceanogr.* **16**, 670–675.
LEWIN, J. C. (1962) In R. A. LEWIN (Ed.), *Physiology and Biochemistry of Algae*. Academic Press, New York and London. pp. 445–455.
LEWIN, J. C. (1963). In C. H. OPPENHEIMER (Ed.), *Marine Microbiology*. Thomas, Springfield, Ill. pp. 229–235.
LEWIN, J. C. and R. L. L. GUILLARD (1963) *Ann. Rev. Microbiol.* **17**, 373–414.
LEWIN, J. C. and R. A. LEWIN (1960) *Can. J. Microbiol.* **6**, 127–134.
LEWIN, R. A. (1962) *Physiology and Biochemistry of Algae*. Academic Press, New York. 929 pp.
LEWIS, E. L. and W. F. WEEKS (1970) *SCAR Syposium on Antarctic Ice and Water Masses*. Tokyo. pp. 23–34.
LEYENDEKKERS, J. V. (1973) *Mar. Chem.* **1**, 75–88.
LIEBIG, J. (1843) *Chemistry in its Applications to Agriculture and Physiology*. Taylor & Walton. London, 3rd ed. 400 pp.
LILLICK, L. C. (1940) *Trans. Am. Phil. Soc.* **31**, 193–237.
LISS, P. S. (1973) *Deep-Sea Res.* **20**, 221–238.
LISS, P. S. (1976) In J. D. BURTON and P. S. LISS (Eds.), *Estuarine Chemistry*. Academic Press, London. pp. 93–130.
LIVINGSTONE D. A. (1963) *Prof. Pap. U.S. Geol. Surv.* **440-G**. 64 pp.
LOCICERO, V. R. (Ed.) (1975) *Proc. First Conference on Toxic Dinoflagellate Blooms*. Mass. Sci. and Tech. Foundation, Wakefield.
LOHMANN, H. (1908) *Wiss. Meeresuntersuch. Abt. Kiel N.F.* **10**, 129–370.
LOHMANN, H. (1911) *Int. Revue Ges. Hydrobiol. Hydrogr.* **4**, 1–38.
LONGHURST, A. R. (1967) *Deep-Sea Res.* **14**, 51–63.
LONGHURST, A. R. (1971) *Oceanogr. Mar. Biol. Ann. Rev.* **9**, 349–385.
LONGHURST, A. R. and V. BAINBRIDGE (1964) *Bull. Inst. Fr. Afr. Noire* sér A, **26**, 337–402.
LORENZEN, C. J. (1976) In D. H. CUSHING and J. J. WALSH (Eds.), *Ecology of the Seas*. Blackwell, Oxford. pp. 173–185.
LUCAS, C. E. (1961) In M. SEARS (Ed.), *Oceanography*. A.A.A.S., Washington, D.C. pp. 499–517.
LUCAS, C. E. and H. G. STUBBINGS (1948) *Hull Bull. Mar. Ecol.* **2**, 133–171.
LUCAS, I. A. N. (1967) *J. Mar. Biol. Ass. U.K.* **47**, 329–334.
LUCAS, I. A. N. (1968) *Br. Phycol. Bull.* **3**, 535–541.
LUI, N. S. T. and O. A. ROELS (1972) *J. Phycol.* **8**, 259–264.
MCALLISTER, C. D. (1970) In J. H. STEELE (Ed.), *Marine Food Chains*. Oliver & Boyd, Edinburgh. pp. 419–457.

McAllister, C. D. (1972) In A. Y. Takenouti (Ed.), *Biological Oceanography of the Northern North Pacific Ocean.* Idemitsu Shoten, Tokyo. pp. 575–579.
McAllister, C. D., T. R. Parsons and J. D. H. Strickland (1960) *J. Cons. Perm. Int. Explor. Mer* **25**, 240–259.
McAllister, C. D., N. Shah and J. D. H. Strickland (1964) *J. Fish. Res. Bd. Can.* **21**, 159–181.
McAllister, C. D., T. R. Parsons, K. Stephens and J. D. H. Strickland (1961) *Limnol. Oceanogr.* **6**, 237–258.
McCarthy, J. J. (1970) *Limnol. Oceanogr.* **15**, 303–313.
McCarthy, J. J. (1972) *Limnol. Oceanogr.* **17**, 738–748.
McCarthy, J. J. and R. W. Eppley (1972) *Limnol. Oceanogr.* **17**, 371–382.
McCave, I. N. (1973) In E. D. Goldberg (Ed.), *North Sea Science.* MIT Press, Cambridge, Massachusetts. pp. 75–100.
McGill, D. A. (1970) In A. R. Miller, P. Tchernia, H. Charnock and D. A. McGill, *Mediterranean Sea Atlas*, Woods Hole Oceanographic Institution, Massachusetts. pp. 95–110.
McGill, D. A. (1973) In B. Zeitzschel (Ed.), *The Biology of the Indian Ocean.* Chapman & Hall, London; Springer-Verlag, Berlin. pp. 53–102.
McIntyre, A. and A. W. H. Bé (1967) *Deep-Sea Res.* **14**, 561–597.
MacIsaac, J. J. and R. C. Dugdale (1969) *Deep-Sea Res.* **16**, 45–57.
MacIsaac, J. J. and R. C. Dugdale (1972) *Deep-Sea Res.* **19**, 209–232.
Mackenzie, F. T. (1975) In J. P. Riley and G. Skirrow (Eds.), *Chemical Oceanography*, Vol. 1, 2nd ed. Academic Press, London. pp. 309–364.
Mackenzie, F. T. and R. M. Garrels (1966a) *Am. J. Sci.* **264**, 507–525.
Mackenzie, F. T. and R. M. Garrels (1966b) *J. Sediment. Petrol.* **36**, 1075–1084.
McLachlan, J. and J. S. Craigie (1966) In H. Barnes (Ed.), *Some Contemporary Studies in Marine Science.* Allen & Unwin, London. pp. 511–517.
McRoy, C. P., R. J. Barsdate and M. Nebert (1972) *Limnol. Oceanogr.* **17**, 58–67.
McRoy, C. P., J. J. Goering and W. E. Shields (1972) In A. Y. Takenouti (Ed.) *Biological Oceanography of the Northern North Pacific Ocean.* Idemitsu Shoten, Tokyo. pp. 199–216.
Machta, L. (1972) In D. Dyrssen and D. Jagner (Eds.), *The Changing Chemistry of the Oceans*, Wiley-Interscience, New York; Almquist & Wiksell, Stockholm. pp. 121–145.
Maddux, W. S. and R. F. Jones (1964) *Limnol. Oceanogr.* **9**, 79–86.
Mague, T. H., F. C. Mague and O. Holm-Hansen (1977) *Mar. Biol.* **41**, 213–227.
Mague, T. H., N. M. Weare and O. Holm-Hansen (1974) *Mar. Biol.* **24**, 109–119.
Mahnken, C. V. W. (1969) *Bull. Mar. Sci.* **19**, 550–567.
Malone, T. C. (1971) *Limnol. Oceanogr.* **16**, 633–639.
Manahan, S. E. and M. J. Smith (1973) *Envir. Sci. Technol.* **7**, 829–833.
Mandelli, E. F. (1968) *J. Phycol.* **4**, 347–348.
Mandelli, E. F. and P. R. Burkholder (1966) *J. Mar. Res.* **24**, 15–27.
Mandelli, E. F., P. R. Burkholder, T. E. Doheny and R. Brody (1970) *Mar. Biol.* **7**, 153–160.
Manheim, F. T., R. M. Meade and G. C. Bond (1970) *Science, N.Y.* **167**, 371–376.
Manton, I. (1967) *J. Cell Sci.* **2**, 265–272.
Manton, I. and G. F. Leedale (1969) *J. Mar. Biol. Ass. U.K.* **49**, 1–16.
Manton, I. and M. Parke (1960) *J. Mar, Biol. Ass. U.K.* **39**, 275–298.
Manton, I. and M. Parke (1962) *J. Mar. Biol. Ass. U.K.* **42**, 565–578.
Manton, I. and M. Parke (1965) *J. Mar. Biol. Ass. U.K.* **45**, 743–754.
Manton, I., D. G. Rayns, H. Ettl and M. Parke (1965) *J. Mar. Biol. Ass. U.K.* **45**, 241–255.
Mantoura, R. F. C. and J. P. Riley (1975) *Analyt. Chim. Acta* **78**, 193–200.
Mantoura, R. F. C., A. Dickson and J. P. Riley (1978) *Estuarine Coastal Mar. Sci.* **6**, 387–408.
Margalef, R. (1958) In A. A. Buzzati-Traverso (Ed.), *Perspectives in Marine Biology.* University California Press, Berkeley. pp. 323–349.
Margalef, R. (1967) *Oceanogr. Mar. Biol. Ann. Rev.* **5**, 257–289.
Marker, A. F. H. (1965) *J. Mar. Biol. Ass. U.K.* **45**, 755–772.
Marshall, H. G. (1968) *Limnol. Oceanogr.* **13**, 370–376.
Marshall, N. (1970) In J. H. Steele (Ed.), *Marine Food Chains.* University of California Press, Berkeley. pp. 52–66.
Marshall, P. T. (1958) *J. Cons. Perm. Int. Explor. Mer* **23**, 173–177.
Marshall, S. M. (1933) *Gt. Barrier Reef Exped. Sci. Rpt.* B.M. (N.H.), **2**, 111–158.
Marshall, S. M. (1947) *Proc. R. Soc. Edinb.* B. **63**, 21–33.
Marshall, S. M. and A. P. Orr (1927) *J. Mar. Biol. Ass. U.K.* **14**, 837–868.
Marshall, S. M. and A. P. Orr (1928) *J. Mar. Biol. Ass. U.K.* **15**, 321–360.
Marshall, S. M. and A. P. Orr (1930) *J. Mar. Biol. Ass. U.K.* **16**, 853–878.
Marshall, S. M. and A. P. Orr (1948) *J. Mar. Biol. Ass. U.K.* **27**, 360–379.

MARTIN, J. H. (1970) *Limnol. Oceanogr.* **15**, 413–418.
MARUMO, R. (1967) *Inf. Bull. Plankt. Japan. Dr. Matsue Commem. No.* pp. 115–122.
MARUMO, R., A. SANO and M. MURANO (1974) *La Mer.* **12**, 145–156 (English Summary only).
MAURER, L. G. (1976) *Deep-Sea Res.* **23**, 1059–1064.
MEASURES, C. I. and J. D. BURTON (1978) *Trans. Am. Geophys. Un.* **59**, 307.
MEEUSE, B. J. D. (1962) In R. A. LEWIN (Ed.), *Physiology and Biochemistry of Algae*. Academic Press, New York. pp. 289–313.
MEGURO, H. (1962) *Antarctic Records* **14**, 1192–1199.
MEGURO, H., K. ITO and H. FUKUSHIMA (1966) *Science N.Y.* **152**, 1089–1090.
MENZEL, D. W. (1967) *Deep-Sea Res.* **14**, 229–238.
MENZEL, D. W. (1970) *Deep-Sea Res.* **17**, 751–764.
MENZEL, D. W. (1974) In E. D. GOLDBERG (Ed.), *The Sea*, Vol. 5. *Marine Chemistry*. Wiley-Interscience, New York. pp. 659–678.
MENZEL, D. W. and J. H. RYTHER (1960) *Deep-Sea Res.* **6**, 351–367.
MENZEL, D. W. and J. H. RYTHER (1961a) *J. Cons. Perm. Int. Explor. Mer* **26**, 250–258.
MENZEL, D. W. and J. H. RYTHER (1961b) *Deep-Sea Res.* **7**, 276–281.
MENZEL, D. W. and J. H. RYTHER (1970) In D. W. HOOD (Ed.), *Symposium on Organic Matter in Natural Waters*. University of Alaska. pp. 31–54.
MENZEL, D. W. and J. P. SPAETH (1962a) *Limnol. Oceanogr.* **7**, 159–162.
MENZEL, D. W. and J. P. SPAETH (1962b) *Limnol. Oceanogr.* **7**, 151–154.
MENZEL, D. W., E. M. HULBURT and J. H. RYTHER (1963) *Deep-Sea Res.* **10**, 209–219.
MIGITA, S. (1967a) *Inf. Bull. Plankt. Japan* **14**, 13–22.
MIGITA, S. (1967b) *Bull. Jap. Soc. Scient. Fish.* **33**, 392–395.
MILLER, J. D. A. (1962) In R. A. LEWIN (Ed.), *Physiology and Biochemistry of Algae*. Academic Press, London. pp. 357–370.
MILLERO, F. J. (1971) *Geochim. Cosmochim. Acta* **35**, 1089–1098.
MILLERO, F. J. (1974) *Ann. Rev. Earth Planet. Sci.* **2**, 101–150.
MIYAZAKI, T., E. WADA and A. HATTORI (1973) *Deep-Sea Res.* **20**, 571–577.
MONOD, J. (1942) *Recherches sur la Croissance des Cultures Bacteriennes*. Hermann, Paris. 211 pp.
MOORE, E. and F. SANDER (1976) *Estuarine Coastal Mar. Sci.* **4**, 589–607.
MOORE, E. and F. SANDER (1977) *Ophelia* **16**, 77–96.
MOORE, R. M. and J. D. BURTON (1976) *Nature, Lond.* **264**, 241–243.
MORRIS, A. W. and J. P. RILEY (1963) *Analyt. Chim. Acta* **29**, 272–279.
MORRIS, A. W. and J. P. RILEY (1966) *Deep-Sea Res.* **13**, 699–705.
MORRIS, I. (1967) *An Introduction to the Algae*. Hutchinson, London. 189 pp.
MORRIS, I. and P. SYRETT (1963) *Arch. Mikrobiol.* **47**, 32–41.
MORRIS, I., C. M. YENTSCH and C. S. YENTSCH (1971a) *Limnol. Oceanogr.* **16**, 854–858.
MORRIS, I., C. M. YENTSCH and C. S. YENTSCH (1971b) *Limnol. Oceanogr.* **16**, 859–868.
MOBERG, E. G. (1926) *Proc. 3rd Pan-Pacif. Sci. Congr. Tokyo*. 233–236.
MOTODA, S. and T. KAWAMURA (1963) In C. H. OPPENHEIMER (Ed.), *Marine Microbiology*. Thomas, Springfield, Ill. pp. 251–259.
MOTODA, S., T. KAWAMURA and A. TANIGUCHI (1978) *Mar. Biol.* **46**, 93–99.
MOTODA, S., A. TANIGUCHI and T. IKEDA (1974) *Indo-Pacific Fisheries Council Proc.* **15** (III), Bangkok.
MULLIN, M. M. and E. R. BROOKS (1970) *Bull. Scripps. Inst. Oceanogr.* **17**, 89–103.
MUNK, W. H. and G. A. RILEY (1952) *J. Mar. Res.* **11**, 215–240.
MURRAY, C. N., J. P. RILEY and T. R. S. WILSON (1969) *Deep-Sea Res.* **16**, 297–310.
MYKLESTAD, S. (1974) *J. Exp. Mar. Biol. Ecol.* **15**, 261–274.
MYKLESTAD, S. and A. HAUG (1972) *J. Exp. Mar. Biol. Ecol.* **9**, 125–136.
NAKAMURA, K. and C. S. GOWANS (1964) *Nature, Lond.* **202**, 826–827.
NAKAYAMA, T. O. M. (1962) In R. A. LEWIN (Ed.), *Physiology and Biochemistry of Algae*. Academic Press, New York. pp. 409–420.
NELSON, D. M. and J. J. GOERING (1977) *Deep-Sea Res.* **24**, 65–73.
NELSON, D. M. and J. J. GOERING (1978) *Limnol. Oceanogr.* **23**, 508–517.
NORRIS, R. E. (1965) *J. Protozool*, **12**, 589–602.
NORTH, B. B. (1975) *Limnol. Oceanogr.* **20**, 20–27.
NORTH, B. B. and G. C. STEPHENS (1971) *Biol. Bull.* **140**, 242–254.
ODUM, W. E. (1970) In J. H. STEELE (Ed.), *Marine Food Chains*. University California Press, Berkeley. pp. 222–240.
OFFICER, C. B. (1976) *Physical Oceanography of Estuaries*. John Wiley, New York. 465 pp.
OGURA, N. (1970) *Deep-Sea Res.* **17**, 221–230.
OGURA, N. (1974) *Mar. Biol.* **24**, 305–312.
OGURI, M., D. SOULE, D. M. JUGE and B. C. ABBOTT (1975) In V. R. LOCICERO (Ed.), *Proc. 1st International Conf. on Toxic Dinoflagellate Blooms*. Mass. Sci. and Technol. Foundation, Wakefield. p. 41.

O hEocha, C. (1962) In R. A. Lewin (Ed.), *Physiology and Biochemistry of Algae*. Academic Press, New York. pp. 421–435.
Ohwada, M. (1972) In A. Y. Takenouti (Ed.), *Biological Oceanography of the Northern North Pacific Ocean*. Idemitsu Shoten, Tokyo. pp. 145–163.
Okada, H. and S. Honjo (1973) *Deep-Sea Res.* **20**, 355–374.
Outka, D. E. and D. C. Williams (1971) *J. Protozool.* **18**, 285–297.
Owen, R. W. and B. Zeitzschel (1970) *Mar. Biol.* **7**, 32–36.
Paasche, E. (1960) *J. Cons. Perm. Int. Explor. Mer* **26**, 33–48.
Paasche, E. (1964) *Physiol. Plant. Suppl. III.* 82 pp.
Paasche, E. (1966a) *Physiol. Plant.* **19**, 271–278.
Paasche, E. (1966b) *Physiol. Plant.* **19**, 770–779.
Paasche, E. (1967) *Physiol. Plant.* **20**, 946–956.
Paasche, E. (1968a) *Physiol. Plant.* **21**, 66–77.
Paasche, E. (1968b) *Limnol. Oceanogr.* **13**, 178–181.
Paasche, E. (1968c) *Ann. Rev. Microbiol.* **22**, 71–86.
Paasche, E. (1973) *Mar. Biol.* **19**, 262–269.
Paffenhofer, G. A. and J. D. H. Strickland (1970) *Mar. Biol.* **5**, 97–99.
Palmer, J. D. and F. E. Round (1965) *J. Mar. Biol. Ass. U.K.* **45**, 567–582.
Pamatmat, M. M. (1968) *Int. Rev. Ges. Hydrobiol.* **53**, 211–298.
Parke, M. (1961) *Br. Phycol. Bull.* **2**, 47–55.
Parke, M. (1966) In H. Barnes (Ed.), *Some Contemporary Studies in Marine Science*. Allen & Unwin, London. pp. 555–563.
Parke, M. and I. Adams (1960) *J. Mar. Biol. Ass. U.K.* **39**, 263–274.
Parke, M. and P. S. Dixon (1964) *J. Mar. Biol. Ass. U.K.* **44**, 499–542.
Parke, M. and P. S. Dixon (1968) *J. Mar. Biol. Ass. U.K.* **48**, 783–832.
Parke, M. and P. S. Dixon (1976) *J. Mar. Biol. Ass. U.K.* **56**, 527–594.
Parke, M. and I. den Hartog-Adams (1965) *J. Mar. Biol. Ass. U.K.* **45**, 537–557.
Parke, M. and I. Manton (1965) *J. Mar. Biol. Ass. U.K.* **45**, 525–536.
Parke, M., I. Manton and B. Clarke (1955) *J. Mar. Biol. Ass. U.K.* **34**, 579–609.
Parke, M., I. Manton and B. Clarke (1956) *J. Mar. Biol. Ass. U.K.* **35**, 387–414.
Parke, M., I. Manton and B. Clarke (1958) *J. Mar. Biol. Ass. U.K.* **37**, 209–228.
Parke, M., I. Manton and B. Clarke (1959) *J. Mar. Biol. Ass. U.K.* **38**, 169–188.
Parke, M., G. T. Boalch, R. Jowett and D. S. Harbour (1978) *J. Mar. Biol. Ass. U.K.* **58**, 239–276.
Parsons, T. R. (1966) *Monographs on Oceanographic Methodology. The determination of photosynthetic pigments in sea water*. UNESCO, Paris. pp. 21–36.
Parsons, T. R. (1972) In A. Y. Takenouti (Ed.), *Biological Oceanography of the Northern North Pacific Ocean*. Idemitsu Shoten, Tokyo. pp. 275–278.
Parsons, T. R. (1975) In J. P. Riley and G. Skirrow (Eds.), *Chemical Oceanography*, Vol. 2, 2nd ed. Academic Press, London. pp. 365–383.
Parsons, T. R. (1976) In D. H. Cushing and J. J. Walsh (Eds.), *The Ecology of the Seas*. Blackwell, Oxford. pp. 81–97.
Parsons, T. R. and R. J. LeBrasseur (1970) *Calif. Mar. Res. Comm. CALCOFI Rept.* **12**, 54–63.
Parsons, T. R. and J. D. H. Strickland (1962) *Deep-Sea Res.* **8**, 211–222.
Parsons, T. R. and M. Takahashi (1973a) *Biological Oceanographic Processes*. Pergamon Press, Oxford. 186 pp.
Parsons T. R. and M. Takahashi (1973b) *Limnol. Oceanogr.* **18**, 511–515.
Parsons, T. R., R. J. LeBrasseur and W. E. Barraclough (1970) *J. Fish. Res. Bd. Can.* **27**, 1251–1264.
Parsons, T. R., K. Stevens and J. D. H. Strickland (1961) *J. Fish. Res. Bd. Can.* **18**, 1001–1016.
Parsons, T. R., M. Takahashi and B. Hargrave (1977) *Biological Oceanographic Processes*, 2nd ed. Pergamon Press, Oxford. 332 pp.
Parsons, T. R., R. J. LeBrasseur, J. D. Fulton and O. D. Kennedy (1969) *J. Exp. Mar. Biol. Ecol.* **3**, 39–50.
Parsons, T. R., K. von Brockel, P. Koeller, M. Takahashi, M. R. Reeve and O. Holm-hansen (1977a) *J. Exp. Mar. Biol. Ecol.* **26**, 235–247.
Parsons, T. R., W. H. Thomas, D. Seibert, J. R. Beers, P. Gillespie and C. Bawden (1977b) *Int. Revue Ges. Hydrobiol.* **62**, 565–572.
Pasciak, W. J., and J. Gavis (1974) *Limnol. Oceanogr.* **19**, 881–888.
Pasciak, W. J., and J. Gavis (1975) *Limnol. Oceanogr.* **20**, 605–617.
Patriquin, D. G. (1972) *Mar. Biol.* **15**, 35–46.
Patriquin, D. G. and R. Knowles (1972) *Mar. Biol.* **16**, 49–58.
Patten, B. C. (1962) *J. Mar. Res.* **20**, 57–75.
Patterson, G. W. (1967) *J. Phycol.* **3**, 22–23.

Percival, E. (1968) *Oceanogr. Mar. Biol. Ann. Rev.* **6**, 137–161.
Pérès, J. M. (1967) *Oceanogr. Mar. Biol. Ann. Rev.* **5**, 449–533.
Peruyeva, Y. G. (1977) *Oceanology* **16**, 617–619.
Peterfi, L. S. and I. Manton (1968) *Br. Phycol. Bul.* **3**, 423–440.
Petersen, C. G. J. (1918) *Rep. Dan. Biol. Stn.* **25**, 1–62.
Petipa, T. S., E. V. Pavlova and Y. Sorokin (1973) In M. E. Vinogradov (Ed.), *Life Activity of Pelagic Communities in the Ocean Tropics*. Israel Program for Scientific Translations, Jerusalem. pp. 135–165.
Petipa, T. S., A. V. Monakov, A. P. Pablyutin and Y. Sorokin (1974) Translation in *Naukova Dumka*, Kiev. pp. 153–160.
Philippon, M. R. (1972) Doctorate Thesis. Université de Provence.
Pickard, G. L. (1963) *Descriptive Physical Oceanography*. Pergamon Press, Oxford. 199 pp.
Pienaar, R. N. (1976) *J. Mar. Biol. Ass. U.K.* **56**, 1–11.
Pingree, R. D. (1975) *J. Mar. Biol. Ass. U.K.* **55**, 965–974.
Pingree, R. D., G. R. Forster and G. K. Morrison (1974) *J. Mar. Biol. Ass. U.K.* **54**, 469–479.
Pingree, R. D., P. M. Holligan, G. T. Mardell and R. N. Head (1976) *J. Mar. Biol. Ass. U.K.* **56**, 845–873.
Pintner, I. J., and L. Provasoli (1963) In C. H. Oppenheimer (Ed.), *Marine Microbiology*. Thomas, Springfield, Ill. pp. 114–121.
Plankton Atlas, Edinburgh Oceanographic Laboratory (1973) *Bull. Mar. Ecol.* **7**, 1–174.
Platt, T. (1969) *Limnol. Oceanogr.* **14**, 653–659.
Platt, T., K. L. Denman and A. D. Jassby (1977) In E. D. Goldberg, I. N. McCave, J. J. O'Brien and J. H. Steele (Eds.), *The Sea*, Vol. 6. Wiley-Interscience, New York. pp. 807–856.
Pomeroy, L. R. (1959) *Limnol. Oceanogr.* **4**, 386–397.
Pomeroy, L. R. (1960) *Bull. Mar. Sci. Gulf Caribb.* **10**, 1–10.
Posner, G. S. (1957) *Bull. Bingham Oceanogr. Coll.* **16**, 106–155.
Postma, H. (1973) In E. D. Goldberg (Ed.), *North Sea Science*. MIT Press, Cambridge, Massachusetts. pp. 326–334.
Poulet, S. A. (1973) *Limnol. Oceanogr.* **18**, 564–573.
Prakash, A. (1971) In J. D. Costlow (Ed.), *Fertility of the Sea*, Vol. 2. Gordon & Breach, New York. pp. 351–368.
Pratt, D. M. (1965) *Limnol. Oceanogr.* **10**, 173–184.
Prescott, G. W. (1969) *The Algae: A Review*. Nelson, London. 436 pp.
Provasoli, L. (1963) In M. N. Hill (Ed.), *The Sea*, Vol. 2. Interscience, New York. pp. 165–219.
Provasoli, L. and J. J. A. McLaughlin (1963) In C. H. Oppenheimer (Ed.), *Marine Microbiology*. Thomas, Springfield, Ill. pp. 105–113.
Pugh, P. R. (1969) Ph. D. Thesis, University of Southampton.
Pugh, P. R. (1971) *Mar. Biol.* **11**, 118–124.
Pugh, P. R. (1975) *Mar. Biol.* **33**, 195–205.
Putter, A. (1909) *Die Ernährung der Wassertiere und der Stoffhaushalt der Gewässer*. Fischer, Jena. 168 pp.
Putter, A. (1925) *Arch. Hydrobiol.* **15**, 70–117.
Pytkowicz, R. M. (1972) *Geochim. Cosmochim. Acta* **36**, 631–633.
Pytkowicz, R. M. and J. E. Hawley (1974) *Limnol. Oceanogr.* **19**, 223–234.
Qasim, S. Z. and P. M. A. Bhattathiri (1971) *Hydrobiologia* **38**, 29–38.
Qasim, S. Z. and C. K. Gopinathan (1969) *Proc. Indian Acad. Sci.* **69B**, 336–348.
Qasim, S. Z., P. M. A. Bhattathiri and S. A. H. Abidi (1968) *J. Exp. Mar. Biol. Ecol.* **2**, 87–103.
Qasim, S. Z., P. M. A. Bhattathiri and V. P. Devassy (1972a) *Mar. Biol.* **16**, 22–27.
Qasim, S. Z., P. M. A. Bhattathiri and V. P. Devassy (1972b) *Mar. Biol.* **12**, 200–206.
Qasim, S. Z., S. Wellershaus, P. M. A. Bhattathiri and S. A. H. Abidi (1969) *Proc. Indian Acad. Sci.* **69B**, 51–94.
Ramamurthy, V. D. (1970) *Mar. Biol.* **6**, 74–76.
Raudkivi, A. J. (1967) *Some Boundary Hydraulics*. Pergamon Press, Oxford. 331 pp.
Rashid, M. A. and L. A. King (1970) *Geochim. Cosmochim. Acta* **34**, 193–201.
Raymont, J. E. G. (1971) In J. D. Costlow (Ed.), Fertility of the Sea, Vol. 2. Gordon & Breach, New York. pp. 383–399.
Raymont, J. E. G. and M. N. E. Adams (1958) *Limnol. Oceanogr.* **3**, 119–136.
Raymont, J. E. G. and R. S. Miller (1962) *Int. Revue Ges. Hydrobiol.* **47**, 169–209.
Rayns, D. G. (1962) *J. Mar. Biol. Ass. U.K.* **42**, 481–484.
Redfield, A. C. (1934) In R. J. Daniel (Ed.), *James Johnstone Memorial Volume*. Liverpool University Press. pp. 177–192.
Redfield, A. C. (1942) *Pap. Phys. Oceanogr. Met., Mass. Inst. Tech. Woods Hole Oceanogr. Inst.* **9**, (2), 1–22.
Redfield, A. C. (1948) *J. Mar. Res.* **7**, 347–361.

Redfield, A. C. (1958) *Am. Scient.* **46**, 205–221.
Redfield, A. C., B. H. Ketchum and F. A. Richards (1963) In M. N. Hill (Ed.), *The Sea*, Vol. 2. Interscience, New York. pp. 26–77.
Reeve, M. R., J. C. Gamble and M. A. Walter (1977) *Bull. Mar. Sci.* **27**, 92–104.
Reid, Jr., J. L. (1965) *Intermediate Waters of the Pacific Ocean*. John Hopkins Press, Baltimore. 85 pp.
Remsen, C. C. (1971) *Limnol. Oceanogr.* **16**, 732–740.
Revelante, N. and M. Gilmartin (1976) *Nethl. J. Sea Res.* **10**, 377–396.
Rheinheimer, G. (1967) *Helgolander Wiss. Meeresunters.* **15**, 243–252.
Rheinheimer, G. (1974) *Aquatic Microbiology*. Wiley, London. 184 pp.
Richards, F. A. (1958) *J. Mar. Res.* **17**, 449–465.
Richards, F. A. (1965) In J. P. Riley and G. Skirrow (Eds.), *Chemical Oceanography*, Vol. 1, 1st ed. Academic Press, London. pp. 197–225.
Richards, F. A. with T. G. Thompson (1952) *J. Mar. Res.* **11**, 156–172.
Richardson, P. (1976) *Oceanus* **19**, 65–68.
Richman, S. and J. N. Rogers (1969) *Limnol. Oceanogr.* **14**, 701–709.
Ricketts, T. R. (1966) *Phytochemistry* **5**, 67–76.
Riley, G. A. (1942) *J. Mar. Res.* **5**, 67–87.
Riley, G. A. (1946) *J. Mar. Res.* **6**, 54–73.
Riley, G. A. (1947) *J. Mar. Res.* **6**, 104–113.
Riley, G. A. (1952) *Bull. Bingham Oceanogr. Coll.* **13**, 40–64.
Riley, G. A. (1956) *Bull. Bingham Oceanogr. Coll.* **15**, 324–344.
Riley, G. A. (1957) *Limnol. Oceanogr.* **2**, 252.
Riley, G. A. (1959) *Bull. Bingham Oceanogr. Coll.* **17**, 83–85.
Riley, G. A. (1963) In M. N. Hill (Ed.), *The Sea*, Vol. 2. Interscience. New York. pp. 438–463.
Riley, G. A. (1967) In G. H. Lauff (Ed.), *Estuaries*. A.A.A.S., Washington, D.C. pp. 316–326.
Riley, G. A. and D. F. Bumpus (1946) *J. Mar. Res.* **6**, 33–47.
Riley, G. A. and S. A. M. Conover (1956) *Bull. Bingham Oceanogr. Coll.* **15**, 47–61.
Riley, G. A. and S. Gorgy (1948) *J. Mar. Res.* **7**, 100–121.
Riley, G. A., H. Stommel and D. F. Bumpus (1949) *Bull. Bingham Oceanogr. Coll.* **12**, 1–169.
Riley, J. P. (1965) *Deep-Sea Res.* **12**, 219–220.
Riley, J. P. and R. Chester (1971) *Introduction to Marine Chemistry*. Academic Press, London. 465 pp.
Riley, J. P. and R. Chester (1976) (Eds.), *Chemical Oceanography*, 2nd ed., Vols. 5 and 6. Academic Press, London. 401 and 414 pp.
Riley, J. P. and R. Chester (1978) (Eds.), *Chemical Oceanography,* 2nd ed., Vol. 7. Academic Press, London. 502 pp.
Riley, J. P. and D. A. Segar (1969) *J. Mar. Biol. Ass. U.K.* **49**, 1047–1056.
Riley, J. P. and G. Skirrow (1975) (Eds.), *Chemical Oceanography*, 2nd ed., Vols. 1–4. Academic Press, London. 606, 647, 564 and 363 pp.
Riley, J. P. and M. Tongudai (1967) *Chem. Geol.* **2**, 263–269.
Riley, J. P. and T. R. S. Wilson (1965) *J. Mar. Biol. Ass. U.K.* **45**, 583–591.
Riley, J. P. and T. R. S. Wilson (1967) *J. Mar. Biol. Ass. U.K.* **47**, 351–362.
Robinson, G. A. (1961) *Bull. Mar. Ecol.* **5**, 81–89.
Robinson, G. A. (1965) *Bull. Mar. Ecol.* **6**, 104–122.
Robinson, G. A. (1970) *Bull. Mar. Ecol.* **6**, 333–345.
Roman, M. R. (1977) *Mar. Biol.* **42**, 149–155.
Rosenberg, R., I. Olssen and E. Olundh (1977) *Mar. Biol.* **42**, 99–107.
Round, F. E. (1965) *The Biology of the Algae*. Edward Arnold, London. 269 pp.
Royal Society (1965) *I.I.O.E., R.R.S. Discovery Cruise 3, Oceanographic Work in the Western Indian Ocean, 1964*. 55 pp.
Rubey, W. W. (1951) *Bull. Geol. Soc. Am.* **62**, 1111–1147.
Russell, F. S., A. J. Southward, G. T. Boalch and E. I. Butler (1971) *Nature, Lond.* **234**, 468–470.
Ryther, J. H. (1954) *Bull. Mar. Biol. Lab., Woods Hole* **106**, 198–209.
Ryther, J. H. (1956a) *Limnol. Oceanogr.* **1**, 61–70.
Ryther, J. H. (1956b) *Nature Lond.* **178**, 861–862.
Ryther, J. H. (1963) In M. N. Hill (Ed.), *The Sea*, Vol. 2. Interscience, New York. pp. 347–380.
Ryther, J. H. (1969) *Science N.Y.* **116**, 72–76.
Ryther, J. H. and W. M. Dunstan (1971) *Science, N.Y.* **171**, 1008–1013.
Ryther, J. H. and R. R. L. Guillard (1959) *Deep-Sea Res.* **6**, 65–69.
Ryther, J. H. and D. W. Menzel (1959) *Limnol. Oceanogr.* **4**, 492–549.
Ryther, J. H. and D. W. Menzel (1960) *Deep-Sea Res.* **6**, 235–238.
Ryther, J. H. and D. W. Menzel (1961) *Bull. Mar. Sci., Gulf Caribb.* **11**, 381–388.
Ryther, J. H. and D. W. Menzel (1965) *Limnol. Oceanogr.* **10**, 490–492.

RYTHER, J. H. and C. S. YENTSCH (1957) *Limnol. Oceanogr.* **2**, 281–286.
RYTHER, J. H. and C. S. YENTSCH (1958) *Limnol. Oceanogr.* **3**, 327–335.
RYTHER, J. H., D. W. MENZEL and N. CORWIN (1967) *J. Mar. Res.* **25**, 69–83.
RYTHER, J. H., J. R. HALL, A. K. PEASE, A. BAKUN and M. M. JONES (1966) *Limnol. Oceanogr.* **11**, 371–380.
RYTHER, J. H., D. W. MENZEL, E. M. HULBURT, C. J. LORENZEN and N. CORWIN (1970) *Anton Bruun. Rpt.* **4**, 3–12.
SAIJO, Y., and S. ICHIMURA (1962) *J. Oceanogr. Soc. Japan, 20th Ann. Vol.* pp. 687–693.
SAMUEL, S., N. M. SHAH and G. E. FOGG (1971) *J. Mar. Biol. Ass. U.K.* **51**, 793–798.
SANDER, F. and D. M. STEVEN (1973) *Bull. Mar. Sci.* **23**, 771–792.
SAVAGE, P. D. V. (1969) *C.E.R.L. Laboratory Memorandum, No. RD/L/M 269.* pp. 44–49.
SCAGEL, R. F. (1966) *Oceanogr. Mar. Biol. Ann. Rev.* **4**, 123–194.
SCHELSKE, C. L. and E. P. ODUM (1962) *Proc. Gulf Caribb. Fish. Invest.* **14**, 75–80.
SCHLEGEL, H. G. (1975) In O. KINNE (Ed.), *Marine Ecology* **2**, (1). Wiley, New York. pp. 9–49.
SCHOENER, A. and G. T. ROWE (1970) *Deep-Sea Res.* **17**, 923–925.
SCHONE, H. K. (1970) *Int. Revue Ges. Hydrobiol.* **55**, 595–677.
SCHONE, H. K. (1972) *Mar. Biol.* **13**, 284–291.
SCLATER, F. R., E. BOYLE and J. M. EDMOND (1976) *Earth Planet. Sci. Lett.* **31**, 119–128.
SEATON, D. D. (1970) *J. Mar. Biol. Ass. U.K.* **50**, 97–106.
SEGUIN, G. (1966) *Bull. Inst. Fr. Afr. Noire* **28**, sér A, 1–90.
SEGUIN, G. (1970) *Bull. Inst. Fond. Afr. Noire* **32**, sér A, 607–663.
SEILER, W. and U. SCHMIDT (1974) In E. D. GOLDBERG (Ed.), *The Sea*, Vol. 5. *Marine Chemistry*. Wiley-Interscience, New York. pp. 219–243.
SEMINA, H. J. (1972) *Int. Revue Ges. Hydrobiol.* **57**, 177–205.
SEMINA, H. J., and I. T. TARKHOVA (1972) In A. Y. TAKENOUTI (Ed.), *Biological Oceanography of the Northern North Pacific Ocean*. Idemitsu Shoten, Tokyo. pp. 117–124.
SHARP, J. H. (1977) *Limnol. Oceanogr.* **22**, 381–399.
SHELDON, R. W. (1972) *Limnol. Oceanogr.* **17**, 494–498.
SHELDON, R. W., W. H. SUTCLIFFE and A. PRAKASH (1973) *Limnol. Oceanogr.* **18**, 719–733.
SHIMADA, B. M. (1958) *Limnol. Oceanogr.* **3**, 336–339.
SIEBURTH, J. McN. (1968) *Adv. Mar. Microbiol.* **1**, 63–94.
SIEBURTH, J. McN. and A. JENSEN (1970) In D. W. HOOD (Ed.), *Symposium on Organic Matter in Natural Waters*. University of Alaska. pp. 203–223.
SILLEN, L. G. (1961) In M. SEARS (Ed.), *Oceanography*, A.A.A.S., Washington, D.C. pp. 549–581.
SILLEN, L. G. (1967) *Science, N.Y.* **156**, 1189–1197.
SKIRROW, G. (1975) In J. P. RILEY and G. SKIRROW (Eds.), *Chemical Oceanography*, Vol. 2, 2nd ed. Academic Press, London. pp. 1–192.
SKOPINTSEV, B. A., F. A. GUBIN, R. V. VOROB'EVA and O. A. VERSHININA (1958) *C.R. Acad. Sci. U.R.S.S.* **119**, 121–124.
SLOWEY, J. F., L. M. JEFFREY and D. W. HOOD (1967) *Nature, Lond.* **214**, 377–378.
SMAYDA, T. J. (1957) *Limnol. Oceanogr.* **2**, 342–359.
SMAYDA, T. J. (1958) *Oikos* **9**, 158–191.
SMAYDA, T. J. (1963a) *Int.-Am. Tropical Tuna Comm. Bull.* **7**, 193–253.
SMAYDA, T. J. (1963b) In C. H. OPPENHEIMER (Ed.), *Marine Microbiology*. Thomas, Springfield, Ill. pp. 260–274.
SMAYDA, T. J. (1964) *Proc. Symposium Exp. Mar. Ecol., Occasional Publ.* **2**. Graduate School, Oceano graphy University, Rhode Island. pp. 25–32.
SMAYDA, T. J. (1965) *Int.-Am. Tropical Tuna Comm. Bull.* **9**, 467–531.
SMAYDA, T. J. (1966) *Int.-Am. Tropical Tuna Comm. Bull.* **11**, 355–612.
SMAYDA, T. J. (1969) *J. Phycol.* **5**, 150–157.
SMAYDA, T. J. (1970a) *Oceanogr. Mar. Biol. Ann. Rev.* **8**, 353–414.
SMAYDA, T. J. (1970b) *Helgolander Wiss. Meeresunters.* **20**, 172–194.
SMAYDA, T. J. (1971) In J. D. COSTLOW (Ed.), *Fertility of the Sea*, Vol. 2. Gordon & Breach, New York. pp. 493–511.
SMAYDA, T. J. (1973) *Norweg. J. Bot.* **20**, 219–47.
SMAYDA, T. J. (1976) In B. MACOWITZ (Ed.), *Effects of Energy-related Activities on the Atlantic Continental Shelf*. Proc. Conf. Brookhaven Nat. Lab. pp. 70–94.
SMAYDA, T. J. and B. J. BOLEYN (1965) *Limnol. Oceanogr.* **10**, 499–509.
SMAYDA, T. J. and B. J. BOLEYN (1966a) *Limnol. Oceanogr.* **11**, 18–34.
SMAYDA, T. J. and B. J. BOLEYN (1966b) *Limnol. Oceanogr.* **11**, 35–43.
SMAYDA, T. J. and B. MITCHELL-INNES (1974) *Mar. Biol.* **25**, 195–202.
SMETACEK, V., B. VON BODUNGEN, K. VON BROCKEL and B. ZEITZSCHEL (1976) *Mar. Biol.* **34**, 373–378.
SMITH, R. L. (1968) *Oceanogr. Mar. Biol. Ann. Rev.* **6**, 11–46.

SMITH, S. L. and T. E. WHITLEDGE (1977) *Deep-Sea Res.* **24**, 49–56.
SMITH, W. O., R. T. BARBER and S. A. HUNTSMAN (1977) *Deep-Sea Res.* **24**, 25–33.
SOLI, G. (1963) In C. H. OPPENHEIMER (Ed.), *Microbiology*. Thomas, Springfield, Ill. pp. 122–126.
SOLORZANO, L. and J. D. H. STRICKLAND (1968) *Limnol. Oceanogr.* **13**, 515–518.
SOROKIN, Y. (1960) *J Cons. Perm. Int. Explor. Mer* **26**, 49–56.
SOROKIN, Y. (1964) *J. Cons. Perm. Int. Explor. Mer* **29**, 41–60.
SOROKIN, Y. (1971) *Int. Revue Ges. Hydrobiol.* **56**, 1–48.
SOROKIN, Y. (1977) *Mar. Biol.* **41**, 107–117.
SOROKIN Y. and A. M. TSVETKOVA (1972) *Oceanology* **12**, 870–878.
SOURNIA, A. (1968) *Int. Revue Ges. Hydrobiol.* **53**, 1–76.
SOURNIA, A. (1969) *Mar. Biol.* **3**, 287–303.
SOURNIA, A. (1970) *Ann. Biol.* **9**, 63–76.
SOURNIA, A. (1972) *J. Mar. Biol. Ass. India* **14**, 139–147.
SOURNIA, A. (1974) *Adv. Mar. Biol.* **12**, 325–389.
SPENCER, C. P. (1954) *J. Mar. Biol. Ass. U.K.* **33**, 265–290.
SPENCER, C. P. (1975) In J. P. RILEY and G. SKIRROW (Eds.), *Chemical Oceanography*, Vol. 2, 2nd ed. Academic Press, London. pp. 245–300.
SPOEHR, H. A. and H. W. MILNER (1949) *Plant Physiol.* **24**, 120–149.
STARODUBTSEV, Y. G. (1970) *Oceanology* **10**, 533–537.
STEELE, J. H. (1956) *J. Mar. Biol. Ass. U.K.* **35**, 1–33.
STEELE, J. H. (1958) *Rapp. P-v. Réun. Cons. Perm. Int. Explor. Mer* **144**, 79–84.
STEELE, J. H. (1962) *Limnol. Oceanogr.* **7**, 137–150.
STEELE, J. H. (1974) *The Structure of Marine Ecosystems*. Blackwell, Oxford. 128 pp.
STEELE, J. H. (1976) In D. H. CUSHING and J. J. WALSH (Eds.), *The Ecology of the Seas*. Blackwell, Oxford. pp. 98–115.
STEELE, J. H. and I. E. BAIRD (1961) *Limnol. Oceanogr.* **6**, 68–78.
STEELE, J. H. and I. E. BAIRD (1962) *Limnol. Oceanogr.* **7**, 42–47.
STEELE, J. H. and I. E. BAIRD (1965) *Limnol. Oceanogr.* **10**, 261–267.
STEELE, J. H. and M. M. MULLIN (1977) In E. D. GOLDBERG, I. N. MCCAVE, J. J. O'BRIEN and J. H. STEELE (Eds.), *The Sea*, Vol. 6. Wiley-Interscience, New York. pp. 857–890.
STEELE, J. H. and C. S. YENTSCH (1960) *J. Mar. Biol. Ass. U.K.* **39**, 217–226.
STEEMANN NIELSEN, E. (1935) *Meddr. Komm. Danm. Fisk, -og Havunders. Serie Plankton* **3**(1), 1–93.
STEEMANN NIELSEN, E. (1937) *J. Cons. Perm. Int. Explor. Mer* **12**, 147–153.
STEEMANN NIELSEN, E. (1940) *Meddr. Komm. Danm. Fisk. -og Havunders. Serie Plankton* **3**(4), 1–55.
STEEMANN NIELSEN, E. (1951) *Meddr. Komm. Danm. Fisk. -og Havunders. Serie Plankton* **5**(4), 1–114.
STEEMANN NIELSEN, E. (1952) *J. Cons. Perm. Int. Explor. Mer* **18**, 117–140.
STEEMANN NIELSEN, E. (1958a) *J. Cons. Perm. Int. Explor. Mer* **23**, 178–188.
STEEMANN NIELSEN, E. (1958b) *Rapp. P-v. Réun Cons. Perm. Int. Explor. Mer* **144**, 92–95.
STEEMANN NIELSEN, E. (1958c) *Rapp. P-v Réun. Cons. Perm. Int. Explor. Mer* **144**, 38–46.
STEEMANN NIELSEN, E. (1962a) *Physiol. Plant.* **15**, 161–171.
STEEMANN NIELSEN, E. (1962b) *Rapp. P-v. Réun. Cons. Perm. Int. Explor. Mer* **153**, 178–181.
STEEMANN NIELSEN, E. (1963) In M. N. HILL (Ed.), *The Sea*, Vol. 2. Interscience, New York. pp. 129–164.
STEEMANN NIELSEN, E. (1964a) *J. Ecol. Suppl.* **52**, 119–130.
STEEMANN NIELSEN, E. (1964b) *Meddr. Komm. Danm. Fisk. -og Havunders, NS* **4**, 31–77.
STEEMANN NIELSEN, E. (1972) *Int. Revue Ges. Hydrobiol.* **57**, 513–516.
STEEMANN NIELSEN, E. (1974) In N. G. JERLOV and E. STEEMANN NIELSEN (Eds.), *Optical Aspects of Oceanography*. Academic Press, London. pp. 361–387.
STEEMANN NIELSEN, E. and V. K. HANSEN (1959a) *Physiol. Plant.* **12**, 353–370.
STEEMANN NIELSEN, E. and V. K. HANSEN (1959b) *Deep-Sea Res.* **5**, 222–233.
STEEMANN NIELSEN, E. and E. AABYE JENSEN (1957) *Galathea Rept.* **I**, 49–136.
STEEMANN NIELSEN, E. and L. KAMP-NIELSEN (1970) *Physiol. Plant.* **23**, 828–840.
STEEMANN NIELSEN, E. and T. S. PARK (1964) *J. Cons. Perm. Int. Explor. Mer* **29**, 19–24.
STEEMANN NIELSEN, E. and M. WILLEMOES (1971) *Int. Revue Ges. Hydrobiol.* **56**, 541–556.
STEEMANN NIELSEN, E. and S. WIUM-ANDERSEN (1970) *Mar. Biol.* **6**, 93–97.
STEEMANN NIELSEN, E., V. K. HANSEN and E. G. JORGENSEN (1962) *Physiol. Plant.* **15**, 505–517.
STEFANSSON, U. and F. A. RICHARDS (1963) *Limnol. Oceanogr.* **8**, 394–410.
STEIDINGER, K. A. (1975) In V. R. LOCICERO (Ed.), *Proc. First International Conf. on Toxic Dinoflagellate Blooms*. Mass. Sci. and Tech. Foundation, Wakefield. pp. 153–162.
STENGEL, E. (1976) In O. DEVIK (Ed.), *Harvesting Polluted Waters*. Plenum Press, New York. pp. 281–297.
STEVEN, D. M. (1966) *J. Mar. Res.* **24**, 113–123.
STOMMEL, H. (1957) *Deep-Sea Res.* **5**, 80–82.

STOMMEL, H. (1958) *The Gulf Stream*. University California Press, Berkeley. 202 pp.
STOSCH, VON H. A. and G. DREBES (1964) *Helgolander Wiss. Meeresunters*. **11**, 209–257.
STRICKLAND, J. D. H. (1960) *Fish. Res. Bd. Can. Bull.* **122**. 172 pp.
STRICKLAND, J. D. H. (1965) In J. P. RILEY and G. SKIRROW (Eds.), *Chemical Oceanography*, Vol. 1, 1st ed. Academic Press, London. pp. 477–610.
STRICKLAND, J. D. H. and T. R. PARSONS (1965) *Fish. Res. Bd. Can. Bull.* **125**, 2nd ed., 203 pp.
STRICKLAND, J. D. H. and L. SOLORZANO (1966) In H. BARNES (Ed.), *Some Contemporary Studies in Marine Science*. Allen & Unwin, London. pp. 665–674.
STRICKLAND, J. D. H., L. SOLORZANO and R. W. EPPLEY (1970) *Bull. Scripps Inst. Oceanogr.* **17**, 1–22.
STRICKLAND, J. D. H., O. HOLM-HANSEN, R. W. EPPLEY and R. J. LINN (1969) *Limnol. Oceanogr.* **14**, 23–34.
STUERMER, D. H. and G. R. HARVEY (1974) *Nature, Lond.* **250**, 480–481.
STUERMER, D. H. and J. R. PAYNE (1976) *Geochim. Cosmochim. Acta* **4**, 1109–1114.
STUMM, W. and J. J. MORGAN (1970) *Aquatic Chemistry*. Wiley-Interscience, New York, 583 pp.
SUBBA RAO, D. V. (1969) *Limnol. Oceanogr.* **14**, 632–634.
SUBRAMANYAN, R. (1959) *Proc. Indian Acad. Sci.* B **50**, 113–187.
SUBRAMANYAN, R., and V. A. H. SARMA (1965) *J. Mar. Biol. Ass. India* **7**, 406–419.
SUGIMURA, Y., Y. SUZUKI and Y. MIYAKE (1976) *J. Oceanogr. Soc. Japan* **32**, 235–241.
SUKHANOVA, I. N. (1969) *Oceanology* **9**, 243–247.
SUNDA, W., and R. R. L. GUILLARD (1976) *J. Mar. Res.* **34**, 511–529.
SVERDRUP, H. U. (1953) *J. Cons. Perm. Int. Explor. Mer* **18**, 287—295.
SVERDRUP, H. U. and W. E. ALLEN (1939) *J. Mar. Res.* **2**, 131–144.
SVERDRUP, H. U., M. W. JOHNSON and R. H. FLEMING (1942) *The Oceans, Their Physics, Chemistry and General Biology*. Prentice-Hall, Inc., New York. 1087 pp.
SWIFT, E. and E. G. DURBIN (1971) *J. Phycol.* **7**, 89–96.
TAKAHASHI, M. and S. ICHIMURA (1972) In A. Y. TAKENOUTI (Ed.), *Biological Oceanography of the Northern North Pacific Ocean*. Idemitsu Shoten, Tokyo. pp. 217–229.
TAKAHASHI, T. (1961) *J. Geophys. Res.* **66**, 477–494.
TAKANO, H. (1967) *Inf. Bull. Plankt. Japan* **14**, 1–12.
TALLING, J. F. (1957) *Physiol. Plant* **10**, 215–223.
TALLING, J. F. (1960) *Limnol. Oceanogr.* **5**, 62–77.
TANADA, T. (1951) *Am. J. Bot.* **38**, 276–283.
TANIGUCHI, A. and T. KAWAMURA (1972) In A. Y. TAKENOUTI (Ed.), *Biological Oceanography of the Northern North Pacific Ocean*. Idemitsu Shoten, Tokyo. pp. 231–243.
TAYLOR, F. J. R. (1972) *Phycologia* **11**, 47–55.
TAYLOR, P. A. and P. J. LEB. WILLIAMS (1975) *Can. J. Microbiol.* **21**, 90–98.
TAYLOR, W. R., H. H. SELIGER, W. G. FASTIE and W. D. MCELROY (1966) *J. Mar. Res.* **24**, 28–43.
TEAL, J. M. and J. KANWISHER (1966) *J. Mar. Res.* **24**, 4–14.
TEIXEIRA, C. (1963) *Bolm. Inst. Oceanogr., S. Paulo* **13**, 53–60.
TEIXEIRA C. (1973) *Bolm. Inst. Oceanogr., S. Paulo* **22**, 49–58.
TEIXEIRA, C. and M. B. KUTNER (1963) *Bolm. Inst. Oceanogr., S. Paulo* **11**, 41–73.
TEIXEIRA, C., J. TUNDISI and Y. J. SANTORO (1969) *Int. Revue Ges. Hydrobiol.* **54**, 289–301.
THEIS, T. L. and P. C. SINGER (1974) *Envir. Sci. Technol.* **8**, 569–573.
THOMAS, J. P. (1971) *Mar. Biol.* **11**, 311–323.
THOMAS, L. P., D. R. MOORE and R. C. WORK (1961) *Bull. Mar. Sci. Gulf Caribb.* **11**, 191—197.
THOMAS, W. H. (1966a) *Deep-Sea Res.* **13**, 1109–1114.
THOMAS, W. H. (1966b) *Limnol. Oceanogr.* **11**, 393–400.
THOMAS, W. H. (1966c) *J. Phycol.* **2**, 17–22.
THOMAS, W. H. (1969) *J. Fish. Res. Bd. Can.* **26**, 1133–1145.
THOMAS, W. H. (1970) *Limnol. Oceanogr.* **15**, 380–394.
THOMAS, W. H. and D. L. R. SEIBERT (1977) *Bull. Mar. Sci.* **27**, 23–33.
THORRINGTON-SMITH, M. (1970a) *Br. Phycol. J.* **5**, 51–56.
THORRINGTON-SMITH, M. (1970b) *Nova Hedwigia* **31**, 815–835.
THORRINGTON-SMITH, M. (1970c) M.Phil. Thesis, University of Southampton.
THORRINGTON-SMITH, M. (1971) *Mar. Biol.* **9**, 115–137.
TIMONIN, A. G. (1969) *Oceanology* **9**, 686–694.
TOWLE, D. W. and J. S. PEARSE (1973) *Limnol. Oceanogr.* **18**, 155–159.
TRANTER, D. J. (1973) In B. ZEITZSCHEL (Ed.), *The Biology of the Indian Ocean*. Chapman & Hall, London; Springer-Verlag, Berlin. pp. 487–520.
TRANTER, D. J. and S. ABRAHAM (1971) *Mar. Biol.* **11**, 222–241.
TRANTER, D. J. and B. S. NEWELL (1963) *Deep-Sea Res.* **10**, 1–9.
TRÉGOUBOFF, G. and M. ROSE (1957) *Manuel de Planctonologie Méditerranéenne*. Vols. 1 and 2. Centre National Rech. Sci., Paris.

TUNDISI, J. G. (1971) In J. D. COSTLOW (Ed.), *Fertility of the Sea*, Vol. 2. Gordon & Breach, New York. pp. 603–612.
TURNER, J. T. (1977) *Mar. Biol.* **40**, 249–259.
UMEBAYASHI, O. (1972) *Bull. Tokai Reg. Fish. Lab.* **69**, 55–61.
UNESCO (1966) *International Oceanographic Tables*, Vol. 1. National Institute of Oceanography, Wormley, and UNESCO, Paris. 118 pp.
UNESCO (1973) *International Oceanographic Tables*, Vol. 2. National Institute of Oceanography, Wormley, and UNESCO, Paris. 141 pp.
VACCARO, R. F. (1962) *J. Cons. Perm. Int. Explor. Mer* **27**, 3–14.
VACCARO, R. F. (1963) *J. Mar. Res.* **21**, 284–301.
VACCARO, R. F. (1965) In J. P. RILEY and G. SKIRROW (Eds.) *Chemical Oceanography*, Vol. 1, 1st ed. Academic Press, London. pp. 365–408.
VACCARO, R. F. and H. W. JANNASCH (1966) *Limnol. Oceanogr.* **11**, 596–607.
VACCARO, R. F. and J. H. RYTHER (1960) *J. Cons. Perm. Int. Explor. Mer* **25**, 260–271.
VACCARO, R. F., F. AZAM and R. E. HODSON (1977) *Bull. Mar. Sci.* **27**, 17–22.
VARGO, G. (1968) *Bull. Mar. Sci.* **18**, 5–60.
VARGUES, M. and J. BRISOU (1963) In C. H. OPPENHEIMER (Ed.), *Marine Microbiology*. Thomas, Springfield, Ill. pp. 415–425.
VEDERNIKOV, V. I. and A. A. SOLOVYEVA (1972) *Oceanology* **12**, 559–565.
VENRICK, E. L. (1974) *Limnol. Oceanogr.* **19**, 437–445.
VINOGRADOV, A. P. (1953) *Sears Foundation Mar. Res. Mem.* **2**, 647 pp.
VINOGRADOV, M. E. (1962) *Rapp. P.-v. Réun. Cons. Perm. Int. Explor. Mer* **153**, 116.
VINOGRADOV, M. E. (1966) *Oceanology* **6**, 894–900.
VINOGRADOV, M. E. (1973) *Life Activity of Pelagic Communities in the Ocean Tropics*. Israel Program for Scientific Translations, Jerusalem. 298 pp.
VINOGRADOV, M. E. and E. G. ARASHKEVICH (1969) *Oceanology* **9**, 399–409.
VINOGRADOV, M. E. and V. V. MENSHUTKIN (1977) In E. D. GOLDBERG, I. N. MCCAVE, J. J. O'BRIEN and J. H. STEELE (Eds.), *The Sea*, Vol. 6. Wiley-Interscience, New York. pp. 891–921.
VINOGRADOV, M. E. and N. M. VORONINA (1962) *Rapp. P-v. Réun. Cons. Perm. Int. Explor. Mer* **153**, 200–204.
VISHNIAC, H. S. and G. A. RILEY (1961) *Limnol. Oceanogr.* **6**, 36–41.
VORONINA, N. M. and I. N. SUKHANOVA (1977) *Oceanology*, **16**, 614–616.
WADA, E. and A. HATTORI (1971) *Limnol. Oceanogr.* **16**, 766–772.
WALL, D. (1970) *Proc. North American Paleontological Convention. 1969*. **G**, 844–866.
WALL, D. and B. DALE (1967) *Rev. Palaeobotan. Palynol.* **2**, 349–354.
WALL, D. and B. DALE (1970) *Micropalaeontology* **16**, 47–58.
WALL, D., R. R. L. GUILLARD and B. DALE (1967) *Phycologia* **6**, 83–86.
WALL, D., R. R. L. GUILLARD, B. DALE, E. SWIFT and N. WATABE (1970) *Phycologia* **9**, 151–156.
WALLEN, D. G. and G. H. GEEN (1971a) *Mar. Biol.* **10**, 34–43.
WALLEN, D. G. and G. H. GEEN (1971b) *Mar. Biol.* **10**, 44–51.
WALLEN, D. G. and G. H. GEEN (1971c) *Mar. Biol.* **10**, 157–168.
WALSH, J. J. (1975) *Deep-Sea Res.* **22**, 201–236.
WALSH, J. J. and R. C. DUGDALE (1971) *Inv. Pesq.* **35**, 309–330.
WANGERSKY, P. J. and R. R. L. GUILLARD (1960) *Nature, Lond.* **185**, 689–690.
WARNER, T. B. (1971) *Deep-Sea Res.* **18**, 1255–1263.
WATSON, S. W. (1965) *Limnol. Oceanogr., Suppl.* **10**, R274–R289.
WATSON, S. W. and I. B. WATERBURY (1971) *Arch. Mikrobiol.* **77**, 203–230.
WATT, W. D. (1966) *Proc. R. Soc. Lond.* B **164**, 521–551.
WEISS, R. F. (1970) *Deep-Sea Res.* **17**, 721–735.
WHEELER, P. A., B. B. NORTH and G. S. STEPHENS (1974) *Limnol. Oceanogr.* **19**, 249–259.
WHITFIELD, M. (1973) *Mar. Chem.* **1**, 251–266.
WHITFIELD, M. (1975a) *Mar. Chem.* **3**, 197–213.
WHITFIELD, M. (1975b) *Geochim. Cosmochim. Acta* **39**, 1545–1557.
WHITFIELD, M. (1975c) In J. P. RILEY and G. SKIRROW (Eds.), *Chemical Oceanography*, Vol. 4, 2nd ed. Academic Press, London. pp. 1–154.
WHITTINGHAM, C. P. (1976) In T. W. GOODWIN (Ed.) *Chemistry and Biochemistry of Plant Pigments*, Vol. 1, 2nd ed. Academic Press, London. pp. 624–654.
WIEBE, P. H. and S. H. BOYD (1978) *J. Mar. Res.* **36**, 119–142.
WILLE, N. (1908) In K. BRANDT and C. APSTEIN (Eds.), *Nordisches Plankton, Botanischer Teil.* **20**, Lipsius & Tischer, Kiel and Leipzig. pp. 1–29.
WILLIAMS, D. B. (1971) In B. M. FUNNELL and W. R. RIEDEL (Eds.), *The Micropalaeontology of Oceans*. Cambridge University Press, London. pp. 91–95.

WILLIAMS, P. J. LEB. (1973) *Limnol. Oceanogr.* **18**, 159–165.
WILLIAMS, P. J. LEB. (1975) In J. P. RILEY and G. SKIRROW (Eds.), *Chemical Oceanography*, Vol. 2, 2nd ed. Academic Press, London. pp. 301–363.
WILLIAMS, P. J. LEB. and C. S. YENTSCH (1976) *Mar. Biol.* **35**, 31–40.
WILLIAMS, P. J. LEB., T. BERMAN and O. HOLM-HANSEN (1972) *Nature, Lond.* **236**, 91–92.
WILLIAMS, P. M. (1969) *Limnol. Oceanogr.* **14**, 156–158.
WILLIAMS, P. M. and L. I. GORDON (1970) *Deep-Sea Res.* **17**, 19–27.
WILLIAMS, P. M., H. OESCHGER and P. KINNEY (1969) *Nature, Lond.* **224**, 256–258.
WILSON, T. R. S. (1975) In J. P. RILEY and G. SKIRROW (Eds.), *Chemical Oceanography*, Vol. 1, 2nd ed. Academic Press, London. pp. 365–413.
WIMPENNY, R. S. (1936a) *J. Mar. Biol. Ass. U.K.* **21**, 29–60.
WIMPENNY, R. S. (1936b) *Fishery Invest. London,* Ser. II, **15**(3), 1–59.
WIMPENNY, R. S. (1938) *J. Cons. Perm. Int. Explor. Mer* **13**, 323–336.
WIMPENNY, R. S. (1946) *J. Mar. Biol. Ass. U.K.* **26**, 271–284.
WIMPENNY, R. S. (1973) *J. Mar. Biol. Ass. U.K.* **53**, 957–974.
WINTER, D. F., K. BANSE and G. C. ANDERSON (1975) *Mar. Biol.* **29**, 139–176.
WOLGEMUTH, K. (1970) *J. Geophys. Res.* **75**, 7686–7687.
WOLLAST, R. (1974) In E. D. GOLDBERG (Ed.), *The Sea*, Vol. 5. *Marine Chemistry*. Wiley-Interscience, New York. pp. 359–392.
WONG, C. S., D. R. GREEN and W. J. CRETNEY (1974) *Nature, Lond.* **247**, 30–32.
WONG, G. T. F. and P. G. BREWER (1977) *Geochim. Cosmochim. Acta* **41**, 151–159.
WOOD, E. J. F. (1963) In C. H. OPPENHEIMER (Ed.), *Marine Microbiology*. Thomas, Springfield, Ill. pp. 28–39 and 275–285.
WOOSTER, W. S., T. J. CHOW and I. BARRETT (1965) *J. Mar. Res.* **23**, 210–221.
WOOSTER, W. S., A. J. LEE and G. DIETRICH (1969) *Limnol. Oceanogr.* **14**, 437–438.
WUST, G. (1930) *J. Cons. Perm. Int. Explor. Mer* **5**, 7–21.
WUST, G., W. BROGMUS and E. N. NOADT (1954) *Kieler Meeresforsch.* **10**,137–161.
WYRTKI, K. (1966) *Oceanogr. Mar. Biol. Ann. Rev.* **4**, 33–68.
WYRTKI, K. (1973) In B. ZEITZSCHEL (Ed.), *The Biology of the Indian Ocean*. Chapman & Hall, London, and Springer-Verlag, Berlin. pp. 18–36.
YAMAZI, I. (1956) *Publs. Seto Mar. Biol. Lab.* **5**, 157–196.
YAMAZI, I. (1959) *Records Oceanogr. Works, Japan, Special No.* **3**, 23–30.
YENTSCH, C. S. (1962) In R. A. LEWIN (Ed.), *Physiology and Biochemistry of Algae*. Academic Press, New York. pp. 771–797.
YENTSCH, C. S. (1965a) *Deep-Sea Res.* **12**, 653–666.
YENTSCH, C. S. (1965b) *Mem. Ist. Ital. Idrobiol. Suppl.* **18**, 323–346.
YENTSCH, C. S. (1974) *Oceanogr. Mar. Biol. Ann. Rev.* **12**, 41–75.
YENTSCH, C. S. and R. W. LEE (1966) *J. Mar. Res.* **24**, 319–337.
YENTSCH, C. S. and D. W. MENZEL (1963) *Deep-Sea Res.* **10**, 221–231.
YENTSCH, C. S. and C. A. REICHERT (1963) *Limnol. Oceanogr.* **8**, 338–342.
YENTSCH, C. S. and J. H. RYTHER (1959) *J. Cons. Perm. Int. Explor. Mer* **24**, 231–238.
YENTSCH, C. S. and R. F. SCAGEL (1958) *J. Mar. Res.* **17**, 567–583.
YENTSCH, C. S. and R. F. VACCARO (1958) *Limnol. Oceanogr.* **3**, 443–448.
YOSHII, O., K. HIRAKI, Y. NISHIKAWA and T. SHIGEMATSU (1977) *Bunseki Kagaku* **26**, 91–96.
ZAJIC, J. E. and Y. S. CHIU (1970) In J. E. ZAJIC (Ed.), *Properties and Products of Algae*. Plenum Press, New York. pp. 1–47.
ZEITZSCHEL, B. (1970) *Mar. Biol.* **7**, 315–318.
ZEITZSCHEL, B. (1973) (Ed.), *The Biology of the Indian Ocean*. Chapman & Hall, London; Springer-Verlag, Berlin. 549 pp.
ZENKEVITCH, L. A. (1963) *Biology of the Seas of the USSR*. Allen & Unwin, London. 955 pp.
ZERNOVA, V. V. (1970) In M. W. HOLDGATE (Ed.), *Antarctic Ecology*, Vol. 1. Academic Press, London. pp. 136–142.
ZERNOVA, V. V. (1970) In M. W. HOLDGATE (Ed.), *Antarctic Ecology* **1**. pp. 136–142.
ZERNOVA, V. V. (1974) *Oceanology* **14**, 882–887.
ZIRINO, A. and S. YAMAMOTO (1972) *Limnol. Oceanogr.* **17**, 661–671.

Index